MILLIMETER WAVE RADAR

THE ARTECH RADAR LIBRARY
David K. Barton, Editor

Laser Radar Systems and Techniques
Christian G. Bachman

MTI Radar
D.C. Schleher

Radar Anti-Jamming Techniques:
A Translation from the Russian of *Zashchita Ot Radiopomekh*
M.V. Maksimov *et al.*

Phased Array Antennas
Arthur A. Oliner and George H. Knittel

Radar Detection and Tracking Systems
Shahen A. Hovanessian

Radar Signal Simulation
Richard L. Mitchell

Radar System Analysis
David K. Barton

Radar Technology
Eli Brookner

RADARS — in seven volumes

I. Monopulse Radar
II. The Radar Equation
III. Pulse Compression
IV. Radar Resolution and Multipath Effects
V. Radar Clutter
VI. Frequency Agility and Diversity
VII. CW and Doppler Radar

David K. Barton

Significant Phased Array Papers
Robert C. Hansen

Synthetic Aperture Radar
John J. Kovaly

Introduction to Synthetic Array and Imaging Radars
Shahen A. Hovanessian

Radar Electronic Counter-Countermeasures
Stephen L. Johnston

Radar Detection
J.V. DiFranco and W.L. Rubin

Introduction to Monopulse
D. Rhodes

Automatic Detection and Radar Data Processing
D.C. Schleher

MILLIMETER WAVE RADAR

Edited by
Stephen L. Johnston

Copyright © 1980

ARTECH HOUSE, INC.
610 Washington St.
Dedham, MA 02026

Printed and bound in the United States of America.

All rights reserved. No part of this book may be reproduced or utilized in any form or by any means, electronic or mechanical, including photocopying, recording, or by any information storage and retrieval system, without permission in writing from the publisher.

Standard Book Number: 0-89006-095-9

Library of Congress catalog card number: 80-53388

CONTENTS

FOREWORD

CHAPTER 1: Introduction and Overview 1

1.1 "Some Aspects of Millimeter Radar" 5
 S.L. Johnston, *Proceedings International Conference on Radar*,
 Paris, France, Dec. 4-8, 1978, pp. 148-159.

1.2 "Millimetre Waves -- The Much Awaited Technological Breakthrough?" 17
 G.S. Sundaram, *International Defense Review*, Vol. 11, No. 2,
 Feb. 1979, pp. 271-277.

1.3 "Millimeter Waves — The Coming of Age" 25
 L.R. Weisberg, *6th DARPA-Tri Service Millimeter Wave Conference*, 1977, pp. 256-265.

1.4 "The Unresolved Problems and Issues" 31
 L.D. Strom, *6th DARPA-Tri Service Millimeter Wave Conference*, 1977, pp. 94-103.

1.5 "The Potential Military Applications of Millimeter Waves" 49
 L.R. Whicker and D.C. Webb, *AGARD Conference Proceedings*, CP 245,
 "Millimeter and Submillimeter Wave Propagation and Circuits," 1978,
 pp. 1-1 through 1-6.

1.6 "Millimeter Wave Technology and Applications" 55
 J.C. Wiltse, *Microwave Journal*, Vol. 22, No. 8, Aug. 1979, pp. 39ff.

1.7 "Millimeter Wave Radar Applications to Weapons Systems" 59
 V.W. Richard, *U.S. Army Ballistic Research Labs Report No. 2631*,
 June 1976, DDC AD B012103 (excerpts).

CHAPTER 2: Millimeter Wave Propagation, Targets and Clutter 117

2.A Propagation

2.A.1 "A Radar System Engineer Looks at Current Millimeter–Submillimeter
 Atmospheric Propagation Data" 127
 S.L. Johnston, *IEEE EASCON-79 Conference Record*, Vol. 1, pp. 27-35.

2.A.2 "Millimeter Wave Propagation" 137
 K.L. Koester, L. Kosowsky, and J.F. Sparacio -- Appendix A to "Millimeter
 Wave Radar Applications to Weapons Systems," V.L. Richard (Paper 1.7).

2.A.3 "Propagation Effects for mm Wave Fire Control System" 167
 T.N. Patton, J.J. Petrovic, and J. Teti, *6th DARPA-Tri Service Millimeter
 Wave Conference*, 1977.

2.A.4 "Environmental Effects on Millimeter Radar Performance" 177
 F.B. Dyer and N.C. Currie, *AGARD Conference Proceedings*, CP 245,
 "Millimeter and Submillimeter Wave Propagation and Circuits," 1978,
 pp. 2-1 through 2-9.

2.A.5 "Atmospheric Attenuation of Millimeter and Submillimeter Waves" 187
 V.J. Falcone, Jr. and L.W. Abreu, *IEEE EASCON-79 Conference Record*,
 Vol. 1, pp. 36-41.

2.A.6 "Atmospheric Turbulence Effects on Millimeter Wave Propagation" 193
 R.W. McMillan, J.C. Wiltse, and D.E. Snider, *IEEE EASCON-79 Conference
 Record*, Vol. 1, pp. 42-47.

2.A.7 "Millimetre Wave Propagation in Smoke" 199
 J.E. Knox, *IEEE EASCON-79 Conference Record*, Vol. 2, pp. 357-361.

2.A.8 "140 GHz Multipath Measurements over Varied Ground Covers" 205
 H.B. Wallace, *IEEE EASCON-79 Conference Record*, Vol. 2, pp. 256-260.

2.A.9 "Millimeter Wave Propagation Measurements over Snow" 211
D.T. Hayes, U.H.W. Lammers, R.A. Marr, and J.J. McNally, *IEEE EASCON-79 Conference Record*, Vol. 2, pp. 362-368.

2.B Targets

2.B.1 "93-GHz Radar Cross Section Measurements of Satellite Elemental Scatterers" 219
R.B. Dybdal and H.E. King, *IEEE Trans.*, Vol. AP-25, May 1977, pp. 396-402.

2.B.2 "95 GHz Pulsed Radar Returns from Trees" 227
R.D. Hayes, *IEEE EASCON-79 Conference Record*, Vol. 2, pp. 353-356.

2.C Clutter

2.C.1 "Fluctuating Clutter" 231
K.L. Koester, L. Kosowsky, and J.F. Sparacio -- Appendix B to "Millimeter Wave Radar Applications to Weapons Systems," V.L. Richard (Paper 1.7).

2.C.2 "mm-Wave Reflectivity of Land and Sea" 241
R.N. Trebits, R.D. Hayes, and L.C. Bomar, *Microwave Journal*, Vol. 21, No. 8, Aug. 1978, pp. 49ff.

2.C.3 "Radar Sea Clutter at Millimeter Wavelengths — Characteristics and Pivotal Unknowns" 245
G.W. Ewell, R.N. Trebits, and J. Teti, *6th DARPA-Tri Service Millimeter Wave Conference*, 1977.

2.C.4 "Multifrequency Millimeter Radar Sea Clutter Measurements" 251
R.N. Trebits, N.C. Currie, and F.B. Dyer, *IEEE EASCON-79 Conference Record*, Vol. 2, pp. 261-264.

CHAPTER 3: Millimeter Wave Radar RF Sources and Components 255

3.A Solid State Sources

3.A.1 "Millimetre Wavelength IMPATT Sources" 265
J.J. Purcell, *The Radio and Electronic Engineer*, Vol. 49, No. 7/8, July/August 1979, pp. 347-350.

3.A.2 "Advances in mm-Wave Components and Systems" 269
B. Adelseck, H. Barth, H. Hoffmann, H. Meinel, and B. Rembold, *AGARD Conference Proceedings*, CP 245, 1978, "Millimeter and Submillimeter Wave Propagation and Circuits," pp. 25-1 through 25-17.

3.A.3 "Solid-State Millimeter-Wave Sources" 287
H.J. Kuno and T.T. Fong, *IEEE EASCON-77 Conference Record*, pp. 24-2A through 2D.

3.A.4 "EHF Solid State Transmitters for Satellite Communications" 291
J.E. Raue, *Proc. AIAA 8th Communications Satellite Systems Conference*, April 1980, pp. 107-111.

3.A.5 "Advanced Solid-State Components for Millimeter Wave Radars" 295
P.M. Schwartz, R.F. Lohr, K.P. Weller, and R.L. Zimmerman, *IEEE-MTTS International Microwave Symposium Digest*, 1975, pp. 261-263.

3.A.6 "A 90 GHz FM-CW-Radar Transmitter" 299
H. Barth and M. Bischoff, *IEEE-MTTS International Microwave Symposium Digest*, 1979, pp. 75-78.

3.A.7 "Millimeter Solid-State Pulsed Coherent Radar" 303
T. Duffield, C. Smith, P. Pusateri, D. English, and F. Bernues, *IEEE-MTTS International Microwave Symposium*, 1979 (late paper).

3.A.8 "High Power Millimeter Wave Radar Transmitters" 311
G.W. Ewell, D.S. Ladd, and J.C. Butterworth, *Microwave Journal*, Vol. 23, No. 8, Aug. 1980, pp. 57ff.

3.B   Electron Tube Sources

3.B.1 "Recent Progress and Future Performances of Millimeter-Wave BWO's" 319
       B. Epsztein, *AGARD Conference Proceedings*, CP-245, 1978, "Millimeter
       and Submillimeter Wave Propagation and Circuits," pp. 36-1 through 36-11.

3.B.2 "The Laddertron — A New Millimeter Wave Power Oscillator" 331
       K. Fujisawa, *IEEE Trans.*, Vol. ED-11, No. 8, Aug. 1964, pp. 381-391.

3.C   Gyrotrons (Fast Wave Devices)

3.C.1 "Millimeter Wave Tubes With Emphasis on High Power Gyrotrons" 343
       R.H. Chilton, *6th DARPA-Tri Service Millimeter Wave Conference*, 1977, pp. 53-62.

3.C.2 "Prospects for High Power Millimeter Radar Sources and Components" 353
       T.F. Godlove, V.L. Granatstein, and J. Silverstein, *IEEE EASCON-77
       Conference Record*, pp. 16-2A through F.

3.C.3 "Gyrotrons for High Power Millimeter Wave Generation" 359
       H.R. Jory, F. Friedlander, S.J. Hegji, J.F. Shivley, and R.S. Symons,
       IEEE International Electron Device Meeting Digest, Dec. 1977, pp. 234-237.

3.D   Mixers

3.D.1 "Review of Mixers for Millimeter Wavelengths" 363
       M. McColl and D.T. Hodges, *6th DARPA-Tri Service Millimeter Wave Conference*,
       1977, pp. 123-127.

3.E   Antennas

3.E.1 "Millimeter Wave Antennas" 373
       A.F. Kay, *Proceedings IEEE*, Vol. 54, No. 4, April 1966, pp. 641-647.

3.E.2 "Performance of Reflector Antennas with Absorber-Lined Tunnels" 381
       R.B. Dybdal and H.E. King, *IEEE APS International Symposium Digest*,
       1979, Vol. 2, pp. 714-717.

3.E.3 "A Millimeter-Wave Scanning Antenna For Radar Application" 385
       O.B. Kesler, W.F. Montgomery, and C.C. Liu, *Proceedings of the 1979 Antenna
       Applications Symposium*, University of Illinois/Rome Air Development Center,
       Sept. 1979 (Session II, first paper) DDC AD A077167.

CHAPTER 4: K_a-Band Radars 409

4.1   "A Millimeter Wave Pseudorandum Coded Meteorological Radar" 411
       M.S. Reid, *IEEE Trans.*, Vol. GE-7, No. 3, July 1969, pp. 146-156.

4.2   "TRAKX: A Dual-Frequency Tracking Radar" 423
       D. Cross, D. Howard, M. Lipka, A. Mays, and E. Ornstein, *Microwave
       Journal*, Vol. 19, No. 9, Sept. 1976, pp. 39-41.

4.3   "An FM/CW Radar with High Resolution in Range and Doppler; Application for
       Anti-Collision Radar for Vehicles" 427
       G. Neininger, *Radar 77*, IEE Conference Publication No. 155, London,
       Oct. 25-28, 1977, pp. 526-530.

4.4   "mm Radar for Highway Collision Avoidance" 433
       Y.K. Wu and C.P. Tresselt, *Microwave Journal*, Vol. 20, No. 11, Nov. 1977, pp. 39ff.

4.5   "Solid State mm-Wave Pulse-Compression Radar Sensor" 439
       J.B. Winderman and G.N. Hulderman, *Microwave Journal*, Vol. 20, No. 11,
       Nov. 1977, pp. 45ff.

4.6   "SEATRACKS — A Millimeter Wave Radar Fire Control System" 443
       G.E. Layman, *IEEE EASCON-78 Conference Record*, 1978, pp. 211-216.

4.7   "A Millimeter Wave Radar for the U.S. Army Helicopters in the 80's" 449
       T.R. Holmes and E.A. Flick, *IEEE NAECON-80 Proceedings*, 1980,
       Vol. 2, pp. 712-716.

CHAPTER 5: Millimeter Wave Radars . . . 455

5.1 "Combat Surveillance Radar AN/MPS-29 (XE-1)" . . . 461
M.W. Long, W.K. Rivers, and J.C. Butterworth, *6th Tri-Service Radar Symposium Record*, 1960, pp. 230-244.

5.2 "A 94-GHz Radar for Space Object Identification" . . . 479
L.A. Hoffman, K.H. Hurlburt, D.E. Kind and H.J. Wintroub, *IEEE Trans.*, Vol. MTT-17, No. 12, Dec. 1969, pp. 1145-1149.

5.3 "Vehicle Mounted Millimeter Radar" . . . 485
F.B. Dyer and R.M. Goodman, Jr., *18th Tri-Service Radar Symposium Record*, 1972, pp. 473-495.

5.4 "Development and Test of a 95 GHz Terrain Imaging Radar" . . . 507
Millimeter Waves Techniques Conference, 1974, Naval Electronics Laboratory Center, Rpt. NELC/TD 308, Vol. 2, pp. F1-1 through 14, DDC AD A020273 (also *Goodyear Aerospace Report* GERA-1987, February 28, 1974).

5.5 "A Solid State 94 GHz Doppler Radar" . . . 521
F.B. Bernues, H.J. Kuno, and J. McIntosh, *IEEE 1975 MTTS International Symposium Digest*, pp. 258-260.

5.6 "Millimeter Wave Monopulse Track Radar" . . . 525
L.H. Kosowsky, K.L. Koester, and R.S. Graziano, *AGARD Conference Proceedings*, CP-197, "New Devices, Techniques and Systems in Radar", June 1976, pp. 35-1 through 35-18.

5.7 "Combined Electro-Optical/Millimeter Wave Radar Sensor System" . . . 543
W.A. Holm, W.S. Foster, and G.R. Loefer, *6th DARPA-Tri Service Millimeter Wave Conference*, 1977.

5.8 "Radar Tracking of an M-48 Tank at 94 and 140 GHz" . . . 553
R.A. McGee and J.M. Loomis, *6th DARPA-Tri Service Millimeter Wave Conference*, 1977.

5.9 "94 GHz Active Missile Seeker Captive Flight Test Results" . . . 559
R. Yoshitani and M.E. Beebe, *6th DARPA-Tri Service Millimeter Wave Conference*, 1977.

5.10 "Potential Applications of Millimeter Wave Radar to Ballistic Missile Defense" . . . 571
G.B. Jones, *IEEE EASCON-77 Conference Record*, 1977, pp. 16-3A through C.

5.11 "Millimeter Radar Application to SOI" . . . 575
R.B. Dybdal, *IEEE EASCON-77 Conference Record*, 1977, pp. 16-4A through I.

5.12 "mm-Wave Instrumentation Radar Systems" . . . 585
N.C. Currie, J.A. Scheer, and W.A. Holm, *Microwave Journal*, Vol. 21, No. 8, Aug. 1978, pp. 35ff.

5.13 "A Millimeter Wave Radar for the Mini-RPV" . . . 593
L. Kosowsky, K. Koester, B. Gelernter and W. Johnson, *AIAA/DARPA Conference on Smart Sensors*, Nov. 1978.

5.14 "Model Simulation of Target Characteristics and Engagement Situations Employing Millimetre Wave Radar Systems" . . . 611
W. Gabsdil and W. Jacobi, *AGARD Conference Proceedings*, CP-245, 1978, "Millimeter and Submillimeter Wave Propagation and Circuits," pp. 7-1 through 7-9.

5.15 "HOWLS Radar Development" . . . 621
V.L. Lynn, *AGARD AVP/GCP Joint Symposium — Avionics/Guidance and Control for RVP's*, Florence, Italy, 4-8 Oct. 1976, Paper No. 31.

5.16 "Millimeter Airborne Radar Target Detection and Selection Techniques" . . . 635
L.M. Novak and F.W. Vote, *IEEE NAECON-79 Conference Record*, Vol. 2, 1979, pp. 807-817.

5.17 "USAF Millimeter-Wave Seeker Development" 647
A.N. Disalvio, *IEEE NAECON-79 Conference Record,* Vol. 1, 1979, pp. 363-370.

5.18 "A Terminal Guidance Simulator for Evaluation of Millimeter Wave Seekers" 655
K.L. Wismer and A.J. Witsmeer, *Proc. Military Microwaves Conference, MM-80,*
London, Oct. 1980.

FOREWORD

The topic "millimeter wave radar" is hardly new, having originated during World War II. Radar activities above 35 GHz (K_a-band) are fairly new, however; experimental 70 GHz radars are about 20 years old. Radars at 95 GHz are more recent.

The first Millimeter-Submillimeter Conference was held in 1970. The first open technical symposium session on Millimeter Wave Radar was at EASCON-77, with a group of sessions on millimeter wave propagation at EASCON-79. The Department of Defense Advanced Research Projects Agency (DARPA) and the three serivces have conducted several millimeter wave workshops since 1975. The NATO Advisory Group for Aerospace Research and Development (AGARD) has sponsored one conference on millimeter waves.

The first comprehensive survey of millimeter wave radar was published in *Microwave Journal* in 1977. The first issue of a technical journal devoted to "short millimetric waves" was published in Great Britain in 1979 by the Institution of Radio and Electronic Engineers. In the last three years, a number of separate papers on millimeter wave radars have been presented at various U.S. and international conferences and have been published in several technical journals.

Up to now, these papers have been widely scattered. This volume is a selected collection of papers from EASCON-77 and 79, the Sixth DARPA Conference, two AGARD Conferences and several technical journals, primarily *Microwave Journal*.

It is hoped that this volume will enable new workers in the field to stay abreast of work by others, and to assist in the exchange of knowledge among all workers in millimeter wave radar.

The author expresses his appreciation to his many colleagues for their cooperation in making these papers available, and for their valuable suggestions. The cooperation of several publishers is gratefully acknowledged. The author's wife again provided vital assistance in typing the manuscript and in securing authors' and publishers' permissions.

Assistance of the Price Gilbert Memorial Library of the Georgia Institute of Technology in obtaining photocopies of some of the papers in this volume is also acknowledged.

CHAPTER 1
INTRODUCTION AND OVERVIEW

Radar originally operated in the "high frequency," e.g., 10 MHz, portion of the electromagnetic spectrum. As technology developed during World War II, radar operating frequencies continually increased through very high frequency, then utilized the cavity magnetron to leap over ultra high frequency to the microwave region (1 - 10 GHz). Although most radar applications today still employ microwave frequencies, technology has, for a number of years, permitted operation at higher frequencies: 15 (Ku), 24 (K) and 35 GHz (K_a-band). Some persons refer to 35 GHz radar as being "millimeter" although the current IEEE Standard [1, 2] defines millimeter radar as 40 - 300 GHz. In the last few years, radar has been developed for experimental applications at 70, 95 and 140 GHz.

Radar technology surveys, such as those referenced in [3 - 5], have extensively treated microwave radar. Although a modest number of radars operating above X-band have been manufactured in production quantities (and hence meet the "Barton Criterion" [3]); these radars have not been included in the above surveys.

While a number of K_u, K, and K_a-band radars have been produced in "quantity", as of this writing no "millimeter" radars have been produced in quantity; however, this may happen in the near future.

Radars operating at K_u-band are generally similar in application to those at X-band. Several military K-band radars were built near the end of World War II. Typical applications included counter-mortar operations and airport surface detection equipment (ASDE). The original U.S. ASDE operated at K-band with the new ASDE at K_u-band.[6] European ASDE, having different environmental conditions, operates at K_a-band. With the exception of the original U.S. ASDE, development of RF components and sources for K_a-band effectively ended most radar activities at K-band.

Recent developments have rekindled radar activities at K_a-band and at even higher frequencies. Review of the literature indicates that a large number of papers describing work in these higher frequencies have been published in the last two or three years; however, these papers are presently scattered in a number of journals and conference proceedings. Several papers were presented at limited government conferences and were not previously available to the general public. This volume presents in one place the most important papers on various aspects of radar in this new frequency region. Radars operating at K_a-band and the millimeter region will be treated.

In some respects, K_a-band radars resemble those at X-band; however, some workers consider K_a-band to be "millimeter" because its wavelength is less than l cm. As a result of recent new frequency allocations for radiolocation (radar) by the World Administrative Radio Conference (WARC-79 [7]), there is a probability that the IEEE Radar Letter Bands Standard [1] may be revised, raising the possibility that "millimeter" might be redefined as frequencies of 30 - 300 GHz, thus then including K_a-band. Accordingly, it is appropriate to include K_a-band radar in this volume.

Since many of the papers in this volume contain extensive bibliographies, the bibliographies for each chapter will contain only items which are not cited in a paper reprinted here.

The first chapter is devoted to several survey papers on millimeter wave radar, and to papers which discuss problems and issues concerning the utilization of this portion of the spectrum.

SYNOPSIS OF REPRINT PAPERS IN CHAPTER 1

1.1 "Some Aspects of Millimeter Radar" — Johnston

Papers on "millimeter waves" date back at least three decades. Most of these are of a general nature. Interestingly, these papers seem to have appeared at cyclic intervals. There were a few millimeter radar papers interspersed in this literature. In 1970, Skolnik published a very comprehensive review of millimeter and submillimeter wave applications.[8, 9] This includes an extensive treatment of radar. Probably the most complete survey of specific millimeter radars appeared in *Microwave Journal* in 1977.[10] This paper lists radars chronologically. It followed the first technical session devoted to millimeter wave radar (IEEE EASCON-77).

This paper, which was presented at the International Conference on Radar held at Paris, France, Dec. 4-8, 1978, is an update of the *Microwave Journal* paper. Its extensive bibliography (140 items) covers three decades. Topics covered include reasons for millimeter radar, possible millimeter applications, past and present radar applications, future millimeter radar systems and issues affecting utilization of millimeter radars. The tables of suggested applications are from the previously cited work of Skolnik. The several tables of representative radar applications are given separately for K_a-band, 70 GHz and 95/140 GHz. Civilian and military applications are shown individually.

Figure 1 in this paper is unique in that it relates atmospheric attenuation as a function of wavelength and of frequency to both the IEEE Radar Letter Bands and to the ITU Radio-location Bands. Photographs of nearly 20 civilian and military K_a and millimeter wave radars are included.*

Of the seven issues affecting millimeter wave radars, only the first two (component capability and radar performance — *i.e.*, atmospheric attenuation) are generally cited by other survey authors. The others (system integration, design tradeoffs, investigation of novel guidance schemes and lack of available millimeter radar frequency allocation) are unique to this paper. These have been given insufficient attention in the literature to date. The last issue has recently been addressed by the previously cited WARC-79[7] and has now been resolved. If WARC-79 had not so acted, this could have prevented utilization of millimeter wave radars.

1.2 "Millimetre Waves — The Much Awaited Breakthrough?" — Sundaram

This recent survey paper from the widely read *International Defense Review* is somewhat in the vein of the preceding paper and its predecessor *Microwave Journal* paper.[10] Certainly millimeter waves have been highly touted at cyclic intervals for over three decades, as a number of references in the preceding paper indicate.

Applications discussed in this paper include radiometry, communications, and missile terminal guidance, in addition to radar. Millimeter aspects of electronic warfare are also treated. Most of them are in K_a-band, however. The radar applications described cite recent European and U.S. work and complement the preceding paper. Further information on the HSA Flycatcher (a dual band X/K_a radar) was published in [12]. Technical characteristics of some of the mm radars, *e.g.*, STARTLE, are contained in [11]. TRAKX is described in Paper 4.2, Chapter 4, of this volume.

Although this paper cites many advantages of millimeter waves and indicates extensive current mm radar work, its conclusions on the future of mm are very reserved, similar to the preceding paper and [10]. Further comments on this paper are in [13].

1.3 "Millimeter Waves — The Coming of Age" — Weisberg

For several years the U.S. Department of Defense Advanced Research Projects Agency (DARPA) and the three military departments have periodically conducted conferences on military applications of millimeter waves. A number of papers in this volume were presented at the sixth such conference (1977).

This introductory paper recognizes the previous cyclic nature of military millimeter wave activities. Three principal topics of this paper are: Why Not Before?, Why Today?, and What Should We Do Now?

Previous unavailability of millimeter wave technology, poor atmospheric propagation of mm waves, and lack of requirements are discarded by the author in favor of "thinking differently". Current thinking emphasizes millimeter waves as an adjunct to laser/EO systems. The table "Applications for Millimeter-Wave Technology" is essentially a refinement of Skolnik's application table given in Paper 1.1 of this chapter. Millimeter waves are considered important in that they can be useful in smoke and haze, which seriously degrade laser/EO systems.

One factor not stressed by the author is that many current millimeter wave applications operate at fairly short ranges, *e.g.*, 1-3 km, and as a result suffer less atmospheric attenuation than at longer ranges, such as 10 - 20 km.

Regrettably, the last topic — What Should We Do Now? — is only briefly addressed. The author's recommendation is essentially that the "separated and entrepreneurial ventures must be brought together and focused to achieve synergism".

---

*Technical characteristics of some of the mm wave radars are given in [11].

1.4 "The Unresolved Problems and Issues" — Strom

This is a companion paper to the preceding paper, and was presented at the same conference. Some of the issues cited in Paper 1.1 of this chapter are treated by this paper. It considers several representative applications of millimeter wave radar to a greater extent than the preceding paper. Interestingly, the author compares a 3 mm radar with a cm (X-band) radar for short range usage, and claims the former is to be less weather-limited due to the increased antenna gain. It will be recalled from the introductory remarks that a U.S. Department of Transportation study [6] concluded that K_u band was more cost-effective for an ASDE radar than K-band. (Presumably this would be even more so when compared to a 3 mm radar).

Review of the previous DARPA/Tri-Service Millimeter Wave Workshops and their objectives should be of great interest to new workers in this field, as should the Workshop results (Figure 5). It is noteworthy that adverse weather effects (recommendation 25) are to be treated in terms of hours of outage per year instead of in terms of rainfall rates. This will be addressed in Chapter 2 of this volume.

1.5 "The Potential Military Applications of Military Waves" — Whicker and Webb

This AGARD (1978) paper draws on the previously cited works by Skolnik [8, 9], Johnston [10] and Papers 1.3 and 1.4 of this chapter. It has a limited review of millimeter wave propagation (cf. Chapter 2 of this volume).

It is pointed out that systems operating in Western Europe will be confronted by ground fog frequently during the fall and winter. The authors do not mention that, accordingly, rain, attenuation would not present a problem in that area. In [6], rainfall is shown to be a major problem in the U.S.

Comparison is made between microwaves, millimeter waves and optical systems for radar. In most of the attributes considered, millimeter wave radar is stated to be between microwave and optical systems: in some ways, it is better than optical systems, and inferior to microwave radar; while in others, it is inferior to optical systems yet superior to microwave radar. Certainly this is different from Paper 1.4 in this chapter. In some cases, the authors propose millimeter wave radars to augment or replace optical systems, and for other applications they propose to augment or replace microwave radars with millimeter radars.

1.6 "Millimeter Wave Technology and Applications" — Wiltse

This very recent survey paper (Aug. 1979) has an extensive bibliography of 65 items. Some of the papers cited are published in this volume. It has strong emphasis on millimeter wave RF sources and components and should be considered with Chapter 3 of this volume. Applications discussed are not confined to radar.

1.7 "Millimeter Wave Radar Applications to Weapon Systems" — Richard

Topics presented in this 1976 government report are millimeter wave propagation characteristics (cf. Chapter 2), millimeter wave radar characteristics, millimeter wave radar and component status (cf. Paper 1.6 and Chapter 3), and technology advances required (cf. Figure 5 of Paper 1.4 in this chapter).

As will be shown in Chapter 2, attenuation due to rainfall is frequently viewed in terms of "average" rainfall rate, whereas the instantaneous rainfall rate actually determines the attenuation. Experimental rain backscatter data in this paper is very interesting. Results of further analysis of this data are presented in Paper 2.C.2 in Chapter 2.

Extensive performance data is presented on the U.S. Army Ballistic Research Labs experimental 70 GHz tracking radar and the Norden 70 GHz radar (see Paper 5.6 in Chapter 5).

The list of existing millimeter radars is more comprehensive than that of Papers 1.4 and 1.6, but not as complete as that of Paper 1.1. The discussion of radar components is very brief (cf. Paper 1.6 and Chapter Three). The technology advances required are essentially consistent with those of Figure 5 in Paper 1.4.

REFERENCES FOR CHAPTER 1

1. "IEEE Standard Letter Designations for Radar Bands", *IEEE Standard* 521-1976, Nov. 30, 1976.

2. "IEEE Radar-Frequency Designations," *Microwave Journal*, Vol. 20, No. 6, June 1977, p. 68.

3. D.K. Barton, "Real-World Radar Technology," *IEEE Int. Radar Conf. Record*, April 1975, pp. 1-22.

4. D.K. Barton, "Radar Technology for the 1980's", *Microwave Journal*, Vol. 21, No. 11, Nov. 1978, pp. 81ff.

5. T.E. Walsh, "Military Radar Systems: History, Current Position and Future Forecast," *Microwave Journal*, Vol. 21, No. 11, Nov. 1978, pp. 87ff.

6. P.J. Bloom, J.D. Kuhn, and J.W. O'Grady, "ASDE-3 — A New Airport Surface Detection Equipment Surveillance Radar," *IEEE ELECTRO-78 Rec*, pt 34, pp. 34/4 - 1 thru 13.

7. R.D. Tompkins, "Report on World Administrative Radio Conference," *IEEE Int. Radar Conf.*, April 1980.

8. M.I. Skolnik, "Millimeter and Submillimeter Wave Applications," *Proc. Symp. on Submillimeter Waves*, Polytechnic Press of the Polytechnic Inst. of Brooklyn, New York, 1970, pp. 9-26.

9. M.I. Skolnik, "Millimeter and Submillimeter Applications," *NRL Memo Rpt. 2159*, Aug. 1970, DDC AD 712055.

10. S.L. Johnston, "Millimeter Radar," *Microwave Journal*, Vol. 20, No. 11, Nov. 1977, pp. 16ff.

11. S.L. Johnston, "Radar Systems for Operation at Short Millimetric Wavelengths," *The Radio and Electronic Engr.*, Vol. 49, No. 7/8, July/Aug. 79, pp. 361-9.

12. L. Klaver, "Combined X/K_a-band Tracking Radar," *Proc. Military Microwaves Conf.*, London, Oct. 1978, pp. 147-56. Reprinted in S.L. Johnston, *Radar Electronic Counter-Countermeasures*, Dedham, Mass.: Artech House, 1979.

13. S.L. Johnston, "Comments on 'Millimetre Waves — the Much Awaited Technological Breakthrough?'", *International Defense Review*, Vol. 13, No. 1, 1980, p. 133.

BIBLIOGRAPHY TOPICS FOR CHAPTER 1

General Millimeter Radar

SUPPLEMENTAL BIBLIOGRAPHY FOR CHAPTER 1

General Millimeter Radar

1. Bibliographies

   Barton, D.K., *International Cumulative Index on Radar Systems*, IEEE Pub. No. JH-4675-5, 1978. (Papers on applications of millimeter waves to radar are entered under various subject categories; e.g., instrumentation, tracking radar, weather radar. Radar operating frequency often is not included in the title of the paper.) Selected entries are given below.

2. Journal Articles

   Blore, W.E., Robillard, P.E., and Primich, R. I., "35 and 70 Gc Phase-Locked CW Balanced-Bridge Model Measurement Radars," *Microwave Journal* Vol. 7, No. 9, Sept. 1964, pp. 61-5.

   Musal, H.M. *et al.*, "Millimeter Radar Instrumentation," *IEEE Trans.* Vol. AES-1 No. 3, Dec. 1965, pp. 225-34.

   Haroules, G.C., and Brown, W.E., "A 60 GHz Multi-Frequency Radiometric Sensor for Detecting Clear Air Turbulence in the Troposphere," *IEEE Trans.* Vol. AES-5 No. 5, Sept. 1969, pp. 712-23.

   Millimeter Waves Technical Conference, Naval Electronics Laboratory Center, San Diego, Ca., March 1974, Rpts. NELC TD-308 Vols. 1 and 2, DDC AD A009512 and A020273. (Contains papers on mm radars, components, and propagation.)

   "mm Wave Technology: New Criteria or Comeback," *Countermeasures Magazine* May 1977, p. 28.

   Johnston, S.L., "A Survey of Millimeter-Submillimeter Radar," *Fourth International Conference on Infrared and Millimeter Waves and Their Applications*, IEEE, Miami Beach, Fl., Dec. 1979.

   Williams, D.T. and Boykin, W.H., Jr., "Millimeter-Wave Missile Seeker Aimpoint Wander Phenomenon," *Jnl. of Guidance and Control*, Vol. 2, May-June 1979, pp. 196-203.

   Howard, D.D., "Comments on 'Millimeter-Wave Missile Seeker Aimpoint Wander'" (with author's reply), *Jnl. of Guidance and Control*, Vol. 3, No. 3, May-June 1980, pp. 284-86. (Points out extensive errors in the paper and cites prior work of others.)

Some Aspects of Millimeter Radar

S.L. Johnston

Proceedings International Conference on Radar, Paris, France, Dec. 4-8, 1978
Pages 148-159
Reprinted by permission.

Ce papier donne un aperçu des radars des bandes K_a et millimétrique, y compris l'asservissement des missiles. Les sujets traités comprennent les fréquences du radar, les raisons pour l'emploi du radar millimétrique et les applications possibles, les applications passées et présentes des radars K_a et mm, les applications futures des radars K_a et mm, et les points à discuter qui affectent l'emploi des radars mm. De nombreuses photos des radars K_a et mm sont présentées avec une longue bibliographie des systèmes et de la technologie du radar mm. Les radars américans et étrangers sont tous les deux compris.

Les radars discutés dans ce papier sont d'une classe que D. K. Barton n'a pas traitée dans son papier pre'dominant du Radar-75, <<Real World Radar Technology>>(<< La technologie pratique du radar>>). Les radars de cette classe ont les caractéristiques communes suivantes: en général, on les emploie pour les cibles autres que les aions aéronefs; ils fonctionnet aux fréquences plus élevées que les radars discutés par Barton; ils ont des systèmes d'asservissement; et dans beaucoup de cas, ils n'ont pas été Fabriquès en plus de deux exemplaires.

Bien qu'on ait Fabriqué de nombreux radars de la bande K_a pour plusieurs applications, et qu'il y ait actuellement un programme modéré de recherche et de dévelopement du mm (surtout à 95 GHz aux États-Unis), on conclut que la production de radars millimétriques n'est pas encore assurée.

This paper is a survey of K_a band and millimeter radars including missile guidance. Topics treated include radar frequencies, why millimeter radars, possible mm applications, past and present K_a and mm radar applications, future K_a and mm radar applications, and issues affecting utilization of mm radars. Numerous photographs of K_a and mm radars are presented with an extensive bibliography of both mm radar systems and technology. Both US and foreign radars are covered.

The radars in this paper are a class of radars not treated by D. K. Barton in his Radar-75 keynote paper "Real World Radar Technology". Radars in this class have the following common characteristics: are generally used for target other than aircraft, operate at higher frequencies, are often not "pure" radars, and in many cases have not been produced in more than two systems.

It is concluded that while a number of K_a band radars have been produced for several applications, and while there is a moderate research and development program today, primarily at 95 GH_z in the USA, production of mm radars is not yet assured.

INTRODUCTION

At Radar-75, D. K. Barton [1] gave an excellent review of radar systems manufactured in "production" in the USA since World War II. Study of his paper indicates some very interesting aspects which he did not mention:

a. With some notable exceptions, most of the radars he described were intended for detection/tracking of aircraft.

b. With some other exceptions, most of the radars described detected/tracked targets at ranges of say 20 Km or more.

c. No specific mention was made of trends in operating frequencies of those radars. A few radars operated in the UHF region but most of those are no longer operational. Otherwise, they operated in L, S, C, or X-Bands. (Band delegations herein are in accordance with the new IEEE Standard Letter Designations for Radar Bands [2], [3]).

d. Only "pure" radars were considered, i.e., only target detections and tracking radars, not radar guidance systems.

In the same vein, the "Barton Criterion" of manufacture of two or more radars had some negative effects:

a. Some very interesting prototype radars which never entered production because of change of military situation were unfortunate casualties.

b. Some radars which were then in development were filtered out.

c. Some "mediocre" radars were produced but are no longer in operational use.

These comments are not intended as criticism of Barton's very informative work but rather to set the stage for this paper as an extention of his, primarily into a higher frequency domain -K_a band (27-40 GHz) and mm (40-300 GHz) [2], [3]. In passing, it should also be noted that the new IEEE definition of radar - "an electromagnetic means for target detection and location [4] now includes so-called "laser radars", "laser rangefinders", laser aids to microwave radar [5], etc. The IEEE Radar Systems Panel now includes a Laser Radar Committee. Operating frequency of laser radars is usually in terrahertz (THz) with wavelengths in micrometers (um). These should be contrasted to microwave radars, e.g., X-band (9 GHz). An HCN laser radar operated at 0.89 THz (890 GHz) with a wavelength of 337 um for a factor of 100 increase in frequency over X-band.

Although not addressed by Barton, considerable work has been devoted to K_a band radars. A number of these were produced in "moderate" quantities. Admittedly the K_a band radar work has been much less than L, S, C. and X-band radars. Additionally, at present a significant effort is being devoted to the development of mm radars, primarily at 95 GHz. Both K_a and mm radars are generally characterized by the previously noted "exceptions" to Barton's paper - targets other than aircraft, operating ranges of less than 20 Km, operating frequencies as noted, limited employment and/or current developmental status.

Millimeter waves have been proposed for a plethora of applications [6], [7] and have been the subject of countless overviews, survey papers, forecasts, etc. over the past three decades appearing in various professional and trade journals as well as many technical reports and symposia [6-53]. These include a 1977 IEEE EASCON special session on mm radar organized by this author [49], a 1977 mm survey paper by this author [50], a 1978 conference paper by this author [51], several US Advanced Research Projects Agency mm workshops [39, 52] and a very recent overview by Wiltse [53]. Comparison of the "real world of mm radar" today as reported herein later, reveals the over-optimism of a number of the overviews and forecasts cited above.

As will be seen later, frequently both K_a and mm radar applications include one further extension of Barton's paper: Missile guidance which employs electromagnetic radiation for guidance purposes whether active, passive, (e.g., "radiometer") or beamrider is actually a radar in consonance with the new IEEE definition of radar [4]. Indeed, such applications appear to represent a major portion of new K_a and mm "radar" applications. Accordingly, this paper will include such uses.

Because of the previously noted absence of K_a band radars from Barton's paper, they will be included here. In several cases K_a band serves as a bridge between say X-band and mm radars.

Finally, this paper is an extension and updating of two previous mm papers and a conference session of this author [49-51].

Radar Frequencies

Most of the current radars described by Barton operate in L, S, C, or X-bands in accordance with the International Telecommunications Union (ITU) radiolocation assignments as listed in [2] and [3]. Most surface based search radars operate in L or S-bands. Tracking radars generally operate in C or X-bands. The widely used World War II SCR-584 automatic tracking radar operated in S-band although a later modification converted it to X-band operation.

A lesser number of radars operate in $K\mu$ band. Above this frequency, atmospheric attenuation begins to increase significantly with several peaks due to water vapor and oxygen absorption at 22, 60, 118 and 183 GHz as shown in Figure 1 [54]. For convenience, IEEE radar letter bands and ITU radiolocation bands for Region 2 have been added. Rainfall attenuation is additional to the one way attenuation shown in Figure 1. Theoretical rainfall attenuation as a function of rain rate is shown in Figure 2 [55] at various frequencies. As Medhurst indicated [55] experimentally measured rainfall attenuation has exceeded theoretical values.

Radars have been built at K-band (24 GHz) (not specifically addressed in this paper), at 35 GHz (K_a band) and at 70, 95, 140, 220 GHz in the mm band.

Why Millimeter Radar

The microwave region has been used for almost innumerable applications. Principle advantages of millimeter waves over microwaves (not necessarily in order of importance) are as follows:

a. Smaller antenna diameter required for same antenna gain/beamwidth.

b. Increased angular resolution for the same antenna diameter.

c. Increased bandwidth (a bandwidth of 1% at 300 GHz is equal to all frequencies below S-band center frequency (3 GHz)).

d. Increased immunity to friendly interference [Electromagnetic Compatibility (EMC)] and hostile interference [Electromagnetic countermeasures (ECM)] by virtue of smaller beamwidth [55] and relatively low present employment density.

e. Increased immunity to unwanted detection by hostile forces.

f. Increased sensitivity to Doppler velocity efforts. These advantages have differing importance in various applications of millimeter waves to radars. They are not free, however, There is no such thing as a free

Introduction and Overview

lunch. For ground based applications, atmospheric absorption, rain backscatter and rain attenuation may pose severe limitations in some radar applications. As Skolnik [6], [7] pointed out, some of these advantages may present problems themselves. As one example, use of very narrow antenna beamwidths may present problems in target acquisition, e.g. a microwave system may be required for this purpose.

To a lesser degree these same advantages apply to K_a band radar. As will be seen later, it is difficult to choose between K_a band and say 95 GHz for some application. Both K_a band and mm have been and are considered for the same application in some cases.

Possible Millimeter Applications

Skolnik [5], [6] presented an extensive list of possible applications of millimeter waves.* Those are repeated here in Tables 1 and 2. For completeness, nonradar applications are included. Some of the nonradar applications listed are already important to millimeter radar, e.g., model radar cross section measurements. Other applications listed could be important in future millimeter radars. The asterisks in these tables denote those applications in which Skolnik believed that millimeter waves may offer some special or unique attraction as compared with microwave frequencies. It is interesting to view Skolnik's tables and his asterisks eight years after his papers were published. Many of his asterisked items are important in the scientific area, e.g., spectroscopy (especially in the submillimeter region) and plasma diagnostics (presently of great importance to ERDA in the field of energy research). Some are important to space, e.g., space object identification yet only a very few, if any, of those application have met or are likely to meet Barton's production criteria. So far all those meeting Barton's criteria were at the K_a band.

It is noteworthy that presently the largest potential commercial application of millimeter radar - automobile braking - was not directly mentioned by Skolnik.

Some of the millimeter radar applications cited by Skolnik [5], [6] seem to have reached their zenith, e.g., weather radar. Employment of some other millimeter radars has not come yet, but may come in the future.

Past and Present Radar Applications

Approximately 60 references [35, 41, 59-118] dealing with past and present K_a and millimeter radar applications were identified. As expected most of them pertain to K_a band radars. Identification data on most of those radars were very difficult to obtain because many of them were for military application. Unfortunately, the most complete open literature radar bibliography [120] does not index by radar frequency band.

For convenience, these past and present applications will be considered by frequency band. In each frequency band, applications will be subdivided into civilian and military applications. These will generally be presented in chronological order within each group. Dates shown are approximate. Some dates are those of nomenclature assignment. Other dates are those of reports on the radars. It is not known whether each radar was produced in substantial quantity although several K_a band radars were produced in limited quantity.

Photographs of a number of K_a and mm radars are included at the end of this paper.

Past and Present Civilian K_a Band Radars

Table 3 lists past and present civilian K_a band radars with references for most entries. The earliest civilian K_a band application appears to have been the Decca (UK) airport surveillance radar [9, 28]. A US Airport Surface Detection Radar (ASDE) operating at 24 GHz (K-band) was built for military use (AN/FPN-31) and installed at several US civilian airports [57] in the mid 1950's. Because of the greater occurrence of rain in many US metropolitan areas, US ASDE has tended to operate at lower frequencies than in Europe. In fact, the new US ASDE [58] will sacrifice angular resolution and will operate in K_u band (16 GHz) rather than K-band (24 GHz) as the ASDE-1.

The first three applications in Table 3 are very similar. All use antennas with rather high rotation speeds (typically 40 RPM), a PPI (plan position indicator) radar display and pulse lengths of 20 ns thus enabling increased range resolution.

Automobile braking is a fairly new application of K_a band radar and will not be discussed further here due to excellent recent coverage elsewhere [70-72].

NASA Wallops Island is adapting a military K_a band radar to the observation of insects for the Department of Agriculture. [NASA, Wallops Island, 1978 private communication.] That same installation has modified a US Army Nike tracking radar into a K_a band tracker for aircraft ground instrumentation. Westinghouse applied their military AN/APQ-97 side looking radar for earth resource exploration [67]. One other civilian application is noteworthy: airborne oil slick detection by the US Coast Guard using a K_a band radiometer - passive radar - in conjunction with an active X-band radar and ultraviolet sensors [69]. A German K_a band radar for safe distance monitoring in industrial applications such as crane operations is also interesting [73].

Past and Present Military K_a Band Radars

Tables 4a and 4b list representative past and present military K_a band airborne and surface radars. Military K_a band radar applications form a contrast with civilian applications. Most of the civilian K_a band applications were developed in countries other than the US. The opposite is true of military K_a band radars. Only two airborne and one ground based non-US military radar developers were identified. K_a band provides increased Doppler resolution over X-band. This permits improved azimuth resolution in synthetic aperture radars [67] and in ground target detection. The airborne AN/APQ-137 has detected a man walking along the ground at velocity of 2 mph [76, 77]. That performance was also enabled by the smaller illumination volume at K_a band which improves the signal=to=clutter ratio.

Recent airborne military applications include missile and bomb guidance. The later results from weather-limited performance of laser guided systems [85]. Dual band (X/K_a) has been used in several military radars, e.g., AN/APN-161, and AN/APQ-122 [78], TRAKX [91], and FLYCATCHER [92]. This permits operation in rain; in tracking radars, it facilitates target acquisition by the K_a band narrower beam.

The LACROSSE tracker [90] represents the first known application of K_a band to ground based missile tracking and guidance. Many of the civilian and military K_a band applications were experimental or were built in small numbers. Quantity production was achieved for the LACROSSE tracker, the AN/SPN-10, -42/GSN-5 family, the AN/TPQ-11 [88][89] and the AN/APD-7 [75].

The Westinghouse Splash Detection Radar [93] is noteworthy in that it was used to detect the water splash ensuing from an intercontinental ballistic missile (ICBM) impact and thus accurately locate the impact point. The Westinghouse WX-50 airborne radar was a private development for military applications [79] and [80].

Past and Present mm Radar Applications

Representative past and present civilian mm band radar applications are shown in Table 5. In the harbor speed monitoring radar, use was made of the increased Doppler sensitivity. For this application, it was required to determine whether ships moving in the harbor were complying with the speed limit of 5 miles per hour. This low velocity was readily detected with a 95 GHz radar.

Table 6 lists representative 70 GHz mm military radar applications while Table 7 lists representative military 95 and 140 GHz mm radar applications. The VEH radar is noteworthy due to its use of a geodesic scanner which employed a mechanical rotating feed movement of 7 feed sections. The 0.5 degree antenna beamwidth covered a 45 sector, typically at a rate of about 20 degrees per second. It used a micro-B display.

Comparison of Tables 5-7 with Tables 3 and 4 offers several contrasts:

 a. There is only one non-US mm entry.

 b. There are fewer civilian mm applications.

 c. There is no known mm radar production although several K_a band radars were produced in "modest" quantities.

 d. The mm applications operate at shorter ranges than the K_a band radars except for the space object identification.

 e. Most of the 95 GHz applications are fairly recent. Most of the 70 GHz applications have now been superceded by the 95 GHz systems.

 f. There is a greater proportion of land applications at 95 GHz.

There is a greater diversity of military mm radar applications e.g., submarine radars, arctic vehicles, and remotely piloted vehicles (RPV).

In one unique mm application, EMI limited (UK) measures both target radar cross section amplitude and glint at 140 GHz [119]. The RF source is an extended interaction oscillator (EIO) [120].

Applications which employ operation at short ranges (a few kilometers) suffer less from rain attenuation and permit use of lower power, enabling all solid state radars [96][121, 122] using IMPATTS or GUNN devices. For applications where average power a watt or more is required, the previously cited EIO is available. The gyroklystron [123, 124] [140] appears to offer potential for tens of kilowatts of average power in the mm region and could greatly extend the range of mm ground based aircraft tracking radars. Jory's Curves of gyrotron CW power versus frequency [140] are very impressive.

Fortunately, some missile guidance applications do not require long ranges and hence do not require large RF power.

Future Millimeter Radar Systems

Godlove [123], Jones [125], Dybdal [126], Green [86], Scheer [95], Tresselt [70, 71], described or implied "coming" millimeter radars at the recent EASCON millimeter radar session organized by the author. Only the presentations of Green and Tresselt appear to have an employment density to meet the Barton production criterion.

Millimeter radiometric seekers similar in concept to those of Green are also under study by other organizations such as the US Army Armament Research and Development Command [82] and the US Army Ballistic Research Laboratories [127].

In both applications, the number of millimeter RF devices per system will probably be small but the total number of systems may be large. Each application is short range and will use small low power solid-state RF sources. The systems of Green may operate at K_a band or in the millimeter region or both.

Some of the other suggested applications of Skolnik in Table 1 are still possible candidates. Studies on their development are currently under way under the sponsorship of several military organizations. There are several forces for these activities: military concern over operation in smoke and dust environments together with recent and projected improvements in millimeter RF sources and components.

At least two technology forecasts relating to millimeter waves are known to the author. One presently under way is by the IEEE Radar Systems Panel (only one item is millimeter waves).

In a recent millimeter wave technology forecast (unpublished) it was the concensus that technology would permit beginning military system engineering development of several kinds of millimeter systems in another five years.

Issues Affecting Utilization of Millimeter Radars

As indicated previously, there are no millimeter radars in production and only a few K_a band radars in use today. There is a fairly sizable millimeter research and development effort now. This includes construction and testing of experimental hardware, system studies, etc. A number of issues must be resolved before millimeter radars can enter production. Major issues will now be discussed, but not necessarily in order of importance. Some of these issues are of greater concern in some potential applications than others.

 a. Component Capability

 Millimeter component capability and availability has long been of concern. Numerous millimeter component surveys have been made [11, 14, 38, 96, 121-123, 128-135]. Most of the millimeter component surveys are primarily for nonradar applications, e.g., communications, especially satellite communications. Accordingly, many radar requirements are not represented. Present component capability is sufficient for a few K_a band radar applications, e.g., the automobile braking radar [70-72]. Godlove [123] pointed out that the coming of very high millimeter RF power sources requires components capable of handling those power levels. The Georgia Tech K_a band instrumentation radars [95, 96] employs some very sophisticated radar techniques. These radars are intended as diagnostic tools especially in target and clutter signatures and signal processing techniques. It is only logical that some sophisticated signal processing techniques utilized in these radars may be found suitable for use in actual millimeter radars. Such techniques will probably require use of polarization/frequency agility as now incorporated into the Georgia Tech radars. Unfortunately, suitable components are only just now being developed to permit building such a radar at 95 GHz [136]. This makes taking those special signatures at 95 GHz difficult now and would make difficult utilization of those processing techniques at 95 GHz in radars. This does not say that "simple" 95 GHz systems are not possible.

Most millimeter component work has been done at 70 and 95 GHz. Component availability at 140 GHz, especially RF sources, is far more limited at 140 GHz. In turn, component availability at 220 GHz is even further limited.

 b. Radar Performance

 As Green [86] and Scheer [95] indicated, there is presently insufficient information available on millimeter radar performance in clutter, rain, fog, haze, etc. Several sophisticated signal processing techniques have been proposed. While preliminary data are encouraging, further work is necessary. It is believed that work presently planned will provide the necessary answers.

c. System Integration

Nearly all of the millimeter radar work today has been confined to the radar or missile guidance unit itself. Very little effort has been devoted in the Army to consideration of incorporating these into a complete missile (weapon) system. The complete weapon system must include means for target acquisition, identification, guidance track initiation, etc., as well as a payload. Fundamental to this is the creation of a missile system concept. Such studies are now in process in several military organizations.

d. Design Tradeoffs

An essential part of system integration is the design tradeoff process, e.g., comparison of capture and tracking antenna beamwidths. Another tradeoff is guidance accuracy and warhead kill radius. Warhead kill radius is a function of warhead size (weight). Warhead weight affects missile size and weight. Millimeter guidance accuracy has not yet been fully determined. Another tradeoff is selection of the optimum radar frequency, e.g., K_a or 95, 140, etc. These design tradeoffs are now in an early phase. In the previously cited US ASDE-3, Bloom [58] compared radar operation at 16 and 24 GHz (Ku and K bands) and showed operation at 16 GHz to be superior.

e. Investigation of Novel Guidance Schemes

Employment of radar against ground/low altitude targets requires tracking targets at very low elevation angles. Several guidance schemes which could reduce the effect of elevation tracking errors on missile miss distance have been proposed. These must be carefully evaluated.

f. Comparison With Other Approaches

The complete millimeter weapon system must be compared with alternate weapon systems, e.g., a laser guided missile. In simple terms, is the added performance of a millimeter system worth its cost differential? Skolnik [6, 7] stated that for most applications, a microwave system could do that job better than a millimeter system. For example: Would a microwave missile guidance system which uses a guidance technique that does not require precise elevation tracking be cheaper or better than a millimeter system?

For another example, Skolnik [7] presented the design of a 95 GHz search radar. Unfortunately, he did not compare its performance and cost with a comparable microwave search radar. Skolnik [7] also mentioned the necessity for economic analyses of millimeter systems.

In a nonradar millimeter application known to the author, a millimeter system was designed for a certain application as a replacement/supplement to an existing microwave system. Subsequent investigation showed that the capability of the microwave system could be significantly improved by the utilization of certain low frequency techniques. The addition was much less expensive than the introduction of a new millimeter system, making the millimeter system unnecessary now.

In this regard, it must be noted that seven years after his initial question of value of millimeter waves for most applications, Skolnik today [48] feels that millimeters may not be the best approach. Events have not yet disproved him. Millimeter missile guidance and auto-mobile braking radar could well be very notable exceptions.

g. Lack of Available Millimeter Radar Frequency Allocation

Initial consideration of the new IEEE Standard letter band designations for radar frequency bands [3, 4] causes one to puzzle over why it lumped together the entire region of 40 to 300 GHz without regard to the windows at 70, 95, 140, and 220 GHz. Careful study of Table I of the Standard reveals the reason: At present there is no ITU frequency allocation for radiolocation (radar/missile guidance) above K_a band. Currently in the continental US it is possible to obtain permission to operate a production system at 95 GHz (for example) on a noninterference basis only. Testing of a US 95 GHz system in Europe (for example) might not be approved by that country if it chooses not to allow operation within its borders.

Experimental operations (radiation) at a contractors plant or at a military test range, e.g., to collect clutter measurements requires a frequency assignment. A frequency assignment must be in accordance with approved allocations.

Requirements for a frequency allocation for production systems are prescribed in the Office of Telecommunications Policy Manual of Regulations and Procedures for Radio Frequency Management.

As many people know, there is a WARC meeting scheduled for 1979. Several radiolocation frequency allocations in the 40 to 300 GHz region are being considered; however, the US position has not yet been established. Even if the US position proposes frequency allocations for radiolocation in the millimeter windows, adoption by the WARC cannot be assumed. Codding [137] recently discussed the US and the ITU.

For millimeter radar proponents, this, if unchanged, could be a case of "all dressed up, but nowhere to go". The importance of frequency allocations should not be confined to millimeter radar proponents. Frequency allocations are of paramount importance to all users of the electromagnetic spectrum whether they be radio, microwave radar, or laser radar specialists.

These technical issues may be illustrated in summary by the country character who was observed dragging a rope through a small town. When asked what he was doing, he replied that he had either found a rope or lost a horse! For most millimeter radar applications today, that is the situation - the jury is still out. Fortunately, there are some very favorable signs. Work which is now under way should provide very useful answers in determining QUO VADIS.

ACKNOWLEDGEMENT

The author gratefully acknowledges many helpful discussions with his colleagues both in government and industry.

DISCLAIMER

Views expressed herein are those of the author and not necessarily those of the Department of the Army or the Department of Defense.

REFERENCES

1. D. K. Barton, "Real-World Radar Technology," *IEEE 1975 International Radar Conference Record*, April 1975, pp. 1-22.

2. "IEEE Standard Letter Designations for Radar Bands,"*IEEE Standard*, 521-1976, November 30, 1976.

3. "IEEE Radar-Frequency Designations," *Microwave Journal*, Vol. 20, No. 6, June 1977, p. 68.

4. "IEEE Standard Radar Definitions," *IEEE Standard*, 686-1977, November 1977.

5. A. V. Jelalian, et. al., OADARS (Optical Aids to Detection and Ranging Systems), *IEEE Electro 1976*, Rec, pt. 30 pp. 30-3-1 to 5.

6. M. I. Skolnik, "Millimeter and Submillimeter Wave Applications," *Proc. Symp on Submillimeter Waves Polytechnic Press of the Polytechnic Inst. of Brooklyn, New York*, 1970, pp. 9-26.

7. M. I. Skolnik, *Millimeter and Submillimeter Wave Applications*, NRL Memorandum Report 2159, 12 August 1970, AD 712055.

8. J. R. Pierce, "Millimeter Waves," *Physics Today*, Vol. 8, November 1950, p. 24.

9. J. R. Christopher, "Advances in Millimetric Radar," *British Comm and Electronics*, Vol. 2, September 1955, p. 48.

10. R. G. Fellers, "Millimeter Waves and Their Applications," *Electr. Eng.* Vol. 75, October 1956, p. 914.

11. P. D. Coleman and R. C. Becker, "Present State of the Millimeter Wave Generation and Technique Art - 1958," *IRE Trans.* Vol. MTT-7, January 1959, p. 42-61.

12. I. P. French, et. al., *The Radio Spectrum Above 10 Kmc/s*, RCA Victor Co., LTD Research Laboratories, Montreal, Report No. 7-400, 1 July 1959, AD228167.

13. A. W. Straiton and C. W. Tolbert, "Millimeters - The New Radio Frontier," *Microwave Journal*, Vol. 3, No. 1, January 1960, pp. 37-39.

14. J. Lurye, Survey of the Literature on *Millimeter and Submillimeter Waves*, TRG-127-SR-2, TRG Inc., Report, 30 June 1960, AD243242.

15. W. Culshauv, "Millimeter Wave Techniques," *Advances in Electronics and Electron Physics*, Vol. 15, 1961, pp. 197-263.

16. F. G. R. Warren, et. al., *The Radio Spectrum From 10 Gc to 300 Gc in Aerospace Communications*, Technical Report ASD-TR-61-589, (7 vols), RCA Victor Co., LTD Research Laboratories, Montreal, 1962 (AD 276274, 294340, 294341, 294452, 336171).

17. D. D. King, "Long Waves or Short-Perspectives on the Millimeter Region," *Spectrum*, Vol. 2, May 1963, pp. 64-68.

18. J. W. Meyer, *Millimeter Wave Research, Past, Present, and Future*, Report ESD-TDR 65-72, (MIT LL TR-389), AD 620727, May 1965.

19. J. W. Dees, V. E. Derr, J. J. Gallagher, and J. C. Wiltse, "Beyond Microwaves, *International Science and Technology*, No. 47, November 1965, pp. 50-56.

20. M. Cohn and R. S. Littlepage, *Implications of Millimeter Wave Research and Technology on Naval Problems*, Advanced Technology Corporation, Contract No. N00014-66-C0166, Final Report, July 1967, AD 819737, 813462.

21. Frank Brand, "Millimeter Waves? You Decide," *Microwave Journal*, Vol. 10, No. 1, November 1967, p. 10.

22. Ted Saad, "Millimeter Waves," *Microwave Journal*, Vol. 10, No. 12, November 1967, p. 14.

23. Ben Lax, "Millimeter - The Step Child of Technology," *Microwave Journal*, Vol. 10, No. 12, November 1967, pp. 16-18.

24. Bill Bazzy, "Market Outlook - A Time for Being - Millimeter Waves," *Microwave Journal*, Vol. 10, No. 12, November 1967, pp. 20-22.

25. D. D. King, "Millimeter Wave Prospectus," *Microwave Journal*, Vol. 10, No. 12, November 1967, pp. 24-29.

26. L. C. Tillotson, Theme Editorial: "Millimeter Waves - Can They Provide Relief From Spectrum Crowding?" *Microwave Journal*, Vol. 10, No. 12, November 1967, p. 32.

27. A. J. Simmons, Theme Article: "Millimeter Wave Components for Use in Systems," *Microwave Journal*, Vol. 10, No. 12, November 1967, pp. 102-107.

28. Staff Report - "Millimeter and Submillimeter Investigation in the United Kingdom," *Microwave Journal*, Vol. 10, No. 12, November 1967, p. 112-117.

29. Marvin Cohen, Business Editorial - "Millimeter Waves: An Appraisal of Their Status," *Microwave Journal*, Vol. 10, No. 12, November 1967, p. 128.

30. Frank Brand, "Millimeter Waves? Yes - But When?," *Microwave Journal*, Vol. 11, No. 11, November 1968, pp. 20-22.

31. Bill Peyser, "Millimeter Waves on the Threshold," *Microwave Journal*, Vol. 11, No. 11, November 1968, p. 128.

32. E. E. Altshuler, "New Applications at Millimeter Wavelengths," *Microwave Journal*, Vol. No. 11, November 1968, pp. 38-42.

33. W. O. Copeland, J. R. Ashwell, G. P. Keflas, and J. C. Wiltse, "Millimeter-Wave Systems Applications," *1969 IEEE GMTT International Microwave Symposium*, Dallas, Texas, May 5-7, 1969, pp. 485-488.

34. M. J. Foral, "Millimeter Wave Radar Investigations," *1969 IEEE GMTT International Microwave Symposium*, Dallas, Texas, May 5-7, 1969, pp. 489-492.

35. M. Lazarus, "Millimetric Microwave Activities in the United Kingdom," *Microwave Journal*, Vol. 12, No. 11, November 1969, pp. 28ff.

36. J. W. Dees and J. C. Wiltse, "An Overview of Millimeter Wave Systems," *Microwave Journal*, Vol. 12, No. 11, November 1969, pp. 42-49.

37. Proceedings of the 1974 Millimeter Waves Techniques Conference, 26-28 March 1974, Naval Electronics Laboratory Center.

38. H. B. Gershman, *A Review of Millimeter Wave Technology*, ESD-TR-73-354, April 1974.

39. Report of the ARPA/Tri Service Millimeter Wave Workshop JHU/APL Report QM 75-009, January 1975.

40. William D. Stuart, "Status of Millimeter Wave Research," Report of the ARPA/Tri-Service Millimeter Wave Workshop APL/JHU QM 75-009, January 1975, pp. 279-309.

41. V. A. Tarulis, "Recommendations from the Systems and Applications Panel," Report of the ARPA/Tri-Service Millimeter Wave Workshop, APL/JHU QM 75-009, January 1975, pp. 271-275.

42. N. B. Kramer, "State of Millimeter Waves," *Microwave Journal*, Vol. 18, No. 12, December 1975, p. 12.

43. C. E. White and E. E. Altshuler, "Millimeter-Wave Progress 1972-75," *Microwave Journal*, Vol. 18, No. 12, December 1975, p. 14 ff.

44. G. S. Kariotis, "Millimeters and Manana," *Microwave Journal*, Vol. 19, No. 11, November 1976, pp. 15-16.

45. C. G. Thornton, "Millimeter Technology - The Future Looks Promising," *Microwave Journal*, Vol. 20, No. 3, March 1977, pp. 14-16.

46. J. D. Montgomery, "Millimeter Market Forecast," *Microwave Journal*, Vol. 20, No. 3, March 1977, p. 30.

47. E. E. Lee and B. S. Franklin, "Millimeter Wave Technology: New Criteria or Comeback?" *Electronic, Electro-Optic and Infrared Countermeasures Magazine*, Vol. 3, No. 5, May 1977, p. 28 ff.

48. M. I. Skolnik, "Millimeter Radar - Introduction to the Session," *EASCON-77 Proc.*

49. *Millimeter Radar, Session No. 16, IEEE EASCON-77*.

50. S. L. Johnston, "Millimeter Radar, *Microwave Journal*, Vol. 20, No. 11, November 1977, p. 16ff.

51. S. L. Johnston, "Radar Applications of Millimeter Waves, Proc," SOUTHEASTCON 78 IEEE Reg III, Conference, April 10-12, 1978, pp. 409-412.

52. Sixth ARPA/Tri-Service Millimeter Wave Conference, 29-30 November 1977, Harry Diamond Laboratories, Adelphi, Maryland.

53. J. C. Wiltse, Jr., "Millimeter Waves - They're Alive and Healthy," *Microwave Journal*, Vol. 21, No. 8, August 1978, p. 16ff.

54. E. E. Altshuler et. al., "Atmospheric Effects on Propagation at Millimeter Wavelengths, *IEEE Spectrum*, Vol. 5, No. 7, July 1968, pp. 83-90.

55. R. G. Medhurst, "Rainfall Attenuation of Centimeter Waves: Comparison of Theory and Measurement," *IEEE Trans.*, Vol. AP-13, July 1965, pp. 550-564.

56. S. L. Johnston, "Radar Electronic Counter-Countermeasure Techniques, *IEEE Trans.* Aerospace and Electron Systems, Vol. AES-14, No. 1, January 1978, pp. 109-117.

57. M. I. Skolnik, *Introduction to Radar Systems*, McGraw-Hill, NY, 1962, pp. 579-581.

58. P. J. Bloom, J. E. Kuhn and J. W. O'Grady, "ASD-3 — A New Airport Surface Detection Equipment Surveillance Radar, *IEEE Electro-78*, Conference, Proc. pp. 34 4-1 to 34 4-4.

59. Seppen, J. M. G. and Verstraten, J., "An 8 mm High Definition Radar Set," *Philips Telecommunication Review*, Vol. 20, No. 1, September 1958, pp. 5-15.

60. Gawron, T. and Jankowski, B., "The High Resolution Radar TOR," *Prace Przemslowego Instutu Telekomunikocji*, No. 57, 1967, pp. 19-25, in Polish (Translation: FTD-ID(RS)1-1116-76, AD A039199).

61. Rzeszowski, M., "(TOR) The New Polish Radar," *Skrzydlata Polska*, No. 6, 1964, p. 5, in Polish (Translation: FTD-TT-1163/1+2+3+4, AD 475267).

62. "The All Seeing Window at the Airport," *Grazhdanskaya Aviatsiya*, No. 4, 1966, pp. 24-25, in Russian.

63. G. Mickam, "Q-Band River Radar," *Nachrichtentechnik*, Vol. 18, No. 4, April 1968, pp. 151-153 (in German).

64. D. Atlas, "Activities in Radar Meteorology, Cloud Pyysics and Weather Modification in the Soviet Union," *Bull American Meteorological Society*, Vol. 11, June 1965, pp. 696-706.

Introduction and Overview

65. M. S. Ried, "Millimeter Wave Pseudorandom Coded Meteorological Radar," *IEEE Trans GE-7*, No. 3, July 1969, pp. 146-156.

66. J. J. G. McCue and E. A. Crocker, "A Millimeter-Wave Lunar Radar," *Microwave Journal*, Vol. 11, No. 11, November 1968, pp. 59-63.

67. R. E. Powers, "Side-Look Radar Provides a New Tool for Topographic and Geological Surveys," *Westinghouse Engineer*, November 1972.

68. "Radar Rides the Rails," *Microwave System News*, October 1977, p. 34.

69. A. T. Edgerton and J. J. Bommarito, *Development of a Prototype Airborne Oil Surveillance System*, US Coast Guard, May 1975, Report No. CG-D-90-75, DDC AD A011275.

70. C. Tresselt and Y. K. Wu, "Millimeter Radar for Highway Collision Avoidance," *Microwave Journal*, Vol. 20, No. 11, November 1977, p. 39ff.

71. G. Neininger, "An FM/CW Radar with High Resolution in Range and Doppler; Application for Anti-Collision Radar for Vehicles," *Radar 77*, IEE Conference Publication, No. 155, pp. 526-530, 25-28 October 1977.

73. AEG Telefunken Brochure, 1978.

74. "Radars for Air Traffic Control," *Interavia*, 11/1970, pp. 1393-1395.

75. "Radars - 35 Years of Airborne Radar Production," Westinghouse Defense and Electronic Systems Center Brochure.

76. "MOTARDES, Moving Target Detection System," Emerson Electric Company Brochure (undated).

77. "Flight Test Report for Moving Target Detection System (MOTARDES)," Emerson Electric Report 2103, 19 October 1966, DDC AD 801549.

78. R. T. Pretty, ed., *Jane's Weapon Systems*, Jane's Yearbooks, London, 1977.

79. "The WX-50 Close Support Radar," International Defense Review, No. 2, 1976.

80. "Compact Radar Adds Punch to Attack Aircraft," *Electronic Warfare* Magazine, Vol. 9, No. 1, January/February 1977, pp. 61-62.

81. J. E. Adair, "Millimeter Wave Radiometry," *Microwave Journal*, Vol. 20, No. 11, November 1977, p. 32ff.

82. Richard Davis, "Millimeter Waves: A Solution That's Finally Found a Problem," *Microwave System News*, Vol. 7, No. 9, September 1977, pp.63-67.

83. "Filter Center," *Aviation Week and Space Technology*, 20 June 1977, p. 65.

84. J. B. Winderman and G. N. Hulderman, "Solid-State Millimeter-Wave Pulse Compression Radar Sensor," *Microwave Journal*, Vol. 20, No. 11, November 1977, p. 45ff.

85. Joseph D. Harrop, Ronald Stump and John J. Teti, Jr., *Surface Navy Applications of Millimeter Wave Sensors*, Vol. 1, Design Trade-Off Study, November 1977, NSWC/DL TR-3748.

86. A. H. Green, Jr., "Applications of Millimeter Wave Technology to Army Missile Systems," *EASCON-77 Rec*.

87. R. J. Donaldson, Jr., "The Measurement of Cloud Liquid-Water Content by Radar," *J. Amer. Meteor.*, Vol. 12, 1955, pp. 238-244.

88. *Preliminary Operational Application Techniques for AN/TPQ-11*, Technical Report 180, HQ USAF Air Weather Service, 1964.

89. D. K. Speed, *General Application of Meteorological Radar Sets*, Hq USAF Air Weather Service Technical Report 184, April 1965, AD 466187.

90. "Millimeter Microwave Technology," Martin Marietta Company Brochure, June 1976, OR 11,566B.

91. D. Cross, et. al., "TRAKX: A Dual-Frequency Tracking Radar," *Microwave Journal*, Vol. 19, No. 9, September 1976, pp. 39-41.

92. R. T. Hill, "Trends in European Radar Technology," *Electro Rec Pt 25*, 1976, pp. 25-1-1 — 25-1-10.

93. "Splash Detection Radar," Westinghouse Defense and Space Center Surface Division Brochure (undated).

94. F. B. Dyer et. al., "Review of Millimeter Wave Radar Development at Georgia Tech," Radar and Instrumentation Laboratory Engineering Experiment Station Georgia Institute of Technology, Internal Technical Report 77-01, May 1977.

95. J. A. Scheer, J. L. Eaves, and N. C. Currie, "Modern Millimeter Instrumentation Radar Development and Research Methodology," *EASCON-77 Proc*.

96. N. C. Currie, J. A. Scheer and W. A. Holm, "Millimeter-Wave Instrumentation Radar Systems," *Microwave Journal*, Vol. 21, No. 8, August 1978, p. 35ff.

97. Aumiller, B., "An Airborne 90-GHz Radiometry Receiver," November 1973, DFVLR, DLR-MITT, 74-05, in German (Translation: NASA N76-11410, ESRO TT-163).

98. F. J. Bernues, et. al., "A Solid State 94 GHz Doppler Radar," *1975 IEEE-MTT-S International Microwave Symposium Proc*., pp. 258-260.

99. Field Test of an Experimental 94-GHz Doppler Radar, US Coast Guard Research and Development Center Report, October 1975.

100. M. W. Long, W. K. Rivers and J. C. Butterworth, "Combat Surveillance Radar AN/MPS-29 (XE-1)," *Sixth Tri-Service Radar Symposium*, 1960.

101. L. H. Kosowsky, "Millimeter Radar for Landing Applications," *American Helicopter Society Proc.* 26th Annual National Forum, June 1970.

102. C. L. Wilson, "70 GC Pulse Radar," BRL Tech Note 1637, November 1966.

103. H. C. Webb, Sr., "Surveillance and Target Identification Using HDL Millimeter Wave Radar," HDL TM 66-21, December 1966, AD 391523.

104. F. B. Dyer and R. M. Goodman, Jr., "Vehicle Mounted Millimeter Radar," *18th Tri-Service Radar Symposium*, 1972.

105. K. L. Koester, L. H. Kosowsky and J. F. Sparacio, "An Experimental Millimeter Monopulse Track Radar," *1974 Millimeter Waves Techniques Conference*, Naval Electronics Laboratory Center.

106. L. H. Kosowsky, K. Koester and R. Graziano, "Millimeter Wave Monopulse Track Radar," AGARD Paper No. 197, June 1976.

107. L. H. Kosowsky, "A 70-GHz Monopulse Radar for Low Angle Tracking," Report of the ARPA/Tri-Service Millimeter Wave Workshop, APL/JHU QM-75-009, January 1975, pp. 139-146.

108. L. A. Hoffman, et. al., "A 94 GHz Radar for Space Object Identification," *IEEE Trans. MTT-17*, December 1969, pp. 1145-1149.

109. F. W. Schenkel and A. Finkel, "NOTER (Nap of the Earth), 95 GHz Radar Wire Avoidance System," JHU/APL Report S-3-R-029, QM-75-081, July 1975.

110. F. Schenkel, "NoteR (Nap-of-the-Earth Radar), Helicopter Wire Detection System," Report of the ARPA/Tri-Service Millimeter Wave Workshop, APL/JHQ QM-75-009, January 1975, pp. 149-162.

111. J. F. Heney, "A 94-GHz Imaging Radar," Report of the ARPA/Tri-Service Millimeter Wave Workshop, APL/JHU QM-75-009, January 1975, p. 147.

112. F. P. Wilcox, "Development and Test of a 95 GHz Terrain Imaging Radar," *1974 Millimeter Waves Techniques Conference*, Naval Electronics Laboratory Center.

113. "Guidance Seeker for Missiles Developed," *Aviation Week and Space Technology*, 5 December 1977, p. 63.

114. Balcerak, Ealy, Martin, Hall, "Advanced IR Sensors and Radars for Target Acquisition," *Proceedings Society Photo Optic Instrumentation Engineering*, Vol. 128, 27-29 September 1977, (Effective Utilization of Radar).

115. "Experimental Millimeter Radar System," *Electronics Magazine*, 1 September 1977.

116. L. H. Kosowsky, et. al., "A Millimeter Wave Radar for RPV's," *Army Aviation Electronics Symposium*, April 1976.

117. J. E. Kammer and K. A. Richer, *140 GHz Millimetric Bistatic Wave Measurements Radar*, BRL Memo Report 1730, January 1966, AD 484693.

118. L. A. Crum and J. R. Stanley, "Recent Developments in Scale Modeling of Radar Reflections by Radar and Solar Methods," *Radar 77*, 25-28 October 1977, IEE Conference Publication, No. 155, pp. 473-477.

119. "EIOs GO Samarium Colbalt," *Microwave Journal*, Vol. 20, No. 11, November 1977, p. 66.

120. D. K. Barton, *International Cumulative Index on Radar Systems*, IEEE Catalog NR JH 4675-5, 1978.

121. F. Bernues and R. Ying, "Recent Advances in Millimeter Radar Technology," *IEEE Electro-78 Conference Rec*, pp. 28/4-1 to 28/4-4, May 1978.

122. N. B. Kromer, "Solid State Technology for Millimeter Waves," *Microwave Journal*, Vol. 21, No. 8, August 1978, pp 57-61.

123. T. F. Godlove, "Prospects for High Power Millimeter Sources and Devices," *EASCON-77 Proc*.

124. H. Jory, S. Hegji, J. Shively and R. Symons, "Gyrotron Developments," *Microwave Journal*, Vol. 21, No. 8, August 1978, p. 30ff.

125. G. Jones, "Application of Millimeter Wave Radar to Ballistic Missile Defense," *EASCON-77 Rec*.

126. R. B. Dybdal, "Millimeter Radar Application to Space Object Identification," *EASCON-77 Rec*.

127. K. A. Richer, "Near Earth Millimeter Wave Radar and Radiometry," *1969 IEEE GMTT International Microwave Symposium*, Dallas, Texas, May 5-7, 1969, pp. 470-474.

128. Sessions A-E, *1974 Millimeter Wave Techniques Conference*, 26-28 March 1974, Naval Electronics Laboratory Center (approximately 25 papers on millimeter components).

129. H. Jacobs, "Millimeter-Wave Integrated Circuits," *Report of the ARPA/Tri-Service Millimeter Wave Workshop*, APL/JHU QM-75-009, January 1975, pp. 45-57.

130. J. J. Heckman, et. al., "Survey Results and Suggested Research and Development Programs for Solid State Millimeter Wave Technology," Applied Engineering Laboratory, Georgia Institute of Technology, TR-1725-3, April 1976, for Naval Electronic Systems Command.

131. H. J. Kuno and T. T. Fong, "Solid State Millimeter Sources," *EASCON-77 Proc.*

132. D. C. Hogg, "Some New Approaches in Design of Millimeter-Wave Antennas," *EASCON-77 Proc.*

133. J. E. Rowe, "High Performance Millimeter-Wave Components," *EASCON-77 Proc.*

134. B. Spielman, "Millimeter-Wave Integrated Circuit Media," *EASCON-77 Proc.*

135. P. M. Schwartz, et. al., "Advanced Solid-State Components for Millimeter Wave Radars," *1975 IEEE-MTT-S International Symposium Proc.*, pp. 261-263.

136. J. Eaves, Georgia Tech. (Private Communication).

137. G. A. Codding, Jr., "The United States and the ITU in a Changing World," *Telecommunication Journal*, Vol. 44-V, May 1977, pp. 231-235.

138. "European Radar Systems," *Interavia*, 3/1962, p. 334.

139. G. E. Layman, "SEATRACKS, A Millimeter Wave Radar Fire Control System," *IEEE EASCON-78 Rec.*

140. H. R. Jory, et. al., "Gyrotrons for High Power Millimeter Wave Generation," *International Electron Devices Meeting Digest*, pp. 234-237, December 1977.

---

*It should be noted that Skolnik's definition of millimeter waves included the submillimeter region (wavelengths of 1 mm to at least 30 μm).

TABLES

TABLE 1
SUGGESTED RADAR APPLICATIONS

Low-angle tracking*
"Secure" military radar*
Interference-free radar*
Cloud-sensing radar*
High resolution radar*
Imaging radar
Ground mapping
Map matching
Space object identification
Lunar radar astronomy
Target characteristics
Weather radar
Clear-air turbulence sensor

Remote sensing of the environment
Surveillance
Target acquisition
Missile guidance
Navigation
Obstacle detection
Clutter suppression
Fuzes
Harbor surveillance radar
Airport surface detection radar
Landing aids
Air traffic control beacons
Jet engine exhaust and cannon blast

* Applications in which Skolnik believed that millimeter waves offer more advantage than microwave frequencies.

TABLE 2
OTHER SUGGESTED APPLICATIONS

SUGGESTED COMMUNICATIONS APPLICATIONS

"Secure" military*
Point-to-point extremely wide-band*
Spacecraft communications during blackout*
Interference-free*
Satellite-to-satellite
Intersatellite relays
Earth-to-space
Retroreflector

SUGGESTED RADIOMETRY APPLICATIONS

Remote sensing of the environment*
Radio astronomy
Radio sextant
Ship detection
Ground target detection
Missile detection
Missile guidance
Clear air turbulence sensor

SUGGESTED INSTRUMENTATION APPLICATIONS

Plasma diagnostics*
Rocket exhaust plume measurements*
Spectroscopy*
Remote vibration sensor
Prediction of blast focusing
Model radar cross measurements
Classroom demonstrations of optics

* Applications in which Skolnik believed that millimeter waves offer more advantages than microwave frequencies.

TABLE 3
REPRESENTATIVE CIVILIAN K_a BAND RADAR APPLICATIONS

APPLICATION	IDENTIFICATION	COUNTRY/SOURCE	YEAR	REFERENCES
Airport Surveillance	Mark I	UK Decca	1955	[28]
	8 GR 250	Europe	1958	[59]
	ASDE 40/500/600	Netherlands/Philips	1970	[74]
	ASMI MK5	UK Decca	1970	[74]
	8 GR 260	Netherlands/Philips	1962	[138]
	TOR	Poland	1964 1967	[60, 61]
	RLS-OLP	USSR	1966	[62]
River Navigation	TOR	Poland	1964 1967	[60, 61]
Harbor Surveillance	River Boat Radar	East Germany	1968	[63]
Meteorological Research	MLR-1	USSR	1965	[64]
	—	South Africa	1969	[65]
Lunar Tracking	—	USA MIT	1968	[66]
Earth Resource Mapping	AN/APQ-97	USA Westinghouse	1968	[67]
Railroad Speed, Distance	VSR	West Germany Telefunken	1975	[68]
Airborne Oil Slick Detection	Radiometer	USA Coast Guard	1977	[69]
Entomological Research	—	USA NASA Dept. AG	1977	*
Automobile Braking	—	USA DOT Bendix	1977	[70, 71]
		West Germany SEL	1977	[72]
Instrumentation Radar	—	USA NASA	1977	*
Distance Monitoring	—	West Germany Telefunken	1978	[73]

* NASA, Wallops Island, 1978 (Private Communication)

TABLE 4a
REPRESENTATIVE MILITARY K_a BAND AIRBORNE RADAR APPLICATIONS

APPLICATION	IDENTIFICATION	COUNTRY/SOURCE	YEAR	REFERENCES
Synthetic Aperture	AN/APQ-56	USA/Westinghouse	1953	[75]
	AN/APQ-55	USA/Texas Instruments	1958	
	AN/APD-7,8	USA Westinghouse	1961/1966	[75]
	AN/APQ-97	USA/Westinghouse	1961	[75]
	P-391	UK EMI	1968	[78]
Forward Mapping	AN/APN-161	USA Sperry	1964	
	AN/APS-36	USA Philco	1948	
Ground Target Detection	AN/APQ-137	USA Emerson	1968	[76, 77]
Nuclear Cloud Mapping	LAPQ-1	USA Bendix	—	
Terrain Following	AN/APQ-58	USA Bendix	1954	
(Some Multimode)	AN/APQ-89	USA NAFI	1960	
	SAIGA	France EMD	1968	[78]
	AN/APQ-122	USA Texas Instruments	1973	[78]
	WX-50	Westinghouse	1973	[79, 80]
Weather Recon	AN/APQ-39	USA GE	1956	
	AN/APQ-70	USA	1963	
Solid State Radiometer	—	USA Sperry	1977	[81, 82]
Missile Seeker/Mini RPV Guidance	—	USA General Dynamics	1977	[83, 84]
Strap-on Bomb Guidance	MICRAD	USA NSWC-DL	1977	[85]
Missile Seeker	AMS	USA MIRADCOM	1977	[86]
Passive Radiometer	—	USA Minn. Honeywell	1978	

TABLE 4b
REPRESENTATIVE MILITARY SURFACE BASED K_a BAND RADAR APPLICATIONS

APPLICATION	IDENTIFICATION	COUNTRY/SOURCE	YEAR	REFERENCE
Surface Search	AN/SPS-7	USA Sylvania	1947	
Weather	AN/TPQ-6	USA Bendix	1947	[87]
	AN/TPQ-11	USA Bendix	1957	[88, 89]
Aircraft Landing	AN/SPN-10	USA Bell Aero	1950	
	AN/GSN-5	USA Bell Aero	1960	
	AN/SPN-42	USA Bell Aero	1970	
Missile Tracking	LaCrosse	USA CAL/Martin	1957	[90]
Range Instrumentation	Trakx	USA NRL	1976	[91]
Air Defense	Flycatcher	Netherlands HSA	1976	[78, 92]
ICBM Splash Detection	Splash Detection	USA Westinghouse	—	[93]
Tracking, Backscatter	GT-K	USA Georgia Tech	1976	[94, 95, 96]
Experimental System	Tracking Radiometer	USA NSWC-DL	1977	[85]
Aircraft Tracking	—	USA NSWC-WO	1977	[85]
Fire Control	Seatrack	USA NSWC-WO	1978	[139]
Intrusion Radar	—	USA NAFI	1977	
Reentry Instrumentation	—	USA BMDATC	1980	

Introduction and Overview

TABLE 5
REPRESENTATIVE CIVILIAN MILLIMETER BAND RADAR APPLICATIONS

APPLICATION	IDENTIFICATION	COUNTRY/SOURCE	YEAR	FREQ (GHz)	REFERENCES
Ground Temperature	Airborne Radiometer	West Germany IFFM	1973	90	[97]
Entomological Research	—	USA NASA Dept. Agr.	1977	70	*
Harbor Speed Monitoring	—	USA Coast Guard	1977	95	[98, 99]
Police Radar	—	USA Hughes Aircraft	1977	95	[98]

* NASA, Wallops Island, 1978 (Private Communication)

TABLE 6
REPRESENTATIVE 70 GHz MILLIMETER RADAR MILITARY APPLICATIONS

APPLICATION	IDENTIFICATION	COUNTRY/SOURCE	YEAR	REFERENCE
Airborne Synthetic Aperture	AN APQ-62	USA	—	
Airborne Search Mapping	JR-9	USA Raytheon	1964	
Submarine Surface Search	AN BPS-8	USA Dumont	1958	
Land Search	AN MPS-29	USA Georgia Tech	1960	[94, 100]
Airborne	Aircraft Landing	USA Norden	1960	[101]
Land	Low Altitude Tracker	USA BRL	1966	[102]
Land	Surveillance Target Target Acquisition	USA HDL	1966	[103]
Land Search	VEH	USA Georgia Tech	1966	[94, 96, 104]
Land	Instrumentation, Backscatter	USA Georgia Tech	1966	[96, 104]
Land	Monopulse Tracker	USA Norden	1974	[105-107]

TABLE 7
REPRESENTATIVE MILITARY 95 GHz AND 140 GHz MILLIMETER BAND RADAR APPLICATIONS

APPLICATION	IDENTIFICATION	COUNTRY/SOURCE	YEAR	REFERENCES
Land	Monopulse Tracker	USA NADC	1969	[34]
Land	Space Object Indent.	USA Aerospace	1969	[108]
Land	Instrumentation, Backscatter GTM	USA Georgia Tech	1966	[94, 96]
Land	Arctic Terrain Avoidance	USA APL	—	
Helicopter	Wire Detection (Noter)	USA APL	1975	[109, 110]
Land	Imaging	USA Hughes	1975	[111]
Airborne	Mapping	USA Goodyear	1974	[112]
Land	CW Doppler Radar	USA Hughes	1975	[98]
Land	Noise Modulated Radar	USA BRL	1976	
Missile	Seeker	USA Hughes	1977	[113]
Land	Antitank Radar (STARTLE)	USA Martin Rockwell	1977	[114]
RPV	Battlefield Surveillance (AQUILA)	USA Norden	1977	[115, 116]
Air	Solid State Radiometer	USA Sperry	1977	[81, 82]
Missile	Seeker	USA USAMIRADCOM	1977	[86]
Land	Tracking, RCS Measurement	USA BRL (140 GHz)	1966	[117]
Land	Scale Model RCS Measurement	UK EMI (140 GHz)	1977	[118]
Land	Reentry Instrumentation	USA BMDATC	1980	
Missile	Radiometer	USA Minn. Honeywell	1978	

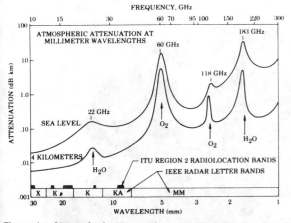

Figure 1. Atmospheric attenuation at millimeter wavelengths with IEEE radar letter bands and ITU Radiolocation bands for Region 2, after Altshuler et. al.

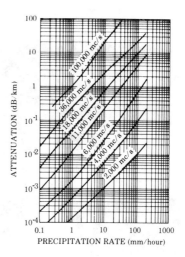

Figure 2. Theoretical rainfall attenuation (various frequencies) after Medhurst.

Figure 3. AN/APQ-137 Airborne Ka Band Ground Target Detection Radar (Courtesy Emerson Electric).

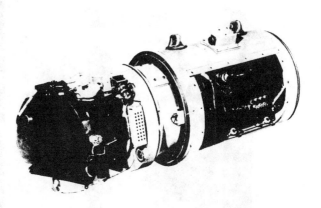

Figure 4. WX-50 Ka Band Airborne Ka Band Multimode Radar (Courtesy Westinghouse).

Figure 5. Ka Band Microwave Radiometric Seeker Subsystem (MRSS) (Courtesy Sperry Rand).

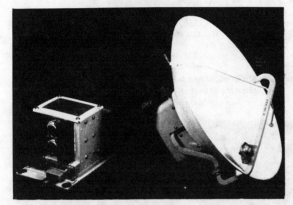

Figure 6. Ka Band Radiometric Area Correlator Radiometric Sensor Subsystem (Courtesy Sperry Rand).

Figure 7. LaCrosse Missile Ground Based Ka Band Tracker (Courtesy Martin Marietta Corporation).

Figure 8. AN/SPN-42 Ka Band Aircraft Landing Radar on Aircraft Carrier (Courtesy US Navy).

Figure 9. TRAKX Dual X/Ka Band Instrumentation Tracking Radar (Courtesy US Naval Research Laboratory).

Figure 10. Ka Band Ground Based Intrusion Radar (Courtesy US Naval Avionics Facility).

Figure 11. AN/MPQ-29 70 GHz Ground Search Radar (Courtesy Georgia Institute of Technology Experiment Station).

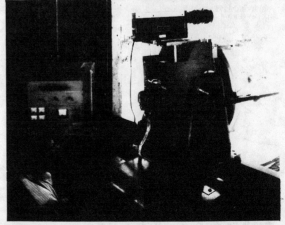

Figure 12. Four Lobe Monopulse 70 Hz Track Radar (Courtesy Norden Systems).

Introduction and Overview

Figure 13. Handheld 95 GHz Police Radar (Courtesy Hughes Aircraft).

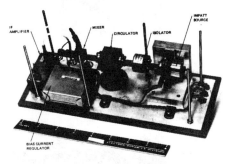

Figure 14. RF Assembly for Police Radar (Courtesy Hughes Aircraft).

Figure 15. 95 GHz Missile Seeker (Courtesy Hughes Aircraft).

Figure 16. 95 GHz Mini-RPV Surveillance Radar Incorporating Polarization Diversity and MTI (Courtesy Norden Systems).

Figure 17. 94 GHz Radar for measuring radar cross-sections (Courtesy Aerospace Corporation).

Figure 18. 15-foot diameter antenna for satellite identification millimeter radar (Courtesy Aerospace Corporation).

Figure 19. 94 GHz Ground Intrusion Radar (Courtesy US Naval Avionics Facility).

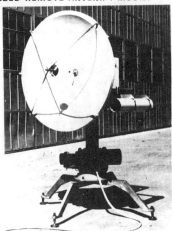

Figure 20. 95 GHz Ground-Based Radar Imagery Antenna Mount (Courtesy Goodyear Aerospace).

Figure 21. 95 GHz radar image of North Island Naval Air Station, San Diego, California (Courtesy Goodyear Aerospace).

Figure 22. 140 GHz radar mixer (Courtesy Hughes Aircraft).

Figure 23. FM-CW Millimeter Seeker Sensor.

Figure 24. Varian Gyrotron VGA 8000, 100 kw cw, 28 GHz Oscillator (Courtesy Varian Associates).

Figure 25. Two-Cavity Gyrotron 28 GHz Amplifier, 65 kw saturated cw output, 40 dB gain, 9% efficiency (Courtesy Varian Associates).

Figure 26. 95 GHz VKB 2443T Pulsed Extended Interaction Oscillator, 1.5 kilowatt peak, 5 watt average (Courtesy Varian Canada).

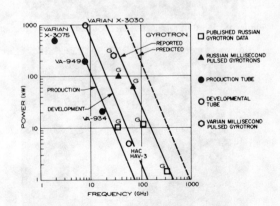

Figure 27. Microwave Tube CW Power State-of-the-Art Summary after Jory et. al. (Reference 140).

Millimetre Waves — The Much Awaited Technological Breakthrough?

G.S. Sundaram

International Defense Review, Vol. 11, No. 2, Feb. 1979
Pages 271-277
© by Interavia S.A., Geneva, Switzerland. Reprinted by permission.

Defense electronics specialists have been placing high hopes on the potential of millimetre waves to improve the operation of many weapons systems. Claimed to provide greater accuracy than conventional microwave (centimetre wave) radars, and to be less susceptible to adverse weather conditions than electro-optical (EO) (wavelength: 0 to 15 microns) sensors, millimetre wave systems should find many military applications in radar, missile guidance and communications. In this article, the special characteristics peculiar to this portion of the electromagnetic spectrum are examined and possible military applications studied in an effort to determine whether the current emergence of millimetre wave techniques and components really heralds a significant technological breakthrough.

Many are the engineers and technicians who, nearly four decades ago, started following the evolution of microwave technology with great fascination, marvelling at each new application found for it. Many technical breakthroughs have since been achieved, and indeed, even the infra-red (IR) and laser regions of the electromagnetic spectrum have been well investigated and a multitude of new applications determined. In fact, several weapons systems employing these frequencies are already in operation. Nevertheless, the same enthusiasm of the early days of 3 and 10 cm microwave radar is being witnessed again now as the millimetre wave era is being brought in, with hopes that all its promise will soon become reality.

The millimetre portion of the electromagnetic spectrum has been the subject of theoretical studies for some years, but it is only since recent advances in semiconductor technology brought about the development of solid-state devices capable of operating at these frequencies, that any practical investigations could be carried out. The current availability of demonstrated millimetre wave techniques and a rapidly increasing range of components 'off-the-shelf', has made it possible for defense specialists to try and incorporate this new technology into weapons systems. The special characteristics of millimetre waves make them very useful for applications involving target tracking, terminal guidance, proximity fuzing, and secure communications where they could either complement or even replace conventional microwave or electro-optical techniques. A host of possible applications has already emerged, and as new components appear, more are being suggested.

What are millimetre waves?

The millimetre wave portion of the electromagnetic spectrum lies between the microwave and far infra-red regions. Frequencies generally attributed to this portion range from 30 to 300 GHz — i.e. corresponding to wavelengths of 10 to 1 mm. At this intermediate position in the spectrum, millimetre waves are not considered to be totally similar to either the microwave or the EO regions. In fact, their importance comes more out of their unique characteristics.

Millimetre waves have three qualities — an interaction with atmospheric constituents and gases; a large bandwidth; and a narrow beamwidth for a small antenna aperture.

Propagation in all parts of the electromagnetic spectrum, ranging from the visible to infra-red and radio frequencies, suffers to a degree from absorption of the electromagnetic energy by atmospheric gases such as water (H_2O), carbon dioxide (CO_2), oxygen (O_2) and ozone (O_3), and from attenuation by atmospheric aerosols such as haze, fog, clouds and rain. This attenuation is sometimes so severe as to render propagation impossible, particularly when using the electro-optic region of the spectrum. Gaseous absorption is a minimum at certain frequencies, and these relatively clearer sections (i.e. regions of superior propagation) are referred to as 'atmospheric win-

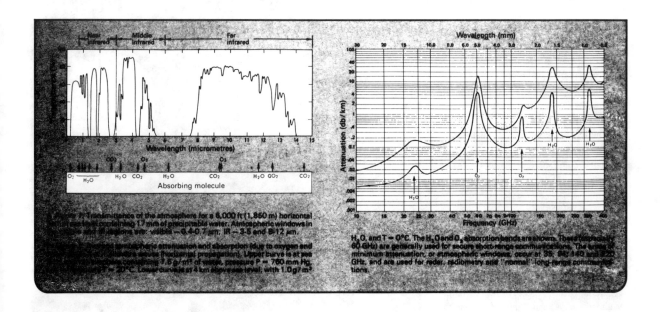

dows. The regions of maximum absorption are referred to as the 'absorption bands'.

The atmospheric windows and absorption bands are shown in Figures 1 and 2. The windows for the visible, IR and RF regions are: visible — 0.4 to 0.7 μm; IR — 3 to 5 μm and 8 to 12 μm; and RF — 10 and 3 cm (main microwave bands). The main millimetre wave windows are — 8.5, 3.2, 2.1 and 1.4 mm corresponding to frequencies of 35, 94, 140 and 220 GHz. The absorption bands in the millimetre wave region are also made use of, with the most important one being around 60 GHz.

As M. I. Skolnik explains in his paper on the subject (presented at the Symposium on Submillimetre Waves, Polytechnic Institute of Brooklyn, March 1970) the electromagnetic energy tends to interact much more with the atmospheric constituents at the millimetric region than at the lower microwave region. In fact, one could virtually consider the RF portion all the way up to 18 GHz as being one continuous window, hence the all-weather capabilities obtained with microwaves. At higher frequencies, however, it has been found that a frequency selective absorption and scattering of the energy takes place — hence the atmospheric windows — and that this is caused by resonances of the atmospheric gases. The millimetre wavelengths are comparable in size to the rain and fog particles, and so similar electrical resonances are caused, without exciting molecular resonance. This results in absorption and scattering of the energy. At the millimetre wave windows, this adverse effect is much less evident, and so most applications are concentrated around these frequencies.

Millimetre waves do not have the same all-weather capability of microwaves, but do have higher resolution. Since component size is related to wavelength, millimetre wave systems are also much smaller. They do not have the extremely high resolution of their EO counterparts, but have superior penetrability through smoke, fog and rain. Millimetre waves thus represent a compromise region where most of the advantageous characteristics of the microwave and EO regions are available while the disadvantageous effects are minimized.

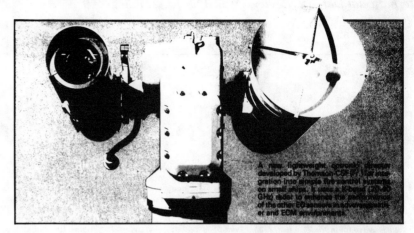

A new lightweight, optronic director developed by Thomson-CSF (F.) for integration into simple fire control systems on small ships. It uses a K-band (35 GHz) radar to measure the range of the other EO sensors in adverse weather and ECM environments.

▲ Sperry Gyroscope (UK) recently presented a model of its *Sea Archer 2* gun fire control system for patrol boats. Along with more conventional EO sensors, it makes use of an M-band (94 GHz) radar (top left) to distinguish small targets from sea clutter in poor weather conditions.

Advantages

● *Small size:* The shorter wavelengths make it possible to reduce the size of components considerably and so build smaller systems. This makes millimetre wave systems of considerable interest for applications where size and weight restrictions are important, as in aircraft and missiles.

● *High bandwidth:* At each millimetre wave window, extremely large bandwidths are available. For example, at the four main windows mentioned earlier — 35, 94, 140 and 220 GHz — the available bandwidths are of the order of 16, 23, 26 and 70 GHz, respectively. This means that the equivalent of all the lower frequencies, including microwaves, can be accommodated in the bandwidth available at any one millimetre wave window. The advantages this presents are considerable. There are many more frequencies which can be used. Thus there is an increased immunity to interference from friendly users (electromagnetic compatibility or EMC). It also makes jamming more difficult, unless the exact frequency to be jammed is known. In communications, more information can be transmitted per unit time. In radar, the range resolution can be increased, while in radiometry, higher sensitivities are possible. The large bandwidths also make radars more sensitive to Doppler frequency shift measurements, since these are much greater.

● *Low beamwidth:* For a given antenna size, smaller radiated beamwidths are possible, providing higher resolution and hence better precision. This is very important in target tracking where the smaller beamwidths can pick out more details and can discriminate better against small targets. They minimize losses due to side-lobe returns — a major problem in microwave radars, and similarly, reduce errors due to multi-path propagation.

A US Army funded competitive study program involving two US companies — Martin Marietta and Rockwell International — is currently underway concerning a means of improving a tank's target acquisition capabilities in battlefield smoke/dust conditions. The Martin Marietta concept is shown here. Called STARTLE (Surveillance and Target Acquisition Radar for Tank Location and Engagement), the system uses a 94 GHz millimetre wave radar. Key to display symbols: a) selected tank thermal sight (TTS) polarity; b) TTS NFOV limits; c) radar angle symbol; d) LED raster; e) FOV indicator; f) track box; g) ballistic reticle.

These narrow beams are very difficult to detect and monitor, and so jamming them presents tremendous problems. Interference from friendly transmissions are also reduced.

Small beamwidths are obtained with comparatively small sized antennas. For example, a 12 cm diameter aperture antenna provides a 1.8° beamdwith at 94 GHz compared to 18° at 10 GHz.

● *Atmospheric losses:* Atmospheric absorption and attenuation losses are relatively low in the transmission windows, compared to the problems faced by laser or IR transmissions in rain, fog and smoke. Millimetre wave sensors are thus more effective than EO ones in adverse weather or battlefield smoke/dust conditions.

maximum gain for that aperture. Similarly, range can be boosted by increasing antenna size, but this also reduces some of the inherent advantages of millimetre waves.

While the large bandwidths available allow better Doppler shift measurements to be made, in ground to space communications this shift can be so large as to go beyond the band limits of the transmission window. The maximum shift to be expected must therefore be known beforehand to ensure that it can be accommodated.

The narrow beamwidths make millimetre wave systems unsuitable for search and target acquisition. Most applications thus tend to be for target tracking and homing.

A major limitation has been the lack of suitable components. This has now been

microwave frequencies for the long-range search and target acquisition roles. Once the target is within the range of the narrow beam millimetre wave radar, it is handed over for more accurate tracking. The smaller millimetre wavelengths also provide better target detail, when the targets are small, such as aircraft, boats and land vehicles. Millimetre wave tracking radars are thus very useful in low-level air defense systems.

The small size, high accuracy, and reasonable operation in adverse weather conditions can make millimetre wave radars very attractive as active seekers for terminal guidance of missiles and smart munitions, as well as for fuzing. Such seekers could complement or replace EO seekers in bad weather and high battlefield dust and smoke

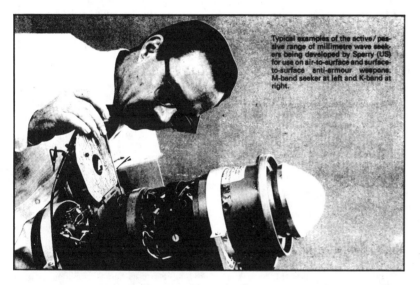

Typical examples of the active/passive range of millimetre wave seekers being developed by Sperry (US) for use on air-to-surface and surface-to-surface anti-armour weapons. M-band seeker at left and K-band at right.

The high attenuation encountered in the absorption bands limit the range so much that, ironically, short range point-to-point secure communications are possible.

Limitations

Needless to say, the millimetre wave region has its limitations. In most cases, however, applications have been found where these are not considered as being particularly restrictive factors.

Because of the atmospheric absorption and attenuation, even in fair weather conditions, the range of millimetre wave radars is limited to 10-20 km. This is further reduced by rain and fog. Major applications therefore involve airborne fire control radars and weapon terminal guidance systems. Atmospheric losses increase with higher frequencies, but at the same time, beamwidths get narrower and components get smaller. Thus, compromise solutions have to be found. At 100 GHz, the atmospheric attenuation is an order of magnitude greater than at 10 GHz. For an average atmosphere containing 7.5 gm of water per cubic metre, this attenuation is approximately 0.06, 0.14, 0.8 and 1 dB/km respectively for the frequencies 10, 35, 94 and 140 GHz.

The small size of the antenna apertures means that less energy is collected, hence reducing the sentivity of the receiving system. If more gain is required, then the aperture must be increased. If this is not possible, then the system should operate at the highest frequency possible to obtain

overcome, at least up to the 94 GHz window. Beyond this frequency, the readily available range of devices is small and users generally have to develop their own. However, high power sources still pose a problem at all millimetre wave frequencies.

Millimetre wave applications

The three main military applications for millimetre wave technology appear to be in the fields of radar, radiometry and communications.

● *Radar:* The microwave region has been used for a multitude of applications, not least of them being radar. However, most of these radars can easily be countered by ECM, and possess some inherent disadvantages which can cause errors or ambiguities. Their wide beamwidths provide spurious returns due to multi-path propagation and side-lobe echos. They can also be cumbersome and heavy particularly for aircraft and missile applications. Such problems can either be minimized or eliminated by using millimetre wave techniques.

The short ranges of millimetre wave radars do not necessarily present major disadvantages since they can usually be offset by the smaller size and greater accuracy. For many aircraft air/air fire control systems, the short ranges are perfectly adequate, and the high angular resolutions obtained with small apertures at millimetre waves are far more important.

In ground-based radars, the range limitation is overcome by using conventional

conditions, when IR/laser/TV-based sensors are virtually useless.

The millimetre wave system's performance is sufficient to be able to distinguish individual tanks, and so appears as the most practical solution in certain scenarios for defense against massed armour. Armoured vehicles have significant radar cross-sections throughout the millimetre wave region and can easily be distinguished on a flat battlefield, although this may be more difficult when the tank is standing, say, next to a tree or a building. This, however, is largely a problem of lack of practical experience in the analysis of such target returns. The next stage is to develop suitable methods of processing the vast amount of fine detail, and study the problems of pattern recognition — what does a tank really look like at these radar frequencies and how does it differ from a haystack?

Millimetre wave radars also have practical applications at sea. Here, they can help the conventional microwave radar to distinguish small ships and other targets from sea clutter and can overcome the multi-path problems of tracking sea-skimming missiles. Again, a sufficiently broad data base (library) of sea clutter and target information is still lacking.

● *Radiometry:* In this application, passive sensors measure the radiometric temperature of objects such as ships or land vehicles. As explained by Skolnik *(see earlier reference on p. 271)*, the radiometer is a passive means of detecting the noise temperature as seen by its antenna. This noise, or so-called radiometric temperature, need not necessarily be the same as the thermal ambient temperature. It is related to the emissivity of the radiating objects as well as to their thermal and brightness (reflection) temperatures. In the case of microwave and millimetre waves, this temperature, when viewing the earth, is approximately the thermal ambient temperature. When viewing metallic targets or the sea surface, the temperature can be much lower than the ambient temperature. This is because the emissivity of such surfaces is low and also because they reflect the brightness temperatures of the cold sky.

The target radiation suffers in practice from interference from undesired background radiation. This can be minimized by having a narrow antenna beam to increase the resolution and target contrast. Millimetre waves are therefore attractive, since the narrow beamwidths can be obtained with small-sized antennas, and the wide bandwidth provides greater sensitivity. They are ideal for use as passive seekers for the terminal guidance of both missiles and smart munitions.

Efforts are being made to develop dual-mode seekers — active / passive millimetre wave, IR / millimetre wave; laser / millimetre wave etc. — in order to make use of the fair and adverse weather performances of each type, and also to provide alternate sensor techniques in case the tactical situation precludes the use of one type.

● *Communications:* Two types of communications links are of particular interest in the millimetre wave region: long-range communications; and short-range secure communications. In these frequency regions, the communications systems are generally of the line-of-sight (point-to-point) type. The propagation is either by free-space radiation, as in conventional microwave relay links, or via special transmission lines and waveguides. The latter are probably limited to about 200 GHz, above which pseudo-optical techniques are needed.

Long range point-to-point communications in free-space require high transmitting gains and large antenna apertures. With millimetre waves, the high gains are possible with small antennas, by operating at the highest frequency possible. An advantage with millimetre wave transmissions is that very wideband communications are possible. In the lower atmosphere, attenuation and absorption can reduce the operating range, but this limitation is not significant in the upper atmospheric regions.

The feasibility of long-range, line-of-sight, wideband communications in free-space may appear dubious, but would seem to be most useful via other propagation media. Moreover, short-range point-to-point links in free-space seem to be attractive for use in the upper atmosphere, say between two satellites, especially because of the small size of the components. Another possible application of millimetre wave links in free-space would be drawn from the ability of the energy at these frequencies to penetrate plasma. This means that communications during the normal 'black-out' period when spacecraft re-enter the lower atmosphere may be possible.

The main attraction of the millimetre wave region for communications is the possibility to make use of the absorption bands to provide secure short-range communications. In these bands, the propagation losses are extremely high, thus limiting the effective ranges to a few kilometres. The narrow beamwidths achieved and the short ranges mean that it is very difficult to detect or monitor, unless the listener is extremely close or actually at the receiving end. The current generation of UHF and VHF equipment radiate detectable energy in all directions. Although wire and cable systems eliminate this, and also the problems of enemy ECM, considerable installation time and logistics are needed, and the equipment is vulnerable to physical attack.

Components, systems, programs

At the present time, a wide range of components is already available for use up to, and including, the 94 GHz window. A number of millimetre wave radar, communications and electronic warfare systems are already in operational use around the world. In addition, a multitude of development

A 6 in (15 cm) diameter, 35 GHz adverse weather millimetre wave seeker developed by Martin Marietta (US). The compact active / passive seeker is a higher powered (4 W) version of a flight-proven design. The antenna has a 4.3° beamwidth, and system employs both shape and amplitude discrimination techniques. *Key:* 1) body-mounted lens; 2) position pickoff; 3) spin coil; 4) permanent magnet ring; 5) spin and torque circuits; 6) circuit cards; 7) RF module; 8) gimbal bearings; 9) torque coil; 10) spin-stabilized gimballed reflector; 11) feed assembly; 12) radome.

▼ One of a family of millimetre wave transceivers brought out by Norden (US). The unit shown operates at around 38 GHz, and was developed for the US Army to provide a moderate range transmission of data and voice communications. It features an 8 in (20 cm) antenna, and can either be tripod or mast-mounted.

programs have been initiated in the last few years, particularly in the US. Brief descriptions of several of these are given below. The list is, however, by no means complete, and includes the names of only some of the companies involved in this field.

● **Radar:** Several radar systems have been developed on both sides of the Atlantic, which make use of millimetre waves to augment the performance of conventional microwave sensors.

— *Flycatcher:* This is a mobile low-level air defense radar developed by Hollandse Signaalapparaten in the Netherlands, to control the fire of both AAA and SAM systems. A dual-frequency approach has been employed, where the target search and rapid acquisition is provided by an I-band (around 10 GHz) system and final tracking by a K-band (around 30 GHz) radar. The higher frequency system is incorporated into the main tracking antenna and automatically controlled by it. This dual mode capability is said to give very good performance down to the lowest altitudes without multi-path propagation problems both on the ground and at sea. Main features include high tracking accuracy, low reaction times, high resistance against ECM (ECCM includes frequency agility and monopulse), and track-while-scan through 360°. (See *IDR* 3/1977, pp. 495–499 for system details.)

— *Blindfire:* Marconi Space and Defence Systems in the UK has developed this tracking radar to provide the British Aerospace *Rapier* low-level missile system with an all-weather fire control capability. This radar reportedly operates at the 35 GHz window (K-band) and is used in conjunction with the *Rapier's* standard lower frequency search radar, replacing the optical sensor to provide accurate night and adverse weather operation. It is operational and has recently been successfully tested with the US Army *Chaparral* air defense system. *Blindfire* employs advanced signal processing techniques to make use of all the target information provided, including 'glint' returns which are normally a hindrance. Thus the system is claimed to offer target classification and identification, sub-clutter visibility without resorting to Doppler or MTI techniques, and minimal effects of multi-path returns at low levels, both on land and at sea.

Introduction and Overview

It is understood that more than one frequency is used, but in the same frequency band.
— *TRAKX:* This radar, developed by the US Naval Research Laboratory, uses both 9 GHz and 35 GHz systems, if necessary simultaneously, to track targets at very low elevation angles while avoiding the weather and propagation problems associated with millimetre wave trackers. A single, dual-feed 8 ft (244 cm) antenna dish allows simultaneous use of the two transmitters and receivers. The system is under full, real-time, computer control, in order to handle the combined data flow. Pulse lengths at both frequencies are the same, while PRF can be varied. Three-channel monopulse tracking techniques are used. Ranging is done using a phase-locked frequency synthesizer, with range data provided by a time interval counter. It is reported that under 'normal' atmospheric conditions, TRAKX is expected to provide good tracking on a 1 m² target out to about 45 km (with the 35 GHz system), with acquisition ranges out to approximately 120 km (at 9 GHz). With 4 mm/h rain, the maximum range at which good track data at 35 GHz can be taken is expected to be about 17 km.

— *STARTLE:* In order to improve target acquisition capabilities for combat vehicles during adverse weather and battlefield conditions, the US company Martin Marietta is developing a prototype millimetre wave system under US Army and DARPA sponsorship. Called STARTLE (Surveillance and Target Acquisition Radar for Tank Location and Engagement), the radar uses the 94 GHz window and provides dual-mode MTI operation. Coherent MTI is used to detect radially moving targets; area MTI allows acquisition of 'creeping' and tangential targets. The spread spectrum waveform used with area MTI processing also provides smoothing of clutter, glint and multi-path effects. The solid-state transmitter operates with a mirror-scanned paraboloid antenna located on the left side of the tank turret. A rotating closure and armoured housing protect the antenna against indirect artillery fire, as well as from damage by trees and brush.

The system avoids undue complexity and increased tank gunner workload by integrating radar information with the thermal sight display. The gunner can consequently locate and fire on targets using either radar or thermal sensors.

The Martin Marietta prototype will compete later this year against another STARTLE prototype being built by Rockwell International. The prototypes will be installed on M60A3 tanks.

— *Sea Archer 2:* This is an advanced, modular gunfire control system developed by Sperry Gyroscope in the UK for use on the fast patrol boat type of craft. Based on the company's *Sea Archer 1* optical fire control concept, it uses improved sensors, including an M-band (94 GHz) tracking radar. The system has an above-deck, remotely-operated, gyro-stabilized, line-of-sight detector to which a range of sensors can be fitted — i.e. a TV tracker, a new IR tracker, a laser rangefinder or the M-band radar — depending on requirements.

The 94 GHz tracker has a 0.5° beamwidth. According to Sperry, the use of M-band allows excellent angle and range resolution which diminish sea clutter and multi-path effects. It also reduces the effects of fog, haze and smoke.

— *TRS 906:* Thomson-CSF, of France, has recently developed this lightweight optronic director for integration with simple fire control systems, such as the company's *Canopus*, for use on fast patrol boats. The TRS 906 can be fitted with any three types of sensors from a range offered by the company which includes: day TV, LLLTV, very-LLLTV, shipborne variant of the SAT *Piranha 1* IR tracker, a CILAS-developed laser rangefinder, or a millimetre wave (K-band) radar tracker/rangefinder. These different sensors can be used to provide simultaneous search, tracking and ranging of targets with high precision even at low altitudes and in adverse weather and ECM conditions.

— *Soviet radars:* Not much is known about this activity. However, it has been reported that Soviet airborne millimetre wave radars exist, and that some Soviet tanks now deploy millimetre wave radars for increased accuracy and a better adverse weather, short-range target acquisition capability. Another possible use of such systems on tanks is to provide an improved IFF capability.

● **Electronic warfare:** Very few systems using millimetre waves have actually been developed in this field. However, it has been recognized that threats do exist in this frequency region and are likely to increase. Hence the US DoD has several ongoing programs for development of suitable ECM capabilities. In addition, many existing systems, such as the F-15's Loral-developed RWR and *Compass Sail* programs are being updated to detect millimetre wave threats. The Westinghouse jamming pods, AN/ALQ-119 and -131, have been designed modularly, so that the higher frequencies can be added when required.

The USAF recently sent out an RFP for a feasibility study of active/passive ECM systems for operation at millimetre waves. AIL, General Instruments, GTE-Sylvania, Hughes and Loral have offered proposals. The US Navy has also made a similar request, but merely for receiver systems. Raytheon has carried out studies of radar warning systems operating up to M-band, while Applied Devices has developed ELINT systems also operating in the M-band.

EW has many limitations in the millimetre wave regions. The jammers are difficult to build, since high powers are not readily available at these frequencies. Until appropriate high power sources are developed, active ECM systems will be hard to come by. The narrow beamwidth and larger bandwidths peculiar to these frequencies also render the jammer's task more difficult. In addition, chaff is not effective above about 20 GHz.

Watkins-Johnson has developed a ground-based countermeasures set, the AN/TLR-31, which provides millimetre wave emitter detection, analysis, and accurate DOA information. It is presently integrated with the AN/MLQ-24 ground-based ECM set, developed by Bogue Electric.

Probe systems is currently producing digital IFM receivers operating from 0.5 to 40 GHz in eight units of which one, the Model IFM-2600, operates in the millimetre wave range 26-40 GHz. EM Systems also offers two IFM receivers, the R3100 18-40

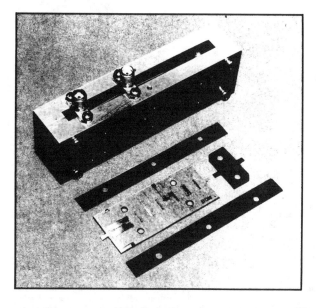 

AIL (US) has built a shipboard mast-mounted millimetre wave *Call Alert* receiver system to provide an all-weather means of alerting the crew of the existence and direction of origin of an incoming signal in the 35 GHz band. System uses the company patented *Fin-line* millimetre wave transmission medium *(left)* on the front end (housed in a radome on mast). The monitor assembly *(right)* is located about 40 m away from the mast-mounted antenna-receiver assembly.

GHz unit and the R3110 18-40 GHz high resolution model.

● **Terminal guidance:** This appears to be the most popular military application for millimetre waves, especially in the US. Much work in this field is also being done in France by Thomson-CSF.

All three Services in the US have initiated development programs for millimetre wave seekers for a variety of applications. As early as 1975, the USAF awarded Honeywell a $2.5 million contract for the design and development (first phase) of an adverse weather, millimetre wave, contrast guidance system for tactical weapons. The Honeywell terminal guidance seeker reportedly employed an active radar and onboard processors which analyzed the returns to distinguish tactical targets. Late last year, the USAF was seeking funded industry studies of a dual-mode seeker for air-to-air missile applications. Plans reportedly called for an active millimetre wave radar and a passive IR sensor. Under its Wide Area Anti-armour Munitions (WAAM) program (see *IDR* 9/1978, pp. 1378-9) the USAF is funding five companies — Boeing, General Dynamics, Hughes, Martin Marietta and Rockwell International — to carry out exploratory development work on an IR or millimetre wave seeker for an air/ground mini missile. Another RFP on this three-phase effort is expected this May. At Eglin AFB, the USAF itself is working on an experimental dual-mode (millimetre/IR) homing head for air-to-ground missiles. It is developing the advanced, onboard signal processor which will detect, identify and locate the targets and accurately aim the missile. The targets will be mainly ground combat vehicles. The USAF has also experimented with active/passive millimetre wave seekers on the GBU-15 glide bomb, to enhance its performance during adverse weather and in smoke/dust conditions.

At the end of 1978, the US Navy was seeking industrial sources to develop and build a dual-mode guidance system for a 7-in diameter air-to-air missile. The gimballed seeker is reportedly required to operate in an active mode at millimetre wave frequencies, as well as in a passive RF receive mode.

In April 1978, the US Army/DARPA awarded Sperry a $1 million contract to demonstrate advanced guidance techniques for air-to-ground anti-armour weapons. Various methods of guidance, using millimetre wave frequencies, will be investigated. Target illumination from a standoff aircraft will be studied, using an illuminator supplied by Norden. The company has also received several US DoD contracts to demonstrate active/passive millimetre wave guidance techniques for both air-to-ground and ground-to-ground missiles for use against massed armour.

Other similar seeker programs include the USAF *Cyclops* and US Army's STAFF (Smart Target Activated Fire and Forget). This latter is a dual-mode EO/millimetre system now undergoing Phase II of its development at ARADCOM.

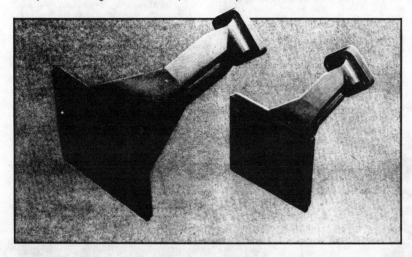

Brief details of three millimetre-wave seekers are given below:
– *Hughes:* One of the leaders in this field, Hughes has developed a 94 GHz (M-band) seeker as a potential terminal guidance unit for short-range missiles, guided projectiles, or longer range weapons equipped with a mid-course guidance system. The prototype has been successfully tested in the laboratory and in captive flight by helicopter. Both active and passive modes were used against tanks and trucks, and better penetration through fog, rain and heavy clouds than with EO seekers was demonstrated. Resolution was greater than that obtained from conventional radars.

The 94GHz seeker will operate in the lock-on-after-launch (LOAL) mode, and will be programmed to detect and destroy tactical targets. A 12.5 cm diameter antenna with a beamwidth of about 1.5° is used. The radiometer employs conical scan guidance techniques. Hughes has developed solid-state power sources of 4 and 10 watts and is working on 100 watt units.
– *Martin Marietta:* A 6 in diameter, 4.3° beamwidth 35 GHz seeker, which is a compact, higher-powered (4W) version of a flight-proven design, has been developed by Martin Marietta. It is compatible with the future needs of direct fire weapons for aircraft, helicopters and RPVs, as well as with terminally-guided sub-munitions, guided bombs, cannon-launched weapons, and tactical nuclear weapons. The active (pulsed)/passive (conical scan) seeker is modular in construction. It consists of three major sub-assemblies — antenna and radome; gimballed, spinning reflector; and electronics package.

The all-solid-state seeker features an automatic target search sequence, multiple range-gated target acquisition and an adaptive threshold detection process.
– *Sperry:* This company is working in both the K and M bands under several US DoD contracts. Sperry active/passive seekers are typically composed of a protective radome, a gimballed tracking antenna system, the millimetre wave transmitter, receiver, IF and video circuits, and signal processing electronics. The systems are all-solid-state, including the low-power CW transmitter, for enhanced reliability. A unique FM/CW modulation technique is used to obtain range resolution without the high peak power (and attendant lower reliability,

▲ Two types of antenna assemblies developed by AEL (US) for operation up to 40 GHz. Shown at left are two horn antennas which provide moderate gain, circularly polarized performance in the 18-26.6 *(left)* and 26.5-40 GHz *(right)* bands respectively. Main use is in air, sea, ground DF systems. Shown at right is a cavity backed spiral antenna providing 2-40 GHz coverage (in a single unit). It is useful, for example, in airborne amplitude comparison DF systems as a broadband dish feed.

it is claimed) of a pulsed signal. Company studies show that production-line seekers with diameters of 4-7 in (10-18 cm), lengths of 12-16 in (30-40 cm) and weights of 10–14 lb 4.5–6.5 kg) will soon be achieved.

● **Communications:** A few communications links have been developed in the millimetre wave region. Many are still in the experimental stages, both in the US and Europe. Under US Defense Communications Agency sponsorship, the Rand Corp. is to investigate the feasibility and advantages of shifting military satellite communications into higher frequency bands (from 20 to 50 GHz). In the UK, Plessey is reportedly experimenting with a short-range 60 GHz communications link.
– *Norden radios:* Norden has developed a range of 5 millimetre wave radios. Four of these operate in the 36.0-38.6 GHz band, while the fifth uses the signal hiding capabilities of the 60 GHz absorption band.

The four lower frequency radios use the same basic transceiver module — a unique design that uses a single varactor-tuned, Gunn oscillator as both the transmitter source and receiver local oscillator. The five radios are: a 38 GHz hand-held radio providing full, duplex voice at ranges up to 40 km and half duplex video (TV) or data transmission up to 13 km; a 60 GHz version of the 38 GHz hand-held radio especially suited to

Introduction and Overview

short range (2 km), highly secure links; a 38 GHz radio, developed for the US Army to provide moderate range transmission of data and voice communications; a 38 GHz tripod or mast-mounted, full duplex transceiver providing high quality voice, wideband data and TV transmissions; and a 38 GHz mast-mounted model developed to provide an economical, longer range information transfer capability, which uses an 18 in (46 cm) parabolic reflector (1.3° beam) to handle full duplex voice and data transmissions at ranges up to 33 km.

Applications for these radios include: command post, local information distribution; remote surveillance information transfer; artillary battery communications; tank-to-tank, ship-to-ship, ship-to-shore communications; aircraft-to-aircraft voice link; aircraft carrier flight deck communications; and patrol aircraft data dump.

— *Call Alert receiving system:* This is a shipboard, mast-mounted system to provide an all-weather means of alerting personnel of the existence and direction of origin of an incoming signal in the 35 GHz frequency band. Developed by AIL, the *Call Alert* receiving system uses the company patented *Fin-Line* millimetre wave transmission medium. Key components of the system are: narrow-beam horn antenna rotating in the azimuth plane; K-band receiver; visual display and audio alarm indicating presence and direction of origin of received signal. The antenna has a beamwidth of 5° in azimuth and 20° in elevation.

— *AIL transceiver:* Under a US Army program, AIL is developing a 60–75 GHz transceiver which can be tuned through the absorption band. Two such transceivers, forming a communications link, will provide transmission and reception of FM signals containing voice, data and video information. Full duplex operation is possible. Antenna beamwidth is 5° and transmitter power is 100 mW.

● **Components:** This is, in fact, the most important development area in this frequency region, and governs the progress made in

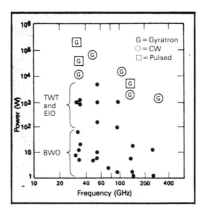

▲ *Figure 3:* Best reported gyrotron results compared with reported results for travelling wave tubes (TWT), extended interaction oscillators (EIO), and backward wave oscillators (BWO) in the millimetre wave region. (*Ref:* Paper presented by T.F. Godlove et al at EASCON 77).

the design and building of millimetre wave systems. Most of the companies which develop systems tend to build their own major components, since these are not always available during these early stages of development. However, certain companies, such as Microwave Associates, AEL, EM Systems, AIL, and AEG-Telefunken (in the FRG) have built up a range of components ranging from diodes, mixers, and power sources to specialized antennas. A variety of components are available for 35 GHz and 94 GHz operation, but at higher frequencies there is still a major requirement to be satisfied.

The main limiting factor in the millimetre wave components field has been the lack of high power sources. At these frequencies, the available power from current TWTs is only 50-100 W, and values comparable to those obtained at lower frequencies appear very improbable using conventional-type sources. For this reason, much attention is being given to the electron cyclotron maser, or gyratron, both in the US and the USSR. This type of power source has now become the prime candidate for high power millimetre wave radar and ECM applications, following the results obtained from a long-term experimental program at Gorki (USSR), a research program at the US Naval Research Laboratories (NRL), and an ERDA sponsored program carried out by Varian Associates in the US. While work in this field was being conducted as 'blue-sky' research in the early 1970s in the US, a team at Gorki State University conducted an intense development effort with electron cyclotron masers, to which it gave the generic name gyrotron. Work was done on both CW and long-pulsed operations, and quasi-CW powers in the 1–60 kW region were obtained at millimetric frequencies. The Soviet achievements spurred the US effort and at NRL, experimental equipment yielded peak powers in the gigawatt region. Varian is currently developing, under ERDA sponsorship, a tube at 28 GHz with a CW power level of 200 kW. In contrast to the Soviet tube, which used a single oscillating cavity and so is correctly called a gyromonotron oscillator, the Varian effort is of the gyroklystron type using resonant cavities separated by drift space.

All these efforts, producing remarkably high power levels, are still very much in the experimental and early development stages, and it will be some years before such equipment becomes readily available.

Conclusions

It is clear that millimetre wave systems will never completely replace both microwave and electro-optical equipment. However they do offer some extremely useful advantages over existing techniques.

Their narrow beam and large bandwidth characteristics provide for greater accuracies than microwaves, eliminating ambiguities due to multi-path, backscatter and clutter phenomena. At the same time, their resolution is less than that of the EO region, although they have better propagation characteristics in adverse weather and battlefield smoke and dust conditions.

A considerable number of the applications mentioned earlier for millimetre wave technology will, no doubt, soon become reality. The real benefits to be reaped from these new advances, however, will come from multi-mode systems, employing the most advantageous characteristics of each of the three regions of the electromagnetic spectrum. Once a useful range of components are readily available for all the millimetre windows, and adequate power sources can be produced, there is no doubt that integrated, multi-mode sensor systems will form a very powerful asset indeed.

Considering the characteristics of millimetre waves, and their potential applications, and on reviewing the recent technical advances, it cannot be denied that we are now on the threshold of a possible technological leap forward. It has taken nearly forty years, since the microwave era began, to achieve this major breakthrough, and go into another region of the spectrum which holds similar promise. Admittedly, twenty years ago, a similar technological jump occurred, when the IR regions were investigated. But this was considerably less spectacular, since the applications were somewhat more restricted. The next decade will surely show whether the hopes of the many engineers and scientists are well-founded.

✦✦

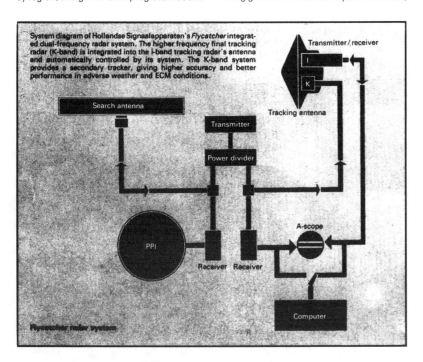

Millimeter Waves — The Coming of Age
L.R. Weisberg

6th DARPA-Tri Service Millimeter Wave Conference, 1977

ABSTRACT

Potential applications for millimeter waves are identified and reviewed with respect to their origin and their key features. A variety of needs for this technology has now been established so that this field is now coming of age. Accordingly, thrust areas must be clearly defined and the many programs must be better focussed and coordinated.

1. INTRODUCTION

Just a few years ago support of millimeter-wave technology was an unpopular cause, shunned by managers and championed by just a few scientists and engineers. Seldom has a subject had more births and deaths than millimeter waves, although reports of its death were exaggerated. Today we are in the somewhat embarrassing position of the 45-year-old confirmed bachelor who is about to get married, and is contemplating three issues: why didn't it seem like a good idea ten years before, is it really a good idea today, and what he should do differently afterwards. Let's consider each of these for millimeter waves.

2. WHY NOT BEFORE?

It is a subtle and interesting question as to why millimeter waves had not emerged as an important thrust ten years ago. In some fields, growth is coupled to a single key invention or discovery such as the laser, microwave oscillations in a semiconductor, or planar technology for integrated circuits. But this is not the case for millimeter waves. No major single advance had occurred that opened up new vistas.

One factor could have been that millimeter-wave technology was not developed enough ten years ago to allow application to systems. However, this was not the primary factor. If the need was sufficiently urgent, we could have developed the technology as we did with lasers. For low-power applications millimeter-wave tubes were close to the needed performance, and the knowledge existed to extend their performance.

Perhaps the reason was the very poor atmosphere propagation in inclement weather for frequencies above X-band. This was particularly important in the past since radar was primarily important for a search function. Systems operating at even Ku-band were rare. However, even this could not have been the gating function. For example, the need for short range tactical battlefield target acquisition under heavy fog was apparent even then.

Another possibility is that we had too many other problems to solve first, so that millimeter waves had too low a priority. Clear needs had not yet emerged. This certainly was a factor. For example, we did not then have passive night vision imaging devices in our inventory.

I believe the reason for the emergence of millimeter waves at this time and not before is that we're thinking differently. We've entered a new era of precision weapons. We could be facing an enemy with a considerably greater weapons numerical superiority, and to counteract this our own weapons must have an unprecedented probability of kill. This has created important new demands, many of which can be met best with millimeter-wave systems.

A second way we are thinking differently is in our approach. One might have expected millimeter waves to have emerged by going up the frequency ladder from L-band to X-band to Ku-band, etc. Instead, we jumped from RF to the optical domain. The new era of precision weapons was ushered in with the invention of lasers and its application to "smart" bombs. But the Achilles' heel of our optical systems such as lasers and FLIRs is that they will not operate in poor weather or through smoke. So we've jumped from the optical domain back to millimeter waves to circumvent this problem in optical systems. Incidentally, you can have some fun by telling an expert in search radar that an important application of millimeter waves is to provide an all-weather capability. He'll probably walk away shaking his head sadly wondering if the laws of nature have changed recently. Yet the statement is true, but referenced to laser systems rather than microwave systems.

A third way we're thinking differently is to include a new-found respect for our potential enemy's capability of disrupting our C^3 networks and finding countermeasures for our weapons. Communication links and guidance systems based on lasers have the virtue of very low probability of intercept (LPI). With the use of millimeter waves, this capability can be preserved and even extended.

3. WHY TODAY

The punchline is that we now have many well-defined needs for millimeter waves which did not really exist before. Let's review some of these and their origins.

As indicated in Table I, the largest group of applications of millimeter waves is to complement electro-optic systems including lasers, FLIRs, and IR seekers. In most cases, in clear weather the electro-optic system will outperform the millimeter-wave system. But as commented above, they perform poorly in inclement weather and smoke.

An important point to remember, and which is sometimes overlooked, is that millimeter-wave systems such as designators and beam riders are useless in heavy fog unless you know what to aim at. That is, the target must first be acquired, classified, and ranged. Of course, this information is needed for any gunfire, but the need for this information becomes even more acute for precision weapons. Because of this need, the Army has initiated a program called STARTLE, and is described in a later paper.

A very interesting possibility with millimeter waves considered in STARTLE would be to image vehicles in heavy fog at a distance of over 1 km. With antennas with a one meter diameter, imaging is just not possible at any wavelength because of either poor resolution or propagation losses, but it misses by only slightly. However, we should not limit our thinking here. One can conceive of larger antennas, perhaps 4x2 meters in a covered truck, which would allow reasonable images to be formed. Additionally, one can consider moving the truck and then applying synthetic aperture or doppler beam sharpening techniques for even better imaging. One can further conceive of using a mini-RPV equipped with millimeter-wave radar and applying synthetic aperture techniques to achieve imaging. Very high speed data links to the ground would be necessary.

Achievement of good imagery through fog or smoke is an exciting possibility. However, even in the absence of imaging, ground vehicles can still be detected and tracked with millimeter waves, and by analysis of the radar returns including the glint characteristics, there is a good possibility of classifying ground objects such as distinguishing between a truck and a tank. Such a system would be a powerful addition to our inventory.

Since the origin of these millimeter-wave applications is from the optical domain, we should also learn some important lessons from optical systems. One area that was not properly explored was optical propagation. We are catching up now, and pinning down key parameters concerning system tradeoffs between 3-5 microns and 8-12 microns. Similarly, for millimeter waves, there are very important tradeoffs still to be determined in choosing among 94 GHz, 140 GHz, and 220 GHz. With respect to multipath, 220 GHz is the clear choice. However, calculations have shown that the required power for a target acquisition system in a humid hot day jumps from a fraction of a watt for 94 GHz to tens of kilowatts for 220 GHz for a range of 3 km. These factors must be accurately determined at an early stage in the millimeter-wave field. We cannot ignore them.

The second group in Table I indicates applications of millimeter waves which are dominated by the ability to use smaller antennas. There are three main platforms for which this is an important factor: satellites, missiles, and mini-RPVs. For example, one future payload for mini-RPVs will be a radar, but to fit it into the small platform and to meet stringent weight requirements, one must go up to at least 94 GHz. For satellites, electro-optic technology is now ahead such as for satellite-to-satellite communications and for detection of objects in space. However, millimeter waves should not be counted out. They could provide some cost advantages, allow faster tracking, and provide better decoy discrimination for reentry vehicle detection. Missiles could employ radiometric sensors operating at 35 GHz for area correlation guidance. Small tactical missiles could employ active seekers operating at 94 GHz.

The third grouping in Table I includes an important possible shipboard application of detection and tracking of low altitude cruise missiles. In this case, the improved multipath capability of millimeter waves compared to lower radar frequencies is the main advantage. With respect to LPI, narrow beams can be achieved with millimeter waves with reasonably sized antennas. An additional LPI capability arises by using the high atmospheric absorption due to oxygen centered at 60 GHz. Although operation at this frequency limits the range of either radar or communications links, it minimizes the danger of detection.

Another less obvious application for millimeter waves is the measurement of radar cross-section. In dealing with large ships, aircraft, or even large ground vehicles, scale models can be quite large and costly even at about 100 GHz when X-band information is desired. By extending the frequency to the sub-millimeter range, smaller scale models can be used.

In all of the above applications, we should not make the mistake of considering millimeter waves to be in competition with other systems. Instead they must be viewed as complementary. For example, millimeter waves can augment FLIR systems or satellite IR detectors, or provide EW protection such as with dual mode seekers for missiles.

Besides the many applications listed in Table I, the other major factor that makes today the ripe time for millimeter waves is the advance of device technology. Many of these advances are discussed in other papers in this Conference. These are really the pacing activities for system developments. A major need is to have a ready supply of sources with a broad range of output powers. In another paper, some advances are presented in a relativistic electron-beam millimeter-wave tube source with which the normal power-frequency squared limits are considerably surpassed. Many other advances will be needed to optimize millimeter-wave systems such as improved detectors and mixers, phase shifters, wave guides, and power combining circuits. As we go higher in frequency we enter a quasi-optical regime, and we must adjust our thinking about circuits accordingly.

4. WHAT SHOULD WE DO NOW?

Here we consider the third of the bachelor's considerations, namely future actions. It is difficult to determine our present level of effort in millimeter-wave technology not only because it is broken up into many bits and pieces, but many of the efforts have not been given visibility perhaps because the technology has been met with a jaundiced eye in the past. In a recent Workshop on millimeter waves organized by the Army, it was estimated that several million is being spent in FY 78 by the three military Services and DARPA on this technology, perhaps as great as $10-million.

If we are to be successful, these separated and entrepreneurial ventures must be brought together and focussed to achieve synergism. This must be a push-pull effort, with management working closely with bench scientists and engineers to define programs. Now is the time for you workers in the field to come out of hiding and be counted! All is forgiven! In this respect, the Army has recently made an excellent step forward by defining near-millimeter-wave technology to be a major thrust, identifying its on-going programs, and coordinating them through the Harry Diamond Laboratories, under Dr. Ed Brown's cognizance. The fact that this Conference is being held at the Harry Diamond Laboratories is an indication of its role in this field. Another factor has been DARPA's early interest in and funding of millimeter-wave technology under the leadership of Dr. Jim Tegnelia.

This Conference promises to be most interesting and is certainly timely. You all know the story of the Phoenix - a legendary bird that burns itself to death to rise young and vigorous from the ashes. Millimeter waves has certainly had its share of births and deaths, and has gone through trial by fire. I believe we have reached the last phase in this cycle, and I regard this Conference as a confirmation event celebrating the coming of age of millimeter waves. It is up to you to see that it properly reaches maturity.

TABLE I. Applications of Millimeter-Wave Technology

<u>Replacement for Lasers and IR Devices</u>
- Target designators
- Beam riders
- Range finders
- Passive seekers
- Detection/classification of ground vehicles
- Imaging

<u>Smaller Size/Weight</u>
- High resolution radar
  (Mini-RPVs/spaced-based)
- Active fuzes
- Radiometers
  (G&C/detection/fuzes/spaced-based)
- Active missile seekers
  (Terminal guidance)

<u>Other</u>
- Narrow beam search and track radar
  (Shipboard low angle)
- LPI Communications
  (Ship-to-ship/battlefield/sat-to-sat)
- LPI radar
- Radar cross section measurements

The Unresolved Problems and Issues
L.D. Strom
6th DARPA-Tri Service Millimeter Wave Conference, 1977

   The advantages and limitations of short wavelength radars and communication sets are identified and the applications which appear to be most suited to the use of millimeter wave equipments are listed in this paper. Typically, millimeter wave radars are used at ranges of 5 to 10 kilometers (or less) and communication sets at 10 to 20 kilometers. Within these limits, a properly designed 3 mm radar can provide adverse weather performance equal or superior to that of an equally well designed 3 cm radar, but it is almost impossible to extend the long range adverse weather operation of millimeter wave equipments to ranges of 20 km or more.

   The recommendations made in workshops held during prior meetings of this group are summarized, and some of the resulting actions are discussed. This paper ends with a reaffirmation of past conclusions concerning the developments needed to demonstrate the utility of millimeter wave equipment.

1. INTRODUCTION

   There are diverse views of the future of short wavelength communication equipments and radar sets. The enthusiast claims that we have not heretofore been able to build useful equipments and systems because the essential building blocks have not been available at a reasonable cost, but that the deficiencies have now been removed. At the other extreme the critic points out that we are simply moving with the flow or a cycle of 15 or 20 years duration. This cycle is characterized during the waxing phase by numerous papers proclaiming that a new and useful frequency band is about to become available. The cycle will wane (says the skeptic) when the realities of an exponential atmosphere are understood and the bill for the pretty little RF circuits comes due. When that comes about, we will not concern ourselves with the impossible dream until another generation of engineers comes along.

   This, then, is the most fundamental of our unresolved problems--which advantages justify the cost resulting from the use of the millimeter wave portion of the electromagnetic spectrum?

Admittedly, the problem is not new. We have searched for answers to it since the early work in millimeter waves during WWII, but the need for a conclusive answer is becoming urgent, and I welcome the opportunity to address the issue. I give some credence to the argument that we haven't heretofore had the right building blocks (but don't believe that to be the complete story); alternately I doubt the skeptical view that work in millimeter waves will ebb and flow throughout the foreseeable future; I estimate that we have only three or four years to solve the problem.

The remarks which follow relate to three distinct topics.

> First, I propose to identify representative applications which can be served by millimeter wave equipments.
>
> Second, I will review the plans and accomplishments of the last few years. In this part of the paper I will skip over the plans and accomplishments made by specific projects and programs since the speakers who follow me to this rostrum may wish to make their own reports. Instead, I will devote my comments to the plans and accomplishments which relate to the ARPA/Tri-Service Millimeter Wave Workshops.
>
> Third, I will present my notions relative to the tasks and developments which will establish the validity of millimeter wave equipments for tactical applications. Since I know nothing about communications, these suggestions will necessarily be restricted to radars.

2. REPRESENTATIVE APPLICATIONS

A number of investigators have looked into the characterisitics of equipments which might advantageously employ millimeter wave carriers. Lists such as Figure 1 typically result from these inquiries, and I assume that most of the points listed there are sufficiently familiar that they do not require discussion. However, to make certain that there is no confusion about the limitations of millimeter waves, I emphasize that these short wavelengths are not suitable for large volume search, broadcasting, or any other application which involves long paths through the lower atmosphere.

To belabor the point a bit, let's consider the matter of large volume search. We will start by pretending that we don't know that one doesn't use a high power telescope to provide fast search over large fields and proceed to calculate the power required in a 3 mm radar designed to provide fast search over large fields and proceed to calculate the power required in a 3 mm radar designed to provide hemispheric search in a dry atmosphere. It turns out that 10 watts of average power is more than enough to provide coverage to 10 km at reasonable frame rates. However, if we provide the power margin needed to burn through a moderate rain, we find that 10 megawatts average power is required. To achieve 20 km range in hemispheric search 160 terawatts are required. Since the terrawatt is not an everyday unit, it may be useful to state that it is very large. If all of the automobiles in use in the United States were efficiently coupled to a single shaft and operated at full throttle, the average power output would be about 5 terawatts.

MILLIMETER RADAR

SUITABLE FOR APPLICATIONS REQUIRING

- HIGH RESOLUTION/BANDWIDTH
- GOOD ADVERSE WEATHER PERFORMANCE
- NEARLY COVERT OPERATION
- LOW SUSCEPTIBILITY TO JAMMING
- LOW AVERAGE/PEAK RF AND PRIME POWER

BUT NOT FOR

- LARGE VOLUME SEARCH/BROADCASTING
- LONG RANGE OPERATION IN LOWER ATMOSPHERE

FIGURE 1. GENERAL CHARACTERISTICS OF APPLICATIONS FOR MILLIMETER WAVE EQUIPMENTS

As indicated by the above, the barrier produced by a soggy atmosphere turns out to be very resistant to attack. This barrier can be pushed back a bit by pouring in the joules, but it cannot be penetrated as can be established by examining Figure 2. The curves are for the exponential equation included on the chart and relate the nominal (no-loss) range, R_n, to the range achieved with atmospheric losses of 1, 2 and 3 dB/km (one way). The dashed curves indicate losses which are typical of a summer day in Washington and a moderate rain of 2 mm/hr. The conclusion I draw from this simple exercise is that (high altitude air-to-air, air-to-ground and exoatmospheric applications excepted) mm radars have maximal ranges of 10 km. At less than 5 or 10 km, atmospheric losses are only an annoyance, but beyond about 10 km, they are all important and decisively limiting.

These conclusions concerning the limitations imposed by atmospheric attenuation also apply to millimeter wave communication links although the effects are less pronounced since the signal propagates over a one-way rather than two-way path.

While I am still on the subject of the limitations of millimeter waves, I should mention the effects of atmospheric scatterers and inhomogeneities. Those which produce time varying refraction include

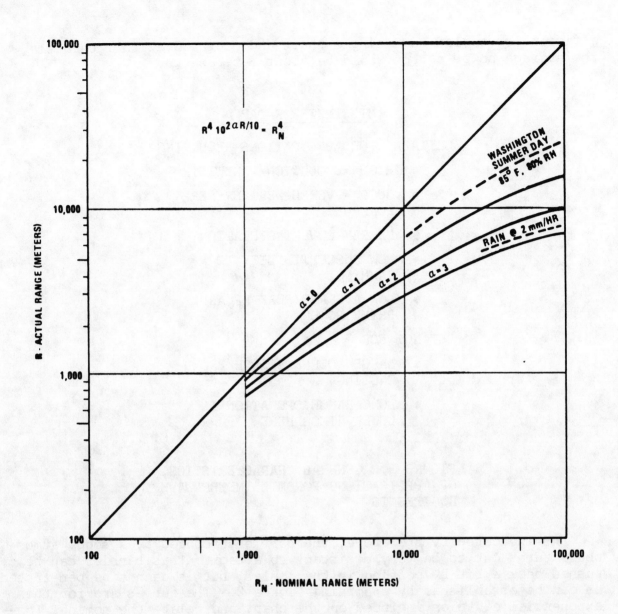

FIGURE 2. RANGE DEGRADATION
DUE TO ATMOSPHERIC ATTENUATION

differentials of temperature, humidity and rainfall. The effect of these is to produce beam wander which results in short term fades. Unless space diversity can be employed, time varying refraction limits the maximum antenna gain which can be used. This time varying refraction is not a significant problem to a radar, but backscatter from rain drops can obscure targets. The use of circular polarization can reduce rain return by 10 dB or more, but the ultimate remedy to clutter backscatter is to reduce the size of the resolution cell. It is interesting that nature limits the magnitude of gain which can be used in a communication link if fading is a problem but demands high gain in a radar to reduce rain clutter.

Turning now to a discussion of the operational advantages provided by millimeter equipments, we note that even the worst of weather produces no more than a few dB per kilometer of attenuation and the clutter from rain varies with both fall rate and wavelength of observation. If we compare a 3 mm radar and a 3 cm radar which are designed for short range usage and are equivalent in all other respects, we note that the equivalence in aperture provides 20 dB more gain at the short wavelength to offset atmospheric attenuation and the equivalence in signal bandwidth (when considered as a percentage of operating frequency) yields a 10-fold improvement in range resolution. Thus, the combined effect of antenna gain and range resolution is to reduce the clutter cell size by 30 dB. It is astonishing to find that under these conditions, the millimeter radar is never as weather limited as the 3 cm radar.

Millimeter waves are totally unaffected by tactical smoke. This is one of the more important points to be made relative to the use of millimeter wave equipments since it is with respect to adverse weather performance that equipments operating at millimeter wavelengths are uniquely superior to those at thermal IR wavelengths. They also provide a limited, but useful, search capability. Equipments which operate at millimeter wavelengths are inferior to their thermal counterparts in all other respects. These include covertness, jam resistance, RF power required, RF bandwidth available, angular resolution, and doppler resolution. However, millimeter wave equipments are superior in all of these latter respects relative to microwave equipments. In short, millimeter waves lie at the geometric mean between the VHF band and the thermal wavelengths. We find that the preferred applications as summarized in Figure 3 are those which require a compromise between the features of either extreme.

To head off the accusations of sophistry that are sure to come during the question period, let me go back to the place where I slipped in that one about equivalence in bandwidth as a percentage of operating frequency. Although there certainly are applications where this is the proper way to think about equivalence, it is also possible to define such as equal time-bandwidth products. In this case the range resolution of the radars are equal and the clutter cell is 20 dB smaller at 3 mm than at 3 cm. At heavy fall rates (e.g.- greater than 25 to 50 mm/hr), the short wavelength radar is less weather limited than the 3 cm radar, but at lower fall rates the clutter effects are up to 10 dB more severe, and of equally great importance.

One noteworthy aspect of Figure 3 is that almost all of the desirable attributes of millimeter wave operation are shared by almost all of the preferred applications. Thus, covertness, jam resistance, and good adverse weather performance are common to all, and all but one or two of these require the wide bandwidth and high angular resolution that can be achieved. The low RF power needed by short wavelength equipments is important in man-portable and forward-area equipments, and fine-grain doppler resolution is useful in several applications.

APPLICATION	COVERTNESS	JAM RESISTANCE	LOW RF POWER RQMT.	GOOD ADV. WEATHER PERF.	WIDE RF BEAMWIDTH	HIGH ANGULAR RES.	DOPPLER RESOLUTION	NOTES & COMMENTS
PT TO PT COMM	●	●	*	*	●			* VALID FOR SHORT LINKS
CHECK PT IDENT.	●	●		*	●	●		* AT SHORT RANGE
TERRAIN FOLLOWING TERRAIN AVOIDANCE	●	●		●	●	●		
LANDING AID	*	*		●	●	●		* IF USED BY HELICOPTERS IN FORWARD AREAS
A-TO-G SURVEILLANCE	●	●		*	●	●	●	* RANGES GREATER THAN 10 KM POSSIBLE. MUST NOT USE SPOILED BEAM
G-TO-A SURVEILLANCE	●	●	●	●			●	BEST USED FOR LIMITED SEARCH (e.g. HORIZON SEARCH)
G-TO-G SURVEILLANCE	●	●	●	●	●	●	●	NOT AFFECTED BY SMOKE
FIRE CONTROL LOW θ GROUND	●	●		●	●	●	●	GOOD ANGLE & RANGE TRACKING
FIRE CONTROL HIGH CLUTTER	●	●	●	●	●	●	●	NEARLY UNIQUE CAPABILITY
MSL SEEKER LOW θ GROUND	●	●	●	●	●	●	●	LOW SCINTILLATION & GLINT EXC FOR MULTI-MODE SKR
MSL SEEKER HIGH CLUTTER	●	●	●	●	●	●	●	NEARLY UNIQUE CAPABILITY
FUZING	●	●	●	●	●	●		IDEAL FOR SLANT RANGE FUZE

Figure 3. Applications for Millimeter Waves

A feature common to all of the preferred applications listed in Figure 3 is that each of them is currently satisfied (to some degree) by microwave radar equipments, thermal IR equipments or human vision. Further, the tactical forces have acquired such equipments in substantial quantity. It appears to me that if (1) all of the projected applications for millimeter wave equipments are satisfied (to some degree) by operational equipments and (2) the numbers of such equipments in the current inventory are large, millimeter communication links and radar sets can supplant the operational hardware only if they can either prove to be substantially superior in performance or provide a major cost advantage. Indeed, they may have to qualify on both counts to succeed.

Despite my doubts, I note that of the twelve applications listed in Figure 3, only the helicopter landing aid is a soft requirement. For the remainder, we can be certain that:

- The insatiated demand for data and other communications will eventually open a market for point to point communication gear at millimeter wavelengths. The combination of features as cited ranging from covertness to channel capacity is surely needed.

- Check point identification by penetrating aircraft (including unmanned cruise missiles) is the key to many strategic missions, and is performed poorly by present equipments. This would appear to be a prime application for early development.

- The need for millimeter wave terrain following/terrain avoidance hardware is not so much based upon the fact that the present microwave equipments do not perform well enough, but rather that they may perform too well. A simple calculation shows that the present radars can be seen by aircraft equipped with intercept receivers for hundreds of miles. (In this application the characteristic of millimeter waves in performing well over limited distances and very poorly over long distances is useful.)

- Although it is probable that most air-to-ground surveillance will continue to be performed at visual or thermal IR wavelengths and much of the remainder by microwave equipments (synthetic aperture and MTI microwave radars), there may be a need for high resolution, real beam millimeter wave radars. The obstacle that I see in fulfilling this application is that such radars will be weather limited by a heavy dew if they use a spoiled beam. The alternative is an agile beam, but we do not yet have the means to rapidly steer beams at millimeter wavelengths.

- Most ground-to-air surveillance applications will be served by visual search, electrooptical equipments or microwave radars. However, one function that can be effectively provided by millimeter wave radar is that of horizon search in which application it will not suffer the multipath limitations of longer-wavelength radars and the weather restrictions of visual/thermal IR wavelengths do not apply.

- The use of millimeter radar for fire control at ranges up to 10 km is a natural application. As noted before, smoke is not a factor and weather is not a limitation at ranges of 10 km or less, and the superior tracking produced by high angular, range and/or doppler resolution radars is a major benefit.

- The advantages cited for millimeter radar in short range fire control apply to both high and low clutter backgrounds. It is clearly evident that the resolution of a millimeter wave radar in angle, range, and doppler will be a decisive advantage in tracking targets which are obscured by clutter.

- Used as a seeker for missiles, an active millimeter radar produces superior tracking relative to active microwave radar since target scintillation and glint are greatly reduced. It is expected that a millimeter radar guided missile is more likely to produce a hit than a microwave radar guided missile.

- The ability of a millimeter wave radar to selectively detect, lock-on to, and track man-made targets in a high clutter environment has been partially demonstrated, but more field measurements are needed to conclusively prove this capability. The incomplete data now available cannot, however, be disputed relative to their demonstration of the unique capability of millimeter wave radars to track targets in high clutter--the major uncertainty is whether such tracking is sufficiently reliable that weapon systems to capitalize upon this capability can be justified.

- Improved Weapon fuzing (and in particular slant range fuzing) is a current need of several developmental weapon systems. At an operating frequency of 300 GHz, an aperture of about one inch produces a beam resolution of about 2° which is adequate for most slant range fuzes. The excellent EC^2M attributes of millimeter wave radars will also be of significant utility in active fuzes.

2. PLANS AND ACCOMPLISHMENTS

As indicated in the introduction to this paper, it would be overly ambitious and somewhat presumptive to attempt a comprehensive review of the accomplishments of the millimeter wave activists during the past three years. We can, however, obtain a good measure of the achievements of this group by reviewing the objectives, recommendations, and series of Millimeter Wave Workshops which have helped promote the common cause of those assembled here.

This series of meetings originated at the ARPA/Tri-Service Millimeter Wave Workshop held at the Applied Physics Laboratory of Johns Hopkins University in December, 1974.

The objective of this first workshop was to "define an integrated program for the development and military utilization of the millimeter (1 mm to 1 cm) portion of the electromagnetic spectrum." I believe we succeeded in this, and many of the remarks of the preceding section were directly drawn from this meeting.

The APL workshop was divided into five topical sessions: Phenomenology chaired by Dr. Bob Hayes; Components chaired by Dr. Harold Jacobs; Communications chaired by Dr. Ed Altschuler; Radar chaired by Dr. Bernie Kulp; and Systems chaired by Mr. V. T. Tarulis. I know that many share with me the opinion that this was a very valuable meeting, and the credit for the success achieved belongs to the panel chairmen along with Dr. Al Stone and Mr. Fletcher Paddison of APL who handled all the arrangements and produced a comprehensive record of the proceedings promptly after the close of the meeting.

In the Fall of 1975 it became evident that a meeting to coordinate planned and on-going projects would be highly beneficial. Dr. Ebeoglu offered to host this meeting at Eglin, and about 40 interested government program managers and engineers convened in November, 1975 at Eglin AFB. I am quite certain that this meeting (#1 in the sequence) produced a high benefit/cost ratio in that duplicative projects were eliminated and replaced with joint projects.

Two of the problems which emerged at the Eglin Meeting were those of adverse weather testing and the measurement of meterological conditions. We accordingly sought out the experts and held the next meeting at Hanscom Field so that staff members from the Air Force's Cambridge Research Center could easily attend. Dr. Ed Altschuler was the host of an exceptionally useful meeting. I know that I was disabused of several notions regarding meterological measurement accuracy.

As indicated in the summary, Figure 4, Meeting #3 was held at the Naval Weapons Center, China Lake thanks to Mssrs Moore and Hooper. The purpose of this workshop was to present the results of work performed during the winter and spring and to coordinate projects. A pattern was now beginning to emerge wherein alternate meetings emphasized project coordination and problems of special interest. The problem which surfaced at China Lake was the prompt communication of results, with special attention given to standardization in instrumentation and data collection. The need for better uniformity being perceived as urgent, a meeting for the Fall of 1975 was accordingly scheduled, and Dr. Kihm of NELC volunteered to be host.

The NELC meeting included sessions devoted to components, phenomenology, and systems. An excellent exchange of recent data resulted from the meeting, and in the follow-on workshop about 20 investigators made their opinions relating to the collection and publication of test data on clutter, atmospherics and like sugjects known. I regret to say the recommendations made at this workshop have not yet appeared in a report or other document. The fault is mine, but I do not know when I will prepare and publish the report. Fortunately, the exchange among the participants appears to have been useful.

The last meeting prior to this was called by Dr. Tegnelia and held in Washington in April of this year. Thanks to the hospitality of Systems Planning Corporation, forty or so attendees assembled across the street from ARPA at the SPC facility. The multiple objectives of this meeting were to discuss the kinds of tests we believed necessary to validate our

WORKSHOP LOCALES AND OBJECTIVES

CHRONOLOGY & TOPICS OF MEETINGS 0 - 5

NOTE: ALL MEETINGS INCLUDED SESSIONS DEVOTED TO THE READING OF TECHNICAL PAPERS PLUS ONE OR MORE WORKSHOP SESSIONS.

MTG	PLACE	HOST	DATES	WORKSHOP OBJECTIVE
0	APL/JHU	DR. AL STONE	16-18 DEC '74	DEFINE INTEGRATED PLAN TO DEVELOP AND UTILIZE mm WAVE PART OF SPECTRUM
1	AFATL	DR. D. EBEOGLU	19 NOV '75	PROGRAM COORDINATION AND TEST PLANNING *
2	AFCRL	DR. E. ALTSHULER	22 JAN '76	ADVERSE WEATHER TESTING
3	NWC	MR. R. MOORE MR. J. HOOPER	22-23 JULY '76	PROJECT PROGRAM COORDINATION *
4	NELC	DR. R.T. KIHM	16-18 NOV '76	STANDARDS FOR DATA COLLECTION
5	ARPA	DR. J. TEGNELIA	26-27 APR '77	1. VALIDATION OF THEORY 2. FORMATION OF STANDING COMMITTEES 3. PROGRAM COORDINATION *

* ATTENDANCE LIMITED TO PERMIT FREE DISCUSSION OF PROCUREMENT SENSITIVE TOPICS.

FIGURE 4. WORKSHOP LOCALES AND OBJECTIVES

Introduction and Overview

expectations for millimeter equipments, re-synchronize projects among the various government laboratories, and establish standing committees for future meetings of this group. The government and industry co-chairmen of the five standing panels which were established at this meeting are identified in your meeting agenda.

The next four charts, Figures 5-A - 5-D, list selected recommendations which have come out of the meetings and workshops. The list is not complete, and some of the recommendations have been consolidated with others of like nature. The order of listing is essentially the order of appearance in my notes and records and does not imply priority or importance. Since there are 28 entries on these charts, I will not discuss each item.

A number of the recommendations on this list are typified by the first, in that work was in progress (in this case on propagation, target/clutter characterization, etc) at the time of the initial meeting, is on-going at this time, and will still be incomplete long after this group has ceased to function. I don't think the recommendations of the ARPA/Tri-Service Workshops have done much to increase the support of this basic activity, but we must, of course, continue to promote measurements programs since a comprehensive understanding of propagation and scatter is fundamental to good communication and radar system design.

Item 4, the development of low cost, high reliability, mm wave integrated circuits is of pivotal importance to the long-term health of mm wave programs of all types. This subject will be addressed again later in this paper.

It is my understanding that the call for the development of mm wave, solid-state combiners, Item 6, has been answered. It is pleasing to note that in December, 1974 designers were working on combiners, that the Workshop reconfirmed the need for such, and that now combiners with the requisite characteristics are becoming available.

A recommendation to develop coherent radars, Item 15, along with Items 17, "clutter model", and Item 18, "improve methods of target acquisition," are all embodied in the ARPA/Army tank fire control radar project, STARTLE. The same co-sponsors are about to award a contract to demonstrate semi-active weapon guidance as called for in Item 21.

Item 25 concerns the specification of equipments in terms of the locale of usage and permissible outage times per year. This is an excellent way to define performance since the user knows what he can expect of the equipment, and the designer is relieved of the burden of ambiguity, as it now stands. One suspects the designer must play twenty questions with the program engineer at the procuring agency to find out enough data to proceed with his work. (Did you say 2 mm/hr? ...Good, but what drop size distribution? ...Well, if you don't have a model, let me use Laws and Parsons distribution ...Now, I need to know what relative humidity to expect ...and the horizontal extent of the rain cell ...and the vertical extent ...and, Oh yes, what temperature did you have in mind?) I'm chagrined that on the projects with which I have been most intimately involved, STARTLE and the Radar Guidance Project, the specification is in terms of mm/hr of rainfall.

RECOMMENDATION	NO ACTION	STARTED OR STARTING	ON-GOING	COMPLETE	COMMENT
1 AUGMENT KNOWLEDGE OF mm WAVE PROPAGATION, TARGET/CLUTTER STATISTICS & ATMOSPHERIC SCINTILLATION			●		
2 PRODUCE mm HANDBOOK SIMILAR TO IR HANDBOOK	●				A SPECIALIZED REPORT LIBRARY HAS BEEN STARTED
3 INVESTIGATE InP mm WAVE DEVICES			●		
4 DEVELOP LOW COST, HIGHLY RELIABLE mm WAVE INTEGRATED CIRCUITS			●		COST CANNOT BE PROVEN UNTIL A QUANTITY "BUY" IS MADE
5 DEVELOP 3 mm TWT WITH 5 KW PEAK, 500 W AVERAGE OUTPUT	●				
6 DEVELOP 3 mm SOLID-STATE COMBINER WITH 1 W OUTPUT				●	
7 DEVELOP MONOLITHIC CONFORMAL ARRAYS	●				

FIGURE 5-A SUMMARY OF WORKSHOP RESULTS (Page 1)

RECOMMENDATION	NO ACTION	STARTED OR STARTING	ON-GOING	COMPLETE	COMMENT
8 DEVELOP ESM RECEIVERS	●				
9 DETERMINE O_2 BAND PROP., VARIATIONS WITH METEOROLOGY MEASURE SCINTILLATION RATES AND MAGNITUDES					
10 OBTAIN MORE DATA ON UPPER ATMOSPHERE ATTENUATION					
11 OBTAIN MORE DATA ON DEPOLARIZATION EFFECTS BELOW 60 GHz					
12 SURVEY CHANNEL IMPAIRMENT ABOVE 60 GHz					
13 COMPARE PROPOGATION AT 35 AND 95 GHz					
14 DEVELOP OBSTACLE AVOIDANCE RADARS AT mm WAVELENGTHS	●				

FIGURE 5-B SUMMARY OF WORKSHOP RESULTS   (Page 2)

RECOMMENDATION	NO ACTION	STARTED OR STARTING	ON-GOING	COMPLETE	COMMENT
15 DEVELOP COHERENT SYSTEMS		● AND	●		
16 IMPROVE STANDARDS AND TEST EQUIPMENT			●		
17 ESTABLISH CLUTTER MODEL			●		PRESENT MODEL NEEDS IMPROVEMENT
18 IMPROVE METHODS OF TARGET ACQUISITION AND IDENTIFICATION			●		
19 DEVELOP mm WAVE SENSORS FOR MAP-MATCHING GUIDANCE	●				
20 DEVELOP mm WAVE EQUIVALENTS TO EO GUIDED WEAPONS	●				FIRST FIELD MEASUREMENTS OF A SEEKER FOR MAVERICK HAVE BEEN MADE BY HAC AS 6.2 LEVEL PROJECT
21 MAKE FULL SYSTEM DEMONSTRATION OF SEMI-ACTIVE GUIDANCE			●		

FIGURE 5-C SUMMARY OF WORKSHOP RESULTS (Page 3)

Introduction and Overview

RECOMMENDATION	NO ACTION	STARTED OR STARTING	ON-GOING	COMPLETE	COMMENT
22 MAKE FULL SYSTEM DEMONSTRATION OF AIRCRAFT TRACKER	●				
23 DEVELOP PHASE SHIFTERS AND BEAM AGILE ANTENNAS			●		
24 EMPHASIZE TARGET AND CLUTTER TESTS WITH WET BACKGROUNDS					
25 SPECIFY ADVERSE WEATHER IN HOURS PER YEAR OF OUTRAGE	●				100 HRS/YEAR INCLUDES ALL RAINS OF 4 mm/HR OR MORE IN MANY TEMPERATE LOCALES (e.g. CENTRAL FRG, WASHINGTON D.C)
26 DO NOT USE MODELS TO PREDICT ADVERSE WEATHER PERFORMANCE	●				CANNOT COMPLY - THEREFORE MODELS MUST BE IMPROVED
27 LOBBY FOR ALLOCATION OF 3 GH_z OR MORE AT 1979 GWARC			●		SEE BOB MOORE, NWC OR HUGO HARDT, TRACOR
28 DEVELOP STANDARD FIELD TEST INSTRUMENTS, PROCEDURES AND DATA SHEETS		●			GOOD WORKSHOP ON THIS SUBJECT AT NELC. NO CURRENT FOLLOW-THROUGH

FIGURE 5-D   SUMMARY OF WORKSHOP RESULTS   (Page 4)

Even though we should not specify our equipments in terms of an incomplete data set and model as suggested in Item 26, I for one do not see how we can avoid using models to judge between competitive proposals. Thus, I believe we must improve our models.

One final comment upon the recommendations which have come from the ARPA/Tri-Service Workshops. We have been advised that unless an allocation is made at the meeting of the General World Administrative Conference in Geneva in 1979, we will definitely and unequivocally be out of business. The merit of Item 27 is therefore self-evident.

TASK & PROJECT SUGGESTIONS: UNRESOLVED PROBLEMS & ISSUES

For the dual reasons that I am not qualified to probe deeply into the requirements for mm wave communications and that time is short, I shall restrict my remarks to mm wave radar. I believe that my suggestions may, however, be applicable to communications equipments as well as to radar sets.

In preparing my paper I cast my thoughts into several patterns, but I kept coming back to two questions as underlying all other problems and issues. Specifically

"Is a mm wave equipment the best possible radar or the worst possible electrooptical sensor? and

Is a mm wave equipment the worst possible radar or the best possible electrooptical sensor?"

A good argument can be presented for any of the theses implied in these questions.

The trade-off topics that must be considered in choosing among the wavelength options are effectiveness in search and track, smoke and adverse weather performance, capability to classify and identify, covertness, jam resistance and cost. By-passing the cost issue for the time being, these trade-offs are arrayed in Figure 6. It is seen that a millimeter wave radar would be preferred to a microwave radar is one or more of the attributes of tracking accuracy, classification/identification, covertness and jam resistance were more important than either (or both) volume search or adverse weather performance.

TRADE-OFF ISSUES

mm-Wave relative to µ-Wave	Thermal IR relative to mm-Wave
Tracking accuracy	Tracking accuracy
Classification/Identification	Classification/Identification
Covertness	Covertness
Anti-jam	Anti-jam

µ-Wave relative to mm-Wave	mm-Wave relative to thermal IR
Volume search	Volume search
Adverse weather performance	Adverse weather performance

FIGURE 6. TRADE-OFF ISSUES

Turning now to the choice between thermal IR and millimeter wavelengths, we find that exactly the same criteria reappear and the same conclusions must be drawn. It may therefore seem that there is no role for millimeter wave radars--nor, for that matter, for X- or K_u-Band radars.

We have looked for attributes of millimeter wave radars which are unique and found none. Thus, if there is a role for millimeter wave radar in tactical warfare, it will be found in a combination of features which make a millimeter wave radar more effective than a counterpart microwave or IR radar. Such combinations of attributes do appear to exist, and we must concentrate upon proving these within the next two or three years before the patience of the sponsors of millimeter wave research and development is exhausted.

Cost is a major factor since we can supplant proven and familiar equipments only if our candidate radar is substantially superior, lower in cost, or both. A 3 dB improvement will not be enough, and this may be the limit of the performance advantage of the first generation hardware, but if we can cut the cost in half and be twice as effective, we achieve a 6 dB gain and this may suffice.

Low cost implies low power since a large cost factor in a conventional radar is the power supply and the transmitter/ modulator. However, the effective ranges of a millimeter radar are only a few kilometers and require only a watt or so of RF power. We can therefore anticipate that the transmitter cost will be modest.

Let's suppose that we have a low cost, short range radar. What will we do with it?

First of all, I would use it in bad weather since, as we have seen, a short range millimeter wave radar is the most effective of all sensors for use in adverse weather.

Second, I would use it to equip men and vehicles for direct fire engagements since the 10 km range performance of a millimeter wave radar is more than sufficient for this use.

Third, I would use it to maximize those attributes which differ most from those provided by a microwave radar. (Since I have qualified my usage to be in adverse weather, the only alternative to a millimeter wave radar is a microwave radar*). The way in which I propose we do this is to concentrate upon the extraction of target classification (and sometimes identification) features from the return.

---

*An adverse weather sensor designed for use in forward areas must also be effective in fair weather since only one set of equipments will be carried into the direct fire engagement region.

Four, I would search out system applications for forward area usage and apply a variety of sensors to the solution of these system requirements. This is not a "cop out". Many of our most successful systems are built upon this principal, and it is only in recent years that we have tried to field universal sensors. HAWK and HERCULES Air Defense Systems as well as the radar suites aboard ships are examples of multi-sensor systems and PATRIOT and AEGIS are examples of systems dependent upon the performance of a universal sensor. In the context of this discussion, a millimeter wave radar might be used cooperatively with CO_2 laser radar to gate the latter and thereby substantially improve its adverse weather performance.

Referring to an earlier list of preferred applications for millimeter radars (Figure 3), the ones which best satisfy the criteria as stated above are:

- A helicopter radar for air-to-ground surveillance, target acquisition and fire control,
- A horizon search radar for detecting low-flying aircraft,
- A fire control radar for ground-to-ground use in a high clutter background,
- A missile seeker for use in a high clutter background, and
- A slant range fuze.

Projects to develop some of the above are in progress, but not all elements of the ultimate design are as yet available. These include:

- A control concept which allows for the adaptation of the radar to the local environment,
- A means of providing beam agility (at an affordable cost) to facilitate target classification and provide track-while-scan operation,
- A means of economically and efficiently using the tens or hundreds of milliwatts available from solid-state sources in coherent, high resolution radars.

In addition to the design elements, we need to understand more precisely how the world appears to a millimeter wave radar. Thus, a systematic investigation of clutter and target characteristics, millimeter radars must be a part of our overall millimeter wave program.

I believe we are right in advocating the development of millimeter wave equipments and systems. Their characteristic of working very well at intermediate ranges in adverse weather and smoke is a sufficient condition for the development and test of millimeter wave hardware.

The Potential Military Applications of Millimeter Waves
L.R. Whicker and D.C. Webb

SUMMARY

This paper reviews the propagation characteristics of millimeter waves, considering effects of rain, clouds and fog. The fundamental limitations of microwave, millimeter and optical systems are discussed and the strengths and weaknesses of each class of system are outlined. Based on these considerations, the most promising application areas for millimeter waves are outlined. Applications for radar, communication and electronic warfare are discussed. Additionally needed component research and development activities are considered.

1.0 INTRODUCTION

Within the past few years, there has been a resurgence of interest in millimeter waves. (SKOLNIK, M.I., 1970), (JOHNSTON, S.L., 1977), (WEISBERG, L.R., 1977), (STROM, L.D., 1977) This is partly attributable to advances in component technology but also reflects the changing priorities of military systems. There is increased emphasis on employing as well as detecting small platforms [(e.g., missiles and RPV's (Remotely Piloted Vehicles)]. The necessity of achieving high resolution as well as being able to penetrate fog and smoke leads to the use of millimeter waves either as the primary system or in complementary equipment to be integrated with present microwave or optical equipment.

This paper first reviews the propagation factors of millimeter waves. The effects of rain, clouds and fog are considered. The fundamental limitations of microwave, millimeter wave and optical systems are described. The strengths and weaknesses of each class are outlined. Based upon these considerations, promising application areas for millimeter waves are outlined. The areas of application are subdivided into radar, communication and electronic warfare. A final section addresses component needs to effectively meet future system requirements.

2.0 CHARACTERISTICS OF MILLIMETER WAVE PROPAGATION

2.1 Weather

A prime consideration in the design and performance of a millimeter wave system is the atmospheric attenuation under the range of anticipated weather conditions. In general the system employs one of two modes: 1) a low attenuation mode for maximum range or 2) an LPI (low probability of intercept) mode for short range secure operation. Figure 1 indicates propagation loss in clear weather and in various types of inclement weather.

For reasonably low values of propagation attenuation in the millimeter wavelength portion of the spectrum the system must operate within one of the atmospheric windows. As is indicated in Figure 1, these windows are centered at 94, 140, and 220 GHz respectively. All have a percentage bandwidth of roughly 20%. The "clear air" attenuation between 94 GHz and the far-infrared is dominated by water-vapor absorption. Under hot humid conditions the attenuation in the windows may increase by as much as a factor of five over the values noted in the figure. Attenuation in the infrared windows centered at 10 and 4 microns is comparable to that of the 94 GHz window and is lower than that of the 140 and 220 GHz windows.

For LPI applications a few dB/km is usually desired and frequencies near the absorption peak centered at 60 GHz are commonly employed. Since this peak is a result of absorption by the oxygen molecule it is relatively stable under changes in temperature and relative humidity. Furthermore, componentry is considerably simpler here than at the higher frequencies where comparable attenuation can be achieved.

Attenuation due to a moderately heavy rain, 10 mm/hr., is also indicated in Figure 1. In general it depends both on particle size and particle density but remains roughly constant from 94 GHz to the visible portion of the spectrum. Below 94 GHz the attenuation due to rain decreases monotonically. Backscatter from raindrops is not a significant problem at frequencies of 94 GHz and higher. (RAINWATER, J.H., 1977)

Figure 1 also shows attenuation due to a heavy fog and a moderately dense cloud. For both the loss at optical wavelengths is prohibitive but is not excessive at millimeter wavelengths and is negligible at microwave frequencies. Characteristics for commonly used smokes are similar, i.e., an attenuation of 100 dB/km is common at infrared wavelengths while there is no measurable attenuation in the millimeter wave portion of the spectrum.

An important point to note is that it is frequently impractical to compensate for the degradation of system performance caused by inclement weather by simply increasing transmitter power. For example, if the propagation attenuation is 5 dB/km, the power must be increased by a factor of ten for each additional kilometer of range.

2.2 Resolution

A prime motivation for employing millimeter wavelengths rather than microwaves is the need for a narrow beam without an excessively large antenna aperture. The diffraction limited angular resolution, Ω, is given by the expression:

$$\Omega = 1.2\lambda/D$$

where λ is the operating wavelength and D is the antenna diameter. Typical high resolution requirements are $\Omega = 10^{-2}$ to 10^{-3} radians for beam riders and tracking radars and $\Omega = 10^{-4}$ and 10^{-3} radians for target detection, identification and classification. The maximum allowable antenna diameter may be a meter or less for aircraft and ground vehicles and as little as a few tens of centimeters for missiles, satellites, RPV's and hand-held radar. There are a significant number of military applications where the resolution offered by conventional microwave systems is inadequate.

For many applications even millimeter waves provide marginal or sub-marginal resolution. For example, a 1.5 meter diameter antenna operating at 230 GHz provides an angular resolution of only 10^{-3} radians. For higher resolution, synthetic aperture radar techniques (TOMIYASU, K., 1978) can be used from moving platforms, otherwise optical techniques must be employed.

2.3   Comparison of Microwave, Millimeter and Optical Systems

As noted above millimeter waves are employed rather than microwaves either to realize a narrower beam in an aperture limited system or to achieve covertness through moderately high atmospheric attenuation. Other advantages of millimeter wave systems are the increased bandwidth capability hence greater information carrying potential and better immunity to countermeasures. Furthermore, components and circuitry are more compact, an important consideration in missile, RPV and man-pack applications.

Where the above considerations are not overriding, microwave systems are usually preferable. Propagation loss is significantly lower in all weather conditions and high power sources and circuitry are much more readily available. These factors lead to much greater range for microwave systems than for millimeter wave systems. Furthermore, most components are more difficult to fabricate at millimeter wavelengths because of the required dimensional accuracies, and are more expensive than their microwave counterparts.

In high resolution applications where millimeter waves have clear advantages over microwaves, millimeter wave systems must compete with the generally better developed thermal imaging systems. The latter possess a 2 to 3 order of magnitude smaller wavelength and by using the powerful focal plane array technology can achieve a considerably greater clear weather range than millimeter wave systems. However, since they are passive an auxiliary active system is required to derive range and velocity information. This may be either a laser or if the clutter problems are not severe may be a complete microwave radar.

The clear advantage millimeter wave systems have over optical systems is their ability to penetrate clouds, smoke, fog and haze. The frequency and predictability of one or more of these conditions often dictates that military systems be operable in them to prevent the enemy from concealing his movements. For example, in Western Europe in approximately one of three mornings during the fall and winter the visibility is reduced to less than 1 km by ground fog. Also, roughly two-thirds of the time the total cloud cover exceeds 50% in the North Atlantic. Effective surveillance and/or operation under these weather conditions mandates use of microwave or millimeter wave systems as primary systems or as backups to optical systems.

3.0   RADAR APPLICATIONS

From the time of World War II much funding has gone into the development and optimization of microwave radar systems. These systems have been designed for particular functions and missions operating at frequencies from UHF to 35 GHz. Generally the search radar function has been accomplished at frequencies below 4 GHz while functions requiring higher resolution, such as airborne ground mapping, have been realized at frequencies up to and above 10 GHz. Radars at 35 GHz, which are technically in the longer millimeter region, are identified generally with the microwave portion of the spectrum. At frequencies above 35 GHz where the high resolution and large bandwidth capabilities of millimeter wave systems might be exploited, considerably less work has been done. Instead, effort has been concentrated in the infrared and visible portion of the spectrum. Table I which is based on considerations stated earlier, lists some of the tradeoffs that must be considered in selecting a radar's operating frequency. As indicated in this chart both optical and microwave radar systems offer certain performance features superior to that promised by millimeter systems. For example, a microwave system is clearly better for a large volume search while an optical system in clear weather is better for an imaging radar. Thus, to find considerable and important applications, the millimeter systems must either complement microwave or optical systems or exhibit a combination of features which make them more effective than their microwave or optical counterparts. Several examples are given below.

TABLE I
RADAR SYSTEM COMPARISONS

RADAR CHARACTERISTIC	MICROWAVE	MILLIMETER WAVE	OPTICAL
TRACKING ACCURACY	✓	✓✓	✓✓✓
CLASSIFICATION/IDENTIFICATION	✓	✓✓	✓✓✓
COVERTNESS	✓	✓✓	✓✓✓
VOLUME SEARCH	✓✓✓	✓✓	✓
ADVERSE WEATHER PERFORMANCE	✓✓✓	✓✓	✓
PERFORMANCE IN SMOKE	✓✓✓	✓✓✓	✓

✓ POOR
✓✓ INTERMEDIATE
✓✓✓ GOOD

3.1   Millimeter Wave Radar Systems to Augment or Replace Optical Systems

The era of precision weapons has materialized with the invention of the lasers and its utilization with "smart" bombs. However, the poor performance of optical systems using lasers or FLIR's in a fog or smoke environment make millimeter systems attractive for this application either as a replacement for or as a complement to present optical systems. Other areas where lasers and IR systems may be replaced include:

- Target Designators
- Beam Riders
- Range Finders
- Passive Seekers
- Detection and Classification
- Imaging

All these applications have the common requirements that the range is relatively short, 1-5 km and the beamwidth is not prohibitively small, 1-10 milliradians.

Short range target detection, identification and classification in a smoke or fog environment both on land and at sea is an important application for millimeter imaging radars. For such a radar an antenna aperture of approximately 10 square meters is required. While such an antenna is too large to mount on a vehicle such as a tank, such a system could be mounted on an auxiliary vehicle. Received data would then be transmitted to other vehicles within the immediate area. With antennas of one meter diameter or smaller, ground vehicles may still be detected and tracked with millimeter waves. Classification of such vehicles may be possible by analysis of the radar returns including the glint characteristics.

Imaging with smaller antenna apertures is possible on moving objects such as RPV's by employing synthetic aperture techniques. Optimum center frequencies for such systems--94 GHz, 140 GHz or 220 GHz--must be determined from tradeoff studies which consider resolution vs. atmospheric attenuation and available power.

3.2 Millimeter Radar Systems to Augment or Replace Microwave Radar Systems

One important application for millimeter shipboard search and track radars is for the detection and tracking of low altitude cruise missiles. It is well known that when a radar tracks targets at elevation angles of the order of an antenna beamwidth or less, the ground reflections produce signals that can cause erroneous measurements of elevation angle resulting in poor tracking. Thus, there is need for the narrow-beam millimeter system. The millimeter system of short to intermediate range can be used to augment other shipboard radars. Recent developments of gyratron tubes (GODLOVE, T.F., GRANATSTEIN, V.L., 1977) give promise of providing suitable power sources at 35 and 94 GHz. This function can be accomplished, also, with an optical system. However, the all-weather capability of the millimeter wave system makes it the logical choice.

Another potentially important application for millimeter wave systems is the LPI radar mentioned previously. Microwave systems utilize frequency hopping and/or matched filtering to achieve security. Millimeter wave systems gain from their narrower beams and also could take advantage of the 60 GHz oxygen absorption line. Such systems could vary their frequency to adjust for weather conditions. The maximum absorption offered at 60 GHz would limit the range of such radars but minimize the danger of detection at distances larger than the operating range.

Other possible microwave-replacement applications include terrain following/terrain avoidance and fire control. The main advantage millimeter wavelengths offer in the former application is their greater security due to the smaller beamwidth and the greater atmospheric attenuation. Millimeter fire control radars would benefit from the increased angular, range and/or doppler resolution.

In still other millimeter radar applications the size and weight reduction of millimeter wave systems vs. their microwave counterpart is the most important feature. Examples are:

- High resolution radars for mini-RPV's and satellites
- Active fuses
- Active missile seekers
- Hand held radars

4.0 COMMUNICATIONS APPLICATIONS

Over the past several years there has been much discussion of the faults and virtues of millimeter wavelength communication links. (SKOLNIK, M.I., 1970) The broad bandwidth capabilities lead to large channel capacity; however, adverse weather conditions can lead to poorer performance than microwave links. In some commercial applications closed waveguide systems can be used. Such systems use the TE_{01} circular waveguide. Here the waveguide may be filled with gas and avoid the irregularities of propagation experienced within the earth's atmosphere. Closed systems appear to have limited military applications, however.

An area of greater potential utilization for military applications is low probability of intercept (LPI) communication links. (VIGNALI, J.A., 1970) As in the case of LPI radar, an LPI communication link would use a narrow beam antenna at frequencies near 60 GHz where the atmospheric attenuation limits the range of detectability. Above the earth's atmosphere a 60 GHz satellite-satellite link can operate with little attenuation. Such systems would be protected from detection or interference from the earth's surface. The other two areas of LPI communication links are for short range (1-5 kw) ship-to-ship communication and for battlefield communications. In these applications the ability of millimeter waves to penetrate fog or smoke is an essential feature. For battlefield applications where large numbers of terminals might be used by individual foot soldiers the terminals must be quite small and lightweight and offer low cost.

5.0 ELECTRONIC COUNTERMEASURE (ECM) APPLICATIONS

As millimeter wavelength radar and communication systems are developed to augment or replace optical and microwave systems, it becomes necessary to develop electronic countermeasures. Before countermeasures can be utilized, the unfriendly millimeter wavelength signal must be located and identified. Thus a first priority in ECM equipment is in surveillance receivers. Such receivers must exhibit good noise characteristics and must cover the various windows between 35 and 300 GHz. An additional band near 60 GHz is required for intercepting LPI radar and communications signals.

In addition to surveillance receivers, the ECM community needs broadband power sources to provide jamming power to defeat the potential threats. Such power must come from conventional microwave tubes or from new tubes using relativistic electron-beams such as the gyratron.

6.0 COMPONENT RESEARCH AND DEVELOPMENT NEEDS

In order to realize new millimeter wavelength radar, communication and ECM systems, substantial military support at the component or technology level is required. Areas requiring particular attention are listed in Table II.

TABLE II
COMPONENT R&D AREAS

SOURCES

(a) High Power
- Conventional Tubes
- Conventional Tubes with New Slow Wave Structures
- New Relativistic Beam Power Tubes Such as Gyratron
  Broadbanding

(b) Medium and Low Power
- IMPATT Power Combining
- High Frequency Operation

CIRCUITS
- Waveguide Broadbanding
- Integrated Circuits
- Quasi-Optical Techniques

SOLID STATE COMPONENTS
- Low Noise Mixers
- Phase Shifters
- JJ Detectors

This table is broken down into three categories--sources, circuits, and solid state components. Some comments on each of these areas follow.

6.1 Sources

High power in the millimeter wave region as in the microwave region is obtained from vacuum tubes. Certain types of tubes (coupled cavity tubes) have provided 100-200 watts of cw power at frequencies in the 30-94 GHz range. However, such tubes are expensive to fabricate and require the use of extremely high tolerance parts. Along with the continued development of conventional tubes, new approaches to forming slow wave structures for millimeter tubes need be investigated.

Applications such as shipboard low angle tracking radar require considerable power. New tubes using relativistic electron beams such as the gyratron promise to provide tens to hundreds of kilowatts of power at frequencies up to 300 GHz. (GODLOVE, T.F., GRANATSTEIN, V.L., 1977) Further work on such tubes is required to bring the technology to the point where they are deployable in systems. Broadbanding studies in particular need to be pursued to address EW and LPI system requirements.

For many millimeter wave military applications the high power capabilities of the tubes will not be required and solid state sources will be adequate. For short range imaging systems, missile seekers and target designators, typical power requirements are 1-10 watts. In excess of 1 watt cw power has been obtained at 94 GHz (STROM, L.D., 1977) by the use of power combining. Recently IMPATT diode oscillators have been reported that offer a cw power output of 50 mW at 202 GHz. (ISHIBASHI, T., OHMARI,M., 1976) Achieving high power with solid state diodes, especially at the lower millimeter wavelengths, is a high priority requirement

6.2 Circuits

In the circuits area, work is needed to improve the bandwidth capabilities of standard waveguide components such as couplers and magic tees. Component development in integrated circuit format is needed also for both cost and size reduction for anticipated high volume applications such as missile seekers and target designators. Several types of transmission lines are needed to be compatible with the various component configurations which may be employed throughout the millimeter wave spectrum. Conventional microstrip approaches are not attractive for frequencies of 94 GHz and higher and work is being concentrated on other formats, such as dielectric waveguide and suspended microstrip. Quasi-optical techniques appear to be useful for the higher millimeter frequencies. (GUSTINCIC, J.J., 1977)

6.3 Solid State Components

A common requirement for all military millimeter wave systems is the need for a high performance receiver. GaAs Schottky barrier diodes have been the most successful at millimeter wavelengths to date with a 5 dB noise figure reported for a 95 GHz sub-harmonically pumped mixer. (SCHNEIDER, M.V., CARLSON, E.R., 1977) Again, further work is needed to realize low cost, ruggedized receivers suitable for military applications. Additional emphasis on wideband versions is required also.

For extremely low noise operation cryogenic receiver technology is being studied. For example, a mixer employing a Josephson Junction as the nonlinear element exhibited a $55^{\circ}K$ noise temperature at 36 GHz. (TAUR, Y., CLAUSSEN, J.H., RICHARDS, P.L., 1974) These devices are likely to be competitive with Schottky barrier devices at the lower millimeter wavelengths; however, significant utilization in military applications is at least 5 to 10 years away because of the devices present limitations in dynamic range and the need for improved planar (non-point-contact) versions. Improved closed cycle refrigerator technology is also essential.

Electronically steerable beams at 94 GHz and higher frequencies are not currently practical because of limitations in control components. Innovative approaches are needed in phase shifters and modulators to overcome the severe cost and fabrication problems at these frequencies. Similar limitations exist for other commonly used solid state components, e.g., circulators, isolators, switches. Again realization of broadband components is an area which needs particular attention.

7.0 CONCLUSIONS

At this point in time, millimeter waves have found only limited use in military systems. This is true since most system needs could adequately be met with better developed microwave and optical systems. This paper has focussed on a range of applications where millimeter waves exhibit a combination of features which make their use attractive compared to other systems. It is felt that millimeter waves will find military application in many of the areas described in this paper. It is difficult at this time to predict which areas of application will be the most important. However, the increasing emphasis on high resolution, small size, high information capability and all weather performance points toward an important role for millimeter waves in future systems.

REFERENCES

GODLOVE, T.F., GRANATSTEIN, V.L., 1977, "Relativistic electron beam interactions for generation of high power at microwave frequencies," 1977 IEEE MTT-S International Microwave Symposium Digest, p.69.

GUSTINCIC, J.J., 1977, "A quasi-optical receiver design," 1977 IEEE MTT-S International Microwave Symposium Digest, p.99.

ISHIBASHI, T., OHMORI, M., 1976, "200 GHz 50 mw cw oscillators with silicon IMPATT diodes," IEEE Trans. Micro.Theory Tech., Vol.MTT-24, p.85.

JOHNSON, S.L., 1977, "Millimeter radar," Microwave Journal, Vol.20, p.18.

RAINWATER, J.H., 1977, "Weather affects mm-wave missile guidance systems," Microwaves, Vol.16, p.62.

SCHNEIDER, M.V., CARLSON, E.R., 1977, "Notch-front diodes for millimetre-wave integrated circuits," Electronics Letters; Vol.24, p.745.

SKOLNIK, M.I., 1970, "Millimeter and submillimeter wave applications," Proc.Symp.on Submillimeter waves," Polytechnic Press of Poly.Inst.of Brooklyn, NY, p.9.

STROM, L.D., 1977, "The unresolved problems and issues," (U), Proc.of Sixth DARPA/Tri-Service Millimeter Wave Conference, p.10 (SECRET).

TAUR, Y., CLAUSSEN, J.H., RICHARDS, P.L., 1974, "Josephson junctions as heterodyne detectors," IEEE Trans.on Micro.Theory Tech., Vol.MTT-22, p.1005.

TOMIYASU, K., 1978, "Tutorial review of synthetic-aperture radar (SAR) with applications to imaging of the ocean surface," Proc.of IEEE, Vol.$\underline{66}$, p.563.

VIGNALI, J.A., 1970, "Millimeter waves and Naval airborne communications," NRL Report 7165.

WEISBERG, L.R., 1977, "Millimeter waves--the coming of age," (U), Proc.of Sixth DARPA/Tri-Service Millimeter Wave Conference, p.4 (SECRET).

DISCUSSION

R.P.Moore, US
What are the assumptions underlying the requirements for resolution required for tracking and target identification and did they take into account other signal characteristics that can be used for identification such as spectral content and thus reduce required resolution and thus antenna diameter required by several fold? Optical resolution criteria applied to mm-wave systems often lead to erroneous conclusions concerning system feasibility, and effectiveness also. Have the requirements taken into account doppler beam sharpness and other techniques used at lower frequencies to obtain better resolution?

Author's Reply
This paper is general in nature. The resolution requirements listed were based on the equation $\Omega = 1.2\lambda/D$. It is true that based on signal return characteristics somewhat smaller antenna sizes may be utilized. The characteristics of specific targets were not given. As indicated

$$\Omega = 10^{-2} \text{ to } 10^{-3} \text{ radians for beam riders and tracking radars.}$$

$$\Omega = 10^{-4} \text{ to } 10^{-3} \text{ radians for target identification and classification.}$$

Specific returns from targets would allow for the larger Ω and thus the smaller antenna, doppler information, glint, and other specific target returns may be utilized.

Millimeter Wave Technology and Applications

J.C. Wiltse

Microwave Journal, Vol. 22, No. 8, Aug. 1979
Pages 39ff.

© 1979 by Horizon House-Microwave, Inc. Reprinted by permission.

INTRODUCTION

The past several years have seen a strong revival of research and applications in the millimeter-wave region, with related investigations extending into the submillimeter-wave range. Several factors have contributed to this revival: they include the advent of new technology; the evolution of new requirements for sensors; and the general superiority of millimeter-wave systems over optical and infrared systems for penetration of smoke, fog, haze, dust, and other adverse environments. The improved technology includes better sources (such as IMPATTS, Gunn oscillators, gyrotrons, extended interaction oscillators, and magnetrons) which have higher power outputs and/or operate at higher frequencies, and in some cases have longer lifetimes. Lower-noise mixers have also been developed, providing noise temperatures as low as 600°K (uncooled) or 200°K (cooled) near 100 GHz. Component development has progressed, too, particularly in the areas of integrated circuits, image lines, fin-line waveguides, and quasi-optical techniques. Nonetheless, there is room for improvement in components and devices, particularly in extending ferrite or other non-linear devices to frequencies above 100 GHz and in reducing losses in most components.

For low power applications (watts of peak power, milliwatts average) solid state IMPATTS have seen extensive development and now operate to 300 GHz.[1-4] Gunn oscillators are available to 100 GHz,[5-7] klystrons to 200 GHz, and carcinotrons to 1000 GHz.[8,9] In medium power tubes, the magnetron has been improved in lifetime and is available to 95 GHz, and the extended interaction oscillator (EIO) has appeared and is in use up to 140 GHz.[8,10] An EIO has been operated at 260 GHz and pulsed units are being developed for 220 GHz. A breakthrough in extremely high power has been obtained with the gyrotron tube and related devices, which have already achieved megawatts peak at efficiencies greater than 30%, and hundreds of kilowatts CW at efficiencies near 50%.[11,12] The gyrotron has seen application in fusion research and undoubtedly will soon be applied in radars. Optically pumped lasers can also be used to produce significant power (up to tens of kilowatts peak) at specific millimeter and submillimeter frequencies, but efficiencies and pulse repetition rates are low.[13]

In the mixer area, both the microwave structures[14-18] and diode materials[19-22] have been extensively studied. After a decade of effort, mixer noise temperatures have been lowered by an order of magnitude. Another innovation has been the subharmonic mixing, wherein employing two back-to-back diodes permits the use of a local oscillator frequency of 1/2 or 1/4 the normal fundamental LO with excellent performance. This permits building all-solid state receivers and radiometers at frequencies much higher (2 to 4 times) than those available from solid state sources.[17] Of course, another approach is to use harmonic generators as sources for fundamental mixing.[23,24] The various types of diodes investigated in recent years have included superconducting tunnel junctions, point-contact Josephson junctions, Schottky devices, Mott diodes (a low-doped Schottky with an epitaxial layer that is completely depleted at zero bias), and Mottky diodes[20] (Schottky barrier diode with a thin epitaxial layer, approximating a Mott). An additional advantage for some of these mixers is a lower requirement for local oscillator power (as low as the order of 0.1 milliwatt). Some of the best research has been done for radio astronomy applications, but the results carry over directly to other fields. Important considerations for some of these mixers include being ruggedized and being made more producible. Nonetheless, the improvements have led to better receivers for radar, communications, and radiometry (remote sensing and radio astronomy applications).

The recent growth in millimeter wave applications is certainly attributable in part to the availability of improved building blocks, particularly solid state sources. While such sources provide relatively low power, they are smaller in overall size and require much lower voltages and prime powers than vacuum tube types, such as magnetrons, EIOs, and klystrons. The most obvious improvements in solid state sources during the past several years have been the availability of higher power outputs and operation at much higher frequencies. Some of the other improvements are more subtle, but very important for systems; these include development of frequency stabilized or phase-locked sources, injection-locked IMPATT amplifiers, and frequency-doubled Gunn oscillators. (As an aside, frequency stabilization and phase-locking are also being developed for pulsed EIOs.) These types of improvements are permitting the extension of all-solid state pulse-compression and coherent MTI (moving target indication) radars to as high as 94 GHz. Similarly, all-solid state passive radiometers are being extended to at least 140 GHz.

In general, atmospheric propagation efforts dominate considerations relating to applications. This is true even for satellite applications outside the atmosphere, since frequencies may be chosen for which the atmosphere is opaque, thus preventing detection of satellite-to-satellite communications by ground-based receivers. Terrestrial systems desiring to prevent signal "overshoot" in range may similarly operate at a frequency of high atmospheric absorption to gain a degree of covertness. Typical values of atmospheric attenuation have been well-established for frequencies up to 100 GHz, including the attenuation and backscatter due to rain or fog, and programs have now been initiated to obtain better information about atmospheric effects above 100 GHz. Of particular interest are the effects in atmospheric "windows," or attenuation minima, near 35, 94, 140, and 220 GHz. Atmospheric turbulence effects due to time-varying localized temperature and humidity variations in clear air produce amplitude scintillations and angle-of-arrival changes, and these are now being investigated in more detail.[25,26] The effects of smoke and/or dust are also being measured. One example is a test ("DIRT-I") conducted in September, 1978, at White Sands Proving Ground by the Army Atmospheric Sciences Laboratory. Propagation through a dust and oil smoke cloud was measured at 94 and 140 GHz and at infrared wavelengths. In other programs the reflectivities and radiometric signatures of land and sea (i.e., "clutter"), as well as vehicles and other targets, are also being obtained at the "window" frequencies.[27]

APPLICATIONS

One of the areas of greatest activity in millimeter waves is that of guidance for missiles and projectiles.[28] Present precision guided weapons are based upon the use of TV or laser seekers; however, they are not adverse weather systems (and mostly daytime-only types). What is needed in general is a precision tracking or weapon guidance capability that will perform satisfactorily in smoke, fog, dust, or rain, day or night.

The Defense Advanced Research Projects Agency and the three Services are now placing particular emphasis on terminal guidance of tactical air-to-surface missiles. Antennas for such "terminal seekers" are aperture limited to less than 6 or 8 inches. In order to obtain narrow beamwidths from such small antennas, millimeter-wave frequencies are being used, with the choice typically being 35 or 94 GHz. Extensive design and measurement programs are underway to develop seekers optimized for various military requirements.[29-32]

In general, all-solid state design has been emphasized for these seekers, consistent with the need for small size and low voltage. Although transmitter output power is low, it is sufficient because the range needed is short. In addition to active (radar) seekers, passive radiometric systems are also being investigated. Even combined active-passive systems are in use, where an active seeker may switch over to a passive radiometric sensor near the end of missile flight, since the radiometer does not have the problem of aimpoint wander which radar glint may produce.

In addition to these types, semiactive systems are being investigated, with obvious analogy to laser designators and seekers. Similarly, millimeter-wave beam-rider missile guidance is being studied. For these examples, higher power transmitters are needed, which generally rules out solid-state sources.

One of the newer radar systems under development is called STARTLE (Surveillance and Target Acquisition Radar for Tank Location and Engagement), which should provide US tanks with the ability to detect and track other tanks (or other armored vehicles) in adverse environmental conditions such as fog, smoke, dust, or rain.[28,31,33-35] The Defense Advanced Research Projects Agency (DARPA) and the Army are jointly sponsoring the program, which is being carried out with two competing contractors. Prototype all-solid state models are being fabricated with the system parameters shown in **Table I** (from Reference 33). The beamwidth of approximately 2/3 of a degree is not narrow enough to provide an image. One of the contractors is providing a dual-mode radar[31] which uses coherent MTI to detect radially moving targets and an area MTI to acquire stationary, creeping, and tangentially-moving targets. The area MTI uses a wideband spread-spectrum transmitted waveform that provides smoothing of clutter, glint, and multipath effects.[31]

Radiometric sensors are being extensively used by NASA for re-

TABLE I

STARTLE SYSTEM PARAMETERS

Frequency: 94 GHz (3.2mm)
Beamwidth: 11 mrad.
Average power: .1 to .5 W
Antenna aperture: 14 inches
Field of view: 15° x 7.5° wide
　　　　　　　5° x 2.5° narrow
Frame time: <2 seconds
Target detection: 3000 meter
　　　　　　　(100m visibility)
Target tracking: .5 mrad @ 2 km
　　　　　　　(100m visibility)

mote sensing[36-39] and by the Navy for navigation and imaging.[40] For example, the Nimbus 6 satellite carried five superheterodyne radiometers with center frequencies between 22 and 60 GHz. Temperature profiles were successfully measured for a variety of surface locations and atmospheric conditions. Recently, various radiometers, centered at frequencies between 90 and 183 GHz, were flight tested in a Convair 990 aircraft. At 183 GHz, the frequency of an atmospheric water vapor transition, the radiometer provides atmospheric temperature data related to water content. A 93 GHz channel obtains surface or lower atmosphere temperature information. A dual channel (93/183 GHz) system is now installed in an RB-57 aircraft.

Other areas where applications are developing include scale modeling (e.g., measuring radar returns from scale models at millimeter/submillimeter frequencies to simulate returns at microwave frequencies)[41,42], electronic warning receivers[43], auto collision avoidance[44,45], and Coast Guard obstacle avoidance.[46]

Several types of millimeter-wave communications systems have been developed or designed. These include terrestrial point-to-point[47] or ship-to-ship systems whose carrier frequencies were chosen to take advantage of atmospheric propagation characteristics; enclosed, low-loss, TE_{01} mode, circular waveguide for use in wideband, long-haul, heavy-route communication systems; and satellite-to-satellite or satellite-to-ground links.

The most straightforward scheme developed for terrestrial (surface) communications is a shortrange (5 to 20 miles) duplex unit operating at 38 GHz, which is a low atmospheric-attenuation region. The advantages of a millimeter-wave system over a microwave system include the availability of a narrow beam width (high gain) from a small aperture, broader bandwidths (possibility of frequency agility), and link privacy.

In other cases, the choice has been to use a frequency near 60 GHz, where attenuation is high, so signal overshoot is low and covert operation is possible. Examples include a line-of-sight communicator for use between ships.[48] The design employs 3-inch diameter paraboloid reflectors (4.5 degree beamwidth), an IMPATT transmitter/local oscillator and battery-pack power. Transceiver weight is 2 pounds and battery pack 6 pounds. Other systems have been devised which emphasize low cost and civilian applications,[49] as well as advanced dielectric waveguide circuitry.[50,51]

For many years, Bell Laboratories and other organizations developed components and technology in relation to the so-called circular-electric mode (TE_{01}) waveguide, which has been shown to provide extremely low loss (about 2 dB/km) over the frequency range from 33 to 125 GHz, while carrying about 220,000 voice circuits simultaneously. Complete systems have been developed in the US, Japan, and England.[52-55] Several years ago the Bell System installed a 14 km long field test system in New Jersey, and a similar 22.7 km length of experimental waveguide has been installed in Japan. Tests apparently have been highly successful.[55,56]

Early satellite communications in the 30-33 GHz region were carried out between the NASA Applications Technology Satellites (ATS-5 and ATS-6) and ground stations.[57] Satellite-to-satellite millimeter-wave relay links were also designed,[58] leading to the development of needed components at V-band, since most of these links would have taken advantage of the very large atmosphere attenuation near 60 GHz to provide isolation from ground receivers.

More recently a variety of satellite links has been reconsidered, partly for reasons of spectrum congestion and/or cost benefits.[59,60] It seems probable that further millimeter-wave applications will evolve in the near future.[61-63]

CONCLUSION

The examples above represent only a modest cross section of millimeter-wave activities today. More could be said about extensive work in spectroscopy, radio astronomy,[64,65] special uses of sensors, and other fields. Certainly the growth in research and development has become much broader in recent years, and now there are overtones of possible millimeter-wave systems which may be produced in quantity, a long-awaited event, and one which is much-needed if the area is to continue to grow.

Dr. Wiltse is the 1979/80 National Lecturer for the IEEE MTT-S group, and his preliminary schedule is shown below. Additional dates and locations will be published in an upcoming issue.

Date	Chapter (Location)
September 26, 1979	Tampa, FL
September 27, 1979	Orlando, FL
October 3, 1979	Baltimore, MD
October 16, 1979	St. Louis, MO
October 17, 1979	Los Angeles, CA
October 18, 1979	Santa Clara, CA
October 25, 1979	Schenectady, NY
November 15, 1979	Boston, MA

REFERENCES

1. Kuno, H. J., and T. T. Fong, "Solid State mm-Wave Sources and Combiners," *Microwave Journal*, June, 1979, pp. 47-48, 73-75, 85.
2. Midford, T. A., and R. L. Bernick, "Millimeter-Wave CW IMPATT Diodes and Oscillators," *IEEE Trans. on MTT*, Vol. MTT-27, May, 1979, pp. 483-492.
3. Fong, T. T., and H. J. Kuno, "Millimeter-Wave Pulsed IMPATT Sources," *ibid.*, pp. 492-499.
4. Bernues, F. J., R. S. Ying, and M. Kaswen, "Solid State Oscillators Key to Millimeter Radar," *Microwave System News*, May, 1979, pp. 79-86.
5. Kramer, N. B., "Solid State Technology for Millimeter Waves," *Microwave Journal*, August, 1978, pp. 57-61.
6. Kramer, N. B., "Millimeter-Wave Semiconductor Devices," *IEEE Trans. MTT*, Vol. MTT-24, November, 1976, pp. 685-693.
7. Fank, F. B., J. D. Crowley, and J. J. Bernz, "InP Material and Device Development for Millimeter Waves," *Microwave Journal*, June, 1979, pp. 86-91.
8. Kantorowicz, G., P. Palluel, and J. Pontvianne, "New Developments in Submillimeter-Wave BWOs," *Microwave Journal*, February 1979, pp. 57-59.
9. Golant, M. B., et al., "Wide Range Oscillators for the Submillimeter Wavelengths," *Pribory i Teknika Eksperimenta*, 1969, pp. 231.
10. "Introduction to Extended Interaction Oscillators," Data Sheet No. 3445 5M, November, 1975, Varian Associates of Canada, Ltd., Georgetown, Ontario, Canada.
11. Jory, H., S. Heggi, J. Shively, R. Sy-

mons, "Gyrotron Developments," *Microwave Journal*, August, 1978, pp. 30-32.
12. Chu, K. R., A. T. Drobot, V. L. Granatstein, and J. L. Seftor, "Characteristics and Optimum Operating Parameters of a Gyrotron Traveling Wave Amplifier," *IEEE Trans. MTT*, Vol. MTT-27, February, 1979, pp. 178-187.
13. Gallagher, J. J., M. D. Blue, B. Bean, and S. Perkowitz, "Tabulation of Optically Pumped Far Infrared Laser Lines and Applications to Atmospheric Transmission," *Infrared Physics*, Vol. 17, 1977, pp. 43-55.
14. Special Issue on Microwave and Millimeter-Wave Integrated Circuits, *IEEE Trans. on Microwave Theory and Techniques*, Vol. MTT-26, No. 10, October, 1978.
15. Special Issue on Millimeter Waves: Circuits, Components, and Systems, *IEEE Trans. on Microwave Theory and Techniques*, Vol. MTT-24, No. 11, November, 1976.
16. Swanberg, N. E., and J. A. Paul, "Quasi-Optical Mixer Offers Alternative," *Microwave System News*, Vol. 9, May, 1979, pp. 58-60.
17. Cardiosmenos, A. P., "Planar Devices Make Production Practical," *Microwave System News*, May, 1979, pp. 46-56.
18. Kawasaki, R., and K. Yamomoto, "A Wide-Band Mechanically Stable Quasi-Optical Detector for 100-300 GHz," *IEEE Trans. MTT*, Vol. MTT-27, May, 1979, pp. 530-533.
19. Schneider, M. V., "Low-Noise Millimeter Wave Schottky Mixers," *Microwave Journal*, August, 1978, pp. 78-83.
20. Keen, N. J., "The Mottky-Diode: A New Element for Low Noise Mixers at Millimeter Wavelengths," AGARD Munich Conference on Millimeter and Submillimeter Wave Propagation and Circuits, *Proceedings No. 245*, September, 1978, pp. 16-1 to 16-9.
21. Kerr, A., "Noise and Conversion Loss of Two-Diode Subharmonically Pumped and Balanced Mixers," *IEEE Int'l Microwave Symp. Digest*, April 30, 1979, pp. 17-19.
22. Held, D., "An Approach to Optimal Mixer Design at Millimeter and Submillimeter Wavelengths," *ibid.*, pp. 25-27.
23. Takada, T., and M. Ohmori, "Frequency Triplers and Quadruplers with GaAs Schottky-Barrier Diodes at 450 and 600 GHz," *IEEE Trans. on Microwave Theory and Techniques*, Vol. MTT-27, pp. 519-523; May, 1979.
24. Calviello, J. A., "GaAs Schottky Barrier Devices and Components for Millimeter and Submillimeter Application," *Microwave Journal*, August, 1979.
25. Andreyev, G. A., et al., "Intensity and Angle of Arrival Fluctuations of Millimetric Radiowave in Turbulent Atmosphere," Joint Anglo-Soviet Seminar on Atmospheric Propagation at Millimetre and Submillimetre Wavelengths, Institute of Radioengineering and Electronics, Moscow, November, 1977.
26. McMillan, R. W., J. C. Wiltse, and D. E. Snider, "Atmospheric Turbulence Effects on Millimeter Wave Propagation," Proc. IEEE Symp. on Aerospace and Electronic Systems, Washington, DC, October, 1979.
27. Trebits, R. N., R. D. Hayes, and L. C. Bomar, "Millimeter Wave Reflectivity of Land and Sea," *Microwave Journal*, August, 1978, pp. 49-53 and 83.
28. Wiltse, J. C., "Millimeter Waves — They're Alive and Healthy," *Microwave Journal*, August, 1978, pp. 16-18.
29. Currie, N. C., J. A. Scheer, and W. A. Holm, "Mm-Wave Instrumentation Radar Systems," *Microwave Journal*, August, 1978, pp. 35-42.
30. Seashore, C. R., J. E. Miley, and B. A. Kearns, "Radar and Radiometer Sensors for Mm-Wave Guidance Systems," *Microwave Journal*, August, 1979.
31. Sundaram, G. S., "Millimetre Waves: The Much-Awaited Technological Breakthrough?," *International Defense Review*, 2, 1979, pp. 271-277.
32. Williams, D. T., and W. H. Boykin, Jr., "Millimeter-Wave Missile Seeker Aimpoint Wander Phenomenon," *AIAA Journal of Guidance and Control*, Vol. 2, May-June, 1979, pp. 196-203.
33. Balcerak, R., W. Ealy, J. Martins, and J. Hall, "Advanced Infrared Sensors and Radar Systems for Tactical Target Acquisition," Society of Photo-Optical Instrumentation Engineers (SPIE) Symposium on Effective Utilization of Optics in Radar Systems, Huntsville, Alabama, 1977, pp. 172-184.
34. Backus, P. H., "Electro Optics," *Journal of Electronic Defense*, Vol. 2, March/April, 1979, pp. 24-35.
35. Fawcette, J., "Army to Test STARTLE Radar While Navy Readies Gyrotron," *Microwave Systems News*, Vol. 8, July, 1978, pp. 23-24.
36. Rainwater, J. Hank, "Radiometers: Electronic Eyes That 'See' Noise," *Microwaves*, Vol. 17, September, 1978, pp. 59-62.
37. Schuchardt, J. M., and J. A. Stratigos, "Detected Noise Levels Guide Radiometer Design," *Microwaves*, Vol. 17, September, 1978, pp. 64-74.
38. Schuchardt, J., J. Stratigos, J. Galiano, and D. Gallentine, "Dual Frequency Multi-Channel Millimeter Wave Radiometers for High Altitude Observation of Atmospheric Water Vapor," *IEEE Int'l Microwave Symp. Digest*, May 2, 1979, pp. 540-542.
39. Blue, M. D., "Reflectance of Ice and Seawater at Millimeter Wavelengths," *ibid.*, pp. 545-546.
40. Hollinger, J. P., J. E. Kinney, and Ballard E. Troy, Jr., "A Versatile Millimeter-Wave Imaging System," *IEEE Trans. MTT*, Vol. MTT-24, November, 1976, pp. 786-793.
41. Cram, L. A., and S. C. Woolcock, "Review of Two Decades of Experience Between 30 GHz and 900 GHz in the Development of Model Radar Systems," AGARD Munich Symposium on Millimeter and Submillimeter Wave Propagation and Circuits, *Proceedings No. 245*, September, 1978, pp. 6-1 to 6-15.
42. Gabsdil, W., and W. Jacobi, "Model Simulation of Target Characteristics and Engagement Situations Employing Millimeter-Wave Radar Systems," *ibid.*, pp. 7-1 to 7-8.
43. Hartman, R., "Multisensor Warning Receiver Tackles Growing Threat," *Journal of Defense Electronics*, Vol. 2, May, 1979, pp. 79-86. (Also see *Microwave System News*, Vol. 9, May, 1979, p. 32.
44. Wollins, Bob, "Auto Anticollision Radar Accelerates Away from the Lab Backburner to Global Test Sites," *Microwaves*, Vol. 17, January, 1978, pp. 9-12.
45. Moncrief, F., "Car Radars Could be Standard in the 1980s," *Microwave System News*, Vol. 8, April, 1978, pp. 23-26.
46. Bearse, S. V., "94 GHz Radar Undergoing Coast Guard Evaluation," *Microwaves*, Vol. 14, September, 1975, pp. 10.
47. Dudzinsky, S. J., Jr., "Atmospheric Effects on Terrestrial mm-Wave Communications," *Microwave Journal*, December, 1975, pp. 39.
48. Davis, R. T., "Mm Transceiver Provides Covert Communications," *Microwaves*, October, 1974, pp. 9.
49. Matsuo, Y., Y. Adaiwa, and I. Takase, "A Compact 60-GHz Transmitter-Receiver," *IEEE Trans. MTT*, Vol. MTT-24, November, 1976, pp. 794-797.
50. Keitzer, J. E., A. R. Kaurs, and B. J. Levin, "A V-Band Communication Transmitter and Receiver System Using Dielectric Waveguide Integrated Circuits," *IEEE Trans. MTT*, Vol. MTT-24, November, 1976, pp. 797-803.
51. Chang, Y., J. A. Paul, and Y. C. Ngan, "Millimeter Wave Integrated Circuit Modules for Communication Interconnect Systems," *Final Report*, No. DELET-TR-76-1353-F on Hughes Aircraft Contract DAAB07-76-C-1353 with ERADCOM; October, 1976.
52. White, R. W., M. B. Read, and A. J. Moore, "Recent British Work on Millimeter Waveguide Systems," *IEEE Trans. Communication Soc.*, Vol. COM-22, September, 1974, pp. 1378-1390.
53. Miyauchi, K., S. Seki, N. Ishida, and K. Izumi, "W-40G Guided Millimeter-Wave Transmission System," *Review of Electrical Communications Laboratories*, Vol. 23, July/August, 1975, pp. 707-741.
54. "Millimeter Waveguide Systems," *Microwave Journal*, March, 1977, pp. 24-26.
55. "Japanese W-40 G. Guided Millimeter-Wave Transmission System," *Microwave System News*, Vol. 6, February/March 1976, pp. 91-97.
56. Warters, W. D., "Millimeter Waveguide Scores High in Field Test," Bell Laboratories Record, November, 1975, pp. 401-408.
57. Ippolito, L. J., "Effects of Precipitation on 15.3 and 31.65 GHz Earth-Space Transmissions with the ATS-V satellite," *Proc. IEEE*, Vol. 59, 1971, pp. 189-205.
58. Dees, J. W., G. P. Kefalas, and J. C. Wiltse, "Millimeter Wave Communications Experiments for Satellite Applications," *Proc. IEEE Int'l Conf. on Communications*, San Francisco, June, 1970, pp. 22-20 to 22-26.
59. Hilson, N. B., J. J. Gallagher, et al, "Millimeter Wave Satellite Concepts," Vol. I — Technical Report No. NASA CR-135227, NASA Lewis Research Center, Contract NAS3-20110, September, 1977.
60. Feldman, N. E., and S. J. Dudzinsky, Jr., "A New Approach to Millimeter-Wave Communications," Rand Report R-1936-RC, April 1977.
61. Mundie, L. G., N. E. Feldman, "The Feasibility of Employing Frequencies Between 20 and 300 GHz for Earth-Satellite Communications Links," Rand Report R-2275-DCA, May, 1978.
62. Tsao, C.K.H., W. J. Connor, and T. E. Joyner, "Millimeter-Wave Airborne Terminal for Satellite Communications," Int'l Conf. on Communications, 11th, 1975, pp. 36-4 to 36-8.
63. Castro, A., and J. Healy, "Transmitter System for mm-Wave Satellite Communications," Int'l Conf. on Communications, 11th, 1975, pp. 36 9 to 36-13.
64. "300-GHz Radio Telescope Nears Completion," *Microwaves*, Vol. 17, September, 1976, pp. 14.
65. Cong, H., A. R. Kerr, and R. J. Mattauch, "The Low-Noise 115-GHz Receiver on the Columbia-GISS 4-ft. Radio Telescope," *IEEE Trans. MTT*, Vol. MTT-27, March, 1979, pp. 245-248

Millimeter Wave Radar Applications to Weapons Systems
V.W. Richard
U.S. Army Ballistic Research Labs Report No. 2631, June 1976, DDC AD B012103
Pages 9-66
This article was prepared by an employee of the U.S. Government as part of his official duties. Such articles are in the public domain and can be reproduced at will. Reprinted by permission.

I. INTRODUCTION

Applications of millimeter wave radars in Army ground-to-ground, ground-to-air, and air-to-ground weapons systems are discussed in this report. The advantages and limitations of operating at millimeter wavelengths in these applications are defined.

The characteristics of millimeter wave propagation in adverse weather are presented; emphasis is placed on rain backscatter and attenuation theory, experimental data, and rain effects on radar system performance.

The theory of fluctuating clutter is described and a terrain clutter model is presented whose output is characterized by its power spectral density and probability density function. Several clutter discrimination techniques are described.

To extend the range of millimeter wave radars in adverse weather, the use of more complex modulation waveforms, such as FM/CW, along with more sophisticated data processing techniques, is proposed. Extended ranges would result from the increased sensitivity of the radar system derived from the use of reduced, optimized bandwidths commensurate with the data rates required. The extra complication of such radars would be justified in applications where the other unique properties of millimeter wave radars are required, such as improved low-angle tracking, high angular resolution of multiple targets, secure operation, and reduced interference between adjacent radars.

Frequencies in the regions of 16, 35, and 95 GHz are suitable for high-accuracy, low-angle tracking. The tracking accuracy improves with increasing frequency. However, at 35 GHz the rain backscatter is relatively large and, therefore, the use of this frequency might be restricted to moving targets where MTI can be used.

II. MILLIMETER WAVE PROPAGATION CHARACTERISTICS

A. Atmospheric Effects

The maximum range of a millimeter wave radar is limited not only by component performance but also by the propagation characteristics of the atmosphere. Millimeter waves are attenuated by (1) the molecular absorption of water vapor and oxygen in clear weather, (2) the absorption of condensed water droplets in fog and rain, and (3) the scattering from water droplets in rain. In addition, raindrops scatter energy back into the radar antenna that appears as noise at the receiver and can obscure desired targets.

There are propagation windows at nominal frequencies of 35, 94, 140, 240, 360, 420, and 890 GHz, as shown in Figure 1. The atmospheric attenuation at these window frequencies increases with increasing frequency, except in the region of 240 GHz where the attenuation is less than at 140 GHz by about 1.3 dB/km. Only the coarse structure of the attenuation curve is shown in Figure 1, although the major propagation windows are shown. A comparison of the attenuation at optical and radar wavelengths can be conveniently made from Figure 1.

The clear-weather atmospheric attenuation at 35, 70, and 94 GHz is generally small compared with other radar system losses and propagation effects. At 140 and 240 GHz, atmospheric attenuation is appreciable, and at 360 GHz and above, it is usually prohibitively large for all but very short-range system applications.

Figure 1 also shows the rain plus atmospheric attenuation at the propagation window frequencies for 4, 10, and 25 mm/hr rainfall rates. The increase in attenuation from rain over clear weather attenuation can be seen in this figure.

Figure 2 shows calculated rain attenuation versus frequency curves that have been found to be in good agreement with measured attenuation. A curve of calculated fog attenuation is also included in Figure 2 which shows the relatively low attenuation in the millimeter wave region compared with the high attenuation in the optical region. It is this unique characteristic of millimeter waves to penetrate fog, drizzle, dust, and dry snow with very little attenuation that makes them useful where good performance must be maintained through conditions of poor visibility.

As stated previously, in addition to attenuation by the atmosphere and rain, further degradation of radar performance can be caused by the backscatter from condensed water droplets. The theory of millimeter wave rain propagation through rain is discussed in Appendix A. The uncertainty that existed previously between calculated and observed rain backscatter at millimeter wavelengths has been resolved with a study conducted by BRL in which rain backscatter and attenuation were measured at 9.375, 35, 70, and 95 GHz.[1] The quantitative relationship between rain characteristics and backscatter was determined experimentally over a wide range of rainfall rates.

---

[1] V.W. Richard and J.E. Kammerer, "Rain Backscatter Measurements and Theory at Millimeter Wavelengths," BRL Report No. 1838, October 1975 (AD B008173L).

Introduction and Overview

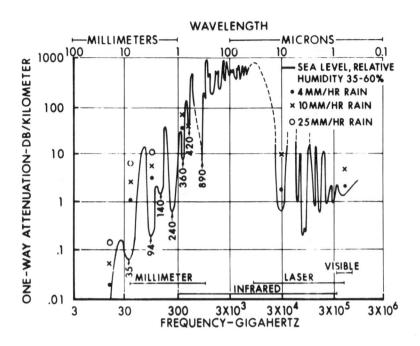

Figure 1. Atmospheric Attenuation vs Frequency

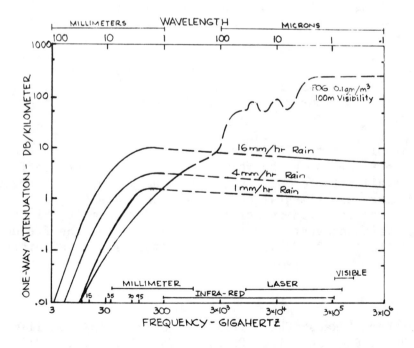

Figure 2. Rain and Fog Attenuation vs Frequency

Figure 3 shows A-scope photographs of rain backscatter measured at 70 GHz for rainfall rates of 1 to 100 mm/hr which are illustrative of the amplitude characteristics of rain clutter. In these photographs, the rain backscatter trace rises to a maximum deflection at a range of about 0.1 km because of the recovery time of the receiver-protector device and then decreases with increasing range because of the normal propagation spreading loss plus rain attenuation. The broad band of noise on the trace is the fluctuation of rain echoes caused by the rapidly changing interference effects between the randomly falling raindrops. The backscatter from light rainfall (Figure 3a, 1 mm/hr) is characterized by a gradually diminishing amplitude versus range. The backscatter from somewhat heavier rainfall (Figures 3b and 3c, 6 and 15 mm/hr) has a greater amplitude at short range and falls off more rapidly with increasing range. For heavy rainfall (Figures 3d, 3e, and 3f, 26-100 mm/hr) the maximum rain return at short ranges is about the same amplitude as medium rainfall, but the backscatter falls off more rapidly with increasing range because of the greater attenuation caused by the heavier rainfall. The reduction in the trihedral target return with increasing rainfall rate is also evident in these photographs.

The maximum range at which the rain backscatter is above the receiver noise is not very large in the photographs in Figure 3, but field radars with higher power and better sensitivity than the experimental radar used for this experiment would show rain backscatter above receiver noise out to much greater ranges.

Figure 3 additionally illustrates the increase in the slope of the rain backscatter trace as the rainfall intensity increases. It has been observed that this slope bears a very definitive relationship to the rainfall rate and, once calibrated, can be used as a very fast response rainfall intensity indicator. This slope technique is particularly useful in evaluating the rainfall rate as a function of range in order to quickly assess the range capability of a millimeter wave radar in rain from moment to moment under rapidly changing rainfall conditions.

The results of the BRL rain backscatter measurement experiment are shown in Figure 4 in the form of rain backscatter coefficient versus rainfall rate for 9.375, 35, 70, and 95 GHz. An extensive review of other experimental and theoretical data on rain backscatter and attenuation is given in Reference 1. For example, Figure 5 shows calculated rain backscatter data by Rozenberg[2] for the frequency range of 10 to 10,000 GHz along with BRL measured data points.

---

[2] V.I. Rozenberg, "Radar Characteristics of Rain in Submillimeter Range," Radio Eng. and Elec. Physics, Vol. 15, No. 12, 2157-2163, 1970.

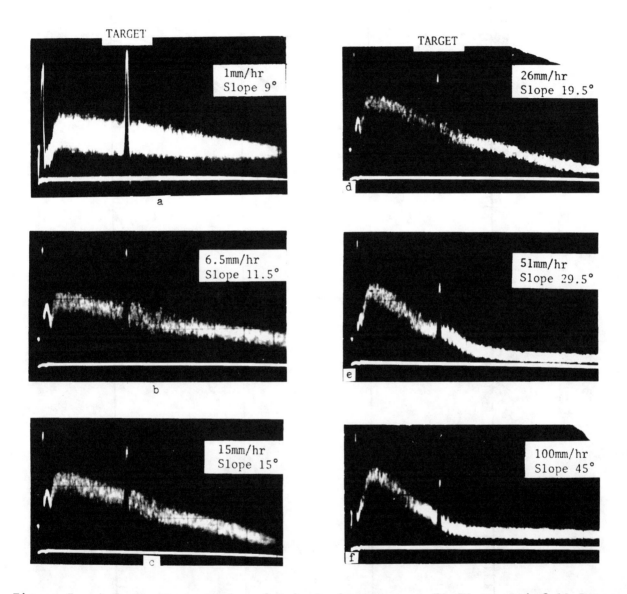

Figure 3. A-Scope Photographs of Rain Backscatter at 70 GHz vs Rainfall Rate

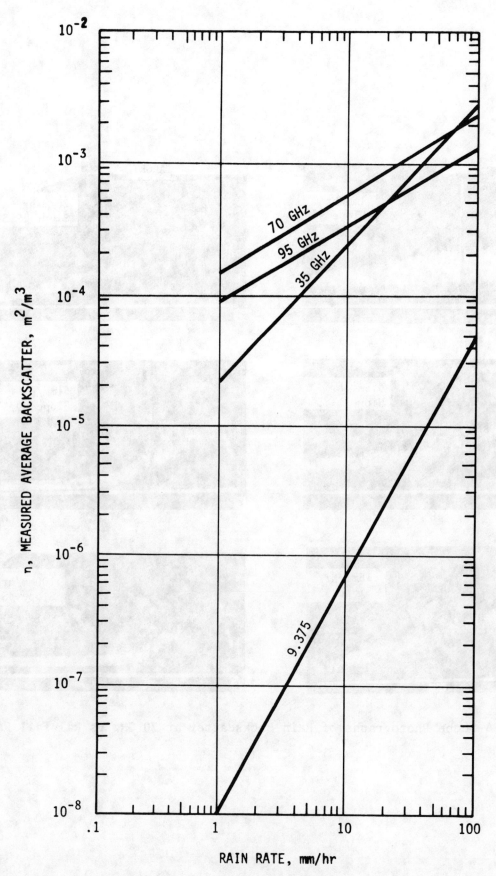

Figure 4. Measured Rain Backscatter Coefficient vs Rainfall Rate

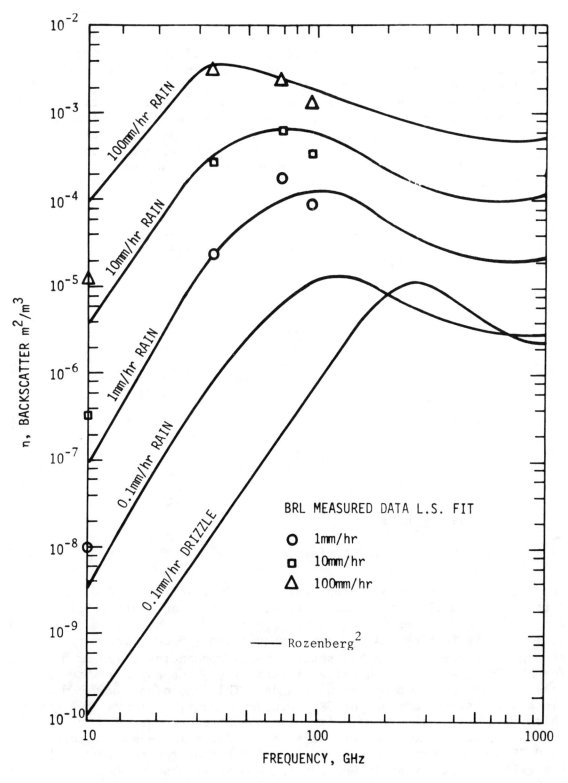

Figure 5. Calculated Rain Backscatter Coefficient vs Frequency

B. Radar Range in Rain

The equations and data in Appendix A and Reference 1 provide basic propagation data from which calculations can be made of the performance of millimeter wave radar systems in rain over a wide range of frequencies and rainfall rates. The combined effects of rain backscatter and attenuation must be taken into account when calculating the maximum radar range in rain. The magnitude of the effect of rain backscatter in reducing range is shown in Figure 6 for radars with a fixed antenna size of 1 meter diameter operating between 10 and 300 GHz. The range at which the rain return is equal to the target return is shown in Figure 6 for targets of 1 and 5 square meters.

The radar range for the condition where the rain return is equal to the target return is given by the following equation which is derived in Reference 1,

$$D = \left(\frac{d^2 \sigma_t}{1.35 \times 10^8 \, \tau \, \lambda^2 \eta} \right)^{1/2}, \qquad (1)$$

where $D \equiv$ range, m (it is assumed that the radar has sufficient power and sensitivity to achieve this range)

$d \equiv$ antenna diameter, m

$\sigma_t \equiv$ target cross section, m^2

$\tau \equiv$ pulse duration, seconds

$\lambda \equiv$ wavelength, m

$\eta \equiv$ rain backscatter coefficient, m^2/m^3.

The rain backscatter coefficient values used are given in Table I (from Reference 1). It should be noted that the backscatter-limited range of Figure 6 goes through a broad minimum, in the region of 20 to 40 GHz, indicating that K_a band frequencies (35 GHz) are more strongly affected by rain backscatter than frequencies lower or higher for a radar that is antenna-size limited; i.e., the antenna size is held constant as the comparison is made. This phenomenon was consistently observed during the BRL rain experiment, as shown in Figure 7 which illustrates the superiority of a 95-GHz radar over 70- and 35-GHz radars (all with the same antenna size) in detecting a sphere target of 0.05 square meter in a heavy rain. The target return is well above the rain backscatter at 95 GHz but is completely obscured at 70 and 35 GHz.

Introduction and Overview

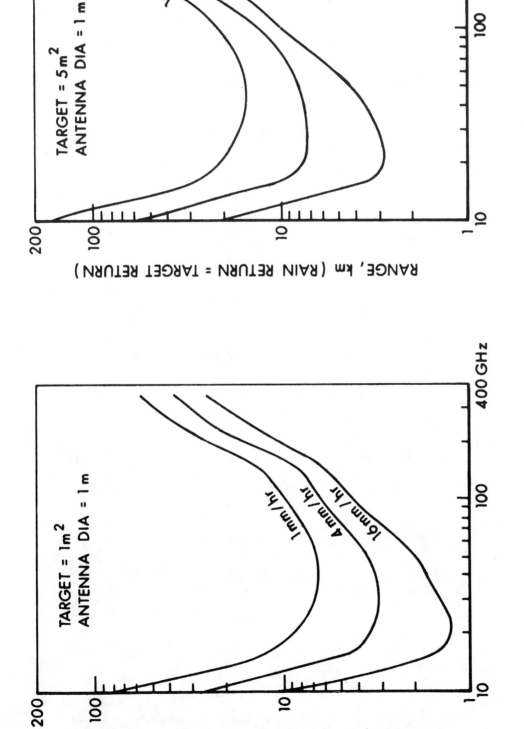

Figure 6. Range at which Rain Return Equals Target Return vs Frequency and Rain Rate with Constant Antenna Size

TABLE 1
RAIN BACKSCATTER COEFFICIENT

Frequency, GHz	Rainfall Rate, mm/hr		
	1	4	16
	η, m^2/m^3		
9.375	1.03×10^{-8}	8.5×10^{-8}	6.9×10^{-7}
16	1.4×10^{-6}	1.1×10^{-5}	9.5×10^{-5}
25	8.5×10^{-6}	5×10^{-5}	3×10^{-4}
35	2.2×10^{-5}	9.2×10^{-5}	4.1×10^{-4}
55	5.5×10^{-5}	1.9×10^{-4}	6.4×10^{-4}
70	1.44×10^{-4}	3.2×10^{-4}	5.52×10^{-4}
95	8.7×10^{-5}	1.9×10^{-4}	4.3×10^{-4}
140	9.5×10^{-5}	2.2×10^{-4}	4.6×10^{-4}
240	4.2×10^{-5}	9.5×10^{-5}	2.2×10^{-4}
360	3.2×10^{-5}	7.2×10^{-5}	1.7×10^{-4}

Introduction and Overview

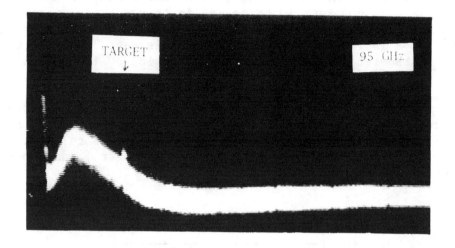

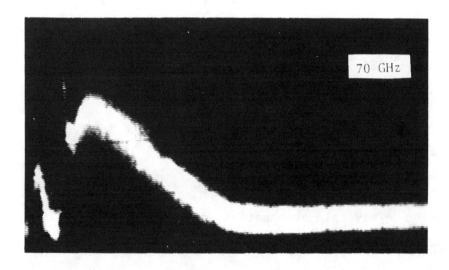

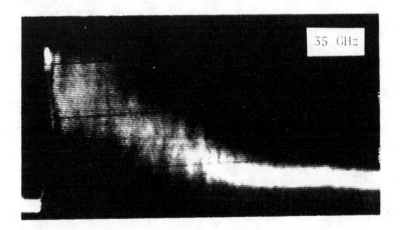

Figure 7. Superiority of 95 GHz over 70 and 35 GHz Radar to Detect Target in Rain

Figure 8 shows the rain backscatter-limited range when the antenna beamwidth is held constant (0.68 degree in this example, the beamwidth for a 1-meter-diameter antenna at 35 GHz) and the antenna size is decreased with increasing frequency. For this case, the rain backscatter-limited range is a minimum in the region of 100 GHz for light rain and 60 GHz for heavy rain.

The magnitude of the range reduction by rain attenuation is illustrated in Figures 9 and 10 for tracking radars operating at 35 and 95 GHz, respectively. Maximum tracking range was calculated using the following equation, radar system parameters, and atmospheric conditions.[1]

$$D = \left(\frac{P_T G^2 \lambda^2 \sigma_t \exp - 0.461\alpha D}{(4\pi)^3 KT (B_{IF} B_{TR})^{1/2} F (S/N)} \right)^{1/4} , \qquad (2)$$

where $D \equiv$ range, m

$P_T \equiv$ peak transmitted power, w

$G \equiv$ antenna gain, power ratio

$\lambda \equiv$ wavelength, m

$\sigma_t \equiv$ target cross section, m^2

$\alpha \equiv$ one-way rain attenuation, dB/m

$KT \equiv 4 \times 10^{-21}$ watts/Hz

$B_{IF} \equiv$ IF bandwidth, Hz

$B_{TR} \equiv$ tracking loop bandwidth, Hz

$F \equiv$ receiving system noise figure, power ratio

$S/N \equiv$ signal-to-noise power ratio.

Tracking ranges were calculated using the following radar system parameters and operating conditions:

Frequency, GHz	35	95
Target cross section, m^2	5	5
Antenna diameter, m	1	1

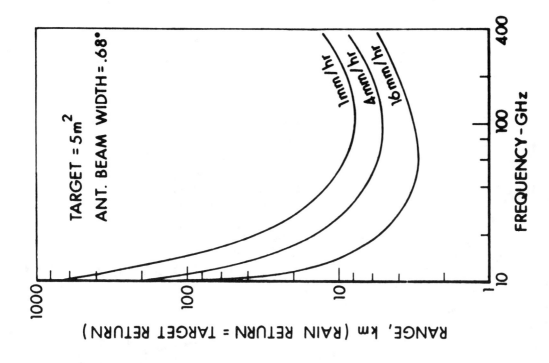
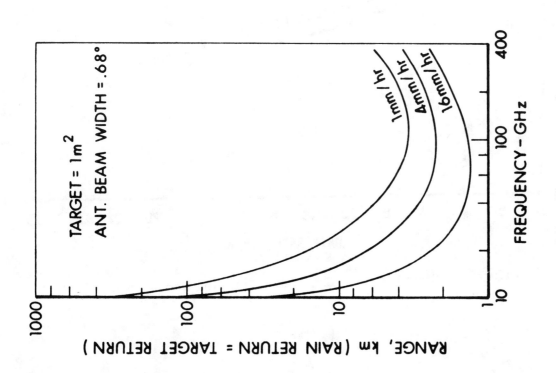

Figure 8. Backscatter Limited Range vs Frequency with Constant Antenna Beamwidth

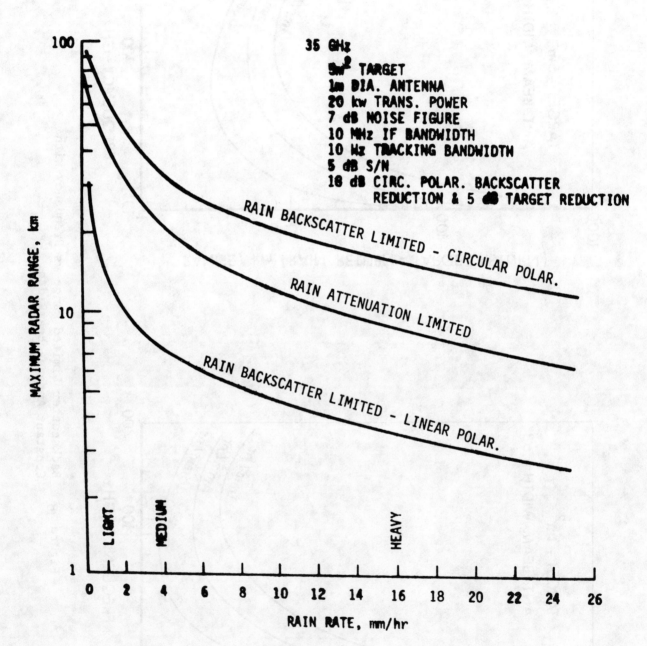

Figure 9. Calculated 35-GHz Radar Tracking Range in Rain

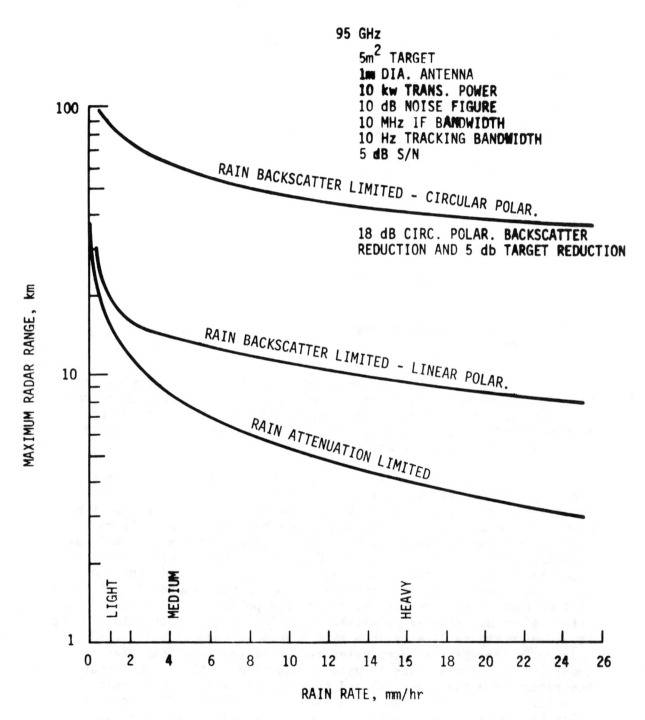

Figure 10. Calculated 95-GHz Radar Tracking Range in Rain

Transmitted power, kw		20		10
Receiver noise figure, dB		7		10
IF bandwidth, MHz		10		10
Tracking bandwidth, Hz		10		10
Signal-to-noise ratio, dB		5		5

Rain attenuation, dB/km, from Figure 2:

Rainfall rate, mm/hr	0	1	4	16
35 GHz attenuation, one-way, dB/km	0.16	0.25	0.95	3.6
95 GHz attenuation, one-way, dB/km	0.5	1.2	3.4	9.2

The backscatter-limited range data for linear polarization from Figures 6 and 8 are also shown in Figures 9 and 10 along with curves for circularly polarized operation based on a rain backscatter reduction of 18 dB with circular polarization and an associated 5 dB reduction of the target cross section.

At 35 GHz, for the radar and target parameters used in Figure 9, the range is severely limited by rain backscatter, with the range being about one-third of what it would be if limited by attenuation only. For moving targets, MTI would possibly reduce the effects of rain backscatter sufficiently to take the system to the rain attenuation limited range in most cases. For non-moving targets, circular polarization can be used to increase the range, provided that the radar has enough power and sensitivity to see the reduced target cross section that usually results when circular polarization is used. The effect of using circular polarization at 70 GHz to reduce rain clutter is shown in Figure 11 where the rain return is reduced below the receiver noise when circular polarization of the same sense for transmitting and receiving is used. Note also in this figure that the return from the trihedral and tree is also reduced when using circular polarization of the same sense. If neither MTI or circular polarization can be used with a 35-GHz radar system, the rain backscatter will impose a significant performance limitation unless a very large antenna can be used.

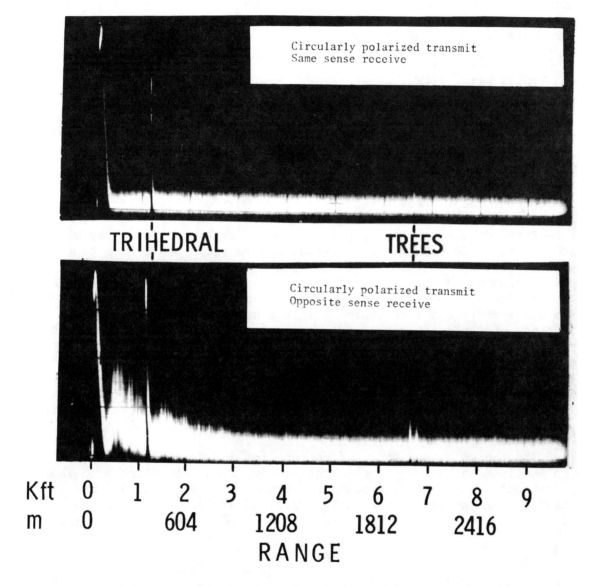

Figure 11. Reduction of Rain Backscatter and Target Return with Circular Polarization at 70 GHz

At 95 GHz, for the radar and target parameters used in Figure 10, the range is attenuation limited, which means that the range could be increased if more power or a more sensitive receiver was used. The use of circular polarization or MTI would not be absolutely necessary for the example chosen in Figure 10 to reduce rain clutter, unless ranges in excess of 10 km were desired.

These calculations of the range of millimeter wave radars operating in rain are given only as examples of the relative magnitudes of rain attenuation and backscatter effects. The calculated ranges do not necessarily represent the maximum performance attainable with higher performance components and the use of optimum modulation and data processing techniques.

C. <u>Optimum Radar Frequencies</u>

Based on atmospheric propagation considerations alone, the choice of an optimum frequency for a weapons system radar would have to take into account the maximum range required under the types of adverse weather specified for the application. For example, rain backscatter can obscure small targets at very short ranges at certain frequencies. The optimum frequency to minimize rain clutter obscuration is a function of the antenna beamwidth and pulse length. In general, the attenuation by rain becomes more of a problem with increasing frequency, although attenuation can be overcome with increased antenna gain and transmitter power, and more sophisticated modulation and signal processing. Thus, it becomes necessary to consider the interaction between the radar system parameters and weather effects to obtain the required performance.

The choice of an optimum frequency for a wide range of radar applications has been very comprehensively treated by Strom.[3] The effects of propagation attenuation, rain and terrain clutter, and radar system parameters typically available at frequencies in the microwave and millimeter region are considered in these references.

III. MILLIMETER WAVE RADAR CHARACTERISTICS

A. General

The millimeter and sub-millimeter wave region of the electromagnetic spectrum between 10 millimeters and 1.0 millimeter wavelength (30-300 GHz) has unique quasi radio-optical propagation characteristics. The extremely high angular resolution and heavily degraded adverse weather properties of optical systems are approached at the high frequency end of this spectrum, and the all-weather propagation and broader resolution properties of radio waves are realized at the low frequency of this spectrum.

The use of very short wavelengths, in the millimeter and sub-millimeter region, and very narrow antenna beamwidths in radar designs offers a number of desirable features for Army weapons systems applications. These features include:

1. Narrow antenna beamwidth with small antenna, which provides

    (a) high tracking and guidance accuracy,

    (b) capability of tracking down to very low elevation angle before ground multipath and ground clutter become appreciable,

    (c) good resolution of closely spaced targets,

    (d) high angular resolution for area mapping and target surveillance,

    (e) high immunity to jamming (narrow beamwidth makes jamming through the main beam difficult),

    (f) high antenna gain, and

    (g) capability of detecting and locating small objects such as wires, poles, and projectiles.

2. Wide frequency spectrum availability, which provides

    (a) high information rate capability for obtaining fine structure detail of target signature with narrow pulses or wideband FM,

    (b) wideband spread-spectrum capability for reduced multipath and clutter,

    (c) high immunity to jamming,

(d) multiple adjacent radar operation without interference, and

(e) very high range resolution capability for precision tracking and target identification.

3. High absorption around transmission windows, which provides

(a) secure operation, if required, by selecting a frequency with higher absorption, and

(b) difficulty of long-range jamming.

4. Low scatter from terrain, which provides

(a) reduced multipath interference, and

(b) reduced terrain clutter.

5. Penetration of dry contaminants in atmosphere, which provides good operation under limited visibility conditions of dust, smoke, and dry snow.

6. Small targets become an appreciable part of a wavelength, which provides good capability of detecting wires, poles, trees, projectiles, birds, and insects.

7. Doppler frequency is high from low radial velocity target, which provides good detection and recognition capability of slowly moving or vibrating target; doppler frequencies are typically conveniently in the audio range.

There are also limitations with millimeter wave radars which include the following:

1. Reduced range in adverse weather

(a) Rain, wet snow, and fog of high moisture content will reduce the range by attenuation that sometimes can be overcome with higher power, greater antenna gain and optimum modulation and data processing.

(b) Rain produces rain backscatter which can mask targets; the use of MTI, narrow antenna beamwidths, narrow pulse widths, and optimum operating frequencies will minimize rain backscatter effects.

2. Poor foliage penetration

   (a) There is little or no penetration of dense green foliage; which is also true at microwaves.

   (b) There is some penetration through the openings between leaves and branches which improves with increasing frequency. Visually obscured aircraft have been tracked behind trees at 70 GHz.

3. High cost of components

   (a) Currently most millimeter wave components are expensive, largely because of lack of high quantity production. Recent advances, such as the millimeter wave integrated circuit developments using microstrip and dielectric waveguides, show promise of greatly reducing the cost of components and assemblies.

   (b) Solid state sources are just becoming available from R&D laboratories, with attendant high cost and some limitations on reliability. Quality-controlled high-quantity production should solve these problems.

4. Lack of some components necessary for rugged, field model millimeter wave radar designs for operation above 35 GHz such as high-power ($\geq$ 10 kw) transmitters, high-power circulators ($\geq$ 1 kw), low-loss radomes, low-loss and compact receiver protectors, and monopulse antenna feeds. Some design and development work has been done on these items, but they have not been developed to a stage where they have been produced in quantity or ruggedized for field use.

B. Ground-to-Air Millimeter Wave Radar Characteristics

1. General. A narrow beamwidth, millimeter wave radar is particularly well suited for tracking low-altitude aircraft since no appreciable ground clutter is intercepted and the sky background gives a high target-to-background contrast. When the aircraft is at tree-top level at very long ranges, the tree background return can present a problem in obscuring the aircraft return. However, the use of a narrow range gate will eliminate the obscuring problem except when the front edge of the tree line is at the range of the aircraft. The use of MTI and automatic tracking circuits with predicted aircraft velocity and acceleration will aid in keeping lock during short periods of high background and tree clutter.

2. **Radar Return from Helicopter.** To better assess the magnitude of the helicopter radar return and the ground and tree clutter problem, measurements were made of the radar return at 70 GHz while tracking a helicopter at low altitudes above and behind trees. The radars used are described in Tables II and III and are shown in Figures 12-14.

A UH-1 helicopter was flown above and behind trees and over water at an altitude of 30 meters in crossing courses and radial flight paths out to a maximum range of about 3 km. A video camera with a zoom telephoto lens was mounted on the radar antenna to record the aircraft position and its surroundings. A second video camera was trained on the A-scope of the radar to record the transmitted pulse and received echoes. The outputs of the two video cameras were mixed and recorded on a ½-inch video tape recorder.

Example of flight records where manual tracking was used are shown in Figures 15 to 23; these are photographs of the video tape operating in the playback stop-action mode. The A-scope trace is shown in the lower portion of the photograph, with the transmitted (T_o) pulse at the extreme left. The narrow, jitter-free quality of the original radar pulse returns is lost in these photographs taken of the video monitor, but the amplitude of the return and the absence of clutter can be seen. While viewing the tapes in motion, the rate and amplitude of the fluctuation of the radar returns can also be seen.

Figures 15 to 17 show the helicopter at a range of about 1.2 km and an altitude of 30 meters flying a crossing course just visible above the trees but beyond the front edge of the tree line, which is at a range of 240 meters. The strong aircraft return echo completely clear of ground and tree clutter can be seen. It is interesting to note that in Figure 17, where the main body of the helicopter is behind the trees and only the tail rotor is exposed, a strong echo is still received. In Figure 18, the entire helicopter is behind the trees and completely visually obscured, but a strong echo is still received by propagation through the openings among the leaves and branches. The tree return is also strong, but it is at a much shorter range since the radar reflection from trees is essentially all from the front edge of the tree line.

Figures 19 and 20 are from radial runs of the helicopter flying at an altitude of 30 meters with the line-of-sight just above the top of the trees. Since the antenna beam is only 6 meters wide at 1 kilometer, the helicopter must be very close to the trees and at the same range before the tree return will compete with the helicopter return. Figure 20 shows the propagation of the radar signal through a small gap between the trees. The helicopter at 0.91 km is giving a strong, clear radar return.

TABLE II

BRL 70-GHz TRACKING RADAR

Power, 10 kw peak, 5 watts average

Pulse width, 50 nanoseconds

PRF, 10 KHz

Antenna, 0.92 m (3-foot) parabolic reflector

Beamwidth, 0.34 degrees

Beam scanning, conical

Polarization, vertical

Antenna Mount, automatic track and manually trainable via optical sight

Transmitter-Receiver, from AN/MPS-29 Search Radar built by Georgia Institute of Technology for the U.S. Army Signal Corps.

TABLE III

UNITED AIRCRAFT CO., NORDEN DIV., 70-GHz RADAR

Power, 500 watts peak, 0.25 watts average

Pulse width, 50 nanoseconds

PRF 2-10 KHz

Antenna, 0.92 m, (3-foot) parabolic reflector

Beamwidth, 0.34 degrees

Beam scanning, none

Polarization, horizontal, vertical, right- or left-hand circular with any combination for transmit and receive

Antenna mount, manually trainable

Figure 12. BRL 70-GHz Tracking Radar Antenna and Mount

Introduction and Overview

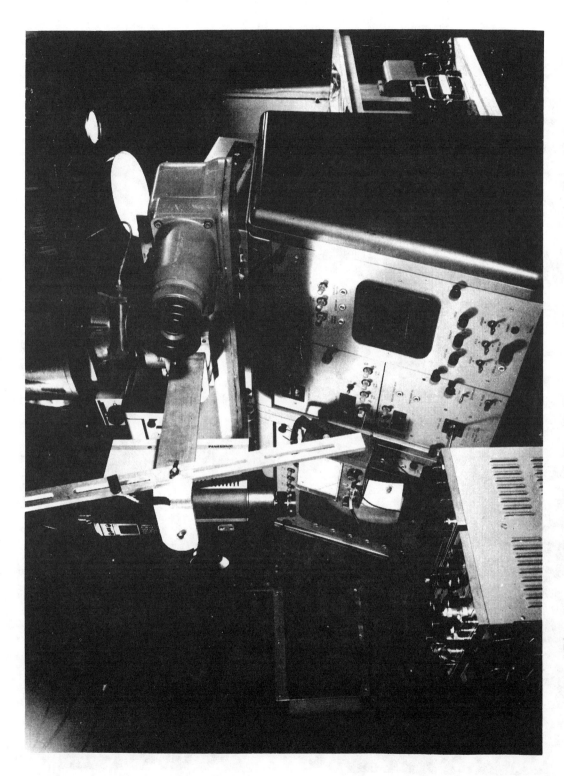

Figure 13. BRL 70-GHz Tracking Radar Interior Showing A-Scope TV Recording

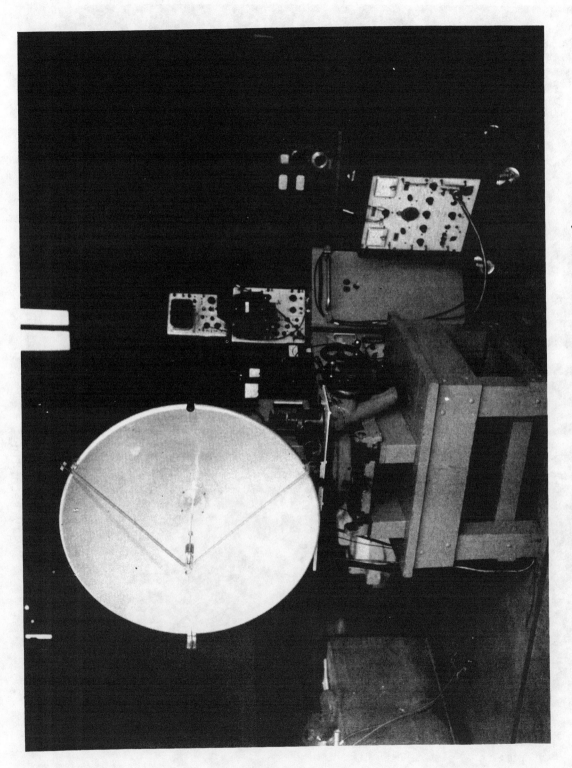

Figure 14. Norden 70-GHz All-Polarization Radar

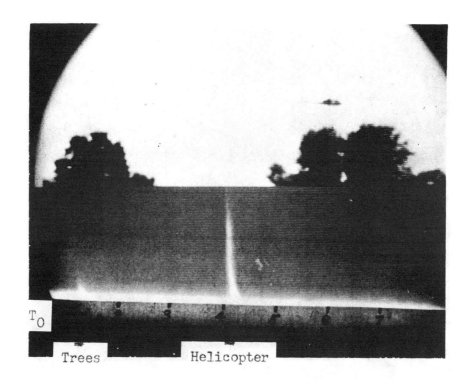

TARGET: UH-1B Helicopter

LOCATION: Flying Cross Course at 1.28 km, 30 meters Altitude, Close to Tree Top Level

RADAR: 70 GHz Norden

Figure 15. Helicopter Flying Crossing Course

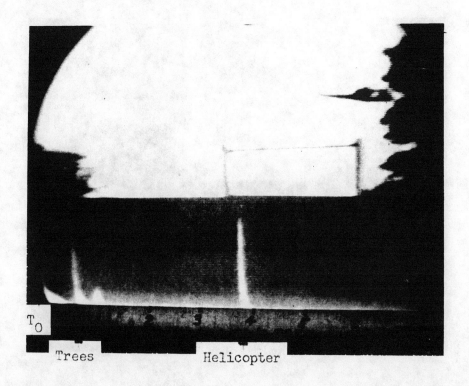

TARGET: UH-1B Helicopter

LOCATION: Flying Cross Course at 1.15 km, 30 meters Altitude, Going Behind Trees

RADAR: 70 GHz Norden

Figure 16. Helicopter Flying Crossing Course, Going Behind Trees

TARGET: UH-1B Helicopter

LOCATION: Flying Cross Course at 1.28 km, 30 meters Altitude, Main Body Behind Trees, Tail Rotor Exposed

RADAR: 70 GHz Norden

Figure 17. Helicopter Flying Crossing Course, Only Tail Rotor Exposed

TARGET: UH-1B Helicopter

LOCATION: Flying Cross Course at 1.28 km, 30 meters Altitude, Completely behind trees.

RADAR: 70 GHz Norden

Figure 18. Helicopter Flying Crossing Course, Completely Obscured by Trees

Introduction and Overview

TARGET: UH-1B Helicopter

LOCATION: Flying North-South Course, Range 455 meters, 30 meters Altitude, Close to Trees

RADAR: 70 GHz Ballistic Research Laboratories

Figure 19. Helicopter Flying Radial Course, Just above Trees

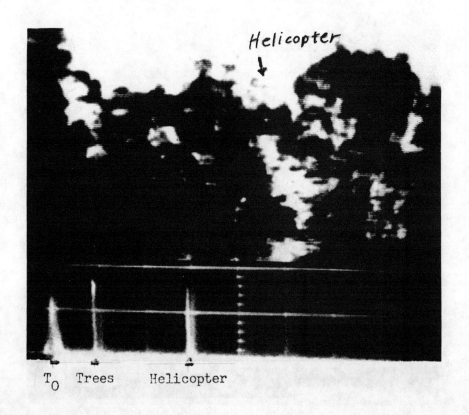

TARGET: UH-1B Helicopter

LOCATION: Flying North-South Course, Range 910 meters, Between Gap in Trees

RADAR: 70 GHz Ballistic Research Laboratories

Figure 20. Helicopter Flying Radial Course, Behind Heavy Tree Cover but Visible Through Gap

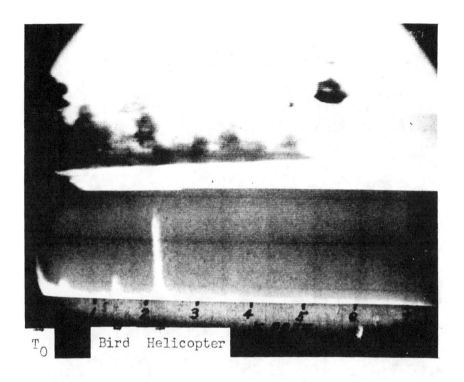

TARGET: UH-1B Helicopter and Bird

LOCATION: Hovering at 670 meters Range, 7.5 meters Altitude. Bird at 400 meters.

RADAR: 70 GHz Ballistic Research Laboratories

Figure 21. Helicopter Hovering 7.5 Meters Above Water

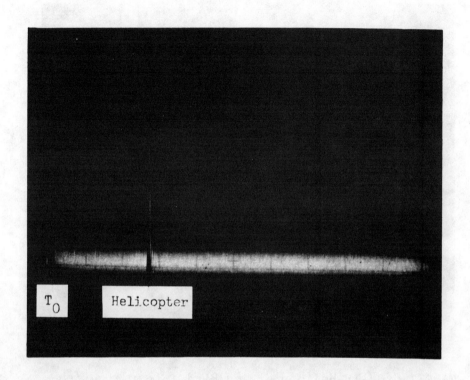

TARGET: UH-1B Helicopter

LOCATION: Hovering at 820 meters Range, 7.5 meters Altitude

RADAR: 70 GHz Norden

SWEEP: 2 microseconds/div; 1000 ft/div; 304 meters/div

Figure 22. Radar Return from Helicopter Hovering, Direct A-Scope Photograph

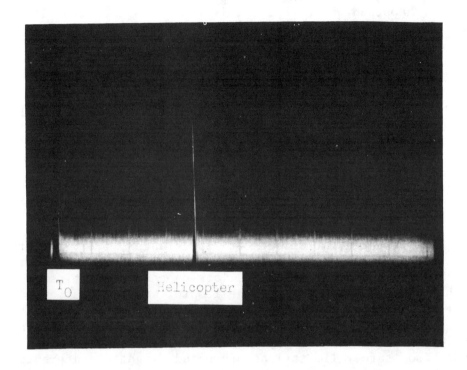

TARGET: UH-1B Helicopter

LOCATION: Flying Cross Course at 1.13 km, 30 meters Altitude just above Tree Top Level

RADAR: 70 GHz Norden

SWEEP: 2 microseconds/div; 1000 ft/div; 304 meters/div

Figure 23. Radar Return from Helicopter Flying Crossing Course over Trees, Direct A-Scope Photograph

Figure 21 shows the helicopter hovering 7.6 meters above the water at a range of 670 meters. The radar return is strong and clear without any clutter. The received signal was recorded as the helicopter slowly rotated in azimuth, presenting all aspects to the radar. A fortuitous circumstance happened while taping this run in that a bird flew across the beam, which shows up clearly as an echo at 460 meters.

Figures 22 and 23 show the radar pulse return from the helicopter as photographed directly from a fast-response oscilloscope. The narrow, jitter-free pulse and clutter-free return is more clearly shown in these photographs than in the previous photographs of the video tape playback. The signal-to-noise ratio is greater than 40 dB for an effective transmitter power to the antenna of about 200 watts peak and 0.1 watt average for this test.

3. <u>Automatic Tracking Tests</u>. Helicopter flight tests were made with the BRL 70-GHz radar operating in a conical scan, automatic tracking mode to assess (1) the tracking accuracy at low altitudes and close to trees, (2) target glint effects at close ranges, i.e., to determine the effective point of track as the aspect angle of the helicopter was changed, and (3) the acquisition and lock-on characteristics of a narrow beamwidth tracking system. A UH-1 helicopter was flown at an altitude of 30 meters in a circle of about 1.5-km diameter with part of the flight behind trees. Unfortunately the foliage was not very dense at the time of these tests and complete visual obscuration could not be obtained; also, safety limitations prevented flights to a greater range.

Figure 24a shows the field and wooded area over which the helicopter was flown. This photograph gives the general perspective of the test area as viewed with the unaided eye. The black horizontal lines on the photographs are transmitter pulses that leaked into the video camera circuits. A telescopic, zoom lens was used for Figures 24b, c and d to more accurately determine the point of track on the helicopter. The aiming circle has a 2-milliradian diameter and is boresighted to the radar antenna.

Figure 24b shows the helicopter being tracked while approaching the radar. The point of track was consistently the main rotor hub when the helicopter flew toward or away from the radar. Figure 24c shows the helicopter turning to make a crossing run; the radar is still tracking the main rotor hub. When the helicopter was exactly broadside to the radar, as shown in Figure 24d, the radar consistently moved back to track the tail rotor. Tracking accuracy was well within the aiming circle on each of the rotor hubs, with the only region of uncertainty occurring when the helicopter aspect was approaching the broadside orientation.

Acquisition and lock-on was not difficult. The radar antenna was pointed toward the helicopter with the aid of an optical sight to within roughly the dimension of the helicopter, and lock-on always occurred.

Introduction and Overview

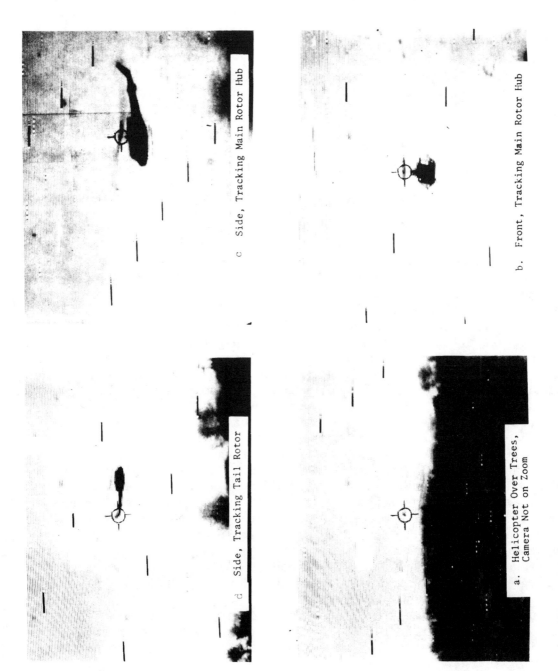

Figure 24. Automatic Tracking of Helicopter with BRL 70-GHz Radar

The radar antenna quickly moved toward the main or rear rotor hub, depending on the orientation of the helicopter.

The helicopter was also tracked while flying behind trees, as shown in Figure 25. Generally good tracking was maintained, although there was considerable modulation of the amplitude of the return caused by foliage obscuration and possibly multipath propagation from the trees. Tracking was much rougher than it was when the helicopter was away from the trees and there was an occassional loss of lock. There was no observable ground clutter multipath effects during any part of these tests.

4. <u>Ground Multipath</u>. When the tracking range is extended or the aircraft altitude is reduced over terrain that has little or no vegetation, a geometric condition will ultimately be reached where the lower portion of the antenna beam will illuminate the earth simultaneously with the target. The radar signal can then reach the target and return to the radar by a direct path and by way of reflection from the earth and/or scatterers on the earth. This multipath situation usually only becomes serious under geometrical conditions where the angular separaion between the direct and reflected wave is comparable to or less than the antenna elevation beamwidth, e.g., a target altitude less than 20 meters at 5-km range when a 1-meter-diameter radar antenna is used at 95 GHz. The reflected wave is then not suppressed by the antenna directivity. The net effect of strong multipath propagation on radar performance is to cause errors in antenna pointing and ranging and to severely modulate the received signal amplitude, causing deep fades and possible loss of automatic track. The effect on tracking is to make the antenna point above and below the true target direction in a cyclic manner as the reflected wave goes alternately in and out of phase with the direct wave. The cyclic rate can vary from near zero to frequencies of many hertz. If the cyclic rate is within the antenna drive servo passband, the antenna will follow the apparently widely changing angle of arrival of the multipath signal. If the reflected signal is large and reaches a critical value, the swing of the antenna can approach the antenna beamwidth. If the reflected signal strength goes above this critical value, the antenna will swing down to the image direction and cycle above and below this point; however, this is not a stable condition and the antenna can swing eratically between the direction of the target and the image, usually causing loss of track.

One of the simplest techniques for reducing low-elevation-angle multipath tracking error is off-boresight tracking.[4] The antenna boresight is maintained at an elevation above the target with the exact

---

[4] W.H. Bockmiller and P.R. Dax, "Radar Low Angle Tracking Study," NASA Contract Report No. 1387, Westinghouse Electric Corp., Baltimore, MD, June 1969.

Introduction and Overview

Figure 25. Automatic Tracking of Helicopter Behind Trees

elevation angle depending upon the radar antenna pattern shape and site conditions. This technique works on the principle that the multipath signal is reduced more than the target signal because of the steep falloff of the antenna pattern. Off-boresight tracking has been effective in preventing gross swings of the antenna and enables continuous automatic track until the target range is reduced enough to operate in the normal on-axis track mode.

Another very effective technique for reducing multipath effects is to change the frequency of the transmitted signal sufficiently to cause the error to go through a complete cycle during the time period of measurement. Thus, the multipath error can be averaged out. This technique is called either frequency agility, swept frequency, frequency diversity, or spread spectrum operation. The transmission frequency can be changed by different waveforms such as sinusoidal, triangular, or in random noise fashion. A theoretical study and experimental investigation of wideband swept frequency methods for reducing multipath effects have been made at BRL.[5] The results are very impressive not only in suppressing multipath but in increasing the target-to-background contrast ratio where the background clutter is caused by scintillation, target vibration, and wind-blown vegetation. Enhanced angular resolution and improved signal-to-noise ratios have been demonstrated.

C. Ground-to-Ground Millimeter Wave Radar Characteristics

1. General. Ground-located microwave radars which operate against ground targets are typically subject to severe performance degradation from multipath propagation, foliage obscuration, terrain masking, clutter, and natural or man-made false targets. If an appreciable portion of the antenna beam intercepts the terrain, the radar performance can be limited because of one or more of the above factors. Ground multipath and clutter do not usually limit the performance of millimeter wave ground-to-ground systems as seriously as foliage obscuration, background clutter, and terrain masking. Operation in the millimeter wavelength region with a narrow beamwidth antenna and a short transmitted pulse or wide bandwidth frequency modulation aids sufficiently in operating successfully close to the ground.

---

[5] R. McGee, "Multipath Suppression by Swept Frequency Methods," Ballistic Research Laboratories Memorandum Report No. 1950, November 1968, AD 682728.

2. **Multipath.** Experimental studies have been made at BRL on millimeter wave multipath pointing errors when propagating very close to the ground.[6-13] Multipath pointing errors have also been measured with a 140-GHz bistatic radar.[14,15] In general, the maximum pointing error in elevation over grassy terrain was observed to be less than plus or minus one-tenth of the 3-dB beamwidth.

---

[6] K.A. Richer, "4.4 Mm Wavelength Near Earth Propagation Measurements (U)," Ballistic Research Laboratories Memorandum Report No. 1403, May 1962 (Confidential), AD 331098.

[7] C.L. Wilson, "Antenna Pointing Errors in Simultaneous Lobing Antenna System," Ballistic Research Laboratories Technical Note No. 1453, June 1962, AD 608997.

[8] C.L. Wilson, "Antenna Pointing Errors in Sequential Lobing Antenna System," Ballistic Research Laboratories Technical Note. 1463, May 1962, AD 609009.

[9] T.W. O'Dell, "Final Report for Missile Guidance Subsystem Feasibility Program - Phase 1A (U)," General Precision Laboratories Report No. P0268 (Confidential).

[10] T.W. O'Dell, "Final Report for Missile Guidance Subsystem Feasibility Program - Phase 2 (U)," General Precision Laboratories Report No. 14990-2, June 1962 (Confidential).

[11] N. Leggett, "Final Report, Breadboard Feasibility Study of the Derringer Weapon Guidance System," Bendix Pacific Div. Report No. 91-139-1 (U), May 1963.

[12] J.E. Kammerer and K.A. Richer, "4.4 Mm Near Earth Antenna Multipath Pointing Errors (U)," Ballistic Research Laboratories Memorandum Report No. 1559, March 1964 (Confidential), AD 443211.

[13] J.E. Kammerer and K.A. Richer, "4.4 Mm Wavelength Precision Antennas and Mount," Ballistic Research Laboratories Memorandum Report No. 1576, July 1964, AD 449726.

[14] J.E. Kammerer and K.A. Richer, "140 GHz Millimetric Bistatic Continuous Wave Measurements Radar," Ballistic Research Laboratories Memorandum Report No. 1730, January 1966, AD 484693.

[15] J.E. Kammerer and K.A. Richer, "Pointing Errors of a 140 GHz Bistatic Radar System Illuminating U.S. Army Targets (U)," Ballistic Research Laboratories Memorandum Report No. 1755, June 1972 (Confidential), AD 375942.

Automatic tracking tests were made with the BRL 70-GHz radar tracking a pickup truck to study multipath and target glint effects. The truck was driven over an irregular course at a range of about 300 meters in a field covered with grass and high weeds. Figure 26a shows the radar tracking the rear edge of the truck as it was driven on a crossing course; this was a consistent tracking point when the truck was broadside to the radar. When the truck turned away from the radar, as shown in Figure 26b, the radar tracked the rear of the cab. Good tracking was maintained when the truck was driven right up to the edge of the woods, as shown in 26c. Figure 26d shows the truck as it is driven on a crossing course. The radar tracked the geometric center of the truck when it was coming toward or going away from the radar, as shown in Figures 26e and 26f.

Lock-on to the truck was easily accomplished either by orienting the antenna toward the truck and switching to automatic or by pointing the antenna ahead of the truck and letting it drive into the beam. Large targets in the field such as a corner reflector and a large wood pole would cause the radar to lose lock on the truck and stay on these targets if the truck was driven past them slowly. If the truck was driven past them at a high speed, the inertia of the antenna carried it past the other targets and the radar remained locked onto the truck. Tracking was generally very smooth and well within the 2-milliradian aiming circle, as can be seen in Figure 25. There was no evidence of multipath within the angular definition of observation using the aiming circle.

The pointing error caused by multipath is dependent upon the type and surface conditions of the terrain, the polarization of the transmitting and receiving antenna, the type of antenna scan, i.e., conical or monopulse, and the type of modulation and signal processing. Propagation over smooth, compacted, flat terrain causes the largest multipath pointing error, which under certain conditions can be as much as the antenna beamwidth; grassy terrain causes the smallest pointing error. The azimuth multipath pointing error is usually random in nature and small; however, if there are large irregularities in the composition or surface conditions in the region of the reflection area, the azimuth errors can also be large.

Monopulse and conical scan multipath pointing errors have been found to be quite similar in magnitude, although the monopulse system tested at 70 GHz had more regions of little or no pointing error as the height above ground was varied.[11]

The use of 140 and 240 GHz for anti-tank beam rider systems looks very promising. Very narrow beamwidths can be obtained with small antennas to obtain high tracking accuracy and avoid ground and background multipath and clutter. The range obtainable in clear and moderately adverse weather appears adequate with the miniaturized, all-solid state transmitters and receivers that are becoming available.

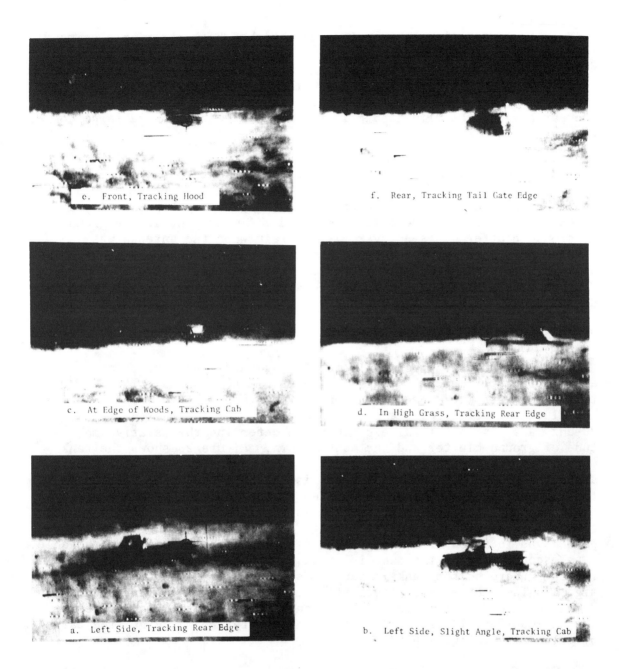

Figure 26. Automatic Tracking of Pickup Truck with BRL 70-GHz Radar

It is interesting to speculate that with a beam rider or a command guidance system, multipath errors might not be serious during the final phase of the flight since both the target and the missile being tracked would be seen by the tracking radar with approximately the same multipath errors. However, one potential problem is that the missile might strike the earth along the flight path if the multipath error has a large downward value.

3. <u>Ground Clutter</u>. When narrow beamwidth antennas are used, the ground clutter return from flat terrain is not usually very large unless the antenna is directed downward to the ground. For example, at 70 GHz with a 0.34-degree antenna mounted 1.4 meters above the ground, no ground clutter was seen over flat terrain when the antenna was pointed horizontally; the antenna had to be pointed downward about a beamwidth to get appreciable ground clutter return. Figure 27 shows the clutter from a field of grass and weeds when the antenna was pointed downward 0.26 degree. The antenna was pointed toward the base of the trees at a range of 304 meters. The clutter has a peak value of about 25 dB above the peak receiver noise level.

The magnitude of ground clutter in a ground-to-ground radar application will depend strongly on the terrain features. If the terrain is flat with low vegetation, targets can be seen well above the clutter level. For example, a tank was manually tracked with a 70-GHz radar as an initial experiment to determine the relative magnitude of ground clutter and the tank return. Figure 28 shows the tank manually tracked with a 70-GHz radar whose antenna is about 1.4 meters above the ground. The A-scope trace is shown in the upper portion of the figure. When the beam was pointed at the junction of the turret and the tank body, the strongest signal was received and very little ground clutter was received at the tank range. In Figure 29, the tank has been moved to a low area among high weeds and the clutter is appreciable. A telescopic view of the tank is shown in Figure 29 where the high vegetation around the tank is causing some clutter close to the tank signal. The clutter signal scintillates rapidly and deeply and generally has a much longer risetime than the tank return pulse, which suggests a technique for discrimination of ground clutter from the desired target by pulse risetime gating.[16]

One of the most effective techniques in separating moving targets from background clutter is to use the doppler shift phenomenon where the frequency difference between frequency spectra of fluctuating clutter and doppler-shifted echoes from a moving target is used. The theory of the statistical nature of fluctuating clutter is given in Appendix B, and a description of MTI techniques is given in Appendix C.

---

[16] M.G. Stansbury and C.L. Wilson, "A Signal Processing Pulse Width Discriminator," Ballistic Research Laboratories Memorandum Report No. 2040, June 1970, AD 710230.

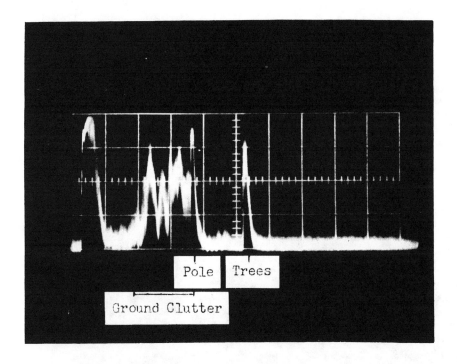

TARGET: Ground in Front of Trees

RADAR: 70 GHz Norden

POLARIZATION: Vertical

SWEEP: 0.5 microsecond/div; 200 ft/div; 76.2 meters/div

Figure 27. Ground Clutter at 70 GHz

TARGET: M 48 Tank

LOCATION: 365 meters Range, Woods Background

RADAR: 70 GHz Norden

Figure 28. 70-GHz Radar Return from Tank in Clear Field

TARGET: M 48 Tank

LOCATION: 260 meters; Telephoto Photograph showing High Grass

RADAR: 70 GHz Norden

Figure 29. 70-GHz Radar Return from Tank in High Grass

Other techniques for clutter rejection and/or target signal enhancement include the use of (1) a swept frequency radar as described earlier under ground-to-air systems, (2) pulse width and slope discrimination circuits,[16] (3) polarization discernment to identify the target by its polarization characteristics, and (4) target vibration and rotation-induced modulation. This fourth technique is particularly useful on stationary targets where either the engine is running or it is windy, since at short wavelengths small physical movement of a reflector causes a large fluctuation in received signal strength at the rate at which the reflector is vibrating or rotating or being blown about.

4. <u>Foliage Obscuration</u>. Foliage obscuration at millimeter wavelengths is quite high; there is essentially no penetration directly through dense green foliage. The propagation through the openings between foliage components has been studied at BRL with a 35-GHz radiometer where a comparison was made between the millimeter wave and the optical blockage.[17,18,19] Millimeter wave obscuration was found to be greater than originally predicted as based on the optical obscuration.

D. <u>Air-to-Ground Millimeter Wave Radar Characteristics</u>

Airborne radars operating against ground targets generally suffer from target acquisition problems, foliage obscuration, and ground clutter. Radars operating at millimeter wavelengths with narrow antenna beams will give high resolution of ground targets. A narrow antenna beamwidth will also reduce the ground clutter, but dense foliage obscuration will not be helped. However, where the foliage is sparse, the target might be seen because of the high resolution of angle and range.

Ground mapping for area surveillance and target acquisition can be done exceptionally well at millimeter wavelengths. The resolution obtained with narrow antenna beams is sometimes high enough for real-time, forward-aspect signal recording which is much simpler than the side-looking, synthetic aperture type system with its complex data processing and lack of coverage forward.

The use of millimeter wavelengths for Instrument Landing Systems (ILS) applications offers advantages of very high resolution and accuracy with acceptable aircraft antenna size.

---

[17] R. McGee, "Millimeter Wave Radiometric Detection of Targets Obscured by Foliage," Ballistic Research Laboratories Memorandum Report No. 1901, January 1968, AD 667962.

[18] R. McGee, W. Sacco and W. Lese, "Effects of Obscuration on Millimeter Wave Signatures of Targets," Ballistic Research Laboratories Memorandum Report No. 1962, February 1969, AD 684900.

[19] R. McGee, "A Foliage Obscuration for Passive Radiometric Detection System," Ballistic Research Laboratories Memorandum Report No. 1999, July 1969, AD 858380.

Given that the target is acquired and designated, a millimeter air-to-ground system should provide excellent gun-aiming accuracy or weapon delivery guidance and terminal homing. The multipath and terrain masking problems will not be as severe as in ground-to-ground applications.

E. Target Acquisition Millimeter Wave Radar Characteristics

The use of millimeter wave sensors for target acquisition holds promise for performing all the functions involved in acquisition such as the volume search; detection of presence of a target; discrimination of target from clutter and false targets; identification by gross category such as a moving tank, truck, rotary-wing or fixed-wing aircraft; location; automatic tracking for determination of range, radial velocity, and azimuth and elevation angles; assignment of target to a weapon for engagement; and assessment of the damage by the indication of cessation of movement or absence of signal return. The sensor could operate in either the pulse, FM/CW, or spread-frequency spectrum modes.

A millimeter wave target acquisition system will, of course, be limited by the handicaps already discussed (reduced performance in rain, foliage obscuration, terrain masking, clutter, and natural and man-made false targets) plus the inherent problem of a longer search time or more rapid scan rate requirement associated with searching a large volume with a narrow beamwidth. Some applications, such as an antiaircraft gun fire control search radar used against low-flying aircraft, require only limited elevation search, primarily at the horizon. These low-altitude search systems reduce the search volume appreciably and make the use of a narrow beamwidth search radar feasible. In addition, multiple-beam search antennas can be used to reduce the time required to search out a large volume. In view of the recent development of potentially low-cost integrated circuit receivers, the multiple-beam concept is becoming more feasible.

A millimeter wave target acquisition sensor would have a number of advantages and possiblities, such as:

1. Small targets can be detected since they will be an appreciable part of a wavelength.

2. Dry contaminants in the atmosphere such as smoke, dust, smog, and dry snow do not affect performance.

3. The absorption in the atmosphere and the narrow beams make detection and jamming by the enemy difficult at long ranges.

4. High angular resolution permits effective separation of the desired target from nearby scatters for effective target discrimination from natural and man-made objects.

5. A very narrow beam allows operation of the sensor close to the ground and acquisition of low-altitude targets with a minimum of ground multipath and ground clutter interference.

6. The combination of narrow beam and narrow transmitted pulse greatly improves the target-to-clutter ratio and permits design of a very effective MTI system for the detection of slowly moving targets hidden in clutter. At millimeter wavelengths, the doppler frequencies resulting from small radial velocities are in the audio frequency range and are convenient to process.

7. Stationary and moving targets have unique detectable signatures if there is some vibration or rotary motion of a component that is a millimeter wave reflector. It is possible to detect very small (millimeter) movements of reflectors since such movements are an appreciable part of a wavelength and cause a large change in the phase of the reflected signal. The phase-modulated reflected signal combines vectorially with the gross reflected signal from the main body of the target, resulting in an amplitude modulation at a rate proportional to the vibration or rotation rate. The success of this technique is dependent upon the difference in the spectra of the target and the clutter.

A millimeter wave radar used in conjunction with an acoustic sensing system (which has a good capability to detect and identify audio noise-emitting targets) would provide accurate target location and tracking information. The acoustic signature of the desired target would serve to aid the radar in acquiring the target by matching the spectrum of the acoustic signal with that of the rotation and/or vibration doppler modulation on the radar signal.

8. Wideband spread-spectrum operation provides good target-to-clutter enhancement by averaging out the clutter as described previously.

9. Extremely high range resolution techniques offer promise for the identification or recognition of spatial features of targets. For example, an FM/CW radar that has 15-cm resolution and a 4.5-mil, 3-db beamwidth has been investigated at BRL.[20] Good depth feature resolution is possible with this type of radar.

---

[20] J. E. Kammerer, "A 94 GHz FM-CW Radar with Six-Inch Range Resolution Capability," Ballistic Research Laboratories Memorandum Report No. 2235, September 1972, AD 908485L.

IV. MILLIMETER WAVE RADAR AND COMPONENT STATUS

A. Existing Radars

A number of radars operating at millimeter wavelengths have been built for military applications which are indicative of component availability and technology status. Most of them have been at the low frequency end of the millimeter wave band in the 33 to 36 GHz region, called the K_a band, and also called bands K-7(32-34 GHz) and K-8(34-36 GHz).

Following is a list of K_a band radars and applications:

Application	Identification
Weather Radars	AN/APQ-39
	AN/APQ-70
Slar Radars (Side-Looking Mapping)	AN/APQ-55, -56, -79, -86, -97
	AN/APQ-7, -8, WX-50
Forward-Scanning Mapping Radars	AN/APQ-113(K_a), -144(K_a)
Navigation Radars	AN/APQ-57, -58
Terrain-Avoidance and Terrain-Following Radars	AN/APQ-89
"MOTARDES" Doppler Airborne Radar	AN/APQ-137
Mapping and Search Radar	AN/APN-61
Carrier Landing Radars	AN/SPN-10, -41, -42
Sea Clutter Measurement	Instrumentation Radar, GIT Georgia Institute of Technology, Atlanta, GA
Cross Section Measurement	Instrumentation Radar, RATSCAT Air Force Special Weapons Command, Holloman AFB, NM
Lunar Radar	Experimental, MIT, Cambridge, MA
Low-Angle Tracking Over Water	Experimental, Naval Research Laboratory, Washington, DC
Anti-Tank Missile Terminal Homing, FM/CW and Noise/FM Solid State Radar	TGSM Project, BRL and Sperry for Missile Command, Redstone Arsenal, AL
Tracking and Cross Section Studies, Pulse/FM Solid State Radar	Experimental, BRL

Fewer military radars have been built for the higher frequency end of the millimeter wave band; most of them are in the experimental or prototype stage. Following is a list of radars operating between 70 and 140 GHz.

Application	Freq.	Identification	Source
Side-Looking Mapping Radar	70 GHz	AN/APQ-62	
Search Mapping Radar	70	JR-9	Raytheon
Search Radar	70	AN/BPS-8	
Search and Surveillance	70	AN/MPS-29	GIT
Search and Surveillance	70	Experimental	Harry Diamond Lab., Wash., DC
Aircraft Obstacle Avoidance Aircraft Instrument Landing Monopulse Tracking	70	Experimental	Norden Div. of United Aircraft, Norwalk, CT
Low-Altitude Aircraft Tracking	70	Experimental	BRL
Obstacle Avoidance Sea Clutter Measurement	95	"	NADC, Warminster, PA
Space Object Identification	95	"	Aerospace Corp., El Segundo, CA
Arctic Terrain Avoidance Airborne Obstacle Avoidance	95	"	Applied Physics Lab., Silver Spring, MD

Airborne Applications
  Terrain Imaging
  Instrument Landing
  Short Range Weapon Delivery
  Sensor Cueing
	95	"	Goodyear Aersopace Corp., Litchfield Park, AZ

FM/CW and Noise/FM Radar for Clutter Suppression and Fine Range Resolution	95	"	BRL
Pulse/FM Solid State Radar for Tracking and Cross Section Measurement	95	"	BRL
Pulse/FM Solid State Radar for Beam Rider	140	"	BRL
Bistatic CW Radar for Cross Section and Propagation Measurements	140	"	BRL
Pulse/FM Solid State Radar Studies	220	"	BRL

A number of organizations outside the United States also have developed millimeter wave radars for military and commercial use.

B. Radar Components

At 35 GHz low-power, solid state transmitters, high-power transmitters up to 150 kw, and all of the other necessary system components and test instrumentation are commercially available to build and test radars.

At 95 GHz, low-power transmitters (1.5 watts peak), solid state sources, and a complete line of components and test instrumentation are available. For high-power transmitters in the 5- to 10-kw range, magnetrons and distributed interaction kylstrons have been built. An extended interaction oscillator type tube has been proposed, but additional development, ruggedization, and reliability improvements are needed before any of these tubes can be put into field use. High-power circulators, compact receiver protectors, low-loss radomes, and monopulse tracking antennas are not now commercially available at 95 GHz.

At 140 GHz, low-power klystrons, solid state sources, Carcinotrons, and extended interaction oscillator type tubes are available, with the latter having an 80-watt CW output capability. In the 220- to 240-GHz region, klystrons, Carcinotrons, and extended interaction type oscillators are available, with the latter having a 10-watt CW output capability.

Mixers with very low noise figures are available; for example, noise figures of 4 dB at 35, 8 dB at 95, and 13 dB at 140 and 220 GHz are available. The Schottky barrier diodes used in these mixers are sensitive to burnout and, therefore, very high isolation receiver protectors are needed when they are used with high power transmitters.

Unfortunately, in general, currently available millimeter wave components are expensive, critical to use and generally have short life when they handle much power or are exposed to transients. This situation exists largely because of the newness of the field, the small volume of usage, and the small amount of reliability engineering being applied. A similar situation prevailed when microwave components were in their infancy. Today we have the advantage, however, that much of the technology for working to extremely small tolerances and with solid state devices is already well advanced.

V. TECHNOLOGY ADVANCES REQUIRED

The areas in which technology advances and basic data not now available are required in order to fully employ millimeter wave sensors are as follows.

A. Weather Effects

Additional experimental studies are needed at frequencies above 95 GHz to supplement the theoretical data that now exist. These data are needed to predict the maximum range of systems which must operate in adverse weather.

B. Target and Background Signatures

These is a great dearth of data on target cross sections and signatures with and without camouflage on the target and general terrain and background clutter data in the millimeter wave region. Particularly needed are power spectral density and probability density function data to optimally design radar systems for effective target acquisition and accurate tracking.

C. Basic Propagation Characteristics

Basic propagation data on foliage obscuration, multipath, clutter, multiple target, and glint effects are needed for very narrow beamwidth, short pulse, high range resolution FM/CW, and spread spectrum millimeter wave radars. Included in this study would be the effects and possible improvements of frequency and/or polarization agility in the conical scan and monopulse modes of operation. These data are needed to determine the tracking accuracy of these special types of radars where the beamwidth is so narrow and the range resolution is so high as to be comparable to the target or clutter source dimensions.

D. Target Acquisition Techniques

These is a need for practical, low-cost, and reliable techniques for the implementation of the complex data processing procedures required to acquire targets in difficult environments.

E. Component Improvement and Development

Improvements and new developments are needed on the following millimeter wave components:

1. Duplexers and mixer diode protectors for high-power ($\geq$ 10 kw) pulse radars that have low insertion loss, adequate isolation to prevent diode burnout, and fast recovery.

2. High-power ($\geq$ 10 kw) sources that have long life and high reliability.

3. Rapid antenna beam scan techniques, electronic or mechanical, with low loss, for target search and acquisition.

Introduction and Overview

4. Radomes with low loss and low boresight shift.

5. High-power (1 to 10 kw) coherent sources or amplifiers for doppler systems.

6. Low-power (1-watt), pulsed, solid state sources at 140 and 240 GHz.

7. Low noise mixers at 140 and 240 GHz.

VI. CONCLUSIONS

There are a number of possible military applications of millimeter wave technology where operation at these short wavelengths offers unique and distinct advantages. Following are suggested applications.

A. Ground-to-Air

1. Target acquisition, particularly effective against low-altitude aircraft.

2. Gun fire control with complete solution radar and possibly closed-loop fire control, particularly effective against low-altitude aircraft.

3. Range-only radar with rapid data acquisition for very short time on the air operation for security.

4. Active seeker for terminal homing on projectile or missile.

5. Tracking radar for missile guidance in beam rider, command, or semi-active mode.

6. Firing error or miss-distance indication of projectiles or missiles with capability of rapidly providing trajectory and precision velocity data.

7. Weather radar, for acquisition of meteorological data with high spatial resolution at short ranges.

B. Ground-to-Ground

1. Ground target acquisition, with best possibilities of working against targets that have some linear or rotary motion or vibration.

2. Anti-mortar radar where the small size and light weight of the antenna is a much needed advantage.

3. Bullet detector and incoming projectile sensor to quickly determine trajectory and source of incoming projectiles.

4. Anti-tank missile or projectile guidance by beam rider or command guidance technique.

5. Artillery rocket tracking immediately after launch to measure and correct for tipoff and ground wind effects.

C. Air-to-Ground

1. Target acquisition.

2. Side-looking and forward-scanning terrain-mapping radar for reconnaissance, target acquisition, and damage assessment.

3. Terrain-following and obstacle-avoidance radar.

4. Instrument landing monitor (ILM) system.

5. Anti-radiation missile (ARM) seeker head.

6. Anti-tank missile guidance by terminal homing of sub-missile (TGSM) or command guidance by sensor cueing.

7. Radar guided bomb.

8. Gun fire control radar against ground targets such as trucks, tanks, and missile launchers.

VII. ACKNOWLEDGEMENTS

The millimeter wave radar study described in this report is the result of group efforts of a number of persons in the Microwave Systems Branch of the Concepts Analysis Laboratory, BRL. In particular, John Kammerer, Harvey LaFon, Lt. Richard Au Hoy and Lt. Bruce Szypot made significant contributions.

The appendices were prepared for BRL by K. L. Koester, Dr. L. Kosowsky, and J. F. Sparacio of the Norden Division of United Aircraft, Norwalk, Connecticut.

REFERENCES

1. V.W. Richard and J.E. Kammerer, "Rain Backscatter Measurements and Theory at Millimeter Wavelengths," BRL Report No. 1838, October 1975, AD 800 8173L.

2. V.I. Rozenberg, "Radar Characteristics of Rain in Submillimeter Range," Radio Eng. and Elec. Physics, Vol. $\underline{15}$, No. 12, 2157-2163, 1970.

3. L.D. Strom, "Applications for Millimeter Radars," System Planning Corp., Arlington, VA, Report No. 108, for ARPA, 31 December 1973 (Confidential), AD 529566; "Adverse Weather Applications for Millimeter Radars on the Battlefield," Paper G.3, NELC/TD 308, 1974 Millimeter Wave λ Technical Conference, Naval Electronics Laboratory Center, San Diego, CA, 26-28 March 1974.

4. W.H. Bockmiller and P.R. Dax, "Radar Low Angle Tracking Study," NASA Contract Report No. 1387, Westinghouse Electric Corp., Baltimore, MD, June 1969.

5. R. McGee, "Multipath Suppression by Swept Frequency Methods," Ballistic Research Laboratories Memorandum Report No. 1950, November 1968, AD 682 728.

6. K.A. Richer, "4.4 Mm Wavelength Near Earth Propagation Measurements (U)," Ballistic Research Laboratories Memorandum Report No. 1403, May 1962 (Confidential), AD 331 098.

7. C.L. Wilson, "Antenna Pointing Errors in Simultaneous Lobing Antenna System," Ballistic Research Laboratories Technical Note No. 1453, June 1962, AD 608 997.

8. C.L. Wilson, "Antenna Pointing Errors in Sequential Lobing Antenna System," Ballistic Research Laboratories Technical Note No. 1463, May 1962, AD 609 009.

9. T.W. O'Dell, "Final Report for Missile Guidance Subsystem Feasibility Program - Phase 1A (U)," General Precision Laboratories Report No. P0268 (Confidential).

10. T.W. O'Dell, "Final Report for Missile Guidance Subsystem Feasibility Program - Phase 2 (U)," General Precision Laboratories Report No. 14990-2, June 1962 (Confidential).

11. N. Leggett, "Final Report, Breadboard Feasibility Study of the Derringer Weapon Guidance System," Bendix Pacific Div. Report No. 91-139-1 (U), May 1963.

12. J.E. Kammerer and K.A. Richer, "4.4 Mm Near Earth Antenna Multipath Pointing Errors (U)," Balllistic Research Laboratories Memorandum Report No. 1559, March 1964 (Confidential), AD 443 211.

REFERENCES (CONT.)

13. J.E. Kammerer and K.A. Richer, "4.4 Mm Wavelength Precision Antennas and Mount," Ballistic Research Laboratories Memorandum Report No. 1576, July 1964, AD 449 726.

14. J.E. Kammerer and K.A. Richer, "140 GHz Millimetric Bistatic Continuous Wave Measurements Radar," Ballistic Research Laboratories Memorandum Report No. 1730, January 1966, AD 484 693.

15. J.E. Kammerer and K.A. Richer, "Pointing Errors of a 140 GHz Bistatic Radar System Illuminating U.S. Army Targets (U)," Ballistic Research Laboratories Memorandum Report No. 1755, June 1966 (Confidential), AD 375 942.

16. M.G. Stansbury and C.L. Wilson, "A Signal Processing Pulse Width Discriminator," Ballistic Research Laboratories Memorandum Report No. 2040, June 1970, AD 710 230.

17. R. McGee, "Millimeter Wave Radiometric Detection of Targets Obscured by Foliage," Ballistic Research Laboratories Memorandum Report No. 1901, January 1968, AD 667 962.

18. R. McGee, W. Sacco and W. Lese, "Effects of Obscuration on Millimeter Wave Signatures of Targets," Ballistic Research Laboratories Memorandum Report No. 1962, February 1969, AD 684 900.

19. R. McGee, "A Foliage Obscuration for Passive Radiometric Detection System," Ballistic Research Laboratories Memorandum Report No. 1999, July 1969, AD 858 380.

20. J.E. Kammerer, "A 94 GHz FM-CW Radar with Six-Inch Range Resolution Capability," Ballistic Research Laboratories Memorandum Report No. 2235, September 1972, AD 908 485L.

CHAPTER 2

MILLIMETER WAVE PROPAGATION, TARGETS AND CLUTTER

Successful operation of a radar is dependent not only on the radar but also on its environment. Propagation characteristics of the medium, target characteristics and clutter characteristics all affect both the design of the radar and its performance. This is especially true for millimeter wave radars. Several papers in the preceding chapter mention millimeter wave propagation. Atmospheric attenuation affects choice of radar operating frequency. The well-known absorption lines generally constrain millimeter wave radar to operation frequencies in the "windows" at 35, 95, 140 and 220 GHz. These windows influenced allocation by WARC-79 of those frequencies for radiolocation (radar). Additionally, some millimeter wave radars may intentionally choose to operate at an absorption line to reduce detection of the radar by hostile intercept receivers. This practice can also reduce unwanted friendly interference.

Papers in this chapter will address these three environmental effects — propagation, targets and clutter.

SYNOPSIS OF REPRINT PAPERS IN CHAPTER 2

2.A Propagation

2.A.1 "A Radar System Engineer Looks at Current Millimeter-Submillimeter Atmospheric Propagation Data" — Johnston

A cursory review of the literature indicates the existence of a plethora of information on "Millimeter" propagation — both theoretical and experimental. Close study, however, reveals a number of shortcomings in this literature for radar purposes, *e.g.:* 1) much of the "millimeter" propagation literature addresses the region around K_a-band, with lesser treatment above K_a-band; 2) a large portion of this literature does not pertain to operating characteristics of millimeter radars (path length, antenna beamwidths, modulation frequencies, antenna polarizations, etc.); 3) few of the measurements were made by/for radar design purposes; and 4) some propagation characteristics of importance to radar have received little attention.

Topics treated in this paper include radar considerations, propagation data requirements, and present propagation data. Characteristics of a "typical" mm radar are presented to indicate propagation measurement system parameters of interest to the mm radar system design engineer. Radar features which are influenced by propagation effects are listed in addition to propagation effects on radar functions and propagation effects on radar/guidance. A table of environmental constraints on microwave radar serves as a "benchmark" leading to the table on propagation data requirements for millimeter wave radar design. The items in this table are then discussed citing available data in the literature and making recommendations for further propagation measurements for millimeter wave radar design. The extensive bibliography (84 items) is a good survey of millimeter propagation literature.

This paper was presented at IEEE EASCON-79 as part of the four sessions on Environmental Effects on Radar/Missile Guidance.

2.A.2 "Millimeter Wave Propagation" — Koester, Kosowsky and Sparacio

This paper, which was originally published as an appendix to Paper 1.7 (Richard) of this volume, is placed here because of its relevance to this chapter.

Propagation of millimeter waves has been investigated for many years. Theory of particle scattering was analyzed by Mie as long ago as 1908. This paper summarizes available (as of 1973) theoretical and experimental data on propagation in clear air, rain and fog at frequencies of 35-94 GHz.

Fundamental definitions and concepts, *e.g.* scattering cross section, absorbing cross section, extinction cross section, raindrop size distributions and fog characteristics, are treated with citations to references for more extensive treatments. These provide the millimeter wave radar systems engineer a good background in millimeter wave propagation. Propagation data are applied in the section "system considerations," where performance of a radar at 35, 70 and 94 GHz is analyzed in clear air, rain and fog.

Several limitations to this paper should be cited:

☐ 1. Frequency band designations of Table A-6 are not in accordance with the current IEEE Standard letter band designations. See discussion in the introduction to Chapter 1 of this volume.

☐ 2. Theoretical analyses do not include the effects of radar polarization, which can be important for non-spherical rain drops. See Paper 2.A.1.

☐ 3. Attenuation data is presented as single valued, *i.e.* as a simple function of rainfall rate. Attenuation due to rainfall has a fluctuation. See Paper 2.A.1 and Reference 2 for this chapter.

☐ 4. Differentiation between instantaneous and average rainfall rates is not made. As Paper 2.A.1 points out, the meteorologist's rainfall rate is generally a value averaged over fairly long periods (minutes or even hours), whereas the much higher instantaneous rainfall rates result in the higher attenuations which are actually experienced.

2.A.3 *"Propagation Effects for mm Wave Fire Control System"* — Patton, Petrovic and Teti

This paper presents results of a survey of propagation effects of the atmosphere on the performance of a sea-based millimeter fire control radar to be used against short range, low altitude targets. It is somewhat similar to the preceding paper. Its conclusion as to the degradation of mm wave propagation for this application, that "...literature is inconclusive on the matter and that experimental work is required," is also similar to the implied results of the first paper. It is interesting to note that the first paper was prepared without knowledge of this paper.

This paper was based on review of a very large number of papers. The "International Radio Index 1975" cited by this paper will be found under its full title [2] in a later edition. Although this paper is oriented towards one specific radar application, much is relevant to other applications.

The consideration of the effects of ducting is noteworthy. Ducting at mm wavelengths is also mentioned by Howard, cited in Paper 2.A.1 of this chapter. It is well known that ducting can occur over land as well as over water.

This paper was presented at the 6th DARPA-Tri-Service 1977 Millimeter Wave Conference.

2.A.4 *"Environmental Effects on Millimeter Radar Peformance"* — Dyer and Currie

This paper complements the preceding three papers in that it contains extensive actual propagation data, whereas the paper 2.A.1 discusses data needs but gives no data. Topics treated include atmospheric effects (attenuation due to gaseous absorption, rain, fog, clouds and snow; scattering and backscatter; background effects; ground cover attenuation) and clutter effects on system parameter choice. The latter topic is extremely important in the design of millimeter wave radar as discussed in the Paper 2.A.1.

This paper was presented at the AGARD Conference on Millimeter and Submillimeter Wave Propagation and Circuits.

2.A.5 *"Atmospheric Attenuation of Millimeter and Submillimeter Waves"* — Falcone and Abreu

Digital computer programs for computation of the atmospheric attenuation of waves having frequencies of 1 - 1000 GHz are described. The program includes both the clear atmosphere and various hydrometeors — fogs, clouds and precipitation (rain). Four fog models and eight cloud models can be used. Liquid water content, rainfall rates, and hydrometeor temperatures can be selected to represent midlatitude propagation conditions. Illustrative resulting data are shown.

Unfortunately, this paper does not address polarization/cross polarization effects which are shown by Paper 2.A.1 to be important. Similarly, attenuation is presented as if it were "steady-state", *i.e.* attenuation fluctuation is not treated here (see Paper 2.A.1 and the following paper).

It would have been helpful if the authors had included comparisons of computer predictions with actual experimental measurements. Such comparisons have been made by a few authors. Comparisons for rainfall attenuation generally show that agreement leaves much to be desired when specific measurements are compared. Lin and Ishimaru [2] address some of the problems in such comparisons.

This paper was presented at IEEE EASCON-79.

2.A.6 "Atmospheric Turbulence Effects on Millimeter Wave Propagation" — McMillan, Wiltse and Snider

This paper considers atmospheric effects on scintillation amplitude (fluctuation) and angle of arrival variation. It is an extension of previous Russian millimeter wave theoretical and experimental measurements.

Angle of arrival variations are actually due to phase front variations in the arriving wave. As Paper 2.A.1 indicates, a moderate amount of angle of arrival work, both theoretical and experimental measurements, has been made at X-band and K_a-band. An overall treatment of angle of arrival would be helpful. Even so, this is the only U.S. paper on this topic.

This paper was presented at IEEE EASCON-79.

2.A.7 "Millimetre Wave Propagation in Smoke" — Knox

The recent attention paid to the adverse effects of some aerosols such as smoke and dust on electro-optical sensors has caused their effects on millimeter wave sensors to be investigated. The U.S. Department of the Army has established a program office to coordinate sensor testing in aerosol obscurants. This paper describes a recent test program, Smoke Week II, and the participation by the U.S. Army Ballistic Research Laboratory.

This paper was presented at IEEE EASCON-79.

Additional data on smoke propagation attenuation is presented in Paper 5.13 of Chapter 5 of this volume.

2.A.8 "140 GHz Multipath Measurements Over Varied Ground Cover" — Wallace

As pointed out in several papers in the preceding chapter, one advantage of millimeter waves is the improved low altitude tracking (reduced multipath effects) as compared to, for example, X-band. Klaver [3] describes a dual X/K_a-band radar and shows the improved low altitude tracking at K_a-band as compared to X-band in the same system. Forward scattered power is the important contributor to multipath effects.

This paper presents results of forward scattered power measurements made at 140 GHz over three types of terrain — high weeds, mowed weeds and very flat asphalt. As a novel departure from forward scattering measurements made at lower frequencies by others, the author simulates the two extreme positions of a conically scanned tracking antenna. Both CW and pulsed IMPATT RF sources are used. The pulsed IMPATT source is actually chirped over 1 GHz. Accordingly, the well-known effects of frequency agility cause decorrelation of the diffuse component of the forward scattered signal. Measurements are presented for only one polarization (vertical).

This paper was presented at IEEE EASCON-79.

2.A.9 "Millimeter Wave Propagation Measurements over Snow" — Hayes, Lammers, Marr and McNally

This Air Force paper may be considered as a companion to and extension of the preceding Army paper. In this paper, simultaneous measurements are made at three frequencies — 35, 98, and 140 GHz. Two types of measurements are made -- backscatter and multipath. Several polarizations are used — HH, VV and VH. Backscatter measurements utilize a tri-band scatterometer. Resulting scatterometer data is shown in σ_o as a function of incident angle. Multipath data are shown as variation of received signal as a function of receiver antenna height. Data are shown for both short dormant grass and for snow cover. Snow greatly increases the depth of the lobe structure. The type of snow is shown to affect multipath.

This paper was presented at IEEE EASCON-79.

2.B Targets

2.B.1 "93-GHz Radar Cross Section Measurements of Satellite Elemental Scatterers" — Dybdal and King

Radar cross section measurements are generally made either in an anechoic chamber with a special RCS measurement radar, or by extracting RCS data from the performance of a normal radar (*e.g.*, see Paper 5.3 in Chapter 5 of this volume). This paper presents actual RCS measurement data made in an anechoic chamber at 93 GHz. Objects reported include solar panels (with and without metallic reference plate), thermal blanket material, plates, corner reflectors (angle stock), UHF dipole antennas, and X-band horn antennas. Only horizontal polarization is used.

All data are presented as a function of aspect angle, showing a complex pattern lobe structure. These resemble RCS patterns taken at microwave (centimeter-wave) frequencies. These patterns clearly show that use of a single number as the value for the radar cross section of a millimeter target is improper (this applies to microwave radars as well).

There is one difference between millimeter wave and microwave radars: a microwave radar will generally illuminate the entire target, thus, the RCS seen by the radar will be the resultant from a number of different elemental scatterers, *e.g.* an antenna alone. As this paper shows, at millimeter waves, a single object, such as those used here, is actually composed of several scatterers and hence has a lobe structure of its own.

This paper is very informative as to the nature of scattering RCS of millimeter wave objects. It clearly indicates the necessity for measuring the RCS of the design target of a mm wave radar and then not using a single number as the radar cross section of a target.

2.B.2 "95 GHz Pulsed Radar Returns from Trees" — Hayes

The preceding paper presents RCS measurements of man-made objects. Trees may be variously considered as a false target (as in vehicular braking radars) or as one type of clutter in low altitude tracking radars. This paper presents the cumulative probability amplitude distributions and frequency spectrum of the RCS of trees at 95 GHz. Data are presented for only one polarization (not stated). The majority are found to be of a log-normal nature, although some follow a Weibull distribution. The Rayleigh distribution appears at very low wind speeds, less than 3 mph. The σ_o is found to be -28 dB.

Fluctuation has two parts, a fixed or slow component some 15 dB in magnitude and about 4 Hz wide, then a fast component approximately 20 dB in magnitude and 200 Hz wide. It is speculated that the former are caused by branches and larger objects, while the latter may be due to moving leaves and small twigs.

These must be considered in the design of a mm wave radar moving target indicator (MTI) for removal of clutter and in the design of detection logic where echoes from trees would be false targets.

This paper was presented at IEEE EASCON-79.

2.C CLUTTER

2.C.1 "Fluctuating Clutter" — Koester, Kosowsky and Sparacio.

This paper, which was originally published as an appendix to Paper 1.7 (Richard) of this volume, is placed here because of its relevance to this chapter.

Publication of clutter reflectivity values (σ_o) may cause some radar system engineers to not properly consider clutter fluctuation in the design of radar signal processors. Clutter fluctuation is treated in this paper with emphasis on millimeter wavelengths. Principal approach is the Rayleigh probability density function. Other fluctuation models treated in papers in this section and in Paper 5.16 include Log-normal, Ricean and Weibull.

This paper provides a background in clutter fluctuation, as well as some experimental data. Other papers in this section provide additional experimental clutter fluctuation data.

2.C.2 "mm-Wave Reflectivity of Land and Sea" — Trebits, Hayes, and Bomar

In most radar system applications, signals reflected by nature — trees, fallen snow, rainfall and sea — are unwanted and hence become "clutter". Reflectivity of several of these are treated somewhat in several preceding papers in this chapter, *e.g.* Papers 2.A.3, 2.A.4, 2.A.9, 2.B.2. This paper addresses reflectivity measurements from nature.

Trees are an important clutter source for radars used to detect or track ground-based military targets. Frequently the target will be surrounded by trees. At other times, the radar must scan tree-covered areas to search for targets. The first four figures in the paper pertain to reflectivity of trees. In Figure 4, foliage attenuation is important. Paper 2.B.2, previously in this chapter, presents an extension of reflectivity from trees at 95 GHz.

Figures 5-7 in the paper pertain to reflectivity from fallen snow. Paper 2.A.9, previously in this chapter, contains additional reflectivity characteristics of snow.

Sea clutter reflectivity measurements are given in Figures 8-10 of the paper. Sea clutter is also treated in Papers 2.C.3 and 2.C.4 to follow.

Rainfall reflectivity data of Figures 11-13 in this paper extend the rain backscatter tables of Paper 2.A.4 previously in this chapter.

2.C.3 "Radar Sea Clutter at Millimeter Wavelengths — Characteristics and Pivotal Unknowns" — Ewell, Trebits and Teti

This paper summarizes measured sea clutter characteristics at frequencies of X-band through 95 GHz reported in the literature through mid 1975. Characteristics discussed include the decidedly non-

Rayleigh amplitude distributions, polarization dependencies, range dependence, auto-covariance and frequency spectra, forward scattering of the sea surface, dependence upon wind direction, dependence upon operating frequency and influence of sea conditions.

As an example, the return using horizontal polarization has a much more "spiky" and irregular appearance than does the return for vertical polarization. The temporal decorrelation (decorrelation time) of sea clutter returns may strongly affect the performance of many radar integration techniques (see Paper 2.A.1).

The author's comments on comparison of data taken by various groups under differing conditions in this paper undoubtedly influenced the extensive clutter measurements program described in the next paper. Difficulty of comparison of experimental measurements is also discussed in [2].

This paper was presented at the Sixth DARPA-Tri-Service Millimeter Wave Conference in 1977.

2.C.4 "Multifrequency Millimeter Radar Sea Clutter Measurements" — Trebits, Currie and Dyer

This paper is an extension of one aspect discussed by Dyer and Currie in Paper 2.A.3, written earlier. It describes results of a sea clutter backscatter measurement program for the U.S. Navy by the Georgia Institute of Technology Engineering Experiment Station. Four frequencies are used simultaneously: 9.5 (X-band), 16 (K_u-band), 35 (K_a-band) and 95 GHz. Polarizations used are HH, VV, and HV. Grazing angles of between 0.5 and 7.8 degrees are used.

Data show a reversal of behavior of HH and VV polarizations between the 35 and 95 GHz measurement frequencies. The spectral width is much greater at 95 GHz than at 35 GHz, with a somewhat greater roll-off slope at 35 GHz than at 95 GHz.

This paper was presented at IEEE EASCON-79.

REFERENCES FOR CHAPTER TWO

[1] D.K. Barton, "International Cumulative Index on Radar Systems," *IEEE Publication Cat. No. JH-4675-5*, 1978.

[2] J.C. Lin and A. Ishimaru, "Propagation of Millimeter Waves in Rain," *Univ. of Washington Rpt. 144*, AFCRL 71-0310, DDC AD 735291, May 1971.

[3] L. Klaver, "Combined X/K_a-band Tracking Radar," *Proc. Military Microwaves Conf.*, 1978, pp. 147-56. Reprinted in S.L. Johnston, *Radar Electronic Counter-Countermeasures.* Dedham, Mass.: Artech House, 1979.

BIBLIOGRAPHY TOPICS FOR CHAPTER 2

Propagation (General)

Propagation (Absorption)

Propagation (Fluctuation)

Propagation (Refraction)

Propagation (Scattering)

Targets

Clutter

SUPPLEMENTAL BIBLIOGRAPHY FOR CHAPTER 2

Propagation (General)

NOTE: Additional propagation references will be found in the bibliographies in Paper 1.6, Chapter 1, and in the bibliographies in papers of Chapters 4 and 5.

Bibliographies and General Publications

Barton, D.K., 1978, *loc cit:* propagation (general) pp. 42-3; propagation (absorption) pp. 43-6; propagation (fluctuation) pp. 46-7; propagation (ionosphere) pp. 47-8; propagation (refraction) pp. 48-9; propagation (scattering) pp. 49-53; (Measurement frequency not always given in title.)

Barton, D.K., "International Cumulative Index on Radar Systems (1977-1979)," *IEEE 1980 International Radar Conference Record*, pp. 486*ff:* propagation (general) p. 21; propagation (absorption), pp. 21-22; propagation (fluctuation) p. 22; propagation (refraction), p. 22; propagation (scattering), pp. 22-23. Selected entries given below. (Measurement frequency not always given in title.)

Spitz, E. and Gachier, G. (Editors), "Millimeter and Submillimeter Wave Propagation and Circuits," *AGARD Conference Proceedings* CP 245, Sept. 1978, Session VIII - Propagation (9 papers).

Kobayashi, H.K., "Atmospheric Effects on Millimeter Radio Waves," *Atmospheric Sciences Lab USAERADCOM Rpt. ASL-TR-0049*, Jan. 1980 (contains 70 references).

Propagation (Absorption)

Zhevakin, S.A. and Viktorova, A.A., "The Waver-Vapor Dimer and its Spectrum," *Soviet Physics-Doklady*, Vol. 11, No. 12, June 1967, pp. 1059-62.

Zhevakin, S.A. and Viktorova, A.A., "Band Spectrum of a Dimer of Water Vapor," *Soviet Physics-Doklady*, Vol. 15, No. 9, May 1971, pp. 836-9.

Zrazhevskiy, A.Yu., "Method of Calculating the Atmospheric Water Vapor Absorption of Millimeter and Submillimeter Waves," *Radio Eng. and Electron. Phys.* Vol. XXI, No. 5, May 1976, pp. 331-6.

Weisback, W., "The Influence of Precipitations on a 35 GHz Short-range Pulse Radar," *Wiss Ber AEG-Telefunken*, Vol. 49 No. 6, 1976, pp. 244-7 (in German).

Zrazhevskiy, A.Yu., Master's Thesis, Institute of Radio Engineering and Electronics, USSR Academy of Sciences, 1977.

McEvan, N.J., *et al.*, "Crosspolarisation from High-Altitude Hydrometeors on a 20 GHz Satellite Radio Path," *Electron. Lett.* Jan. 6, 1977, Vol. 13, No. 1, pp. 13-14.

Crane, R.K., "Prediction of the Effects of Rain on Satellite Communication Systems," *Proc. IEEE*, Vol. 65, No. 3, March 1977, pp. 456-74.

Dintelmann, F. and Rucker, F., "Analysis of an XPD Event at 30 GHz Measured with ATS-6," *ibid.*, pp. 477-9.

Liebe, H.J., Gimmstad, G.G., and Hopponen, J.D., "Atmospheric Oxygen Microwave Spectrum-Experiment vs. Theory," *IEEE Trans.*, Vol. AP-25, No. 3, May 1977, pp. 327-35.

Bergmann, H.J., "Satellite Site Diversity: Results of a Radiometer Experiment at Band 18 GHz,:: *ibid.*, pp. 483-9. (Frequency scaled to 35 GHz)

Troitskii, A.V., "Measurements of Atmospheric Radio-wave Absorption on the Slope of O_2 band $\lambda \approx 5.6$ mm," *Radiophys. and Quant. Electron.* Vol. 20 No. 8, Aug. 1977, pp. 861-3.

Rainwater, J.H., "Weather Affects mm-wave Missile Guidance System," *Microwaves*, Sept. 1977, pp. 62-4, 101-2.

Goldhirsch, J., "Prediction of Slant Path Rain Attenuation Statistics at Various Locations," *Radio Science*, Vol. 12, No. 5, Sept.-Oct. 1977, pp. 741-7.

Moffat, P.H., Bohlander, V.R., and Gebbie, H.A., "Atmospheric Absorption Between 4 and 30 cm^{-1} Measured Above Mauna Kea," *Nature*, Vol. 269, No. 5630, Oct. 20, 1977, pp. 676-7.

Sinitskiy, V.B., Savenko, A.A., and Kivva, F.V., "Attenuation of Millimeter Waves in Rain," *Telecomm. Radio Engrg.*, Pt. 1, Vol. 31, No. 10, Oct. 1977, pp. 53-6.

Delonge, P. and Sobieski, P., "Fine Structure of Microwave Cross-Polarization Due to Precipitation," *Anales des Telecomm.*, Vol. 32, No. 11-12, 1977, pp. 377-85. (in English)

Breuer, L.J. and Keruels, R.K., "Rainfall Drop Spectra Intensities and Fine Structures on Different Time Bases," *ibid.*, pp. 430-6. (in English)

Yamada, M., Ogewa, A., Furuta, O., and Yuki, H., "Rain Depolarization Measurement Using Intelsat-IV Satellite in 4 GHz Band at a Low Elevation Angle," *ibid.*, pp. 524-9. (in English)

Kislykov, A.G., *et al.*, "Investigations of Millimeter and Submillimeter Wave Propagation Through the Whole Atmosphere," paper presented at the Anglo-Soviet Seminar on Atmospheric Propagation at Millimeter and Submillimeter Wavelength, Nov. 28-Dec. 3, 1977, pp. I1-I18.

Gibbens, C.J., Wrench, C.L., and Croom, D.L., "Atmospheric Emission Measurements Between 22 and 150 GHz," *ibid.*, pp. K1-K5.

Fomin, V.V., "Molecular Absorption in Transparent Gaps in the Atmosphere," *12th All Union Conference on the Scattering of Radiowaves*, TOMSK, 1978, Tezsi Doklodov, Vol. 2, pp. 85ff.

Zabolotniy, V.F., *et al.*, "Attenuation of Radiowaves at Wavelengths of 1.25 and 1.9 mm," *ibid.*, pp. 20ff. Also published as "Attenuation of Radiowaves at Wavelengths of 1.25 and 2.0 mm," *Infrared Phys.*, Vol. 18, 1978, pp. 851-7.

Aganbekyan, K.A. and Zrazhevskiy, A. Yu., "Absorption of Submillimeter Waves in Pure Water Vapor in Mixtures of H_2O-H_2O, B_2O-CO_2, H_2O-Ar," *ibid.*, pp. 52ff.

Cverklov, B.A. and Furashchov, N.I., "An Investigation of the Absorption in Pure Water Vapor in Transparent Gaps of 105.6 and 118.9 μm," *ibid.*, pp. 54ff.

Stankevich, V.S., "Evaluation of the Effects of Individual Absorption in the Atmosphere," *ibid.*, pp. 56ff.

Aganbekyan, K.A., *et al.*, "Absorption of Radiowaves by Water Vapor and Oxygen into the Thick Earth Atmosphere in 0.75 - 3.0 mm Waves," *ibid.*, pp. 58ff.

Iskhakov, I.A., *et al.*, "Measurement of the Attenuation of Radiation with 4.1 mm Wavelength in Slant Paths in the Presence of Hydrometeoric Formations," *12th All Union Conference on Radiowave Scattering*, TOMSK, 1978, Part 2, pp. 72ff.

Olsen, R.L., Rogers, D.V., and Hodge, D.B., "The aR^b Relation in the Calculation of Rain Attenuation," *IEEE Trans.*, Vol. AP-26, No. 2, Mar. 1978, pp. 318-29.

Bangham, M.J., "Theoretical Calculations of Atmospheric Emission and Absorption Spectra in the Far Infrared," *3rd Int. Conf. on Submillimeter Waves and Their Applications*, March 27-April 1978, Univ. of Surrey, UK, pp. 106ff. Also published in *Infrared Phys.*, Vol. 18, pp. 799-801.

Kemp, A.J., *et al.*, Calculation of Line Shape Parameters for Water Vapour from Dispersive Data," *ibid.*, pp. 98ff.

Nills, R.E., "Absolute Measurements of Atmospheric Emission and Absorption in the Range 3-40 cm^{-1}," *ibid.*, pp. 105ff.

Wrixon, G.T. and McMillan, R.W., "Measurements of Earth-Space Attenuation at λ = 1.3 mm," *ibid.*, pp. 111ff. (Related paper in *IEEE Trans.*, Vol. MTT-26, June 1978, pp. 434-9.)

John, L. and Nelson, M., "Atmospheric Water Vapor Absorption at 350 and 450 Microns," *ibid.*, pp. 119ff.

Woody, D.P., *et al.*, "New Measurement of the Submillimeter Cosmic Background," *ibid.*, pp. 200ff.

Zrazhevskiv, A. Yu. and Iskhakov, I.A., "Absorption of Radio Waves by Water Vapor with the Atmosphere at Wavelengths of 0.8 - 2.0 mm," *Radio Engrg. and Electron. Phys.*, Vol. 23, No. 7, July 1978, pp. 8-13.

Zravkevskiy, A.Yu. and Iskhakov, I.A., "Absorption of Radiowaves by Water Vapor Within the Atmosphere at Wavelengths of 0.8 - 20 mm," *Radio Eng. and Electron. Phys.*, Vol. 23, No. 7, Aug. 1978, pp. 8-13.

Plambeck, R.L., "Measurements of Atmospheric Attenuation near 225 GHz: Correlation with Surface Water VApor Density," *IEEE Trans.*, Vol. AP-26, No. 5, Sept. 1978, pp. 737-41c.

Ho, K.L., Mavrokoukoulakis, N.D., and Cole, R.S., "Rain-induced Attenuation at 36 GHz and 110 GHz," *IEEE Trans.*, Vol. AP-26, No. 6, Nov. 1978, pp. 873-5c.

Kuznetsov, I.V., *et al.*, "Measurement of Optical Thickness of the Atmosphere in Wavelengths of 0.87, 2.09, and 3.34 mm," *All Union Symposium on MM and SMM Waves*, Vol. 10, 1978, Part P, pp. 157ff.

Iskhakov, I.A., "The Attenuation of Radiation in Wavelengths of 4.1 mm in the Earth's Atmosphere," *ibid.*, pp. 159ff.

Zrazhevskiy, A.Yu. and Strelkov, G.M., "The Shape of the Harmonious Oscillator Line with Conditional Stresses," *ibid.*, pp. 164ff.

Lin, S.H., "Empirical Calculation of Microwave Rain Attenuation Distributions on Earth-Satellite Paths," *IEEE EASCON 78 Rec.*, 1978, pp. 372-8.

Zabolotniy, V.F., *et al.*, "Attenuation of Radiation at Wavelengths of 1.25 and 2.0 mm," *Infrared Phys.*, Vol. 18, Dec. 1978, pp. 815-7.

Fedi, F., "The Eurocop-Cost 25/4 Project on Propagation above 10 GHz," *Alta Freq.*, Vol. 48, No. 4, April 1979, pp. 153-7. (in English)

Fedi, F., "Attenuation Due to Rain on a Terrestrial Path," *ibid.*, pp. 167-84. (in English)

Blomquist, A. and Norbury, J.R., "Attenuation Due to Rain on Series, Parallel, and Convergent Paths," *ibid.*, pp. 185-90. (in English)

VanderVorst, A., "Cross Polarisation on a Terrestrial Path," *ibid.*, pp. 201-9. (in English)

Brussaard, G., "Attenuation Due to Rain on a Slant Path," *ibid.*, pp. 210-4. (in English)

Ho, K.L., Mavrokoukoulakis, N.D., and Cole, R.S., "Propagation Studies on a Line-of-Sight Microwave Link at 36 GHz & 110 GHz," *Jnl. on Microwave, Optics and Acoustics*, Vol. 3, No. 3, May 1979, pp. 93-8.

Lee, W.C.Y., "An Approximate Method for Obtaining Rain Rate Statistics for Use in Signal Attenuation Estimating," *IEEE Trans.* Vol. AP-27, No. 3, May 1979, pp. 407-13.

Efanov, V.A., et al., "Oscillation of the Sun's Millimetric Radioemanations Near the Horizon," *12th All Union Conf. on Radiowave Scattering*, TOMSK, June 1979, Vol. 2, Tezis Dokladov, p. 9, M.

Morita, Kazuo, "Estimation of Rain Attenuation Duration Time Distribution in Microwave and Millimeter Wave Bands," *Electrical Commun. Lab. Tech. Jnl.*, Nipp Telegraph & Telephone, Japan, Vol. 28, No. 4, 1979, pp. 663-70.

Fedoseev, L.I. and Kuznetsov, I.V., "Synchronous Measurements of the Atmospheric Radio Emission in the 2 and 3 Millimeter Wavelength Regions," *IEEE Fourth International Conference on Infrared and Millimeter Waves and their Applications*, Miami Beach, Fl., Dec. 1979 (late paper).

Evans, B.G., "Millimetre-Wave Propagation," *Proc. Military Microwaves Conf.*, 1980 (MM-80).

Gallagher, J.J., et al., "Atmospheric Effects on Near Millimeter Wave Systems Applications: An Overview," *Proc. Military Microwaves Conf.*, 1980 (MM-80).

Zurheiden, D., Kloevehorn, V., and Raudonat, U., "94 GHz Radar Propagation in a Realistic Battlefield Environment," *Proc. Military Microwaves Conf.*, 1980 (MM-80).

Propagation (Fluctuation)

Andreyev, G.A. and Chernaya, L.F., "Fluctuations in Millimeter-wave Beams Propagating in a Turbulent Absorbing Troposphere," *Telecom. Radio Engrg.* Pt. 2, Vol. 33 No. 1, Jan. 1978, pp. 64-74.

Hodge, D.M., Theobold, D.M., and Devasirvather, D.M.J., "Amplitude Scintillation at 2 and 30 GHz on Earth Space Paths," *Union Radio Scientific Internationale Open Symp.*, La Baule, Loire-Atlantique France, April 8-May 6, 1977, pp. 421-5.

Cole, R.S., Ho, K.L., and Mavrokoukoulakis, N.D., "The Effect of the Outer Scale of Turbulence and Wavelength of Scintillation Fading at Millimeter Wavelengths," *IEEE Trans.* AP-26 No. 5, Sept. 1978, pp. 712-5.

Mavrokoukoulakis, N.D., Ho, K.L., and Cole, R.S., "Temporal Spectra of Atmospheric Amplitude Scintillations at 110 GHz and 36 GHz," *IEEE Trans.* Vol. AP-26, No. 6, Nove. 1978, pp. 875-7c.

Matthews, P.A., "Scintillation on Millimeter Wave Radio Links and the Structure of the Atmosphere," papers presented at Anglo-Soviet Seminar on Atmospheric Propagation at Millimeter and Submillimeter Wavelengths, Nov. 28-Dec. 3, 1977, pp. H1-H8.

Andreev, G.A. and Sokolov, A.V., "Millimeter Wave Fluctuations During Scattering in Earth's Atmosphere," *12th All Union Conf. on Radiowave Scattering*, TOMSK, 1978, Part 2, pp. 66*ff*.

Zrazhevskiy, A.Yu., et al., "The Fluctuation Characteristics During Scattering on an Earthly Path," *ibid.*, pp. 69*ff*.

Sharapov, L.I., "Amplitude of Fluctuation of Radiowaves in a Short Range Unit of Millimeter Waves," *All Union Symposium on Millimeter and Submillimeter Waves*, 10, 1978, Part P, pp. 149*ff*.

Propagation (Ionosphere)

Fang, D.J., Kennedy, D.J., and Devieux, C., "Ionospheric Scintillations at 416 GHz and Their System Impact," *IEEE EASCON-78 Rec.*, pp. 385-90.

Propagation (Refraction)

Kolosov, M.A., et al., "Influence of Refraction on the Attenuation of Water Vapor with Millimeter and Submillimeter Waves in the Troposphere," *All Union Symposium on Millimeter and Submillimeter Waves*, 10, 1978, Part P, pp. 147*ff*.

Liebe, H.J. and Hopponen, J.D., "Variability of EHF Air Refractivity with Respect to Temperature, Pressure, and Frequency," *IEEE Trans.* Vol. AP-25, No. 3, May 1977, pp. 336-45.

Propagation (Scattering)

Zhevakin, S.A. and Viktorova, A.A., papers presented at the 9th All Union Conf. on Radiowave Scattering, 1969, Part P, pp. 107ff.

Evans, B.G. and Holt, A.R., "Scattering Amplitudes and Crosspolarisation of Ice Particles," *Electron. Lett.*, Vol. 13, No. 12, 1977, pp. 342-4.

Shimbakuro, F.I., "Bistatic Scattering from Rain at Millimeter Wavelengths," *IEEE EASCON-79 Rec.*, pp. 369-71.

Andreev, G.A., *et al.*, "The Diffusion of Millimetric and Submillimetric Waves in the Atmosphere," *Problemi Sovremenroi Radiotekniki i Elektronika*, Vol. 1, 1978, M, Institute of Radio Engineering and Electronics of USSR Academy of Sciences, pp. 211ff.

Andreev, G.A., *et al.*, "The Diffusion of Millimeter and Submillimeter Radiowaves in the Earth's Atmosphere and Possible Fields of Practical Application of These Waves," *12th All Union Conf. on the Scattering of Radio Waves*, TOMSK, 1978, Tezis Dokladov, Part 2, pp. 62ff, 'NAUKA' M.

Targets

Target radar cross section data is extremely important in the design and performance analysis of all radars. Many treatments of millimeter wave radar system design assume the conventional convenient (but still meaningless) value of one square meter. Very little data has been published on millimeter target radar cross section measurement data. Paper 5.3, Chapter 5 of this volume also includes results of limited millimeter wave Target RCS measurements.

1. **Bibliographies**

    Barton, D.K., 1978, *loc cit;* cross section, pp. 15-18. (Measurement frequency generally not indicated.)

    Barton, D.K., 1980, *loc cit;* cross section, pp. 8-9. (Measurement frequency generally not indicated.)

2. **Journal Articles**

    Weathers, G.D., Emmons, G., and Graf, E.R., "Glint at Millimeter Wavelengths," *Proc. IEEE Southeastcon 78*, Apr. 78, pp. 405-8. (Treatment is very limited; does not reflect extensive glint analyses and measurements made at microwave frequencies.)

Clutter

Barton, D.K., 1978 *loc cit;* clutter (general) pp. 11-12, clutter (atmospheric), pp. 12-13, clutter (land) pp. 13-14, clutter (sea) pp. 14-15.

Barton, D.K., 1980 *loc cit;* clutter (general), pp. 6-7; clutter (atmospheric), p. 7; clutter (land), p. 7; clutter (sea), pp. 7-8.

A Radar System Engineer Looks at Current Millimeter-Submillimeter Atmospheric Propagation Data

S.L. Johnston

ABSTRACT

Numerous reviews have been made of theoretical propagation characteristics in the millimeter and submillimeter regions. Additionally, extensive experimental atmospheric propagation measurements have been made in these regions. These have generally been made for purposes other than radar systems design. It appears that the propagation community does not appreciate the needs of the radar systems design engineer. This paper will set forth the general atmospheric propagation requirements of a millimeter/submillimeter radar systems design engineer. These requirements include much more than mere atmospheric absorption/attenuation. The requirements will be illustrated by discussion of presently existing mm/smm atmospheric propagation data. Needs for additional data will be identified.

INTRODUCTION

The term "millimeter wave" is widely used at present. Although ostensibly obvious is meaning, the terms "millimete wave" and "radar" have different meanings to different people although each term has an official IEEE definition (1). Millimeter wave as used in this paper is the frequency region of 40-300 GHz (2). Figure 1 (3) shows the IEEE radar bands and the ITU radiolocation allocations superimposed on a plot of atmospheric attenuation as a function of frequency adapted from Altshuler et. al (4).

Radar as used in this paper is in accordance with the new IEEE definition (5). Accordingly it includes missile seekers and "radiometers" when used for missile guidance. Millimeter wave radar in general constitutes substantial departure from the well known microwave radar as exemplified by large ground based aircraft surveillance radars.

Possible applications of millimeter wave radar were identified by Skolnik almost a decade ago (6, 7) and are listed in Table I. Extensive attention is currently being devoted to millimeter wave radar as reported in several survey papers (8, 3, 9). Radar in the higher frequency range - .1 - 1 THz has also been surveyed recently (10). Millimeter wave survey papers almost always include a discussion of propagation of millimeter waves. Generally, this is confined to absorption/attenuation by the atmosphere and rainfall effects. Some millimeter wave propagation surveys add line broadening. Other propagation effects of concern to the millimeter wave radar systems design engineer are generally not considered in millimeter wave propagation surveys. This paper will address millimeter wave radar propagation needs and will relate them to presently existing propagation data.

RADAR CONSIDERATIONS

Inspection of the previously cited millimeter wave radar surveys indicates a wide range of mm wave radar applications from very short range tactical missiles to an extremely long range space object identification system. The former, in number, constitute the largest number of potential applications. Characteristics of a "typical" short range mm wave radar shown in Table II were extracted from references in (3). These data will serve to indicate propagation parameters of interest to the mm radar systems design engineer.

Table III from an unpublished work indicates radar features which are influenced by propagation effects. Table IV lists propagation effects on the basic radar functions. Propagation effects which affect radar/guidance are shown in Table V (11). Tables III - V clearly indicate that the mm wave radar system design engineer is concerned with much more than attenuation due to the atmosphere and rain.

PROPAGATION DATA REQUIREMENTS

Most of the techniques used in millimeter wave radar e. g. angle measurement or target detection were obtained from microwave radar. Accordingly, it is instructive to consider the literature on atmospheric effects on microwave radar. Due allowance must be made for frequency effects, however. The ionosphere can greatly affect low frequency radar but can be neglected for millimeter wave radar. Rainfall attenuation has little effect on low frequency radar but may greatly affect millimeter wave radar. Barton (12) recently identified environmental constraints on microwave radar as contained in Table VI. References shown are those cited in (12). This table should serve as a point of departure for mm wave radar system design engineers and mm wave propagation specialists.

With Table VI as a "bench mark" and Tables III-V as inputs, the propagation data requirements for mm radar systems design shown in Table VII were established. This table should be compared with the scope of most mm wave propagation surveys. It should be recognized that most if not all of the mm wave propagation surveys were not prepared for the radar design engineer. Even mm wave communications systems design may be affected by angle arrival variations however.

PRESENT PROPAGATION DATA

It is instructive to briefly review currently existing mm wave propagation data for its suitability for mm wave radar systems design. The ensuing discussions will serve to identify suitability of various data and indicate needs for additional mm wave propagation data for radar systems design. It is appreciated that much of the data cited were not taken for radar systems design thus it should not be expected to find sufficient suitable data for that purpose. It will be noted in some instances that some data cited were not taken at mm frequencies. In such cases the data cited were used to illustrate the phenomena and of course per se indicate need for mm wave propagation measurements. Many of the references cited were not located from propagation surveys; but rather from the extensive radar literature bibliography compiled by the IEEE Radar Systems Panel (13). Unfortunately that bibliography does not include an indication of frequency (unless in the title) and the large number of entries in the bibliography precluded extensive examination other than cursory inspection of titles and selection of interesting ones. Certainly a more thorough search is warranted.

Dyer and Currie (79) presented an excellent survey of existing millimeter wave propagation data on attenuation, backscatter and foliage with data contained in several tables and figures. Unfortunately several of the characteristics required in Table VII herein were not addressed there. Treatment of the effects of clutter on radar system performance was noteworthy.

Actual propagation data will not be included in this paper for several reasons:
1. Space limitations preclude this.
2. The data user should carefully study the full paper which contains referenced data in order to learn of its applicability, limitations, and assumptions rather than attempt to use data based on limited extraction.
3. In some instances microwave data will be cited as being illustrative of the type being considered in absence of millimeter wave data.

4. In other instances only theoretical data have been identified. Theoretical data indicates what the experimenter should expect to observe in experiments based upon the validity of various assumptions. Experimental data may indicate what was actually observed under the conditions reported. Ideally one should have both theoretical and experimental data.

In the following section, propagation data will be discussed in the order shown in Table VII. Most of the references cited should be readily available for further study. Reference (14) described some submillimeter wave propagation measurements techniques and included an extensive bibliography of smm wave propagation measurements.

1. Aerosols (dust, smoke, etc.)- Data on attenuation, influence on angle arrival, etc. on aerosols at mm waves is very limited. Ahmed and Auchterlonie (15) reported measurements of refractive index and loss tangent for clay and for sand at X-band (10 GHz). These data would enable estimation of propagation characteristics in dust-laden atmospheres. Such data is needed at millimeter frequencies.

Downs (80) primarily treated optical transmission-absorption and scattering by clear air, absorption by water vapor, scattering by haze and fog, scattering by rain, absorption by water drops and attenuation by smoke and dust. He indicated that on one smoke test, 3.2 and 2 mm wave radar signals showed no attenuation.

Knox (16) in another EASCON-79 paper presents results of recent experiments on millimeter wave propagation in smoke. Further measurements are needed.

2. Angle of Arrival Variations - The literature contains extensive references on angle of arrival variations (17-28, 36). Most of the references cited (listed chronologically) were at the microwave regions. Several papers in this group are theoretical analyses. One reference (17) reported horizontal angle of arrival measurements (X-band) although most of the microwave measurements concerned vertical angle of arrival. One paper reported measurements at K_a-band (23) and one Soviet paper (27) reported measurements at 2mm wavelength. Further angle of arrival measurements are needed both in the millimeter wave and submillimeter wave regions. Emphasis on the correlation interval and power spectral density are needed.

3. Atmospheric Attenuation -- Most of the references on millimeter wave propagation were concerned with attenuation. Many of them considered only the DC term or the slowly varying component; i. e. fluctuation is not addressed. Falcone (28) in an EASCON-79 paper describes a digital computer model which computes steady state attenuation at wavelengths from L-band to submillimeters as function of various hydrometeors (fog, clouds, rain, etc.). Must of the mm wave propagation

measurements were not made for radar applications hence the parameters of interest to radar e. g. Table II were not employed. Extension of those non-radar measurements to radar applications may not be possible. Reference (14) contains references to a large number of smm wave attenuation; unfortunately, fluctuation was frequently not considered. Theory of attenuation fluctuation was treated in (26, 29-35). Reference (31) is noteworthy in that it analyzed correlation of signals for widely separated frequencies to determine the maximum useable signal bandwidth. References (36-51, 81) contain fluctuation measurement data at various microwave and millimeter wavelengths. Reference (47) reported measurements at several closely spaced mm frequencies and is relevant to the previously cited theoretical paper of reference (31). Unfortunately, the full paper of (47) has not been published however.

Review of these fluctuation papers indicated absence of data at wavelengths less than 3mm. Correlation intervals, fluctuation distribution, and power spectral density when reported were generally not relevant to Table II. These again indicate need for further measurements.

4. Atmospheric Backscatter - Measurements of atmospheric backscatter due to rain and to snow have been reported at wavelengths down to 3mm (52-59, 82). Measurements at shorter wavelengths would probably require more sensitive radars than now exist at these shorter wavelengths. Generally, snow backscatter measurements treat fallen snow rather than while falling. Reference (82) reported measurements on falling snow at 2 mm. Some of the reported measurements were made at only one polarization or sequentially at various polarizations. Sometimes this was done at only one wavelength. Simultaneous measurements at both horizontal and vertical polarizations at several different wavelengths e. g., X, K_u, K_a, 3, 2 and 1 mm are needed.

5. Foliage Penetration - Only two references on foliage penetration measurements (60, 61) were identified at K_a Band and 3 mm. Polarization does not seem to have been considered by (61). Data at both horizontal and vertical polarizations are needed at shorter wavelengths.

6. Phase Variations - Phase variations due to propagation may affect a radar in numerous ways; a very common one is in angle of arrival variations, i. e. angle (direction) measurements. If the phase varies at some indicated rate, an equivalent frequency variation results. This frequency variation can affect coherent radars such as those equipped with moving target indicators (MTI). Almost a dozen references on phase variations were located (26, 32, 38, 48, 49, 62-66). Some of these were cited previously in the discussion of amplitude fluctuations. The references cited cover theory as well as measurements from VHF thru 3 mm. Most of them did not include effect of polarization. Simultaneous measurements are needed using both horizontal and vertical polarizations at several frequency bands.

7. Polarization Effects - Polarization effects result from the behavior at different polarizations of various propagation phenomena. In cross polarization, an electromagnetic wave of one linear polarization when propagated thru a medium may be received as if it were of the orthogonal polarization. This, if of sufficient magnitude, may be very important in those communications systems where both linear polarizations are used, each with separate modulation signals. A cross modulation results. In radar, this could be important for some electronic countermeasures (ECM) situations. References (67-77) addressed polarization and cross-polarization. Nearly all of these were presented in relation to communication systems. No specific mm band (i. e. above 40 GHz) measurements were noted in these references.

Long (83) presented a scattering model for land and sea at microwave frequencies. His model consists of depolarizing diffuse and non-depolarizing quasi-optical scatterers. He indicated that in alternative terms the depolarizing scatterers are incoherent whereas the non-depolarizing scatterers are coherent. Polarization is thus important for some scatterers.

Hayes (84) in another EASCON-79 paper reports back scattering at 3.2 mm wavelength from trees. Although the back-scattering model presented is incomplete (i. e. only one polarization), the spectrum is shown to contain two fluctuation components: a fixed or slow fluctuation component some 15 dB in magnitude and about 4 Hz wide and a fast fluctuation component of approximately 20 dB in magnitude and 200 Hz wide. These must be considered in the design of a mm wave radar moving target indicator (MTI) for removal of clutter.

As in several of the previously discussed propagation effects, additional measurements are needed using the parameters of Table II. Polarization measurements would generally be made in conjunction with studies of the effects of rain, atmospheric turbulence, etc. on attenuation, angle of arrival, etc.

8. Surface Phenomena - This last group includes monostatic backscatter bistatic clutter and forward scatter. The nature of surface roughness may influence them. Decorrelation time and power spectral content are important. Measurements of the effects of surface phenomena at millimeter wavelengths are very sparse. Trebits (78) measured sea clutter at several wavelengths as reported in another EASCON-79 paper.

CONCLUSIONS

This paper has set forth propagation data requirements needed by a millimeter wave radar design engineer and parameters of a "typical" millimeter wave radar. A brief review of existing data on many propagation data requirements has shown fairly extensive data at microwaves below K_a-Band. For most of the propagation requirements the data above 40 GHz is very limited. Even attenuation data which is relatively more plentiful is inadequate. Most of the measurements of all the propagation data cited were made for non-radar applications and much are not directly extendable to the typical radar parameters. Clearly the millimeter radar design engineer is at a disadvantage as compared to the microwave radar designer despite the greater importance of propagation effects to the millimeter wave radar.

A more extensive study of the literature on these propagation data would greatly benefit the millimeter wave radar designer and serve to present better guidance to the propagation specialists.

TABLE I
SUGGESTED RADAR APPLICATIONS

Low-angle Tracking*	Remote Sensing of the Environment
"Secure" Military Radar*	Surveillance
Interference-free Radar*	Target Acquisition, Recognition, Identification
Cloud Sensing Radar*	Missile Guidance
Imaging Radar	Navigation
Ground Mapping	Obstacle Detection
Map Matching	Clutter Suppression
Space Object Identification	Fuzes
Lunar Radar Astronomy	Harbor Surveillance Radar
Target Characteristics	Airport Surface Detection Radar
Weather Radar	Landing Aids
Clear-air Turbulence Sensor	Air Traffic Control Beacons
High Resolution Radar*	Jet Engine Exhaust and Cannon Blast

*Applications in which Skolnik believed that sub-millimeter waves offer more advantage than microwave frequencies.

TABLE II
CHARACTERISTICS OF A "TYPICAL" MM RADAR

Maximum Application Range	3 km
Maximum PRF for Unambiguous Range	50 kHz (Max)
Single Scan Correlation Time	20 u s (Min)
Pulse Width	20 n s (Min)
RF Bandwidth	50 MHz (Max)
Antenna Beamwidth	6-15 mrad (.5-1 deg)
Conical Scan Frequency	30-100 Hz
Angle of Arrival Measurement Time	100 s (K_a Band)
Average RF Power Output	.1-1W
Antenna Diameter	.25-1 m
Radar Size	Breadbox Up
Power Requirements	50 W

*250 MHz for spread spectrum (frequency agility)

TABLE III
RADAR FEATURES INFLUENCED BY PROPAGATION EFFECTS

1. Maximum Range.
2. Choice of Detection Thresholds.
3. Type of Detector.
4. Detection Logic.
5. Signal Processing Technique.
6. Search Scan Rate.
7. Antenna Polarization.
8. Type of Angle Measurement: Monopulse/conical Scan.
9. Pulse Width (RF Bandwidth).
10. Conical Scan Frequency Choice.
11. Doppler Bandwidth.
12. Number of Pulses Integrated.
13. False Alarm Rate.
14. Moving Target Indicator Performance.
15. Receiver Automatic Gain Control Characteristics.
16. Angle Tracking Accuracy.
17. Range Tracking Accuracy.
18. Range Gate Width.
19. Choice of Frequency.

TABLE IV
PROPAGATION EFFECTS ON RADAR FUNCTIONS

RADAR FUNCTION	PROPAGATION EFFECTS
1. DETECTION	Atmospheric attentuation, fluctuation (scintillation), foliage attenuation backscatter, polarization variation fade depths and durations.
2. ANGLE MEASUREMENT	Angle of arrival variation due to meteorological causes, ducting, fluctuation (especially for conical scan), depolarization.
3. SPECIAL FUNCTIONS (including imaging)	spatial coherence same as detection.

TABLE V
PROPAGATION EFFECTS ON RADAR/GUIDANCE

MULTIPATH/FORWARD SCATTER
    Modeling sea/terrain
    Experimental measurements
        - angle errors
        - signal fading

BACKSCATTER (THEORETICAL AND EXPERIMENTAL)
    Rain
    Sea
    Terrain

DUCTING AT mm WAVES

ATMOSPHERIC/FOLIAGE ATTENUATION
    Attenuation in troposphere
        - rain
        - fog
        - humid air
    Near ocean surface moisture layer
    Foliage penetration

TABLE VI
ENVIRONMENTAL CONSTRAINTS FOR MICROWAVE RADAR
(From Barton: in Brookner)

CONSTRAINT	DATA SOURCES
Line of sight	
Horizon range	Equation (1)
Range-height-angle charts	Blake (1, 2)
Terrain masking	Geodetic survey maps
Attenuation*	
Clear atmosphere	Blake (4)
Rain, clouds, snow	Gunn and East (5), Barton (6) Beam, Dutton, and Warner (7)
Clutter*	
Weather	Gunn and East (5), Barton (6), Bean, Dutton, and Warner (7),
Land	Moore (8), Skolnik (9),
Sea	Nathanson (10), Barton (6)
Refraction	
Tropospheric Bias	Barton (11)
Trophospheric fluctuation	Barton and Ward (12),
Ionosphere	Barton (11)
Ducting	Bean and Dutton (13)
Surface reflection	

Lobing (fading) — Blake (2)
Multipath error — Barton and Ward (12), Barton (14)
Diffraction — Blake (2)
Interference (ECM) model — Arbitrary selection

*See all Chapter 1

TABLE VII
PROPAGATION DATA REQUIREMENTS FOR MILLIMETER WAVE RADAR DESIGN

Aerosol (dusts, etc.) Characteristics
   Loss tangent
   Refractive index

Angle of Arrival Variations
   Correlation interval
   Power spectral density

Atmospheric Attenuation
   Correlation interval
   Fluctuation distribution
   Power spectral density

Atmospheric Backscatter
Foliage Penetration
Phase Variations
Polarization Effects
Surface Phenomena
   Bistatic clutter
   Forward scatter
   Nature of surface roughness
   Decorrelation time
   Power spectral content
   Backscatter (monostatic)

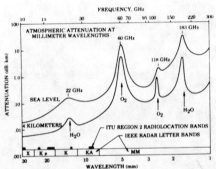

Figure 1. Atmospheric attenuation at millimeter wavelengths with IEEE radar letter bands and ITU Radiolocation bands for Region 2, after Altshuler et. al.,

REFERENCES

1. "IEEE Standard Dictionary of Electrical and Electronic Terms," IEEE Standard 100 - 1977, New York, IEEE Press, 1977.

2. "IEEE Standard Letter Designation for Radar Bands," IEEE Standard 521 - 1976, November 30, 1976.

3. Johnston, S. L., "Some Aspects of Millimeter Radar," Proc. Int. Conference on Radar, Paris, 4 - 8 Dec. 1978, pp 148-59.

4. Altshuler, E. E., et al, "Atmospheric Effects on Propagation at Millimeter Wavelengths," IEEE Spectrum, vol 5, no. 7, Jul 1968, pp 83-90.

5. "IEEE Standard Radar Definitions," IEEE Standard 686 - 1977, Nov 1977.

6. Skolnik, M. I., "Millimeter and Submillimeter Wave Applications," Proc. Symp. on Submillimeter Waves, Polytechnic Press of the Polytechnic Inst. of Brooklyn, N. Y., 1970, pp 9-26.

7. Skolnik, M. I., "Millimeter and Submillimeter Wave Applications, NRL Memorandum Report 2159, 12 Aug. 1970, DDC AD 712055.

8. Johnston, S. L., "Millimeter Radar," Microwave Jnl, vol 20 no. 11, Nov. 1977, pp 16ff.

9. Sundaram, G. S., "Millimetre Waves - The much-awaited technological Breakthrough?" Int. Def. Rev., 2/1979, pp 271-7.

10. Johnston, S. L., "Radar Systems for Operation at Short Millimetric Wavelengths," The Radio and Electron. Engr., vol 49 no 8, Aug. 1979, in press.

11. Private communication, D. D. Howard, NRL, 1979.

12. Barton, D. K., "Philosophy of Radar Design," in: E. Brookner, "Radar Technology," ARTECH House Books, Dedham, Mass., 1977.

13. Barton, D. K., "International Cumulative Index on Radar Systems," IEEE Pub Cat. No JH 4675-5, 1978.

14. Johnston, S. L., "Submillimeter Wave Propagation Measurement Techniques," Second International Conference and School on Millimeter Waves and their Applications, San Juan, Puerto Rico, 6-11 Dec. 1976, Conf. Digest (Summary).

15. Ahmed, I. Y., and Auchiterlonie, L. J., "Microwave Measurements on Dust, Using An Open Resonator," IEE Electron. Lett. vol 12 no 17, Aug. 19, 1976, pp 445-6. (x-band).

16. Knox, J. E., "Millimeter-wave Propagation in Smoke," EASCON-79 Rec.

17. Straiton, A. W., and Gerhardt, "Results of Horizontal Microwave Angle-of-Arrival Measurements by the phase-Difference Method," Proc IRE, vol 36 no 7, July 1948, pp 916-22. (x-band).

18. Fannin, B. M., and Jehn, K. H., "A Study of Radar Elevation-Angle Error Due to Atmospheric Refraction," IRE Trans vol AP 5 Jan 1957, pp 71-7. (theory).

19. Strohbehn, J. W., and Clifford, S. F., "Polarization and Angle-of-Arrival Fluctuations for a Plane Wave Propagated Through a Turbulent Medium," IEEE Trans vol AP-15 no. 3, May 1967, pp 416-21. (theory).

20. Evans, G. C., and Shaw, A. H., "Measurements of the Propagation Component of Radar Tracking Noise at a Wavelength of 3 cm," Proc IEE, Vol 115 no. 10, Oct. 1968, pp 1431-8.

21. Thompson, M. C., Jr., and Janes, H. B., "Measurements of Phase-Front Distortion on an Elevated Line of Sight Path," IEEE Trans. vol AES-6 no. 5, Sept 1970, pp 645-55. (x-band).

22. Bennett, J. A., "Refractive Errors in Angle-of-Elevation Measurements," IEEE Trans vol AES-7 no. 2, Mar 1971, pp 243-7. (theory).

23. Lees, M. L., "High Resolution Measurement of Microwave Refraction on Short Tropospheric Paths," IEEE Trans vol AP 20 no 2, Mar 1972, pp 176-81. (K_a band).

24. Valley, G. C., "Angular Jitter in Amplitude Comparison Monopulse Radar Due to Turbulence," IEEE Trans vol AP-23 no 2, March 1975, pp 274-8. (theory).

25. Vickers, W. W., and Lopez, M. E., "Low-Angle Radar Tracking Errors Induced by Nonstratified Armospheric Anomalies," Radio Science, vol 10 no. 5, May 1975, pp 491-505. (C-band measurements).

26. Andreev, G. A. and Chornaya, L. F., "Amplitude and Phase Fluctuations of Millimeter Wave (MMW) Beam Propagating in a Turbulent Atmospheric Atmosphere," Proc Anglo-Soviet Seminar on Atmospheric Propagation at Millimetre and Submillimetre Wavelengths, Moscow, Nov 28 - Dec 3, 1977, pp P1-10. (theory 0.8-10 mm).

27. Andreev, G. A., et al, "Intensity and Angle of Arrival Fluctuations of Millimetric Radiowave in Turbulent Atmosphere," Proc Anglo-Soviet Seminar loc cit, pp R1-8. (measurement 2mm).

28. Falcone, V. J., Jr., "Atmospheric Attenuation of millimeter and Submillimeter Waves," EASCON-79 Conf. Rec.

29. McMillan, R. W, Wiltse, J. C., and Snider, D. E., "Atmospheric Turbulence Effects on Millimeter Wave Propagation," EASCON-79 Conf. Rec.

30. Tukiz, O, "Theory of the Scintillation Fading of Microwaves," IRE Trans vol AP 5 no 1, Jan. 1957, pp 130-6.

31. Muchmore, R. B., and Wheelon, A. D., "Frequency Correlation of Line-of-Sight Signal Scintillations," IEEE Trans vol AP-11 no 1, Jan 1963., pp 46.51.

32. Izyumov, A. O., "Amplitude and Phase Fluctuations of a Plane Monochromatic Wave in a Near-Ground Layer of Moisture Containing Turbulent Air," Radio Engrg. and Electron. Phys. vol 13, no 7, 1968, pp 1009-13.

33. Gurvich, A. S., "Effects of Absorption on the Fluctuation in Signal Level During Atmospheric Propagation," Radio Engrg and Electron. Phys. Vol 13 no 11, 1968, pp 1687-94.

34. Izyumov, A. O., "Frequency Spectrum of Amplitude Fluctuations of a Plane Electromagnetic Wave in a Surface Layer of Turbulent Atmosphere," Radio Engrg and Electron. Phys Vol 14 no 10, 1969, pp 1609-12.

35. Armand, N., et al, "Fluctuation of Submillimeter Radio Waves in a Turbulent Atmosphere," Radio Engrg and Electron. Phys. vol 16 no 8, 1971, pp 1259-66.

36. Crawford, A. B., and Jakes, C. W., Jr., "Selective Fading of Microwaves," BST J vol 31 no 1, Jan 1952, pp 68-90 (Measurements: angle of arrival 1.25 cm, fading, S-band).

37. Kiely, D. G., "Some Measurements of Fading at a Wavelength of 8MM over a Very Short Sea Path," J. Brit IRE, vol 14, Feb 1954, pp 89-92.

38. Tolbert, C. W.; Fannin, B. M; and Straiton, A. W.; "Amplitude and Phase Difference Fluctuations of 8.6MM and 3.2 cm radio waves on line-of-sight Paths," Univ. of Texas EERL Rpt 78, Mar 1956, DDC AD 08810.

39. Tolbert, C. W.; and Straiton, A. W., "Attenuation and Fluctuation of Millimeter Radio Waves, IRE Nat. Conv Rec vol 5 pt 1, 1957, pp 12-18.

40. Tolbert, C. W.; Britt, C. O.; and Straiton, A. W.; "Antenna Pattern Fluctuations at 4.3 Millimeter Wavelengths Due to Atmospheric Inhomogeneities," Univ of Texas EERL Rpt no 96, 13 Dec 1957.

41. Weibel, G. E., and Dressel, H. O., "Propagation Studies in Millimeter-Wave Link Systems," Proc IEEE vol 55 no 4, Apr 1967, pp 497-513. (measurements, 3.3mm).

42. Lane, J. A., et al, "Absorption and Scintillation Effects at 3mm Wavelength on a Short Line-of-sight Radio Path," IEE Electron. Lett. vol 3 no 5, May 1967, pp 185-6.

43. Keelty, J. M., "Millimeter Investigations, Vol 4, Link, Test Facility and Results," RCA Research Labs, Montreal Quebec Rpt RCA 3576-B/96690-3 vol 4, Jan 1969, DDC AD 857435.

44. Janes, H. B., et al, "Comparison of Simultaneous Line-of-sight Signals at 9.6 and 34.5 GHz," IEEE Trans, vol AP-18 no 4, Jul 1970, pp 447-51.

45. Mandics, P. A.; Lee, R. W.; and Waterman, A. T., Jr.; "Spectra of Short-term Fluctuations of line-of-sight signals: Electromagnetic and Acoustic," Radio Science Vol 8 no 3, Mar 1973, pp 188-201 (Theory and measurement, K_a band).

46. Vilar, E., and Matthews, P. A., "Importance of Scintillation in Millimetric Radio Links Proc 4th European Microwave Conf. Montreux,Switzerland, Sept 10-13, 1974, paper C3.2, pp 202-6.

47. Moffat, P., "Fluctuations in Atmospheric Transmission at about 3mm Wavelength," Inter-Union Commission on Radio Meteorology (IUCRM) Colloquium on Probing of Atmospheric Constituents, Bournemouth, England, May 14-21, 1975.

48. Thompson, M. C., et al, "Phase and Amplitude Scintillations at 9.6 GHz on an Elevated Path," IEEE Trans vol AP-23 no 6, Nov 1975, pp 850-4.

49. Thompson, M. C., et al, "Phase and Amplitude Scintillation in the 10-40 GHz band," IEEE Trans vol AP-23 no 6, Nov 1975, pp 792-7.

50. Ho, K. L.; Mavrokoukoulakis, N. D.; and Cole, R. S.; "Wavelength Dependence of Scincillation Fading at 110 and 36 GHz," IEE Electron Lett. vol 13 no 7, 31 Mar 1977, pp 181-3.

51. Matthews, P. A., "Scintillation on Millimetre Wave Radio Links and the Structure of the Atmosphere," Proc. Anglo-Soviet Seminar loc cit, pp H1-8.

52. Currie, N. C.; Dyer, F. C.; and Hayes, R. D.; "Some Properties of Radar Returns from Rain at 9,375,35,70 and 95 GHz," Int Radar Conf. Rec 1975, pp 215-20.

53. Dyer, F. B.; Currie, N. C.; and Applegate, M. S.; "Radar Backscatter from Land, Sea, Rain, and Snow at Millimeter Wavelengths," Radar 77, London, IEE Conf Pub No 155, Oct 1977, pp 559-63.

54. Richard, V. W., and Kammerer, J. E., Ballistic Research Laboratory, Aberdeen Proving Ground, Md., Oct 1975, private communication.

55. Sackinger, W. M., and Byrd, R. C., "Backscatter of Millimeter Waves from Snow, Ice, and Sea Ice," IEE Conf Pub no 98, "Propagation of Radio Waves Above 10 GHz," 1973, pp 219-28 (K_a band).

56. Currie, N. C., et al, "Radar Millimeter Backscatter Measurements from Snow," Georgia Inst. of Technol. Engineering Experiment Sta. Final Rpt Con F-08635-76-C-0221, Jan 77, (AFAL TR 77-4) DDC AD B021148.

57. Hofer, R., and Schanda, E., "Signatures of Snow in the 5-94 GHz range," Radio Science Vol 13 no 2, Mar-Apr 1978, pp 365-9. (Brightness Temperatures from Radiometer measurements).

58. Hayes, D., "Millimeter-Wave Scattering Measurements over Snow," EASCON-79 Conf Rec.

59. Currie, N. C.; Dyer, F. B.; and Ewell, G. W.; "Characteristics of Snow at Millimeter Wavelengths," IEEE AP-S Int. Symp. Proc., Oct. 1976, pp 579-592. (K_a band and 3mm; backscatter and penetration).

60. Waite, W. P., and MacDonald, H. C., "Vegetation Penetration with K-Band Imaging Radars" IEEE Trans Vol GE-9 no 3, Jul 71, pp 147-155.

61. Currie, N. C.; Dyer, F. B.; and Martin, E. E. "Millimeter Foliage Penetration Measurements," IEEE AP-S Int. Symp Proc Oct 1976, pp 575-81. X-Band to 3mm).

62. Herbstreit, J. W. and Thompson, M. C., "Measurements of the Phase of Signals Received over Transmission Paths with Electrical Lengths Varying as a Result of Atmospheric Turbulence," IRE Electromagnetic Wave Theory Symposium in IRE Trans vol AP-4 no 3, Jul 1956, pp 352-8 (VHF and L-Band).

63. Deam, A. P., and Fannin, B. M., "Phase-Difference Variations in 9350-Megacycle Radio Signals Arriving at Spaced Antennas," Proc. IRE Vol 43, Oct 1956, pp 1402-5.

64. Etcheverry, R. D., et al, "Measurements of Spatial Coherence in 3.2-mm horizontal Transmission," IEEE Trans vol AP-15 no 1, Jan 1967, pp 136-140.

65. Vilar, E., and Matthews, P. A., "Measurement of Phase Fluctuations on Millimetric Radiowave Propagation," IEE Electron. Lett. vol 7 no 18, Sept 9, 1971, pp 566-7. (K_a-band).

66. Seliga, T. A., and Bringi, V. N., "Differential Reflectivity and Differential Phase Shift; Applications in Radar Meteorology," Radio Science, vol 13 no 2, Mar-Apr 1978, pp 271-5.

67. Watson, P. A.; Goodall, F.; and Arbabi, M.; "Linear Cross-Polarisation and attenuation measurements at 11 and 36 GHz," IEE Conf Pub no 98, Propagation of Radio Waves Above 10 GHz, 1973, pp 155-61.

68. Evans, B. G., and Troughton, J., "Calculation of Cross-Polarization due to Precipitation," IEE Conf Pub no 98 loc cit, pp 162-71.

69. Evans, B. G. and Troughton, J., "Linear and Cross-Polarization Statistics," IEE Conf Pub no 98 loc. cit, pp 172-8.

70. Ghobrial, S. I., and Watson, P. A., "Cross-Polarization During Clear Weather Conditions" IEE Conf Pub no 98, loc cit pp 179-81.

71. Evans, B. G., and Thompson, P. T., "Measurement of Cross-Polarization Using the SIRO Satellite," IEE Conf Pub no 98 loc cit, pp 183ff.

72. Turner, D. J. W., "Measurements of Cross-Polarization Discrimination at 22 GHz and 37GHz," IEE Conf Pub no 98, "Propagation of Radio Waves Above 10 GHz," 1973, pp 256ff.

73. Bostian, C. W., et al, "The Influence of Polarization on Millimeter Wave Propagation Through Rain," NASA CR-143686, Dec 73. (K_u-band).

74. Semplak, R. A., "Simultaneous Measurements of Depolarization by Rain Using Linear and Circular Polarizations at 18GHz," BSTJ vol 53 no 2, Feb 1974, pp 400-04.

75. Sobieski, P., "Depolarization Measurements at 12 GHz," Proc 4th European Microwave Conf., Montreux, Switzerland, Sept 10-13, 1974 paper C 3.4, pp 212-6.

76. Oguchi, T., "Rain Depolarization Studies at Centimeter and Millimeter Wavelengths: Theory and Measurement," J. Radio Res Lab of Japan vol 22 no 109, 1975, pp 165-211. (K_u-band).

77. Cox, D. C., et al, "Attenuation and Depolarization by Rain and Ice Along Inclined Paths Through the Atmosphere at Frequencies above 10 GHz," EASCON-79 Conf Rec.

78. Trebits, R. N., et al, "Multifrequency Radar Sea Clutter Measurements," EASCON-79 Conf Rec.

79. Dyer, F. B., and Currie, N. C., "Environmental Effects on Millimeter Radar Performance," AGARD Conf Proc no 245, Millimeter and Submillimeter Wave Propagation and Circuits, 1978, pp 2-1 thru 9.

80. Downs, A. R., "A Review of Atmospheric Transmission Information in the Optical and Microwave Spectral Regions," BRL memo Rpt 2710, Dec 1976.

81. Weintroub, H. J. and Hoffman, L. A., "Space Communications Systems Considerations," AGARD Conf Proc CP 107, Symp on Telecommun. Aspects of Frequencies between 10 and 100GHz, 1972, pp 22-1 thru 17.

82. Richard, V. W.; Kammerer, J.E.; and Reitz, R. G.; "140-GHz Attenuation and Optical Visibility measurements of fog, rain, and snow," ARBRL-MR-2800, Dec 77.

83. Long, M. W., "A Radar Model for Land and Sea," Proc URSI Open Symposium, L. Baule, Loire-Atlantique, France, Apr 28-May 6, 1977 (A78-25823), pp 117-9.

84. Hayes, R. D., "95 GHz Pulsed Radar Returns from Trees," IEEE EASCON-79 Proc.

Millimeter Wave Propagation
K.L. Koester, L. Kosowsky, and J.F. Sparacio
Appendix A to "Millimeter Wave Radar Applications to Weapons Systems"
V.L. Richard (Paper 1.7)
Pages 77-105

This article was prepared by an employee of the U.S. Government as part of his official duties. Such articles are in the public domain and can be reproduced at will. Reprinted by permission.

I. INTRODUCTION

The propagation of microwave energy through the atmosphere results in a reduction in intensity of the primary beam. This attenuation is the result of two phenomena - absorption and scattering. The absorption arises from both free molecules and suspended particles such as dust and water droplets condensed in fogs and rains. In addition to the attenuation, the energy scattered back in the direction of incidence is of prime interest for radar applications since it contributes additional noise to the system.

This section summarizes the available theoretical and experimental data on millimeter wave (35-94 GHz) propagation in rain and fog. In addition, an analysis of clear attenuation in the 60 to 75 GHz region is presented.

II. CLEAR AIR ATTENUATION

The gases that absorb microwave energy in a noncondensed atmosphere are oxygen, which has a magnetic interaction with the incident radiation, and water vapor, which contributes because of the electric polarity of the water molecule. In each case, there are resonance frequency regions where the absorption is abnormally large.[1] For oxygen, the resonance occurs at wavelengths of approximately 0.5 and 0.25 cm. In the case of water vapor, the resonance occurs at approximate wavelengths of 1.35, 0.27, 0.16 and 0.09 cm.

The clear air attenuation coefficient is given as a function of frequency by Rosenblum[2] for sea level and an altitude of 4 kilometers, as reproduced in Figure A-1. Examination of the figure shows regions of relatively low attenuation at wavelengths of approximately 8.6, 3.2, 2.1, and 1.3 millimeters. These clear weather windows are usually chosen as the operating frequency of sensors when optimum range performance is required.

---

*Appendices A-D were prepared for BRL under Contract DAAD05-72-C-3059 by K. L. Koester, L. Kosowsky, and J. F. Sparacio of Norden Division, United Aircraft, Norwalk, Connecticut.

[1] J.H. Van Vleck, "Theory of Absorption by Uncondensed Gases," in D.E. Kerr (ed.), Propagation of Short Radio Waves, pp. 646-664, Boston Technical Publishers, Massachusetts, 1964.

[2] E.S. Rosenblum, "Atmospheric Absorption of 10 to 400 KMCPS Radiation," Microwave Journal, pp. 91-96, March 1961.

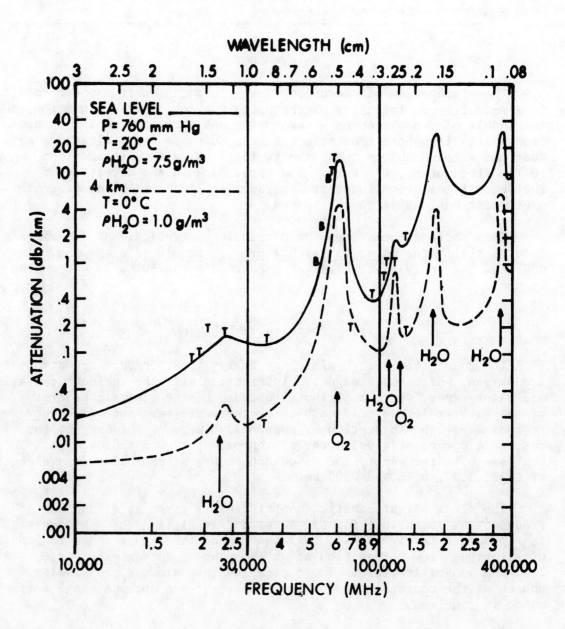

Figure A-1. One Way Attenuation per Kilometer for Horizontal Propagation in Clear Air (Rosenblum)

Further examination of Figure A-1 shows that the attenuation falls off quite rapidly as the frequency increases from 60 to 94 GHz. Because of the rapid variation in attenuation in this frequency region, a study of clear air propagation at sea level was conducted by Norden and is summarized below.

Tables of theoretical values of absorption by oxygen and water vapor at several discrete wavelengths were presented by Van Vleck.[1] Measurements have been made of the clear air attenuation into the 60 to 75 GHz region by several authors.[3-6] Their experimental results were combined with Van Vleck's to obtain the clear attenuation coefficient of 64 to 80 GHz energy shown in Figure A-2. It should be noted that the water absorption is dependent on the water vapor content of the air. In a temperate climate, approximately one percent of the molecules in the atmosphere are H_2O; this corresponds to a vapor content of 7.5 g/m^3.

Figure A-2 is plotted for this vapor content. At saturation at 20°C at sea level, the water vapor content is 17 g/m^3. Under tropical conditions the water vapor content can be even higher.

An examination of the figure indicates that the attenuation decreases rapidly from 6.7 dB/km at 64 GHz to 0.27 dB/km at 76 GHz. In the 76 to 94 GHz frequency range, there is only a slight decrease in attenuation.

---

[3] D.C. Hogg, "Millimeter-Wave Communication through the Atmosphere," Science, pp. 39-46, 5 January 1968.

[4] C.W. Tolbert, and A.W. Straiton, "Experimental Measurement of the Absorption of Millimeter Radio Waves Over Extended Ranges," IRE Transactions on Antennas and Propagation, pp. 239-241, April 1957.

[5] A.W. Straiton, and C.W. Tolbert, "Anomalies in the Absorption of Radio Waves by Atmospheric Gases," Proceedings of the IRE, pp. 898-903, May 1960.

[6] A.G. Kislyakov, V.N. Nikonov and K.M. Strezknew, "An Experimental Study of Atmospheric Absorption on a 4.1 mm Wave as Function of the Height above Sea Level," translated from Russian by the Foreign Technology Division, WPAFB, Ohio, Report FTD-MT-24-339-68, 30 October 1968, (AD 685 996).

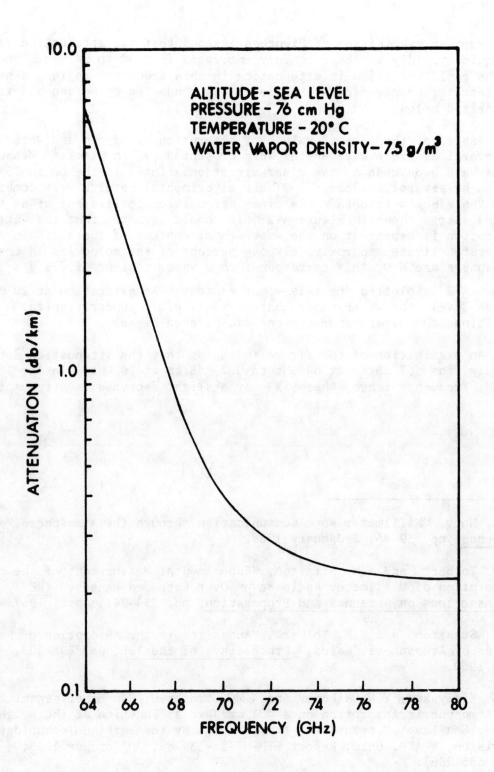

Figure A-2. One Way Attenuation per Kilometer for Horizontal Propagation in Clear Air

III. PROPAGATION IN RAIN

The study of the attenuation and scattering in a medium such as rain is based upon the knowledge of the effects of the individual particles. The complete theory for spherical particles of any material in a non-absorbing medium was developed by Gustav Mie.[7] This work was extended by Stratton[8] and is outlined in Kerr.[9] A comprehensive study of the theory of electromagnetic scattering from small particles is also offered by Van de Hulst.[10] A detailed examination of this theory is beyond the scope of this report. However, a brief discussion of the basic is summarized in Section III-A.

A. Cross Sections of Single Spheres

A single dielectric sphere in the path of a plane wave will scatter and absorb some of the incident energy. These effects are characterized by several quantities called cross sections and have the dimensions of area. The Gunn and East[11] definitions of the scattering, absorption, extinction, and backscatter cross-sections are:

$$\text{Scattering Cross Section } (Q_s) = \frac{\text{Total Power Scattered (over } 4\pi \text{ sterradians)}}{\text{Incident Power Density}} \quad \text{(A-1)}$$

$$\text{Absorption Cross Section } (Q_a) = \frac{\text{Total Power Absorbed (as heat)}}{\text{Incident Power Density}} \quad \text{(A-2)}$$

$$\text{Extinction Cross Section } (Q_a) = \frac{\text{Total Power Lost (to the incident wave)}}{\text{Incident Power Density}} \quad \text{(A-3)}$$

---

[7] G. Mie, Ann Physik, Vol. 25, p. 377 et. seq., 1908.

[8] J.A. Stratton, Electromagnetic Theory, McGraw-Hill, New York, 1941.

[9] H. Goldstein, "Attenuation in Condensed Water," in D.E. Kerr (ed.), Propagation of Short Radio Waves, pp. 641-692, Boston Technical Publishers, Boston, Massachusetts, 1964.

[10] H.C. Van de Hulst, Light Scattering by Small Particles, Wiley, New York, 1957.

[11] K.L.S. Gunn and T.W.R. East, "The Microwave Properties of Precipitation Particles," Quarterly Journal of the Royal Meteorological Society, Vol. 80, pp. 533-545, October 1954.

The term extinction is used to describe the energy lost by the incident wave to a single particle; attenuation is the energy lost to a continuous volume of particles.

It should be noted that the conservation of energy requires that[10]

$$Q_e = Q_a + Q_s \qquad (A-4)$$

$$\text{Backscatter Cross Section} (\sigma) = \frac{\text{Total Power Scattered Backward (along the direction of incidence)}}{\text{Incident Power Density}} \qquad (A-5)$$

The scattering and absorption properties of single particles are complex functions of the size, shape, and index of refraction of the particles as well as the wavelength of the incident energy. The scattering cross sections for water spheres ranging from 0.04 to 6.0 millimeters in diameter in 0.04 millimeter increments were calculated by SRI[12] for various temperatures. The various cross sections at 18°C are shown in Figure A-3.

B. Mie Scattering Theory

Our interest lies in the backscatter and attenuation cross sections associated with a continuous distribution of particle sizes within a given volume. The relationships for a single particle were described in Section III-A. If the particle size distribution is known, the reflectivity and attenuation can be determined, using the appropriate scattering theory.

The scattering theory to be used in the determination of the backscatter and attenuation of rain and fog depends on the size of the drops in the medium and the wavelength of the radiation. Mie scattering theory must be used for drops larger than 0.06 wavelength in diameter. For drops smaller than 0.06 wavelength the Rayleigh theory approximations are applicable.

---

[12] "Study of Atmospheric Propagation Factors for 70 GHz Energy," Stanford Research Institute, 26 August 1969 (Supplied by SRI under Norden PO 0016146 - not available for distribution).

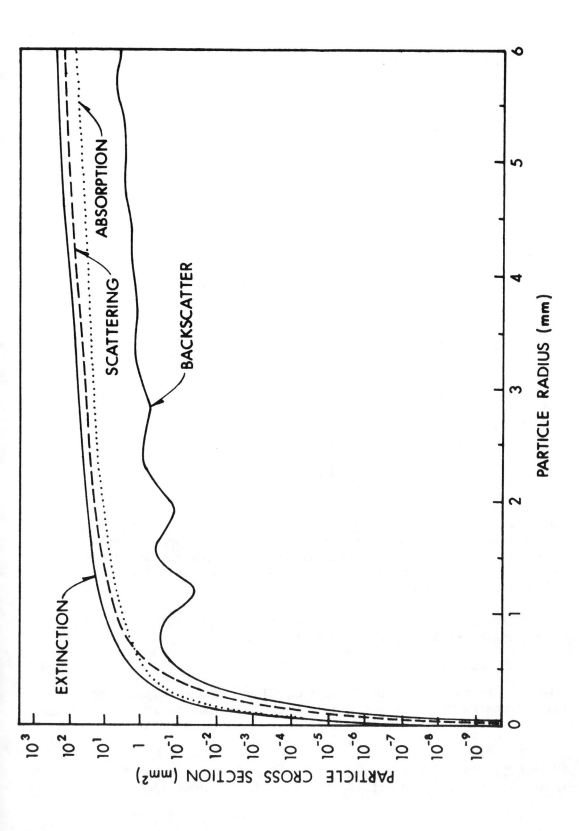

Figure A-3. Cross Sections of Water Spheres at 18°C for 4.3 mm Wavelength Energy [1]

Lukes[13] describes the size of the drops present under various atmospheric conditions. The ranges of drop size diameters for several atmospheric conditions are given in Table A-1.

Table A-1

Drop Diameters for Various Atmospheric Conditions[13]

ATMOSPHERIC CONDITION	DROP SIZE RANGE Micrometers
Haze	0.01 - 3
Fog	0.01 - 100
Clouds	1 - 50
Drizzle (0.25 mm/hr)	3 - 800
Moderate Rain (4.0 mm/hr)	3 - 1500
Heavy Rain (16.0 mm/hr)	3 - 3000

Since rain is comprised of drops 258 micrometers (0.258 mm) and larger, Mie scattering theory must be used for millimeter wavelength radiation. The smaller drops in haze and clouds allow the use of the Rayleigh approximation. Rain is investigated in Section III-D and fog is discussed in Section IV.

The reflectivity or backscatter cross section per unit volume in the Mie Scattering Region is defined by Mitchell[14] as

---

[13] G.D. Lukes, "Penetrability of Haze, Fog, Clouds and Precipitation by Radiant Energy Over the Spectral Range 0.1 Micron to 10 Centimeters," The Center for Naval Analyses of the University of Rochester, Report No. 61, May 1968, AD 847658.

[14] R.L. Mitchell, "Radar Meteorology at Millimeter Wavelengths," Aerospace Corporation, Report TR-669 (6236-46)-9, June 1966.

$$\eta = \int_{D_{min}}^{D_{max}} \sigma(D)N(D)dD, \qquad (A-6)$$

where $\sigma(D)$ = backscatter cross section of particle with diameter D,

$N(D)dD$ = the number of particles with diameter between D and D + dD per unit volume.

Similarly the attenuation is defined as

$$\alpha = 4.343 \int_{D_{min}}^{D_{max}} Q_e(D)N(D)dD, \qquad (A-7)$$

where $Q_e(D)$ = extinction cross section of particle with diameter D.

Note that the integral is multiplied by 4.343 to give the units of decibels per unit length as in standard radar practice.

C. Drop Size Distributions

The relation of raindrop size to intensity was first investigated by Laws and Parsons[15] in 1943. Their experimental study indicated that the median drop size was a fairly strict function of rain intensity. The change in median drop size over the range of intensities covered was quite small, increasing by only three diameters while the intensity increased by a factor of four hundred. Tolbert and Gerhardt[16] presented the drop size distribution for the Laws and Parsons data shown in Figure A-4. Additional raindrop measurements

---

[15] O.J. Laws and D.A. Parsons, "The Relation of Raindrop-Size to Intensity," *Transactions of the American Geophysical Union*, Vol. 24, pp. 452-460, 1943.

[16] C.W. Tolbert and J.R. Gerhardt, "Measured Rain Attenuation of 4.3 Millimeter Wavelength Radio Signals," University of Texas, Report No. 83, 31 May 1956.

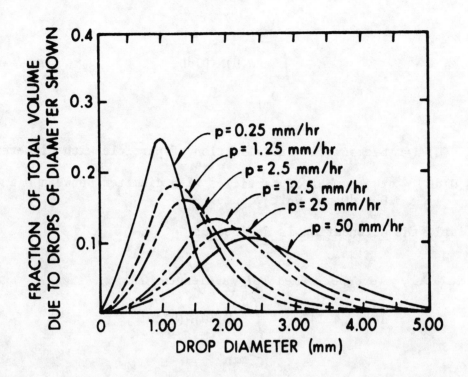

Figure A-4. Rainfall Drop Size Distribution as a Function of Rain Rate for Laws and Parsons Data with 0.25 mm Diameter Interval [16]

were performed by Marshall and Palmer[17] in 1948. The distribution of drops with size was found to be in agreement with that of Laws and Parsons.

Cantaneo and Stout[18] studied raindrop size distributions in several locations and found the distributions to be quite similar. A difference was found between warm and cold frontal rains with the cold rain having smaller drops. Despite the existing differences in rain of a given intensity, the drop size distributions described are valid for a general study of propagation through rain.

D. Attenuation Due to Rain

The first extensive investigation of attenuation in rain was performed by Ryde and Ryde[19] in 1945. Their calculations, which are presented in Kerr,[9] were performed for wavelengths ranging from 3 millimeters to 10 centimeters. A form of equation (A-7) was used with the Laws and Parsons drop size distribution and the extinction cross section determined from Mie scattering theory. The one-way attenuation in rain at 18°C for 4.3 millimeter energy is shown in Figure A-5.

Mie theory was also used to determine the rain attenuation at 70 GHz at 0°C by Crane[20] and at 10° and 30°C by SRI.[12] Their results are also shown in Figure A-5.

Experimental confirmation of these results was provided by Tolbert and Gerhardt[16] and Hogg.[21] The latter's results are also shown in Figure A-5. Examination of this illustration indicates general agreement on the attenuation in rain at 70 GHz.

---

[17] J.S. Marshall and W.McK. Palmer, "The Distribution of Raindrops with Size," *Journal of Meteorology*, Vol. 5, pp. 165-166, August 1948.

[18] R. Cantaneo and G.E. Stout, "Raindrop-Size Distributions in Humid Continental Climates, and Associated Rainfall Rate-Radar Reflectivity Relationships," *Journal of Applied Meteorology*, Vol. 7, pp. 901-907, October 1968.

[19] J.W. Ryde and D. Ryde, "Attenuation of Centimeter and Millimeter Waves by Rain, Fog, and Clouds," British General Electric Co., Report No. 8670, 1945.

[20] R.K. Crane, "Microwave Scattering Parameters for New England Rain," Technical Report 426, Lincoln Laboratories, MIT, 3 October 1966.

[21] D.C. Hogg, "Millimeter-wave Communication through the Atmosphere," *Science*, Vol. 159, pp. 39-46, 5 January 1968.

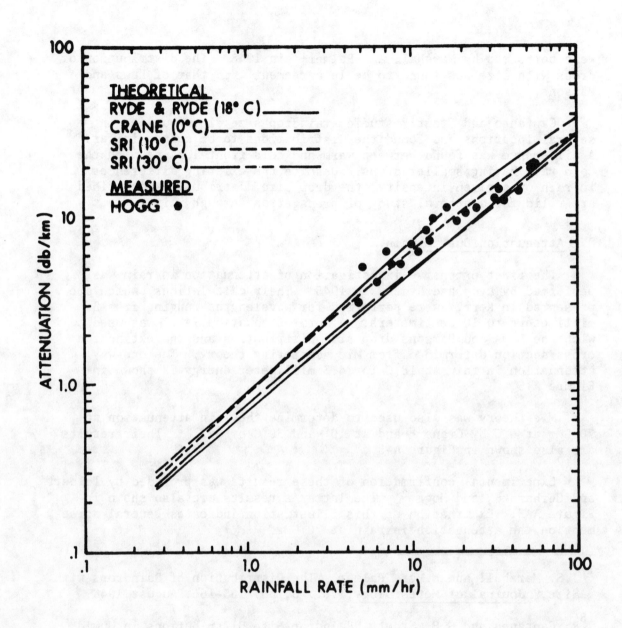

Figure A-5. One-Way Attenuation in Rain at 70 GHz

Figure A-6 shows the attenuation coefficient for 35, 70 and 94 GHz, the "millimeter" frequencies. Also shown for comparison is the attenuation at 15.5 GHz. These curves were derived from data presented by Crane.[20]

E. Backscatter Cross Section of Rain

Until recently, the study of radar meteorology at millimeter wavelengths has largely been ignored. The Rayleigh scattering approximations used at lower frequencies are not applicable above X-band. The exact, but more difficult, Mie theory was not utilized because as Skolnik[22] points out, "...the larger attenuations at the higher frequencies preclude the making of measurements conveniently." As components were developed, interest in millimeter radar systems increased because of their high resolution capability. At the relatively short ranges of less than 5 nautical miles where the resolution would be most useful, the rainfall attenuation is small. Since backscatter data were not available, many people incorrectly extrapolated lower frequency results.

The Rayleigh approximation of the backscatter cross section of rain at 35, 70 and 94 GHz is shown in Figure A-7. Also shown is the backscatter cross section predicted by Crane, using Mie scattering theory. At 70 and 94 GHz the difference between the Mie and Rayleigh values increases with rainfall rate. It is interesting to note the crossover of the 35, 70 and 94 GHz Mie curves at a rain rate of 9 millimeters per hour.

Crane's backscatter curve for 70 GHz energy is repeated in Figure A-8 for comparison with the SRI 70 GHz curves, which show the temperature dependence of the backscatter coefficient. (Both used a form of equation (A-6) to obtain the volume backscatter coefficient in units of km^{-1} $ster^{-1}$. The radar backscatter coefficient, plotted in Figure A-8, is defined as 4π times the volume backscatter coefficient).

The measurement of the backscatter cross section of rain was included in the Norden experimental program. The results for rains of 2 and 3.8 millimeters per hour are shown in Figure A-8. They compare favorably with the predicted values.

---

[22] M. Skolnik, Introduction to Radar Systems, pp. 521-569, McGraw-Hill, New York, 1962.

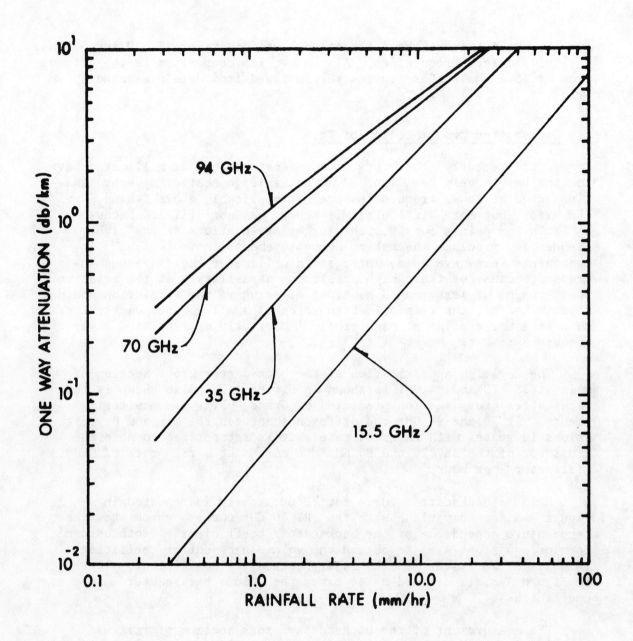

Figure A-6. One-Way Attenuation in Rain [20]

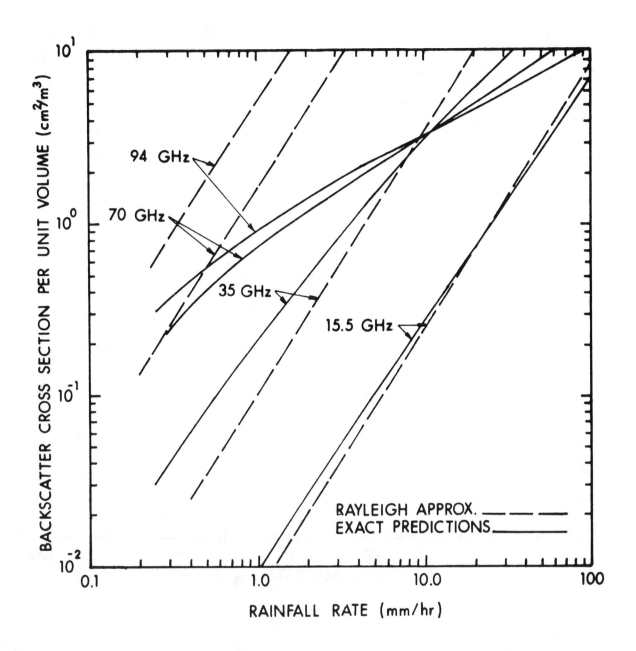

Figure A-7. Backscatter Cross Section Per Unit Volume of Rain at 0°C [20]

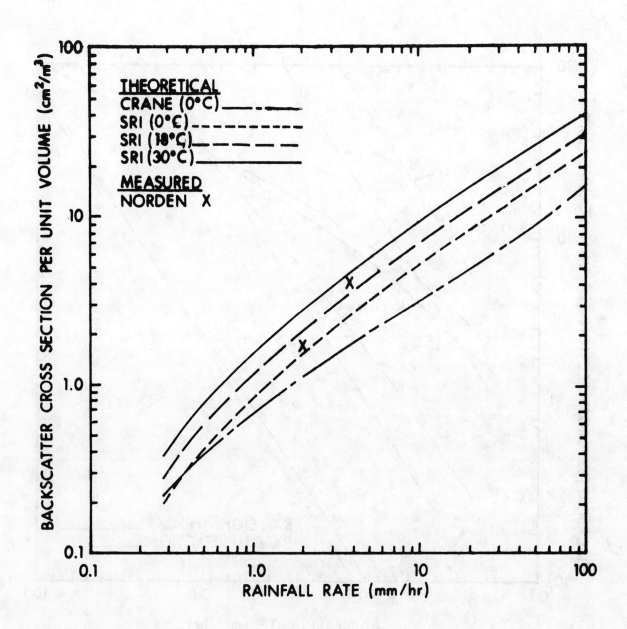

Figure A-8. Backscatter Cross Section Per Unit Volume of Rain at 70 GHz

IV. FOG

A. Characteristics of Fog

Fog results from the condensation of atmospheric water vapor into water droplets that remain suspended in the air.[23] When the resulting cloud or water droplets or ice crystals envelop an observer and restrict his horizontal visibility to one kilometer or less, the international definition of fog has been satisfied.[24] Evaporation and cooling are the principal physical processes which contribute to the formation of fog. Of the various fog classifications used by meteorologists, the two basic types of interest in radar applications are advection fog and radiation fog.

Advection is the horizontal movement of an air mass that causes changes in temperature or other physical properties. An advection (or coastal) fog is one which forms over open water as a result of the advection of warm moist air over colder water.

Radiation (or inland) fog forms in air that has been over land during the daylight hours preceding the night of its formation. Fogs which form in low, marshy land and along rivers on calm, clear nights are also considered radiation fogs.

The characteristics of these two fogs are given in Table A-2:

Table A-2

Fog Characteristics[12]

	RADIATION (INLAND) FOG	ADVECTION (COASTAL) FOG
Average Drop Diameter	10 microns	20 microns
Typical Drop Size Range	5-35 microns	7-65 microns
Liquid Water Content	0.11 g/m^3	0.17 g/m^3
Droplet Concentration	200 cm^{-3}	40 cm^{-3}
Visibility	100 m	200 m

---

[23] J.F. O'Connor, "Fog and Fog Forecasting," *Handbook of Meteorology*, pp. 727-736, McGraw-Hill, New York, 1945.

[24] J.J. George, "Fog," *Compendium of Meteorology*, pp. 1179-1189, American Meteorological Society, Boston, 1951.

Note that the advection fog has a higher liquid water content, but greater visibility than the radiation fog. The correlation of visibility in fog to liquid water content is shown in Figure A-9 for both advection and radiation fogs.

There is considerable variation in the water content of clouds and water fogs, but in general, stratus (or low) clouds and typical radiation or advection fogs have water contents on the order to 0.25 g/m^3 or less.[12] Mason[25] reports that the maximum liquid water content of an advection fog approaches 0.4 g/m^3 when there is a strong temperature inversion. On rare occasions, the liquid water content can become as large as 0.5 to 1.0 g/m^3 in very dense radiation fogs (with 20 to 30 meters visibility).[12,19,26,27]

B. Fog Attenuation

The small size of water droplets comprising a fog allows the use of the Rayleigh approximations in the determination of the reflectivity and attenuation at 70 GHz. Atlas[28] shows that in the Rayleigh scattering region the one-way attenuation coefficient, α, is given by

$$\alpha = \frac{81.86 \, M \, \text{Im}(-K)}{\lambda \rho} \quad \text{dB/km,} \tag{A-8}$$

where M = liquid water content per unit volume of fog in g/m^3,

$\text{Im}(-K)$ = absorption coefficient,

$K = \dfrac{m^2-1}{m^2+2}$,

m = complex index of refraction,

λ = wavelength in mm,

ρ = density of water in g/cm^3.

---

[25] B.J. Mason, *The Physics of Clouds*, p. 97 et. seq., Clarendon Press, Oxford, 1957.

[26] P.N. Tverskoi, "Physics of the Atmosphere," Translated from Russian for the National Aeronautics and Space Administration and the National Science Foundation, NASA TTF-288 TT65-50114, 1965.

[27] Ralph G. Eldridge, "Haze and Fog Aerosol Distributions," *Journal of the Atmospheric Sciences*, Vol. 23, September 1966.

[28] D. Atlas, "Advances in Radar Meteorology," *Advances in Geophysics*, Vol. 10, pp. 317-478, Academic Press, New York, 1964.

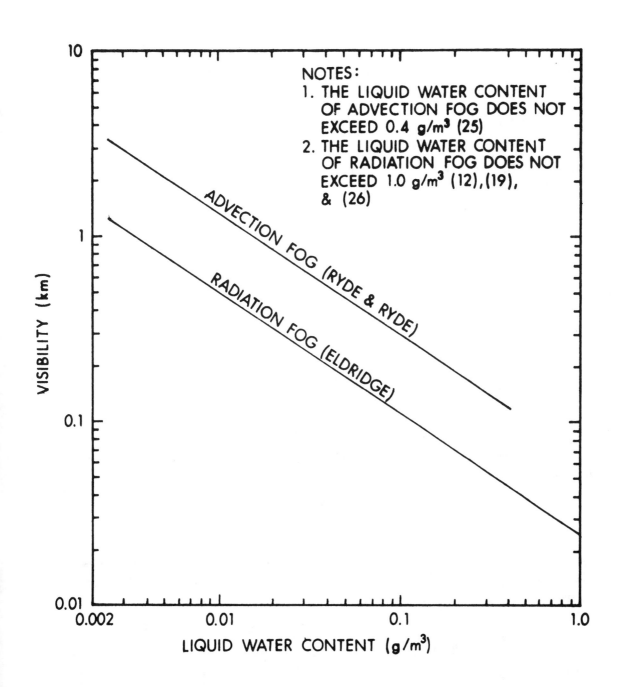

Figure A-9. Correlation of Visibility in Fog to Liquid Water Content

A density of 1 g/cm³ for water is generally assumed for all temperatures, since the density varies no more than 0.78% over the 0°C to 40°C temperature range.[29]

In the Rayleigh scattering region, attenuation is due mainly to absorption. To calculate the absorption coefficient for fog, the index of refraction for water must be determined for the frequencies of interest. The complex index of refraction, m, is given in terms of the complex dielectric constant, ε_c, by[11]:

$$m^2 = \varepsilon_c = \varepsilon_1 - j\varepsilon_2, \qquad (A-9)$$

where ε_1 and ε_2 are the real and imaginary parts of the dielectric constant.

The dielectric constant may be evaluated by the Debye formula[9]:

$$\varepsilon_c = \frac{\varepsilon_o - \varepsilon_\infty}{1 + j\frac{\Delta\lambda}{\lambda}} + \varepsilon_\infty, \qquad (A-10)$$

where ε_o, ε_∞, and $\Delta\lambda$ are empirically derived constants. The Debye constants for water over the 0°C to 40°C temperature range are presented in Table A-3.

Table A-3

Constants for the Debye Formula

T(°C)	ε_o	ε_∞	$\Delta\lambda$ (cm)
0	88.0	5.5	3.59
10	84.0	5.5	2.24
18	81.0	5.5	1.66
20	80.0	5.5	1.53
30	76.4	5.5	1.120*
40	73.0	5.5	0.857*

* These constants have been rederived and differ from those published in Reference 9. The derivation of these new values will be presented in a forthcoming paper.

---

[29] D.E. Gray, American Institute of Physics Handbook," McGraw-Hill, New York, 1957.

The complex indices of refraction of liquid water for 8.6, 4.3, and 3.2 mm energy were evaluated from equations (A-9) and (A-10) and are shown in Table A-4 as a function of water temperature.

Table A-4

Complex Index of Refraction of Water

Temp. (0°C)	8.6 mm 35 GHz	4.3 mm 70 GHz	3.2 mm 94 GHz
0	3.947-j2.367	3.039-j1.603	2.801-j1.302
10	4.802-j2.735	3.543-j2.059	3.173-j1.732
20	5.607-j2.838	4.077-j2.380	3.596-j2.076
30	6.266-j2.733	4.608-j2.574	4.031-j2.323
40	6.748-j2.501	5.108-j2.649	4.457-j2.477

Using the complex indices of refraction listed in Table A-4, the absorption coefficients for 35, 70 and 94 GHz radiation were determined for the 0°C to 40°C temperature range. These values are given in Table A-5.

Table A-5

Absorption Coefficients for Water

TEMPERATURE (0°C)	ABSORPTION COEFFICIENT		
	35 GHz	70 GHz	94 GHz
0	0.114	0.172	0.183
10	0.079	0.137	0.162
20	0.058	0.107	0.133
30	0.044	0.085	0.109
40	0.036	0.069	0.090

Examination of Table A-5 indicates that the absorption coefficient of water is highly sensitive to temperature in the 0°C to 40°C range. Assuming the density of water to be 1 g/m^3, the attenuation coefficient of fog at 35, 70 and 94 GHz may be computed using equation (A-8), and is given by:

$$\alpha_{35} = 9.52 \text{ M Im}(-K) \quad \text{db/km,} \tag{A-11}$$

$$\alpha_{70} = 19.04 \text{ M Im}(-K) \quad \text{db/km,} \tag{A-12}$$

$$\alpha_{94} = 25.58 \text{ M Im}(-K) \quad \text{db/km.} \tag{A-13}$$

The attenuation coefficient has been computed as a function of liquid water content using equations (A-11), (A-12), and (A-13) at 0°C and 40°C. These results are shown graphically in Figure A-10. An examination of the figure shows that the attenuation is substantially greater at 0°C than it is at 40°C.

Fog is generally described by its apparent optical visibility. This, however, is an inadequate characterization of fog in that similar visibility conditions may result from various meteorological conditions.

Visibility restrictions result from fog or low clouds with or without precipitation. These natural conditions occur in wide variety as a result of different weather conditions with seasonal and geographical variations. The detailed physical nature of the phenomena varies considerably even when producing comparable effects on vision. Reduced visibility can result from widely different distributions of water substance in terms of numbers of droplets of different dimensions.

As discussed in Section IV-A advection fogs have a higher liquid water content, but greater visibility than radiation fogs. The small number of large drops in an advection fog attenuates light, and hence visibility, much less than the large number of small drops in a radiation fog. The larger liquid water content of advection fogs causes greater attenuation of millimeter wave energy.

It is possible, however, to obtain representative values of attenuation based on the apparent visibility in each of the two basic fogs by combining the advection and radiation fog visibility curves from Figure A-9 with the attenuation curves of Figure A-10. However, the most meaningful description of the attenuation coefficient is obtained from the liquid water contents.

C. Fog Backscatter

The backscatter cross section of fog can be determined from the Rayleigh scattering theory approximation given by Atlas[28] as

$$\eta = \frac{\pi^5 |K|^2}{\lambda^4} Z, \tag{A-14}$$

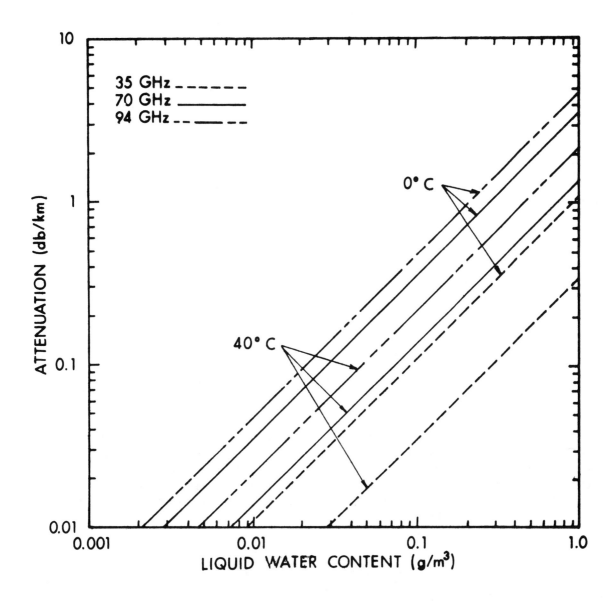

Figure A-10. One-Way Attenuation in Fog As a Function of Liquid Water Content

where

$$|K|^2 = \text{backscatter coefficient,}$$

$$\lambda = \text{wavelength in mm,}$$

$$Z = \text{reflectivity factor.}$$

Atlas[28] gives the reflectivity factor in terms of the liquid water content for radiation fog as

$$Z_R = 0.48M^2 \qquad \text{(A-15)}$$

and for advection fogs as

$$Z_R = 8.2M^2. \qquad \text{(A-16)}$$

For M in units of g/m^3, Z is in units of mm^6/m^3. It is obvious from the above relationships that the largest backscatter cross section would be found in advection fogs. However, in the millimeter wavelength region (35 to 94 GHz) the largest cross section was found to be less than $1.0 \text{ mm}^2/m^3$. Inasmuch as the cross section is more than two orders of magnitude smaller than that of rain, it has a negligible effect on radar system performance. Consequently, it will not be considered further.

V. SYSTEM CONSIDERATIONS

A. Introduction

The primary considerations in the selection of frequency relate to the antenna size, transmission properties of the atmosphere, and the state of component technology. Since radar systems must be designed to operate under adverse weather conditions, such as rain and fog, close examination must be made of attenuation and backscatter in these media, as a function of radar wavelength.

From an installation point of view, selection of a higher frequency would be advantageous in reducing antenna swept volume, with a corresponding reduction in radome size and the attendant structural, cost, and weight problems. Since swept volume varies approximately as the cube of the antenna size, any reduction in antenna size would prove quite beneficial. However, an increase in frequency to achieve this benefit must be carefully weighed against possible deleterious effects of weather.

B. Clear Air Performance

The attenuation of millimeter energy in clear air was discussed in Section II. The clear air attenuation coefficient for the frequencies of interest is shown in Table A-6.

Table A-6

Clear Air Attenuation Coefficient

Frequency GHz	New Band Designation	Old Band Designation	Attenuation Coefficient dB/km
35	K	K_A	0.18
70	M	V	0.41
94	M	W	0.24

If the system is designated for clear air operation only, the frequency at which the attenuation is smallest should be chosen. However, any operational radar would be used under adverse weather conditions, such as rain and fog. The effect of rain and fog on system performance must be analyzed.

C. Rain Performance

The performance of a radar operating in the rain is degraded by the absorption and scattering of the energy by the raindrops. For analysis purposes, the attenuation coefficient for light and heavy rain is given in Table A-7 for the three millimeter frequencies of interest. These values are taken from Figure A-6. (An attenuation coefficient of 1 dB/km would reduce the radar range of a system from 10 nautical miles in clear air to approximately 6.5 nautical miles; 7 dB/km reduces the range to approximately 1.1 nautical miles.)

Table A-7

Rain Attenuation and Backscatter Coefficient

Frequency GHz	Light Rain (1 mm/hr)		Heavy Rain (16 mm/hr)	
	Attenuation (dB/km)	Backscatter (cm²/m³)	Attenuation (dB/km)	Backscatter (cm²/m³)
35	0.24	0.21	4.0	4.9
70	0.73	0.72	6.9	4.1
94	0.95	0.89	7.4	3.9

In addition to the attenuation, the energy reflected back to the radar is of interest since it contributes additional noise to the system. The backscatter coefficients are also presented in Table A-7 based on the curves of Figure A-7.

As can be seen, the backscatter cross section of heavy rain is comparable for V- and W-band radars and is actually less than at Ka-band. Thus, a V- or W-band radar operating in heavy rain would have less rain clutter than a Ka-band system. A W-band system would have slightly less rain backscatter than a V-band system at the heaviest rain rates, but the higher attenuation at all rain rates would tend to discourage the use of frequencies higher than V-band.

Even though rain attenuates V-band more than Ka-band, it does not preclude its choice as the frequency of an operational radar. Consider the target-to-rain clutter ratio for a given radar.

The return from a target is given by

$$P_{TGT} = \frac{P_T G^2 \lambda^2 \sigma_T}{(4\pi)^3 R^4 \exp(2\alpha R)} \quad . \tag{A-17}$$

The return due to rain is given by

$$P_{RAIN} = \frac{P_T G^2 \lambda^2 \sigma_i \, \theta\phi(\frac{c\tau}{2})}{(4\pi)^3 R^2 \exp(2\alpha R)} \quad , \tag{A-18}$$

where P_T = peak power,

G = antenna gain,

λ = wavelength,

σ_T = target cross section,

R = range,

α = one way attenuation coefficient,

σ_i = rain backscatter cross section per unit volume,

θ = azimuth beamwidth,

ϕ = elevation beamwidth,

c = speed of light in free space,

τ = transmitted pulse width.

The target-to-clutter ratio is given by

$$\frac{P_{TGT}}{P_{RAIN}} = \frac{\sigma_T}{R^2 \left(\frac{c\tau}{2}\right) \sigma_i \theta \phi} \; .$$

For a parabolic antenna of dimension ℓ

$$\theta = \phi = \frac{65 \lambda}{\ell} \; .$$

Substituting,

$$\frac{P_{TGT}}{P_{RAIN}} = \left[\frac{\sigma_T \ell^2}{R^2 \left(\frac{c\tau}{2}\right) (65)^2}\right] \frac{1}{\lambda^2 \sigma_i} \; .$$

For a fixed target cross section at a given range, the term in brackets is a constant. It is now possible to define a figure of merit, F_m, for system performance in rain which is a function only of frequency; namely,

$$F_m = \frac{1}{\lambda^2 \sigma_i} \; . \tag{A-19}$$

This factor gives an indication of the effect of rain backscatter on system performance.

Figure A-11 shows the "figure of merit" as a function of radar frequency. An examination of the curve indicates that a frequency in the Ka-band region would provide the poorest performance under all rain conditions. Consequently, a frequency other than 35 GHz should be chosen. Inasmuch as a frequency in the V- or W-band region offers acceptable performance in rain, it will be considered further. Before a frequency is chosen, its performance in fog must be investigated.

D. Fog Performance

The attenuation coefficient for millimeter propagation in fog was derived in Section IV-B. As shown in Figure A-10, the attenuation is linearly dependent on the liquid water content of the fog. In addition, there is a significant temperature dependence. The attenuation coefficient is given in Table A-8 for a liquid water content of 0.1 g/m^3.

Table A-8

Attenuation Coefficient of Fog
Liquid Water Content - 0.1 g/m^3

Frequency GHz	Attenuation Coefficient dB/km	
	0°C	40°C
35	0.11	0.034
70	0.36	0.138
94	0.47	0.22

The attenuation increases with frequency. Thus the lowest frequency (35 GHz) should be chosen for adequate fog performance.

E. Frequency Choice

Consideration of clear air and fog attenuation would lead to the choice of 35 GHz as the frequency for an operational system. However, once the effect of rain backscatter is considered, a frequency in M-band (70 or 94 GHz) would offer the best performance under all weather conditions. A choice between 70 and 94 GHz requires a careful examination of component technology, noise, and ECM. There has been sufficient development of 70 GHz components to allow the design of practical systems at this frequency band.

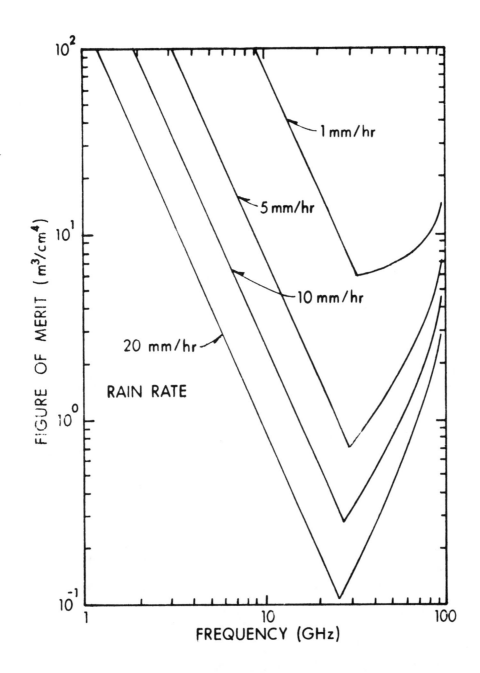

Figure A-11. Figure of Merit of Radar Performance in Target Cross Section = 1 Square Meter

Propagation Effects for mm Wave Fire Control System
T.N. Patton, J.J. Petrovic, and J. Teti
6th DARPA-Tri Service Millimeter Wave Conference, 1977
Reprinted by permission.

ABSTRACT

This paper presents the results of a survey of propagation effects of the atmosphere on a millimeter wave signal used by a sea based fire control radar for use against short range, low altitude, targets. Factors which influence detection range and tracking accuracies are considered. Primary goal was to determine whether normal low altitude conditions of over sea refractivity will seriously degrade the performance of a pencil beam mm wave radar. It is concluded that the literature is inconclusive on the matter and that experimental work is required.

1. INTRODUCTION

As a preliminary step in a support program to Naval Surface Weapons Center, Dahlgren Laboratory, a survey of the literature on millimeter wave propagation in the troposphere was undertaken. We were specifically interested in those effects which might bear on the performance of a fire control radar functioning as part of a surface weapons system. Factors which might influence detection range and tracking accuracies were of particular interest. We hoped that a literature search might resolve the questions:

Will normal, low altitude conditions of over sea refractivity seriously degrade the performance of pencil beam, fire control radars operating at millimeter wavelengths; and is there reason to consider the use of the 72 GHz and 95 GHz transmission windows as well as that at 35 GHz for shipborne fire control radars? The effort was limited to a survey and summary of the literature on the subject, and included no new analyses or experimental work. Propagation effects of interest included those due to refraction in both standard atmospheres and nonstandard atmospheres; and attenuative phenomena due to atmospheric gases as well as the various forms of condensed water vapor. In view of recent concern over refractively induced "holes" in radar coverage, ducting phenomena were emphasized.

Sources of information employed in the survey included (1) Indices and bibliographies, (2) General references, and (3) Selected specific references. In the first category are included the International Radar Index 1975, bibliographies and abstracts on tropospheric propagation from Defense Documentation Center, NASA, and National Technical Information Service, and the extensive bibliographies provided by Nathanson and Skolnik. Of particular importance among the general references is the 1975 ARPA, Tri Service Symposium on Millimeter Waves, and Kerr's Propagation of Short Radio Waves, as well as Nathanson. The literature search was supplemented with a visit to the Electromagnetic Compatibility Analysis Center (ECAC) to discuss millimeter wave propagation. These discussions seemed to point up the lack of generally agreed upon results at present.

The approach taken has been to skim indices and bibliographies identifying likely sources, while digesting the summary presentations provided by the general references. This allowed time for accumulating the more specific sources. Mathematical models, analytical techniques, and experimental results were of primary interest; but system analyses were also reviewed.

General or growing acceptance of a model or of an experimental result has been taken as evidence of its probable validity, although it must be noted that some models, techniques, or experimental data are widely used because, even though they may be questionable, no practical alternative is available. The use of ray tracing rather than the waveguide mode formulation for portraying the trapping of energy in atmospheric ducts is a good example of this sort of thing.

Because the literature in this field is most extensive it was not possible to carry out anything even approaching comprehensive coverage. An attempt was made, in fact, to restrict coverage to documents which bear directly on the issues this report addresses. Not even all of these documents could be reviewed and assimilated in the time available, however. Consequently we are quite conscious of the possible inadequacies of the results. In addition to the simple limitation of coverage it is likely that in some cases we have misread our sources or inadvertently put words into their mouths; and in other cases that we may be guilty of false inferences. It is with these cautions that we present the following results. They are given in three categories, including general results, refractive effects, and attenuative effects.

2. GENERAL RESULTS

The first general conclusion which one might draw from the literature is that while a substantial body of well developed theory is available, there seems to be no comparable, comprehensive base of experimental data. It was generally agreed in 1974 [6]* that no good metrology base exists for millimeter wave propagation. This complicates the production of reliable propagation data. Furthermore, small scale atmospheric inhomogeneities which are insignificant for longer wave lengths may take on great significance in this area, and techniques for instrumenting them do not seem to exist. This lack of a metrology base, and the consequent lack of reliable data has impact on modeling since the modeler is forced to rely heavily on conjecture rather than on hard interpretable data. Emphasis tends to be limited to statistical reduction of scattered data.

One good example of the difficulty of instrumenting millimeter wave experiments is given by the problem of rainfall. The drop size distribution and density are needed if results are to be correlated with theory. But these items must be measured at a point and they are not constant in space. In particular for rain rates greater than about 2 mm/hr they may vary widely [5]. Furthermore, even at a measurement point, reliable measurement of these items is a challenge.

3. REFRACTIVE EFFECTS

Refraction studies are based either on a horizontally stratified atmosphere or on some sort of randomly lumpy atmosphere. The bulk of available results are theoretical and are based on some form of standardized, atmosphere, the refractive index of which varies only with altitude. The most representative of the standard atmospheres seems to be the Central Radio Propagation Laboratory (CRPL) Exponential Reference Atmosphere. (See Figure 1 for comparison) Substantial material on its

---

*Bracketed numbers denote reference sources.

effects has been developed; and its use is fully justified in systems analysis whenever a standard atmosphere is appropriate.

The main refractive effects which are of concern include reduction of signal strength due to trapping or focusing phenomena, and errors in the estimation of range and angle to the target. Within the operational bounds of the shipborne fire control problem, the CRPL Exponential Reference Atmosphere does not perceptibly degrade radar performance. Neither the refractively produced attenuation nor the inaccuracy of range and angle is of significant magnitude. This can be seen in the curves shown in Figures 2 [10] and 3 [1]. For ranges of less than 40 nmi it can be seen that the expected focusing loss is of the order of 0.05 db, while elevation angle and range will be in error by less than 2 mrad and 80 feet respectively. Furthermore, the fluctuation of these values would be expected to be only about 0.7 mrad and 2 feet in the worst case, of heavy cumulus weather.

The probable impact of non-standard atmospheres is by no means so clear. And this is a most important area since over the sea there is almost always the potential for an evaporation duct. (Experimental data published by Hitney [3] suggests that one exists more than 95% of the time.) There is also experimental evidence of partial signal trapping in such ducts at millimeter wavelength [8]. This suggests the existence of a radar hole above the duct but does not demonstrate it. No simple procedures or analytical tools are available for incorporating non-standard atmospheres into systems analyses.

Although considerable discussion of the evaporation duct was discovered, little was reviewed on elevated ducts; and, consequently, no conclusions are drawn about that aspect of the problem.

It is worth noting that there are comments in the literature [9] which call into question the validity of the horizontally stratified model atmosphere for millimeter waves, and some independent experimental results [8] tend to support the allegation that small scale atmospheric irregularities mitigate trapping at these frequencies. At present the question seems to remain open, and is an area which should be given some careful experimental consideration.

To the extent that the horizontally stratified model atmosphere remains valid at millimeter wave length tools exist for computing some trapping effects. The Naval Ocean Systems Center (NOSC) has produced a set of computer programs which implement the analytical technique (wave guide theory) for computing path loss in the presence of an evaporation duct [3]. To be relevant to the fire control problem, however, these programs will have to be modified to incorporate a pencil beam antenna pattern. This would seem to be a worthwhile modification to carry out.

4. ATTENUATIVE EFFECTS

For the cases of oxygen and uncondensed water vapor the general theory appears to be quite stable and widely accepted. On the resonance lines, predictions of absorption are precise. In the transmission windows, however, there are some discrepancies, though not generally serious ones [6, 11]. Currently available data are adequate for system analysis.

In the case of the various forms of condensed water vapor, on the other hand, the matter is not so well defined. Theory is well developed but as I mentioned earlier there are practical difficulties on the experimental side [7]. These difficulties have to do with droplet shape, size distribution, and density, in space. No reliable techniques exist for measuring these as support to propagation measurements. For rain, some excellent correlations have been shown between rate (in mm/hr) and attenuation (in db/km). But unaccounted for variations in measured results demand explanation. This probably will be available when experimental conditions are brought under control.

Several atmospheric pollutants, of the type which might accompany the firing of weapons in a fire fight, are known to have resonance lines in the millimeter wave region [4, 11]. No data on their operational concentration was found, however, and that important point remains unresolved.

While several experiments have been performed to measure the clutter and attenuation characteristics of rain, aerosols, water vapor, and oxygen, their attenuation varies widely and, furthermore, most of this work appears to have been oriented toward specific systems. The data therefore, are not comprehensive and it does not seem presently possible to reliably choose a frequency which offers the best tradeoff between propagation, attenuation and backscatter for some particular application. At least one study of system design (conducted by the Applied Physics Lab of JHU) for a fire control radar capable of supporting a three mile missile, and having a nominal detection range versus a one square meter target at 12 nmi was discovered [6]. It reported that ECCM capability (main lobe resistance to stand off jamming) was markedly enhanced by choosing frequencies of 49 GHz or higher. However, the influence of the rain requirement (4 mm/hr) made it undesirable to go to frequencies beyond 37 GHz. This suggests that any system which must function in even moderate rain must remain at the lower end of the millimeter range even though clear weather advantages would accrue at the higher frequencies; or, alternatively must have an optional frequency for bad weather use.

5. RECOMMENDATIONS

In view of these general findings, and with particular reference to the fire control radar problem, the following efforts seem fully warranted.

Continued survey and organization of the literature of the field.

Analysis to support simplified treatment of non-standard refractive effects, along the lines of the Lens Loss approach, on the assumption that the horizontally stratified model is valid.

Development of measurement techniques to reduce variability of millimeter wave data, and increase the credibility of experimental results.

Coordinated experimental/analytical efforts in the area of propagation effects due to condensed water vapor in its various forms.

Experimentation to test the hypothesis that, at millimeter wavelengths, small scale atmospheric irregularities reduce trapping or nullify it; and to determine how these irregularities would affect the radar beam.

General experimentation to examine thoroughly the impact of tropospheric effects on pencil beam, fire control radar systems.

All of these items are mutually supporting and should be coordinated if they are to achieve their full potential. Experimentation and analysis performed by the millimeter wave community at large seems to be only loosely coordinated at present, although symposia, such as the present one, contribute substantially to its unity.

6. REFERENCES

1. Barton, D.K., and Ward, H.R., *Handbook of Radar Measurement*, Prentice-Hall (1969).

2. Bean, B.R., et. al., "Weather Effects on Radar" in *Radar Handbook*, Skolnik, M.I., Editor, McGraw-Hill (1970).

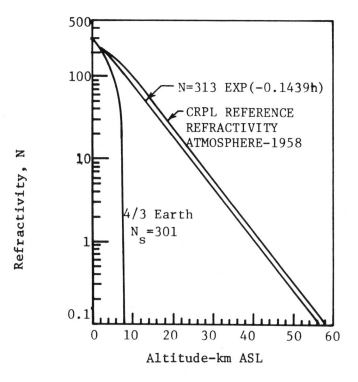

FIGURE 1 RADIO REFRACTIVITY PROFILES

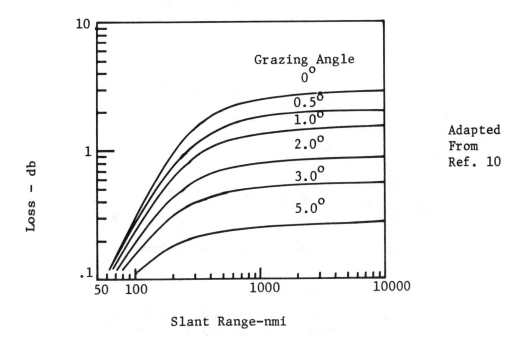

FIGURE 2 LENS LOSS EFFECT

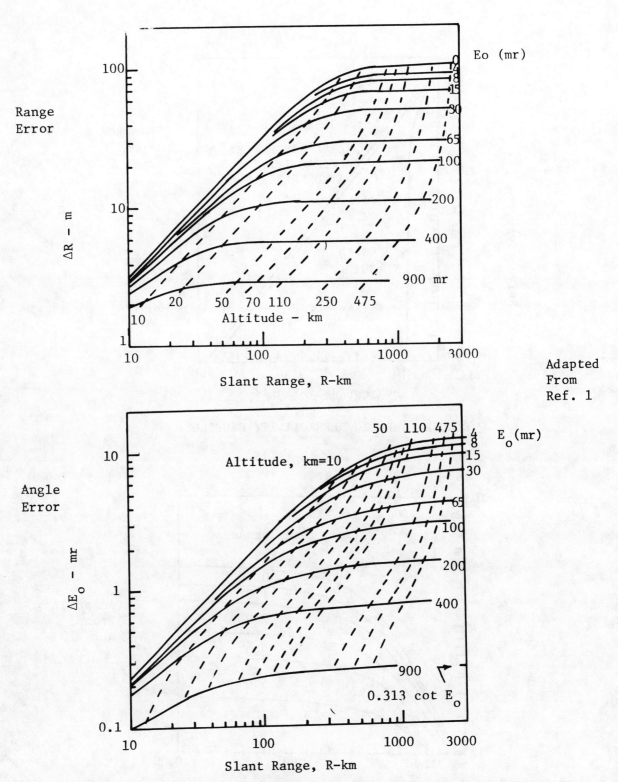

FIGURE 3 RANGE AND ANGLE BIAS ERROR DUE TO REFRACTIVE EFFECTS IN THE CRPL EXPONENTIAL REFERENCE ATMOSPHERE (N_o=313)

Adapted From Ref. 1

3. Hitney, H.V., "Radar Detection Range Under Atmospheric Ducting Conditions", Record of the IEEE, *International Radar Conference*, (April 1975), pp. 241-243.

4. Morgan, L.A., and Ekdahl, C.A., Jr., *Millimeter Wave Propagation*, RADC-TR-66-342. Rome Air Development Center (1966) AD 489 424.

5. Page, P.M., and Williams, M.A., "Preliminary Results of Measurements of Sea Reflectivity and Attenuation Due to Rain at 81 GHz", ASWE Working Paper WP-XAT-7711, Admiralty Surface Weapons Establishment, May 1977.

6. *Report of ARPA/Tri-Service Millimeter Wave Workshop, December 16-18, 1974*, ARPA Report #ARPAT10-75-3, dated January 1975. (Compiled by Applied Physics Laboratory, Johns Hopkins University).

7. Richard, V.W., and Kammerer, J.E., "Rain Backscatter Measurements and Theory at Millimeter Wavelengths", Report No. 1838. USA Ballistic Research Laboratories, Aberdeen Proving Ground, MD (October 1975).

8. Richter, J.H., and Hitney, H.V., "The Effect of the Evaporation Duct on Microwave Propagation", Technical Report No. NELC/TR-1949, (17 April 1975). Naval Electronics Laboratory Center.

9. Straiton, A.W., and Scarpero, D.C., "A Survey of Millimeter Wavelength Radio Propagation Through the Atmosphere, Technical Report TR68-1 (30 November 1968), US Air Force Avionics Laboratory, AD 844 946.

10. Weil, T.A., "Atmospheric Lens Effect, Another Loss for the Radar Range Equation", *IEEE Trans. on Aerospace and Electronic Systems*, AES-9, No.1 (January 1973). pp. 51-54.

11. Zhevakin, S.A., and Naumov, A.P., "Propagation of Centimeter, Millimeter, and Submillimeter Waves in the Earth's Atmosphere", FSTC Translation FSTC-NT-23-18-69. US Army Foreign Science and Technology Center (1969). AD 694 411.

Environmental Effects on Millimeter Radar Performance
F.B. Dyer and N.C. Currie
AGARD Conference Proceedings, CP 245, "Millimeter and Submillimeter Wave Propagation and Circuits," 1978
Pages 2-1 — 2-9
© 1978 by AGARD Conference Proceedings. Reprinted by permission.

SUMMARY

This paper explores the effects of clutter and non-ideal atmospheric conditions on the performance of millimeter radar systems. A series of investigations of radar returns from land, sea, rain, and snow have been performed over the past several years which were designed to define the potential utility of millimeter waves to selected system applications. The data and supporting analyses include results for all of the commonly used radar operating frequencies between 9.5 GHz and 95 GHz. Average values were measured to develop frequency dependencies, and considerations were made of those factors such as polarization, amplitude distributions, and spectra which significantly impact the application of modern signal processing techniques to millimeter radar.

1. INTRODUCTION

The objective of any measurement program lies in the application of data to some real-world problem; the key to this aspect of the program lies in the analysis and presentation of results. Results might include presentation of such average quantities such as radar cross-section, radar cross-section per unit area, and attenuation coefficient. These parameters can be coupled with meteorological, temporal, or environmental parameters in the final presentations of results. Fluctuation characteristics such as amplitude distributions, frequency spectra, or correlation functions are also important for many applications; however, considerable care is needed in the interpretation of such data. Finally, there are data which are related to specific signal processing techniques, such as special polarization or frequency processing, which will require their own specific presentations.

2. ENVIRONMENTAL PARAMETERS AFFECTING MILLIMETER RADAR PERFORMANCE

Millimeter radar performance is affected by the environment in a number of ways including attenuation and backscatter from the atmosphere, backscatter from the background such as sea or land clutter, and attenuation due to natural cover such as snow or foliage.

2.1 Atmospheric Effects

Millimeter waves propagating through the atmosphere are affected by their interaction with atmospheric gases, particulate matter, and hydrometeors. These interactions occur through three primary mechanisms: absorption, scattering, and refraction. The effects of these interaction mechanisms on millimeter waves are summarized in Table I.

Absorption and scattering are the result of direct interaction between the millimeter waves and the atmosphere (gases, particulate matter, hydrometeors). The primary gases that affect millimeter waves are water vapor and oxygen. Gaseous absorption varies with millimeter wave frequency, atmospheric pressure, temperature, and absolute humidity.

The primary particulate matters that effect millimeter waves are dust, smog, and smoke particles. Because these particles have dielectric constants much smaller than the dielectric constant of water droplets, their attenuation cross section is small. Thus particulate matter has a negligible effect on millimeter waves.

The primary hydrometeors that affect millimeter waves are rain, cloud droplets, hail, fog, and snow. Of these, rain has the most significant effect. Suspended water droplets and rain cause absorption of millimeter waves that exceeds that caused by the combination of water vapor and oxygen (gaseous absorption in the non-resonant frequency region). Rain absorption is a function of rain rate, drop size, drop size distribution, temperature, frequency, and path length. Cloud droplets and fog attenuate millimeter waves in the same manner as rain, but the amount of attenuation is generally much less than that due to rain. Dry snow produces very small attenuation except at very heavy snow-fall rates.

In a clear, non-precipitating atmosphere, gaseous water vapor and oxygen are the major absorbers of millimeter waves. The attenuation functional dependencies are determined by the atmospheric parameters of temperature and pressure. In the atmosphere, the values of these parameters will depend on altitude and prevailing weather conditions. For oxygen, the pressure parameter depends not only on altitude, but also on the humidity at the altitude being considered. Figure 1 illustrates typical millimeter wave attenuation characteristics due to the combination of oxygen and water vapor for representative relative humidity values at standard sea level altitude.

In a precipitating atmosphere, liquid water is the major absorber of millimeter waves. A precipitating atmosphere comprises rain, fog, snow, and hail. Of these, rain has the most significant effect on millimeter waves. Over the past 30 years, many people working in both radar and meteorology have made calculations attempting to model rain and determine the properties of electromagnetic absorption and scattering by raindrops. Adjustments have been made in the models employed to account for such parameters as drop shape; the dielectric constant, which is a function of temperature and frequency; frequency of incident power; type of rain (drop size distributions; type and location of rain, such as thunderstorms, frontal systems, maritime or orographic environment); and polarization of incident power. One result of these investigations is a general relationship between the rain rate and the absorption and scattering properties. The usual presentation shows both the radar cross-section and attenuation to increase as the rain rate increases.

Absorption of millimeter waves by water droplets is proportional to the size of the droplets. The absorption cross-section of a water droplet is generally expressed as follows:

$$q_a = \frac{\lambda^2}{\pi} \beta^3 \, \text{Im}\left(-\frac{m^2-1}{m^2+2}\right) \qquad (1)$$

where; $\beta = 2\pi r/\lambda$
λ = wavelength of the millimeter wave
r = radius of the droplet
m = complex index of refraction of water
$\text{Im}(\ )$ = imaginary part of $(\ )$

Equation 1 is valid to within 0.5 dB for droplets having a radius less than 0.06 times the wavelength (i.e., 0.49 mm for 35 GHz and 0.18 mm for 95 GHz). Droplets having a radius less than 0.1 mm are considered non-precipitating, and their absorption cross-section is more than 10 times their scattering cross-section at these frequencies. The absorption of millimeter waves by rain in a particular storm, therefore, depends on the nature of the storm itself and consequently on how the rainfall rates and drop sizes are distributed.

Extensive data (CURRIE, N. C., DYER, F. B. and HAYES, R. D., 1975; GODARD, S. L., 1970; SEMPLAK, R. A. and TURRIN, R. H., 1969; IPPOLITO, L. J., 1970; KERR, E., ed., 1951; RICHARD, V. W., 1976) have been generated concerning rain attenuation of millimeter waves and radar performance in rain. In general, a one way attenuation of 1 dB km^{-1} would reduce the radar range of a system from 10 nautical miles in clear air to approximately 5.4 nautical miles (LO, L. T., FANNIN, B. M. and STRAITON, A. W., 1975). Typical values of rain attenuation range from 4 dB km^{-1} at 35 GHz and 9.1 dB km^{-1} at 95 GHz for a 16 mm hr^{-1} (heavy) rainfall to 0.24 dB km^{-1} at 35 GHz and 0.95 dB km^{-1} at 95 GHz for 1 mm hr^{-1} (very light) rainfall (RICHARD, V. W. and KAMMERER, J. E., 1975).

Typical values of fog attenuation at 35 and 95 GHz, respectively, range from 0.11 and 0.47 dB km^{-1} at 0°C to 0.034 and 0.22 dB km^{-1} at 40°C for a fog with 0.1 gm m^{-3} liquid water (150 - 250 m optical visibility (LO, L. T., FANNIN, B. M. and STRAITON, A. W., 1975).

Attenuation in the millimeter windows by clouds of various types has been thoroughly treated by Lo et al (ALTSHULER, E. E. and EBEOGLU, D. B., 1976). They report average attenuation of 0.18 and 0.61 dB in stratocumulus and 0.2 and 0.34 dB in cumulus at 35 and 95 GHz, respectively.

Limited calculations and measurements of attenuation in falling snow indicate that attenuation by dry snow is much less than that of rain having an equivalent water content, except at the shortest wavelengths in the millimeter region. Attenuation by wet snow is more significant, particularly in the melting region, where water-coated snowflakes yield backscattering 5 to 10 dB greater than that of the rain below. Backscatter measurements by McCormick and Hendry (MCCORMICK, G. C. and HENDRY, A., 1975) indicate that the radar "bright band" at the melting level may be less distinct at 16.5 GHz than at lower microwave frequencies. The absorption and backscatter characteristics of the melting level have not been documented at higher frequencies. Absorption and scattering when hail is present is comparable to or larger than that due to rain alone. The magnitude depends strongly on the sizes and shapes of the hailstones and on the presence and thickness of a water coating.

Particulate matter in the atmosphere has little effect on millimeter waves. Dust, smog, and smoke particles have dielectric constants that are small compared to that of water droplets, hence they have negligible effect on millimeter-wave propagation. Ice crystals in clouds cause attenuation particularly near the melting level; however, few data have been reported.

Atmospheric scattering is associated with atmospheric absorption; both are caused by the interaction of electromagnetic waves with atmospheric matter (i.e., gases, hydrometeors, and particulate matter. Scattering and absorption both cause attenuation of an incident electromagnetic wave, but the mechanisms differ. Absorption is a quantum effect, but scattering can be explained in terms of the wave energy loss due to scattering is a function of wavelength and particle size.

Calculations on the backscattered radiation from fog and clouds reveal that the energy returned to the receiver as noise is more than two orders of magnitude smaller than from rain (LO, L. T., FANNIN, B. M. and STRAITON, A. W., 1975; LUKES, G. D., 1968). Hence, backscatter range reduction from fog and clouds can be ignored.

A study was conducted by EES and the Ballistic Research Laboratories, Aberdeen, Maryland to measure millimeter rain backscatter (CURRIE, N. C., DYER, F. B. and HAYES, R. D., 1975; RICHARD, V. W. and KAMMERER, J. E., 1975). Simultaneous measurements were made of rain backscatter at 10, 35, 70 and 95 GHz for vertical and circular polarizations. The basic results are indicated in Tables 2 and 3. As can be seen from the Tables, the polarization of the incident wave is important in terms of the degree of backscatter due to the differential scattering by nonspherical rain drops. The circularly polarized signals were seen to be 10 to 15 dB lower in backscatter magnitude as was the rain rate. In fact, variations of ±5 dB were obtained in the data when plotted as a function of rain rate without considering drop size.

2.2   Background Effects

Radar performance can be limited by interference from the background near a target of interest in addition to atmospheric effects. The interfering area can include snow-covered ground, trees, grassy fields, and the sea surface. Measurements have been performed by EES to characterize the reflectivity of these types of background over the last several years and the results are summarized in Tables 2 through 5 (DYER, F. B., CURRIE, N. C. and APPLEGATE, M. S., 1977). Tables 2 and 3 give average values of $\sigma°$ for land and sea backscatter respectively while Tables 4 and 5 give standard deviations for the land and sea data assuming log-normal distributions. (Clutter is generally log-normal for the small antenna spot sizes encountered at millimeter waves.) Table 4 shows that the average reflectivity ($\sigma°$) of trees, grass, and snow increases with decreasing wavelength and increasing depression angle. It also shows that crusted snow (that is, snow which has melted and refrozen) gives the highest return of the three types of clutter listed in the table at 8.6 mm and below while at longer wavelengths (3.2 cm) dry trees have the highest reflectivity. Table 5 shows that the trend of increasing reflectivity with decreasing wavelength also holds for sea clutter although the effect is not nearly as pronounced as for land clutter, and for higher sea states the reflectivity at 3 mm wavelength is less than that at 3 cm wavelength. The values for $\sigma°$ do correlate strongly with wave height and to a lesser degree with wind speed. Tables 6 and 7 give typical values of standard deviations for various types of clutter. Briefly, for land clutter, the standard deviation increases with decreasing wavelength. For sea clutter, the standard deviation decreases with decreasing wavelength.

2.3   Ground-Cover Attenuation

In addition to backscattering energy which interferes with detection of a target, certain types of ground cover such as foliage or snow can attenuate millimeter wave energy so that a target is masked by this cover. Measurement programs were conducted to measure the attenuation properties of foliage and snow at millimeter wavelengths and the results are summarized in Tables 8 and 9. (CURRIE, N. C., DYER, F. B. and EWELL, G. W., 1976; CURRIE, N. C., DYER, F. B. AND MARTIN, E. E., 1976; CURRIE, N. C. AND MORTON, T. M., 1977). As can be seen from the figures, very little penetration would be achieved through either foliated trees or through snow at 35 GHz and above.

3.   CLUTTER EFFECTS ON SYSTEM PARAMETER CHOICE

The ultimate weight and size of a radar will depend very strongly on the frequency of operation, as both component size and power consumption vary inversely with frequency (for a given upper bound of performance). The choice of operating frequency depends on a number of factors, including allowable antenna aperture, maximum range, weather performance, anticipated target-to-clutter ratios, and the anticipated processing gains which may be obtained with that choice. The key factor in system sizing considered here is the behavior of the target-to-clutter ratio as a function of frequency.

Although a limited quantity of comparative data is available in the literature on the cross-section of various man-made targets (HAYES, R. D. and DYER, F. B., 1973) a number of independent investigations have been made by EES of target returns at the various frequencies of interest here. The general tendency seen in such data is for the effective cross-sections of vehicles, buildings, and personnel to be approximately independent of frequency above 9 GHz. The assumption used below is that target cross-section is not frequency dependent.

Normally the initial question to be answered in any system study is the effect of frequency scaling on maximum detection range. The overall consideration includes, of course, any physical constraints of the platform, application, etc. It is necessary to establish an acceptable set of ground rules for such a comparison; however, the detailed treatment of such a comparison is beyond the scope of this paper. General considerations for guiding the comparison of frequency effects of system parameters can be found in several radar handbooks (SKOLNIK, M. I. (ed.) 1970; NATHANSON, E. E., 1969). Given a knowledge of available transmitter power, receiver noise figure, and losses as a function of frequency it is possible to make order-of-magnitude performance comparisons which will bound the anticipated system capability.

The frequency dependence of clutter on radar system performance can be illustrated by defining three radar systems differing only in frequency as shown by Table 10. Figures 2 through 4 show the equivalent average signal-to-clutter ratios as functions of range for land clutter, sea clutter, and rain clutter assuming a +10 dBsm constant target, a -10 dBsm constant target, and a 0 dBsm constant target respectively. (These target sizes are typical of those which would be encountered for each type of environment.) The target-to-clutter ratios indicated are independent of transmitted power (assuming that the signals are not in the noise) and are dependent only on aperture size, frequency, pulse length and reflectivity of the clutter (σ or n).

The problem of radar detection in clutter is further complicated by the statistical nature of the returns from targets and clutter. Investigations described in References HAYES, R. D. and DYER, F. B., 1973; DYER, F. B., GARY, M. J. and EWELL, G. W., 1974; HAYES, R. D., DYER, F. B. and CURRIE, N. C., 1976; and CURRIE, N. C., DYER, F. B. and HAYES, R. D., 1975 suggest that often experimental amplitude distributions behave as log-normal distributions with large standard deviation. Example sets of Receiver Operating Characteristic curves are shown in Figures 5 and 6. While these figures are idealized they serve to illustrate the impact on the detection problem which results when the extension is made from a non-fluctuating target in receiver (Rayleigh) noise to the case of log-normally distributed clutter and targets.

For example, consider the difference in performance due only to the clutter background which is implied in the comparison of Figures 5 and 6. For a false alarm rate of 10^{-6}, the required signal-to-background clutter is similar to those discussed above. Figure 6 also illustrates the greatly reduced possibility of achieving very high probabilities of detection when returns from both target and background are log-normally distributed. If a probability of detection of 80% is required with a false-alarm probability of 10^{-6}, a clutter background of trees (for example), and a moderately fluctuating target, then a target cross section of 20 dBsm or more will be required for the example radars illustrated in Table 10.

4. CONCLUSIONS AND RECOMMENDATIONS

It should be concluded from these discussions that operation in the millimeter wave region offers a number of potential advantages, principally small size and weight, but at the same time, systems analysis is complicated by the relatively small and inconsistent data base which exists for systems operating at these frequencies. This is in spite of the systematic efforts of a number of investigators.

A few comprehensive surveys of available data on reflectivity at millimeter wavelengths have appeared--for example, HAYES, R. D. and DYER, F. B., 1973. In addition, some additional data appear in SKOLNIK, M. I. (ed.), 1970; NATHANSON, E. E., 1969; and LONG, M. W., 1975. Unfortunately, examination of these data indicates that the amount of variety of data decreases strongly as frequency increases above 10 GHz.

Much of the data available at frequencies above 10 GHz suffers from being taken under widely varying conditions, using systems having widely varying parameters, thus making meaningful comparisons of the results difficult. Perhaps the most obvious deficiencies in the data involve the lack of measurements taken using coherent systems or short pulse (less than 20 ns) systems, and the lack of exploration of polarization and frequency as variable parameters. Much of the data which are available are presented as averaged, highly processed information; considerably more useful are more "dynamic" or less processed information such as amplitude distributions, power density spectra, or temporal correlation functions.

The data base for ground clutter needs to be expanded to include: trees at grazing incidence, tree lines and fields, roads, buildings, rivers, and lakes. Some significant variable parameters will include the amount of surface water, wind speed, foliage content, bistatic angle (where appropriate), and specific geometries of the target and the antenna axis.

Similarly, the atmospheric propagation path requires additional data to fully describe backscatter, attenuation, and refraction for paths containing fog, clouds, snow, rain, and ice. Some of the more important parameters for these measurements will include characterization over the entire path length of relative (or absolute) humidity, drop-size distribution, water content, and location of water (i.e., whether water coated ice, ice in fog or ice in cloud).

In particular, recent rain attenuation and backscatter measurements have pointed out the need to attempt simultaneous measurements at 70, 95, 140 and 230 GHz in order to resolve some serious modeling questions, particularly the Mie scattering problem. Of particular interest during these measurements would be detailed characterization of drop-size distributions, dielectric properties of the individual drops, and relative humidity--all throughout the transmission path.

The problem of target characterization, particularly of motorized vehicles, is as significant, and no less difficult, as the problem of characterizing the environment. Particularly for the active radar case, the geometrical sensitivity of the backscattering from the target makes a three-dimensional description of the reflectivity desirable. The influence of surface coatings such as snow, ice, or water on the target return requires detailed attention to the dielectric properties of the coating, surface roughness, and possible influence on specular reflection properties. Possibly the exploitation of polarization or frequency properties of reflectivity would prove useful; however, no valid data base exists in the millimeter region. Bistatic behavior of targets and clutter is also a matter which should receive attention since little or no definitive work has been done on this aspect of the reflectivity problem, especially at the millimeter wavelengths.

5. REFERENCES

ALTSHULER, E. E. and EBBOGLU, D. B., 1976, "Second DoD Workshop on Millimeter Wave Terminal Guidance Systems (Adverse Weather Effects)," RADC-TR-76-9, Rome Air Development Center, Griffiss AFB, New York.

CURRIE, N. C., DYER, F. B. and EWELL, G. W., October 1976, "Characteristics of Snow at Millimeter Wavelengths," Proceedings of the IEEE AP-S International Symposium, Amherst, Massachusetts. 13.

CURRIE, N. C., DYER, F. B. and HAYES, R. D., 1975, "Analysis of Radar Rain Return at Frequencies of 9.375, 35, 70 and 95 GHz," Final Technical Report on Contract DAAA25-73-C-0256, Georgia Institute of Technology, February, AD A007254.

CURRIE, N. C., DYER, F. B. and HAYES, R. D., 1975, "Some Properties of Radar Return at 9.375, 35, 70 and 95 GHz," IEEE International Radar Conference, Arlington, Virginia.

CURRIE, N. C., DYER, F. B. and MARTIN, E. E., October 1976, "Millimeter Foliage Penetration Measurements," Proceedings of the IEEE AP-S International Symposium, Amherst, Massachusetts.

CURRIE, N. C. and MORTON, T. M. January 1977, "Comparison of Winter/Summer Microwave Foliage Penetration Measurements," Engineering Experiment Station, Georgia Institute of Technology, Atlanta, Georgia, Final Report on Task 2 of MIT/LL P. O. No. CK-1069.

DYER, F. B., CURRIE, N. C. and APPLEGATE, M. S., October 1977, "Radar Backscatter From Land, Sea, Rain, and Snow at Millimeter Wavelengths," Digest of the 1977 IEEE International Radar Conference, London, England.

DYER, F. B., GARY, M. J. and EWELL, G. W., 1974, "Radar Sea Clutter at 9.5, 16.5, 35, and 95 GHz," IEEE AP-S International Symposium, Atlanta, Georgia.

GODARD, S. L., July 1970, "Propagation of Centimeter and Millimeter Wavelengths Through Precipitation," IEEE Transactions on Antennas and Propagation, Vol. AP-18, No. 4.

HAYES, R. D. and DYER, F. B., 1973, "Land Clutter Characteristics for Computer Modeling of Fire Control Radar Systems," Engineering Experiment Station, Georgia Institute of Technolougy, Atlanta, Georgia, Technical Report No. 1 on Contract DAAA 25-73-C-0256.

HAYES, R. D., DYER, F. B. and CURRIE, N. C., 1976, "Backscatter From Ground Vegetation at Frequencies Between 10 GHz and 100 GHz," IEEE AP-S International Syposium, Amherst, Massachusetts.

IPPOLITO, L. J., July 1970, "Millimeter Wave Propagation Measurements from the Applications Technology Satellite (ATS-V)," IEEE Transactions on Antennas and Propagation, Vol. AP-18, No. 4.

KERR, P. E., (ed.) Propagation of Short Radio Waves, MIT Radiation Lab Series, 13, New York: McGraw-Hill, 1951.

LO, L. T., FANNIN, B. M. and STRAITON, A. W., November 1975, "Attenuation of 8.6 and 3.2 mm Radio Waves by Clouds," IEEE Transactions on Antennas and Propagation, Vol. AP-23, No. 6.

LONG, M. W., 1975, Radar Reflectivity of Land and Sea, D. C. Heath and Company, Lexington, Massachusetts.

LUKES, G. D., May 1968, "Penetrability of Haze, Fog, Clouds and Precipitation by Radiant Energy Over the Spectral Range 0.1 Micron to 10 Centimeters," The Cneter for Naval Analyses, Study No. 61.

MCCORMICK, G. C. and HENDRY, A. 1975, "Principles for the Radar Determination of the Polarization Properties of Precipitation," Radio Science, Vol. 10, pp. 421 - 434.

NATHANSON, E. E., 1969, Radar Design Principles, McGraw-Hill Book Co., New York.

RICHARD, V. W., June 1976, "Millimeter Wave Radar Applications to Weapons Systems," Ballistic Research Laboratories, Memorandum Report No. 2631.

RICHARD, V. W. and KAMMERER, J. E., October 1975, "Rain Backscatter Measurements and Theory at Millimeter Wavelengths," BRL Report No. 1838, AD B008173L.

SEMPLAK, R. A. and TURRIN, R. H., July-August 1969, "Some Measurements of Attenuation by Rainfall at 18.5 GHz," The Bell System Technical Journal, Vol. 48, Pt. 2.

SKOLNIK, M. I. (ed.), 1970, Radar Handbook, McGraw-Hill Book Co., New York.

TABLE 1. ATMOSPHERIC EFFECTS ON MILLIMETER-WAVES

	Atmospheric Absorption	Interaction Scattering	Mechanism Refraction
Attenuation	x	x	
Depolarization	x	x	
Backscatter		x	
Amplitude Scintillation	x	x	x
Phase Scintillation	x	x	x
Ducting			x

TABLE 2. SUMMARY OF AVERAGE RAIN ATTENUATION AND BACKSCATTER SHARACTERISTICS

Wavelength (mm)	Rain Rate (mm/hr)	Attenuation (dB/km)	Backscatter (m^2/m^3)
32*	1	0.01	1.3×10^{-8}
	5	0.05	2×10^{-7}
	10	0.13	6.2×10^{-7}
	20	0.30	2×10^{-6}
8.6*	1	0.27	1.0×10^{-6}
	5	1.33	1.9×10^{-5}
	10	2.64	6.4×10^{-5}
	20	5.22	2.0×10^{-4}
3.2*	1	1.60	1.5×10^{-5}
	5	4.48	9×10^{-5}
	10	6.98	1.9×10^{-4}
	20	10.88	4.0×10^{-4}
32**	1	0.01	1.1×10^{-9}
	5	0.05	9×10^{-9}
	10	0.13	5×10^{-8}
	20	0.30	1.6×10^{-7}
8.6**	1	0.27	3.5×10^{-8}
	5	1.33	8.5×10^{-7}
	10	2.64	3.5×10^{-6}
	20	5.22	1.3×10^{-5}
3.2**	1	1.60	2.2×10^{-6}
	5	4.48	1.3×10^{-5}
	10	6.98	2.7×10^{-5}
	20	10.88	6×10^{-6}

*Vertical polarization
**Circular polarization

TABLE 3 - TYPICAL STANDARD DEVIATION VALUES FOR RAIN RETURN EXPRESSED IN dB ASSUMING LOG NORMAL DISTRIBUTION

Wavelength (mm)	Rain Rate (mm/hr)	Standard Deviation (dB)	
		Vertical	Circular
32	5	7.5	8.0
	20	8.0	8.5
8.6	5	3.0	2.7
	20	3.2	2.5
3.2	5	3.2	3.0
	20	3.0	2.9

TABLE 4 - SUMMARY OF AVERAGE LAND CLUTTER REFLECTIVITY CHARACTERISTICS

Wavelength (mm)	Depression Angle Degrees	Radar Cross-Section per Unit Area (dB)				
		Wet Trees	Dry Trees	Dry Grass	Wet Snow	Crusted Snow
32	2	-	-32	-36	-45	-
8.6	2	-	-22	-28	-	-
3.2	2	-	-22	-	-	-
32	8	-	-28	-	-35	-
8.6	8	-	-22	-	-20	-10
3.2	8	-	-17	-	-	-
32	15	-	-	-	-30	-
8.6	15	-	-	-	-18	-8
3.2	15	-	-	-	<-10	<-10
8.6	30	-15	-19	-	-	-
3.2	30	-11	-8	-	-	-
8.6	45	-14	-18	-	-	-
3.2	45	-7	-5	-	-	-

TABLE 5 - SUMMARY OF AVERAGE SEA BACKSCATTER CHARACTERISTICS

Wavelength (mm)	Wave Height** (ft)	Wind Speed (mph)	Average Cross-Section per Unit Area (dB)	
			H	V
32	0.6	0	<-70.0	<-73.0
	1.0	0	-69.5	-60.2
	1.6	12	-47.0	-42.0
	3.7	11	-42.8	-36.5
8.6	0.6	0	-63.7	-59.8
	1.0	0	-41.8	-43.6
	1.6	12	-44.1	-37.3
	3.7	11	-	-
3.2	-	4	-38.0	-43.0
	-	7	-35.0	-43.0
	-	12	-35.0	-38.0

*Data were acquired at a 1.4° depression angle.
**Significant wave height.

TABLE 6 - TYPICAL STANDARD DEVIATION VALUES FOR LAND CLUTTER EXPRESSED IN dB ASSUMING LOG NORMAL DISTRIBUTIONS

Wavelength (mm)	Standard Deviation (dB)			
	Dry Trees	Dry Grass	Wet Snow	Crusted Snow
32	4.0	1.5	-	-
8.6	4.4	1.7	1.0	1.0
3.2	6.0	2.0	1.5	1.5

TABLE 7 - TYPICAL STANDARD DEVIATION VALUES FOR SEA CLUTTER EXPRESSED IN dB ASSUMING LOG NORMAL DISTRIBUTIONS

Wavelength (mm)	Wave Height* (ft)	Wind Speed (mph)	Standard Deviation (dB)	
			H	V
32	0.8	0	9.0	5.2
	1.6	12	7.7	4.4
	3.7	11	5.7	5.4
8.6	0.8	0	7.2	4.2
	1.6	12	5.7	4.3

*Significant wave height.

TABLE 8 - MEASURED DECIDUOUS FOLIAGE ATTENUATION FOR MILLIMETER WAVE ENERGY

Season	Frequency (GHz)	One-Way Attenuation α* (dB/m)
Summer	10	2.9
Winter	10	1.4
Summer	16	3.8
Winter	16	1.6
Summer	35	8.0
Summer	95	15.0

*Foliage depth was defined as the sum of the diameters of the branches in the path of the signal.

Millimeter Wave Propagation, Targets and Clutter

TABLE 9 - ONE-WAY SNOW ATTENUATION
FOR MILLIMETER WAVE ENERGY

Snow Condition	Frequency (GHz)	One-Way Attenuation α** (dB/m)
Wet	35	200
Dry	35	50
Wet	95	>300
Dry	95	250

**Dry snow was crusted; wet snow was packed into a snow bank.

TABLE 10. EXAMPLE SETS OF RADAR PARAMETERS

Frequency, f	9.4 GHz	35 GHz	94 GHz
Wavelength, λ	3.2 cm	8.6 mm	3.2 mm
Antenna Diameter	50 cm	50 cm	50 cm
Antenna Gain	27.4 dB	33.1 dB	37.4 dB
Beamwidth	4.1°	1.1°	0.4°
Transmitted Power	100 kW	60 kW	6 kW
Pulse Length	50 nsec	50 nsec	50 nsec
Noise Figure $\overline{NF}_o$	8 dB	8 dB	10 dB
Loss (T/R), L	2 dB	4 dB	6 dB
Polarization	HH	HH	HH
Beam Area (σ°)	0.379R	0.101R	0.037R
Beam Volume (η)	$0.23R^2$	$0.00167R^2$	$0.00022R^2$

R is the slant range.

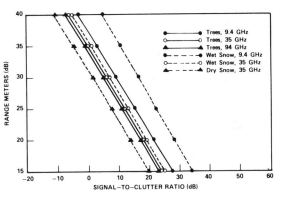

Figure 2. Signal–to–Clutter Ratio for Land Clutter Assuming a +10 dBsm Target.

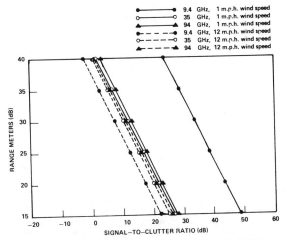

Figure 3. Signal–to–Clutter Ratio for Sea Clutter Assuming a –10 dBsm Target.

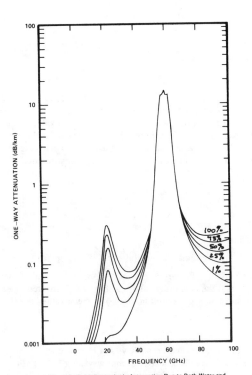

Figure 1. Total Atmospheric Attenuation Due to Both Water and Oxygen for Several Different Relative Humidities.

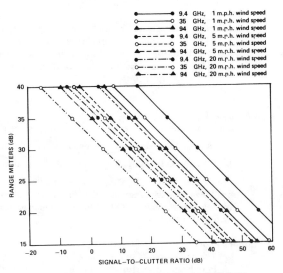

Figure 4. Signal–to–Clutter Ratio for Rain Assuming a 0 dBsm Target.

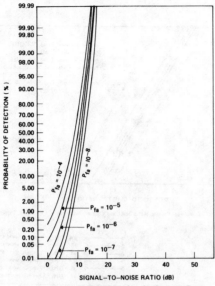
Figure 5. Receiver Operating Curves for a Nonfluctuating Target in a Noise (Rayleigh) Background, Single Pulse Detection.

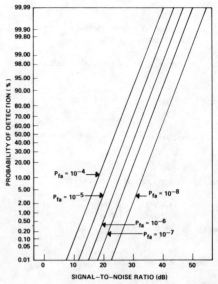
Figure 6. Receiver Operating Curves for a Log–Normal Target and Log–Normal Clutter with Standard Deviations of 4.34 and 6.5 dB Respectively, Single Pulse Detection.

DISCUSSION

E.P.Baars, FRG

Given signal-to-clutter ratios over a range of 30 to 10.000 m as straight lines in log-log scaling (Figs 3 and 4), how are the dimensions of the clutter cell calculated for near field returns?

Author's Reply

The target-to-clutter values shown are intended simply to allow comparison of frequency effects on target-to-clutter ratios. As such, factors such as near field effects and propagation effects were ignored. However, the curves can be used to obtain "ballpark" values for system applications. The tables in the paper contain the accurate values of $\sigma°$ obtained for the various types of clutter and can be used to calculate more accurate estimates of signal-to-clutter ratios. To be of use, however, the standard deviations should be used which are listed in the paper to calculate the clutter distributions. Also, some type of fluctuating target model should be used to achieve realistic estimates of target detectability in clutter.

R.P.Moore, US

Have measurements been made or have you considered:

(1) the structure and inhomogenities in adverse weather atmosphere

(2) the geographical variations of ground clutter affecting both its amplitude and statistical characteristics.

Both of these factors are extremely important in determining effects of environment on performance as they determine the percentage of the time that the systems can operate and the percentage of areas that can be covered as a function of performance. Calculations assuming uniform atmospheres or, as you pointed out, mean values can give descriptive answers.

Author's Reply

(1) The inhomogenity of the atmosphere is known to be a problem. However, we have not been funded to look into the problem. However, during an experiment to measure rain backscatter at mm wavelengths in 1974 (Reference B in the paper) rain drop size distributions were recorded and analyzed for short intervals. It was determined that the drop size distributions varied widely over short time intervals (at least for the type of rain encountered which were mainly localized thunderstorms) and that the reflectivity depended as strongly on the dropsize distribution as on the rain rate.

(2) We are aware of the possible effects on clutter reflectivity characteristics of different geographies. For this reason, present and future measurement programs in which we are participating or expect to participate in include measurement of the reflectivity of snow, land and the sea at several sites both in CONUS and in Europe. Also, attempts are being made to characterize the environment at each site so that reflectivity differences can be explained. These efforts need to be expanded to include other scenarios not included in the present program plans. An example of our interest in this problem is presented in References 5 and 6 in which we define a new approach to the characterisation of foliage path thickness for the purpose of providing a more uniform way of comparing results for foliage penetration measurements obtained in different regions of the world.

Atmospheric Attenuation of Millimeter and Submillimeter Waves

V.J. Falcone, Jr. and L.W. Abreu

AFGL/OPI

ABSTRACT

A computer model of atmospheric attenuation of millimeter and submillimeter waves for frequencies of 1-1000 GHz has been developed for clear, foggy, cloudy and precipitating (rain) atmospheres. The molecular absorption spectra of water vapor, oxygen, ozone, carbon monoxide, and nitrous oxide are calculated under clear conditions by an efficient computer algorithm of the AFGL HITRAN Code, named FASCODE-1. The hydrometeor attenuation of fog, clouds, and rain typical of midlatitude conditions is addended to the FASCODE-1 molecular absorption model. The hydrometer models include 4 fog models (both radiative and advective) with liquid water contents of 0.02 - 0.37 gm^{-3}, 8 cloud types with liquid water contents of 0.01 - 1.4 gm^{-3} and rainfall rates of 0 - 150 mm/hr. The models can accommodate hydrometeor temperatures of $0^{\circ} - 40^{\circ}C$. This operational computer program permits calculation of worldwide atmospheric attenuation of microwave, millimeter and submillimeter radiation for an arbitrary input atmospheric model and geometry (slant range).

INTRODUCTION

The increasing use of microwaves for determination of geophysical parameters, high data rate communication channels, weapons guidance and target acquisition in the past few years has emphasized the requirement for better modeling and more accurate calculations of attenuation through clear and hydrometeor atmospheres. The advantages of millimeter and submillimeter wavelengths are; sensitive to geophysical parameters, provides larger communication bandwidths, secure communications, narrower beamwidths, good angular resolution, and imagery capability.

The disadvantage is atmospheric attenuation caused by gaseous absorption and hydrometeor attenuation.

AFGL has initiated a study to model and calculate atmospheric transmissions and atmospheric attenuation of radio frequency waves in the spectral region of 1-1000 GHz. This research follows in the spirit of the original LOWTRAN[1] by McClatchey, and that is, an operationally directed computer code for user implementation.

The codes are available from National Climatic Center, Federal Building, Asheville, NC 28801 for a service charge.

The models utilized in this study are the following:

a) Clear atmosphere given Valley[2]

b) Fog and Cloud models are from Silverman[3] and the research of the authors.

c) Rain Models are the research of the authors. The rainfall rates considered are from 0.25 mm/hr to 150 mm/hr. The drop size distribution of these rainfall rates are determined from the Marshall-Palmer(M-P) distribution. The models have been chosen to be representative of midlatitude conditions. The computer program will allow each researcher to calculate transmission and/or attenuation for "any" horizontally stratified model which represents real world conditions.

CLEAR ATMOSPHERE

The atmospheric molecular absorption spectra is calculated by a computer efficient algorithm of AFGLs HITRAN Code. This new code was advanced by Clough et al[4]. It is called FASCODE-1, an acronym for Fast Atmospheric Signature Code. The HITRAN code is a compilation of spectral line parameters of the significant atmospheric absorption lines for frequencies from the microwave through the infrared regions of the spectrum. The paper by Clough et al should be consulted for further information about this program and the theory of clear air transmission.

HYDROMETEORS

Fog, clouds and precipitation are the meteorological results of nuclei that began as haze. The only distinction between fog and clouds is their distance from the earth. Similarly the distinction between clouds (or fog) and rain is almost non-existant when they are mixed. Fog and clouds refer to a collection of water drops with diameter less than 100 μm. Rain refers to a collection of water drops with diameter up to 6.5 mm.

It is assumed that the hydrometeors are homogeneous, spherical masses of pure water, with a density slightly less than $1g/m^3$. For fog and clouds the Rayleigh scattering approximation holds, and the attenuation is directly proportional to the liquid water content and inversely proportional to the wavelength considered, see equation 2. On the other hand, rain attenuation requires the full Mie theory calculations.

CALCULATIONS

Clear Atmosphere Transmission/Attenuation

In the frequency range 1-1000 GHz the clear atmosphere progresses from 100% transparent to opaque, as the frequency is increased. This transmission and attenuation is present 100% of the time. Figure 1 was calculated by FASCODE-1.

HYDROMETEORS ATTENUATION

The attenuation of millimeter and submillimeter waves by hydrometeors is appreciable for a signficiant percentage of a year. Consequently estimates of fog, cloud, and rain attenuation are required for the design of various systems. Physically, the removal of energy from an electromagnetic wave is well understood. It is the meteorological statistics that are not well known. This latter fact leads to difficulties in modeling.

In most scattering calculations it is assumed that the shape of hydrometeors is approximately spherical and their temperature is homogeneous. These two assumptions are true for fogs and clouds, but not strictly true for rain. For modeling purposes these assumptions are reasonable for rain.

The absorption and scattering properties of a single dielectric sphere was first advanced by Mie. Fog, clouds, and rain are not composed of a single size drop, but are a composition of a distribution of many sizes. If this distribution of drop size radii is known, then the Mie theory may be used to calculate the exact absorption and scattering properties.

The Mie theory calculations depend upon the wavelength of incident radiation, the size (r) and index of refraction (m) of the particle, and the angular position (θ) around the scatterer. The relative size (x), i.e. ratio of the particle size to wavelength (r/λ), is the physically important parameter. At lower relative size, i.e. x < 1, Mie scattering reduces to Rayleigh scattering. The Mie extinction cross section Q_{ext} and scatter cross section Q_s and absorption cross section Q_a are defined in Deirmendjian [5].

The dielectric behavior of water at radio frequencies has been reported in the literature by many authors. The calculational procedures required to quantify the real and imaginary terms of the index of refraction are such, that there is disagreement amongst the values calculated. In order to avoid these calculational differences, Ray's [6] calculational procedure has been chosen as a standard. Ray's research covers the frequency and temperature range of interest and most importantly is an empirical model.

Meteorologists have studied these drop size distributions in depth and have found them to vary with the type of fog, cloud, and rain, geographical location, and season of the year. The fog and cloud drop size distributions employed in this research are given by Silverman, see Table III.

Rain drop size distributions are the most complex and difficult to analyze. The distributions vary temporally and spacially within a single storm, due to the effects of wind sorting (updrafts), evaporation, accretion, as well as changes in the precipitation process. The droplet distribution even differs in different parts of a single rain cloud, both vertically and horizontally.

The values of liquid water content (M) in Table III are typical clouds. The variability of liquid water content may be as large as a factor of 5 from these stated. Research has shown that the liquid water content of non-precipitating clouds have values from 0.1 g/m^3 - 0.5 g/m^3 whereas precipitating clouds often have L.W.C. greater than 1 g/m^3.

The volume extinction cross section (the integral of the extinction cross section over the distribution of drops) for Rayleigh scatterers is

$$Q_{V_{EXT}} = \int_0^\infty Q_{EXT}\, n(r)\, dr \qquad 1.$$

where n(r) = Deirmendjian modified gamma function. It can be shown that

$$Q_{V_{EXT}} = \frac{6\pi}{\lambda} M\, I_m\left(-\frac{m^2-1}{m^2+2}\right) \qquad 2.$$

where m = complex dielectric constant.
Thus the fog and cloud attenuation is due to the liquid water content of clouds.

For rain, the Marshall-Palmer (M-P) dropsize distribution was chosen because it relates the meteorologically measured parameter R (rain rate mm/hr) to the dropsize distribution

$$n(r) = 16000 \exp[-82R^{-0.21} r] \qquad 3.$$

Adjustment of the amplitude 16000 and exponent $-82R^{-0.21}$ allow the M-P distribution to fit measured data more accurately. The full Mie equations are required for the calculation of equation 1. To perform these calculations for a system would be too time consuming from a computer standpoint. To alleviate this impracticability we have chosen to set up interpolation tables. The attenuation table was calculated in units of dB/km at 20°C temperature for wavelengths from 15. to 0.03 cm and for rainrates of 0.25 to 150 mm/hr. A temperature correction table for the same wavelengths and rainrates was calculated for the temperature range 0-40°C. Abreviated versions of these tables are illustrated, see Table I and Table II. An interpolation and extrapolation

Millimeter Wave Propagation, Targets and Clutter

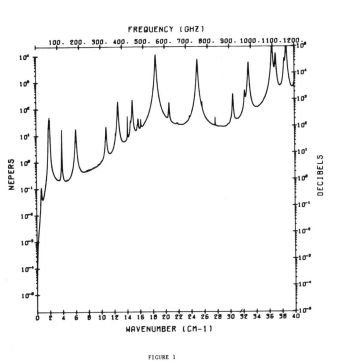

FIGURE 1

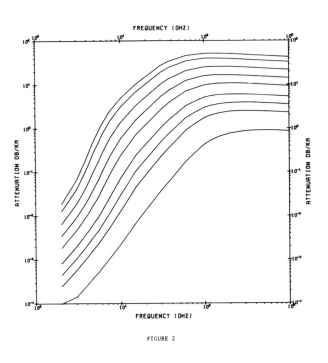

FIGURE 2

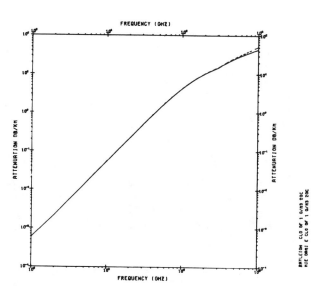

FIGURE 3

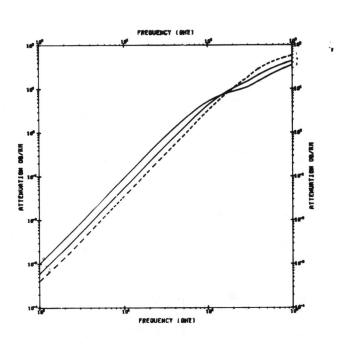

TABLE I ATTENUATION (db/km) at 20°C

PRECIPITATION RATE (MM/HR)	WAVELENGTH (CM)														
	.03	.05	.1	.15	.2	.25	.3	.5	.8	1.0	2.0	3.0	5.0	6.0	15.0
0.25	.867	.900	.874	.773	.656	.539	.434	.179	.0634	.0381	.685	.231	.657	.434	.631
1.15	2.31	2.43	2.51	2.41	2.22	1.99	1.74	.919	.374	.232	.0449 $\times 10^{-2}$	.0134 $\times 10^{-2}$	.304 $\times 10^{-3}$	.191 $\times 10^{-3}$	.249 $\times 10^{-4}$
2.50	3.51	3.71	3.90	3.83	3.63	3.34	3.01	1.77	.783	.497	.104	.0311	.618 $\times 10^{-2}$	.374 $\times 10^{-2}$	.454 $\times 10^{-3}$
5.00	5.35	5.65	6.01	6.02	5.83	5.49	5.08	3.29	1.60	1.05	.239	.0750	.0132	.758 $\times 10^{-2}$	.829 $\times 10^{-3}$
12.50	9.35	9.86	10.59	10.80	10.69	10.33	9.81	7.13	3.94	2.70	.698	.245	.0399	.0209 $\times 10^{-2}$	.186 $\times 10^{-3}$
25.00	14.27	15.03	16.18	16.67	16.70	16.38	15.81	12.36	7.51	5.38	1.52	.591	.100	.0488	.348 $\times 10^{-2}$
50.00	21.78	22.90	24.68	25.61	25.89	25.70	25.14	20.89	13.87	10.37	3.23	1.38	.265	.124	.661 $\times 10^{-2}$
100.00	33.22	34.85	37.55	39.18	39.84	39.96	39.50	34.54	24.83	19.40	6.66	3.09	.706	.338	.0128 $\times 10^{-2}$
150.00	42.48	44.51	47.93	50.16	51.10	51.54	51.22	45.94	34.46	27.59	10.06	4.86	1.24	.613	.0190

TABLE II

TEMPERATURE CORRECTION FACTOR
FOR TABLE I
FOR REPRESENTATIVE RAINS

PRECIPITATION RATE MM/HR	WAVELENGTH CM	0°	10°	20°	30°	40°
0.25	.03	1.0	1.0	1.0	1.0	1.0
	.1	0.99	0.99	1.0	1.01	1.02
	.5	1.02	1.01	1.0	1.0	1.0
	1.25	1.09	1.02	1.0	1.0	0.99
	3.2	1.55	1.25	1.0	0.81	0.65
	10.0	1.72	1.29	1.0	0.79	0.64
2.5	.03	1.0	1.0	1.0	1.0	1.0
	.1	1.0	1.0	1.0	1.0	1.01
	.5	1.01	1.01	1.0	0.99	0.98
	1.25	0.95	0.96	1.0	1.05	1.10
	3.2	1.28	1.14	1.0	0.86	0.72
	10.0	1.73	1.30	1.0	0.79	0.64
12.5	.03	1.0	1.0	1.0	1.0	1.0
	.1	1.0	1.0	1.0	1.0	1.01
	.5	1.02	1.01	1.0	0.99	0.97
	1.25	0.96	0.97	1.0	1.04	1.07
	3.2	1.04	1.03	1.0	0.95	0.88
	10.0	1.74	1.30	1.0	0.79	0.63
50.0	.03	1.0	1.0	1.0	1.0	1.0
	.1	1.0	1.0	1.0	1.0	1.01
	.5	1.02	1.01	1.0	0.98	0.97
	1.25	0.99	0.99	1.0	1.02	1.04
	3.2	0.91	0.96	1.0	1.01	1.01
	10.0	1.75	1.31	1.0	0.78	0.62
150.0	.03	1.0	1.0	1.0	1.0	1.0
	.1	1.0	1.0	1.0	1.0	1.01
	.5	1.03	1.01	1.0	0.98	0.97
	1.25	1.01	1.0	1.0	1.0	1.01
	3.2	0.88	0.95	1.0	1.04	1.06
	10.0	1.72	1.31	1.0	0.78	0.62

scheme allows all wavelengths and temperature ranges to be calculated. The results are shown in figures 2, 3 and 4.

Figure 2 is the attenuation (dB/km) vs. frequency (GHz) for rainrates 0.25, 1.25, 2.50, 5.00, 12.5, 25.0, 50.0, 100. and 150. mm/km.

Figure 3 is the attenuation (dB/km) for a 1 km thick cloud with a L.W.C. of 1 g/m^3 at 20°C. The dash line is the full Mie theory calculation, the solid line the Rayleigh.

Figure 4. Temperature dependence of attenuation for a 1 km thick cloud with L.W.C. of 1 g/m^3 for temperatures 0°, 20°, and 40°C, the top, middle and bottom lines respectively (left side of figure).

MODELS

Table III gives typical fog and cloud models which are representative of midlatitude conditions. These "typical" models are not average values, e.g. model 5, cummulus has a LWC of 1 g/m^3. This type of cloud normally has LWC of 0.5-1g/m^3 with values as high as 4g/m^3 depending on geographical location (Wyoming vs. Florida).

Rain modeling requires knowledge of rainfall rate and drop size distribution with altitude. In addition the models have to be consistent with respect to LWC i.e. the cloud associated with a specific rainrate, must possess a LWC the minimum value of which is the same magnitude as that LWC of the precipitation. For modeling clouds/rain the LWC of the cloud is $M = 0.05R$ and for the precipitation $M = 0.072R^{0.88}$. These equations have been used in Table IV.

Models I-IV were constructed from the following rules.

a) LWC of rain from ground to cloud base is determined by the M-P formula $M(g/m^3) = 0.072R^{.88}$ (mm/hr).

b) LWC of cloud is expressed as $M = 0.05R$; for R 8mm/hr the cloud may touch the earth (Models I and II).

c) At cloud base the sum of a) plus b) is multiplied by a factor 3. Subsequent distances by 4, 2, and 0.6.

TABLE III

Cloud Type	(M^{-3})	Thickness (m) Bottom	Top
Heavy Fog 1	0.37	0	150
Heavy Fog 2	0.19	0	150
Moderate Fog 1	0.06	0	75
Moderate Fog 2	0.02	0	75
Cumulus	1.00	660	2700
Altostratus	0.41	2400	2900
Stratocumulus	0.55	660	1320
Nimbostratus	0.61	160	1000
Stratus	0.42	160	660
Stratus	0.29	330	1000
Stratus - Stratocumulus	0.15	660	2000
Stratocumulus	0.30	160	660
Nimbostratus	0.65	660	2700
Cumulus-Cumulus Congestus	0.57	660	3400

TABLE IV

TYPICAL RAIN/CLOUD MODELS

Name	Altitude Base (km)	Top (km)	M(g/m^3)
Drizzle I R 1.25 mm/hr	0.	0.5	0.55
	0.5	1.0	1.1
	1.0	1.5	0.25
Steady Rain 5mm/hr Model II	0.	0.5	0.55
	0.5	1.0	1.65
	1.0	2.0	2.20
	2.0	3.0	1.10
	3.0	4.0	0.30
Steady Rain 12.5 mm/hr Model III	0	1.0	0.66
	1.0	2.0	3.87
	2.0	3.0	5.16
	3.0	4.0	2.58
	4.0	5.0	0.77
Summer Cumulus 15 mm/hr Model IV	0.	2.0	0.8
	2.0	3.0	4.65
	3.0	4.0	6.2
	4.0	5.0	3.1
	5.0	6.0	0.93

REFERENCES

1. McClatchey R.A., Selby J. "Atmospheric Transmittance from 0.25 to 28.5 μm: Computer Code" AFCRL-TR-75-0255 May 75.

2. Valley S.L., Ed (1965) "Handbook of Geophysics and Space Environments" AFCRL

3. Silverman B.A., Sprague E.D., "Airborne Measurements of In-Cloud Visibility" Second National Conference on Weather Modification of the American Meteorological Society Apr. 6-9, 1970, Santa Barbara, California

4. Clough et al. AFGL Technical Report. To be published.

5. Deirmendjian D. "Electromagnetic Scattering on Polydispersions", Elseviev Publishing Co. 1969.

6. Ray P.S., "Broadband Refractive Indices of Ice and Water" App. Optics, Vol. II, No. 8. pg. 1836-1844, 1972.

Atmospheric Turbulence Effects on Millimeter Wave Propagation

R.W. McMillan, J.C. Wiltse, and D.E. Snider

ABSTRACT

The effects of atmospheric turbulence on the propagation of optical signals have been thoroughly analyzed by other workers; in the case of millimeter microwave radiation less work has been done, and theory predicts a strong dependence of the scintillation amplitude and angle of arrival variations on the humidity structure parameter C_q in addition to the temperature structure parameter C_T. This paper extends the work of several Russian authors to the point of calculating both intensity and angle of arrival fluctuations in the atmospheric windows at 94 and 140 GHz. The calculated results are compared to experiment for both frequencies, and reasonably good agreement is obtained in both cases.

1. INTRODUCTION

Scintillation of electromagnetic energy traversing the turbulent atmosphere is caused by refractive index inhomogeneities in the path that cause phase shifts, giving rise to selective reinforcement or degradation of the energy across the beam. The resulting energy distribution is log normal [1], characterized by a variance σ_ℓ^2 that is a function of the degree of atmospheric turbulence. These inhomogeneities also cause angle of arrival fluctuations, depolarization, frequency shift, and thermal blooming; although the latter three effects are thought to be of minor importance in the millimeter spectral region.

Most of the original work on atmospheric turbulence was done in Russia by Chernov [2] and Tatarski [1], who treated mainly optical fluctuations and neglected the effects of absorption on the fluctuation intensity. This approach has worked well for optical wavelengths, as attested by the large number of turbulence papers which show reasonable agreement between theory and experiment.

More recently, several other Russian workers have examined the problem of millimeter and submillimeter wave fluctuations; which requires that absorption by atmospheric constituents, mainly water vapor, be considered. This approach was apparently first taken by Izyumov [3] who solved the wave equation using a complex index of refraction to account for absorption and thus obtained expressions for amplitude and phase fluctuations valid for mm propagation. This work was refined by Gurvich [4] and by Armand, Izyumov and Sokolov [5], who obtained the reasonably tractable expressions used for calculations in this paper.

2. MILLIMETER WAVE TURBULENCE THEORY

2.1 Intensity Fluctuations

The method used for determining the magnitudes of the amplitude and phase variations of an electromagnetic wave propagating through the turbulent atmosphere consists of solving the scalar wave equation

$$\nabla^2 \psi + k^2 N^2 \psi = 0, \quad (1)$$

under the conditions

$$n = n_0 + \mu$$
$$m = m_0 + \nu \quad (2)$$

where ψ is the wave function, k is the wave number, and N is the index of refraction. To account for index fluctuations, the form of this index is taken to be that of Equation (2), where n and m are the real and imaginary parts of the refractive index, n_0 and m_0 are the mean values of these parts, and μ and ν are the fluctuating parts.

Equation (1) is solved by the method of smooth perturbations which is discussed in detail in Tatarski [1]. The results of solving this equation are the autocorrelation functions R_χ and R_ϕ of the fluctuations of

---

*This work was supported by the U. S. Army Atmospheric Sciences Laboratory and by Battelle Columbus Laboratories under Contract No. DAAG29-76-D-0100.

the amplitude and phase of the wave, as given below:

$$\begin{pmatrix} R_\chi(a,L) \\ R_\phi(a,L) \end{pmatrix} = 2\pi^2 k^2 L \int_0^\infty J_0(qa) \left\{ \phi_\mu(q) \left(1 \mp \frac{k}{q^2L} \sin\frac{q^2L}{k}\right) \right.$$
$$\left. \mp \frac{4k}{q^2L} \phi_{\mu\nu}(q) \sin^2\frac{q^2L}{2k} + \phi_\nu(q) \left(1 \pm \frac{k}{q^2L} \sin\frac{q^2L}{k}\right) \right\} qdq, \quad (3)$$

where the upper sign gives amplitude fluctuations and the lower sign gives phase fluctuations. Previously undefined parameters appearing in these equations are:

- a = correlation distance
- L = transmission path length
- q = wavenumber, the integrations are carried out over all wavenumbers
- J_0 = Bessel function of the first kind of order zero
- $\phi_\mu, \phi_{\mu\nu}, \phi_\nu$ = spectra of fluctuations of real and imaginary parts of index and their cross correlation.

Armand, et al [5] state that these spectra are related to the spectra of the temperature and humidity fluctuations by the relations

$$\phi_\mu(q) = \left(\frac{\partial\mu}{\partial T}\right)^2 \phi_T(q) + \left(\frac{\partial\mu}{\partial\rho}\right)^2 \phi_\rho(q)$$
$$\phi_\nu(q) = \left(\frac{\partial\nu}{\partial T}\right)^2 \phi_T(q) + \left(\frac{\partial\nu}{\partial\rho}\right)^2 \phi_\rho(q) \quad (4)$$
$$\phi_{\mu\nu}(q) = \left(\frac{\partial\mu}{\partial T}\right)\left(\frac{\partial\nu}{\partial T}\right) \phi_T(q) + \left(\frac{\partial\mu}{\partial\rho}\right)\left(\frac{\partial\nu}{\partial\rho}\right) \phi_\rho(q),$$

where ρ is absolute humidity and T is absolute temperature.

Values of μ and ν suitable for use in Equation (4) have been given by Gurvich [4], who states that:

$$\mu = \left(K_1 \frac{P}{T} + K_2 \frac{e}{T} + K_3(\lambda) \frac{e}{T^2}\right) \times 10^{-6}$$
$$\nu = \gamma(P_0, T_0, e_0) \frac{e}{e_0} \frac{T_0^2}{T^2} \frac{P}{P_0} \frac{\lambda}{2\pi} \times 10^{-6}, \quad (5)$$

where previously undefined parameters are defined below.

- K_1 = 78°K/mb
- K_2 = 72°K/mb
- K_3 = 3.7 × 10^5 (°K)^2/mb (only weakly dependent on λ)
- P = atmospheric pressure in mb
- e = partial pressure of water vapor in mb
- λ = wavelength of transmitted radiation
- γ = absorption coefficient in neper/km
- e_0, P_0, T_0 = stationary values of e, P, T.

The forms of the spectral distributions of fluctuations of temperature and humidity $\phi_T(q)$ and $\phi_\rho(q)$ are taken from Gurvich [4], who gives

$$\phi(q) = \frac{5}{36} \left(\frac{2}{\pi}\right)^{2/3} L_0^{11/3} \frac{\Gamma(5/6)}{\Gamma(2/3)} C^2 (1 + q^2 L_0^2)^{-11/6}. \quad (6)$$

The factor C may be either C_T or C_ρ which are the temperature and humidity structure parameters, respectively. However, it will be seen that the integrals corresponding to Equation (3) for angle of arrival calculations do not converge when Equation (6) is used. For this reason a Gaussian "tail" of the form [5]

$$\phi'(q) = \frac{5}{36} \left(\frac{2}{\pi}\right)^{2/3} L_0^{11/3} \frac{\Gamma(5/6)}{\Gamma(2/3)} C^2 e^{-q^2 L_0^2/4} \quad (7)$$

was added to ensure convergence. Since Equation (7) is also a legitimate choice for the fluctuation spectrum, no loss of rigor is inherent in this approach.

Now consider again Equation (3). The log amplitude variance of the fluctuations is related to the autocorrelation function $R_\chi(a,L)$ by the equation

$$\sigma_\chi^2 = R_\chi(0,L).$$

Since $J_0(0) = 1$, the Bessel functions in Equation (3) vanish and the log amplitude variance may be written as

$$\sigma_\chi^2 = \frac{5}{18} 2^{2/3} L_0^{11/3} \pi^{1/2} k^2 L \frac{\Gamma(5/6)}{\Gamma(2/3)}$$
$$\cdot \left\{ I_1 \left[\left(\frac{\partial\mu}{\partial T}\right)^2 C_T^2 + \left(\frac{\partial\mu}{\partial\rho}\right)^2 C_\rho^2\right] - 4I_2 \left[\left(\frac{\partial\mu}{\partial T}\right)\left(\frac{\partial\nu}{\partial T}\right) C_T^2 \right.\right. \quad (8)$$
$$\left.\left. + \left(\frac{\partial\mu}{\partial\rho}\right)\left(\frac{\partial\nu}{\partial\rho}\right) C_\rho^2\right] + I_3 \left[\left(\frac{\partial\nu}{\partial T}\right)^2 C_T^2 + \left(\frac{\partial\nu}{\partial\rho}\right)^2 C_\rho^2\right] \right\},$$

where the I's are the integrals

$$I_1 = \int_0^\infty \frac{1}{(1+q^2L_0^2)^{11/6}} \left(1 - \frac{k}{q^2L}\sin\frac{q^2L}{k}\right) qdq$$
$$I_2 = \int_0^\infty \frac{1}{(1+q^2L_0^2)^{11/6}} \frac{k}{q^2L} \sin^2\frac{q^2L}{2k} qdq \quad (9)$$
$$I_3 = \int_0^\infty \frac{1}{(1+q^2L_0^2)^{11/6}} \left(1 + \frac{k}{q^2L}\sin\frac{q^2L}{k}\right) qdq,$$

and where Equations (3), (4) and (6) are used. These integrations may be performed numerically using parameters measured in field experiments to obtain results for comparison. These comparisons will be discussed in a later section.

It should be noted that these results are obtained subject to the inequality $L \ll \ell_0^4/\lambda^3$, where ℓ_0 is inner scale dimension and λ is wavelength. This criterion may not be well met for the longer millimeter waves, but the results discussed in Section 3 indicate that this limitation is not severe.

2.2 Angle of Arrival Fluctuations

The angle of arrival of an electromagnetic wave may be defined as the normal to the wavefront at any point. Fluctuations in angle of arrival are caused by refractive index inhomogeneities that cause phase shifts resulting in constructive and destructive interference across the wavefront of the beam, causing the localized angle of the wavefront normal to change relative to the line of sight to the transmitter.

Using the above definition, Strohbehn and Clifford [6] have derived an expression for the angle of arrival of an electromagnetic wave in turbulence based on the phase correlation function $R_\phi(a,L)$ given by the lower signs of Equation (3). Consider a wavefront propagating in the direction $\hat{n}$ which is nominally in the x-direction. The angles with the x-axis and with the x-y plane are θ and ϕ respectively. Now consider the placement of two receivers in the plane $x = L$. The angle α is the angle which the line between these receivers makes with the y-axis, and a is the distance between them. Using this geometry, Strohbehn and Clifford show in a very straightforward way that the correlation functions between these two receivers for angle fluctuations of θ and ϕ are, respectively

$$R_\theta(a,L) = -\frac{1}{k^2} R_\phi''(a,L)\cos^2\alpha - \frac{1}{k^2 a} R_\phi'(a,L)\sin^2\alpha ,$$

$$R_\gamma(a,L) = -\frac{1}{k^2} R_\phi''(a,L)\sin^2\alpha - \frac{1}{k^2 a} R_\phi'(a,L)\cos^2\alpha ,$$

(10)

where the primes denote diffrentiation with respect to a. The only a-dependent term in the expression for $R_\phi(a,L)$ is the Bessel function $J_o(qa)$. Using the rules for differentiation of Bessel functions and assuming a point receiver which implies that $a = 0$ gives

$$R_\beta(0,L) = \sigma_\beta^2 = R_\theta(0,L) + R_\gamma(0,L) = -\frac{1}{k^2} R_\phi''(0,L)$$

(11)

Substituting this result into Equation (3) gives

$$R_\beta(0,L) = \pi^2 L \int_0^\infty \left[\phi_\mu(q)\left(1 + \frac{k}{q^2 L}\sin\frac{q^2 L}{k}\right) + \frac{4k}{q^2 L}\phi_{\mu\nu}(q)\sin^2\frac{q^2 L}{2k} \right.$$
$$\left. + \phi_\nu(q)\left(1 - \frac{k}{q^2 L}\sin\frac{q^2 L}{k}\right)\right] q^3 dq .$$

(12)

Using this result, essentially all that must be done in evaluating the angle of arrival variance is to perform the integrals of Equations (9) with q replaced by q^3, as a result of the differentiation with respect to a. These integrals will be evaluated numerically and compared to measured results of other authors in the next section.

3. COMPARISON OF MEASURED AND CALCULATED RESULTS

In the fall of 1978, millimeter wave propagation measurements were made at both 94 and 140 GHz at White Sands Missile Range, New Mexico over a 2 km path. The index of refraction structure parameter C_n was also measured, together with other pertinent meteorological parameters. Strong intensity fluctuations were observed at both frequencies during these tests. A typical calibration run showing fluctuations at 94 GHz is shown in Figure 1.

Unfortunately, equipment was not available during these tests for measurement of the absolute humidity structure parameter C_ρ, so that it was necessary to assume a value for this important quantity before comparisons between theory and experiment could be made. The value assumed was $C_\rho = 1.45 \times 10^{-4}$ m$^{-1/3}$ g/m^3 which was chosen to give good agreement with the level of fluctuation measured during the initial series of propagation measurements at 94 GHz. It will be seen that this choice gives only fair agreement for the other 94 GHz measurements but good agreement for the 140 GHz measurements. Table I summarizes the parameters used in calculations of intensity and angle of arrival fluctuations at both 94 and 140 GHz. The values of C_T used in these calculations were obtained from the measured values of C_n^2 by using the equation

$$C_T = \frac{C_n T^2}{79P} \times 10^{-6} .$$

(13)

Using the parameters defined above, and the approach discussed in the last section, peak-to-peak intensity fluctuations as a function of C_T were calculated for both 94 and 140 GHz. These results are shown in Figures 2 and 3 respectively in which the circles represent measured data and the solid curve shows calculated results. Similarly, angle of arrival fluctuations were calculated, and the results are shown in Figure 4. Since these fluctuations were not measured, the calculations for 140 GHz are compared to the results of Andreyev et al [7], who measured 150 GHz fluctuations over a 5 km path, and obtained peak-to-peak levels of 0.49 mrad. This value compares favorably with the range of 0.34 to 0.53 mrad shown in Figure 4. Unfortunately, the conditions under which these measurements were made are not specified. Furthermore, no 94 GHz data are available, although Figure 4 shows that the frequency dependence is not strong.

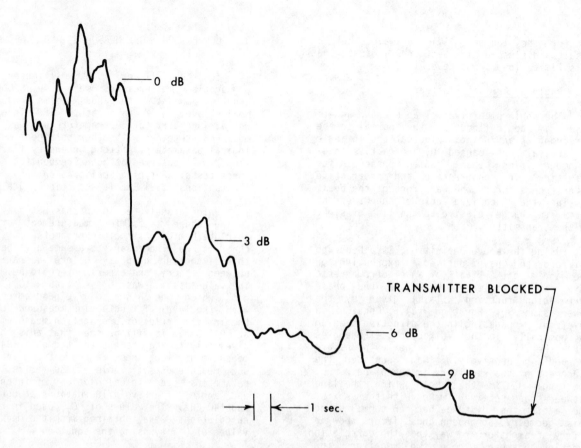

Figure 1. Fluctuations Measured During a Typical Calibration Run. Note the Absence of Fluctuations with the Transmitter Blocked.

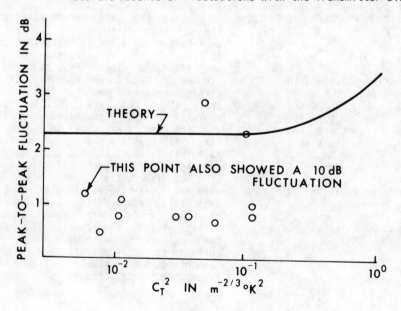

Figure 2. Peak-to-Peak Intensity Fluctuations vs. C_T^2 at 94 GHz. Values of Parameters Used for Calculation are as Follows: L_0 = 1.5 m, T = 288.8 K, ρ = 8.5 g/m^3, and C_ρ = 1.45 × 10^{-4}m$^{-1/3}$g/m^3. Measured Results are Shown as Circles.

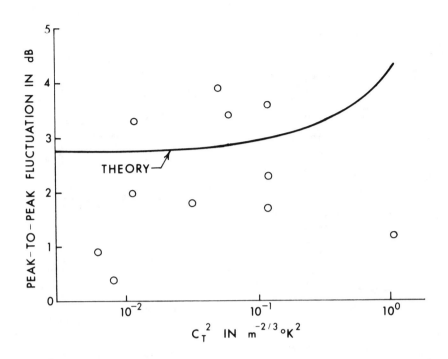

Figure 3. Peak-to-Peak Intensity Fluctuations vs. C_T^2 at 140 GHz. Parameters Used for Calculations are the Same as for Figure 2. Measured Results are Shown as Circles.

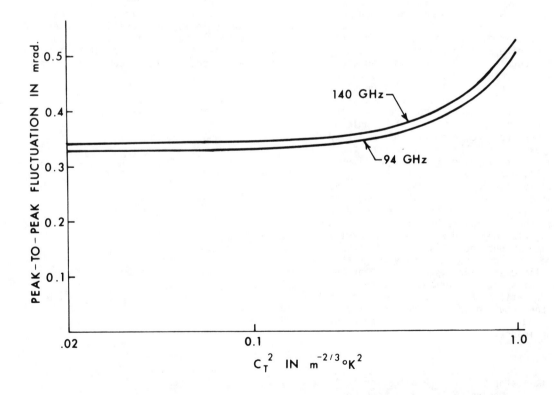

Figure 4. Calculated Angle of Arrival Fluctuations vs. C_T^2. The Conditions are the Same as Those Used for Figures 2 and 3.

TABLE I

Values of Parameters Used in Calculations

PARAMETER	VALUE
P	1013 mb
e	10.7 mb
λ	3.19, 2.14 mm
γ	0.092 neper/km (94 GHz) 0.346 neper/km (140 GHz)
T	289°K
e_o	9.45 mb
P_o	1013 mb
T_o	293°K
C_ρ	1.45×10^{-4} g/m$^3 \cdot$ m$^{-1/3}$

4. CONCLUSIONS

To briefly review the approach, the scalar wave equation was solved under the conditions in which both real and imaginary parts of the index of refraction are fluctuating. Atmospheric parameters measured during the White Sands propagation tests were then substituted into the solutions to these equations to obtain calculated results for both intensity and angle of arrival fluctuations. Finally, the calculated fluctuation intensity results were compared to the measured results, and the calculated angle of arrival errors were compared to those measured by Andreyev, et al [7]. Based on these comparisons, it is concluded that millimeter and submillimeter wave turbulence theory is substantially correct, and is able to give useful anwsers for systems calculations. At the same time, it is conceded that the data base for these comparisons is small, and results should be used with caution until a broader base is obtained. Intensity fluctuations are not considered likely to greatly affect millimeter wave systems except at the extreme limits of performance, but angle of arrival variations are of the same order of size as the required angular accuracy of many systems, and are therefore expected to be of some significance.

5. REFERENCES

1. V. I. Tatarski, Wave Propagation In A Turbulent Medium, McGraw-Hill Book Co., New York, 1961, Chapter 7.

2. L. A. Chernov, Wave Propagation In A Random Medium, McGraw-Hill Book Co., New York, 1960, Chapter 5.

3. A. O. Izyumov, "Amplitude and Phase Fluctuations of a Plane Monochromatic Submillimeter Wave in a Near-Ground Layer of Moisture - Containing Turbulent Air", Radio Engineering and Electronic Physics, Vol. 13, No. 7, 1968, pp. 1009-1013.

4. A. S. Gurvich, "Effect of Absorption on the Fluctuation in Signal Level During Atmospheric Propagation", Ibid., Vol. 13, No. 11, 1968, pp. 1687-1694.

5. N. A. Armand, A. O. Izyumov, and A. V. Sokolov, "Fluctuations of Submillimeter Waves in a Turbulent Atmosphere", Ibid., Vol. 10, No. 8, 1971, pp. 1257-1266.

6. J. W. Strohbehn and S. F. Clifford, "Polarization and Angle of Arrival Fluctuations for a Plane Wave Propagated Through a Turbulent Medium", IEEE Trans. Antennas and Propagation, Vol. AP-15, No. 3, May 1967, pp. 416-421.

7. G. A. Andreyev, V. A. Golunov, A. T. Ismailov, A. A. Parshikov, B. A. Rozanov, and A. A. Tanyigin, "Intensity and Angle of Arrival Fluctuations of Millimetric Radiowaves in Turbulent Atmosphere", Joint Anglo-Soviet Seminar on Atmospheric Propagation at Millimetre and Submillimetre Wavelengths, Institute of Radioengineering and Electronics, Moscow, November, 1977.

Millimetre Wave Propagation in Smoke
J.E. Knox

ABSTRACT

The transmission of millimetre waves was measured through clouds of typical battlefield smokes and obscurants at 35, 94, and 140 GHz.

Dust dispersed by a high explosive detonation was the only obscurant to affect transmissivity. The smallest transmissivity observed in that case was 91%. It is concluded that none of the materials dispersed would have prevented a millimetre wave sensor from performing its mission.

INTRODUCTION

Weapons systems using millimetre wave sensors have been proposed and are currently being evaluated by the Army. Millimetre wave radars have an advantage over longer wavelength radars in situations requiring short-range, low-angle target acquisition and tracking with "tactically sized" radar antennas. The smaller beamwidths associated with the shorter wavelengths alleviate the multipath problem which normally limits low-angle tracking.

Millimetre wave sensors are more attractive than electro-optical (E.O.) devices when the target or the sensor is obscured by dust or smoke. Since E.O. devices operate in, or near, the visible wavelengths, it is generally true that when the target cannot be seen by eye, it cannot be seen by an E.O. sensor. Exceptions to this rule occur with longer wavelength E.O. sensors. For example, infrared imagers can operate through some smokes, but are blocked by others.

Although these advantages of millimetre wave devices have long been known, it is only recently that light-weight, low-power, solid-state, millimetre wave sources have become available. With these sources, and with other millimetre wave components, it is now possible to assemble small and effective radar systems.

The Ballistic Research Laboratory has assembled such systems and has used them to demonstrate low-angle tracking under a variety of conditions. We have also participated in a smoke and obscurant test program to verify experimentally the effectiveness of millimetre wave sensors in typical battlefield smokes and obscurants.

This paper will describe the most recent part of our test program, Smoke Week II, and will present our results.

SMOKE WEEKS

The Army has established a smoke office, headquartered at Aberdeen Proving Ground, Md. The duties of this office include the coordination of sensor testing in smokes and obscurants. Part of their job is to ensure that smoke tests are standardized (as much as possible) and are well instrumented. In this way, the test results can compare competing sensors, or weapon systems. Another benefit of well instrumented and standardized testing is the creation of a data base of smoke effects which will be useful for battlefield modelling.

In November, 1977, and again in November, 1978, the smoke office held test programs which were called Smoke Week I and Smoke Week II. The test programs were occasions for government agencies and contractors to operate their sensors through clouds of typical battlefield smokes and obscurants. Some participants brought laboratory equipment for transmission tests while others brought ready-to-field weapon systems for tracking and acquisition tests. This paper describes our measurements during Smoke Week II.

Smoke Week II was held at Eglin AFB, Fl. The test area was arranged as shown in Figure 1. Range instrumentation formed a grid 300 metres long. Participants located their sensors at distances from the target area which were convenient to them; some were as far away as 3 km from the target area. Within the instrumented grid there were 100 aerosol samplers, 20 aerosol photometers, and three particle size analyzers. These instruments yielded information about the particle sizes within the cloud, cloud density, and cloud thickness (along the axis of the range).

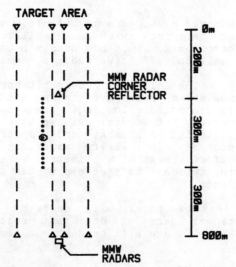

FIGURE 1. LAYOUT OF INSTRUMENTED SMOKE RANGE.

Transmissometers and telephotometers were aligned along the axis of the range to provide information about the time duration of the cloud over the range, its density (measured at various visible and infrared frequencies), and its luminance. Total distance between the transmitters and receivers of these instruments was 800 metres. They were kept far enough from the central 300 metres so that they were never within the smoke cloud.

Meteorological instruments measured air density, barometric pressure, solar radiation, percent of cloud coverage, wind direction, wind speed, air temperature, and humidity. A common time base (range time) was used for all the instrumentation and was available to each participant for recording with his sensor data. After the test program, each participant was sent a package of processed range data so that he could correlate it with his own sensor data.

Table I lists the "typical" battlefield smokes and obscurants which were used. Each of these materials were dispersed at least two different times in order to improve the quality of each participant's data. The method of dispersion was chosen so as to reproduce as closely as possible the condition likely to be encountered on a battlefield.

The materials, along with explosive charges, were positioned on the windward side of the range. The amount of material and the amount and type of explosive with each container of material, and the method of positioning (on the surface of the ground, or buried a few centimetres) were all chosen to simulate fielded smoke ordnance. The ordnance simulated in each trial is listed in the last column of Table I.

The actual ordnance, delivered by artillery, say, could not be used to disperse the obscurants because there is not the required control over shell placement. Safety factors alone would have forced range instrumentation to be farther from the smoke area, and some instrumentation could have been expected to be destroyed. Propagation testing has been done using real artillery and has been reported by J. D. Lindberg (1).

At Smoke Week II, after the munitions were fired, the duration of each trial depended upon the type of smoke being generated. Dust clouds thrown up by high explosives drifted across the range as a unit, and the test was over in about three minutes. On the other hand, chunks of phosphorous, dispersed by small explosive charges, would burn for several minutes, generating a dense cloud. The cloud would then become less dense as the smaller chunks burned out. The phosphorous tests were ended after 10 minutes, even though some pieces of phosphorous continued to burn for half an hour.

THE RADARS

Figure 2 is a block diagram of the radars. There were three radars, one each at 35, 94, and 140 GHz. The pulsed transmitters used in all three radars were IMPATT diodes. The local oscillators in the 35 and 94 GHz radars were GUNN diodes, so that these radars were all solid-state. The local oscillator at 140 GHz was a Klystron tube.

Each radar had an I.F. attenuator immediately after an I.F. pre-amplifier. The attenuator was used to reduce signal level in the radar receiver when it was necessary to ensure that the receiver would not be saturated by the reflected signal from the target corner cube reflector. In this way, we were able to use one reflector for all three radars. The reflector was large enough to give a strong return signal at 35 GHz., and the I.F. attenuators prevented saturation at 94 and 140 GHz.

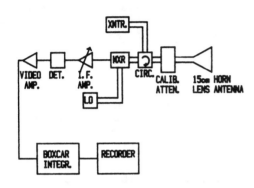

FIGURE 2. MILLIMETER WAVE RADARS USED FOR SMOKE/OBSCURANT MEASUREMENTS

Each radar also had a calibrated R.F. attenuator. These attenuators were used to calibrate the radar immediately prior to each test. When an attenuator was changed from 0 dB to 0.5 dB, the transmitted pulse was attenuated on its way to the antenna and the reflected pulse was attenuated on its way to the receiver, for a total power reduction of 1.0 dB. The reduced radar output was recorded for calibration.

During a test, with the attenuator set at 0 dB, if a cloud which attenuated the signal by 0.5 dB per pass came between the radars and the reflector, the transmitted and the reflected pulse would each have been attenuated by 0.5 dB for a total of 1 dB. Hence a 0.5 dB calibrated attenuator change reduced the radar output by the same amount as 0.5 dB attenuation per pass through the cloud.

Figure 1 (which shows a layout of the test range) also indicates the position of the BRL radar equipment. The three radar units, along with the radar controls and the data recording equipment, were set out at the 800m point on the range. All three radars were aimed at a corner cube reflector which was set on a tripod out in the field just past the instrumented grid, 600 metres from the radars. The radar cross-sectional area of the reflector was 625 square metres at 35 GHz (4500 sq. metres at 94 and 10,000 sq. metres at 140 GHz).

TEST RESULTS

Of the seven obscurants listed in Table I, only one, labeled H.E. Dust, affected transmissivity of any of the radars. Figure 3 provides an example of nearly all the data that were taken during Smoke Week II. It shows a set of radar

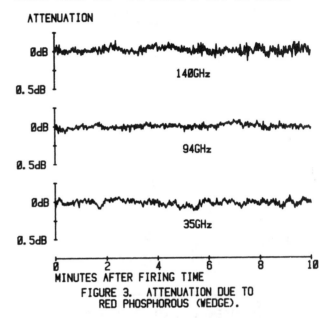

FIGURE 3. ATTENUATION DUE TO RED PHOSPHOROUS (WEDGE).

signals received during a test of red phosphorous (wedge) (RP). Six charges of RP, arrayed in a line 100 m long, 75 m from the center line of the range, were fired simultaneously. The wind was blowing at 3 m/sec across the range. The cloud which was formed was 60 m deep (along our line of propagation), and persisted over the center line for approximately 15 minutes. The median particle size was 0.7 microns, and material concentration was approximately 15 to 30 gm/cu.m.. The received signals shown in figure 3 do not differ from those of the minutes immediately prior to the start of the tests.

Figure 4 shows the radar outputs during one of the H.E. dust trials. The particular trial shown here was the one which attenuated the radar signals the most. It can be seen that the maximum value of attenuation was only 0.4 dB (corresponding to a transmissivity of 91%), and that attenuation persisted for approximately 10 seconds.

Table II shows the variability of the results of the H.E. dust trials. There appears to be no correlation between the values of maximum attenuation and the sizes of the explosive charges. Some trials using the 5 lb. charges generated clouds which affected the radars, while one which used the 15 lb. charges apparently generated a cloud which was transparent to millimetre waves. Attempts to correlate maximum attenuation values with other range data, such as humidity, wind speed, or particle sizes, were not successful. Apparently none of the parameters measured during the trials were those which affected the transmissivity of the dust clouds at millimetre wavelengths.

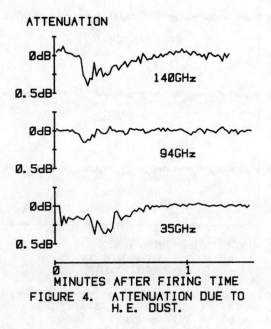

FIGURE 4. ATTENUATION DUE TO H. E. DUST.

DISCUSSION

The test results for the smokes were not unexpected. The sizes of the smoke particles generated during Smoke Week II were 0.5 to 20 microns in diameter, or less than 0.01 times the wavelength. Furthermore, the materials were nonconductive. For these reasons, one would not expect interaction between the radar signals and the smoke.

The radars did see the dust clouds, however. Since range data could not be correlated with radar signal attenuation, we must look to other sources of information for an explanation.

The largest sized particle which could be measured by the range instrumentation was 13 microns. One of the participants was operating a particle counter with a particle size window around 50 microns, and it detected a "significant" number of particles. Calculations show that particles larger than 100 microns could still have been falling through the radar line-of-sight during the time of maximum attenuation, approximately 20 seconds after firing time. It is likely, then, that the attenuation was a result of blockage by larger particles than were sampled by the range instrumentation.

There were no obscurants at Smoke Week II which would have prevented a millimetre wave sensor from operating. There still exists a need to demonstrate the transparency of these obscurants to millimetre waves under other operating conditions, however. For example, phosphorous is anhydrous, and will absorb moisture in amounts which are very sensitive to the relative humidity in the air. It would be worth-while, therefore, to conduct a series of tests under conditions of very high humidity.

BRL plans to continue measuring attenuation through obscurants, especially any which are new or experimental. With this in mind, BRL is presently assembling a set of radars (at 35, 94, and 140 GHz) which will be used solely for propagation measurements, both chamber tests and open air tests.

Table I. Obscurants Used During Tests
November, 1978

TYPE	NUMBER OF TRIALS	NUMBER PER TRIAL	DESCRIPTION
HC	2	15	155mm Cannisters
W.P. Wick	2	12	2.75" Rocket
Fog Oil	2	2	M3A3 Generators
Plasticized W.P.	2	6	5 in. Zuni Warhead
W.P. Wedge	2	3	155mm XM825
R.P. Wedge	2	6	155mm XM803
H.E. Dust	5	6	30-90lb. C4 (105-155mm)

Notes:
H.C. is hexachloroethane
W.P. and R.P. are white and red phosphorous, respectively. The phosphorous are packaged in various ways to achieve desired burning rates.
H.E. Dust is dust dispersed by detonating high explosives.

Table II. Maximum Attenuation
During H.E. Dust Trials.

TRIAL NUMBER	AMOUNT OF H. E.	MAXIMUM ATTENUATION		
		35 GHz	94 GHz	140 GHz
23	6ea. 5lb. C4	0 dB	0 dB	0 dB
23R	6ea. 5lb. C4	0	0.3	0.2
26	6ea. 15lb. C4	0.3	0.2	0.4
29	6ea. 15lb. C4	0	0	0
30	6ea. 5lb. C4	0.3	0	0.3

Reference:

(1) James D. Lindberg (Comp.), "Measured Effects of Battlefield Dust and Smoke on Visible, Infrared, and Millimeter Wavelength Propagation," Report Number ASL-TR-0021, Atmospheric Sciences Laboratory, White Sands Missile Range, N.M., Jan. 1979, (Distrib. Limited to U.S. Gov't Agencies Only).

140 GHz Multipath Measurements over Varied Ground Covers

H.B. Wallace

ABSTRACT

As part of a 140 GHz low-angle tracking program measurements of the forward scattered power were made at 140 GHz over three types of terrain: high weeds, mowed weeds, and asphalt. A theoretical model was developed assuming specular reflection, to determine the forward scattering coefficient. The results tend to indicate that for the terrain with a vegetative cover, the scattering coefficient was less than -0.1 and for the asphalt it was approximately -0.5.

INTRODUCTION

The Ballistic Research Laboratory, U. S. Army Armament Research and Development Command (BRL/ARRADCOM) has been investigating millimetre-wave technology as applied to the near-earth environment. One area of investigation has been the tracking of targets at very low elevation angles. As with all radar systems, multipath was of primary concern. In fixed site radar systems, there are a multitude of methods for alleviating multipath errors: radar fences, offset null tracking and complex angle tracking, to name a few. However, the simplest and perhaps most efficient method has been to avoid letting the tracking beam come in contact with the ground. In mobile systems, it is perhaps the only method. In order to track close to the ground one can either decrease the antenna beamwidth by making the antenna larger or by increasing the frequency of operation. The use of large antennas to reduce the antenna beamwidth sufficiently for very low-angle tracking can have adverse effects on the overall system mobility and characteristics.

The BRL/ARRADCOM is primarily concerned with low-angle tracking of targets which are not far removed from the radar (less than 10 km). This tends to make the use of EHF an effective method of alleviating some multipath problems. In order to evaluate some of the system constraints and parameters in very low-angle tracking, and to advance the state-of-the-art in millimetre-waves, a 140 GHz tracking radar system was constructed with a .91 m conically scanned antenna. This system was not readily usable as a means of measuring the forward scattering coefficient for various terrains thereby requiring the construction of a seperate system which could be transported to various terrains and operated with a minimum amount of effort.

This presentation will describe the test instrumentation for measuring the forward scattering coefficient and a comparison between the experimental results and a theoretical model which was developed.

TEST SCENARIO

A 100 m range was set up to measure the multipath effects on a conically scanning beam at 140 GHz (Figure 1). Three different earth covers were measured: high weeds, mowed weeds, and an asphalt road flat to approximately 1 cm. A conically scanning antenna smaller than 0.91 m was not available at the time for these experiments so single horn-lens antennas were used. The antenna was mounted on a precision altazimuth mount to permit positioning the transmitted beam to simulate the upper and lower lobes. The transmitter was placed 1.82 m above the ground and the beams had a 3-dB crossover parallel to the ground. A 10 mW CW IMPATT oscillator was initially used as a source. Later measurements used a 0.5 W pulsed IMPATT source.

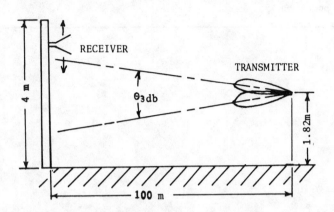

Figure 1. Multipath Range Configuration

At the other end of the 100 m range, a superheterodyne receiver was placed on a vertical positioner. The positioner could move the receiver from just above ground level to a height of approximately 4 m. The receiver was maintained parallel to the ground at all times and aimed in azimuth with a 7X riflescope.

Two different antennas were used for the transmitter: a 50.8 mm and 152.4 mm horn-lens. The receiver used a 50.8mm horn-lens. All of the measurements were made using vertical polarization. Prior to running the test, the one-way beam patterns were measured with the receiver elevated to 12 m above the ground. Figure 2 shows the measured pattern (x's) and computer generated patterns (solid lines) for the 152.4 mm antenna.

Prior to measuring the simulated upper and lower lobes of the transmitter, the receiver was moved from the top of the positioner to ground level while the transmitter continually tracked it. This provided a baseline measurement of power and multipath. Since the positioner used a manually cranked winch, some noise was introduced into the receiver output from oscillation of the positioner. Aiming was difficult above the 3 m position and resulted in improper pointing in some of the pattern measurements.

After the tracking pattern was made, the transmitter was positioned to simulate the upper lobe and a power measurement was taken. The lower lobe was then simulated and measured. A calibrated RF attenuator in the receiver was used to measure received power relative to the received power at the time the transmitter was pointed at the receiver in its uppermost position. Figures 5 through 8 are examples of the patterns measured for the 50.8 and 152.4 mm antennas over the dif-

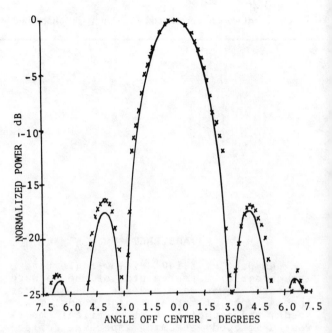

Figure 2. 152.4 mm Antenna Patterns

ferent terrains. Only those measurements simulating the upper and lower conical scan lobes are shown.

THEORETICAL MODEL

The equations used to calculate the multipath will not be discussed here as they can be found in many other references (1,2,3,4). It should be pointed out that the ground reflections were assumed to be specular, i.e., the ground was perfectly smooth.

According to Barton (Ref.1), specular reflection will apply if the RMS height variation in the first Fresnel zone of reflection is

$$\sigma < \frac{1}{8} \frac{\lambda}{\sin\theta}$$

where

σ = RMS surface height variations
θ = reflection angle
λ = wavelength = 2.14 mm

In this case θ = 3.3 degrees (receivers at 4 m height), so if $\sigma < 4.61$ mm, specular reflection can be assumed. This case is met for the asphalt surface but not for the two vegetation covered surfaces.

When the ground is rough the maxima and minima of the multipath returns will be reduced somewhat. An additional component will be added to the return at the

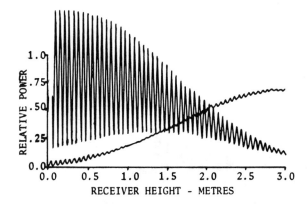

Figure 3. Theoretical Patterns, 50.8 mm Antenna

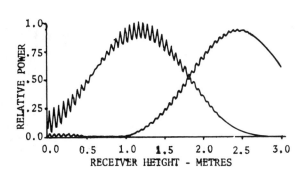

Figure 4. Theoretical Patterns, 152.4 mm Antenna

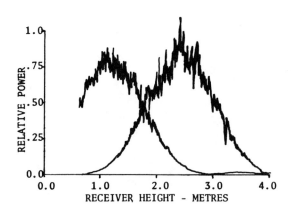

Figure 5. Measured Patterns Over Grass, 152.4 mm Antenna

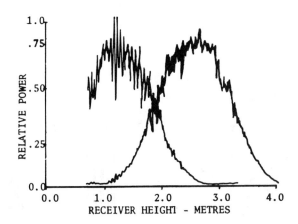

Figure 6. Measured Patterns Over Asphalt, 152.4 mm antenna

lower receiver elevations due to the diffuse scattering from the ground just in front of the receiver. This component tends to produce a noisy and unpredictable signal unless the terrain can be mapped to a fraction of a wavelength. Any vegetation which covers the terrain will produce two effects. Primarily it will act as an attenuating medium covering the terrain surface thus reducing the apparent scattering coefficient. A secondary effect is the noisy modulation of the reflected signal by wind. Since the various parameters of fine grain surface variations and vegetative loss could not be measured with sufficient accuracy at such high frequencies the theoretical model treated the surface as a smooth, somewhat lossy medium.

The antenna patterns were simulated using a $[J_1(x)/(x)]^2$ function for the 50.8 mm antennas and a

$$\frac{2}{3}\left[\frac{\sin x}{x}\right]^2 + \frac{1}{3}\left[\frac{\sin x}{x}\right]^3$$

function for the 152.4 mm antenna. Very good agreement was obtained for both antennas as can be seen in Figures 2. Figures 3 and 4 show the predicted returns for the 152.4 and 50.8 mm transmitter antennas over a terrain with a forward scattering coefficient of -0.5. The relative power indicated is the received power relative to the free space power received when the transmitter and receiver are coaligned.

RESULTS

Figure 5 shows the receiver output over the grassy terrain for the 152.4 mm

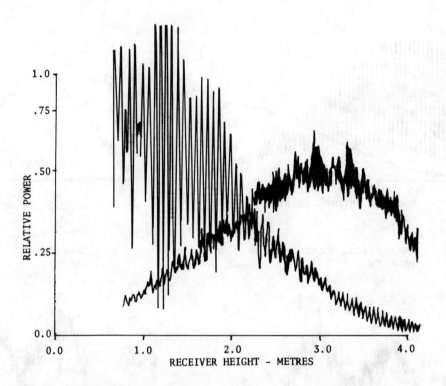

Figure 7. Measured Lobe Patterns Over Asphalt, 50.8 mm Antenna, CW Source

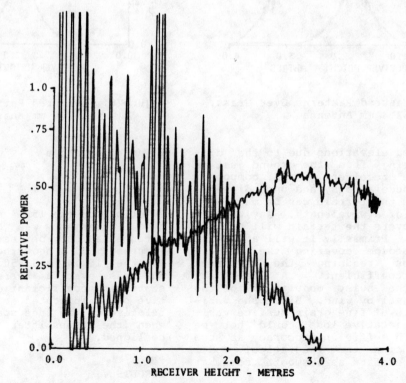

Figure 8. Measured Lobe Patterns Over Asphalt, 50.8 mm Antenna, Pulsed Source

antenna. It is obvious that the multipath lobing that should be present is masked by the system noise and the noise introduced by the vegetative covering. Figures 6 and 7 are the outputs of the receiver over the asphalt road. The multipath cycles overwhelm the signal from the 50.8 mm antenna and are just visible in the 152.4 mm antenna data. A comparison of this data with the theoretical model indicates the forward scattering coefficient is less than -0.1 for the vegetative covered terrain and approximately -0.5 for the asphalt surface.

Some interesting phenomena were noted when the measurements were made using the pulsed IMPATT source. This source had a center frequency of 140 GHz but chirped over 1 GHz during its 50 nsec duration. Figure 8 is the output of the receiver with the 50.8 mm transmitter antenna over asphalt. The first and most notable difference is the reduction of noise on the signal. This is undoubtably due to the decorrelation of the diffuse component by the spread spectrum. The second difference is apparent by examining the number of cycles between the 1 and 2 metre positions. The output for the chirped source shows 13 cycles of multipath. Figure 7, the output for CW source, shows 15 cycles which agrees with the theoretical model. The reason for this anomaly is not yet understood.

REFERENCES

(1) David K. Barton, "Low-Angle Radar Tracking," Proc. IEEE, Vol 62, No. 6, Jun 74.

(2) Richard A. McGee, "Multipath Suppression by Swept Frequency Methods," BRL Memorandum Report 1950, Nov 68. (AD #682728)

(3) Cecil L. Wilson, "Pointing Errors in Sequential Lobing Antenna Systems," BRL Technical Note 1463, May 1463, May 62. (AD #609009)

(4) Miles V. Klein, Optics, New York, John Wiley & Sons, Inc., 1970, pp 184-189.

Millimeter Wave Propagation Measurements over Snow

D.T. Hayes, U.H.W. Lammers, R.A. Marr, and J.J. McNally

ABSTRACT

Advances in millimeter wave technology have made possible the development of radar, guidance and digital communication systems. This has resulted in a need to better understand millimeter wave propagation phenomena. A study has been made of multipath propagation and clutter caused by reflection from terrain, in particular snow-covered terrain. Data are given which show the scattering properties of snow as a function of frequency (35, 98 and 140 GHz), incident angle, polarization, snow type and free water content. Preliminary data are also included to demonstrate the effect of snow cover on multipath propagation.

1. INTRODUCTION

For millimeter wave systems intended to operate close to terrain the designer of millimeter wave radars needs reliable data on propagation phenomena which may interfere with system operation. Background clutter and multipath propagation are caused by snow on the ground which is known to be a strong scatter source at millimeter wavelengths. For this reason scattering from snow has been studied experimentally at the Electromagnetic Sciences Division of RADC.

An intensive measurement program was carried out during the winter of 1977-78. Tests were conducted at 35, 98 and 140 GHz using a scatterometer designed and constructed by RADC for the express purpose of gathering data from snow cover. Backscatter, bistatic scatter and attenuation data were collected for various types of snow cover. Particular emphasis was placed upon obtaining scattering data from metamorphic snow, which is old and granular in structure. These results are presented in Section 2.

During the past winter (1978-79) a new system was designed in-house to determine multipath effects on millimeter wave propagation. Initial measurements over snow-covered terrain are discussed in Section 3.

2. BACKSCATTER MEASUREMENTS FROM SNOW

A photograph of the scatterometer is shown in Figure 1. The transmitters and receivers are mounted at the upper end of the boom. The associated power supplies at the base of the scatterometer act as counterweight to the weight of the boom. The support structures are made of wood to minimize possible interference. The scatterometer is seated on a motor-driven turntable which rotates and sweeps the antenna beams over the snow field. The return signal is recorded on an X-Y recorder. By rotating the upper part of the boom about the boom midpoint, the angle at which the beam intersects the snow surface is varied. In-situ measurements were made at three incident angles (90°, 45°, and 15°) as illustrated in the upper half of Figure 2.

FIGURE 1. MULTI-FREQUENCY SCATTEROMETER

In addition, the system was operated in the configuration shown in the lower half of Figure 2. In this so-called "sample mode" a slab of snow is placed in a basket at the midpoint of the boom. The transmitters are located at the base of the boom and the receivers at the other end. The distance from transmitters and receivers to the snow sample is 1 meter. This assures that measurements are made in the far field of the antennas. Half-power beamwidths are typically 7°. The basket (strung much like a tennis racket with string) was designed to interfere minimally with the transmitted beam. The size of the snow slab is large enough to neglect edge effects. Operation in the sample mode has the advantage that, in addition to backscatter, both transmission and bistatic scatter measurements may be made.

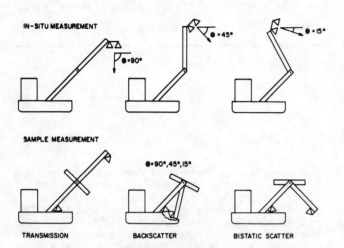

FIGURE 2. GEOMETRIES OF SCATTER MEASUREMENTS

In the in-situ and sample modes operation is possible for vertical, horizontal and crossed polarizations. Relative power measurements are made by comparison of the snow return and that from a large flat-plate reflector. In the in-situ mode the transmitting and receiving antennas are positioned at nadir over the flat plat (see Figure 1) and the peak power return from the plate is recorded. Calibration in the sample mode is achieved by attaching the flat plate to the underside of the basket and recording the peak return at perpendicular incidence. When making transmission measurements the equipment is calibrated by monitoring the received power with the basket empty. Comparison between the received signal levels in the transmission mode without plate and the backscatter mode with plate, confirm the validity of the assumption of an image source behind an infinite, conducting plane.

A block diagram of the scatterometer system set up for the sample mode is shown in Figure 3.

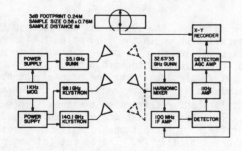

FIGURE 3. SCATTEROMETER BLOCK DIAGRAM

The transmitters consist of 1 kHz-square-wave modulated oscillators connected to standard gain horn antennas. A Gunn oscillator is employed at 35.1 GHz and klystrons at 98.1 and 140.1 GHz. Standard gain horn antennas are also used for reception. The receiver consists of a single-ended K_a-band mixer and a K_a-band Gunn local oscillator. The oscillator is tuned to provide fundamental mixing at 35 GHz, third-order harmonic mixing at 98 GHz and fourth-order harmonic mixing at 140 GHz. The resulting 100 MHz IF signal is then amplified and detected. Narrowband post detection amplification at 1 KHz is performed before the signal is recorded on the Y-axis of an X-Y recorder. The dynamic range of the system exceeds 50 dB at all frequencies. Automatic gain control is provided in the receiver circuit to accommodate the wide dynamic range of the input signal. A voltage proportional to the angular position of the turntable is fed into the X-axis when in-situ measurements are made. When the basket is used to hold a snow slab the grazing angle of the incident radiation on the slab is recorded on the X-axis.

The scattering cross-section (σ^0) normalized to unit area is determined in the following manner. The return from the snow sample is given by the standard radar equation

$$P_r = \frac{P_t \, G_t \, \sigma^0 \, A A_e}{(4 \pi R^2)^2}$$

where P_r = power received

P_t = power transmitted

G_t = transmitting antenna gain

A_e = receiving antenna effective area

R = distance from transmitter and receiver to snow sample.

A = area of snow illuminated between the 3 dB points of the transmitted beam.

To a good approximation

$$A = \frac{\pi \beta_E \beta_H R^2}{4 \sin \theta}$$

where β_E and β_H are the 3 dB E and H-plane beamwidths, and θ is the incident angle.

For calibration, the power P_{rp} received from the flat plate at range R_p is given by

$$P_{rp} = \frac{P_t \, G_t \, A_e}{4\pi \, (2R_p)^2} \, .$$

Dividing Eq. (1) by Eq. (3) and substituting Eq. (2) for the footprint area A results in

$$\sigma^0 = \frac{P_r}{P_{rp}} \, \frac{4 \sin \theta}{\beta_E \, \beta_H} \, (\frac{R}{R_p})^2 .$$

The system was operated in the sample mode throughout January and February 1978. During this period the scattering properties of dry snow only were examined. Early in March daytime temperatures exceeded 0° C by a sufficient amount to cause appreciable melting. At this time the system was reconfigured to make in-situ measurements. Initial measurements were begun before daybreak. This allowed data to be collected from dry snow. The measurements were continued throughout the day. At the same time, the free-water content of the snow was measured using a hot-water calorimeter. During all measurements extensive ground-truth data were collected. These included: snow depth, type, density, temperature, air temperature and other weather parameters.

The backscatter signal scintillated rapidly as the antenna beams swept across the snow field. As expected, the scintillation frequency was directly proportional to the signal frequency. At 98 and 140 GHz the scintillations were so rapid that only the local peaks of the recorded signal could be resolved on the X-Y plots. The scintillation amplitudes were always in excess of 20 dB for the return from dry metamorphic snow. Only in the case of very wet snow was the amplitude less than 20 dB.

At all frequencies the local peak scatter returns were analyzed to determine the statistics and the mean of the peak backscatter coefficient. At 35 GHz, the statistics of the complete backscatter signal were established. Mean peak backscatter cross-sections were determined as a function of grazing angle, polarization and frequency. The effect of free water on the scattering cross-section was also determined as a function of the same parameters. To establish the validity of measurements made in the sample mode, the scattering cross-section calculated from sample data was compared with that calculated from in-situ measurements from similar snow.

At all frequencies the peak backscatter coefficient was found to obey lognormal statistics. On the other hand, at 35 GHz the statistics for the complete signal were best fitted by a Rayleigh distribution. An example of this situation is shown in Figure 4. The cumulative probability for the peak backscatter coefficient and the total scintillation backscatter coefficient are plotted on log-normal probability paper for vertical polarization and perpendicular incidence at 35 GHz. Note that the median value of σ^0 differs by approximately 3 dB between the two curves. Although this relationship could not be determined at the two higher frequencies it is reasonable to assume that it still holds. In the remainder of this section only mean values of the peak backscatter coefficient will be presented.

Figures 5, 6 and 7 show the mean peak backscatter coefficient determined for dry and wet snow for 35, 98 and 140 GHz, respectively. In each figure the solid lines connect the mean σ^0 determined by averaging all data collected over dry snow at each grazing angle. The dashed lines connect the mean σ^0 determined from similar data collected over wet snow. The free-water content varied from a low of 3% to a high of 15% by weight. Both like-polarized (VV) and cross-polarized (HV) backscatter coefficients are shown.

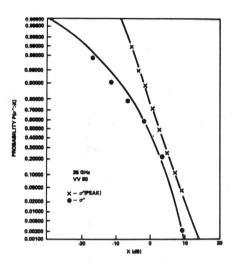

FIGURE 4. CUMULATIVE PROBABILITY FOR σ^0 (PEAK) AND σ^0: 35 GHz. σ^0 (PEAK) - x, σ^0 - ●. Curve through σ^0 data is for Rayleigh distribution fitted to median value of measured σ^0.

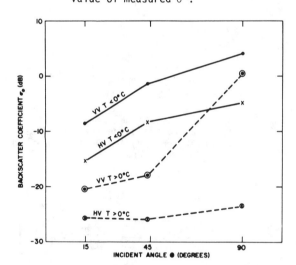

FIGURE 5. σ^0 VS INCIDENT ANGLE: 35 GHz. Vertical Polarization (VV) and Cross Polarization (HV) for Dry (T < 0° C) and Wet (T > 0° C) Snow.

Examining the results for 35 GHz (Figure 5) it is seen that σ^0 for dry snow is in excess of 0 dB at 90° and is lower by about 10 dB at 15°. The trend for the cross-polarized σ^0 is similar, being approximately 10 dB lower than the corresponding VV return at each incident angle. When free water is present there is a dramatic change in the angular dependence of σ^0. While the like-polarized coefficient drops only by about 3 dB at 90°, the decrease is in excess of 10 dB at 45° and 15°. The cross-polarized coefficient drops to a constant level almost 20 dB below that for dry snow. The angular dependence for the return from wet snow appears to be almost specular by nature. It may be caused by a water film forming at the surface which then appears smooth at 35 GHz.

Examining the data for 98 GHz (Figure 6) and for 140 GHz (Figure 7) one sees a different situation. At these frequencies σ^0 is higher for dry snow than was the case at 35 GHz. At 90°, σ^0 is approximately 7 dB for 98 GHz and 8 dB for 140 GHz. For dry snow the dependence of σ^0 on incident angle is similar to the 35 GHz data for both the like-polarized and cross-polarized data. However, for these higher frequencies, there is no change in the character of the angular dependence for wet snow. The decrease of σ^0 is no more than 6 dB at 98 GHz and 4 dB at 140 GHz. This may be due to two effects. First, at these two higher frequencies the snow surface still looks rough in spite of the water film. Second, the relative change in dielectric constant between ice and water is decreasing as the frequency increases from 35 to 140 GHz.

3. MULTIPATH MEASUREMENTS OVER SNOW

The scatterometer was modified this past winter to permit multipath measurements. The multipath geometry is shown in Figure 8. The three-frequency receiver was removed from the scatterometer and mounted on a movable platform attached to a vertical column. The platform acts as a motor-driven elevator. It travels from its lowest position of 0.2 meters above the surface to a maximum height of 4 meters. The transmitter remains at a fixed height of 2 meters at the top of the scatterometer boom.

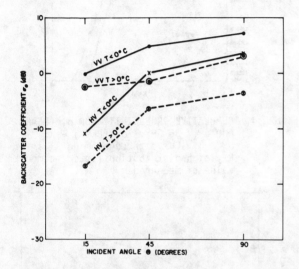

FIGURE 6. σ^0 VS INCIDENT ANGLE: 98 GHz. Vertical Polarization (VV) and Cross Polarization (HV) for Dry (T < 0°) and Wet (T > 0° C) Snow.

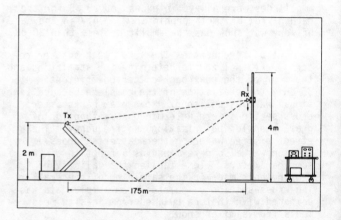

FIGURE 8. GEOMETRY OF MULTIPATH MEASUREMENTS

While varying the height of the receiver continuously over its full range the interference pattern produced by the direct signal from the transmitter and that part of the transmitted signal scattered off the terrain surface toward the receiver is recorded. The beamwidth of the transmitting and receiving antennas is wide enough to encompass both the direct and reflected rays. At the conclusion of a run the receiver is positioned at the peak of the strongest interference lobe. The signal is then calibrated with a precision attenuator inserted behind the first receiver IF-amplifier.

Multipath measurements were performed with the transmitter and receiver located at the opposite ends of a flat, grassy field. The transmitter-to-receiver distance was 175 meters. Initial baseline measurements were made at a time when the field was not covered by snow. The ground cover consisted only of short dormant grass. The measurements were repeated over the same propagation path under varying snow conditions. These included new crystalline snow, old metamorphic snow and deposited sleet particles.

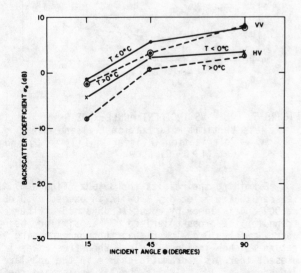

FIGURE 7. σ^0 VS INCIDENT ANGLE: 140 GHz. Vertical Polarization (VV) and Cross Polarization (HV) for Dry (T < 0° C) and and Wet (T > 0° C) Snow.

Figures 9 through 12 show the relative signal intensity vs receiver height measured over bare ground for vertical and horizontal polarization. At each frequency it is seen that the differences in the multipath interference structure for both polarizations are minimal. This is to be expected. The specular angle for the wave reflected from the ground varies from a minimum value of 0.5° to a maximum of 2° as the receiver moves from its lowest to its highest position. This angular range is well below the Brewster angle and thus the reflection coefficients for horizontal and vertical polarization are roughly equal.

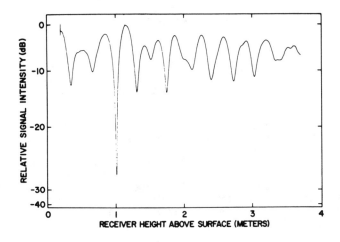

FIGURE 9. RELATIVE SIGNAL INTENSITY VS RECEIVER HEIGHT: 35 GHz. Vertical Polarization. No Snow Cover.

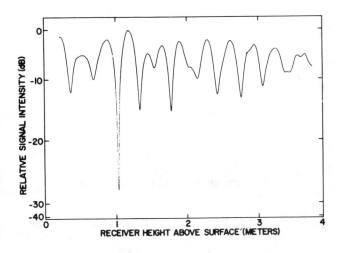

FIGURE 10. RELATIVE SIGNAL INTENSITY VS RECEIVER HEIGHT: 35 GHz. Horizontal Polarization. No Snow Cover.

The number of lobes increases approximately proportional with frequency. This is reasonable since the change in receiver height necessary to cause a path difference of one wavelength decreases with frequency. Also the lobes appear more irregular at the higher frequencies. Terrain inhomogeneities would have a greater effect on the reflected wave as the wavelength of the transmitted signal decreases. Over bare ground one cannot discern any trend in the lobing intensity as a function of frequency.

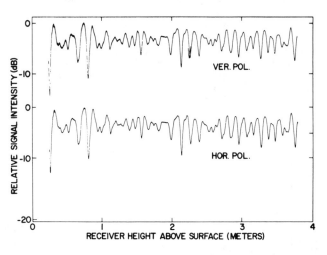

FIGURE 11. RELATIVE SIGNAL INTENSITY VS RECEIVER HEIGHT: 98 GHz. Vertical and Horizontal Polarization. No Snow Cover.

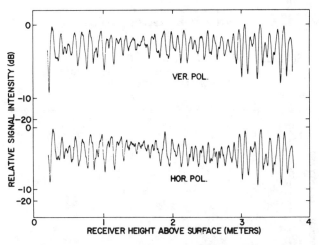

FIGURE 12. RELATIVE SIGNAL INTENSITY VS RECEIVER HEIGHT: 140 GHz. Vertical and Horizontal Polarization. No Snow Cover.

This past winter was almost void of snow in the Boston area. Consequently, only a limited amount of data was obtained. Results are preliminary in nature, intended only to demonstrate that substantial snow multipath propagation is possible at millimeter wavelengths. A thorough investigation requires that more measurements be conducted.

The following plots show data obtained over snow for vertical polarization only. In all cases results for horizontal polarization were quite similar. Figures 13 through 15 present the multipath interference measured at 35, 98 and 140 GHz for propagation over a 5 to 8 cm layer of new snow. The snow (density 0.1 gm/cm^3) was still in its pristine crystalline state when the measurements were conducted. At all frequencies one sees an increase in depth of the lobe structure when snow is present. At 35 GHz (Figure 13) there is a 3 to 4 dB increase over the bare ground lobing. The lobing at 98 GHz (Figure 14) is about 10 dB on the average. Randomly interspersed through the data are nulls varying from 15 to 25 dB in depth. The lobes at 140 GHz (Figure 15) almost always exceed 10 dB. At some receiver heights the depth of the nulls approach 20 dB.

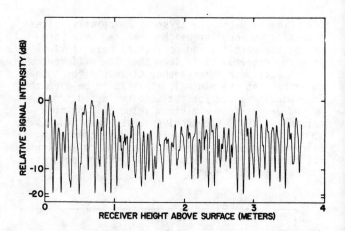

FIGURE 15. RELATIVE SIGNAL INTENSITY VS RECEIVER HEIGHT: 140 GHz. Vertical Polarization. Dry New Snow (5 to 8 cm).

The multipath intensity increases dramatically when the ground is covered with snow that has metamorphosed from its initial crystalline form. Figure 16 shows the multipath lobing measured at 35 GHz over a very light cover of old granular snow. The snow depth was less than two centimeters. The diameter of the granules varied from one to two millimeters. Over much of the range the depth of the nulls is in excess of 15 dB.

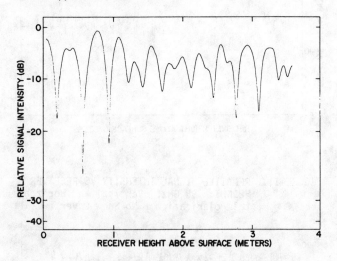

FIGURE 13. RELATIVE SIGNAL INTENSITY VS RECEIVER HEIGHT: 35 GHz. Vertical Polarization. Dry New Snow (5 to 8 cm).

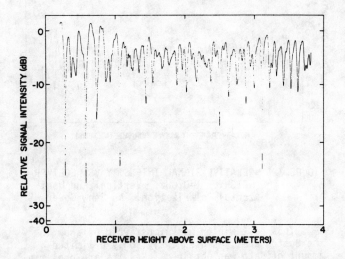

FIGURE 14. RELATIVE SIGNAL INTENSITY VS RECEIVER HEIGHT: 98 GHz. Vertical Polarization. Dry New Snow (5 to 8 cm).

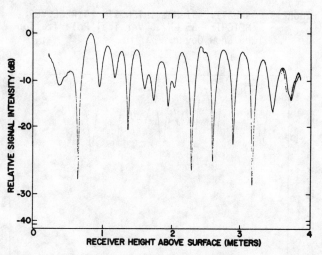

FIGURE 16. RELATIVE SIGNAL INTENSITY VS RECEIVER HEIGHT: 35 GHz. Vertical Polarization. Light Snow Cover (2 cm).

No multipath data were collected at 98 and 140 GHz over granular snow. However, data were obtained following a sleet storm which deposited a layer approximately 8 cm deep of almost spherical ice particles. The diameter of the particles varied from one to two millimeters. The sleet layer differed from the granular snow only in the layer depth and in the regularity of the particle shape. The signal measured at 98 GHz is shown in Figure 17 and at 140 GHz in Figure 18. At both frequencies the majority of the nulls are 20 dB below the peaks of the received signal.

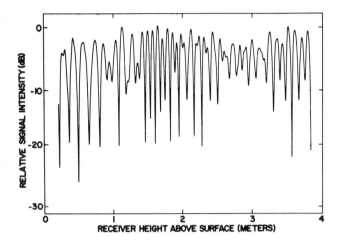

FIGURE 17. RELATIVE SIGNAL INTENSITY VS RECEIVER HEIGHT: 98 GHz. Vertical Polarization. Sleet (8 cm).

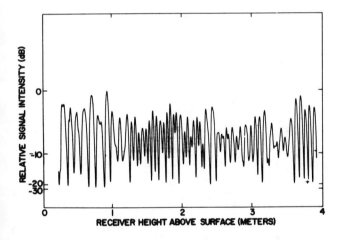

FIGURE 18. RELATIVE SIGNAL INTENSITY VS RECEIVER HEIGHT: 140 GHz. Vertical Polarization. Sleet (8 cm).

The multipath measurements program will be continued through next winter. During the intervening months measurements of multipath propagation are being made over varied terrain, e.g. asphalt, grassy fields, croplands, etc.

93-GHz Radar Cross Section Measurements of Satellite Elemental Scatterers

R.B. Dybdal and H.E. King

Abstract—The radar cross section (RCS) of representative spacecraft materials and components was measured at 93 GHz. Measurements of solar cells, thermal blanket material, structural components, and sensors are included. These measurements indicate the highly specular nature of scattering at millimeter wavelengths. The measurements generally agree with results predicted on the basis of common asymptotic formulations.

I. INTRODUCTION

Millimeter wavelength radar systems are enjoying more consideration as the technology in this regime develops. The feasibility and performance evaluation of these radars require knowledge of the radar cross section (RCS) properties of the targets, and at present, little measured RCS data are available at these frequencies. This succinct paper presents RCS data measured at 93 GHz on typical satellite elemental scatterers. These measurements provide data to determine the RCS response of satellites as well as other targets.

The data presented here were obtained from measurements performed in an anechoic chamber. Only a portion of an entire satellite could be measured in the anechoic chamber in order that the far-field criteria be satisfied. Measurements of the elements of a target are useful since superpositon may be used to obtain the response of the entire satellite and the returns from individual portions of the target are important for high resolution radars. The available bandwidth and high Doppler sensitivity achievable at millimeter wavelengths allow the resolution of individual target components. High resolution systems using a chirp waveforms may achieve 6-in range resolution at these frequencies [1], and the responses of the individual components comprising the radar visible portion of the target are distributed in the separate range cells of the radar. The high Doppler sensitivity available at these frequencies allows a display of individual portions of the target in different Doppler bins. The contents of the range and Doppler bins depend on the RCS levels of the individual components of the target.

II. EQUIPMENT

The measurements were performed in The Aerospace Corporation 90-ft quasi-tapered anechoic chamber. RCS measurements at 93 GHz were initiated in 1968 with the performance and description of the chamber documented in [2]. This frequency is the highest ever used in a microwave anechoic chamber, to the best of the authors' knowledge. This chamber has been used for scale model measurements and typical results for a simple sphere-cone reentry shape are presented in [2]. Later documentation [3] described measurements of simple flat plate targets. The results obtained indicate good correspondence with those predicted on the basis of common asymptotic formulations.

The separation between the target and the radar in these measurements is 50 ft. This target range restricts the maximum target dimension to 6 in in order to satisfy the $2D^2/\lambda$ far-field criteria. Horizontal polarization is generally used in these measurements since the peak specular return of the monofilament lines used for target support is -20 dBsm for vertical polarization. Since the targets are large in terms of wavelengths, the returns, particularly in the specular regions, are not strongly polarization dependent. A sphere was used as a calibration standard in these measurements and the returns from spheres of several different diameters were used to establish an absolute calibration level. The phase-locked electronics for the radar are described in [4].

III. RCS MEASUREMENTS

A typical satellite has two types of targets, structural and sensor, that are visible to a radar. The structural portion of the target includes metal components, fasteners, solar-array-covered panels, and thermal blanket material. Satellites generally include sensors that would be visible to an interrogating radar, and, accordingly, various antenna and telescope RCS returns were measured. Additional RCS patterns can be found in [5].

A. Solar-Array Material

Solar-array cells are present over a large part of the visible surface of a typical satellite, and, therefore, the RCS characteristics of this material are important. A solar cell sample mounted on honeycomb material is shown in Fig. 1(a). The RCS response of this target was taken by rotating the target with a conducting plate the same size as the solar cell area affixed to the back side. The measured RCS response of this target rotated about each axis of its rectangular shape is given in Figs. 1(b) and 1(c). The grating lobes present in Fig. 1(b) are attributed to scattering from the metal portions of the honeycomb grid since they appear on both sides of the pattern. For the pattern in Fig. 1(c), the grating lobes do not appear since the metal edges that yield the grating lobes are cross polarized with respect to the incident field and have a very low return under that condition. Measurements without the reference plates indicate the honeycomb material is relatively transparent at millimeter wavelengths since the RCS return was similar for both sides of the plate. Measurements performed at X band have indicated the honeycomb behaves as a conducting plate.

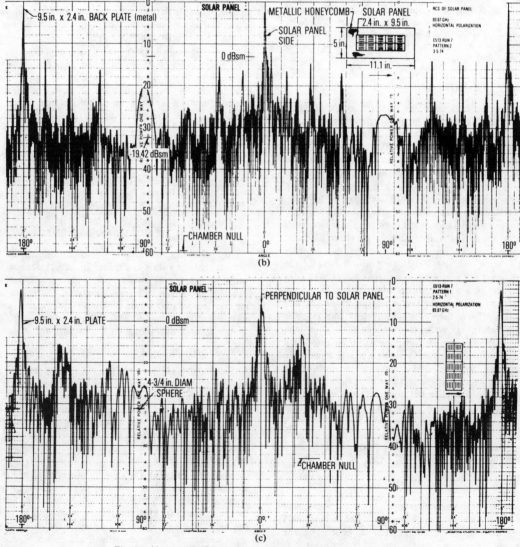

Fig. 1. Measured RCS of solar panel with and without back metallic reference plate. (a) Solar cell sample mounted on honeycomb material and reference plate. (b) RCS of solar cell rotated about long direction with reference plate. (c) RCS of solar cell rotated about narrow direction with reference plate.

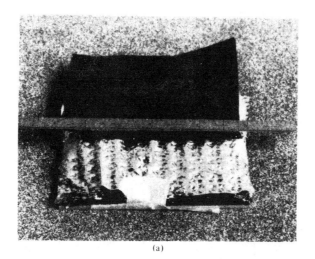

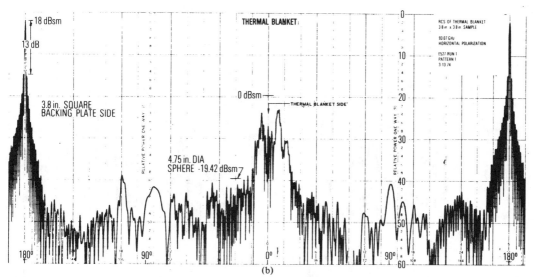

Fig. 2. RCS of thermal blanket material with metal reference plate. (a) Thermal blanket material sample. (b) Thermal blanket before handling.

B. Conducting Plates

Earlier measurements reported on flat plate targets [3] were performed on relatively small sizes. Two plates constructed from precision stock were fabricated to yield a +25 dBsm peak specular return; the resulting dimensions are 5 in square and 5.7 in diameter circular. The measured response from these plates agreed closely with predicted values as derived from asymptotic formulations [6].

Another material that is generally abundant on a satellite is thermal blanket material. A small sample of this material was obtained and taped to a metal plate of the same dimensions shown in Fig. 2(a). Thermal blanket material is composed of a layer of sheets with crinkles on the inner layers and a smooth gold foil outer layer. The smoothness of the thermal blanket material varies with the amount of handling required to place it on the satellite.

The measured response in Fig. 2(b) indicates the effects of this handling. The back surface is a square aluminum plate 3.8 in on a side and exhibits a normal response. The peak specular value for this plate should be +20 dBsm ($4\pi A^2/\lambda^2$) versus the +18 dBsm which was measured. The angular variation in the principal planes should be $(\sin X)/X$ as predicted from physical optics considerations and the first lobe removed from specular should be 13.3 dB below the peak value. The measured values correspond well with the predicted envelope. The thermal blanket covered side indicates the effects of surface distortions. The response on this side has a lower amplitude and different angular behavior. This response, of course, varies as the surface undergoes further handling.

The photo in Fig. 2(a) indicates the thermal blanket construction; the outer layer is a smooth gold foil and the inner layers are formed in a waffle-like pattern. The measured response of the inner layer material [5] yields a pattern which appears to be the envelope of an individual bump modulated by the grating lobe spectra for an array of bumps. Again the level is lowered by the surface roughness and has a peak level

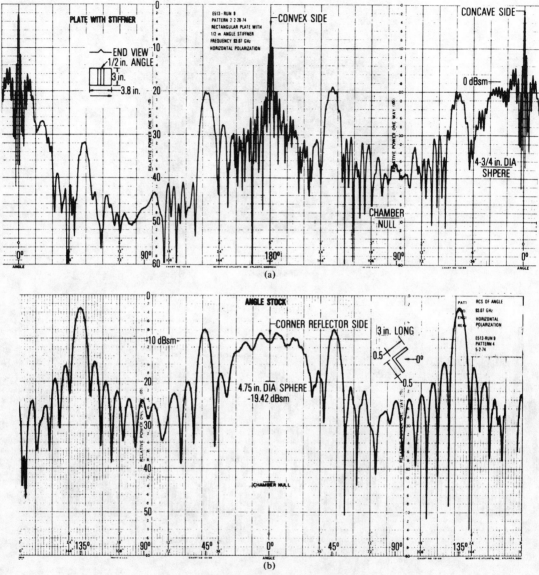

Fig. 3. Measured RCS of plates with structural components. (a) Plate with stiffner. (b) Angle stock.

on the order of −10 dBsm. Surface roughness effects were also illustrated by covering one side of a 2.3 in diameter plate with aluminum foil which were crumpled with varying degrees of roughness. A sample with a peak roughness height of ≈3/16 in had sufficient roughness to destroy the normal specular return.

A set of plates was constructed to illustrate the effects of some structural components. The response of a plate having four rivet heads was quite similar to one without the rivet heads [5]. A plate with a piece of 1/2 in angle stock in it was constructed to illustrate the effects of a panel with a stiffner ridge. The RCS response in Fig. 3(a) illustrates the effects of such a ridge. The RCS response of the angle stock by itself is shown in Fig. 3(b), which shows the specular return from the flat portions of the stock and indicates the concave portion of the stiffner focuses like a dihedral corner reflector.

C. Antenna Scattering

Dipole antennas are commonly used on satellite system. The response of an available dipole antenna was measured. The antenna is shown in Fig. 4(a) and includes a sleeve that broadbands its antenna performance [7]. The measured response is given in Fig. 4(b). An available commercially made AEL Model ALN108B conical log spiral antenna was measured to illustrate scattering from fiberglass structures. The specular returns from the fiberglass are identifiable in the measured response given in Fig. 5. A set of measurements were performed on X band (Narda 640) and E band (60–90 GHz waveguide, FXR E 638A) standard-gain horn antennas. Both E- and H-plane cuts for horizontal polarization were measured and the E-plane patterns are given in Figs. 6 and 7. The radar frequency is ten times the normal operating frequency for the

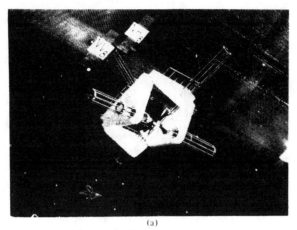

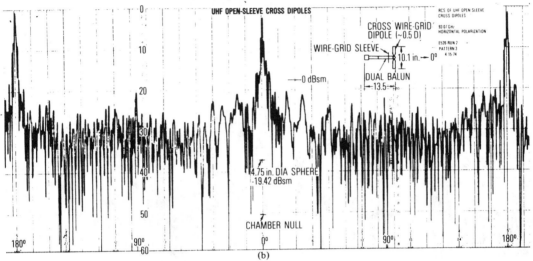

Fig. 4. 93-GHz RCS characteristics of UHF open-sleeve dipoles without metallic reflector. (a) Open-sleeve dipole. (b) RCS of cross dipoles including sleeve.

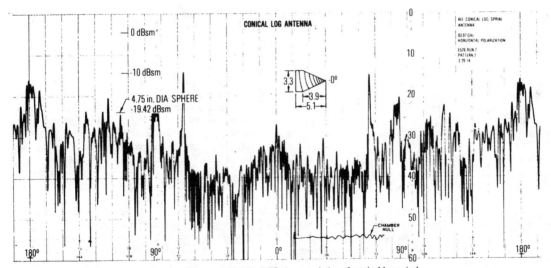

Fig. 5. Measured 93-GHz RCS characteristics of conical log-spiral antenna.

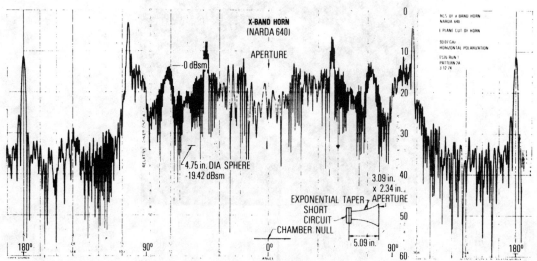

Fig. 6. Measured 93-GHz RCS characteristics of X band horn (E-plane cut).

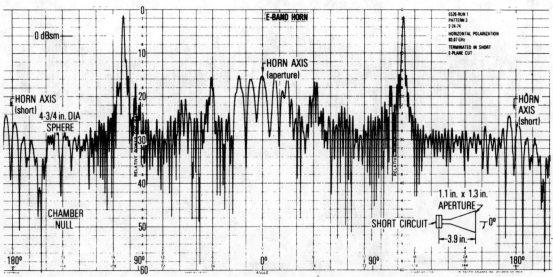

Fig. 7. Measured 93-GHz RCS characteristics of E-band (60 to 90 GHz) horn (E-plane cut).

X band antenna and multimode propagation effects are seen in the scattered return. The return from a small telescope has also been measured [5]; the return from the optics (1.5 in diameter) has a noticeable lobe structure with peak levels on the order of −20 dBsm.

IV. CONCLUSIONS

The RCS response measured at 93 GHz from representative spacecraft materials and components has been described. The scattered responses from these targets are generally what might be predicted on the basis of common asymptotic formulations; however, some important differences were noted. The measured RCS responses exhibit a wide dynamic range; a minimal level is perhaps on the order of −30 dBsm; while the peak specular return for the targets investigated is approximately +25 dBsm.

In contrast to returns associated with lower frequencies for the same targets, the millimeter response is characterized by narrower specular returns with higher peak values, yielding greater spacial separation between scattering centers. This behavior should allow the details of the target to be examined on a more individual basis as the aspect angle with respect to the radar changes, as compared with responses measured with lower frequency radars. Solar cell and thermal blanket materials, which form a major portion of the visible surface of a typical satellite, were measured. Honeycomb material is transparent to some degree, and the metal portions of the honeycomb can behave as an array of dipoles possessing a grating lobe spectra. Surface roughness plays a more dominant role in influencing scattering at higher frequencies; this effect becomes more important as the size of the target increases. The surface roughness initially lowers the peak specular amplitude and finally provides sufficient decorrelation to destroy the normal behavior.

ACKNOWLEDGMENT

The authors gratefully acknowledge the technical support of L. U. Brown and J. Fedarka in testing the various targets.

REFERENCES

[1] L. A. Hoffman, K. H. Hurlbut, D. E. Kind, and H. J. Wintroub, "A 94-GHz Radar for Space Object Identification," *IEEE Trans. on Microwave Theory and Techniques,* vol. MTT-17, pp. 1145–1149, December 1969.

[2] R. B. Dybdal and C. O. Yowell, "VHF to EHF Performance of a 90-ft Quasi-Tapered Anechoic Chamber," *IEEE Trans. Antennas and Propagation,* vol. AP-21, pp. 579–581, July 1973; also, TR-0073 (3230-40)-2, The Aerospace Corporation, El Segundo, CA, December 28, 1972.

[3] —, "Millimeter Wavelength Radar Cross Section Measurements," in *IEEE 1973 G-AP International Symposium,* Boulder, CO, August 22-24, 1973.

[4] R. D. Etcheverry, G. R. Heidbreder, W. A. Johnson, and H. J. Wintroub, "Measurements of Spatial Coherence in a 3.2-mm Horizontal Transmission," *IEEE Trans. Antennas and Propagation,* vol. AP-15, pp. 136–141, January 1967.

[5] R. B. Dybdal and H. E. King, "93-GHz Radar Cross Section Measurements of Satellite Elemental Scatterers," The Aerospace Corporation Technical Report TR 0075(5230-40)-4, SAMSO TR 75-127, April 10, 1975.

[6] G. T. Ruck, D. E. Barrick, W. D. Stuart, and C. K. Krichbaum, *Radar Cross Section Handbook.* New York: Plenum, 1970, vol. 2, ch. 7.

[7] H. E. King and J. L. Wong, "225-400 MHz Antenna System for Spin-Stabilized Synchronous Satellites," Aerospace Corporation Technical Report TR-0172(2162)-1, SAMSO TR 72-77, March 15, 1972. DDC Document No. AD 893 125L.

95 GHz Pulsed Radar Returns from Trees

R.D. Hayes

ABSTRACT

The foliage region of trees have been observed with a narrow beam, short pulsed 95 GHz radar to determine the spectral content and amplitude distribution of the signal. Depression angles and selected foliage regions were used to minimize the effects of exposed tree trunks in the radar cell-of-resolution.

INTRODUCTION

Field measurements of the radar backscatter from trees and other vegetation have been a continuing program at the Engineering Experiment Station of the Georgia Institute of Technology for over two and a half decades. The radars employed have operated in the frequency region from 6 to 100 GHz with the majority of the data collected at X-band. In recent years, a series of measurements have been conducted at 9.4, 16, 35 and 95 GHz (1) in order to relate the radar backscatter from trees at microwave and millimeter wavelengths. General features of the radar cross-section have been reported in previous articles (2) and at previous conferences (3,4). The purpose of this article is to look specifically at 95 GHz data in more detail to determine the spectral content from zero hertz to 200 hertz and to characterize the amplitude distribution of the same data. The data used in this analysis were selected from the large data bank and remain a portion of the existing general depository.

The radar was located on a mountain some 700 feet above surrounding terrain. The radar antenna was depressed 8 degrees below the horizon and the trees were located past a cleared area so that there was a clear line-of-sight between the radar and the tree canopy under investigation. Basic parameters of the radar system are given in Table I.

A 10 inch trihedral was placed in an adjacent radar cell-of-resolution and used as a calibrated reference for the radar. An anemometer was also located in the area for the purpose of recording wind speed. Data were recorded for periods of several minutes to insure constant environmental and meteorological conditions. After visual observation of the magnetically recorded data, short runs of 5 to 10 seconds were chosen for evaluation on the CYBER 74 computer. For consistency the only data analyzed for this study were under the conditions of clear skies and wind speeds in the range of 7 to 15 mph.

TABLE I

Frequency	95 GHz (Nominal)
Peak Power	6 kw
Pulsewidth	50 ns
PRF	2000 pps
Azimuth Beamwidth	0.70 degrees
Elevation Beamwidth	0.65 degrees
IF Center Frequency	60 MHz
IF Bandwidth	20 MHz
IF Response	Logarithmic
Dynamic Range	70 dB
Recorder	Magnetic

Previous analysis of these data was made using a NOVA computer, which at the time was limited in storage and processing. The purpose of a recent study was to retain all frequency components from dc to 1000 Hz and amplitudes of the recorded data.

The data were converted to digital format; linearized with calibration from the corner reflector; exponentiation to account for logarithmic receivers; and square rooted to place the data in a true linear voltage format. After this preparation of the recorded data, processing and analysis was performed. After analysis, the data were prepared for presentation with typical units as generally used by engineers in the radar community.

AMPLITUDE DISTRIBUTION

Cumulative probability distributions were made of the amplitude of the radar return. A majority of these amplitude distributions have a log-normal nature, with standard deviations typically 5 $\pm$ 1 dB.

The log-normal probability function of the radar cross-section, σ, is defined by

$$p(\sigma) = \frac{1}{s\sigma\sqrt{2\pi}} \exp\left\{-\frac{1}{2s^2}\left[\ln\left(\frac{\sigma}{\sigma_m}\right)\right]^2\right\}$$

where

    s = standard deviation
    σ_m = median value of σ

and the average cross-section is given by

$$\bar{\sigma} = \exp\left[\frac{s^2}{2} + \ln \sigma_m\right].$$

There were a number of measurements where the power return followed the Weibull distribution. The Weibull probability function of the radar cross-section, σ, is

$$p(\sigma) = \frac{b\sigma^{b-1}}{\alpha} \exp\left(-\frac{\sigma^b}{\alpha}\right)$$

where

    b is the shape factor
    α is the scale factor

and the average cross-section is given by

$$\bar{\sigma} = \alpha^{\frac{1}{b}} \Gamma\left(1 + \frac{1}{b}\right)$$

where

    Γ is the Gamma function.

For the data following the Weibull function, the shape factor was usually between 0.4 and 0.6. There were less than 1 percent of the data where b = 1, which is the condition for a Rayleigh distribution. The Rayleigh distribution appeared at very low wind speeds, less than 3 mph.

A typical amplitude distribution for deciduous trees is shown in Figure 1. The wind speed was between 7 and 15 mph and the recording time was 5 seconds. A Weibull cumulative distribution, with a shape parameter b = 0.5, has been drawn on the figure and presents a reasonable match to the recorded data. The Weibull curve was set to match the data at the 0.1 percent point. The average radar return value can be calculated in a straight forward manner from the equation of the Weibull function. Then considering the antenna beamwidth, depression angle, range between the radar and the trees, it is found that the value of radar cross-section per area, σ_o, was -28 dB for this particular data batch. This value was typical of that recorded over many observations at a depression angle of 8 degrees.

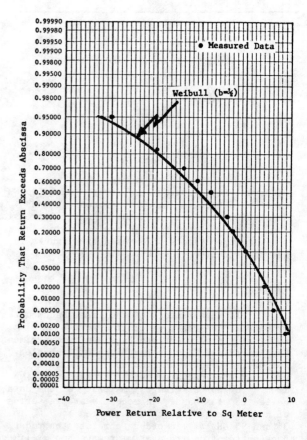

Figure 1. Cumulative Distribution 95 GHz Tree Echoes

SPECTRUM

The data were processed on a CDC 6400 computer to determine the frequency content of the 5 seconds of recording. The result of this processing is shown in Figure 2. The power spectral density is divided into two parts, a fixed or slow fluctuation component some 15 dB in magnitude and about 4 Hz wide; then, a fast fluctuation component of approximately 20 dB in magnitude and 200 Hz wide. These two components can be presented by

$$P_1(f) = \frac{A}{1+\left(\frac{f}{.8}\right)^2} \quad \text{low frequency}$$

$$P_2(f) = \frac{B}{1+\left(\frac{f}{30}\right)^2} \quad \text{high frequency}$$

and curves representing these functions are shown in Figure 2. For the example of data shown in Figure 2, A = 150 and B = 2.0. When evaluating the area under the density curve for P_1 and P_2, it is found that P_1 contains 70 percent of the total power and P_2 contains 30 percent of the total power. These values are typical and have been found at K_a- and X-band for both ground clutter (5) as well as sea clutter under moderate winds, not calm conditions.

The significance of the large low frequency component appears when one attempts to design Doppler filters to remove fixed and low frequency clutter components and still detect slowly moving targets.

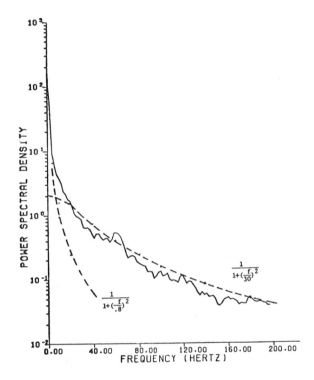

Figure 2. 95 GHz Tree Echo Spectrum

AMPLITUDE HISTOGRAM

In an attempt to better describe the mechanism which cause fixed or slowly moving components in the spectral return, and to account for the probability amplitude distributions, amplitude histograms were generated. An example of such amplitude histograms is shown in Figure 3. The quantifying steps are such that the first unit is some 60 dB below the reference corner reflector. This first width contains all the system noise.

From Figure 3, it is observed that there are two major regions and one minor region where the amplitude distribution is not a monotonically decreasing function with increased magnitude of return signal. These three regions are marked by arrows. These reversals in the slope of the amplitude distribution reveal that there is more than one type of scattering mechanism present. From the target geometry, it is speculated that there is a collection of randomly oriented scatterers such as leaves and small twigs which cause a Rayleigh type response, and a collection of fixed (or very slowly moving) scatters caused by branches and larger objects causing discrete returns.

This concept of several discrete plus random scatterers is consistent with both the power spectrum data and the amplitude histograms presented here for a radar operating at 95 GHz. The histogram in Figure 3 is typical and occurred for most all of the data collected at 95 GHz. Previous data collected at X-band (6) also revealed, on occasions, this same phenomenon of two independent mechanism in the radar echo.

Assume that two independent echo mechanisms, whose signals add, are simultaneously present. Then the observed probability density function is given by the convolution of the probability density functions of the two mechanisms

$$p_o(P) = \int_0^P p_1(P-x) p_2(x) dx$$

Now assume that P_1 is a Rayleigh density function and P_2 is unknown, then it can be shown that

$$p_1(P) = p_o(P) + \bar{P} \frac{dp_o(P)}{dP}$$

This mean that the density function of the unknown mechanism can be determined from the observed density function and its derivative, when the average power, $\bar{P}$, of the Rayleigh component is known. Although it is difficult to obtain the average of the Rayleigh component when only the total signal is known, this can be done by iterating the signal processing procedure.

The importance of this model is that there will be a random scatter mechanism and a collection of fixed or slowly moving echoes. Now when these fixed scatterers occur with sufficient amplitude then there will appear false targets, and care must be taken in designing CFAR circuits.

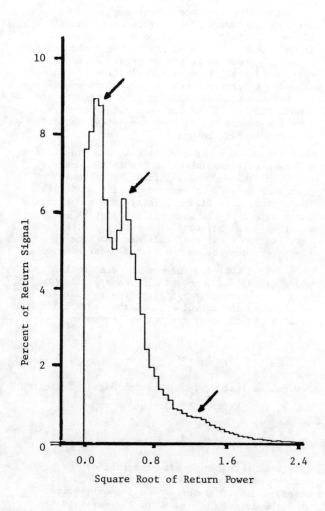

Figure 3. Amplitude Spectra for 95 GHz Tree Echoes

References

1. N.C. Currie, F.B. Dyer and R.D. Hayes, "Radar Land Clutter Measurements at Frequencies of 9.5, 16, 35, and 95 GHz," 2 April 1975.
2. R.N. Trebits, R.D. Hayes and L.C. Bomar, "MM Wave Reflectivity of Land and Sea," Microwave Journal, August 1978.
3. R.D. Hayes, "Fluctuations in Radar Backscatter from Rain and Trees," DARPA/Tri-Service MM Wave Workshop, NELC, San Diego, Nov. 1976.
4. R.D. Hayes, et al "Backscatter from Ground Vegetation at Frequencies Between 10 and 100 GHz," IEEE AP-S International Symposium, 1976.
5. M.W. Long, Radar Reflectivity of Land and Sea, Lexington Books, 1975.
6. R.D. Hayes, et al "Study of Polarization Characteristics of Radar Targets," AD 304 957, Oct. 1958.

CONCLUSIONS

Probability density functions reveal that 95 GHz radar echoes from trees are not Rayleigh distributed but are sometimes log-normal and sometimes Weibull. Power spectral density diagrams and amplitude histograms reveal both discrete fixed returns and fast fluctuations in the radar echoes. It is speculated that the fast fluctuations are caused by moving leaves and small twigs while the fixed (slowly moving) returns are caused by branches and larger objects. The discrete echoes could appear as false targets.

Acknowledgement: This work was in part sponsored by the Martin-Marietta Aerospace Co. of Orlando, Florida and in part by the U.S. Army ARRADCOM.

Fluctuating Clutter

K.L. Koester, L. Kosowsky, and J.F. Sparacio
Appendix B to "Millimeter Wave Radar Applications to Weapons Systems"
V.L. Richard (Paper 1.7)
Pages 107-115

I. INTRODUCTION

In this section we shall be concerned with the reradiation characteristics of terrain at millimeter wavelengths where the short wavelength permits operation of small lightweight radar systems capable of operation in adverse weather over ranges of several nautical miles.

Because of the short wavelength, the clutter environment plays a very significant role in obstacle detectability in that small variations in clutter position cause large fluctuations in the strength of the background signal. A terrain clutter model is presented whose output is characterized by its power spectral density and its probability density function. A comparison with experimental data indicates that the model may confidently be used for signal detection studies at these wavelengths.

II. STATISTICAL NATURE OF CLUTTER

The signal radiated by a large group of scatterers has been studied on a statistical basis for many years. The theoretical models which have served to describe observed phenomena generally consist of an infinite number of independent scatterers, each free to move at a different velocity. The description of the signal emanating from such an array forms a branch of the general theory of random processes dating back to Rayleigh.[1]

The received electric field strength can be shown to be of the form[2]

$$E(t) = R(t) \cos [w_o t + \phi(t)], \qquad (B-1)$$

---

[1] Lord Rayleigh, *Theory of Sound*, Second Edition, Volume 1, MacMillian, 1894.

[2] D. Kerr, *Propagation of Short Radio Waves*, Radiation Laboratory Series, Volume 13, McGraw-Hill Book Company, 1951.

where w_o is the transmitted frequency and $R(t)$ and $\phi(t)$ are functions of scatterer motion. If we define a vector, $R(t)$, whose magnitude is equal to R and whose phase is ϕ, then the probability density function of R, i.e., the probability that R lies between R and R + dR, is given by the well-known Rayleigh distribution[2]

$$P(R) = \frac{2}{P_o} R \exp\left(-\frac{R^2}{P_o}\right), \tag{B-2}$$

where P_o is a parameter of the distribution. A plot of equation (B-2) is given in Figure B-1. The mean value of R is $\frac{\sqrt{\pi P_o}}{2}$, and the rms value is $\sqrt{P_o (4 - \pi)/2}$.

Extending the discussion to the case of a fixed target return plus clutter, Schwartz[3] has shown that the probability density function of R, $q(R)$, is given by

$$q(R) = \frac{2R}{P_o} \exp\left[-\left(\frac{R^2}{P_o}\right) + m^2\right] I_o\left(\frac{2mR}{\sqrt{P_o}}\right), \tag{B-3}$$

where I_o is the modified Bessel function, m^2 is the signal-to-noise power ratio. For $m^2 \gg 1$, equation (B-3) reduces to the Gaussian function,

$$q(R) = \frac{1}{\sqrt{\pi P_o}} \exp\left[-\left(\frac{R}{P_o} - \left(\frac{R}{\sqrt{P_o}} - m\right)^2\right)\right]. \tag{B-4}$$

Thus, for high signal-to-noise ratios, the probability density function of the magnitude of the composite return approaches a Gaussian distribution with mean, $m\sqrt{P_o}$, and standard deviation, $\sqrt{P_o/2}$.

---

[3] M. Schwartz, <u>Information Transmission, Modulation, and Noise</u>, McGraw-Hill Book Company, 1959.

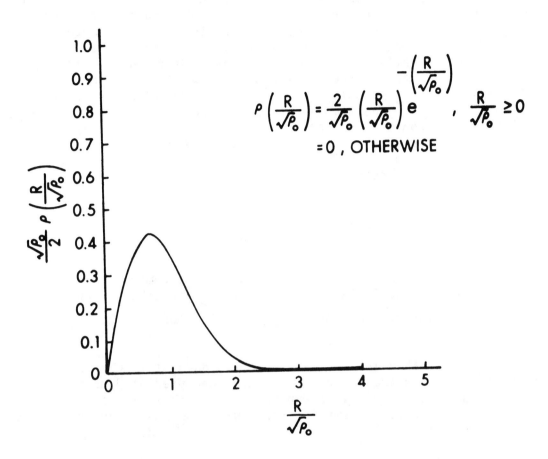

Figure B-1. The Rayleigh Probability Density Function

Up to this point we have been concerned with the long-term statistics of the clutter which essentially determine the "betting curve" of the noise process. Now let us examine the correlation between samples of the clutter signal spaced at second intervals, i.e., the autocorrelation function of the clutter process. This problem has not been solved analytically for the model discussed previously, although much experimental data has been obtained for a wide variety of scatterer mobilities and forcing functions, such as wind. The Fourier transform of the autocorrelation function is the power spectral density, and it is with this function that we shall be concerned. Figure B-2 is a plot of experimental data obtained for various types of terrain when observed at the frequency 70 GHz. Barlow[4] suggests that a good approximation to the power spectral density is given by

$$W(f) = \exp\left[-a\left(\frac{f}{f_o}\right)^2\right], \quad (B-5)$$

where "a" is a spectrum parameter dependent on the composition of the area being observed and f_o is the frequency of operation. Shown on Figure B-2 is a plot of equation (B-5) for values of "a" obtained by Barlow at 1 GHz.

It is evident that in each case observed data at 70 GHz has a broader spectrum than that which would have been predicted using Barlow's parameters. A possible explanation for this discrepancy is that at low frequencies the smaller structures such as leaves and grass play a more prominent role.

In Figures B-3 and B-4 are shown the probability density functions for tree-covered terrain with the mean value removed. For the cases considered, the mean values were large compared to the fluctuating component and produced a Gaussian distribution as predicted by equation (B-4), where noise is replaced by the fluctuating component and m>>1.

III. PROBABILITY DENSITIES OF FLUCTUATING CLUTTER

The problem of describing the radar video return from a fixed target immersed in clutter is analogous to that of describing a signal composed of a sine wave plus narrow band noise. This problem has been extensively treated in the literature.[5,6]

---

[4] E.J. Barlow, "Doppler Radar," Proc. IRE, Vol. 37, April 1949.

[5] S.O. Rice, "Mathematical Analysis of Random Noise," BSTJ, Vol. 24, pp. 98-107.

[6] D. Middleton, Introduction to Statistical Communication Theory, McGraw-Hill Book Company, 1960.

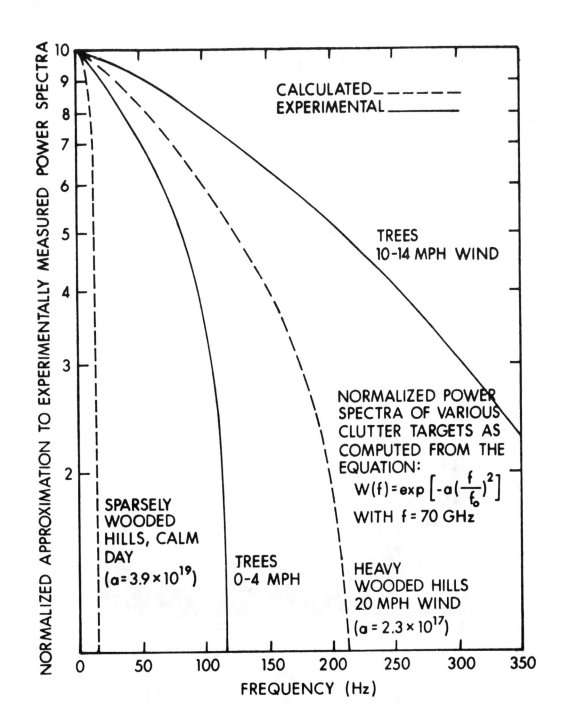

Figure B-2. Clutter Spectrum of Wooded Ground at Millimeter Wavelengths

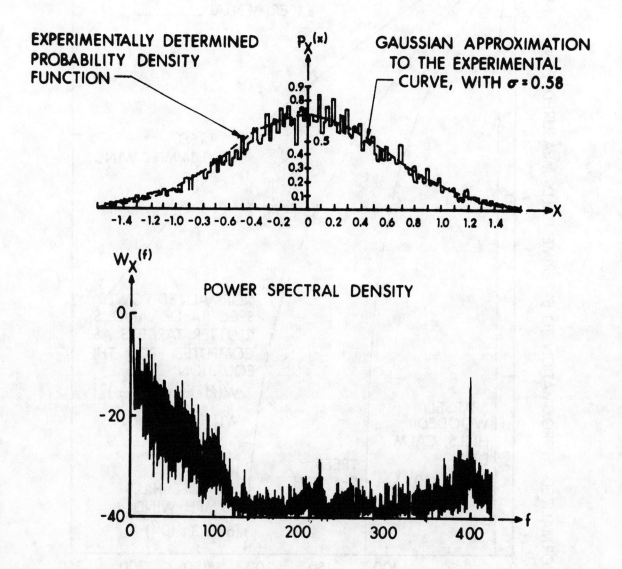

Figure B-3. Characteristics of the Radar-Video Return From A Fluctuating Target

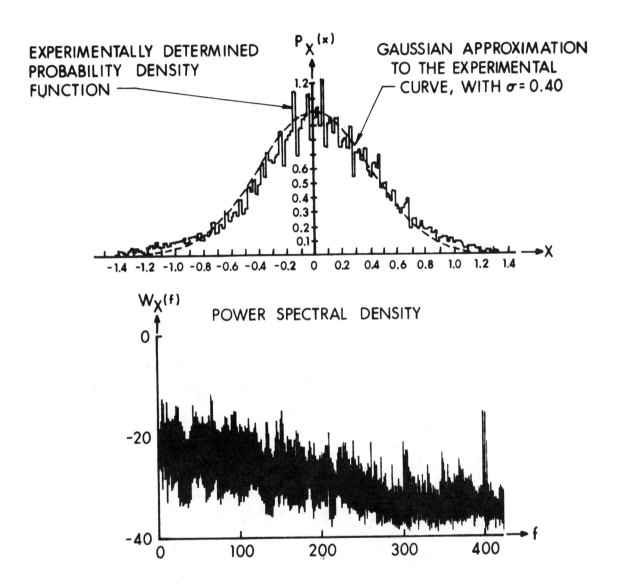

Figure B-4. Characteristics of the Radar-Video Return From A Fluctuating Target

The instantaneous signal voltage, $v(t)$, in the receiver prior to detection can be written as the superposition of the return from a large number of independent scatterers, representing clutter, and an unmodulated sine wave at the radar frequency, f_o, corresponding to the return from the fixed target. Rice[5] has shown that

$$v(t) = [x_c(t) + A] \cos 2\pi f_o t - x_s(t) \sin 2\pi f_o t, \qquad (B-6)$$

where x_c and x_s are normally distributed random variables related to the clutter process. Their probability density functions are given by[5]

$$p(x_c) = \frac{1}{2\pi C} e^{-\frac{x_c^2}{2C}}, \qquad (B-7)$$

$$p(x_s) = \frac{1}{2\pi C} e^{-\frac{x_s^2}{2C}}, \qquad (B-8)$$

where C, the variance of the distributions, is equal to the mean clutter power, and A is the constant amplitude of the return from the fixed target. As the clutter spectrum is narrow band,[4] it is appropriate to write

$$v(t) = r(t) \cos [2\pi f_o t + \theta(t)] \qquad (B-9)$$

with

$$\theta = \tan^{-1} \frac{x_s}{x_c + A} \qquad (B-10)$$

and

$$r^2 = x_s^2 + (x_c + A)^2. \qquad (B-11)$$

For a radar receiver employing a linear envelope detector, envelope $r(t)$ of instantaneous voltage $v(t)$ is equal to the radar video return. It is known that the probability density function of r, $p(r)$, is given by[5]

$$p(r) = \frac{1}{C} e - s^2 \; re - \frac{r^2}{2C} \; I_o \; (rs \, \frac{\sqrt{2}}{C}), \; r \geq 0$$

$$= 0 \qquad\qquad\qquad r < 0 \tag{B-12}$$

where $p_R(r)$ is zero for negative values of r as the radar video is not allowed to make negative excursions. I_o is the modified Bessel function of the first kind and zero order,[7] and

$$s^2 = \frac{A^2}{2C} \tag{B-13}$$

is the target-to-clutter power ratio.

---

[7] Handbook of Mathematical Functions, U.S. Department of Commerce, National Bureau of Standards, Applied Mathematics Series 55, 1964, pp. 374-377.

mm-Wave Reflectivity of Land and Sea

R.N. Trebits, R.D. Hayes, and L.C. Bomar

INTRODUCTION

The current interest in radar applications at millimeter wave frequencies, especially on the part of the military, has presented system designers with a number of new conceptual hardware applications. Unfortunately, there is only a limited data base from which probabilistic models or system performance can be evaluated. Very few calibrated radar reflectivity studies have been conducted to examine the phenomenology of land and sea clutter at frequencies above 16 GHz. In addition, the conventional measures of characterizing radar reflectivity may well not adequately quantify the effects of discrete clutter contributors in light of the improved angular resolution in the radar cell possible at millimeter wavelengths.

The intent of this paper is to present a synopsis of the currently available amplitude and power spectral characteristics of land clutter, sea clutter, and rain backscatter above K_u-band. Emphasis in the data presented is placed on those environmental and radar parameters which can affect the statistical properties of the radar backscatter. In addition, clutter characteristics in the millimeter wave region are compared with those for the lower radar frequencies. References are included which will lead the interested reader into the bulk of the available millimeter reflectivity data.

TERRAIN CLUTTER[1-6]

Radar backscatter data from trees and grass fields were collected by Georgia Tech in a series of experiments to determine the characteristics of ground clutter at millimeter wavelengths. The radars were located on hills in north Georgia and were operated in a searchlight mode for these measurements. The antennas were selected to produce about a 1-degree half-power beamwidth, and the transmitted pulses were usually 100 ns, so that a well defined section of vegetation could be observed. The radar cross section per unit area ($\sigma°$) was observed to increase with increasing radar frequency and to decrease as the radar beam approached the horizon (low grazing angle). The relationship follows the function

$$\sigma° = -20 + 10 \log \frac{\theta}{25} - 15 \log \lambda ,$$

where

$\sigma°$ is in dB,
θ is the depression angle in degrees below the horizon,
λ is radar wavelength in centimeters.

The results of many measurements at X- and K_a-band and at 95 GHz are summarized in **Figure 1**. There did not appear to be much difference in $\sigma°$ between dry deciduous trees and foliage or coniferous trees. Vertically polarized backscatter was usually 3 to 4 dB higher than horizontally polarized backscatter. The return from wet foliage was usually about 5 dB stronger than the return from dry foliage. Grassy fields produced radar returns about 10 dB below that for foliated trees at all frequencies from 10 to 100 GHz. The grass was 2 to 3 feet tall for these measurements, so that the data were not influenced by moisture in the ground.

The spectral density of the radar return from wind-blown trees was dependent on the radar frequency, as shown in **Figure 2**. These data represent the intermodulation of the moving leaves and branches in winds from 6-to-15 mph. As the wind decreased below 5 mph, the spectral content decreased drastically to the extent that there was essentially no intermodulation at millimeter waves on a calm day.

Autocovariance functions generated from these radar signals show that the radar returns decorrelated in less time as the windspeed increased. The time required for the normalized correlation coefficient to reach a value of 0.25 was also noted to decrease as the radar frequency was increased from 10 to 100 GHz. **Figure 3** shows how the decorrelation time was influenced by windspeed and, thus, tree movement as well as the radar frequency.

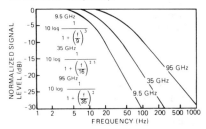

Fig. 2 Normalized Frequency Spectra of Deciduous Tree Clutter.

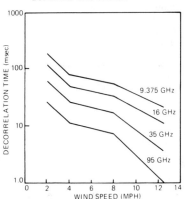

Fig. 3 Decorrelation Time of Tree Clutter as a Function of Wind Speed.

Penetration of the radar signals through the foliage decreased as the radar frequency was increased. The results of penetration measurements through both deciduous and coniferous tree foliage are summarized in **Figure 4**. Attenuation of the radar signals was observed to vary as a function of the foliage density. The data indicate that attenuation varied by a factor of 2 to 1 as a function of seasonal variation of foliage density.

There is only a limited amount of reflectivity data from snow covered

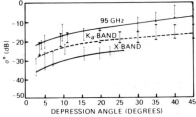

Fig. 1 Measured Normalized Radar Cross Section of Deciduous Tree Clutter, for Dry Conditions, as a Function of Depression Angle.

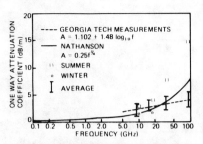

Fig. 4 Median Values of Measured Attenuation Constant for Tree Foliage as a Function of Frequency.

terrain at frequencies above 35 GHz. The data at X-, K_u-, and K_a-bands are somewhat more plentiful and have been calibrated. However, even at these lower radar bands, measurement programs are continuing. The data presented here are considered reliable, but as additional data are collected under differing environmental conditions, some modifications to the σ° dependencies presented can be expected. Power level changes of 10 dB in snow backscatter are to be expected between radar frequencies of 10 GHz and 35 GHz. **Figure 5** is a composite of data from the Georgia Institute of Technology and the University of Kansas depicting this frequency effect. Data were collected in Colorado during the spring when the snow was well frozen and very dry at night. As the sun heated the surface in the morning, the snow would melt and produce free water just below the surface.

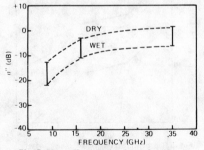

Fig. 5 Normalized Radar Cross Section of Snow Clutter at 40° Incidence Angle.

Reflectivity from metamorphic, aged snow was observed to be strongly influenced by the free water near the surface, as shown in **Figure 6**. This has also been noted by many researchers when observing the power level changes between night and sunny days.

The critical free water content, which caused appreciable changes in signal levels, appeared to be about 2 percent. Appreciable changes in the signal level were not apparent for free water contents between 5 and 20 percent, and the signal level was reasonably stable for free water contents of 1 percent or less. Uncalibrated obser-

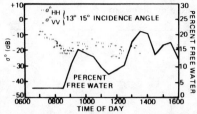

Fig. 6 Normalized Radar Cross Section of Smooth Snow at 35 GHz as a Function of Time of Day and Percent Free Water in Snow.

vations with a passive radiometer demonstrated the same trend in the effective temperature of wet and dry snow, indicating that free water at the snow surface decreases the snow emissivity.

The radar cross section of snow per unit area has been observed to change with grazing angle. **Figure 7** is a composite of several investigations at 35 GHz; note the general trend for the value of σ° to decrease as the line of sight becomes more horizontal to the surface (decreased grazing angle).

Measurements made at the Rome Air Development Center (RADC) indicate a grazing angle dependency not only at 35 GHz, but also at 94 and 140 GHz. The lower grazing angles below 10° result in values of σ° that were about 10-to-15 dB less than the values of σ° measured near normal incidence. The data presented by RADC represent variations in peak values, and the effects of radar frequency on average values of σ° have not been interpreted. Data reduction is still in process, and average values will be obtained in the near future. The trends observed in the RADC data, however, are consistent with data reported by other researchers.

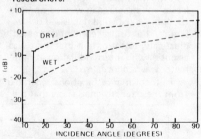

Fig. 7 Normalized Radar Cross Section of Snow Clutter at 35 GHz for Linear Polarizations.

SEA CLUTTER[7-10]

The general characteristics of sea clutter differ significantly from those of land clutter, precipitation, and receiver noise. These differences include decidedly non-Rayleigh amplitude distributions, polarization dependencies, range dependency, frequency spectra, wind/wave vector dependency, and operating frequency dependency, among others. Furthermore, analyses of radar sea clutter data at frequencies above X-band indicate that the general frequency dependencies observed in the microwave region may not necessarily be extended throughout the millimeter wave region.

The general frequency dependency of σ° for sea clutter at near grazing incidence, sea states 2 and 3, and the upwind look direction is shown in **Figure 8**, which represents a compilation of data points collected by Georgia Tech and several other investigators for both horizontal and vertical polarizations. One of the more significant observations of these data is that σ°_{HH} does not tend to increase as rapidly above X-band as it does below, while σ°_{VV} appears to decrease slightly near 95 GHz. In addition, σ°_{HH} exceeds σ°_{VV} by about 5 dB, on the average, at 95 GHz; this relationship is opposite that observed at microwave frequencies.

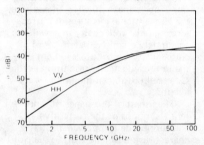

Fig. 8 Normalized Radar Cross Section of Sea Clutter as a Function of Frequency, for Near Grazing Incidence.

These observations imply that it may be advantageous from a sea clutter viewpoint to use vertical, rather than horizontal polarization for 95 GHz radar systems. Conventional microwave radar systems, however, use horizontal polarization in a marine environment.

The amplitude distributions of 95 GHz sea backscatter signals are observed to be approximately log-normal, with standard deviations for horizontal and vertical polarizations near 7 dB and 5 dB, respectively. These distributions at 95 GHz exhibit statistical characteristics which are essentially the same as those observed for microwave sea clutter.

The dependency of σ° on incidence angle (θ) is shown in **Figure 9** for X-, K-, and K_a-bands, vertical polarization, 15 to 20 knot winds, and incident angles between 90° (normal incidence) and 10°. At 95 GHz, the angle dependency near grazing is between θ^0 and θ^{+1} for rough water and the upwind look direction, but it can be inversely proportional to θ for low sea state conditions or the cross-wind look direction. As a rule, σ° tends to be larger for upwind than for downwind look directions.

Millimeter Wave Propagation, Targets and Clutter

Sea backscatter at millimeter wave frequencies, as at microwave frequencies, increases with increasing wind speeds. The dependency on wind speed W is approximately proportional to $W^{1.5}$ at $1.4°$ incidence and to W^4 at $2.8°$ incidence, as calculated from 9.4 to 95 GHz data collected by Georgia Tech at Boca Raton, Florida.

The representative autocorrelation dependency for 95-GHz sea clutter, shown in **Figure 10**, indicates regions associated with the decorrelation due to random scatterers, the rearrangement of the surface, and the passage of waves through the resolution cell. The width of the initial spike is directly related to operating frequency, but other characteristics tend to be independent of frequency. There is, however, a tendency for the decorrelation to vary with look direction relative to the sea wave propagation vector.

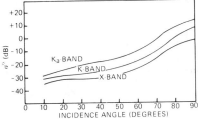

Fig. 9 Normalized Radar Cross Section of Sea Clutter as a Function of Incidence Angle.

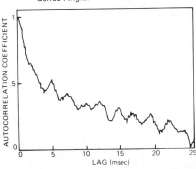

Fig. 10 Autocorrelation Coefficient of Sea Clutter Return at 95 GHz.

RAIN BACKSCATTER[11]

Rain radar backscatter has been measured for many years by many researchers, but the first observation made at several frequencies above 16 GHz, simultaneously, was a joint project between Frankford Arsenal, Balistics Research Laboratories, and Georgia Tech. The volumetric radar cross section was observed to have an exponential relationship with rain rate, as shown in **Figure 11**. Note that the relationships at 9.375 and 35 GHz have the same slope, but the slopes change at 70 and 95 GHz, and the backscatter at 95 GHz is less than that at 70 GHz.

The backscatter at X- and K_a-bands follows the general relationship described by Rayleigh, since the drop diameter is small relative to the radar wavelength. As the radar wavelength is decreased, the scattering approaches that described by Mie, and at 95 GHz there is perhaps even some forward scattering.

Backscatter from rain was observed to depend strongly on polarization. The magnitude of backscatter for circular polarization was from 10 to 15 dB below the backscatter observed for linear polarization. This difference appeared to be true at all frequencies from 10 GHz to 100 GHz.

Although not displayed in **Figure 11**, the backscatter was found to be dependent on the drop size distribution (for a given rain rate) as well as on the rain rate, or total water content. Drop size spectrometers were employed during these experiments, and the mean values plotted in **Figure 11** had ± 5 dB spread. In every case investigated, the volumetric backscatter for rain containing drops of 2 to 6 mm in diameter was some 10 dB higher than for rain containing drops 1 mm in diameter or less, even though the rain rates were the same.

Intermodulation of the radar signal by the rain drops in the cell of observation creates high frequency components in the return signal. As can be seen in **Figure 12**, the higher the radar frequency, the higher was the frequency content in the return signal. Note also that the rate of roll-off of the high frequency content is slower at 95 GHz than at 9.4 GHz. It is postulated that this higher frequency content is due to (1) the 3.2 mm radar wavelength being

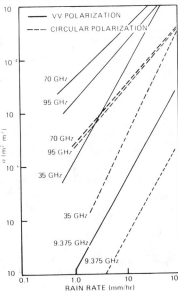

Fig. 11 Least Mean Square Fit to Radar Backscatter Data from Rain.

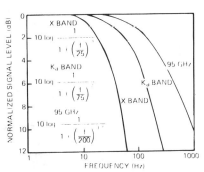

Fig. 12 Normalized Frequency Spectra of the Backscatter from Rain.

approximately the same as the rain-drop diameter and (2) the multi-bounce scattering between resonant size drops in the cell of resolution. A comparison of the harmonic content in the backscatter from rain with backscatter from trees indicated that the signal content for rain at a 5 mm/hr. rain rate was eight times that for trees in winds of 6 to 15 mph.

Autocovariance functions were calculated from the radar returns, and when the correlation coefficient decreased to a value of 0.25, the returns were considered to be decorrelated. **Figure 13** shows graphically that the return decorrelates more rapidly with increasing rain rate as well as with increasing radar frequency. At millimeter wavelengths the decorrelation time is short and almost constant with respect to rain rate, having a change of only 2 to 1 between rainfall rates of 5 and 90 mm/hr., whereas the slope of the decorrelation time with rain rate is greatly affected by the magnitude of the rain rate for microwave radar opera-

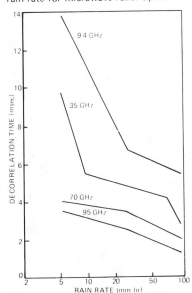

Fig. 13 Decorrelation Time of Radar Backscatter from Rain.

short time period. The mixers have a single sideband noise temperature of a few hundred degrees at 100 GHz and about 1000°K near 200 GHz. Further improvements can be expected at the higher frequencies using thinner epitaxial layers and new junction geometries for the mixer diodes.

REFERENCES

1. Armstrong, E. H., "A new system of short-wave amplification," *Proc. IRE*, Vol. 9, Feb. 1921, pp. 3-27.
2. Poss, H. L., "The patent system—one scientist's story," *Physics Today*, Vol. 31, Jan. 1978, pp. 9-11.
3. Linke, R. A., M. V. Schneider and A. Y. Cho, "Cryogenic millimeter-wave receiver using molecular beam epitaxy diodes," *IEEE Trans. MTT*, Vol. MTT-26, Dec. 1978.
4. Cong, H., A. R. Kerr, P. Thaddeus and R. J. Mattauch, "The low-noise 115-GHz receiver front-end on the Columbia-GISS four-foot telescope," *IEEE Trans. MTT*, to be published.
5. Schottky, W., and E. Spenke, "Zur quantitativen Durchfuhrung der Raumladungs-und Randschichttheorie der Kristallgleichrichter," *Veröffentichungen aus den Siemens-Werken*, Vol. 18, 1939, pp. 225-291.
6. Sze, S. M., *Physics of Semiconductor Devices*, New York, Wiley, 1968.
7. Schneider, M. V., R. A. Linke and A. Y. Cho, "Low-noise millimeter-wave mixer diodes prepared by molecular beam epitaxy (MBE)," *Appl. Phys. Letters*, Vol. 31, August 1, 1977, pp. 219-221.
8. Schneider, M. V., and E. R. Carlson, "Notch-front diodes for millimeter-wave integrated circuits," *Electronics Letters*, Vol. 13, Nov. 1977, pp. 745-747.
9. Kerr, A. R., "Low-noise room-temperature and cryogenic mixers for 80-120 GHz," *IEEE Trans. MIT*, Vol. MTT-23, Oct. 1975, pp. 781-787.
10. McMaster, T. F., M. V. Schneider and W. W. Snell, Jr., "Millimeter-wave receivers with subharmonic pump," *IEEE Trans. MTT*, Vol. MTT-24, Dec. 1976, pp. 948-952.
11. Carlson, E. R., M. V. Schneider and T. F. McMaster, "Subharmonically-pumped millimeter-wave mixers," *IEEE Trans. MTT*, Vol. MTT-26, Oct. 1978.
12. Goldsmith, P. F., and R. L. Plambeck, "A 230-GHz radiometer system employing a second-harmonic mixer," *IEEE Trans. MTT*, Vol. MTT-24, Nov. 1976, pp. 859-861.
13. Torrey, H. C., and C. A. Whitmer, *Crystal Rectifiers* (M.I.T. Radiation Lab. Ser. Vol. 15). New York, McGraw-Hill, 1948.
14. Dragone, C., "Analysis of thermal shot noise in pumped resistive diodes," *Bell Syst. Tech. J.*, Vol. 47, 1968, pp. 1883-1902.
15. Saleh, A. A. M., *Theory of Resistive Mixers*, Cambridge, MA, M.I.T. Press, 1971.
16. Barber, M. R., "Noise figure and conversion loss of the Schottky barrier mixer diode," *IEEE Trans. MTT*, Vol. MTT-15, Nov. 1967, pp. 629-635.
17. Held, D. N., and A. R. Kerr, "Conversion loss and noise of microwave and millimeter-wave mixers," *IEEE Trans. MTT*, Vol. MTT-26, Feb. 1978, pp. 49-61.
18. Kerr, A. R., R. J. Mattauch and J. A. Grange, "A new mixer design for 140-220 GHz," *IEEE Trans. MTT*, Vol. MTT-25, May 1977, pp. 399-401.

tions. When comparing radar returns from rain with those from trees, rain returns decorrelate in about 1/8 the time.

CONCLUSIONS

As broad a subject as radar clutter cannot be extensively covered in just a few pages, even when restricted to just the millimeter wave region. This discussion was, therefore, intended to acquaint the reader with some of the more pertinent characteristics of terrain, sea, and rain backscatter at millimeter wave frequencies, and to compare them to the perhaps more familiar signal characteristics at microwave frequencies.

The amount of millimeter wave backscatter data is far less extensive than that documented at microwave frequencies. The need for expansion of this millimeter wave data base has been recognized, and Georgia Tech is presently participating in several research programs involving measurement of the reflectivity characteristics of terrain, snow, and sea clutter at 35 and 95 GHz. These investigations should lead to significant improvements in the characterization and understanding of millimeter wave clutter signals.

REFERENCES

1. Hayes, R. D., F. B. Dyer, and N. C. Currie, "Backscatter from Ground Vegetation at Frequencies Between 10 and 100 GHz," *1976 Int IEEE/AP-S Symp.*
2. Currie, N. C., F. B. Dyer, and E. E. Martin, "Millimeter Foliage Penetration Measurements," *1976 Int'l IEEE/AP-S Symp.*
3. Currie, N. C., et al., "Radar Millimeter Backscatter Measurements Volume I. Snow and Vegetation," *Report No. AFATL-TR-77-92*, July, 1977.
4. Currie, N. C., F. B. Dyer, and G. W. Ewell, "Characteristics of Snow at Millimeter Wavelengths," *1976 Int'l IEEE/AP-S Symp.*
5. Ulaby, F. T., et al., "Microwave Remote Sensing of Snow," The University of Kansas Center for Research, Inc., *RSL Technical Report 340-1*, June, 1977.
6. Currie, N. C., and T. P. Morton, "Comparison of Microwave Winter/Summer Foliage Penetration Measurements," *Final Technical Report, Georgia Tech Project A-1858*, January 1977.
7. Grant, C. R., and B. S. Yaplee, "Reflectivity of the Sea and Land at Centimeter and Millimeter Wavelengths," *NRL Report 4787*, July, 1956.
8. Dyer, F. B., M. J. Gary, and G. W. Ewell, "Radar Sea Clutter at 9.5, 16.5, 34 and 95 GHz," *1974 Int'l IEEE/AP-S Symp.*
9. Rivers, W. K., "Low-angle Radar Sea Return at 3-mm Wavelength," *Final Technical Report, Georgia Tech Project A-1268*, November, 1970.
10. Trebits, R. N., et al., "Millimeter Radar Sea Return Study," *Technical Report, Georgia Tech Project A-2013*, to be published.
11. Currie, N. C., F. B. Dyer, and R. D. Hayes, "Some Properties of Radar Returns from Rain at 9.375, 35.70 and 95 GHz," *1975 Int'l IEEE Radar Conf.*

Radar Sea Clutter at Millimeter Wavelengths — Characteristics and Pivotal Unknowns
G.W. Ewell, R.N. Trebits, and J. Teti
6th DARPA-Tri Service Millimeter Waves Conference, 1977
Pages 199-205
Reprinted by permission.

2.C.3

ABSTRACT

Principal characteristics of millimeter radar sea clutter which impact surface Navy requirements are reviewed. Available data are briefly surveyed, and additional data requirements addressed.

SUMMARY

An accurate knowledge of the behavior of radar sea clutter at millimeter wave frequencies is essential if a meaningful assessment of the applicability of millimeter waves to the surface Navy's applications is to be accomplished. The general characteristics of radar sea clutter differ significantly from those of many other forms of interfering signals, such as those from land clutter, precipitation, and receiver noise. Some of the more startling features of radar sea clutter return involve the decidedly non-Rayleigh amplitude distributions, polarization dependencies, range dependence, auto-covariance and frequency spectra, forward scattering of the sea surface, dependence upon wind direction, and dependence upon operating frequency, to name only a few.

In recent years researchers at the Georgia Tech Engineering Experiment Station have conducted a number of programs involving measurement of radar sea clutter reflectivity at X-band frequencies and above. Results of these programs are scattered throughout the literature and some data are in reports which are no longer available or whose distribution is limited. This paper collects a selected subset of these data in order to make these results more generally available. These data are particularly significant since they indicate the frequency dependences which have been observed at lower frequencies may not necessarily be extended to millimeter wavelengths.

When one observes the backscattered energy from the sea using a microwave radar system, one of the more startling features of the return involves its change in character with transmitted and received polari-

---

*Dr. Ewell is with the Engineering Experiment Station, Georgia Tech, Atlanta, GA. Dr. Trebits is with the Engineering Experiment Station, Georgia Tech, Atlanta, GA. Mr. Teti is with the Naval Surface Weapons Center, Dahlgren, VA. This work is related to efforts conducted for NSWC under Contract No. N60921-77-C-A168.

zations. For many conditions, the return using horizontal polarization has a much more "spiky" and irregular appearance than does the return for vertical polarization; the return using horizontal polarization has a nature more like that resulting from a number of isolated point scatterers than does the signal for vertical polarization. While vertical polarization is more nearly noise-like, substantial differences still exist between the fluctuation behavior of sea clutter return and that of receiver noise. For instance, the width of the amplitude distributions is considerably wider for either horizontal or vertical polarization than it is for receiver noise. In fact, system models based on Rayleigh distributed sea clutter will be decidedly in error and polarization dependence must form an integral part of any radar sea clutter modeling effort.

Both the average value and statistical behavior of the sea clutter return are affected by the local grazing angle. For a mixed homogeneous atmosphere, range dependence of sea clutter return approaches R^{-3} at short ranges, decreasing to R^{-7} at longer ranges; this varying range law is due to the interference between direct and reflected rays, and the range at which the range dependence changes is commonly referred to as the "transition range," which is a function of antenna height, operating frequency, and sea surface characteristics (1).

The look direction relative to the wind/sea direction affects the sea clutter return. The degree of dependence is a function of polarization, operating frequency, and sea state. Data indicate up sea/down sea ratios as high as 20 dB, with a fairly weak frequency dependence.

The behavior of σ^o with frequency is a matter of primary concern in system planning. The theoretically predicted frequency dependence is a function of the specific models assumed, and only fragmentary data for a limited range of environments are available (2-7). The temporal decorrelation of sea clutter return may strongly affect the effectiveness of many system integration techniques; returns may exhibit substantial decorrelation over time periods from 10-100 milliseconds.

These are only a few of the major features of radar sea clutter at millimeter wavelengths. When one examines the data available describing dependences, one finds very little data readily available above 10 GHz (1,2,3,4,5,6,7). When one further confines oneself to pulsed systems near grazing incidence, the body of available data is even smaller. In recent years, Georgia Tech has conducted several experiments which have yielded data describing radar sea clutter return near grazing incidence at millimeter wavelengths. Principal experiments were a simultaneous 9.4 and 95 GHz program conducted at Georgia Tech Field Site, Boca Raton, Florida, and 9.5, 16.5 and 35 GHz experiments conducted at the Tech Field Site and from the NUC Oceanographic Research Tower located off Mission Beach, San Diego, California.

The 9.4 and 95 GHz sea return measurements (2) show trends with changes in geometry and environmental variables which are similar to those which are generally observed at lower frequencies. The average cross-section per unit area, σ^o, increases with increasing wind and is generally larger for upwind than for crosswind look directions. The dependence on wind speed, W, is approximately proportional to $W^{1.5}$ at an incident angle of 1.4 degrees, and proportional to W^4 at 2.8 degrees. Only low and moderate wind conditions were experienced so it is not known specifically how saturation of σ^o occurs for high sea states. The angle dependence near grazing is between ψ^o and ψ^{+1} for rough water and upwind look direction, but it can be inversely proportional to angle for low sea state or cross wind look direction. In contradistinction to most experience at low frequencies, σ^o_{HH} is greater than σ^o_{VV} by about 5 dB, average (see Figure 1). Comparison of values measured at 95 GHz with those measured at 9.4 GHz suggests that σ^o does not increase with frequency above X-band, and that for the higher sea states some decrease is evident, particulary for vertical polarization (see Figure 2). Other experiments tend to indicate sea clutter return may be maximized at frequencies near 16 GHz (3). Statistical distributions were again approximately lognormal, with standard deviations for horizontal and vertical polarizations near 7 and 4.7 dB respectively.

A representative autocorrelation function for 95 GHz sea clutter return is given as Figure 3, clearly showing regions associated with the decorrelation due to random scatterers, due to re-arrangement of the surface, and due to passage of waves through the resolution cell (2). The width of the initial spike is directly related to operating frequency, but other characteristics tend to be independent of frequency. There is, however, a decided dependence of decorrelation upon look direction relative to the sea wave propagation vector.

As was the case of the two-frequency experiment, the three-frequency data showed some trends with changes in geometry not dissimilar to those observed at microwave frequencies. Again, σ^o increased with increasing wind speed and was largest in the up wind/sea direction; information summarizing variation with polarization and look direction is given in Figure 4. Some representative cross section data are summarized in Table I. Returns exhibited amplitude fluctuations which were approximately lognormally distributed; standard deviations were functions of depression angle, frequency, polarization, sea state, and look direction. Correlation was a function of frequency, polarization, and look direction.

When one examines the body of sea clutter data available at millimeter wavelengths, much of it taken with systems having substantially different characteristics, acquired for geometries which vary widely, and in environments whose characterization is sketchy at best, it is difficult to quantify the principal dependencies of millimeter radar sea clutter. It would be of considerable benefit if a carefully instrumented and calibrated multifrequency experiment could be conducted with particular emphasis placed on characterization of the environment, and identification of sea clutter variations with look direction, depression angle (or range), polarization, environmental factors, system resolution and, of course, operating frequency.

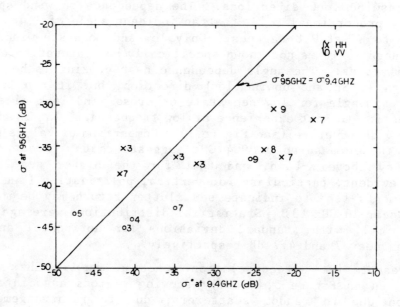

FIGURE 1. σ^0 AT 9.4 AND 95 GHz (2).

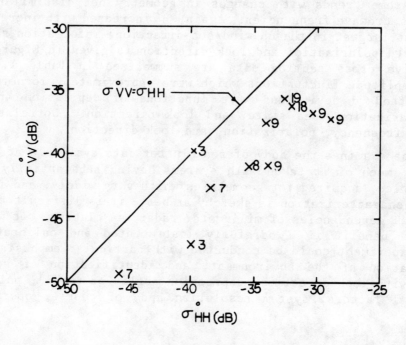

FIGURE 2. SCATTER DIAGRAM FOR HORIZONTALLY AND
VERTICALLY POLARIZED RETURN AT 95 GHz (2).

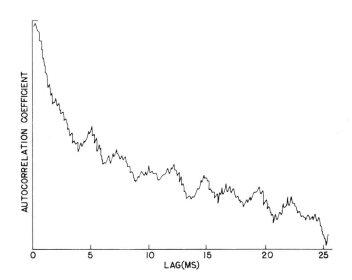

FIGURE 3. AUTOCORRELATION FUNCTION OF 95 GHz
SEA CLUTTER RETURN (2).

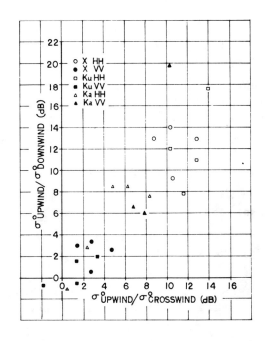

FIGURE 4. VARIATION OF RETURN WITH LOOK DIRECTION
RELATIVE TO THE LOCAL WIND/WAVE DIRECTION.

Table I

σ^o

FOR SEA CLUTTER IN dB (3)

Significant Wave Height (ft)	Wind Speed (mph)	σ^o (dB) Frequency/Polarization					
		9.5 GHz		16.5 GHz		35 GHz	
		HH	VV	HH	VV	HH	VV
0.64	0	<-70.0	<-73	<-70	<-70	-63.7	-59.8
0.8	0	-69.5	-60.2	<-70	-65.6	-41.8	-43.6
0.96	2	-64.0	-50.0	-43.4	-45.8	-	-
1.1	11	-49.5	-41.0	-49.3	-44.1	-	-
1.6	5	-47.4	-45.4	-37.3	-33.8	-42.8	-38.2
1.6	12	-47.0	-42.0	-38.2	-38.6	-44.1	-37.3
1.7	14	-48.4	-54.4	-30.0	-24.7	-	-
3.7	11	-42.8	-36.5	-	-	-	-
3.7	11	-41.2	-39.8	-	-	-	-

REFERENCES

1. "Modeling for Radar Detection (U)," W. K. Rivers, S. P. Zehner, and F. B. Dyer, Final Technical Report, Contract N00024-69-C-5430, Georgia Institute of Technology, 31 December 1969, SECRET.

2. "Low-Angle Radar Sea Return at 3 mm Wavelength," W. K. Rivers, Final Technical Report on Contract N62269-70-C-0489, Georgia Institute of Technology, 15 November 1970, UNCLASSIFIED.

3. "Testing of Radars to Determine Characteristics for Swimmer Defense Programs (U)," G. W. Ewell, Final Recommendation Report, Contract N00024-73-C-5415, October 1973, CONFIDENTIAL.

4. "Backscattering from Water and Land at Centimeter and Millimeter Wavelengths," C. R. Grant and B. S. Yaplee, Proc IRE 45:976, July 1957.

5. "Backscattering Characteristics of the Sea in the Region from 10 to 50 KMC," J. C. Wiltse, S. P. Schlesinger, and C. M. Johnson, Proc IRE 45:220, February 1957.

6. "OWEX II Radar Tests," G. W. Ewell, Final Technical Report, Contract N00123-74-C-2048, Georgia Institute of Technology, June 1975, UNCLASSIFIED.

7. "Backscattering Cross-Sections at 4.3 mm Wavelengths of Moderate Sea Surfaces," C. W. Tolbert, C. O. Britt, and A. W. Straiton, University of Texas EERL Report No. 95, 8 November 1957, UNCLASSIFIED.

Multifrequency Millimeter Radar Sea Clutter Measurements

R.N. Trebits, N.C. Currie, and F.B. Dyer

ABSTRACT

A recent radar sea clutter backscatter measurement program at 9.5, 16, 35, and 95 GHz is described; parallel polarization data were recorded for a variety of wave height, wind speed, wind direction, and geometric conditions. Preliminary analyses of selected data are presented for horizontal and vertical transmitted polarizations.

INTRODUCTION

A set of radar sea clutter measurements was undertaken during the winter of 1978-79 by the Georgia Tech Engineering Experiment Station under contract to the Naval Surface Weapons Center at Dahlgren, Virginia. The major objective of this measurement program was to create a radar sea clutter backscatter data base in the millimeter wave region over a wide range of environmental conditions which are characteristic of the open sea. The subsequent analytical characterization of millimeter wave sea clutter returns will permit more accurate modeling of sea reflectivity and calculation of radar system performance in this frequency region. Especially important features of this experiment include the simultaneous measurement of sea return across the microwave and millimeter wave regions and the measurement of sea return in both the plateau and interference regions in a single experiment, thereby reducing the uncertainty of combining data from different experiments having different radar systems, geometries, and environments.

Simultaneous measurement of sampled sea clutter backscatter signals for frequencies between 9.5 and 95 GHz permits parametric characterization of these returns for nearly identical sea state, meteorological, and geometric conditions. The measurements made at 9.5 GHz provide reference to the rather extensive radar sea clutter data base documented at that frequency, while the measurements made at 16, 35, and 95 GHz will significantly extend the presently very limited radar sea clutter data base at the higher microwave freqencies into the millimeter wave region. [1-3]

FACILITY DESCRIPTION

The site used for this measurement exercise was the Stage II Research Tower, which is operated by the Naval Coastal Systems Center at Panama City, Florida. Stage II is located approximately three kilometers off-shore Panama City Beach in reasonably open sea conditions and contains onboard power generators, laboratory space, and marginally adequate living accommodations. The water depth beneath Stage II is twenty meters, and the sea bottom is essentially flat and featureless. Southwest to southeast of Stage II the nearest significant land masses are farther than 500 km away.

Radar backscatter data were recorded under a variety of sea state conditions, including sea states 1 through 3. Radar look directions included upwind, crosswind, and downwind aspects over a range of grazing angles between 0.5 and 7.8 degrees. Additional data were recorded for conditions where the wind direction differed significantly from the wavefront direction, i.e., decoupled wind/wave conditions which may precede the development of a fully arisen sea. Radar data recording was augmented by documentation of local wind speed/direction, significant wave height (highest 1/3 and 1/10), wavefront direction, atmospheric temperature, barometric pressure, and relative humidity. The sensors, for most cases, were located on Stage II itself. One hundred and forty-six (146) useful data runs were recorded, each of which consists of three minutes of sampled radar video sea return, accompanied by an additional 46 calibration runs.

RADAR DESCRIPTIONS

The relevant parameters of the four radar systems used in this measurement program whose data are described in this paper are listed in Table 1. All four radar systems transmit either horizontal or vertical polarization (selectable) and simultaneously receive both the parallel- and cross-polarization. Polarization isolation exceeded 20 dB for all radar systems, and all intermediate frequency amplifiers used had logarithmic transfer characteristics. Simultaneous sampling of all eight radar video signals (parallel-and cross-polarized components for each of four radar systems) was performed for recording of the sea return data on FM tape recorders. These sampled signals were also monitored on CRT displays and on strip chart recorders.

Table 1
Radar System Parameters

Frequency (GHz)	9.5	16	35	95
Peak Power (kW)	50	50	5	1
Pulse Length (ns)	50	50	100	100
Beamwidth (deg)	1.5	1.5	0.7	0.7

CALIBRATION PROCEDURES

The system amplitude calibration for the 9.5, 16, and 35 GHz radar systems was effected using appropriate signal generators and sets of calibrated attenuators. A second, reference calibration of these radars was made using the signal backscattered from a trihedral corner reflector of known radar cross section and calculating the received signal level from the two-way radar range equation and measured system parameters. Since a signal source calibration was not feasible for the 95 GHz radar, this radar system's response function was established using the return from the trihedral corner reflector and a calibrated attenuator. For most of this measurement program, calibration of the 95 GHz cross-polarization channel was not performed.

The vertical multipath field was probed for all four radar frequencies over a three meter vertical extent. A trihedral corner reflector was mounted on a motor driven carriage, which allowed a relatively constant radar cross section over the entire range of target heights. In addition, propagation loss (i.e., path loss) was also monitored at the lower three frequencies, to be used to correct the sea backscatter signal amplitudes at all four frequencies for excessive path loss effects.

DATA ANALYSIS

For the purposes of presenting this paper, a limited subset of the data was analyzed. The data set chosen consisted of three minute recordings of sea return for upwind, downwind, and crosswind conditions at 9.5, 16, 35, and 95 GHz. The wave height for this data set was approximately two to five and a half feet, while the wind speed was 12 to 15 knots. The data recordings were processed by minicomputer to yield amplitude distributions, average values, and frequency spectra. Some of these results are discussed below.

The average backscatter values were normalized to radar cross section per unit area and are given in Figures 1 and 2. The solid lines shown on the figures are the predicted return generated by a computer program which was previously developed at Georgia Tech utilizing the millimeter sea data then available in the literature. [1] The figures show that at 35 GHz the VV downwind data are larger than the values predicted by the model, and at 95 GHz all of the data are larger than the values predicted by the model.

One major difference between the analyzed data and the model is that the model predicts 95 GHz returns to be smaller than 35 GHz returns, while the analyzed data show that 95 GHz returns are about 5 dB higher than the 35 GHz returns.

Figure 3 summarizes a set of data points collected and recorded during one day, for all four radar systems. Each point in Figure 3 represents horizontally polarized backscatter for several depression angles within the plateau region. The X- and Ku-band data are consistent with formerly reported data. The overall frequency dependence of normalized radar cross section appears to be a weak one.

Figures 4 and 5 illustrate typical amplitude distributions for the data. A large difference in the distribution width between horizontal polarization and vertical polarization is observed at 35 GHz while much less difference is observed at 95 GHz. Typical standard deviation values were 5-6 dB for 35 GHz, VV polarization, 10-11 dB for 35 GHz, HH polarization; and 10-12 dB for VV and HH polarizations at 95 GHz.

Previous data at 95 GHz have indicated that the relative magnitude of VV polarization versus HH polarization is different at 95 GHz as opposed to 35 GHz and lower frequencies. That is, while VV polarization is normally larger than HH polarization at frequencies up to 35 GHz, at 95 GHz horizontal polarization tends to be larger than vertical in magnitude. The present data support this conclusion. Figure 6 is a scatter diagram of the values of HH and VV for the same beam positions for 35 and 95 GHz. Values to the left of the diagonal dotted line indicate that $\sigma^{0}(VV)$ is larger than $\sigma^{0}(HH)$ while values to the right of the dotted line indicate that $\sigma^{0}(HH)$ is larger than $\sigma^{0}(VV)$. As can be seen, at 35 GHz most of the values fall to the left of line, while at 95 GHz most of the values fall to right of the line, indicating quite different polarization behavior between the two frequencies.

Typical frequency spectral plots at 35 GHz and 95 GHz are given in Figures 7 and 8. The figures show that the spectral width is much greater at 95 GHz than at 35 GHz, while the roll-off slope is somewhat steeper at 35 GHz than at 95 GHz. It has been found that the spectra for other types of clutter, such as backscatter from rain and trees, tend to roll-off with a characteristic given by $1/(1 + (f/f_c)^n)$ at millimeter wavelengths. The limited data analyzed so far indicate that f_c should be between 100-200 Hz at 35 GHz and 350-500 Hz at 95 GHz and that n should be three for both frequencies.

CONCLUSIONS

Since the data presented here represent only a portion of one day's data and are not supported by a full analysis and evaluation, it is perhaps premature to draw any major conclusions. Clearly the results of the initial analysis are encouraging in that it appears that the data are both internally consistent and generally support the anticipated behavior of sea return from 9.5 to 95 GHz. The data of the scatter diagram (Figure 6) suggest that a careful analysis of the influence of the discrete (i.e., spikey) returns is needed to develop a satisfactory physical model of the scattering mechanism. An initial assessment of the data relative to the model suggests that the present model will continue to serve as a useful basis for system analysis with only modest adjustment.

REFERENCES

1. R. N. Trebits, et al., "Millimeter Radar Sea Return Study," Contract No. N60921-77-C-A168, Interim Technical Report, Georgia Institute of Technology, July 1978.
2. M. W. Horst, et al., "Radar Sea Clutter Model," 1978 International IEEE/AP-S URSI Symposium, May 1978.
3. M. W. Long, Radar Reflectiivity of Land and Sea, (Lexington, Massachusetts: D.C. Heath and Company, 1975).

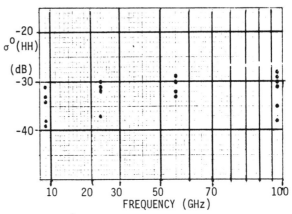

Figure 3. σ^0 (HH) Versus Frequency for Various Depression Angles.

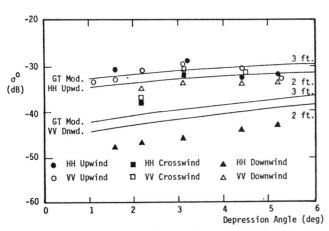

Figure 1. 35 GHz Sea Return Data for 2.6 Foot Wave Height.

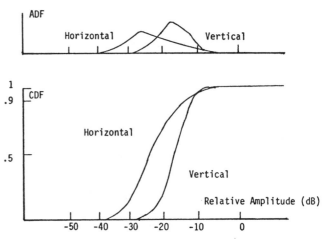

Figure 4. 35 GHz Sea Return Amplitude Distributions.

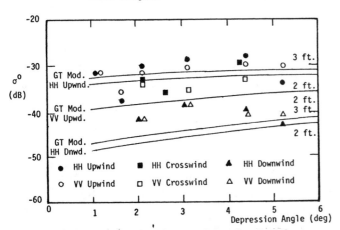

Figure 2. 95 GHz Sea Return Data for 2.6 Foot Wave Height.

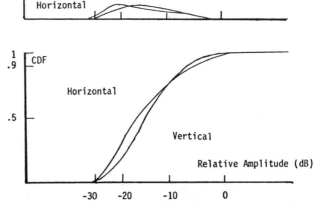

Figure 5. 95 GHz Sea Return Amplitude Distributions.

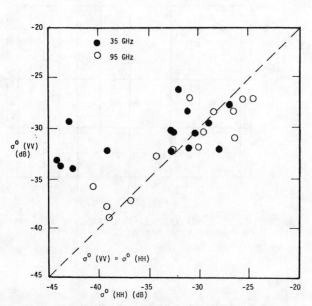

Figure 6. Comparison of HH and VV Radar Cross Section Data for 35 GHz and 95 GHz.

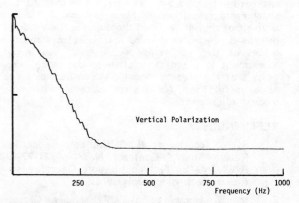

Figure 7. Power Spectral Density of 35 GHz Sea Return.

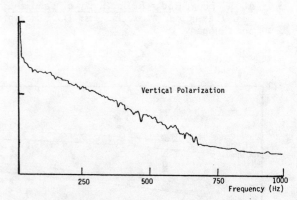

Figure 8. Power Spectral Density of 95 GHz Sea Return.

CHAPTER 3

MILLIMETER WAVE RADAR RF SOURCES AND COMPONENTS

Practically every survey paper on millimeter wave applications (*cf.* Chapter 1) indicates that strong factors in the ultimate employment of millimeter wave systems are the availability, cost, and performance of RF sources and components. This is especially true for radar applications. Such applications will vary in complexity from simple systems to highly complex ones which hopefully would employ sophisticated waveforms, antennas, and signal processing already used in microwave radar. Availability for many applications will require large quantities of components, not merely "ones and twos" as in some functions, *e.g.* spectroscopy. Reliability will also be extremely important for millimeter wave radar.

Papers in this chapter treat millimeter wave RF sources, wave guide, mixers and antennas. Some millimeter wave radar source and component requirements are similar to those of some other millimeter wave applications, *e.g.* communications. This is partially true for RF sources. For example, local oscillators in superheterodyne receivers are generally similar in both radar and in communications. Solid state CW RF sources are ideally suited for local oscillators. Solid state local oscillators are smaller, lighter, operate at lower voltages and are more reliable than, for example, klystrons. For some millimeter wave radar applications, solid state RF sources may not have adequate performance such as power output, or operating frequency capability. Accordingly, electron sources (thermonic devices) such as the backward wave oscillator (BWO), the extended interaction oscillator (EIO), and the gyrotron have roles in millimeter wave radar. Some millimeter wave radars operating at very short ranges are ideally suited for solid state RF sources (both CW and pulsed), of course. Solid state millimeter wave radar transmitters will be described. One is a hybrid unit.

SYNOPSIS OF REPRINT PAPERS IN CHAPTER 3

3.A Solid State RF Sources

3.A.1 "Millimetre Wavelength IMPATT Sources" — Purcell

The IMPact ionization Avalanche Transit Time (IMPATT) or avalanche diode is now an important solid state RF source for low power millimeter wave radar transmitters. This paper, from a recent special issue of *The Radio and Electronic Engineer*, describes the theory of operation, diode design and fabrication, present status, and future prospects. Both CW and pulsed operation are treated. Discussion of IMPATT reliability is noteworthy. Interestingly, IMPATT life expectancy follows the classical "bath-tub" pattern in which mortality rates are high during infancy and senility, and are at a low constant level between the two. Data points showing results from various companies are very interesting.

3.A.2 "Advances in mm-Wave Components and Systems" — Adelseck, Barth, Hoffman, Meinel, Rembold

This *AGARD 1978* paper treats state-of-the-art and availability of mm-wave semiconductors (RF sources, mixer/converter diodes, and PIN diodes) mm-wave waveguides, and examples of mm-wave systems for military applications. Many survey paper authors do not qualify their data as to its applicability. Data in this paper contain curves labeled "laboratory" and labeled "commercial". Data for both IMPATTs and Gunn diodes show power output increases of about 3 dB per year.

Comparison of Figures 2 and 3 of this paper with Figures 3 and 4 of the preceding paper seems to bear out the extrapolations for IMPATT diodes in Figure 1. This paper seems to highlight U.S. and German work, in comparison to the preceding paper which cites two Japanese, U.K. and one U.S. work for IMPATTs.

Fin-line structures are attractive alternatives to the commonly used rectangular waveguide. A number of interesting mm-wave fin-line components are described. These components are usable below 100 GHz at present. Other work on waveguides for mm-wave systems is listed in the bibliography for this chapter.

MM-wave systems for military applications recently developed at AEG-Telefunken described here include a K_a-band coherent-pulse radar, a 64 GHz radio-link using a parametric up-converter, and a 90 GHz radiometer receiver front end.

3.A.3 "Solid-State Millimeter-Wave Sources" – Kuno and Fong

Noise characteristics and frequency stability of RF sources are important in many radar applications, both in the transmitter and the receiver local oscillator. This IEEE *EASCON-77* paper addresses both aspects of solid state millimeter wave RF sources. Figure 2 is instructive in its comparison of the AM noise characteristics of IMPATTS, Gunn diodes and klystrons. FM noise of free running and phase locked IMPATT oscillators (Figure 7) show that for frequencies of less than about 400 kHz from the carrier, phase locking provides excellent noise reduction.

Four methods of frequency and phase stabilization are described: temperature control, the AFC loop, injection locked oscillator and the phase locked oscillator. Frequency stabilization is important in radar applications to insure that the resulting IF signal remains in the IF amplifier pass band. An additional technique for accomplishing this is the homodyne or zero IF frequency radar described in Paper 5.5 of Chapter 5 of this volume.

3.A.4 "EHF Solid State Transmitters for Satellite Communications" – Raue

Although IMPATT technology has advanced greatly in the last several years, even double drift techniques will not yet permit sufficient power output from a single device for some system applications. Combiners have been successfully used for power transistors up thru L-Band. This paper describes amplifiers developed by TRW for operation at 37, 41, and 60 GHz. Although intended for satellite communications systems, they could be used in coherent CW Doppler radar systems. Amplifier power outputs of 5 W at 37 GHz and 1 W at 60 GHz have been achieved. Units under development are to have outputs of 10 W at 41 GHz and 2 W at 60 Hz.

Combining circuits which are the heart of such amplifiers are treated extensively in this paper. Another unique feature is a network analyzer having a 10 GHz instantaneous bandwidth at a 60 GHz center frequency.

This paper was presented at the AIAA 8th Communications Satellite Systems Conference, Orlando, Fl., Apr. 1980.

3.A.5 "Advanced Solid-State Components for Millimeter Wave Radars" – Schwartz, Lohr, Weller and Zimmerman

Although the title indicates "components," this paper actually describes a novel hybrid millimeter wave radar transmitter-receiver. The transmitter uses a linear FM chirp bandwidth of 1.5 GHz. This waveform is generated at a low frequency, then upconverted to 95 GHz. The received signal is then downconverted. In both the transmitter and the receiver, the offset frequency is a klystron oscillator which is injection locked to the output of a varactor diode frequency multiplier chain. A two-stage IMPATT amplifier boosts the upconverted transmitter signal to required power level (30 mW). Various parts of this apparatus are described in the paper.

3.A.6 "A 90 GHz FM-CW Radar Transmitter" – Barth and Bischoff

This second solid state millimeter wave radar transmitter paper describes a 90 GHz radar transmitter which uses an FM-CW waveform. Here the RF source is two synchronized and power combined IMPATT oscillators. The synchronizing source consists of a 45 GHz Gunn oscillator which is simultaneously swept in frequency and frequency doubled by the same varactor diode. Outputs of the two IMPATT oscillators are combined in a 90° hybrid. Combined power output is 240 mW.

3.A.7 "Millimeter Solid-State Pulsed Coherent Radar" – Duffield, Smith, Pusateri, English and Bernues

The third solid state millimeter wave transmitter paper illustrates employment of two of the methods of frequency/phase stabilization described previously in Paper 3.A.3. The block diagram resembles that of a microwave coherent radar MOPA (master oscillator-power amplifier) transmitter. A phase-locked Gunn oscillator is used as the STAMO (stable master oscillator). This in turn injection locks a pulsed IMPATT oscillator.

Two operating modes of the pulsed IMPATT oscillator are provided in this transmitter: coherent operation for MTI Doppler receiver processing, and also spread spectrum. In the latter mode, the oscillator operates as a free running pulsed oscillator with a controlled intrapulse thermal frequency chirp.

Performance measurements on the coherent mode are included in the paper.

This paper was presented at the 1979 IEEE MTT Symposium.

3.A.8 "High Power Millimeter Wave Radar Transmitters" – Ewell, Ladd, Butterworth

A more appropriate title for this paper might be "Millimeter Wave Radar Transmitters," since this paper addresses millimeter wave radar transmitters with average powers from milliwatts to kilowatts. In peak powers these are watts to hundreds of kilowatts.

This paper considers both solid-state and thermionic RF sources. The discussion on modulators for pulsing of RF sources (primarily thermionic) is noteworthy. Many of the points made are frequently not properly considered in modulator design for millimeter-wave devices.

Other papers in this chapter and in this bibliography should be used to supplement the bibliography for this paper.

3.B   Electron Tube Sources

3.B.1 "Recent Progress and Future Performances of Millimeter-Wave BWO's" — Epsztein

Although solid-state millimeter wave RF power sources have interesting capabilities as illustrated by the preceding eight papers, there are some applications which cannot be met by these devices. The backward wave oscillator (BWO) was created a few years before the IMPATT diode, and has oscillated up to 850 GHz in France and 1300 GHz (1.3 THz) in Russia. Power output of 5 W at 280 GHz has been achieved in a narrow band tube.

This paper describes BWO principle of operation and tube structure, lists tubes currently in production at Thomson-CSF (in France), reviews new developments, and defines objectives for further BWO evolution. Figure 1 in the paper, which shows output power and frequency of various millimeter-wave and submillimeter-wave sources, is very informative. It includes the principal types of RF sources — klystrons, TWT, EIO, magnetron, IMPATT diodes, gyrotrons, TEO lasers and FIR lasers.

Although not specifically stated, the main thrust of this paper is CW operation of the BWO as a local oscillator signal source for a superheterodyne receiver utilizing voltage tuneability for wide frequency coverage. The BWO has been used in an experimental FM radar in the U.S. Figure 1 shows data for pulsed BWO operation, but the paper does not address this mode. The bibliography at the end of these summaries includes references to an experimental BWO FM millimeterwave radar and to pulsed BWO operation.

This paper was presented at the 1978 AGARD Conference on Millimeter and Submillimeter Wave Propagation and Circuits.

3.B.2 "The Laddertron — A New Millimeter Wave Power Oscillator" — Fujisawa

The Backward Wave Oscillator (e.g., CARCINOTRON) of the preceding paper is only one of several types of electron tubes for millimeter wave RF sources. (Certainly the BWO is the most developed one, however.) Others are the Laddertron described in this paper, the Ledatron [2], and the Extended Interaction Oscillator/Amplifier (EIO/EIA) [3].

The Laddertron, which was invented in Japan in 1954, is a multi-gap klystron using ladders. Two types of Laddertrons have been developed. One uses a series of identical slots cut regularly on the conducting plates. Its dominant mode has the same field and hence is called the O-mode Laddertron. The other Laddertron type uses two kinds of slots with different lengths interleaved on the conducting plates, with its dominant mode having the same field on alternate slots and mutually reverse fields on adjacent slots. This is the π-mode Laddertron.

A demountable experimental version of the π-mode Laddertron gives a power output of 10 watts at the 50 GHz band with an electronic tuning range of 300 MHz and a mechanical tuning range of 2.5 GHz. This paper describes the theory and experimental results obtained in 1964 on the Laddertron.

The Extended Interaction Oscillator (EIO), which is similar to the O-mode Laddertron, was developed in Canada [3]. The EIO uses a magnetic field to focus the beam as it passes through the tube. Both CW and Pulsed EIOs have been built. Several experimental millimeter radars at 95 GHz have been built using an EIO. Recently, an Extended Interaction Amplifier (EIA) has been developed. It has a peak power output of 2.3 KW at 95 GHz with a 10% duty cycle, 33 dB gain, a mechanical tuning range of 1 GHz, and a bandwidth of 200 MHz [4].

3.C   Gyrotrons (Fast Wave Devices)

3.C.1 "Millimeter Wave Tubes With Emphasis on High Power Gyrotrons" — Chilton

Thermionic millimeter-wave RF sources discussed in the preceding section have average power output capabilities on the order of 100 W at 100 GHz. For higher average powers and/or very high peak powers, fast wave devices must be used. This paper from the 6th DARPA-Tri-Service Millimeter

Wave Conference describes the Hughes HAW-7 500 W (average) 94 GHz coupled cavity TWT as a starting point.

DARPA and the U.S. Air Force Rome Air Development Center sponsored work on two fast wave RF sources — the gyrotron and the Ubitron are discussed. This paper includes principles of operation, and performance measurements on experimental tubes.

Unfortunately, this paper does not contain a bibliography. Limited bibliographies are provided in the next paper and in the bibliography at the end of these summaries.

3.C.2 "Prospects for High Power Millimeter Radar Sources and Components" — Godlove, Granatstein, and Silverstein

The gyrotron is sometimes called the electron cyclotron maser. There are several forms of the gyrotron; three of these are the gyroklystron, the gyromonotron oscillator, and the gyrotravelling wave amplifier (Gyro-TWA). They are sometimes also called cyclotron resonance masers (CRM).

This paper, which was presented at IEEE *EASCON-77*, describes the CRM theory and early developments from the studies of Twiss, of Schneider, and of Gapanov. It also discusses CRM present status. A very useful table of representative CRMs operated to date (1977) or presently being designed is included in the paper (Table 2). This list with references pertains to CRMs which use thermionic cathode guns as the source of the electron beam.

Although the CRM is an interesting device, and very impressive results have been achieved, there are a number of technical issues which affect its successful employment. These are treated by the authors. These issues, which involve auxiliary components and system implications, include susceptibility to spurious oscillation, tuneability, phase stability, size and weight (especially of the magnet, power supply, and modulator), harmonic operation and components themselves (vacuum windows, mode converters, duplexers, various ferrite devices, rotary joints and waveguide).

Unfortunately, this paper does not include detailed consideration of radar system applications of the gyrotron. Certainly it has a number of potential radar applications, (*e.g.* see Paper 5.10 in Chapter 5); however, gyrotron power capabilities cannot overcome rain attenuation of millimeter waves at "long" ranges.

3.C.3 "Gyrotrons for High Power Millimeter Wave Generation" — Jory, Friedlander, Hegji, Shively, and Symons

This paper describes in more detail the Varian Associates' several gyrotron development programs, which are listed in a table in the preceding paper. Some of this is also briefly mentioned in Paper 3.C.1. Figure 5 in this paper, which is a "microwave" tube CW power state-of-the-art summary, is noteworthy. It lists both development tubes and production tubes. The first curve actually pertains to tubes other than CRM devices, *e.g.* TWTs and klystrons. Care must be taken in interpretation of the Russian gyrotron results, since sometimes peak power of pulsed devices is misinterpreted as average (CW) power.

3.D Mixers

3.D.1 "Review of Mixers for Millimeter Wavelengths" — McColl and Hodges

RF sources described in the previous three sections provide for generation of the millimeter wave radar signal to be transmitted, and for the local oscillator in the accompanying superheterodyne receiver. The receiver requires the use of a mixer which, with local oscillator, converts the received signal to IF frequency for amplification prior to detection. Diode mixers (sometimes called crystal rectifiers) for operation in the microwave region were highly developed during World War II. Necessity for operation at millimeter-wave frequencies has led to several new types of mixer devices.

This excellent DARPA-Tri-Service Conference survey paper discusses the conventional Schottky diode, Josephson junction, superconducting-semiconductor (super-Schottky) diode, varactor down converter and InSb hot electron bolometric mixers in terms of their performance, physical operating principles and limitations. Frequency of concern is 30 - 300 GHz. Table 1, which pertains to Schottky barrier diodes, contains reported results of 22 different workers with references to each in the extensive bibliography.

Although not emphasized in the text, some of the references actually discuss harmonic mixing. This can be very beneficial if a local oscillator frequency operating at the received frequency is not available.

Figure 3 is an overall comparison of reported performance of the four types of mixers as a function of frequency. This figure also includes the quantum limit (the quantity hf) described in the paper.

In the submillimeter region, this quantity can exceed kT and become the theoretical performance limit of mixers.

Both room temperature and cryogenic devices are treated in this paper. Several of the references address mixer performance in millimeter wave radio astronomy receivers. These results are generally extendable to radar receivers; however, the latter require consideration of transmitter signal leakage (for single antenna operation with TR device) or transmitter signal reflected from nearby objects (for two antenna operation).

Frequency agile operation and wide transmitter bandwidth may also present mixer problems.

3.E Antennas

3.E.1 "Millimeter Wave Antennas" — Kay

Most practical millimeter wave antennas, like many microwave antennas, are based on geometrical optics designs, *e.g.* parabolic reflectors, Cassegrain systems, and lenses. Comparative merits and the few design features peculiar to millimeter waves are discussed in this survey paper. A few experimental arrays and ferrite scanners at millimeter wave frequencies have been built. These are treated together with millimeter wave absorbers and radomes. Radiation patterns are included for several types of antennas.

Interestingly, the author shows that in terms of gain per dollar (unit cost), millimeter wave antennas can be cheaper than microwave antennas.

Although antenna sidelobes are mentioned in this paper, their treatment is much less detailed than that of Rudge and Foster [1]; however, this paper was prepared over a decade earlier, and very low sidelobe antennas were not of that great concern then.

Most microwave radar antennas operate in the far field and also illuminate the entire target. A millimeter wave radar antenna may operate in both near and far fields as the target range changes. Antenna sidelobe levels may be of concern in millimeter wave radars, due to their interaction with the lobe structure of a target radar cross section pattern (*e.g.* Paper 2.B.1), or in a multi-target situation.

Millimeter wave antennas may be focussed to a spot, as in plasma diagnostic systems and in certain radar applications. Focussing antennas are briefly described in this paper.

The extensive bibliography is a good survey of the literature on millimeter wave antennas up to the time of this paper (1966).

3.E.2 "Performance of Reflector Antennas with Absorber-lined Tunnels" — Dybdal and King

This paper applies a twenty-year-old technique to the sidelobe level reduction of a 92 GHz millimeter wave antenna. Measured sidelobe level of about 39 dB is achieved for the close-in sidelobes, with much greater reduction for the far-out lobes.

3.E.3 "A Millimeter-Wave Scanning Antenna for Radar Application" — Kesler, Montgomery and Liu

Very narrow beamwidth of millimeter wave radar antennas is an excellent attribute for tracking radars. This narrow beamwidth can, however, present problems to the tracking radar. The radar must, of course, acquire its intended target before tracking can be initiated. This problem is frequently glossed over in millimeter wave radar design (*cf.* Paper 1.1, Chapter 1).

Several solutions are possible: 1) searching a large volume by mechanical motion of the antenna; 2) temporary spoiling the antenna beam (and hence also lowering antenna gain during search); 3) employment of a separate auxiliary acquisition radar; 4) utilization of electronic scanning (phased array); 5) mechanically scanning the flat plate of Cassegrain antenna.

This paper describes an implementation of the last approach. This technique has been previously demonstrated on a microwave radar antenna. It has many advantages for a millimeter wave tracking radar.

The antenna described permits 45 degrees of scan by moving the lightweight special flat plate through only 22½ degrees. A four-horn monopulse angle measuring system is included. Although not clearly stated in the paper, this plate scanning technique is used only in target acquisition. Conventional antenna drive servos to move the entire antenna assembly in azimuth and elevation would be used for monopulse tracking.

REFERENCES FOR CHAPTER 3

[1] A.W. Rudge and P.R. Foster, "Low Sidelobe Radar Antennas," *Proc. Military Microwaves Conf.*, 1978. Reprinted in S.L. Johnston, *Radar Electronic Counter-Countermeasures.* Dedham, Mass.: Artech House, 1979.

[2] K. Mizuno and S. Ono, "The Ledatron," Chapter 5 in K.J. Button (Editor), *Infrared and Millimeter Waves*, Vol. 1: *Sources of Radiation*, New York: Academic Press, 1979.

[3] Varian Associates of Canada, "Introduction to Extended Interaction Oscillators," undated brochure.

[4] Varian Associates, private communication.

BIBLIOGRAPHY TOPICS FOR CHAPTER 3

RF Sources and Components, General

Solid State Sources

Electron Tube Sources

Gyrotrons

Circuits and Components

Mixers/Receivers

Antennas

SUPPLEMENTAL BIBLIOGRAPHY FOR CHAPTER 3

NOTE: Additional references on millimeter wave RF sources and components are contained in the bibliographies in Paper 1.6 Chapter 1, and in the bibliographies of papers in Chapters 4 and 5.

RF Sources and Components, General

Reports

Devyatkov, N.D. and Golant, M.B., "Development of Electron Devices for the Millimeter and Submillimeter Wavelength Ranges," *Radio Engrg. and Electron. Phys.* Vol. 12, No. 11, Nov. 1967, pp. 1835-46 (contains 60 references).

Chiron, B., Mahieu, J.R., and Fache, M., "Nouveaux Ensembles Hyperfrequence Emission — Reception Pour Radars Ondes Millimetriques ("New Devices, Techniques and Systems in Radar" — in French), *AGARD Conf. Proceedings CP-197*, 1976, pp. 7-1 through 7-15, DDC AD A040144, NASA N 77 - 22346.

Spitz, E. and Gachier, G. (Editors), "Millimeter and Submillimeter Wave Propagation and Circuits", *AGARD Conference Proceedings CP-245*, 1978, Session II: Solid State Sources (5 papers), Session III: Submillimeter Receivers (5 papers), Session IV: New Technologies and Integration Techniques (12 papers), Session V: Components and Circuits (4 papers), Session VI: Tubes (2 papers), Session VII: Special Devices (3 papers).

Guenther, B.D. and Carruth, R.T., "Millimeter and Submillimeter Wave Sources for Radar Applications," *USAMICOM Technical Rpt. H-78-6*, May 1978, DDC AD A056768 (contains 65 references).

Solid State Sources

1. Books

   Kuno, H.J., "IMPATT Devices for Generation of Millimeter Waves," Chapter 2 in Button, K.J. (Editor), *Infrared and Millimeter Waves; Vol. I: Sources of Radiation*, New York: Academic Press, 1979.

2. Journal Articles

   Kuno, J., "Devices Ready for Millimeter Systems," *Microwave System News*, May 1979, Vol. 9, No. 5, pp. 71ff. (Summarizes 8 papers at IEEE International Solid State Circuits Conference, Feb. 1980 session "Millimeter-wave Technology for the 80s".)

   Ondria, J., "The Noise Properties of Millimetre Wave Gunn and IMPATT Diode Local Oscillators," *Proc. 9th European Microwave Conf.*, Sept. 1979, Session II, "Millimetre Wave Devices."

   Midford, T.A., "Advances in Millimetre Wave IMPATTS," *Proc. 9th European Microwave Conf.*, Sept. 1979, Session II: "Millimetre Wave Devices."

   Amboss, K., "The Current Art of Millimeter Wave Solid State and Tube Type Power Sources," *Proc. Military Microwaves Conf.*, 1980 (MM-80)

Electron Tube Sources

1. Books

   Convert, G. and Yeou, T., "Backward Wave Oscillators," Chapter 4 in Benson, F.A. (Editor), *Millimetre and Submillimetre Waves*, London: ILIFFE, 1969.

   Kantorowicz, G. and Palluel, P., "Backward Wave Oscillators," Chapter 4 in Button, K.J. (Editor), *Infrared and Millimeter Waves; Vol. I: Sources of Radiation*, New York: Academic Press, 1979.

2. Journal Articles

   Favre, M., "Results Obtained on Cross Field Carcinotrons under Pulsed Operations," *Proc. IEE* Vol. 105 Pt. 5 Sup. 10, pp. 533-7 and 542-3, 1958 (*Proc. Int. Conf. on Microwave Valves*, May 1958).

Chodorow, M. and Wessel-Berg, T., "A High Efficiency Klystron with Distributed Interaction," *IEEE Trans.* Vol. ED-8-9 No. 1, Jan. 1961, pp. 44-55.

Bacon, L.C., Enderby, C.E., and Phillips, R.M., "V-Band Ubitron Amplifier Development," General Electric Co., Palo Alto, CA, AFWC Rpt. No. TDR 62 295, Contract AF 33 601 2818, July 196, DDC AD 331446.

Phillips, R.M., "Study and Investigation of Millimeter Wave Generator Using the Fast Wave Undulating Beam Principle," General Electric Co., Palo Alto, CA Rpt. R 61 ELM 209, Contract No. AF 33 616 7474, Dec. 1961, DDC AD 325089.

Enderby, C.E. and Phillips, R.M., "V-Band Ubitron Amplifier Development," Final Report, General Electric Co., Palo Alto, CA, Contract AF 33-616-8356, Jan. 1964, ASD TDR 63-847, DDC AD-347628.

Preist, D.H. and Leidigh, W.J., "A Two-Cavity Extended Interaction Klystron Yielding 65 Percent Efficiency," *IEEE Trans.* Vol. ED-11 No. 8, Aug. 1964, pp. 369-73.

Phillips, R.W. and Enderby, C.E., "Millimeter Wave Generator Fast Wave Undulation Principle," Final Report, General Electric Co., Palo Alto, CA, Contract AF 33-657-11390, Nov. 1964, GE Rpt. R64 ELM 220-12, AF Rpt. AL TDR 64-249, DDC AD 452588.

Chodorow, M. and Kulke, B., "An Extended-Interaction Klystron: Efficiency and Bandwidth," *IEEE Trans.* Vol. ED-13 No. 4, April 1966, pp. 439-47.

Sprangle, P., "Fast Wave Interaction with a Relativistic Electron Beam," *Jnl. of Plasma Phys.*, Vol. 11, April 1974, pp. 299-309. (Pertains to Ubitron.)

Gorshkova, M.A. and Smorgonskii, A.V., "The Theory of Ubitron – 'O' Type Amplifier with Ultra-Relativistic Electron Beam," *Radio Phys. and Quant. Electron*, Vol. 18, No. 8, Sept. 1975, pp. 888-891.

Krementsov, S.I., Raizer, M.D., and Smorgonskii, A.V., "Ubitron Oscillators with a Relativistic Electron Beam", Pis'ma v Zhurnal Tekhnicheskoi Fiziki, Vol. 2, May 26, 1976, pp. 453-7. (In Russian, abstract trans 76 A 37620.)

Baird, J.M., Sensiper, S., Amboss, K. and Heney, J.F., "Millimeter Wave Ubitron Development Phase I, Final Tech. Rpt. 8 Mar.- 31 Dec. 1976," Hughes Aircraft Co. Contract F 30602 - 76 - C - 0215, DDC AD-A040043.

Tkach, Iu. V., et al., "Ion Acceleration in the Interaction of a Strong-Current Relativistic Electron Beam with a Spatially-Periodic Magnetic Beam," *JETP Letters*, Vol. 28, No. 5, 1978, pp. 535-8. (Translated from Russian.)

Puri, M.P. and Handy, R.A., "EHF Mini-TWT for Broadband Operation at 20 Watts CW," *10th European Microwave Conf. Proc.*, Sept. 1980, Paper p. 9.

"mm Magnetrons from 15 - 96 GHz," *Microwave System News* Vol. 9, No. 5, May 1979, p. 78.

Bratman, V.L., Ginzburg, N.S., and Petelin, M.I., "Common Properties of Free Electron Lasers," *Optics Communications*, Vol. 30, Sept. 1979, pp. 409-12. (General theory of microwave devices, in particular Ubitron and gyrotron.)

Mourier, G., "Research and Development on High Power Millimeter Wave and Submillimeter Wave Electron Tubes," *Proc. Military Microwaves Conf.*, 1980 (MM-80).

Gyrotrons

1. Books

Hirshfield, J.L., "Gyrotrons," Chapter 1 in Button, *loc cit.*

2. Journal Articles

Loshakova, I.I., "Megawatt Relativistic Klystron With Extended Interaction," *Radioelectron. and Commun. Syst.* Vol. 20 No. 16, 1977, pp. 85-6.

Goldenberg, A.L., "The Gyrotrons – High Power Sources at Millimeters and Submillimeter Wavelengths," *Proc. 10th European Microwave Conf.*, Sept. 1980, Paper INV 5.1.

Circuits and Components

1. Journal Articles

Cohen, L.D. and Meier, P.J., "E-Plane mm-Wave Circuits," *Microwave Jnl.* Vol. 21, No. 8, Aug. 1978, pp. 63ff.

Goldie, H., "A High-Power Broadband Millimeter Wave Switch and Receiver Protector," *IEEE MTT Int. Microwave Symp.* June 1978, pp. 345-6.

Goldie, H., "Plasma Switch Protects 95 GHz Radar," *Microwave Syst. News* Vol. 8 No. 9, Sept. 1978, pp. 71-2.

(Various authors) Session PC-4: "Millimetre and Submillimetre Components" (six papers), *Proc. 9th European Microwave Conf.*, Sept. 1979, pp. 441-72. Also poster sessions pp. 721-30. (For Table of Contents, see *Microwave System News*, May 1979, pp. 89ff.)

IEEE/MTT-S International Symposium and Workshops Digest, May 1980: millimeter wave papers in sessions on millimeter receivers and components, millimeter wave ICs, high power devices and techniques (*e.g.*, gyrotron), guides and components.

Inggs, M. and Williams, N., "Dielectric Waveguide Technology and its Implications for MM Wave Integrated Circuits and Antennas," *Proc. Military Microwaves Conf.*, 1980 (MM-80).

Scarman, R.S. and Oxley, T.H., "Millimetre-Wave Components of Hybrid-Open Microstrip Form," *Proc. Military Microwaves Conf.*, 1980 (MM-80).

Meindel, H., Adelseck, B., and Gallsen, H., "A Survey of Planar Integrated MM-Wave Components," *Proc. Military Microwaves Conf.*, 1980 (MM-80).

Lemke, M. and Hoppe, W., "Hexaferrite Components — Tuneability at Millimetre-Waves," *Proc. Military Microwaves Conf.*, 1980 (MM-80).

Bates, R.N. and Coleman, M.D., "Millimetre Wave E-Plane MICs for Use up to 100 GHz," *Proc. Military Microwaves Conf.*, 1980 (MM-80).

Sudbury, R.W., "Monolithic Ga As Circuits for Millimeter-Wave Radar Application," *Proc. Military Microwaves Conf.*, 1980 (MM-80).

Mixers/Receivers

1. Journal Articles

Vystavkin, A.N. and Migulin, V.V., "Receivers for Millimeter and Submillimeter Waves," *Radio Engrg. and Electron. Phys.* Vol. 12 No. 11, Nov. 1967, pp. 1847-60 (extensive survey paper).

Dryagin, Yu A. and Lubyako, L.V., "Mixer for the Millimeter Range of Waves with the Selection of a Mirror Channel Using An Interferometer," *Radio Phys. and Quant. Electron.* Vol. 20 No. 4, April 1977, pp. 447-8.

Averin, S.V. and Popov, V.A., "A Mixer Employing a Schottky Barrier Diode for the Short-Wavelength Portion of the Millimeter and Submillimeter Wavelength Bands," *Radio Engrg. and Electron. Phys.* Vol. 22 No. 8, Aug. 1977, pp. 126-8.

Vystavkin, A.N., *et al.*, "A Josephson Junction as a Signal Mixer — Multiplier at Millimeter Wavelengths," *Radio Engrg. and Electron. Phys.* Vol. 22 No. 7, July 1977, pp. 142-3.

Cardiasmenos, A.G., "Planar Devices Make Production Practical," *Microwave System News* Vol. 9 No. 5, Aug. 1979, pp. 46ff.

Swanberg, N.E. and Paul, J.E., "Quasi-Optical Mixer Offers Alternative," *Microwave System News* Vol. 9 No. 5, Aug. 1979, pp. 58ff.

Bernues, F.J. and Pusateri, P.H., "Millimeter-Wave Solid State Radar Front-Ends," *IEEE WESCON-79*, Session 20/Paper 5, pp. 5-1 thru 5-5, 1979 (includes also RF sources.)

Keen, N.J., "Millimetre Wave Mixer Diodes," *Proc. 9th European Microwave Conf.*, Sept. 1979, Session II — Millimetre Wave Devices.

Kollberg, E., "Low Noise mm-Wave Receivers," *Proc. 10th European Microwave Conf.*, Sept. 1980, Paper INV 5.2.

Raiisanen, A., *et al.*, "A Cooled Schottky-Diode Mixer for 75 - 120 GHz," *Proc. 10th European Microwave Conf.* Sept. 1980, Paper SS 5.1.

Knoechel, R. and Schlegel, A., "Octave-Band Double-Balanced Integrated fin-line Mixers at mm-wavelengths," *Proc. 10th European Microwave Conf.* Sept. 1980, Paper SS 5.2.

Cardiasmenos, A.G., "Future Trends in Millimeter Low Noise Receivers," *Proc. Military Microwaves Conf.*, 1980 (MM-80).

Antennas

Note: Paper 5.3, Chapter 5 has an extensive discussion of geodesic Luneberg lens antennas for millimeter wave radar.

1. **Bibliographies**

    Barton, D.K., "International Cumulative Index on Radar Systems," *IEEE Pub. Nr JH 4675-5*, 1978; antennas (general) p. 1; antennas (gain and pattern considerations) pp. 1-2; antennas (lenses) pp. 2-3; antennas (radomes) p. 3; antennas (reflector systems) pp. 3-4; antennas (scanners) pp. 4-5; arrays (general) pp. 5-6; arrays (feed systems) p. 7; arrays (gain, pattern, accuracy considerations) pp. 7-8; arrays (phase shifter, frequency scanning and control) pp. 8-9; array radar systems, pp. 9-10. Some K_a-band and millimeter antennas contained in the above entries. Operating frequency sometimes not included in paper title.

2. **Journal Articles**

    Semplak, R.A., "100 GHz Measurements on a Multiple-beam Offset Antenna," *Bell Syst. Tech. Jnl.*, Vol. 56 No. 3, March 1977, pp. 385-98.

    Kooi, P.S. and Leong, M.S., "Q-Band Short Backfire Antenna Arrays," *Proc. 10th European Microwave Conf.*, Sept. 1980, Paper AN 4.6.

    Williams, N. and Adiata, N.A., "Millimetre Wave Antennas," *Proc. Military Microwaves Conf.*, 1980 (MM-80).

    Carter, M., "Millimetric Aerials for Full Illumination Radars," *Proc. Military Microwaves Conf.*, 1980 (MM-80).

    Cashen, E.R. and Carter, M., "Linear Arrays for Centimetric and Millimetric Radars," *Proc. Military Microwaves Conf.*, 1980 (MM-80).

Millimetre Wavelength IMPATT Sources

J.J. Purcell

1 Introduction

The rapid technological development of solid-state millimetre-wave components has resulted in a significant increase in interest in the exploitation of the frequency spectrum above 90 GHz. There are two mechanisms which are currently exploited to produce solid-state oscillators at frequencies as high as 100 GHz; the Gunn-effect and the Avalanche Transit Time effect. The Gunn effect describes the N-shaped current–voltage characteristic exhibited by Group III–V semiconductors such as GaAs or InP. The negative differential resistance decreases, however, with increasing modulation frequency owing to energy relation effects, and imposes a practical upper-frequency limit of about 100 GHz. The present 'state-of-the-art' of about 100 mW at 94 GHz from both InP and GaAs devices is probably within 3 dB of the Gunn diode's potential performance. The avalanche diode, on the other hand, has produced power at fundamental frequencies as high as 400 GHz, and exhibits an almost constant conversion efficiency from 40 to 100 GHz.

The advent of reliable, small sized, impatt sources producing continuous power levels of hundreds of milliwatts and pulsed powers of watts at frequencies around 100 GHz has allowed the atmospheric 'windows' at 94 to 140 GHz to be used in system applications such as:

Radar,
Communications,
Missile guidance,
Imaging,

whilst in the laboratory, the availability of relatively low-cost and reliable sources has extended the spectroscopists' field of interest into the 'microwave' spectrum of 90 to 300 GHz.[1]

Military applications are predominantly concentrated upon guidance and imaging, where the high resolution conferred by the millimetre wavelengths, and the relative immunity to smoke and fog can offer an all-weather capability superior to both microwave and far-infra-red frequencies.

2 Microwave Generation by Impatts

The IMpact ionization Avalanche Transit Time diode (or avalanche diode) was first proposed as a solid-state microwave source by W. T. Read in 1958.[2] Read proposed that an antiphase voltage/current relationship could be sustained at microwave frequencies by the combined delay of avalanche multiplication in a reverse-biased diode, coupled with the drift of generated carriers through the depletion layer of the junction. In 1965,[3] Johnston, De Loach and Cohen at Bell Laboratories demonstrated the concept with a commonplace silicon computer diode embedded in a C-band (4–8 GHz) waveguide circuit.

In the relatively short period of 14 years, the silicon impatt has matured, to become a commercially available solid-state alternative to thermionic devices such as klystrons and carcinotrons over the spectrum from 8 to 300 GHz. Manufacturers' estimates of life-times are of the order of tens of thousands of hours for continuous operation with junction temperatures of about 200°C. Recent developments have been in the fields of high power operation (1–10 watts) over short pulse lengths for imaging and guidance applications, mainly in the United States,[4] and in extending the upper frequency limit as high as 400 GHz through operation in harmonic modes, in both England[1] and Japan.[5]

3 Diode Design and Fabrication

In theory, almost any semiconductor which could be doped to appropriate levels could be used as a p–n junction avalanche diode oscillator. However, the available power from a particular material is determined simply by a figure of merit, F, where

$$F = V_s E_g$$

and V_s and E_g are the saturated drift velocity of carriers and the semiconductor bandgap.

Maximizing F and confining the choice to practical materials with well-developed growth and processing technologies at present restricts the designer to either GaAs or Si. Comparison of the values of F ($E_g = 1.44$ eV, $V_s = 0.6 \times 10^7$ cm s^{-1} for GaAs, $E_g = 1.1$ eV and $V_s = 1.0 \times 10^7$ cm s^{-1} for Si) would appear to offer comparable power performance. However, a more detailed examination of the avalanche mechanism indicates an upper frequency limit of about 60 GHz for GaAs, beyond which Si diodes offer an increasingly marked advantage.

Millimetre-wave impatts are produced from a single-crystal layer which is deposited by vapour phase epitaxy onto a high conductivity substrate. Small quantities (a few parts in 10^6) of impurities are included within the lattice to produce either acceptor or donor-type semiconductor material of a prescribed carrier-density level. The diode structure is commonly of one of three forms:

p^+–n–n^+	'n-type'	single drift
p^+–p–n^+, or n^+–p–p^+	'p-type'	single drift
p^+–p–n–n^+		double drift

labelled in the direction of surface to substrate.

The double-drift diode is, in effect, two back-to-back single-drift diodes, and has advantages of higher power and higher impedance when compared with a single drift diode at the same frequency. However, technological limitations tend to confine use of this structure to frequencies below 100 GHz at the present time.

Theoretical analysis of the alternative 'n' and 'p'-type single-drift diodes suggests that higher output powers and higher frequency limits could be realized through use of the 'complementary' hole-drift device, though in practice, most laboratories favour the production of 'n'-type diodes owing to the practical difficulty of achieving low parasitic losses with p-type material. Power dissipation is predominantly confined to the undepleted epitaxy, where the current is conducted by carriers under the influence of low electric fields. It is important, therefore, that the low-field mobility of the semiconductor should be high, maximizing the conductivity ($Nq\mu$), otherwise power dissipation will counteract the power generation of the avalanche process. The relative mobilities, at room temperature, at impurity levels of 10^{17} cm^{-3} are about 250 cm^2 V^{-1} s^{-1} for holes, and 650 cm^2 V^{-1} s^{-1} for electrons.

The impatt is a 'transit-time' device and hence, assuming a constant conversion efficiency, follows the customary $PZF^2 = $ const. power decrease with increasing frequency, where P is the available power and Z is the minimum circuit impedance which can be realized. This expression is a direct result of the output power being proportional to the chip volume, the cross-sectional area and length being each inversely proportional to frequency. Peak power is generated at a frequency determined by the mean depletion layer width, and this width in turn is determined by the layer doping level. A typical 100 GHz single-drift diode might have an optimum space-charge layer width of only 0·6 μm and a breakdown voltage of about 12 V. Diodes designed for operation at frequencies as high as 300 GHz require layer widths as narrow at 0·2 μm, which is probably close to the technological limit of current fabrication techniques.

A 100 GHz impatt has a cross-sectional area of about 10^{-5} cm^2 and an active-layer width of 6×10^{-5} cm. The dissipation of 2 watts of input power therefore represents a remarkably high power density of 3×10^9 W cm^{-3}! Clearly the diode must be in contact with a heat sink presenting a sufficiently low thermal

Fig. 1. Typical millimetre-wave impatt device, showing quartz 'stand-offs', bond tape and plated heat sink.

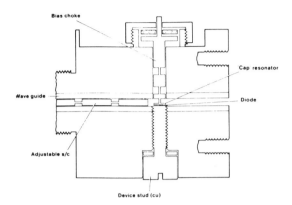

Fig. 2. Schematic diagram of typical 100 GHz impatt oscillator.

impedance (deg C/W) to restrict the temperature rise to within safe limits. In practice, manufacturers either electroplate a high conductivity integral heat sink of gold, copper or silver onto the junction contact, or for highest powers, bond the diode directly to a diamond heat sink. (Type IIA diamonds have a thermal conductivity, at room temperatures, of 2 to 5 times that of copper.)

A typical millimetre wave silicon impatt with a gold heat sink is shown in Fig. 1. The device would be mounted on a post by either wafer-bonding (solder) or ultrasonic-bonding.

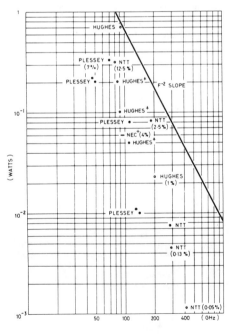

Fig. 3. 'State-of-the-art (1978)' impatt performance (c.w.).

At millimetre wavelengths, it is necessary to minimize the parasitic reactances associated with the diode encapsulation and commercially available device packages are usually unsuitable. Most manufacturers use a stand-off or quartz-ring 'encapsulation' of their own design.

The diode is embedded in a microwave circuit tuned to resonate at the appropriate frequency. The resonator is usually either of the 'radial cap' or 'coupled post' designs. Manufacturers pay particular attention to the cavity surface finish and mechanical stability in order to produce sources with low noise and stable output.

Figure 2 is a cross-sectional schematic drawing of a 100 GHz impatt oscillator of the radial cap design.

4 Present Status

In Figs. 3 and 4 are shown the present performance of millimetre-wave impatts from laboratories in the US, UK and Japan. Most activity is centred around the low atmospheric attenuation windows of 94, 140 and 225 GHz. At 94 GHz, c.w. impatt sources can produce output powers as high as 850 mW (Hughes) whilst powers of 50 to 100 mW are available commercially. High-powered, short pulse length operation enables the diode to be operated at current densities limited purely by space-charge effects. With pulse lengths of only 300 ns, output powers as high as 5 to 10 watts have been reported at 94 GHz (Hughes).[4]

The upper frequency for operation in a fundamental mode is about 400 GHz, determined by physical limitations which include diffusion, the rate of change of ionization coefficients with electric field, energy relaxation and the onset of tunnelling at high electric fields. However, microwave power is available, at low levels, at frequencies as high as 600 GHz by harmonic selection.[1]

As far as their suitability for systems' usage is concerned, the output power level of impatt sources is primarily determined by the device lifetime required. Impatts generally either fail in a very short time known as the 'infant mortality', which can be encompassed within the manufacturers' burn-in period, or at a constant rate following a log-normal temperature dependence. The life expectancy follows the classical 'bath-tub' pattern in which mortality rates are high during infancy and senility and at a low constant level between the two. At a junction temperature of 250°C, for example, an m.t.t.f. of greater than 10^6 hours would normally be expected. The activation energy is such that a change in junction temperature of 10°C has the effect of changing the m.t.t.f. by a factor of about 2 times. A typical plot of diode reliability with junction temperature is shown in Fig. 5. Clearly in a practical situation, the user gains dramatically increased reliability at the expense of either relatively modest reductions in output power, or the use of a cooled environment.

Analysis of failure modes indicates that the diode's

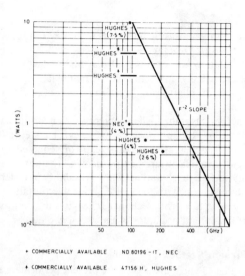

Fig. 4. 'State-of-the-art (1978)' peak impatt power.

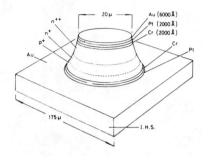

Fig. 6. Schematic diagram showing the construction of a '100 GHz' single drift N-type impatt.

demise is associated with an incursion of the contact metallization into the bulk of the semiconductor. High reliability figures have been achieved through the use of intermediate refractory layers such as Pt or W between the surface metallization of Cr or Ti and the Au wire bond. A schematic diagram showing a typical metallization pattern is shown in Fig. 6.

5 Future Prospects

Industrial activity in solid-state millimetre-wave source development is a direct reflection of the level of interest of mainly military applications in the atmospheric windows at 94 to 140 GHz. These frequencies are being utilized for tracking, guidance, radiometry and communications systems. Source requirements tend to fall into two distinct categories: either low-power, stable, c.w. outputs for local oscillators, parametric amplifier pumps and sources for Doppler radars, radiometers and transponders, or high-power pulsed oscillators for such applications as missile seekers, terminal homing and imaging. The low-power requirement is within the capabilities of established sources. However, those applications using short (50–300 ns) pulse operation tend to require output powers as high as 10 watts. State-of-the-art laboratory results[4] have shown that 10 watts is achievable from double-drift silicon impatts at 94 GHz. The frequency 'chirp' is high (1 GHz/100 ns) owing to the large temperature excursion over the pulse length; however, Hughes workers successfully controlled the frequency variation by shaping the applied bias pulse. No information is currently available on the effects on reliability of pulsing diodes at these power levels but it is likely that diode or circuit combining techniques will be necessary in order to achieve such operation with useful reliabilities.

As far as higher frequencies are concerned, theoretical work has indicated that tunnelling injection rather than avalanching may extend operation as high as 1000 GHz.[6] To date, however, Tunnel injection Transit Time (tunnett) diodes have produce only small amounts of power (1 mW) at frequencies up to 278 GHz.

7 Acknowledgment

This paper has been published with the kind permission of the Directors of Plessey Research (Caswell) Limited.

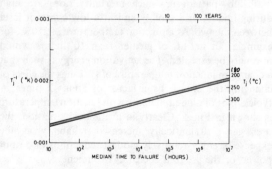

Fig. 5. Typical impatt reliability plot.

6 References

1. Llewellyn-Jones, D., 'Spectroscopy at frequencies near 180 GHz utilising the harmonic content of a millimetre-wave impatt oscillator', Proc. 4th European Microwave Conference, Montreux, 1974, pp. 86.
2. Read, W. T., 'A proposed high frequency negative resistance diode', Bell Syst. Tech. J., 37, pp. 401–46, March 1958.
3. Johnston, R. L., De Loach, B. C. and Cohen, G. B., 'A silicon diode microwave oscillator', Bell Syst. Tech. J., 44, pp. 369–72, February 1965.
4. Kramer, N. B., 'Solid state technology for millimetre waves', Microwave J., August 1978, pp. 57–61.
5. Ino, M., Ishibashi, T. and Ohmori, M., 'Submillimetre wave Si p^+–p–n^+ IMPATT diodes', Jap. J. Appl. Phys., 16, pp. 89–92, 1966.
6. Nishizawa, J., Motoya, K. and Okuno, Y., 'Tunnel injection oscillator over 200 GHz', Proc. 8th European Microwave Conference, Paris (1978), pp. 780–4.

Advances in mm-Wave Components and Systems

B. Adelseck, H. Barth, H. Hoffmann, H. Meinel, and B. Rembold

SUMMARY

In the first part this paper gives an overview on state-of-the-art and availability of mm-wave semiconductors. Laboratory results are compared with the data of commercially available diodes. The second part is dealing with new active and passive mm-wave components. Because of different applications, very distinguished problems have to be solved leading to different technologies. While for high Q and high performance rectangular waveguide devices are in use, in recent time new combinations of waveguide and planar structures (fin-lines) has been developed to reduce cost and weight and to increase reliability and reproductivity. Typical examples of components will be demonstrated: High Q and high stable Gunn oscillators as well as swept Impatt sources up to 90 GHz, frequency converters as parametric up converters at 64 GHz, mixers at 35, 60 and 90 GHz and doublers at 90 GHz. In addition, fin-line components, for instance detectors, PIN-attenuators and double-throw-switches will be shown. Using these components some mm-wave-appliances have been developed and will be presented.

1. INTRODUCTION

The rise in mm-wave development observed in recent years is caused by the availability of semiconductors like Gunn devices, Impatt- and Schottky-diodes as well as by the increasing activities on mm-wave circuit design. Both trends are at least promoted by the broad spectrum of applications on civil market and by military employments like radar, radiometry, and communications.

Though well known since many years, the breakthrough in mm-wave radarsystems at 8 mm wavelength came only recently as now the system dimensions can be minimized using exclusively semiconductors. Mm-wave radars operate at frequencies determined by the atmospheric attenuation minima about 35 and 90 GHz. Due to the small dimensions mm-wave radars mounted on mobil carriers are well suited for short distance applications (< 10 km). Typical radar employments are terrain surveillance, defence against low-flying aircrafts, signification of aircrafts, helicopter radar, and active terminal guidance. Coherent as well as incoherent-puls devices and FM-CW-radars find their application in these tasks.

Radiometer receivers used for military purposes in general are established at frequencies about 35 and 90 GHz. Radiometers are utilized for terrain observation, terminal homing and radiomapping. Main topics of research are to decrease the noise figure of the front ends and to improve the computer interpretation of the measured contrast.

Using mm-wave frequencies mobile radio links with small demensions can be established which almost are not sensitive against jamming. Making use of the narrow antenna beams and the high atmospheric attenuation for instance at 60 GHz these links have a likewise low probability of intercept (LPI). Typical applications are battle field communications, datalink between radarstations, and microwave ampling employments in a convoy.

2. STATE OF THE ART AND AVAILABILITY OF MM-WAVE SEMICONDUCTORS

Because of the rapid development of semiconductor technologies mm-wave techniques today mostly use diodes for power generation, signal detection, and frequency conversion.

Fig. 1 shows a graph of the power increase of Impatt- and Gunn oscillators respectively at the example of 35 GHz diodes. The data are the highest catalogue specifications which has been available in recent years. The annual increase of power observed up to now is about 3 dB/year.

Impatt diodes mainly are used as transmitter diodes in radar or communication systems. Compared with Gunn oscillators, Impatt devices are easier to sweep by changing the current. Sweep ranges up to 10 GHz at 90 GHz center frequency are achievable. In the question of available power, the silicon technology today delivers up to 10 watt peak power at 90 GHz using double drift Impatts on diamond heat sink, fig. 2. These values are the best reported data /1/. The best commercially available data, today are about 5 W peak power and 200 mW CW-power. Because of the price, however, these diodes are not suited for massproduction up to now.

Gunn devices today are available up to 100 GHz, fig. 3. The well known advantage of these elements is the better noise measure as compared to Impatt diodes. Therefore Gunn elements are used for local oscillators and for transmitters of self-mixing doppler systems for instance. Furthermore Gunn-elements are appropriate for puls-oscillators especially at frequencies in Ka-band. The disadvantage of Gunn-elements, however, is the smaller obtainable power which today is about 10 mW at 90 GHz. Standard material of Gunn devices is GaAs. The InP-technology, theoretically promising better efficiency, recently renders better results /2/.

The package of Gunn- or Impatt diodes in most cases consists of a quartz or ceramic ring surrounding the chip, which is soldered on a copper or diamond heatsink. The chip-top is contacted with a metal coverdisk using a small ribbon in a single or crossover version.

Schottky-Barrier-diodes on GaAs today are the standard semiconductors used for downconverting mm-wave signals in the lower IF-bands below X-band. Compared to Impatt- or Gunn-diodes the variety of commonly used package types is obvious, which is caused by different problems to be solved. Some examples of typical mixer diodes are shown in fig. 4.

Best results in question of noise figure can be obtained using honeycomb-whisker-diodes. The diode chip containing thousands of single Schottky-diodes is contacted by a whisker. The disadvantage of poor mechanical stability can be avoided using planar structures (slot-line or fin-line on quartz-substrat). Experiences concerning with this technique have shown similar noisefigures compared to the conventional whisker method /3/.

Whisker-diodes encapsulated in a ceramic package deliver quite good electrical values as well as good mechanical performance /4/. The built-in procedure is easier compared to the pure whisker-technologie, but the price of these diodes is very high. For lower frequency application, bonded diodes incorporated in a similar package give good results. For lower frequencies also beam-lead diodes or "chip+ wire"-devices are well suited again. Beam-lead devices - Schottky as well as PIN - become more and more important at all especially in mm wave integrated circuits (MMIC). These devices are less expensive than whisker contacted diodes. The mechanical performance is better. The electrical properties though somewhat worse compared to whisker diodes are sufficient for many applications.

3. MM-WAVE CIRCUITS, DIFFERENT TECHNOLOGIES FOR DIFFERENT DEMANDS

Mm-wave circuits are not only an accumulation of semiconductors. Circuit and technology are rather more important if thermic, mechanic or low cost requirements must be observed. Concerning a mm-wave mass production fin-line structures obviously are the most sucessfull alternative to rectangular wave-guide. Nearly all microwave components can be realised using fin-lines unless high Q-factors are required. For high Q and high frequency devices about 90 GHz and above, however, today conventionally tooled block mounts using rectangular waveguides still give better results.

3.1 Oscillators and mixers up to 90 GHz

Gunn- or Impatt-oscillators as well as Schottky mixers are the most important components for mm-wave-equipments at all. Concerning oscillators results can be predicted very well. Using a test oscillator-mount unknown values of diode parameters, e.g. negative resistance, junction capacitance, and post reactance are derived from measured behaviour of oscillator frequency and power depending of iris-diameter /5/. In conjunction with producer data of package capacitance and ribbon inductance the oscillator performance can be calculated including power output, frequency, temperatur stability, quality-factor, pushing, and pulling.

Fig. 5 shows an approved oscillator device well suited for cw- or pulse applications up to 60 GHz. The oscillator does not need a reduced hight waveguide, thus for mass production a short piece of a commonly used waveguide embedded in a roughly tooled block mount can establish the cavity /6/. The power obtained is in general higher than that noted on the producer's diode data sheet. Temperatur stability as well as mechanical tuning range are sufficient for most applications in radar, radiometry, and communications.

For frequencies above 60 GHz oscillators containing a coaxial low pass filter as shown in fig. 6 are easier to fabricate. Due to the higher frequency this mount is mainly used for Impatt diodes. To obtain good results the low pass filter ("choke") must be designed and tooled very carefully to achieve reproductivity and to avoid rf-leakage. Depending on the employed diodes cw-power of more than 100 mW at 90 GHz is a typical value.

Mixer devices at mm-wavelength in general exhibit more variety compared with oscillators. This is caused by different diode packages as well as by different mixer applications. Fig. 7 shows a very simple but rugged mixer device, which exhibit excellent low conversion loss and noise figure at frequencies up to 60 GHz /7/. Similar to the oscillator as described above the mixer's cavity is not more than a piece of the related rectangular waveguide incorporating a quartz-substrat in cross plane. A slot resonator carries a beam-lead diode on it's backside thus separating IF and RF circuits. The diode's IF-circuit is floating in respect to ground so that diode polarities can be chosen arbitrarily. This is important for combining units to balanced mixers. A tuning screw enables to compensate deviations of diode parameters or to change the center frequency up to 500 MHz. Depending on diodes, best mixer data are 4.6 dB conversion loss at 7 dB SSB noise figure including 2 dB IF amplifier noise contribution.

At frequencies about 60 GHz good results have been obtained using encapsulated whisker diodes. To achieve the right diode position a small collet is used to pick up the diode /8/. This important detail is responsible for a reliable mixer operation at good electric values: Conversion losses of 4.5 dB at 60 GHz and 6.5 dB at 90 GHz can be reproduced.

3.2 Fin-line structures

As mentioned above, the fin-line is a very successfull alternative to the commonly used rectangular waveguide. The most important fin-line cross sections are shown in fig. 8. The simplest type of line is the wellknown unilateral fin-line which is best suited for almost all fin-line components /9/. The bilateral fin-line carries a metalisation on both substrate sides. This provides a lower transmission loss as well as more freedom in biasing active components, for instance matched PIN-attenuators or mixers with antiparallel switched diodes. The smallest wave impedance this line is delivering is about 100 Ω. Lower wave-impedances down to 10 Ω however, can be realized using antipodal fin-lines, which are best suited for wave transformers with very high transformation factors ($>$ 20:1). Therefore the antipodal fin-line is a good transition medium between rectangular waveguide and microstrip. Nevertheless great care must be taken to design a well matched broadband transition from rectangular waveguide to fin-line. Extensive experiments testing a lot of transition contours lead to an unstepped device incorporating a curved shape similar to a parabola /10/. As a result transitions with better than 20 dB return loss covering entire waveguide bands have been realized. In addition, because of no broadband demands for most applications, narrow band return losses better than 40 dB are possible. This of cause only can be carried out using computer aided design and automatic plotted layout generation. Fig. 9 gives a summary of realized passive components. Regarding fin-line mixers, detectors and PIN-diode-components the inherent broadband characteristics overcome all conventional waveguide devices. In most cases the band limits are not given by the circuit but by the measurement equipment. The conversion loss of balanced mixers can be reduced down to 7,5 dB broadband using a tuned LO. Otherwise losses do not exceed 11 dB over the entire Ka-band incorporating a fixed LO in midband as an integrated part of the front end. These oscillators can quite easily be built up by heatsinking the diode at the waveguide wall.

PIN-attenuators and switches are developed in manyfold versions: Broadband-devices with insertion loss less than 1 dB and isolation better 25 dB over the entire Ka-band have been realized. Narrow band resonator-typ attenuators only need one diode to achieve similar values as the broadband attenuators. Even at 90 GHz 2 dB loss and 13 dB isolation have been obtained using only one low cost beam-lead PIN-diode.

For some applications, in the turned-off case no power should be reflected. Providing a seperately biased matching PIN-diode in front of the others 15 dB return loss at all transmission values has been realized.

PIN-attenuators furthermore can be expanded to multi-port-pin-switch-components using fin-lines. These devices in general exhibit the same data as compared to PIN-attenuators. Designing power dividers, the choice of impedance relation between outgoing arms determins the power division.

Fin-line couplers have shown good results. Remarkable properties are high coupling factors up to cross-over operation as well as excellent broadband directivity. Finally the main advantages of fin-lines compared with the rectangular waveguide, should be summarised:

- o concentration of field and waves on small diode dimensions
- o multiband devices are possible
- o almost all components can be realized
- o a high integration level is possible
- o more flexible than monolithic devices
- o suited also for small size mass production
- o low cost circuit design and development

4. MM-WAVE SYSTEMS FOR MILITARY APPLICATIONS

As mentioned in the introduction the most important applications of mm-wave components are radar, communications and radiometry. In the following some examples recently developed at AEG-TELEFUNKEN are given.

4.1 A coherent pulse-radar at 35 GHz

In fig. 10 a block diagramm of a coherent 35 GHz pulse-radar is shown /11/. The heart of the system is a 34.5 GHz CW Gunn-oscillator which is locking a 500 MHz amplitude-modulated Gunn source by means of the lower sideband. This delivers a quartz stabilized IF-reference to provide a coherent down conversion. In addition the overlayed pulse modulation avoids spurious signals at 35 GHz which may decrease mixer's sensitivity. After passing a 35 GHz bandpass the pulse is amplified using a pulsed synchronised Gunn-oscillator. The antennas output power is about 1.5 W.

The received signal is attenuated by means of a PIN-STC using fin-lines. The insertion loss is 0.4 dB, the isolation 28 dB. Noise figure of the mixer is about 6 dB including 1.4 dB IF amplifier contribution.

Fig. 11 shows the radar without cover. The microwave circuit is embedded in the upper part. The right and left sides contain the transmitter- and receiver electronic circuits. Inputs and outputs as well as controll equipment is situated on the backside.

4.2 A 64 GHz radio-link using a parametric up-converter

Another example of mm-wave applications is a broadband radio-link at 64 GHz which is under construction at present, fig. 12. Most important part of the transmitter (on left) is a parametric up-converter, which renders a broadband modulation up to 500 MHz bandwidth at a conversion gain of 7 dB, /12/. The mixer's SSB noise figure is less than 6 dB. Fig. 13 shows on left the parametric up-converter containing two waveguide-filters for signal- and pump-power respectively and the varactor double-tee-mount in between. On the right, the down - converter is shown, containing the directional filter, the LO, the mixer mount, and an adjustable high precision tunable backshort.

4.3 A 90 GHz radiometer receiver frontend

A well known application of mm-wave components at higher frequencies are radiometer receivers. A typical block diagram of a frontend is shown in fig. 14. Although today sufficient Gunn-LO power at 90 GHz is available, varactor doubled sources at 45 GHz are less expensive and deliver more power. This is usefull if encapsulated mixer diodes are utilized requiring somewhat more power than whisker contacted diodes without package. The LO decoupling of the signal input considerably can be increased by compensating the input circulator by means of a sliding load. More than 40 dB is reproducible. The total system DSB-noise figure which can be reproduced including circulator loss and IF-noise contribution is about 9 dB.

REFERENCES

/1/ Kramer, N.B. — Solid State Technology for Millimeter Waves.
Microwave Journal 8 (21) (1978) S. 57-61.

/2/ Kennedy, J.K.; Lessoff, H.; Lile, D.L. — InP - An Assessment of United States Activities.
Symposium on Micr. Comp. for the Frequ. Range above 6 GHz, Bruxelles (1978).
Symp. Digest.

/3/ McMaster, T.F.; Carlson, E.R.; Schneider, M.V. — Subharmonically pumped mm-wave mixers built with notch-front and beam-lead diodes.
IEEE MTT-S Conf. (1977) Symp. Digest pp 389-392.

/4/ — Texas Instruments Diode data sheet MDX 623.

/5/ **Barth, H.; Bischoff, M.; Schroth, H.:** — 64/4-GHz-Empfangsmischer für ein 60 GHz-Transpondersystem. Abschlußbericht Vertr. Nr. 01 TI 077A-AK/RT/WRT 2077 AEG-TELEFUNKEN, Ulm (1978).

/6/ Lindner, K.; Wiesbeck, W. — Die Mikrowellenbaugruppen eines 35 GHz Abstandswarnradars für Kraftfahrzeuge.
Mikrowellenmagazin 5 (1977) pp 398-403.

/7/ Meinel, H. — Aufbau eines Ka-Band-Mischers in hybrider Hohlleitertechnik.
Interner Bericht AEG-TELEFUNKEN, Ulm (1978).

/8/ Barth, H.; Rembold, B. — 90 GHz Radiometrie-Empfänger.
Interner Bericht AEG-TELEFUNKEN, Ulm (1978).

/9/ Hofmann, H.; Meinel, H.; Adelseck, B.
New integrated mm-wave components using fin-lines.
IEEE MTT-S (1978). Symp. Digest pp 21-23.

/10/ Adelseck, B. e.a.
Neue Wellenleiter und Schaltkreise für mm-Wellensysteme. 2. Zwischenbericht zum Vertrag T/RF 31/60007/61010.
AEG-TELEFUNKEN, Ulm (1977).

/11/ Lindner, K.
Ka-Band Radar (Gera). Abschlußbericht zum Vertrag T/RF 31/71418/71310.
AEG-TELEFUNKEN, Ulm (1978).

/12/ Barth, H.; Meinel H.
Design of a 4 GHz to 64 GHz parametric upper-sideband up-converter.
Symp. on Microwave Components for the Frequency Range above 6 GHz, Bruxelles (1978) Symp. Digest.

/13/ Barth, H.; Rembold, B.
Design and performance of 90 GHz radiometer front ends.
AGARD-Symposium on Millimetre and Submillimetre wave propagation. Munich (1978).

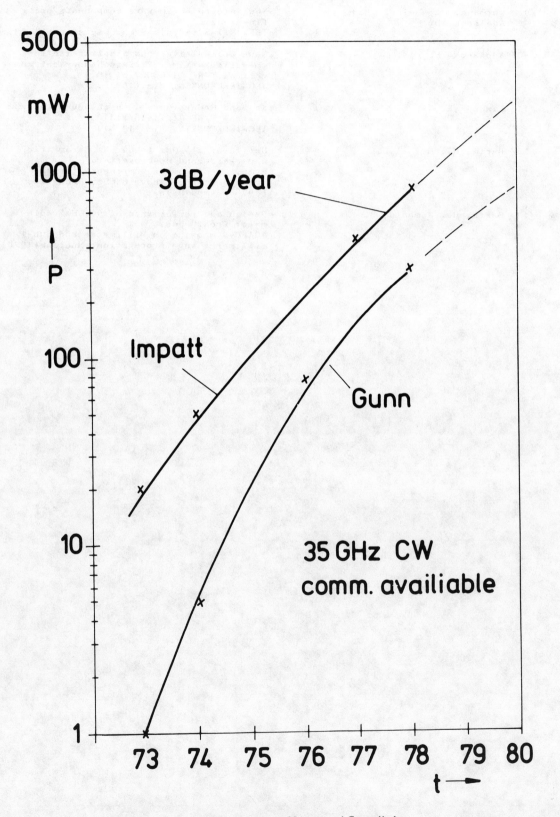

Fig.1 Power increase of Impatt and Gunn diodes

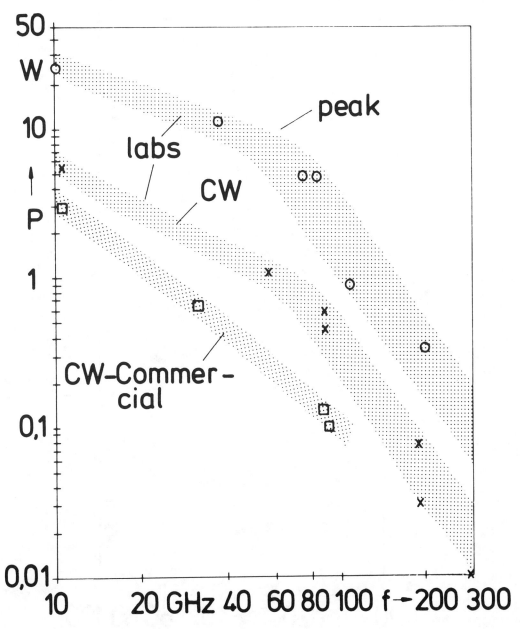

Fig.2 Impatt-diode power vs. frequency

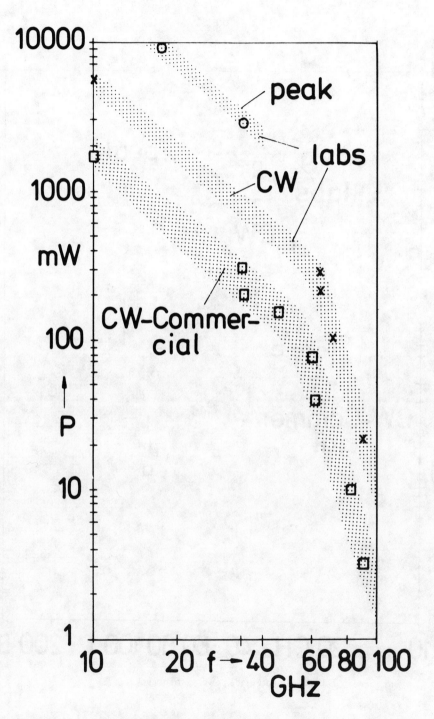

Fig.3 Gunn-element power vs. frequency

Millimeter Wave Radar RF Sources and Components

diode	whisker	notchfront + whisker	ceramic package + whisker	ceramic package + gold ribbon	beamlead	chip + wire
	100 μm	Quartz 100 μm	600 μm	1000 μm	1000 μm	800 μm
C_j/fF	6 – 12	6 – 12	30	150	50	60
C_p/fF	–	–	60		15	
R_s/Ω	5 – 15	5 – 15	10	10	5	10	25
conversion-loss/dB	5–6 (90 GHz)	8 (90 GHz)	5 (60 GHz) 7,5 (90 GHz)	5 (35 GHz)	5 (35 GHz) 8 (90 GHz)	6 (35 GHz)
noise figure/dB SSB, without IF	4 (90 GHz)	4,6 (98 GHz)	5–6 (90 GHz)	5–6 (35 GHz)	8 (98 GHz)	6–7 (35 GHz)
advantages	best values at f > 60 GHz	good values and rugged performance	good values and easy to build in	low cost, high reliable, suited up to 60 GHz	suited for planar structures up to 100 GHz	low cost suited up to 40 GHz

Fig. 4 mm-wave mixer diodes

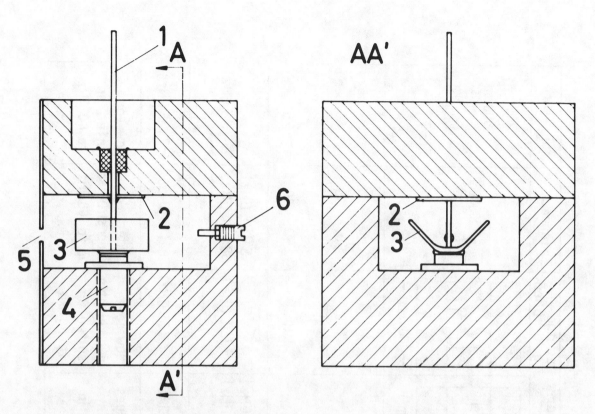

1 bias connector
2 capacitive blocking choke
3 matching wings
4 gunn element
5 iris
6 tuningscrew

Operation: CW or pulsed
Frequency: 26-60 GHz
Power: 100 mW$_{cw}$, 2 W$_{peak}$
Pulse: down to 5 nsec
Stability: 10^{-6}/°C
Mechan. tuning: >1 GHz

Fig.5 Gunn-oscillator

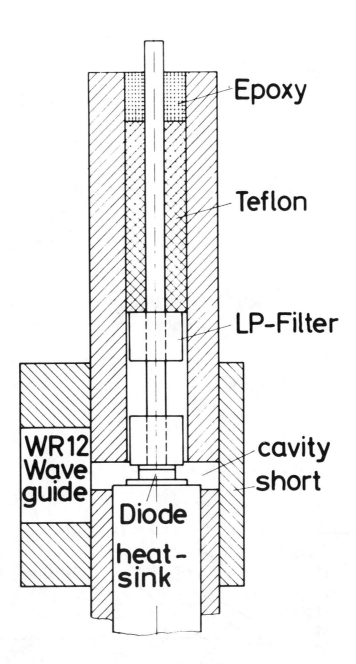

Operation:	CW or pulsed
Frequency:	60 - 100 GHz
Power:	100 mW$_{cw}$, 2 W$_{peak}$
Pulse:	$\geq$ 100 nsec
electr. tuning:	up to 10 GHz

Fig.6 Impatt-oscillator

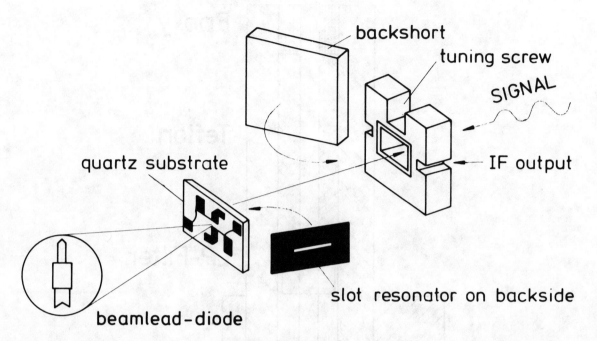

Fig.7 mm-wave-mixer

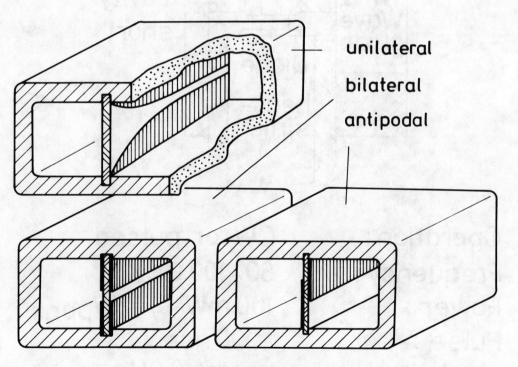

Fig.8 Finline-structures

component		frequency, GHz	properties
balanced mixer		26–40	L_c <8dB broadband
oscillator		26–40	Gunn devices for integrated LO's
detector		26–40	Sensitivity >.5mV/µW broadband
PIN-attenuator		18–60 60–90	loss <1dB, isol. >30dB entire band loss <2dB, isol. >13dB/diode 90GHz
PIN-switch power divider		18–60	loss <2dB, isol. >20dB broadband power division adjustable, ripple ±.2dB broadb.
coupler		26–40	k down to 0dB (crossover) directivity: >15dB entire band; >40dB narrowband

Fig.9 Finline components

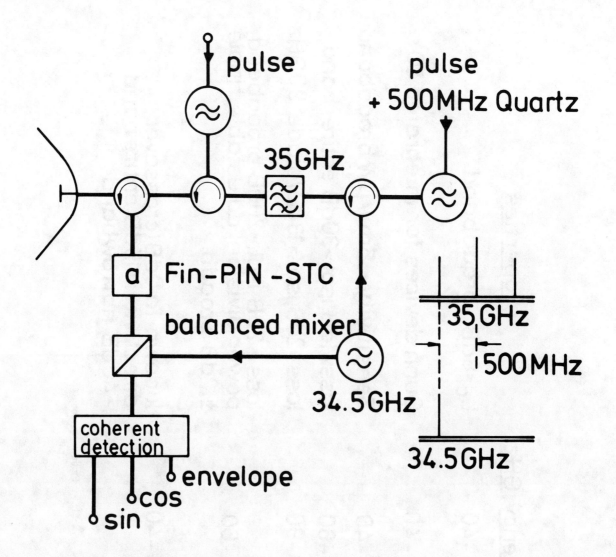

Fig.10 35 GHz pulse doppler radar principle block diagram

Millimeter Wave Radar RF Sources and Components

Fig.11 A coherent 35 GHz pulse radar

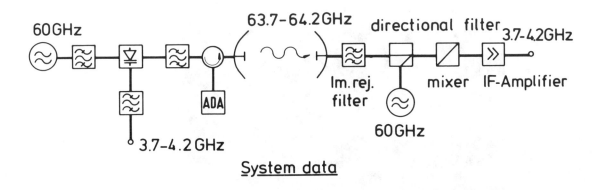

System data

transmitter		receiver	
power	10 mW (100 mW)	noise figure	6 dB
bandwidth	500 MHz	LO decoupling	> 30 dB
ripple	± 0.2 dB	gain	30 dB
gain	7 dB		

Fig.12 Broadband radio-link at 64 GHz

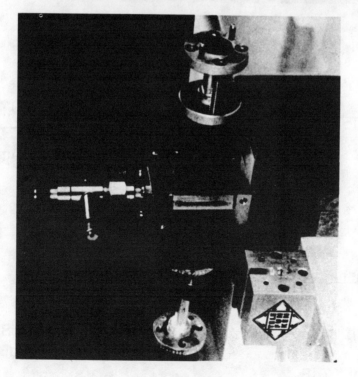

Fig.13 Up- and down-converter of a 64 GHz radio link

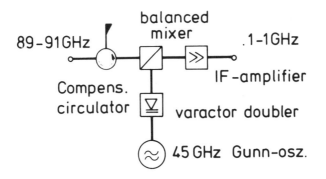

Systemdata

noise figure (dsb)	9 dB
LO decoupling of signal input	>40 dB
input matching	18 dB
RF/IF-gain	54 dB
IF-bandwith	0.9 GHz
ripple	±0.8 dB

Fig.14 A 90 GHz radiometer receiver front end

DISCUSSION

H.G.Oltmam, US
   (1) What was the instantaneous and tunable bandwidth of your example 35 GHz coherent radar?

   (2) Is the local oscillator tunable?

Author's Reply
   (1) Bandwidth is about $\frac{1}{10 \times 10^{-9}}$ Hz, given by the transmitter pulse.

   (2) The local oscillator is not tunable, but may be replaced by another at a different frequency.

M.C.Carter, UK
   Were there any problems due to leakage through the circulator.

Author's Reply
   The attenuation of the isolated circulator path combined with the PIN-STC-attenuation is sufficient for our application.

Solid-State Millimeter-Wave Sources

H.J. Kuno and T.T. Tong

Abstract

Recent advances in solid-state millimeter-wave sources are reviewed. Specifically, the development of IMPATT and Gunn oscillators and amplifiers are presented with emphasis placed on system applications.

Introduction

Recently the development effort on millimeter-wave tracking radars and missile seekers has increased at a rapid rate. Millimeter-wave systems offer many advantages over both microwave systems and electro-optical systems. In comparison with microwave systems, millimeter-wave systems offer smaller size, lighter weight, improved accuracy, greater resolution, as well as narrower beamwidth for a given antenna aperture. In comparison with electro-optical or infrared systems, millimeter-wave systems provide improved penetration through cloud, smoke, and dust. The development activities for millimeter-wave tracking radars and missile seekers are currently centered around 35 GHz and 94 GHz, where atmospheric attenuation is low.

Motivated by the system requirement, the development of solid-state millimeter-wave devices has progressed rapidly in recent years. In this paper solid-state sources, such as IMPATT and Gunn sources, are reviewed. Pulsed and CW power sources for transmitters, and local oscillators for receiver front ends are also discussed.

CW Oscillators

Perhaps the most commonly available power sources are free running CW oscillators. Shown in Figure 1 are state-of-the-art output power of millimeter-wave IMPATT and Gunn sources as a function of frequency. IMPATT oscillators can provide CW power output of 1.5 W at 35 GHz, 1 W at 60 GHz, 700 mw at 94 GHz, 100 mw at 140 GHz, and 25 mw at 220 GHz. The recent improvement in output power from the millimeter-wave IMPATT oscillators are primarily due to the development of double-drift silicon IMPATT diodes with diamond heat sink package. Gunn devices, on the other hand, provide 500 mw at 35 GHz, 150 mw at 60 GHz, and 25 mw at 94 GHz.

In addition to the output power, noise characteristics are important properties for system applications. Shown in Figure 2 are AM noise characteristics of millimeter-wave oscillators. Typically Gunn oscillators provides about 10 dB less noise than IMPATT oscillators. Thus, Gunn oscillators are better suited for LO applications while the IMPATT oscillators are suited for high power applications. At frequencies higher than 100 GHz, where Gunn LO's are not available, IMPATT oscillators must be used as LO's. IMPATT LO's can be used effectively with mixers when filters or balanced mixers are used to properly suppress the LO noise. In general a noise suppression factor of 30-40 dB can be achieved. In fact IMPATT local oscillators are used in many millimeter-wave receivers with noise figures as low as those using a klystron.

Pulsed Oscillators

For many system applications high peak power pulsed oscillators are required. Significant progress has recently been achieved in the development of pulsed millimeter-wave sources. Peak output power levels of 10 W at 35 GHz and 5 W at 94 GHz can be achieved on a reproducible basis with double drift IMPATT diodes. With Gunn devices peak output power of 1-2 W was achieved at 35 GHz. In general more than 10 times as high peak power output as CW power output can be achieved with pulsed IMPATT and Gunn oscillators (See Figure 3). In operating pulsed sources, maximum pulse width and duty factor are important parameters which determine the achievable peak power output. Since millimeter-wave solid-state devices have small thermal time constant, the device temperature rises rapidly within a pulse. Shown in Figure 4 are calculated temperature transient resistance of IMPATT diodes as a function of time under pulsed operation. In order to achieve the maximum peak output power, pulse width should be kept below 100 ns. For operations with longer pulse widths peak output power will be reduced. In addition, trade-off between peak output power and pulse duty factor must be made.

Another important property associated with the pulsed operation is the frequency chirping effect. As the device temperature increases with time within a pulse cycle, the diode impedance changes which results in a decrease of oscillation frequency with time. Since the frequency of operation is also a function of bias current, the amount of the frequency chirp can be controlled to meet specific systems requirements by shaping the bias pulse current waveform (as shown in Figure 5). Typically, frequency chirp greater than 1 GHz can be achieved.

Power Amplifiers

Both IMPATT and Gunn devices have effectively been used as millimeter-wave power amplifiers. For power amplification both stabilized amplifier and injection locked oscillators have been developed. Injection locked oscillators are suited for high-gain, narrow bandwidth operations, while stabilized amplifiers are for low gain (~10 dB per stage), broader bandwidth applications. The maximum output power achievable from an amplifier is approximately the same as that obtainable from the same device operated as an oscillator.

In order to achieve higher output power, a number of devices can be combined in an amplifier cavity or a number of amplifiers can be combined by means of hybrid coupler. The feasibility of power combining techniques has also been demonstrated at millimeter-waves. The power combiner/amplifier approach should prove useful in many future systems development.

Frequency and Phase Stabilizations

Free running oscillators, both IMPATT and Gunn, typically have frequency stability of $-0.5 \times 10^{-4}/°C$. This means that the frequency drift rate is approximately -2 MHz/°C at 35 GHz and -2 MHz/°C at 94 GHz. For applications where temperature variations cause excessive frequency drifts, a number of techniques have recently been developed for controlling the frequency. The simplest method is to control the oscillator cavity temperature by means of a small heater and a control circuit. Since solid state oscillator cavities have small masses, it is relatively easy to control its temperature in this way. The temperature variation can be kept to less than 1°C.

Another approach to the frequency stability is to use a frequency discriminator such as an invar cavity filter with an AFC loop (See Figure 5). This method does not require heater power but requires a more complex control circuitry.

In addition to the long term frequency stability, many systems require phase stabilities of crystal controlled oscillator quality. For such applications two basic approaches have recently been developed for millimeter-wave sources. One is a phase lock loop approach and the other is an injection locking approach using a multiplier chain as shown in Figures 6c and 6d. In the phase locked oscillator, sampled power of the millimeter-wave frequency is converted down to an IF by a harmonic mixer, the phase is compared with the phase of the reference crystal oscillator, and the phase error is then corrected by means of a feedback loop to the millimeter-wave oscillator. This technique has successfully been applied to both IMPATT and Gunn oscillators. Shown in Figure 7 is a comparison of phase noise of a free running and phase locked millimeter-wave IMPATT oscillators. Significant reduction in phase noise can be seen within the locking band. The locking bandwidth is limited by the locking loop bandwidth, which is typically 1-10 MHz.

Similar improvement in phase stability can be achieved by means of an injection-locked oscillator with a frequency multiplier chain. Since trade-off between locking gain and bandwidth can be made, a broader locking bandwidth can be achieved in an injection locked oscillator than in a phase locked oscillator. However, an injection locked oscillator using multiplier chain is considerably more complex than a phase locked oscillator at millimeter-wave frequencies.

Summary

Solid-state device technology has advanced to the point where millimeter-wave sources are ready for system applications. Along with small size and light weight, solid state devices also offers high reliability. Extensive reliability studies of millimeter-wave devices have been made in the past several years so that systems can now be designed not only for performance but also for reliability. It appears that the key solid-state devices necessary for systems such as tracking radars and missile seekers are ready up to 100 GHz and that the frequency coverage is rapidly extending into 140 GHz and beyond.

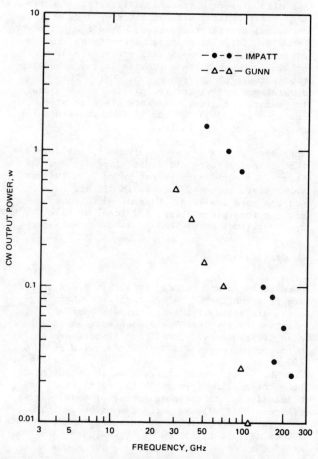

Figure 1   State-of-the-art of cw IMPATT and Gunn oscillators.

Millimeter Wave Radar RF Sources and Components

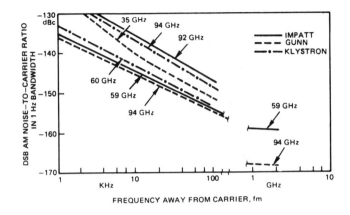

Figure 2 AM noise characteristics of millimeter-wave oscillators.

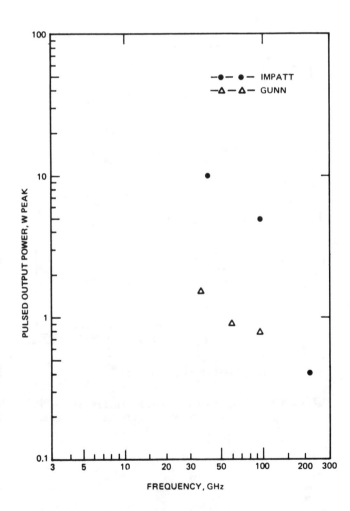

Figure 3 State-of-the-art of pulsed IMPATT and Gunn oscillators.

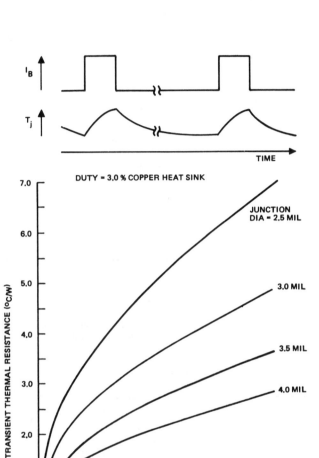

Figure 4 Transient thermal resistance of a pulsed 94 GHz IMPATT diode on a copper heat sink.

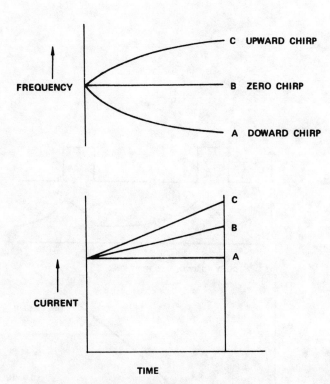

Figure 5 Frequency chirp characteristics of a double drift IMPATT diode and the response to a current ramp.

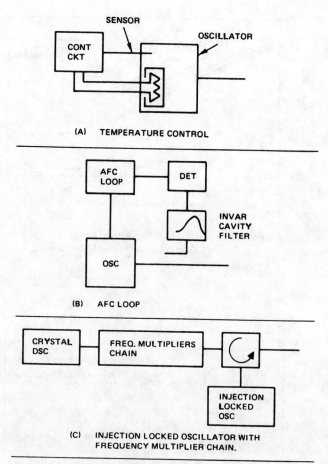

Figure 6 Frequency and phase stabilizations of millimeter-wave solid-state sources.

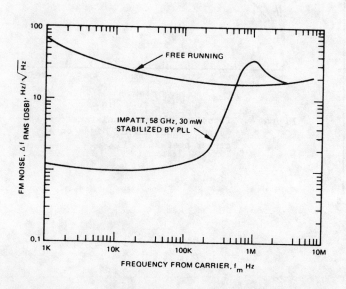

Figure 7 FM noise of free running and phase locked IMPATT oscillator.

EHF Solid State Transmitters for Satellite Communications
Jorg E. Raue
Proc. AIAA 8th Communications Satellite Systems Conference, April 1980

Since 1970 TRW has been active in all phases of millimeter component, subsystem and system development, including passive high performance components and antenna systems, active semiconductor power and low noise receiver devices, and active circuits. The intense 1970-75 efforts led to a series of technological breakthroughs, including the first successful application of network analysis and computer aided design at millimeter wave frequencies, and resulted in a series of components that clearly advanced the state-of-the-art. (References 1-6). This work then formed the basis for subsequent successful millimeter wave subsystem and system development. One of these is an on-going concentrated effort to develop high power (greater than 1 watt CW) solid state amplifiers suitable for space applications. This paper describes these development efforts for 2 to 10 watt communication amplifiers in the frequency ranges near 35, 40 and 60 GHz.

The two technical challenges that have to be overcome in order to produce reliable multi-watt solid state amplifiers are: (1) solid state sources with reasonable output power and efficiency are required, and (2) diode combining circuitry with good combining efficiency and small size are essential. Significant progress has been made in both areas in the past two years. Hughes, Raytheon and TRW have produced IMPATT power sources with output power exceeding 1 watt at 40 GHz and 0.5 watt at 60 GHz with reasonable junction temperatures. In the circuits area, TRW is developing a 10 watt solid state amplifier at 41 GHz and a 2 watt amplifier at 60 GHz. Table 1 summarizes these and other recent developments.

The success of these technology development efforts, all of which are directed toward space communication applications, is primarily attributable to the successful application of advanced design techniques to millimeter wave frequencies. This includes detailed device and circuit analysis and modeling with heavy reliance on computer aided design. In addition, great strides have been made recently in advancing the state-of-the-art of EHF waveguide circulators - components of fundamental importance in the design of high performance reflection amplifiers.

This paper describes the aforementioned amplifiers, including device and circuit design aspects, design techniques, gain and bandwidth tradeoffs, efficiency and reliability considerations, etc. For example, two of these amplifiers are depicted in Figures 1 and 2.

Our in-house developed line of breadboard laboratory network analyzers, which covers the 26 to 40, 33 to 50, 55 to 65 and 90 to 100 GHz band, each with 10 GHz instantaneous bandwidth, is typified by the V-band unit shown in Figure 3. These units, used together with the proper RF circuits, allow accurate device terminal impedance measurements and circuit evaluation. For example, the circuit impedance seen by a diode in a Ka-band circuit is shown in Figure 4. Of particular value here is the 10 GHz instantaneous bandwidth which eliminates the time consuming measurements typical of commercial units available up to 40 GHz. These units have an instantaneous bandwidth of 2 GHz.

Also evident is the extraordinary simplicity of the RF circuit, which allows analytical treatment.

The starting point of the driver amplifier circuit design was the measurement of diode impedance. From this data and from the measured impedance of other diodes from the same lot, it was determined that the reactive component of the

Table 1. Summary of TRW's High Power EHF Solid State Amplifier Developments

Amplifier Description				
Frequency	Output Power	Gain	Bandwidth	Status
37.3 GHz	5 Watts	33 dB	700 MHz	Completed 1977
38.1 GHz	5 Watts	30 dB	700 MHz	Completed 1978
41 GHz	10 Watts	30 dB	100 MHz	Under Development
60 GHz	1 Watt	14 dB	600 MHz	Completed 1978
60 GHz	2 Watts	17 dB	200 MHz	Under Development

*The author is now with the California Polytechnic State University, San Luis Obispo, CA, 93407, as Head of the Department of Electronic and Electrical Engineering.

small signal impedance under bias, clustered principally around $-j0.5$ normalized to 50 ohms. Based on this assessment, the reactive component of the circuit impedance was targeted to $+j0.5$ ohms (normalized to 50) to complement the diode impedance.

A coupled coaxial circuit was selected as the primary circuit for development because of its potential for analytical characterization and because of its similarity to the combiner coaxial segment. Figure 4 depicts this circuit. Of particular interest are the thick iris coupling from the coaxial line to the waveguide port and the dumbbell type bias line. This type of circuit had previously been successfully employed with Gunn devices. The pin, however, has the drawbacks of being relatively difficult to modify and presents stability problems when implemented with an IMPATT device. As a consequence, a bias line similar to that used in the combiner, was implemented. This line consists of a pin extending from the diode package, past the waveguide coupling iris, through a lossy termination element to a spring tensioner and bias connection. In addition, a transformer may be attached to the end of the line to facilitate impedance matching to the diode. The advantages of this type of line are high ease of modification and the stable terminated environment presented to the diode to suppress out of band spurious oscillations. This circuit is characterized by exceptionally low Q's - typically less than 10.

Various forms of power combining at the circuit level were considered for the power stage. A previous successful effort at 37 GHz involving a circular TM_{010} cavity combining scheme provided for substantial impetus to employ the same technique at 41 GHz. Due to the increased frequency, the diameter of the basic TM_{010} cavity is restricted to a point where the maximum number of diodes that can be accommodated is reduced. A minimum of twelve diodes will be required on the assumption that 1 watt RF is available per device and a projected combining efficiency of 80% is realized. This large number of devices at a higher frequency dictates that a higher order mode be utilized, specifically the TM_{020} mode. The experimental breadboard combining circuit for the final power stage of the 10 watt, 41 GHz amplifier is seen in Figure 5, together with a schematic representation of the combining structure. All stages of these various amplifiers are designed and built in a fully deterministic fashion, i.e., no tuning elements are used.

High performance circulators are imperative for successful quality amplifier design. For this reason, we developed, from scratch, an entirely new line of exceptionally high performance circulators. The new design is compared with the well-known previous approach in Figure 6. The cylindrical junction, mechanically interlocked without the use of any adhesives within the waveguide housing, and consisting of simple shapes which are relatively easy to fabricate, inspect and assemble, results in excellent repeatability of performance and unlike the commonly used triangular, rigidly epoxy-bonded structures which are susceptible to thermal, shock and vibrational stresses, this junction is soft-mounted and impervious to these stresses. Similar junctions, constructed at lower frequencies for space applications, were thoroughly tested to spacecraft level specifications without a single failure. The ferrites of the junction are designed to operate with a high level of magnetic bias. This requires higher quality, more expensive, rare earth permanent magnets, but the added expense is easily justified considering that the need for a precise adjustment of the magnetic bias is eliminated saving labor, and that the thermal stability of the circulator and its power handling capability are significantly enhanced.

In the microwave frequency ranges below 30 GHz, the circulator designer may select the ferrite material from an available range of compositions to satisfy the specification requirements. Above 30 GHz, this freedom of choice vanishes; there is only one material suitable and even this ferrite with the highest saturation magnetization of 5250 gauss becomes marginal as we approach 50 GHz. This limitation of saturation magnetization poses difficult problems to the wideband junction design. Narrowband, or Faraday rotation components can be constructed, but the narrowband circulators are not suitable for use with solid state devices with wideband negative resistance characteristics, and the Faraday rotation circulators are characterized by prohibitively high insertion loss.

The wideband performance of the circulators is obtained by the design of the junction ferrites, which allow a simultaneous propagation of two adjacent dielectric resonator modes. With an adequate level of saturation magnetization for the given frequency range, this design approach insures good balance of the amplitude of the VSWR and isolation ripple. After a proper selection of the ferrite dimensions, the length of the matching dielectric spacer, the length and the diameter of the metallic transformer is calculated to complete the circulator design. The task is relatively simple and easy when the ferrite material has adequate magnetic properties and when a relatively narrow bandwidth is acceptable, or when the circulator is used in a system where the insertion loss is not a crucial consideration. The present design (patent issued in 1979) evolved from a thorough and logical consideration of the whole problem. A detailed evaluation of the electrical, environmental and manufacturing requirements pointed out the way to the cylindrical geometry, which combines the structural simplicity with the best potential for analytical optimization of the electrical parameters. A representative sample of the full line of circulators is shown in Figure 7. Typical performance data for Ka- and Q-band circulators comprises Figures 8 and 9. Typical insertion loss per junction in the 7 to 40 GHz range is 0.07 dB to 0.1 dB per pass, with bandwidth of 25 to 30 percent, and CW power handling capability of 200 Watts at X-band and 20 Watts at 45 GHz.

References

[1] Bayuk, F. J. and J. E. Raue, "Large Signal Device Characterization for Broadband Ka-Band Avalanche Diode Amplifier Design", IEEE-MTT International Microwave Symposium Digest, pp. 210-212, June, 1976.

[2] Bayuk, F. J. and J. E. Raue, "Ka-Band Solid State Power Amplifier", IEEE-MTT International Symposium Digest, June 1977.

[3] J. E. Raue, "High Performance Millimeter Wave Receiver and Transmitter Component Development", EASCON-77, pp. 24-4A-E, September, 1977 (invited paper).

[4] Yuan, L.T., et al, "A V-Band Network Analyzer/Reflection Test Unit", IEEE-MTT International Microwave Symposium Digest, pp. 221-223, June, 1976.

[5] U.S. Patent No. 4,145,672, March, 1979.

[6] Piotrowski, W.S. and J. E. Raue, "Low Loss High Power Latching Waveguide Switch", International Symposium Digest, pp. 103-104, June 1978.

Acknowledgements

The author gratefully acknowledges the key contributions of Franklin Bayuk who is responsible for the power amplifier development, Wieslaw Piotrowski, who is responsible for the ferrite development, and Dale Mooney who has made significant contributions in the area of computer aided EHF amplifier design and development. This work was supported in part by the Air Force Avionics Laboratory, under contracts F 33615-77-C-1184 and F 33615-77-C-1185. R.T. Kemerly is the project engineer.

Advanced Solid-State Components for Millimeter Wave Radars

P.M. Schwartz, R.F. Lohr, K.P. Weller, and R.L. Zimmerman

Abstract

Low power components for use in W-band (75-100 GHz), chirp radars have been developed. The development was based on a chirp bandwidth of 1.5 GHz. The components included resistive mixers for frequency conversion and generation of the chirp waveform at the radar output frequency and for bandwidth compression in the receiver front end, varactor multipliers, phase stabilized power sources to provide basic RF power for radar processing, and IMPATT amplifiers to boost the output power from the low power levels available from the mixers.

A. Introduction

The components described in this talk were developed to show the feasibility of operating a millimeter-wave radar with a linear FM chirp waveform used for pulse compression. In this project we were concerned only with the design of the millimeter wave components and were able to assume that all the signals with frequencies in Ku-band and lower are derived from a single, stable master oscillator. We did, however, impose limitations on the "low" frequency power demanded by the millimeter wave components so that it would have been practical for all the power to be supplied from a single solid-state source at each frequency. The chirp bandwidth was chosen to be 1.5 GHz; this sweep could reasonably be generated at lower frequencies and represented a potentially useful bandwidth.

The block diagram for a basic implementation is shown in Fig. 1. In the transmitter there is a waveform generator to create the FM chirp and an amplifier to increase the output power level. The waveform generator operates inside a feedback loop in order to improve the frequency linearity of the chirp waveform. This loop can contain the amplifier as well as the waveform generator so that the major portion of the phase distortion in the transmitter is cancelled. The receiver contains a mixer followed by a low noise amplifier and second IF conversion stage; there is also another waveform generator to generate an FM chirp for the local oscillator to the input mixer so that bandwidth compression is performed at the receiver input.

In order to maintain the frequency stability and phase coherence that was assumed for the low frequency driving signals, it was necessary to provide each of the waveform generators with millimeter wave power that was suitably stabilized in phase and frequency. This stabilization was accomplished by designing circuits that could injection lock the output of a klystron and simultaneously control the natural frequency of oscillation of the klystron to be coincident with the frequency of the injected signal.

B. Component Design and Performance

1. Waveform Generator

The waveform generators were required to create the FM chirp signals for the transmitter and receiver. The generation of the millimeter wave chirp was done with an upconverter to translate a chirp generated at a low frequency to the output frequency. In addition, there had to be some provision for detecting the frequency linearity of the output chirp waveform so that a signal could be obtained to control the VCO providing the chirp. It was decided that this function could be performed most efficiently by sampling the output and translating back to the original frequency. The major benefit derived from this approach was that it allowed the linearity of the chirp to be measured at lower frequencies where compact nondispersive delays are available to allow the comparison of the chirp waveform with a delayed version of itself.

Because the two translation operations were so similar there was very little difference in the design of the up and downconverters. Of the two options available for use in the frequency converters, varactor converters or nonlinear resistive mixers, the latter type were universally chosen for this demonstration because of the easier task of matching to a resistive load rather than a reactive load. The millimeter wave Schottky diodes that were used in the frequency converters had a planar, honeycomb array structure, and their basic characteristics have been described previously.[1]

The typical converter circuit used is shown in Fig. 2. The converter is single ended, requiring only one diode. The diode is mounted in a modified Sharpless wafer package. This package has one port for the injection of bias and IF signals which is isolated with a low pass filter from two RF waveguide ports. The waveguide ports are used for the injection of millimeter wave pump power and in the case of an upconverter for the extraction of the converted signal. (With the downconverter the second waveguide port is used to inject the input signal, and the converted signal is extracted from the IF port). Isolation between the two waveguide ports was achieved with bandpass filters that were coupled to the diode package with two-section transformers. The transmission and reflection of the filters had a major influence on the final output characteristics of the converter circuits. Special low loss filters using rectangular TE_{101} cavities were developed with tightly controlled characteristics. The filters in conjunction with a multislug tuner at the IF port were used to achieve the flat in-band response that would be needed in this type of application. Most of the waveguide filters were based on a

---

*Hughes Aircraft Company, Research Laboratories Division, Malibu, California.

**Hughes Aircraft Company, Electron Dynamics Division, Torrance, California.

ACKNOWLEDGEMENT: This work sponsored by the Air Force Systems Command's Rome Air Development Center, Griffiss AFB, NY

three-section, .1 dB Chebyshev ripple design and had an excess insertion loss of ~.5 dB.

The most important performance measure for the upconverters is the converted output power. Fig. 3 shows a typical upconverter output where the minimum peak power is 1 mW and the output ripple is less than 1 dB peak-to-peak across the band. For the downconverters the important factor is conversion loss; the typical conversion loss achieved with the downconverters developed in this project was 10 dB.

The output power from the upconverter in the transmitter chain was raised to the 30 mW level by a two stage IMPATT amplifier. The two stage, circulator coupled amplifier is provided with a broadband isolator at the output to prevent out of band oscillation when the unit is connected to a high VSWR load. No isolation is required between stages. The overall gain of this unit for a 1 mW input is 15 dB with less than 2 dB gain variation over a 1.5 GHz bandwidth centered at 94 GHz.

2. Stabilized Millimeter Wave Power Sources

The millimeter wave power used to create the transmitter and receiver local oscillator waveforms was supplied by klystron oscillators. The output of the klystrons was stabilized in phase and frequency by injection of a low level signal into the tube through a circulator as shown in Fig. 4. This low level locking signal was derived from a stabilized Ku-band signal with a x5 varactor multiplier. The bulk of the circuitry in Fig. 4 was needed to compensate for the narrow locking bandwidth of the klystrons. This bandwidth was so narrow for the locking power available ($\sim$16 dB below the output power) that normal temperature and power supply variations would have been sufficient to unlock the klystron. To counter this effect the reflector voltage of the klystron was used to control the free running frequency of the klystron to be equal to the frequency of the injection locking signal. The control signal for the klystron reflector voltage was derived from the phase difference between the locking signal and the klystron output. Any change in the klystron's free-running frequency changes the relative phase of the output so this parameter could be used to control the klystron reflector voltage. The use of frequency converters with an IF frequency of 1.6 GHz made it possible to use the same component design for this control circuitry as was used for the waveform generator converters. This design choice also made it possible to implement the required phase bridge with off-the-shelf, commercial microwave components.

The x5 multiplication needed to provide the injection locking power for the klystron was performed with specially designed snap-action millimeter wave varactors. These diodes were similar to the varactors reported earlier[1] but were made with a lower doped epitaxial layer. The multiplier circuit used did not contain any special provision for idler tuning and had 7 mW output with 8% efficiency at 80 GHz.

3. Receiver

The millimeter wave receiver had a mixer front end that could accept local oscillator and input signals over the 1.5 GHz chirp bandwidth. Its construction was essentially the same as that used for the frequency converters in the waveform generators. The conversion loss of the mixer with 1 mW of local oscillator power $\sim$9 dB at the peak of the IF response with a bandwidth of $\sim$0.4 GHz. The first IF amplifier was a 4 GHz paramp with a 1.6 dB NF. The noise figure measured with hot and cold loads was 11-12 dB although the experimental uncertainty was high because the hot load noise temperature was only $\sim$390K.

C. Summary

The feasibility of operating a linear FM chirp radar at frequencies in W-band (75-110 GHz) has been demonstrated by developing the solid-state components needed to produce the needed waveforms and process the return signal at the receiver input. This demonstration effort relied on semiconductor devices that were developed primarily for use in V-band (50-75 GHz) and was mainly directed at developing the high frequency circuitry needed for the different circuit functions. These developments included high quality waveguide bandpass filters, and improved IMPATT amplifier circuitry. The components developed for the project included a bandwidth compressing mixer with a conversion loss of 9 dB for the receiver, chirp generators for both the transmitter and the local oscillator for the mixer in the receiver, a two stage IMPATT amplifier with 15 dB gain at a 1 mW input level, and moderate power (20-30 dBm), phase stabilized millimeter wave power sources.

D. Acknowledgement

We would like to thank R. M. Madden for supplying the silicon varactors used in the x5 multiplier circuits.

Reference

1. H.L. Stover, et al., "Solid-State Components for a 60-GHz Receiver Transmitter" 1973 IEEE International Solid State Circuits Conference, Philadelphia, Pennsylvania.

Millimeter Wave Radar RF Sources and Components

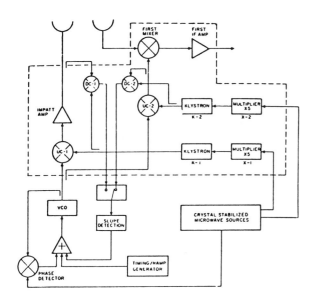

Fig. 1. Basic Block Diagram

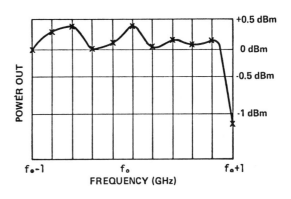

Fig. 2. Frequency Converter Circuit

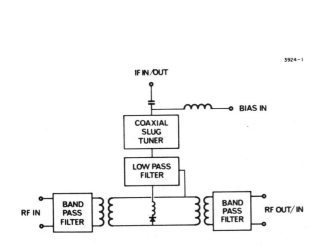

Fig. 3. Upconverter Output Characteristic

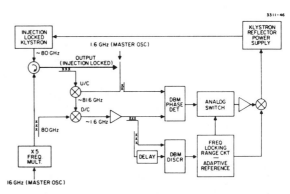

Fig. 4. Klystron Stabilization Circuit

A 90 GHz FM-CW-Radar Transmitter

H. Barth and M. Bischoff

ABSTRACT

The design and the performance of an FM-CW-RADAR transmitter for 90 GHz set up with two synchronized and powercombined Impatt-oscillators are presented in this paper. The synchronising source consists of a 45 GHz-Gunn-oscillator which is simultaneously swept and doubled by the same varactor. The outputpower of the entire unit is 240 mW and the achievable sweep range amounts to 1 GHz.

Introduction

This paper describes the performance of an FM-CW-RADAR transmitter in the 90 GHz-range. In order to achieve sufficient power while exhibiting high spectral purity, the outputs of two synchronized Impatt-oscillators are combined using a 90°-hybrid. The master oscillator incorporates a 45 GHz-Gunn-diode and one Varactor-diode mounted in a common waveguide resonator. The varactor simultaneously doubles the frequency from 45 to 90 GHz and operates as a sweep device when its bias voltage is changed. The output power of the entire system is 240 mW, having the spectral purity of the Gunn-oscillator.

IMPATT-Oscillator

The simple, low-cost set up of the IMPATT-oscillator can be seen in fig. 1, showing the cross-section of the oscillator block mount.

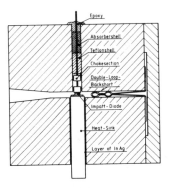

Fig. 1
IMPATT-oscillator

The diode is soldered on a copper heat sink covered with a thin layer of In Ag solder which is press fitted in the precisely machined hole. Moving this heat sink up and down and simultaneously optimizing the position of the sliding short, the diode can be matched to the reduced height waveguide.

In order to achieve excellent mechanical stability and good electrical performance, the choke has to be designed very carefully. The structure, centered by a teflon support shell, consists of two quarterwave long silver rolls which are stringed on a thin hardened steel needle. The attenuation of the choke is better than 30 dB at 90 GHz. An additional shell of absorber material gives defined conditions in the choke section and prevents residual RF-leakage. The front plane of the choke and the diode cap are covered with solder.

After optimizing the output power and adjusting the frequency, the unit can be fixed by heating up. Small mismatches caused by this solder procedure can be compensated by changing the position of the short and the bias current. Finally the choke is fixed by a small drop of epoxy. Some comments concerning the design of the sliding short.

The well known sliding short (i.e. [1]) is improved by use of a second loop one half wavelength behind the first; fig. 2.

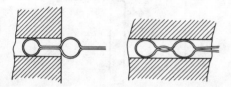

Fig. 2a   Fig. 2b
Double loop sliding short.
Compression of the second loop leads to high contact pressure of the first one.

The diameter of the first loop is exactly the waveguide height whereas the diameter of the second loop is slightly bigger; the compression of the second loop leads to a high contact pressure of the first loop. The material used is fine rolled silver, which is strong enough to compensate some roughness occuring at the milled inner waveguide walls.

The output power of each IMPATT-oscillator is about 150 mW. Fig. 3a shows a spectrogramm of the unsynchronized oscillators, fig. 3b the spectrogramm under synchronisation

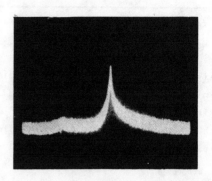

Fig. 3a
Spectrogramm of the powercombined not synchronized IMPATT-oscillators

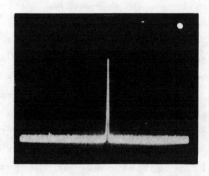

Fig. 3b
Spectrogramm of the powercombined synchronized IMPATT-oscillators
IF-Bandwith 300 kHz; Scan Width 10 MHz/div

Master Oscillator

In fig. 4 the design used for the master oscillator is shown.

Millimeter Wave Radar RF Sources and Components

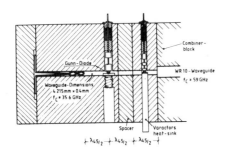

Fig. 4
Basic dimensions of the master-oscillator

Varactor- and Gunn-mount are built up using reduced height waveguide technique. The cut-off-frequency of the waveguide is chosen in such a way, that one guide-wavelength at 45 GHz amounts to three wavelength at 90 GHz. The basic dimensions are shown in fig. 4. The distance between the diodes is varied by spacers to compensate parasitics.

Power output can be optimized by a spacer-iris combination. The diodes are soldered in the same way as shown above. The time-dependent elastance $S(\omega_0 t)$ of the 45 GHz pumped varactor can be expanded in a fourier series, see [2], giving the following normalized fourier coefficients S_k / S_{max}

$$S_k / S_{max} = \frac{1}{2\pi} \int_0^{2\pi} \left(\frac{\emptyset + V_0 + 2V_1 \cos\omega_0 t}{V_B + \emptyset} \right)^\gamma \cos k\omega_0 t \, d\omega_0 t$$

$$k = 0, 1, 2$$

with

$V_1^2 \sim P_{pump}$; $\emptyset$ = contact potential;
V_B = Breakdown voltage
S_{max} = maximum elastance (at V_B)

The doubler efficiency $\mathcal{E}$ can be calculated from:

$$\mathcal{E} = \frac{\frac{f_c}{2f_0} \cos\Theta - 2\frac{S_2}{S_1}}{\frac{f_c}{2f_0} \cos\Theta + \frac{S_{max}}{2S_2}}$$

with

$$\Theta = \tan^{-1} \frac{S_0}{(R_2 + R_S) 4\pi f_0}$$

f_c = cut-off frequency of the varactor,
f_0 = pump frequency (45 GHz),
R_2 = impedance seen by the varactor at 90 GHz, and
R_S = series resistance of the varactor.

Because of the bias voltage dependence of S_0, S_1 and S_2 the following three effects can be explained:
firstly:
frequency change by changing S_0, which is desired,
secondly:
an undesired efficiency modulation by changing S_0, S_1 and S_2, and
thirdly:
variations of the inner matching conditions caused by frequency changing, which are undesired, too.

In a tuning range of 1 GHz the output power of the master oscillator varies about 3 dB. But together with the power addition the effective output power modulation of the entire system amounts to only .1 dB.

The data achieved are
Input power at 45 GHz : 120 mW
Output power at 90 GHz : 20 mW
Efficiency of the doubler: 15 %

A significant feature should also be pointed out: Due to the frequency doubling the virtual quality factor of the oscillator, measured by synchronisation or pushing, is extremely high (more than 20,000). Such a VCO is obviously predestined for synchronisation applications.

Power-Combiner

Fig. 5 shows the coupler system for power-combining and a connection to a balanced mixer.

Fig. 5
Power-combiner, dismantled
1   3 dB-coupler for power combining
2   3 dB-coupler for the balanced mixer
3   15 dB-coupler, couples out the LO-Power
a   IMPATT port   b   IMPATT port
c   master oscillator port   d   mixer connection   e   receiver input
f   transmitter output

Coupler 1 is a short-slot coupler working as a 90°-hybrid and makes possible the synchronizing and the power-combining [3]. Coupler 2 supplies the mixer with the necessary driving power and achieves decoupling of the signal and LO-branches. The coupler 3 supplies the mixer diodes with about -15 dB of the emitted power via coupler 2.

To suppress interactions of the IMPATT-oscillators with each other by reflections, the master oscillator is connected to coupler 1 via an isolator.

The hybrids have a coupling attenuation of 3.5 dB and a directivity of more than 20 dB at 90 GHz.

The output power of the entire system is 240 mW. The driving power per mixer-diode amounts to + 6 dBm giving a SSB-conversion loss of the self biased mixer of 6 dB.

The maximum sweep range is about 1 GHz.

Acknowledgement

This work has been supported by the German Government (Grant T/RF 31/714 29/71 327).

References

/1/ T.T. Fong, Kenneth P. Weller, David L. English: "Circuit Characterization of V-Band IMPATT-oscillators and Amplifiers", IEEE Transactions on Microwave Theorie and Techniques Vol. MTT-24, No. 11, November 1976, pp. 752-755.

/2/ Penfield/Rafuse: "Varactor Applications", The M.I.T. Press.

/3/ H.J. Riblet: "The Short Slot Hybrid Junction", Proc. IRE Vol. 40, Febr. 1952, No. 2, pp. 180/184.

Millimeter Solid-State Pulsed Coherent Radar
C. Smith, P. Pusateri, D. English, F. Bernues, and T. Duffield
IEEE-MTTS International Microwave Symposium, 1979 (late paper)
Reprinted by permission.

ABSTRACT

This paper describes the performance of a 94 GHz radar transmitter designed to demonstrate the feasibility of operating a solid state high power pulsed IMPATT diode oscillator as a coherent and spread spectrum power source. The radar system requirements are discussed, and performance characteristics of phase-locked oscillators and pulsed IMPATT diode oscillators are analyzed. Feasibility hardware measurement techniques and results are included. Overall performance is analyzed on the basis of the measured data.

Introduction

Detection of stationary and moving targets in clutter can be improved by the incorporation of both coherent and noncoherent techniques in a radar system. This approach includes the capability for both doppler and area MTI processing, but can create complex hardware configurations unless careful consideration is given to optimum hardware implementation concepts. The hardware described is used in a radar employing both coherent and noncoherent processing. This paper describes the performance of a transmitter output stage that operates in both modes. In the spread spectrum (noncoherent) mode, the IMPATT oscillator operates as a free-running pulsed oscillator with a controlled intrapulse thermal frequency chirp. This same IMPATT oscillator is used as the final power amplifier driven by a low noise stable master oscillator (STAMO) and operates in the injection locked oscillator (ILO) mode. The STAMO consists of a VHF crystal controlled reference oscillator followed by two phase-locked oscillators (PLOs) which provide amplification and frequency multiplication to 94 GHz. Coherency is evaluated by separate measurements on the STAMO and ILO and, also, on the combination. The test results indicate a signal with pulse-to-pulse coherency consistent with system requirements.

Pulsed Coherent Transmitter

The feasibility transmitter hardware is shown in Figure 1. It consists of:

1. A phase-locked Gunn oscillator
2. An injection-locked CW IMPATT oscillator
3. A pulsed IMPATT oscillator which operates in two different modes:
   a. As an injection-locked coherent pulsed oscillator
   b. As a free-running chirped pulsed oscillator.

The phase-locked system shown in Figure 1 is a "two-loop" system: a 100 MHz reference signal (supplied by a low FM noise crystal-controlled oscillator) is used to phase-lock a transistor oscillator/multiplier at 7.8 GHz, which provides LO power for a harmonic mixer. The 12th harmonic of the 7.8 GHz signal is mixed with the sampled RF output of the Gunn oscillator, and the resultant IF signal is amplified and phase-compared with the reference signal. The resultant error signal (output of the phase detector) is amplified and applied to bias-tune the Gunn oscillator.

The Gunn PLO injection-locks a CW IMPATT oscillator with 240 mW output power with 17 dB locking gain, without degrading the FM noise of the resultant output signal.

In the coherent mode, the CW ILO injection-locks a pulsed IMPATT oscillator. In the spread-spectrum mode, the CW ILO is turned off and the pulsed IMPATT oscillator exhibits a linear frequency chirp across the pulse.

Two different pulse modulators are used for the two modes. Peak power output in both cases is 6 watts.

Test Data

The test data of the feasibility hardware transmitter is shown in Figures 2 through 5. Figure 2 shows the measured FM noise of the Gunn PLO. The same

FM noise performance is exhibited by the IMPATT CW ILO. The pulse-to-pulse coherency of the pulsed IMPATT ILO is shown in Figure 3, which is the spectrum of the output pulse train from the phase detector in the phase bridge shown in Figure 4. The equivalent signal-to-phase noise ratio is 90 dBc/Hz. Figure 5 shows the first PRF lines of the downconverted pulse coherent 94 GHz spectrum. The signal-to-noise ratio is over 70 dBc/Hz.

Conclusions

This paper demonstrates the feasibility of solid state pulsed coherent radar transmitters at 94 GHz. The performance shown establishes a new state-of-the-art for solid-state components.

ACKNOWLEDGEMENTS

The results reported in this paper represent the contributions of many individuals. Acknowledgements are due to D. Ryan, D. Bowyer, and Steven Gray of Martin Marietta, and R. S. Ying, J. Moyer, E. Nakaji, K. Lee, M. Simonutti, M. Shimizu, H. Bell, J. Rachal, and C. Ross of Hughes Aircraft Company, Electron Dynamics Division.

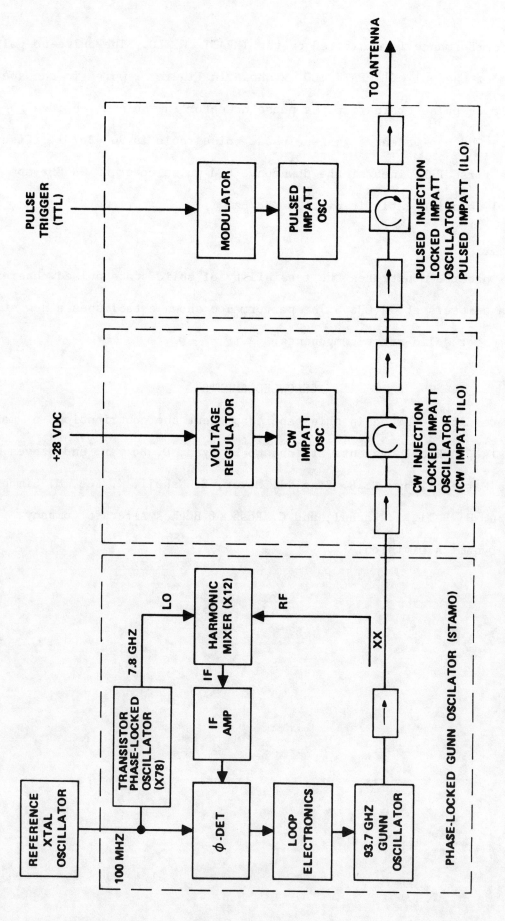

Figure 1 Block diagram of 94 GHz solid state pulsed coherent transmitter.

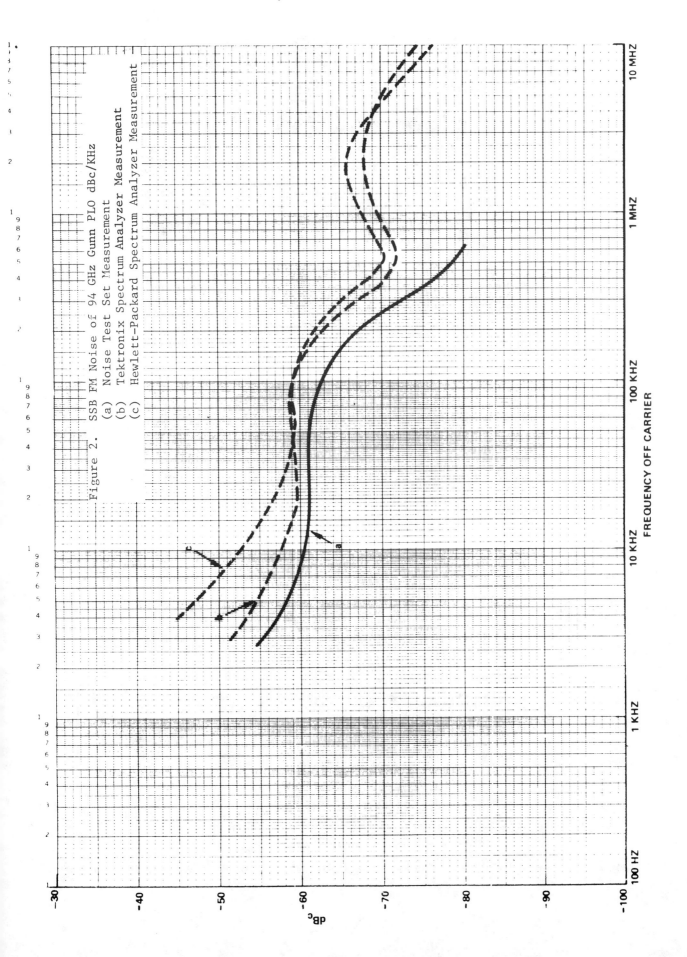

Figure 2. SSB FM Noise of 94 GHz Gunn PLO dBc/KHz
(a) Noise Test Set Measurement
(b) Tektronix Spectrum Analyzer Measurement
(c) Hewlett-Packard Spectrum Analyzer Measurement

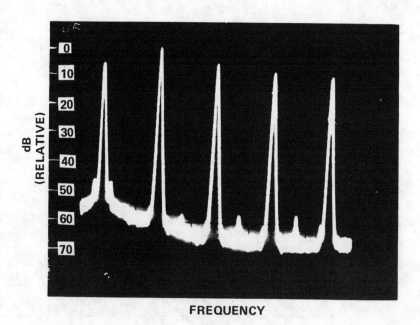

Figure 3. Spectral Display of Pulse
Bandwidth: 1 KHz
Vert: 10 dB/Div

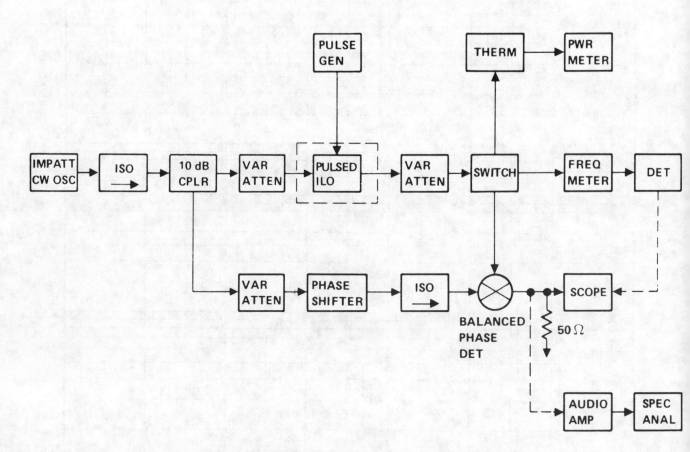

Figure 4. Test Setup for Pulsed ILO (coherent mode and spread spectrum mode)

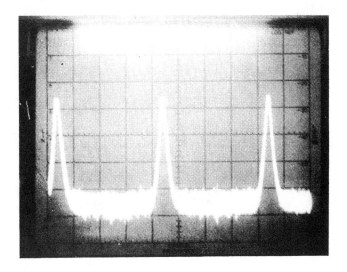

Figure 5. Spectrum of Downconverted (10GHz) Output of 94 GHz Pulsed Coherent Transmitter

High Power Millimeter Wave Radar Transmitters
G.W. Ewell, D.S. Ladd, and J.C. Butterworth
Microwave Journal, Vol. 23, No. 8, August 1980
Pages 57ff
© 1980 by Horizon House-Microwave, Inc. Reprinted by permission.

The selection of a radar transmitter for a millimeter-wave radar system (35 GHz and above) is governed by many of the same factors which influence the selection of any other radar transmitter. Such factors may include peak and average power, pulsewidth, PRF, stability, bandwidth, and tunability, to name only a few. The small physical size of the RF structure, transmission line losses and breakdown, and the limited variety of devices available at millimeter wavelengths, however, makes implementation of a millimeter-wave radar transmitter a somewhat specialized effort.

This article discusses the selection and implementation of a millimeter-wave radar transmitter with emphasis on available devices, modulator techniques, and interaction of the modulator with the RF generation device.

RF SOURCES

RF sources[1,2] for millimeter-wave radar transmitters may be either solid-state or thermionic devices. The solid-state devices are primarily the IMPATT and the Gunn devices. Thermionic sources include magnetrons, TWT's, klystrons, extended interaction oscillators (EIO's), BWO's, and gyrotrons. **Figure 1** shows several moderate to high-power sources, including EIO's, magnetrons, and an IMPATT diode oscillator.

Most of the high-power, pulsed solid-state sources at millimeter wavelengths are IMPATT diodes and they may be used as amplifiers or oscillators. A cross sectional view of a millimeter-wave IMPATT diode package is shown in **Figure 2**. This package may be mounted in the waveguide in two different manners as shown in **Figure 3**, providing tuning of the mount in both the waveguide and the coaxial sections so that the device may be well matched. Figure 4 shows output power, frequency, and efficiency for a 94 GHz IMPATT oscillator as a function of diode or bias current. Because both current changes and thermal heating during the pulse may affect the frequency of oscillation of the pulse, the pulse shape is sometimes controlled as shown in **Figure 5** to provide a degree of control of the transmitted pulse spectrum. Even with such compensation, the output of IMPATT devices sometimes tends to be broadband with considerable phase noise; these tendencies can be reduced by injection locking the diode. Typically injection locking with about 13 dB gain considerably improves the output spectral shape.[4] **Figure 6** gives power output versus frequency for both CW and pulsed IMPATT oscillators, and the outputs of several amplifiers may be combined to increase the total output power.[6] It may not be unreasonable to expect to see hundreds of watts produced by pulsed, combined IMPATT diodes in the next few years.

Fig. 1 Photograph of several commonly used high power millimeter-wave transmitter sources. In the middle is a 5 watt Gunn oscillator, surrounded by (from left to right) an EIO with a samarium-cobalt magnet, and EIO with an Alnico magnet, a 500 watt 70 GHz magnetron, a 8 kW 95 GHz magnetron, and a 1 kW 96 GHz magnetron.

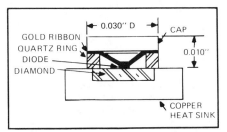

Fig. 2 Cross section view of a millimeter-wave IMPATT diode package.[3]

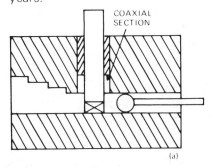
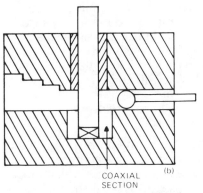

Fig. 3 Two methods of mounting an IMPATT diode in a waveguide structure, with provision for matching included.[3]

At the present time, 5 to 10 watts are available from pulsed IMPATT's at 95 GHz,[4] laboratory development models have demonstrated 3 watts at 140 GHz,[7] and devices at 225 GHz are under development.[3] Both IMPATT and Gunn oscillators are suitable as CW sources and may be quite useful where a few milliwatts of output power are desirable for use as an FM-CW or a biphase coded transmitter.

Magnetrons may be used as sources for higher power, pulsed transmitters. At millimeter-wave frequencies, most magnetrons are the "rising sun" type, although experimental inverted coaxial devices have been fabricated. Simplified cross sectional views of such configurations are given as **Figure 7**. In a magnetron, electrons are emitted from the cathode and under influence of crossed electric and magnetic fields pass to the anode. During this period, energy is coupled to the RF field, which is subsequently transferred to the output. One of the major problems with the operation of magnetrons, particularly when operated at short pulse lengths, is that more than one mode of RF oscillation is possible and obtaining correct oscillation in the desired mode may be difficult. One of the major factors influencing this mode selection is the rate of rise of voltage (rrv) applied to the tube. This, coupled with specifications on pulse shape and ripple, are often the dominant factors in the tube/modulator interface.[1] Typical peak powers available are 125 kW at 35 GHz, 10 kW at 70 GHz, 1 to 6 kW at 95 GHz, to 1 kW at 140 GHz. Typical duty cycles are 0.0002 to 0.0005, and efficiencies are on the order of ten percent.

One of the problems with millimeter-wave magnetrons is the small cathode sizes which must be used; this results in high cathode current densities which may result in short tube life. The extended interaction oscillator (EIO) was developed to circumvent this problem. **Figure 8** shows a cross sectional view of an EIO; the cathode is separated from the RF interaction region, to permit a substantial increase in cathode area. **Table I** summa-

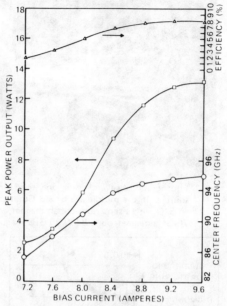

Fig. 4 Output power, frequency, and efficiency for a 94 GHz IMPATT oscillator as a function of current.[4]

rizes the characteristics of some pulsed millimeter-wave EIO's. Connections to an EIO are typically as shown in **Figure 9**. There are anode, cathode, and filament voltages which must be provided for proper operation of the tube. The anode may be operated with either a negative dc or pulsed voltage, the cathode voltage is pulsed negative with respect to the anode and ground, and the filament is operated at cathode potential. Variation in anode or beam voltage can provide a degree of electronic tuning, but larger amounts or tuning require mechanical adjustments of the extended interaction oscillator.

When comparing EIO operation with that of a pulsed millimeter-wave magnetron, a number of significant differences are soon apparent. The most obvious difference is the ease of operating an EIO; arcing and moding problems often associated with magnetrons are not evident. While one prototype tube has exhibited too low voltage, low power modes as beam voltage was varied, the modes were well separated and presented no problems, and final versions of this tube only evidenced a single mode. Another striking difference is that the EIO (being a space-charge limited diode for which $I = KV^{3/2}$) behaves like an almost resistive load-significantly different from the biased diode characteristics of magnetrons. An EIO, however produces no RF output until the beam voltage reaches approximately 70% of its final value. Finally, unlike a magnetron, the EIO operating frequency is a quite sensitive function of electrode voltages, requiring extremely flat pulses to avoid excessive frequency shifting. The EIO has proven to be a rather easy device to pulse; nevertheless, satisfactory operation (particularly for short pulses) requires careful selection and design of the modulator. The extreme voltage sensitivity and large amounts of stray capacitance of the EIO often necessitate special modulator techniques. Plans are underway for development of gridded EIO's and extended interaction amplifiers, but none are available at this time.

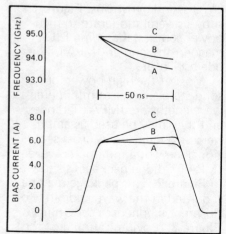

Fig. 5 IMPATT device frequency as a function of pulse shape showing compensation for frequency change during

TABLE I

CHARACTERISTICS OF SOME PULSED MILLIMETER EIO's

Tube Type	Mech. Tuning Range (GHz)	Power Output (Peak, Watts)	Beam Voltage WRT Cathode	Anode Voltage WRT Cathode	Electronic Tuning Range MHz
VKF 2443	92.7-96.0	950-1770	21 kV	12 kV	300-360
VKT 2419	139.7-140.3	270	20 kV	7.6 kV	370
VKY 2429	225.5	70	21.3 kV	8.2 kV	400

The development of millimeter-wave TWT's has been rather limited until recently, but Hughes has operated an experimental 1 kW peak, 250 watt average power tube at 95 GHz,[10,11] and a 60 GHz, 5 watt tube for communications purposes has been developed.[12]

Backward wave oscillators (BWO's) are another source of CW power in the millimeter-wave region; they have been operated at frequencies as high as 1300 GHz. However, more representatives results are 10 milliwatts over the 325 to 390 GHz band, or 5 watts at 280 GHz.[13]

In recent years, a new class of microwave and millimeter-wave oscillators and amplifiers has been developed. These devices, called gyrotrons or electron cyclotron masers, show promise of providing peak millimeter wave power outputs considerably higher than that obtainable using previous techniques.[14-20]

Gyrotron devices typically utilize a relativistic electron beam and convert constant electron energies to microwave energies in an intense electro-magnetic field. Initial results typically involved operation at megavolt levels, and super-conducting magnets were necessary to obtain the extremely high magnetic fields required. Devices having much more modest voltage and magnetic field requirements can be built, although they operate at somewhat lower power levels.

Figure 10 is a presentation of the achievable and predicted gyrotron power levels over a range of frequencies, and **Table II** presents additional details on a number of Soviet devices. Although relativistic devices are capable of substantially higher peak powers, their large size and weight make them unsuitable for many radar applications. Thus, the data presented in **Figure 10** and **Table II** appear to be representative of present and projected gyrotron capability.

Up to this point, discussions have centered about the utilization of gyrotrons as oscillators; however, it is possible to configure gyrotrons as amplifiers by providing appropriate input and output couplings to initially bunch the electron beams and to

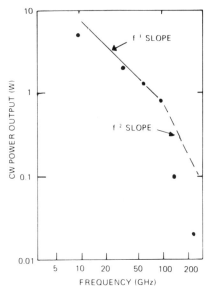

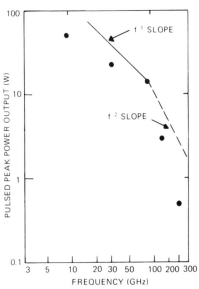

Fig. 6 Power output as a function of frequency for IMPATT oscillators.[3]

TABLE II

REPORTED GYROTRON OPERATING CONDITIONS AND OUTPUT PARAMETERS
(Adapted from Reference 16)

Model No.	Mode of Oscillation	Wavelength mm	CW or Pulsed	Harmonic Number	B-field kG	Beam Volts kV	Beam Amps.	Output Power kW	Measured Eff. %	Theoretical Eff. %
1	TE_{021}	2.78	CW	1	40.5	27	1.4	12	31	36
2	TE_{031}	1.91	CW	2	28.9	18	1.4	2.4	9.5	15
	TE_{231}	1.95	Pulsed	2	28.5	26	1.8	7	15	20
3	TE_{231}	0.92	CW	2	60.6	27	0.9	1.5	6.2	5

extract energy from the resulting beam. To date, most activities have concentrated about the implementation of a gyro-klystron as an amplifier in the lower portion of the millimeter-wave spectrum, and 200 kW has been achieved at 28 GHz.

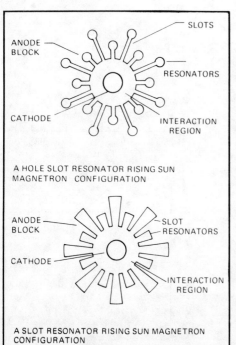

Fig. 7 Cross sectional view of two possible configurations of rising sun magnetron interaction regions. The slot resonant structure is most commonly used at millimeter wavelengths.

MODULATORS

The device used to provide proper voltages and currents to the transmitter RF source is usually called the modulator or pulser. Modulators[1,21,22] can be built over a wide range of powers, pulsewidths and efficiencies, and a number of techniques may be used in such devices. IMPATT diodes operate at relatively low voltages and currents, and conventional solid-state circuit design techniques may be used in such cases; SCR discharge circuits are often employed for pulsing IMPATT diodes. Since the pulser for IMPATT diodes is often supplied with the diode, they will not be discussed further. Modulators for magnetrons and EIO's may require relatively high voltages and currents, necessitating special design techniques. These higher power modulators may be either of the line-type or hard tube variety, and each has its own advantages and disadvantages. The line-type modulator utilizes discharge of an energy storage network (called a PFN or pulse forming network) to generate the output pulse. Special forms of the line-type modulator include the Blumlein, "Pedestal," and SCR-magnetic modulators. Such modulators are small and lightweight, but lack flexibility in controlling of pulse shape. Hard tube modulators use a grid controlled vacuum tube to control formation of the output pulse; they tend to be larger and more complex than line-type modulators but provide more control of the output pulse.

LINE-TYPE MODULATORS

Conventional line-type modulators[1] use a pulse-forming network (similar to an artificial transmission line), a switch (such as a hydrogen thyratron, a SCR,

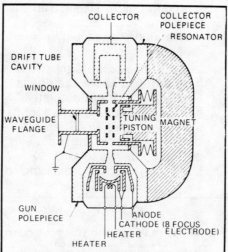

Fig. 8 Cross section view of a millimeter-wave extended interaction oscillator (EIO). Note the removal of the cathode from the RF interaction region, permitting increased cathode area.[8] (Courtesy Varian Associates of Canada).

or an RBDT), and a pulse transformer to step-up the output voltage supplied to the magnetron or EIO. A simplified schematic diagram of such a line-type modulator is shown as **Figure 11**.

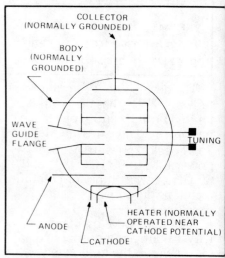

Fig. 9 Schematic diagram of an EIO with various electrodes identified.[9]

As is the case with most high power tubes, occasional momentary arcs may develop in the tube structure. With millimeter-wave tubes, the small physical size of the device makes serious damage to the tube likely if large amounts of energy can be dissipated in the tube during such arcs. The use of a line-type modulator is quite desirable to limit the energy dissipated in the RF oscillator/amplifier on a per-pulse basis.

Many millimeter-wave applications require high range resolution, necessitating the use of short pulses. The relatively high operating impedance and the large stray cathode capacitance of the EIO, however, make it difficult to achieve satisfactory operation for short pulse lengths. In particular, there are several criteria which must be satisfied by the transformer if reasonable pulse fidelity is to be achieved. If good waveform fidelity without excessive overshoot is to be achieved, then the condition

$$z \approx \frac{L_\ell}{\sqrt{C + C_D}}$$

should be satisfied[1,23], where

Millimeter Wave Radar RF Sources and Components

Z is the ratio of operating voltage to operating current

L_ℓ is the leakage inductance of the transformer

C is the tube element capacitance of the EIO

C_D is the distributed capacitance of the transformer.

If one assumed $C_D = 0$, C = 15 pF, and that Z = 8 kV/.65A, then one can solve for the desired value of L_ℓ as 2.3 mH. The rise time achievable with such a transformer is as in Reference 1.

$$t_r = 1.78 \sqrt{L_\ell (C + C_D)}$$

or the minimum rise time is approximately 0.33 µs.

There are several methods by which rise time may be reduced while still preserving reasonable

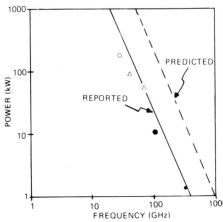

Fig. 10 Gyrotron power levels as a function of frequency (adapted from reference 19).

pulse shape; one straightforward approach being to parallel the cathode with a low resistance to reduce the operating impedance, permitting lower values of L_ℓ and reduced rise time. Unfortunately, this approach results in the consumption of appreciable amounts of power. A second approach is to eliminate the transformer entirely and pulse the tube directly; unfortunately, this approach necessitates the use of rather high voltages in the modulator.

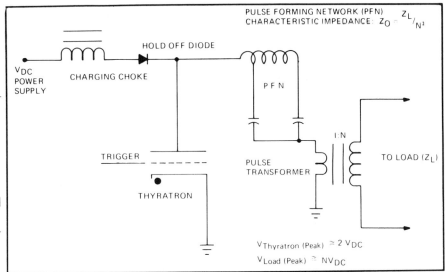

Fig. 11 Simplified schematic diagram of a line-type modulator for use with an EIO or a magnetron. The switch may be either a hydrogen thyratron, SCR, SCR-magnetic core, or a spark gap, depending upon the application. The output transformer may be bifilar wound to carry filament current to the tube.

The use of the Blumlein modulator configuration[1,24] permits the use of lower voltages in the modulator; a schematic diagram of such a modulator is shown as **Figure 12**. The PFN's may be lumped constant networks, or open-circuited coaxial cable sections may be utilized for short pulses.

When operating magnetrons at short pulse lengths, so-called "pedestal" modulator techniques are sometimes utilized.[1,25] This approach involves the application of a slowly rising voltage pedestal with a short pulse superimposed on top to place the tube at its operating point. When using magnetrons, this approach helps

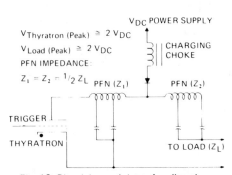

Fig. 12 Blumlein modulator for directly pulsing an microwave tube. The pulser may be capacitively coupled to the cathode but a dc recharge path must be provided.

the magnetron to oscillate in the proper mode. EIO's exhibit no such moding problems, nevertheless such an approach does have some merit when pulsing EIO's, since reduction of the required voltage swing of the short pulse reduces the effective load impedance, thus simplifying modulator design problems.

The use of an SCR-magnetic modulator could certainly be considered for millimeter wave applications. In such a modulator, SCR's and magnetic cores are used as switches to discharge the PFN. The short pulses and the high powers required and the limited numbers of units produced, however, argues against widespread use of such techniques, even though they result in small size and weight. SCR modulators have been extensively used to pulse millimeter-wave IMPATT oscillators.

When using the various line-type modulators at short pulse lengths (less than 10 ns), the modulator switching time may become significant. The switching time of hydrogen thyratrons (the most commonly used switch) varies with tube type and among thyratrons of the same type. Representative switching times of from ten to fifteen ns have been achieved with conventional thyratrons, permitting generation of pulses approximately ten nanoseconds in width. Other candidate rapid switching devices include multiple grid thyratrons and triggered spark gaps.

When an EIO is pulsed, there is only a limited portion of the beam voltage over which appreciable RF energy is generated, thus producing RF pulse rise times substantially shorter than the rise time of the applied voltage pulse. A delay of from 15 to 20 ns between application of full voltage and generation of the RF output pulse has been observed. A 3 to 5 ns jitter of the leading edge of the RF pulse, with respect to the voltage pulse, was observed. The termination of RF pulse was coincident with the removal of the applied voltage.

HARD TUBE MODULATORS

The use of a grid controlled vacuum tube to control formation of the output pulse is quite attractive if detailed control of the output pulse shape is desired. One example of such a modulator was fabricated for a series of millimeter wave EIO's.[22]

EIO's typically exhibit a large frequency change for a relatively small change in cathode voltage; thus, an extremely flat top on the cathode voltage pulse is required to minimize frequency changes during the pulse. A hard-tube modulator operated in the unsaturated region can be used to precisely control the pulse applied to the EIO.

The modulator can be conveniently divided into three sections: a pulse shaper, a driver, and an output stage. The output stage, which is shown in **Figure 13**, consists of a Y690 switch tube operating in the unsaturated mode. The output stage is driven by the driver, consisting of bifilar-wound inverting transformer which is driven from the plates of three 5687 tubes. The grids of the 5687 tubes are driven through a bifilar-wound inverting transforming by a power FET, which in turn is driven by the pulse shaper.

The pulse shaper is shown in **Figure 14** in simplified form. An input trigger generated the "start" pulse which is longer than the desired transmitter pulse, and which is used to drive a tapped delay line. The output of each delay line tap is buffered and connected to a NAND gate and through an inverter to an adjacent NAND gate.

The drawing of a single combination of these gates shown in **Figure 15** illustrating the NAND gating of the delayed and undelayed waveforms causing the output to go low for a time interval equal to the delay line spacing. This action causes current to flow through transformer 1 to

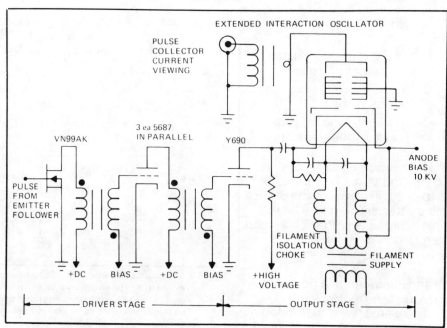

Fig. 13 Simplified schematic diagram of the driver and output stage of the hard-tube modulator used to pulse a millimeter-wave EIO.

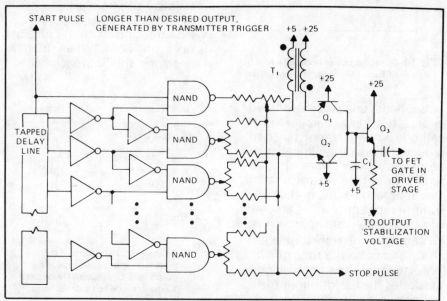

Fig. 14 Simplified schematic diagram of pulse shaper circuit used with EIO modulator.

Millimeter Wave Radar RF Sources and Components

transistor Q_1, and directly through transistor Q_2. The ratio of these currents is adjusted by the potentiometers, and the difference in the two currents either charges or discharges capacitor C_1 during the appropriate interval. In this manner, the pulse shaper generates an output pulse whose shape is adjustable in increments equal to the spacing of the delay line taps. The capacitor voltage on C_1 is buffered by the emitter follower Q_3 and sent to gate of the FET's in the driver stage. A voltage which is the average of the peak current through the EIO is fed back to the emitter power supply of Q_3 to stabilize the peak EIO current. In the particular version that was constructed, the spacing of the delay line taps was 10 nanoseconds, so each adjustment potentiometer controlled a 10 nanosecond section of the driver pulse. The outputs of the delay lines were multiplexed and a selected tap was monitored to terminate the pulse; when the signal reached this preselected tap, the "stop" pulse was generated, discharging the capacitor C_1 and terminating the driver pulse. Selection of the tap to be multiplexed permitted ready adjustment of the output pulsewidth.

A relatively low value resistor for the plate load for the Y690 triode switch tube was chosen to time. Unfortunately, this resulted properly discharge the stray capacitance associated with the EIO to achieve the desired rapid fall in a reduction in modulator efficiency, but for this application pulse fidelity and pulse shape control were more important than efficiency. Capacitive coupling to the cathode was utilized, and a bifilar-wound choke provided heater current for the EIO from a filament supply floating at the –10 kV bias level.

Using this design approach, it was possible to generate a voltage waveform to pulse an EIO with a high degree of flatness; **Figure 16(a)** shows the voltage waveform on the cathode of an EIO operating at 140 GHz. The extreme flatness achieved is shown in **Figure 16(b)**, an expanded view of the same waveform, illustrating that a voltage flatness of less than 50 volts was achievable by this method.

Another example of a hard-tube modulator is a so-called blocking oscillator modulator. A series of these were built to moderate 70 GHz magnetrons operating at the 500-1000 W level. A simplified schematic diagram of such a modulator is shown in **Figure 17**.

Operation of the circuit of **Figure 17** was initiated by a positive input trigger which turned on V_1, and through transformer T_1, turned on V_2. Feedback through T_1 kept V_2 saturated, generating an output pulse, until either T_1 saturated or C_1 discharged sufficiently to increase the plate voltage of V_1. Once the plate voltage of V_2 began to increase, feedback through T_1 rapidly turned the tube off, terminating the output pulse. If C_1 is made variable and used to control output pulse width, simple control of the output pulse is achieved. A continuously variable pulse width of 20 to 45 ns has been demonstrated, as was a switched capacitor generating 15, 30, and 45 ns output pulses.

FUTURE TRENDS

Probably the most active area of development in millimeter wave radar transmitters is in the IMPATT device area. As discussed earlier, the improvement in material and development of combining techniques may allow several hundred watts to be obtained from IMPATT arrays at 95 GHz within the next few years.

Another area for development is in the gyrotron area. Here efforts are centering about development of coherent amplifiers, primarily of the gyro-klystron type.

Due to technical problems, development of millimeter wave magnetrons does not appear to be an active area at this time; no

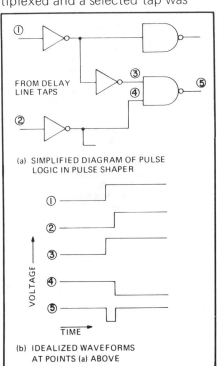

Fig. 15 Simplified circuit diagram and waveforms for one section of pulse shaping logic.

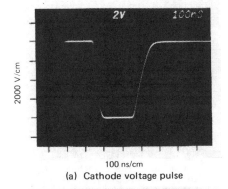

(a) Cathode voltage pulse

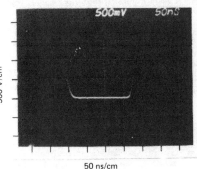

(b) Expanded cathode voltage pulse

Fig. 16 Modulator output waveforms at cathode of EIO.

major new developments are anticipated in the immediate future.

While millimeter wave TWT transmitters have not been an area for intense high visibility development, significant powers have been achieved at frequencies as high as 95 GHz; status of future development programs in this area are not clear at this time.

Current research efforts centered about the EIO include grid controlled devices and the development of an extended interaction amplifier (EIA). Current efforts are underway to determine the feasibility of injection or phase locking an EIO and frequency modulating the device for use in pulse compression systems. Coupled with these efforts are development of techniques for the precise control of the voltages and currents to achieve the phase and amplitude stabilities required for such applications.

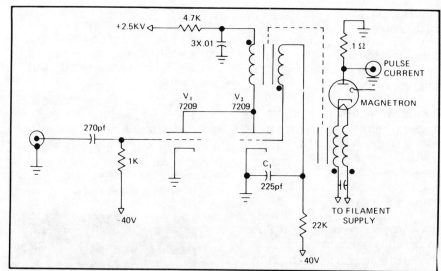

Fig. 17 Schematic diagram of a parallel-triggered blocking oscillator used to pulse a millimeter magnetron.

REFERENCES

1. Ewell, G. W., *Radar Transmitters*, to be published by McGraw-Hill Book Co., 1980.
2. Reedy, E. K., and G. W. Ewell, "Millimeter Radar" in *Infrared and Millimeter Waves Volume IV — Millimeter Wave Techniques and Systems*, to be published by Academic Press, 1980.
3. Kuno, H. J., and T. T. Fong, "Hughes IMPATT Device Work Above 100 GHz," in *Millimeter and Submillimeter Wave Propagation and Circuits, AGARD Conference Proceedings*, No. 245, pp. 14-1, 14-2, September 1978.
4. Chang, K., et. al., "High Power 94 GHz Pulsed IMPATT Oscillators," in *1979 IEEE MTT-S International Microwave Symposium Digest*, IEEE Catalog No. 79CH1439-9MTT, pp. 71-72.
5. Kramer, N. B., "Solid State Technology for Millimeter Waves," *Microwave Journal*, Vol. 21, No. 8, pp. 57-61, August, 1978.
6. Quine, J., J. McMullen, and D. Khandelwal, "K_u-Band Amplifiers and Power Combiners," in *1978 IEEE Int'l. Microwave Symposium Digest*, IEEE Catalog No. 78CH1355-7MTT, pp. 346-348.
7. Ygan, Y. C., and E. M. Nakaji, "High Power Pulsed IMPATT Oscillator Near 140 GHz," in *1979 IEEE MTT-S Symposium Digest*, IEEE Catalog No. 79CH1439-9MTT, pp. 73-74.
8. "Introduction to Extended Interaction Oscillators," published by Varian Associates of Canada, Ltd., Georgetown, Ontario, Canada.
9. "Preliminary Data Sheet for VKB-2443T Pulsed Extended Interaction Oscillator," published by Varian Associates of Canada, Ltd., Georgetown, Ontario, Canada.
10. Henry, J. F., "A 94 GHz Imaging Radar," in *Report of the ARPA/Tri-Service Millimeter Wave Workshop*, Applied Physics Laboratory, Johns Hopkins University, p. 149, 1974.
11. Arnold, K., "3.2 Millimeter Wave Transmitter Tube," R & D Technical Report DELET-TR-78-3015-1, U.S. Army ERADCOM, Ft. Monmouth, NJ, 1980.
12. Panter, N., "Development of a 5 Watt Traveling Wave Tube for 60 GHz," in AGARD Conference Proceedings No. 245, *Mm and Submm-Wave Propagation and Circuits*, Chap. 37, 1978.
13. Epszstein, B., "Recent Progress and Future Performance of Millimeter-Wave BWO's," in AGARD Conf. Proceedings No. 245, *Mm and Submm-Wave Propagation and Circuits*, Chap. 36, 1978.
14. Granatstein, V. L., P. Sprangle, R. K. Parker, and M. Herndon, "An Electron Synchrontron Maser Based on an Intense Relativistic Electron Beam," *Journal of Applied Physics*, Vol. 46, No. 5, pp. 2021-2028, May, 1975.
15. Hirshfield, J. L., and V. L. Granatstein, "The Electron Cyclotron Maser — An Historical Survey," *IEEE Trans on MTT*, Vol. MTT-25, No. 6, pp. 522-527, June, 1977.
16. Flyagin, V. A., A. V. Gaponor, M. I. Petelin, and V. K. Yulpatov, "The Gyrotron," *IEEE Trans. on MTT*, Vol. MTT-5, No. 6, pp. 514-521, June, 1977.
17. Zapevalov, V. Ye., G. S. Korablev, and Sh. Ye. Tsimring, "An Experimental Investigation of a Gyrotron Operating at the Second Harmonic of the Cyclotron Frequency with an Optimized Distribution of High Frequency Field," *Radio Engineering and Electronic Physics*, pp. 86-94, 1977.
18. Ahn, Saeyoung, "Design Study of Gyrotron TWA: Initial Consideration on Low Magnetic Field," in *Proceedings 1978 International Electron Devices Conference*, pp. 394-395.
19. Jory, H. R., F. Friedlander, S. J. Hegji, J. F. Shively, and R. S. Symons, "Gyrotrons for High Power Millimeter Wave Generation," in *Proceedings 1977 IED Conf.*, pp. 234-237.
20. Jory, H., et al., "Gyrotron Developments," *Microwave Journal*, Vol. 21, No. 8, pp. 30-32, August 1978.
21. Ewell, G. W., D. S. Ladd, and J. C. Butterworth, "Operation of Pulsed Millimeter Wavelength Extended Interaction Oscillators," paper presented at 1979 IEEE/MTT-S Symposium, Orlando, Florida, May, 1979.
22. Ladd, D. S., G. W. Ewell, and J. C. Butterworth, "A Modulator Technique for Cathode-Pulsed Millimeter Wave Extended Interaction Oscillators with Fine Control of Pulse Shape," paper to be presented at the 14th Pulse Power Modulator Symposium, to be held at Orlando, FL, June 1980.
23. Fenoglio, P., et. al., "High Power Voltage Pulse Transformer Design Criteria and Data, Final Report, Volume 1," p. 13, Final Report on Contract DA36-039-sc-117, General Electric Company, February 1, 1953.
24. Glasoe, G. N., and J. V. Lebacqz, ed., *Pulse Generators*, p. 465, McGraw-Hill Book Company, Inc., 1948.
25. Parker, T. J., "A Modulator Technique for Producing Short Pulses in High Powered Magnetrons," *IRE Nat'l. Convention Record*, pp. 142-151, 1954.

Recent Progress and Future Performances of Millimeter-Wave BWO's
B. Epsztein

SUMMARY

Up to now, backward-wave oscillators have been the most powerful and reliable millimeter wave generators. Among microwave tubes, they have also reached the highest frequencies (840 GHz for a CSF tube, 1200 - 1300 GHz for a Russian tube).

The present communication will be devoted to the recent progress achieved on these tubes.

- A modification of the slow-wave structure has resulted in a considerable increase in the electronic tuning range : having initially a few percent bandwidth, a tube operates now between 325 and 390 GHz, delivering more than 50 mW over most of the band and more than 10 mW over the whole band.*
- A narrow-band tube delivers more than 5 W at 280 GHz.
- Improvements are under way to reduce the weight and improve focusing.*

I. INTRODUCTION

As evidenced by this meeting, the millimeter-wave area, after a long period of low-level activity, is again becoming very much alive, since a number of new applications are coming up.

New military applications have already been described during this conference and will not be touched upon here. But new scientific applications have also recently appeared, among which one may mention far-infrared spectroscopy. This method of analysis constitutes a new approach to the study of molecular structure and promises to yield basic information on the structure and composition of the universe and, perhaps, the origin of life, since molecules can now be detected and identified at extra-galactic distances. Beside their use as a diagnostic tool, either in interferometry or in radiometry, millimeter waves are also now under serious consideration for heating of plasmas in thermonuclear fusion reactors.

Most applications, if not all, require millimeter-wave sources, whether it is for transmission or for reception. Some of the needs at low power levels and at the lower frequencies in the millimetric range are beginning to be filled by solid-state oscillators, but for high powers and high frequencies, millimeter-wave tubes are still without competition and will remain so for some time.

Among these tubes, the backward-wave oscillator (BWO), otherwise known as the "O" type carcinotron, is particularly outstanding for several reasons :

Created in 1951 at CSF and independently at Bell Labs, this tube has reached the highest frequencies ever obtained with a microwave tube : 850 GHz at CSF in 1963 [1] and 1300 GHz by a Russian team led by Golant in 1969 [2]. Thus it has closed the far infrared gap in coherent generation between low-frequency devices such as microwave tubes and the optical frequencies obtained by lasers. For example, the HCN laser delivers its power at a frequency of 890 GHz. Figure 1 shows a plot of power versus frequency for most millimeter-wave sources described in the literature.

On the other hand, this tube is continuously voltage tunable over a wide frequency range, which can reach one octave in the UHF region and is still greater than 20 % in the millimeter wave region.

Backward-wave oscillators are the only commercially available sources at frequencies above 150 GHz. Up to now, tubes in current production are available at up to 300 GHz, but it is to be expected that the available frequency range will be extended far beyond that value.

And last but not least, these tubes have shown a surprising long life, the average being 4000 hours for 70-GHz tubes and 1000 hours for 300 GHz tubes, when one takes into account the very difficult conditions under which they operate, such as the small size of the tube parts as well as the very large current densities which are required.

II. PRINCIPLE OF OPERATION AND STRUCTURE OF THE BWO

Since BWO's have been mostly replaced in the UHF frequency band by solid-state devices, such as varactor-tuned or YIG - tuned solid-state oscillators, and have therefore fallen somewhat out of fashion, it is perhaps useful to give a brief reminder of their principle of operation.

Basically, a BWO (Figure 2) is a traveling-wave tube, the beam of which interacts with an electromagnetic wave supported by a periodic slow-wave structure[4]. The main difference with ordinary traveling-wave tubes lies in the fact that the electron beam and the electromagnetic wave propagate in opposite directions, cumulative interaction still taking place due to a stroboscopic effect. Thus a continuous feedback process is set up, the beam current modulated by the wave giving up energy to the wave which flows toward the electron gun [1,2], thus enhancing the beam modulation. The RF power is extracted by means of a coaxial or waveguide output section [5] located near the gun.

The spent beam strikes a collector [6]. The slow-wave structure is usually attenuated toward the collector to prevent unwanted extra feedback due to multiple reflections.

The electron beam is part of the feedback loop described above. Thus, by modifying its velocity, it is possible to continuously change the feedback delay, so that the tube's oscillating frequency can be continuously varied as a function of the beam voltage.

*Studies supported by ESTEC, a division of ESA.

There principles have been applied, in particular at THOMSON-CSF, to the generation of millimeter-wave power. For this application, three difficult conditions have to be met :
- First, a very high current density in the beam is required in order to achieve acceptable efficiency.
- Second, the slow-wave structure, which includes a number of very minute parts, has to be machined extremely accurately, with tolerances of one micrometer.
- Third, this structure should have excellent thermal properties, in particular a high thermal conductibility and a large thermal capacitance.

The first condition is fulfilled by using a highly converging Pierce-type electron gun (area convergence on the order of 100), including an impregnated cathode operating at very high current densities (15 A/cm² for the 1 millimeter wavelength tube).

The second and third conditions are satisfied by using a vane-type slow-wave structure (Figure 3), whose transverse dimensions are on the order of a quarter of the free-space wavelength. Its pitch, which is twice the vane thickness, is about half this value. This structure is obtained by milling a solid copper block. Each vane includes a hole (the diameter being 1/10 of the wavelength) to allow the passage of the beam in a region where the high-frequency electric field is strong and fairly uniform.

III. TUBES IN PRODUCTION

Several carcinotrons operating at wavelengths in the range of 1 cm to 1 mm have been developed at THOMSON-CSF. At the time of their development, the primary goal was to obtain as much power as possible, even at the expense of bandwidth. There is a trade-off between these characteristics due to the fact that the higher the coupling impedance of the slow-wave structure, the higher the efficiency but the smaller the bandwidth.

Table I gives the performances of the millimeter-wave carcinotrons presently in production at THOMSON-CSF.

TABLE I

Tube type	Center Frequency (GHz)	Bandwidth (GHz)	Power out (W)	Voltage (V)	Current (mA)	Average life (hours)
CO. 80	40	1	min. 10 max. 40	3000 6900	60 80	5000
CO. 40	70 or 74	3	min. 3 max. 15	3000 6000	60	4000
CO. 20	136 or 154	2 to 4	min. 1.5 max. 3	3000 6000	max. 60	2000
CO. 10	282	15	min. 0.2 max. 1	5000 11000	max. 30	1000

IV NEW DEVELOPMENTS

As new requirements have recently appeared, a development effort has been made and is still in progress to respond to them. This effort is oriented toward several directions and includes :
- Increase of the bandwidth by redesign of the slow-wave strucure.
- Increase of the efficiency and output power as well as bandwidth by redesign of the electron optics and beam transmission improvement.
- Reduction in size and weight by use of samarium-cobalt magnets.
- Increase of life by use of new cathodes.

IV. 1 Wideband Tubes

Recently, the European Space Agency has initiated a program of research and development of components required for space borne submillimeter-wave spectroscopy. As a part of this program, a general study of submillimeter-wave sources tunable over a wide band for use as a local oscillator in heterodyne detection has been launched.

The general requirements for such sources are listed in Table II.

TABLE II - Requirements for the Submillimeter-Wave Local Oscillator

REQUIREMENTS	REASONS FOR
. Extended frequency range (wavelengths between 0. 5 and 1 mm) . Frequency stability	Heterodyne detection with high spectral resolution
. Broadband tunability	As few models as possible to cover the spectral range of interest
. Sufficient output power	Not to restrict the choice of the detector
. Reduced mass, volume and power consumption	Space-borne equipment
. Continuous tunability . Limited operational constraints	Easier use on board a space platform

Millimeter Wave Radar RF Sources and Components

To meet there requirements, ESA decided to support the development by THOMSON-CSF's Electron Tube Division of a new carcinotron model, with extended bandwidth together with sufficient output power in the 0.75-1 mm range [3]. The specifications for this carcinotron are listed in Table III.

TABLE III - Extended-Bandwidth Carcinotron Specifications

Central wavelength	$0.8 \text{ mm} < \lambda_o < 0.9 \text{ mm}$
Tunable bandwidth	20 %
Output power	>10 mW
Power consumption	< 150 W

In order to meet these specifications, an extensive computer evaluation of the interaction was first made, followed by a completely new design of the slow-wave circuit. Its dimensions were chosen so that, compared to the circuit of a regular production tube, it would operate in a region allowing much wider frequency tuning, at the expense of a reduced coupling impedance, which results in a decrease of output power. In view of the requirements, this reduction could well be afforded in this particular case.

A tube was built whose wavelength versus voltage characteristics are shown in Figure 6. The measurement accuracy is ± 0.5 %. The tube covers a continuous spectrum from 0.95 mm at 4000 V to 0.75 mm at 9000 V. Depending on the operating frequency, the voltage frequency sensitivity varies between 8 and 40 MHz/volt.

Figure 7 shows output power, beam power and frequency versus voltage. The main result is that more than 10 mW are obtained above 330 GHz. The ripples exhibited by the power curve are explained by combined reflections at the line terminals, output window and output load. They never exceed a slope of 0.01 dB/MHz, characteristic value for a well designed and well matched X-band TWT.

As shown in Figure 8, the efficiency is about 0.07 %, of the same order of magnitude as for previous narrow-band carcinotrons. The beam power is always less than 100 W.

Voltage tunability might seem to preclude the use of a BWO when a stable frequency is needed. However, experiments conducted at ESA [4] on this tube have shown that it can be frequency-locked to an external stable reference source of lower frequency. It has thus been possible to phase-lock a carcinotron operating at 244 GHz with the 24th harmonic of a frequency synthesizer. The measured linewidth is 750 Hz at 40 dB below the peak value and is determined by the resolving power of the spectrum analyser.

The same studies were also concerned with the noise properties of the carcinotron. Experimental results show that the noise temperatures measured on three carcinotrons between 230 GHz and 380 GHz lie between 1000 K and 3000 K, which corresponds to a signal-to-noise ratio approximately equal to 120 dB/MHz.

In 1977, radiometers using carcinotrons were flown on NASA's CV-990 and C-141 aircraft . System performance was as good as in laboratory conditions [4].

These results show that carcinotrons are particularly well suited as tunable local oscillators for low-noise heterodyne detection.

IV. 2   Further Improvements

For the spaceborne application planned by ESA, it is important that the bulk and weight of the tube plus its impedimenta, that is its power supply and cooling system, be as limited as possible. In order to achieve this goal, one approach is to use new magnetic materials in order to reduce the size and weight of the focusing magnet.

A study to achieve this, sponsored by ESA, has been initiated, leading to the results shown in Figure 9. Shown on the left is the new magnet, made of samarium-cobalt (SmCo5). On the right, the usual magnet made of Ticonal 600 (this material is similar to Alnico) is shown for comparison. Both magnets produce an induction of 0.6 tesla (6 kilogauss) over a focusing length of 30 mm. The total weight of the tube plus its magnets is thus reduced to 8.5 kg. This is to be compared to 30 kg for the tube with a Ticonal magnet.

Another approach to reduce size and weight is to try to improve the beam transmission which would allow, for the some output power, having reduced power consumption, hence a smaller mass for the power supply and the cooling circuit. Furthermore, an improvement in beam transmission should result in a widening of the bandwidth, since it is the factor which limits the tube operation toward lower frequencies, at low voltages. For these reasons, a new gun structure has been designed with the help of computer programs, the main difficulty, beside the sheer size of its elements, being that it should be able to operate correctly with widely varying beam voltages.

Preliminary experiments made on a beam tester have shown a large increase in total as well as in relative current transmission at lower beam voltages, below 6 kV, while the transmission at higher voltages is correspondingly decreased. The comparison is, however, somewhat pessimistic since the drift sections are different. A valid conclusion will be reached when these improvements are tested on an operating tube.

IV. 3   Higher-Power Tube

As a result of the know-how acquired during these studies and the improvements achieved in the accuracy of carcinotron fabrication, a narrow-band production tube has been built that delivers 5 watts CW at a frequency of 280 GHz (Figure 10).

V   FUTURE EVOLUTION

The achievement just mentioned shows that the evolution of the millimeter-wave BWO is by no means terminated. Among others, the following objectives for further development of this tube can be defined :

Increase in power output and efficiency, a goal of 10 watts CW at 1 mm wavelength appearing perhaps feasible. This, of course, is well below what one could expect from the gyrotron, but it should be remembered that a gyrotron operating at this frequency would unavoidably require superconducting magnetic coils and the liquid helium equipment that goes with them.

Development of tubes capable of being put into production, either narrow band, high power, or wideband, in the submillimetric region, down to 0.5 - 0.3 mm.

Increase of life by use of new cathodes : for instance, it should be possible to obtain the same current density as with an impregnated tungsten matrix cathode at a temperature 100 °C lower by using an osmium-coated tungsten matrix.

Even better results may be obtainable with barium-scandate cathodes, for which current densities of 100 A/cm^2 at 1050 °C with a life of many thousands of hours have recently been mentioned [5].

Last but not least, more widespread use of these tubes, which are now produced in very small quantities and are therefore expensive, should result in an improvement of the production efficiency and a reduction in their cost.

REFERENCES

[1] Convert, G., Yeou, T. (1964) "Millimeter and submillimeter waves" - Chap. 4. Benson Ed., Iliffe Books, London.

[2] Golant, M.B.; Alekceenko, Z.I.; Korotkova, Z.S.; Lunkind, L.A.; Negerev A.A.; Petrova, O.P.; Rebrova, T.B.; Saveleva, V.S. (1969). Pribory i Tekhnika Eksperimenta p. 231.

[3] Kantorowicz, G.; Palluel, P.; Pontvianne, M. (1978) 3^d Int. Conf. on Submillimeter Waves, University of Surrey, Guilford.

[4] de Graauw, Th.; Anderegg, M.; Fitton, B.; Bonnefoy, R.; Gustencic, J.J., (1978) 3^d Int. Conf. on Submillimeter Waves, University of Surrey, Guilford.

[5] Van Ostrom, A. (1978) 1978 Tri-Service Cathode Workshop, Naval Research Laboratory, Washington (D.C.).

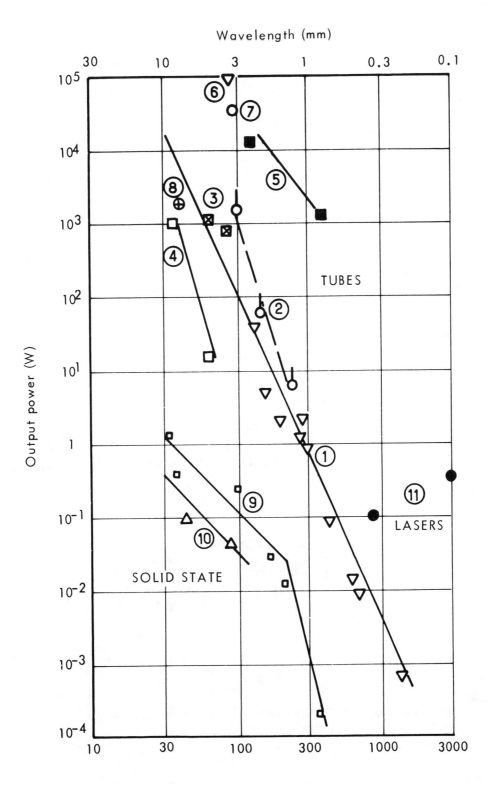

Figure 1. Output power of millimeter-wave and submillimeter-wave sources: (1) CW BWO, (2) pulsed EIO, (3) CW solenoid TWT, (4) CW ppm TWT, (5) CW gyrotron, (6) pulsed BWO, (7) pulsed magnetron, (8) CW klystron, (9) IMPATT diodes, (10) TEO, (11) CW FIR lasers.

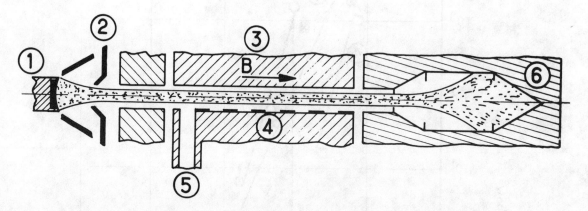

Figure 2. Cross-sectional diagram of a BWO showing (1) the cathode, (2) the anode, (3) the focusing magnet, (4) the slow-wave structure, (5) the RF output and (6) the collector.

Figure 3. Construction of a vane-type slow-wave structure for a BWO (greatly enlarged).

Figure 4. CO.10 carcinotron.

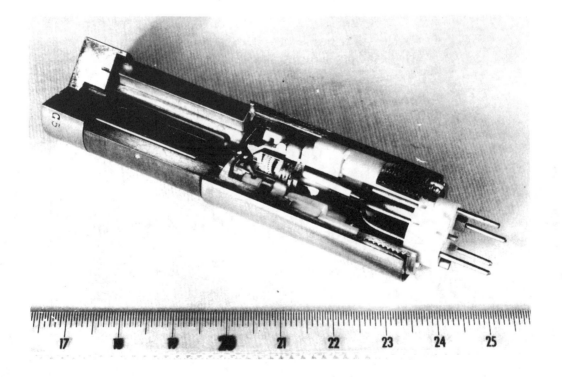

Figure 5. Cutaway view of a production Carcinotron.

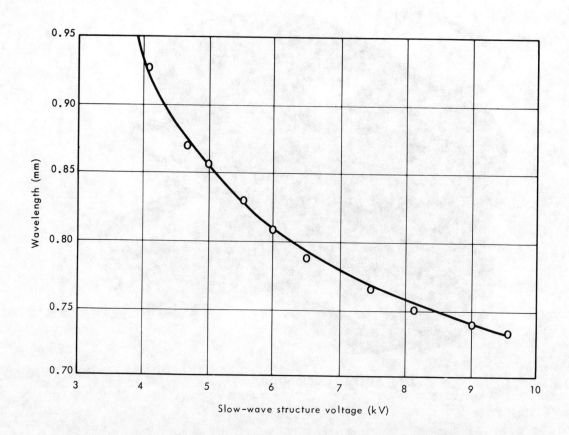

Figure 6. Wavelength versus voltage characteristics for the new broadband carcinotron.

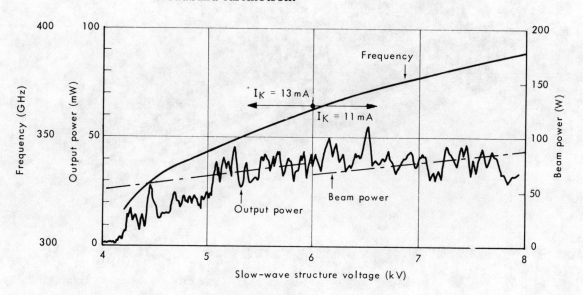

Figure 7. Characteristic curves for the new broadband submillimeter-wave BWO.

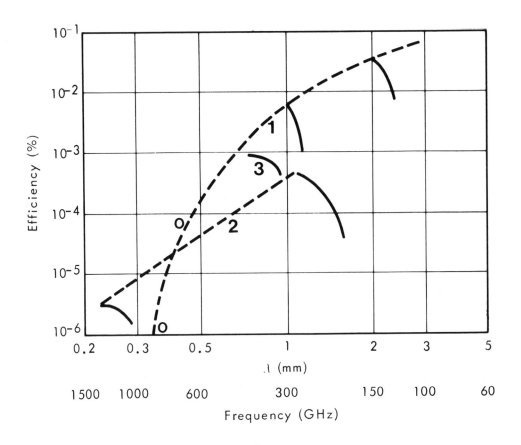

Figure 8. Efficiency of the new broadband carcinotron (3) as compared to earlier narrow-band models (1 and 2).

Figure 9. The new lightweight SmCo$_5$ focusing magnet (left) and a standard Ticonal® 600 magnet, for comparison.

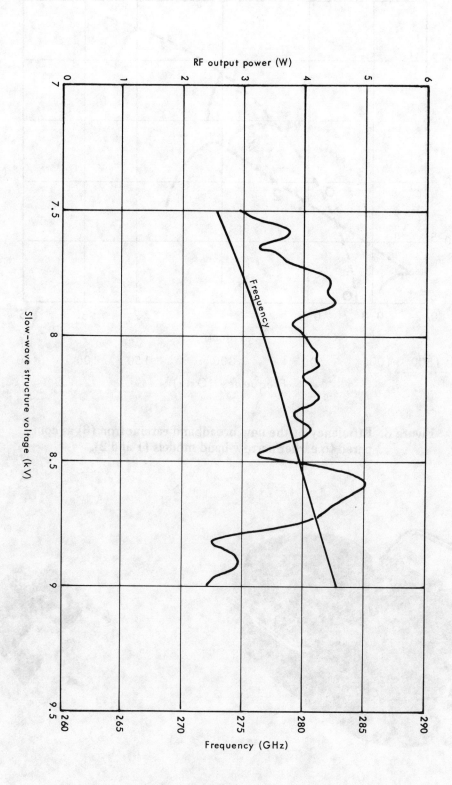

Figure 10. Characteristics of an improved narrow-band BWO in production.

DISCUSSION

N. Pranter, FRG
How high is the cathode emission density of your BOW's?

Author's Reply
Emission density is 15 A/cm^2.

S. Kulpa, US
For the tube with

$$\lambda = 0.80 - 0.9 \text{ mm}$$
$$\Delta f/f \sim 20\%$$

For > 10 mm.

What are the anticipated $\dfrac{\Delta f}{\Delta v}$?

How stable a power supply can Thomson provide?

Author's Reply
$\dfrac{\Delta f}{\Delta v}$ is larger than 8 MHz/V and smaller than 40 MHz/V depending on the frequency of operation.

Thomson-CSF does not, for the time being, manufacture the power supply for this tube but can help to obtain it at given specifications. Usual power supplies have a 10^{-4} stability but 10^{-5} and even 10^{-6} can be obtained.

The Laddertron — A New Millimeter Wave Power Oscillator

K. Fujisawa

Summary—The Laddertron is a multi-gap klystron using ladders. This paper describes the theory and experimental results of the tube. Two types of Laddertrons are distinguished, namely the O-mode Laddertron and the π-mode Laddertron. The operation of each is analyzed and the equivalent circuit representations determined theoretically, whereby one can compute the resonant frequencies within an error of several per cent. A demountable experimental version of the π-mode Laddertron gave an output power of 10 watts for the 50-Gc band, an electronic tuning range of 300 Mc, and a mechanical tuning range of 2.5 Gc.

I. Introduction

THE LADDERTRON is a kind of multi-gap klystron invented by the author in 1954.[1] It uses a flat electron beam flowing through a drift section between two parallel conducting plates with several slots cut on them. Two types of Laddertrons are now developed. One uses a series of identical slots cut regularly on the conducting plates. Its dominant mode has the same field on every slot, so this mode is called the O-mode and the tube, the O-mode Laddertron. The other one uses two kinds of slots with different lengths interleaved on the conducting plates, and its dominant mode has the same field on alternate slots and mutually reverse fields on adjacent slots. This mode is called the π-mode, and the tube, the π-mode Laddertron. In this paper, the oscillation modes of the two types of Laddertrons are theoretically analyzed and the interaction of the slot fields with the beam is treated under the assumption of small signals, whereby the operation of the Laddertron is completely clarified.

II. The O-Mode Laddertron

In Fig. 1, the cavity of the O-mode Laddertron is shown. Let us assume that the slot field is perpendicular to the slot and is distributed sinusoidally along the slot

---

Manuscript received March 23, 1964. This work was supported by the Asahi Science Research Fund in 1960 and also by the Tōyōreiyon Science Research Fund in 1961.

K. Fujisawa is with the Faculty of Engineering Science, Osaka University, Toyonaka, Osaka, Japan.

[1] K. Fujisawa, Japanese Patent No. 234547, applied on December 28, 1954.

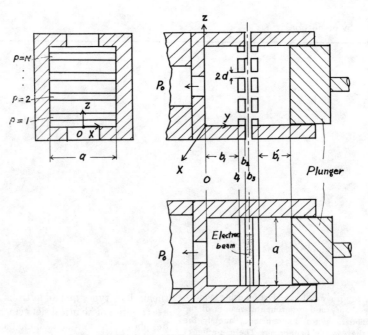

Fig. 1—The cavity of the O-mode Laddertron.

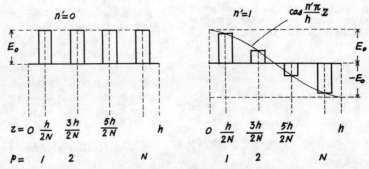

Fig. 2—The distribution of slot fields of the O-mode Laddertron.

with its maximum at the center and its nodes at each end of the slot. Taking the coordinate axis, as shown in Fig. 1, the electric field component E_x disappears by the assumption, and the remaining E_y and E_z are related with H_x by the following equations:

$$\left.\begin{aligned}\frac{\partial^2 E_y}{\partial x^2} + k^2 E_y &= -j\omega\mu_0 \frac{\partial H_x}{\partial z}, \\ \frac{\partial^2 E_z}{\partial x^2} + k^2 E_z &= j\omega\mu_0 \frac{\partial H_x}{\partial y},\end{aligned}\right\} \quad (1)$$

$$\frac{\partial^2 H_x}{\partial x^2} + \frac{\partial^2 H_x}{\partial y^2} + \frac{\partial^2 H_x}{\partial z^2} + k^2 H_x = 0. \quad (2)$$

Here, let us assume the following distribution of the slot fields: on the pth slot,

$$E_{z,p} = E_0 \cos \frac{\pi x}{a} \cos \left(\frac{2p-1}{2N} n'\pi\right),$$
$$\text{for} \quad p = 1, 2, \cdots N, \quad n' = 0, 1, 2 \cdots, \quad (3)$$

where N is the total number of slots. As shown in Fig. 2, this distribution of the slot fields corresponds to the sinusoidal field distribution in the cavity that has n' nodes in between. Clearly, n' gives the mode number. In the following, let us put the case of $n' = N$ out of consideration, because in this case all the slot fields disappear and this mode cannot be excited.

By the above distribution of slot fields, we have the following boundary condition on the plane $y = b_1$:

$$[E_z]_{y=b_1} = \begin{cases} E_0 \cos \frac{\pi x}{a} \cos \left(\frac{2p-1}{2N} n'\pi\right), \\ \qquad \text{for } \frac{2p-1}{2N} h - d \leqq z \leqq \frac{2p-1}{2N} h + d, \\ \qquad\qquad\qquad\qquad p = 1, 2, \cdots N, \quad (4) \\ 0, \text{ elsewhere.} \end{cases}$$

We can solve (1) and (2) under the above boundary condition, and obtain for $0 \leq y \leq b_1$,

$$H_z = \sum_{n=0}^{\infty} C_{0n} \cos \frac{\pi x}{a} \cosh (\gamma_{0n} y) \cos \frac{n\pi z}{h},$$

$$E_z = j\omega\mu_0 \sum_{n=0}^{\infty} C_{0n} \cos \frac{\pi x}{a} \frac{\gamma_{0n} \sinh (\gamma_{0n} y)}{k^2 - (\pi/a)^2} \cos \frac{n\pi z}{h},$$ (5)

where

$$\gamma_{0n}^2 = \left(\frac{\pi}{a}\right)^2 + \left(\frac{n\pi}{h}\right)^2 - k^2,$$ (6)

and

$$C_{0n} = \begin{cases} \dfrac{(-1)^r E_0}{j\omega\mu_0} \dfrac{1 + \delta_{n',0}}{1 + \delta_{n,0}} \dfrac{2Nd}{h} \dfrac{\sin (n\pi d/h)}{n\pi d/h} \dfrac{k^2 - (\pi/a)^2}{\gamma_{0n} \sinh (\gamma_{0n} b_1)}, \\ \quad \text{for } n = (2rN \pm n') \geq 0, \quad r = 0, 1, 2, \cdots \\ 0, \quad \text{otherwise.} \end{cases}$$ (7)

In the above equation, δ is given by

$$\delta_{i,j} = \begin{cases} 1, & \text{for } i = j \\ 0, & \text{otherwise.} \end{cases}$$ (8)

Let us then consider the region between the two slotted conducting plates—the interaction space of the beam with the fields—and assume the case of symmetrical cavity ($b_1 = b_1'$). Then the magnetic wall comes to the center of the interaction space and the fields are obtained from (5) by substituting $\sinh [\gamma_{0n}(y - b_3)]$ and $\cosh [\gamma_{0n}(y - b_3)]$ for $\cosh (\gamma_{0n} y)$ and $\sinh (\gamma_{0n} y)$, respectively, and by substituting $\cosh [\gamma_{0n}(b_3 - b_2)]$ for $\sinh (\gamma_{0n} b_1)$ in the expression of C_{0n} given by (7).

Now, the input admittance $Y_{c,p}$ as seen from the pth slot toward the cavity is defined as

$$Y_{c,p} = -\frac{\iint_{p\text{th slot}} [E_z H_x]_{y=b_1} \, dx \, dz}{[2dE_{z,p}]_{z=0}^2}.$$ (9)

Similarly, the input admittance $Y_{i,p}$ as seen from the same slot toward the interaction space is given by carrying the integration on the plane $y = b_2$ and omitting the minus sign in (9). This definition means that the input admittance as seen from the slot is given by the power radiated from the slot divided by the square of the maximum slot voltage. By using the above values of field quantities, we can compute the values of $Y_{c,p}$ and $Y_{i,p}$ as follows:

$$\left.\begin{array}{l} Y_{c,p} \\ Y_{i,p} \end{array}\right\} = j\sqrt{\frac{\epsilon_0}{\mu_0}} \frac{Na}{2h} \frac{k^2 - (\pi/a)^2}{k} \sum_{\substack{n=(2rN\pm n') \\ \geq 0}}^{\infty} \frac{1 + \delta_{n',0}}{1 + \delta_{n,0}}$$

$$\cdot \left(\frac{\sin n\pi d/h}{n\pi d/h}\right)^2 \frac{1}{\gamma_{0n}} \begin{cases} \coth (\gamma_{0n} b_1) \\ \tanh [\gamma_{0n}(b_3 - b_2)] \end{cases}.$$ (10)

The slot itself is considered as a short section of a rectangular waveguide having width a, height $2d$, and length $(b_2 - b_1)$, and a propagating TE_{10} mode. Its characteristic admittance Y_{s0} and guide wavelength λ_g are given, respectively, by

$$Y_{s0} = \frac{a}{4d} \sqrt{\frac{\epsilon_0}{\mu_0}} \frac{\sqrt{k^2 - (\pi/a)^2}}{k},$$ (11)

$$\frac{2\pi}{\lambda_g} = \sqrt{k^2 - (\pi/a)^2}.$$ (12)

The value of Y_{s0} given above is derived from a definition similar to (9). Then the above two input admittances on both sides of the slot are matched at the slot by the following equation:

$$-Y_{c,p} = Y_{s0} \frac{Y_{i,p} + jY_{s0} \tan [2\pi(b_2 - b_1)/\lambda_g]}{Y_{s0} + jY_{i,p} \tan [2\pi(b_2 - b_1)/\lambda_g]}.$$ (13)

Noting that $(b_2 - b_1)/\lambda_g \ll 1$ and $|Y_{i,p}| \ll Y_{s0}$ in the above equation, we can obtain the following approximate equation:

$$Y_{c,p} + Y_{i,p} + jY_{s0} \sqrt{k^2 - (\pi/a)^2}(b_2 - b_1) \simeq 0.$$ (14)

Clearly, this is the admittance matching condition at the pth slot. It is to be noted that (14) is really independent on p, as is seen from (10). So, if the admittance matching condition is satisfied at any one slot, it also applies to every other slot, and thus it assures the existence of the n'th mode.

Now, let us consider the various resonant modes in detail. Here, we assume the following frequency range:

$$\left(\frac{\pi}{a}\right)^2 < k^2 < \left(\frac{\pi}{a}\right)^2 + \left(\frac{\pi}{h}\right)^2.$$ (15)

Then for the higher modes which correspond to $n' \geq 1$, γ_{0n}'s in (6) are all real numbers; so $Y_{c,p}$ and $Y_{i,p}$ both become capacitive susceptances, and hence (14) cannot be satisfied. Namely, it is concluded that all higher modes are eliminated for the frequency range given by (15).

As for the dominant mode or the O-mode, which is given by $n' = 0$, we have the following relations:

$$\left.\begin{array}{l} \gamma_{00} = j2\pi/\lambda_g, \quad \coth(\gamma_{00} b_1) = -j \cot(2\pi b_1/\lambda_g), \\ \tanh[\gamma_{00}(b_3 - b_2)] \simeq \gamma_{00}(b_3 - b_2), \\ \gamma_{0n} \simeq n\pi/h, \quad \text{for } r \geq 1, \\ \coth(\gamma_{0n} b_1) \simeq \tanh[\gamma_{0n}(b_3 - b_2)] \simeq 1. \end{array}\right\}$$ (16)

By using (10) and (16), we can rewrite (14) in the following form:

$$-jY_0 \cot(2\pi b_1/\lambda_g) + Y_i + Y_f + Y_s \simeq 0,$$ (17)

where

$$\left.\begin{array}{l} Y_0 = \dfrac{Na}{2h} \sqrt{\dfrac{\epsilon_0}{\mu_0}} \dfrac{\sqrt{k^2 - (\pi/a)^2}}{k}, \quad Y_i = j\omega C_i + \dfrac{1}{j\omega L_i}, \\ Y_f = j\omega C_f + \dfrac{1}{j\omega L_f}, \quad Y_s = j\omega C_s + \dfrac{1}{j\omega L_s}, \end{array}\right\}$$ (18)

and

$$C_i = \epsilon_0 \frac{a}{2} \frac{b_3 - b_2}{h/N}, \qquad L_i = \frac{\epsilon_0 \mu_0}{C_i} \left(\frac{a}{\pi}\right)^2,$$
$$C_f = \epsilon_0 \frac{a}{2} \frac{4d}{h/N} \sum_{r=1}^{\infty} \frac{\sin^2 (2rN\pi d/h)}{(2rN\pi d/h)^3}, \quad L_f = \frac{\epsilon_0 \mu_0}{C_f} \left(\frac{a}{\pi}\right)^2, \quad (19)$$
$$C_s = \epsilon_0 \frac{a}{2} \frac{b_2 - b_1}{2d}, \qquad L_s = \frac{\epsilon_0 \mu_0}{C_s} \left(\frac{a}{\pi}\right)^2.$$

Clearly, Y_0 is the characteristic admittance of the partial cavity, which is formed by dividing the total cavity into N sections each containing one slot, and Y_i is the admittance of one half of the partial interaction space. Y_f is the admittance corresponding to the fringing fields on both sides of one slot, and Y_s is the slot admittance. Considering the physical meaning of these constants, (17) is very easily understandable by a transmission line model.

Until now, we considered the special case of the symmetrical cavity ($b_1 = b_1'$), but for more general case of $b_1 \neq b_1'$, we can easily generalize (17) by a physical reasoning and obtain

$$-jY_0\left(\cot \frac{2\pi b_1}{\lambda_g} + \cot \frac{2\pi b_1'}{\lambda_g}\right) + 2(Y_i + Y_f + Y_s) \simeq 0. \quad (20a)$$

From this equation, we obtain the equivalent circuit representation for the O-mode Laddertron, as shown in Fig. 3. In this figure, $Y^{(e)}$ is the electronic admittance derived later. Eq. (20a) is rewritten in the following form:

$$\frac{Na}{2h}\left(\cot \frac{2\pi b_1}{\lambda_g} + \cot \frac{2\pi b_1'}{\lambda_g}\right) \simeq \frac{4\pi}{\lambda_g} \frac{C_i + C_f + C_s}{\epsilon_0}, \quad (20b)$$

by which the resonant frequency of the O-mode Laddertron is easily calculable. In Fig. 4, a numerical example of the resonant frequency of the O-mode Laddertron is shown. The calculated values of the resonant frequency by (20b) agree quite well with experimental values. Also, it is seen from this figure that the resonant frequency

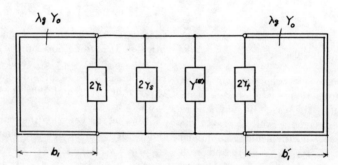

Fig. 3—The equivalent circuit of the O-mode Laddertron.

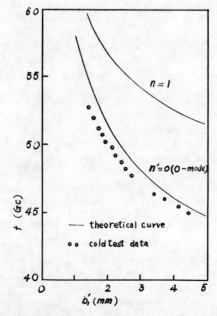

Fig. 4—The resonant frequencies of a O-mode Laddertron cavity.
$a = 3.8$ mm, $h = 5.76$ mm, $b_1 = 1.5$ mm, $N = 12$.

is widely tunable by changing the cavity thickness b_1'. This is one of the important characteristics of the O-mode Laddertron.

III. THE π-MODE LADDERTRON

In Fig. 5, the cavity of the π-mode Laddertron is shown. It differs from the O-mode Laddertron in that it uses two kinds of slots interleaved on two conducting plates. Also, in this case, it is assumed that there exist sinusoidal distributions of electric fields along both kinds of slots with the maxima at the centers and the nodes at both ends of the slots. Then, we have the following boundary condition on the plane $y = b_1$:

$$[E_z]_{y=b_1} = [E_z^{(1)}]_{y=b_1} + [E_z^{(2)}]_{y=b_1}, \quad (21)$$

where

$$[E_z^{(1)}]_{y=b_1} = \begin{cases} E_0 \cos\frac{\pi x}{a} \cos\left(\frac{p}{N}n'\pi\right), \\ \quad \text{for } \frac{ph}{N} - (1-\delta_{p,0})d \leq z \leq \frac{ph}{N} + (1-\delta_{p,N})d \\ \quad\quad p = 0, 1, 2, \cdots N, \quad (22) \\ 0, \quad \text{elsewhere}, \end{cases}$$

$$[E_z^{(2)}]_{y=b_1} = \begin{cases} -E_1 \cos\frac{\pi x}{a_1} \cos\left(\frac{2q-1}{2N}n'\pi\right), \\ \quad \text{for } -\frac{a_1}{2} \leq x \leq \frac{a_1}{2}, \\ \quad \frac{2q-1}{2N}h - d \leq z \leq \frac{2q-1}{2N}h + d, \\ \quad\quad q = 1, 2, \cdots, N \quad (23) \\ 0, \quad \text{elsewhere}. \end{cases}$$

The boundary condition given by (22) corresponds to the case where, in the cavity shown in Fig. 5, all the shorter slots are removed and there remain only the longer slots. The boundary condition given by (23) corresponds to the case where, in the same cavity, all the longer slots are removed and there remain only the shorter slots. The distribution of slot fields given by (21) is schematically shown in Fig. 6. Clearly, the fundamental mode corresponding to $n' = 0$ is the so-called π-mode.

As for the solutions of this boundary value problem, it is to be noted that Maxwell's Equation is linear, so its solution is obtained by simply superposing the two solutions that are obtained by imposing each boundary condi-

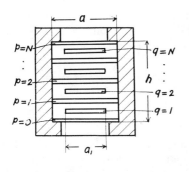 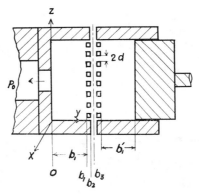

Fig. 5—The cavity of the π-mode Laddertron.

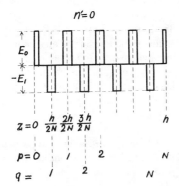 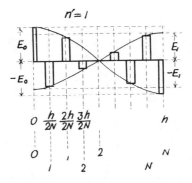

Fig. 6—The distribution of slot fields of the π-mode Laddertron.

tion of (22) and (23) separately. Let us designate the solutions under the boundary condition of (22) alone by the upper suffix (1), and those under the boundary condition of (23) alone by the upper suffix (2). Then we have, for $0 \leq y \leq b_1$,

$$H_x^{(1)} = \sum_{n=0}^{\infty} C_{0n}^{(1)} \cos \frac{\pi x}{a} \cosh (\gamma_{0n} y) \cos \left(\frac{n\pi z}{h}\right),$$

$$E_z^{(1)} = j\omega\mu_0 \sum_{n=0}^{\infty} C_{0n}^{(1)} \cos \frac{\pi x}{a} \frac{\gamma_{0n} \sinh (\gamma_{0n} y)}{k^2 - (\pi/a)^2} \cos \left(\frac{n\pi z}{h}\right), \quad (24)$$

$$H_x^{(2)} = \sum_{m=0}^{\infty} \sum_{n=0}^{\infty} C_{mn}^{(2)} \cos \left(\frac{2m+1}{a} \pi x\right) \cdot \cosh (\gamma_{mn} y) \cos \left(\frac{n\pi z}{h}\right),$$

$$E_z^{(2)} = j\omega\mu_0 \sum_{m=0}^{\infty} \sum_{n=0}^{\infty} C_{mn}^{(2)} \cos \left(\frac{2m+1}{a} \pi x\right) \cdot \frac{\gamma_{mn} \sinh (\gamma_{mn} y)}{k^2 - [(2m+1)\pi/a]^2} \cos \left(\frac{n\pi z}{h}\right), \quad (25)$$

where

$$\gamma_{mn}^2 = \left(\frac{2m+1}{a}\pi\right)^2 + \left(\frac{n\pi}{h}\right)^2 - k^2, \quad (26)$$

and

$$C_{0n}^{(1)} = (-1)^r C_{0n}, \quad \text{for} \quad n = (2rN \pm n') \geq 0$$
$$r = 0, 1, 2, \cdots, \quad (27)$$

$$C_{mn}^{(2)} = \begin{cases} -\dfrac{(-1)^r}{j\omega\mu_0} \alpha_m E_1 \dfrac{1+\delta_{n',0}}{1+\delta_{n,0}} \dfrac{2Nd}{h} \\ \quad \cdot \dfrac{\sin(n\pi d/h)}{n\pi d/h} \dfrac{k^2 - [(2m+1)\pi/a]^2}{\gamma_{mn} \sinh(\gamma_{mn} b_1)}, \\ \quad \text{for} \quad n = (2rN \pm n') \geq 0, \\ \quad r = 0, 1, 2, \cdots, \\ 0, \quad \text{otherwise}, \end{cases} \quad (28)$$

where

$$\alpha_m = \frac{\dfrac{4\pi}{aa_1} \cos\left[(2m+1)\dfrac{a_1\pi}{2a}\right]}{\left(\dfrac{\pi}{a_1}\right)^2 - \left(\dfrac{2m+1}{a}\pi\right)^2}, \quad \text{for } m = 0, 1, 2, \cdots \quad (29)$$

and

$$\alpha_0 \simeq 1 - \frac{a-a_1}{2a} \simeq \sqrt{\frac{a_1}{a}}. \quad (30)$$

The resultant fields are obtained by

$$\begin{aligned} H_x &= H_x^{(1)} + H_x^{(2)}, \\ E_z &= E_z^{(1)} + E_z^{(2)}. \end{aligned} \quad (31)$$

Substituting these fields into (9), we can calculate the input admittances as seen from each slot toward the cavity as follows: on the pth longer slot,

$$Y_{c,p} = j\sqrt{\frac{\epsilon_0}{\mu_0}} \frac{Na}{2h} \frac{k^2 - (\pi/a)^2}{k} \sum_{\substack{n=(2rN \pm n') \\ \geq 0}}^{\infty} \frac{1+\delta_{n',0}}{1+\delta_{n,0}}$$
$$\cdot \left[1 - (-1)^r \frac{\alpha_0 E_1}{E_0}\right] \left[\frac{\sin(n\pi d/h)}{n\pi d/h}\right]^2 \frac{\coth(\gamma_{0n} b_1)}{\gamma_{0n}}, \quad (32)$$

and on the qth shorter slot,

$$Y_{c,q} = j\sqrt{\frac{\epsilon_0}{\mu_0}} \frac{Na}{2h} \frac{k^2 - (\pi/a)^2}{k} \alpha_0^2 \sum_{\substack{n=(2rN \pm n') \\ \geq 0}}^{\infty} \frac{1+\delta_{n',0}}{1+\delta_{n,0}}$$
$$\cdot \left[1 - (-1)^r \frac{E_0}{\alpha_0 E_1}\right] \left[\frac{\sin(n\pi d/h)}{n\pi d/h}\right]^2 \frac{\coth(\gamma_{0n} b_1)}{\gamma_{0n}}$$
$$+ j\sqrt{\frac{\epsilon_0}{\mu_0}} \frac{Na}{2h} \sum_{m=1}^{\infty} \sum_{\substack{n=(2rN \pm n') \\ \geq 0}}^{\infty} \frac{1+\delta_{n',0}}{1+\delta_{n,0}} \alpha_m^2$$
$$\cdot \frac{k^2 - [(2m+1)\pi/a]^2}{k} \left[\frac{\sin(n\pi d/h)}{n\pi d/h}\right]^2 \frac{\cot(\gamma_{mn} b_1)}{\gamma_{mn}}. \quad (33)$$

The input admittances, as seen from the same slot toward the interaction space, are similarly calculated, and the results for the symmetrical cavity ($b_1 = b_1'$) are given by substituting $\tanh[\gamma_{0n}(b_3 - b_2)]$ and $\tanh[\gamma_{mn}(b_3 - b_2)]$ for $\coth(\gamma_{0n} b_1)$ and $\coth(\gamma_{mn} b_1)$ in (32) and (33), respectively.

Then, the input admittances on both sides of each slot are related through the slots as follows:

$$Y_{c,p} + Y_{i,p} + j\sqrt{\frac{\epsilon_0}{\mu_0}} \frac{k^2 - (\pi/a)^2}{k} \frac{a(b_2 - b_1)}{4d} \simeq 0 \quad (34)$$

$$Y_{c,q} + Y_{i,q} + j\sqrt{\frac{\epsilon_0}{\mu_0}} \frac{k^2 - (\pi/a_1)^2}{k} \frac{a_1(b_2 - b_1)}{4d} \simeq 0. \quad (35)$$

The above two equations give the admittance matching conditions at each slot, and they are really independent on p and q, as is clear from (32) and (33). So, if the admittance matching condition is satisfied at any one pair of the longer and the shorter slots, then it is automatically satisfied at every other slot, and thus the solution is obtained. Hence, (34) and (35) give the existence condition of the n'th resonant mode, and from these we can determine the resonant frequency and the slot field strength ratio E_1/E_0 of this mode. It is to be noted here that the above conclusions result from assuming the same value of n' in (22) and (23). If we assign different values of n' in (22) and (23), the admittance matching conditions at each slot differs from each other and we cannot find solutions that satisfy such complex boundary conditions.

Now, let us consider in detail the dominant mode ($n' = 0$) or the π-mode. Substituting $n' = 0$ in (32) and (33) and the corresponding expressions for $Y_{i,p}$ and $Y_{i,q}$, we can rewrite (34) as follows:

$$E_0\left[Y_{e,p} + Y_{i,p} + j\sqrt{\frac{\epsilon_0}{\mu_0}}\frac{k^2 - (\pi/a)^2}{k}\frac{a(b_2 - b_1)}{4d}\right]$$

$$\simeq (E_0 - \alpha_0 E_1)\left(-jY_0\cot\frac{2\pi b_1}{\lambda_g} + Y_i\right)$$

$$+ E_0(Y_f + Y_s) - E_1 Y_t \simeq 0, \quad (36)$$

where Y_0, Y_i, Y_f and Y_s are given by (18), and Y_t is the transfer admittance between one pair of adjacent longer and shorter slots, and is given by

$$\left.\begin{array}{l} Y_t = j\omega C_t + \dfrac{1}{j\omega L_t}, \\[6pt] C_t = \epsilon_0\,\dfrac{\sqrt{aa_1}}{2}\,\dfrac{4d}{h/N}\sum_{r=1}^{\infty}(-1)^r\,\dfrac{\sin^2(2rN\pi d/h)}{(2rN\pi d/h)^3}, \\[6pt] L_t = \dfrac{\epsilon_0\mu_0}{C_t}\left(\dfrac{a}{\pi}\right)^2. \end{array}\right\} \quad (37)$$

Similarly, applying (16) and the following approximations

$$\left.\begin{array}{l} k^2 - \left(\dfrac{2m+1}{a}\pi\right)^2 \simeq \left(\dfrac{\pi}{a}\right)^2 - \left(\dfrac{2m+1}{a}\pi\right)^2, \\[6pt] \gamma_{m0} \simeq \dfrac{2\pi}{a}\sqrt{m(m+1)}, \quad \coth(\gamma_{m0}b_1) \simeq 1, \\[6pt] \tanh[\gamma_{m0}(b_3 - b_2)] \simeq \gamma_{m0}(b_3 - b_2), \text{ for } m \geq 1, \\[6pt] \gamma_{mn} \simeq \dfrac{n\pi}{h}, \text{ for } r \geq 1 \end{array}\right\} \quad (38)$$

in the expression of $Y_{e,q}$ and $Y_{i,q}$, we can rewrite (35) as follows:

$$E_1\left[Y_{e,q} + Y_{i,q} + j\sqrt{\frac{\epsilon_0}{\mu_0}}\frac{k^2 - (\pi/a_1)^2}{k}\frac{a_1(b_2 - b_1)}{4d}\right]$$

$$\simeq \alpha_0(\alpha_0 E_1 - E_0)\left(-jY_0\cot\frac{2\pi b_1}{\lambda_g} + Y_i\right) + E_1\alpha_0^2 Y_f - E_0 Y_t$$

$$+ E_1 j\sqrt{\frac{\epsilon_0}{\mu_0}}\frac{k^2 - (\pi/a_1)^2}{k}\frac{a_1(b_2 - b_1)}{4d} - E_1 j\sqrt{\frac{\epsilon_0}{\mu_0}}\frac{Na_1}{2h}\left(\frac{2\pi}{a}\right)^2\frac{1}{k}$$

$$\cdot\left\{\sum_{m=1}^{\infty} m(m+1)\left(\frac{\alpha_m}{\alpha_0}\right)^2\left(\frac{1}{\gamma_{m0}} + b_3 - b_2\right)\right.$$

$$\left. + 4d\sum_{m=1}^{\infty} m(m+1)\left(\frac{\alpha_m}{\alpha_0}\right)^2\sum_{\substack{n=2rN\\r\geq 1}}^{\infty}\frac{\sin^2(n\pi d/h)}{(n\pi d/h)^3}\right\}$$

$$\simeq \alpha_0(\alpha_0 E_1 - E_0)\left(-jY_0\cot\frac{2\pi b_1}{\lambda_g} + Y_i\right)$$

$$+ E_1(Y'_f + Y'_s) - E_0 Y_t \simeq 0, \quad (39)$$

where

$$Y'_f = \alpha_0^2 Y_f, \quad (40)$$

$$\left.\begin{array}{l} Y'_s = j\omega C'_s + \dfrac{1}{j\omega L'_s}, \\[6pt] C'_s = \alpha_0^2 C_s, \quad L'_s = \dfrac{\epsilon_0\mu_0}{C'_s}\left(\dfrac{a'_1}{\pi}\right)^2. \end{array}\right\} \quad (41)$$

In the above equation, a'_1 is the effective length of the shorter slot, and is given by

$$\left(\frac{\pi}{a'_1}\right)^2 \simeq \left(\frac{\pi}{a_1}\right)^2 + \frac{2Nd}{h(b_2 - b_1)}\left(\frac{2\pi}{a}\right)^2$$

$$\cdot\left\{\frac{a}{2\pi}\sum_{m=1}^{\infty}\sqrt{m(m+1)}\left(\frac{\alpha_m}{\alpha_0}\right)^2 + \sum_{m=1}^{\infty} m(m+1)\left(\frac{\alpha_m}{\alpha_0}\right)^2\right.$$

$$\left.\cdot\left[b_3 - b_2 + 4d\sum_{r=1}^{\infty}\frac{\sin^2(2rN\pi d/h)}{(2rN\pi d/h)^3}\right]\right\}. \quad (42)$$

Clearly, Y'_f is the equivalent admittance corresponding to the fringing fields on both sides of the shorter slot, and Y'_s is the admittance of the shorter slot.

Here, (36) and (39) are easily understandable by a coupled transmission line model. While these equations can be applied for the special case of the symmetrical cavity ($b_1 = b'_1$), it is very easy to modify them so as to be applicable for the general case of $b_1 \neq b'_1$ by physical reasoning, and we obtain

$$(E_0 - \alpha_0 E_1)\left[-jY_0\left(\cot\frac{2\pi b_1}{\lambda_g} + \cot\frac{2\pi b'_1}{\lambda_g}\right) + 2Y_i\right]$$

$$+ 2E_0(Y_f + Y_s) - 2E_1 Y_t \simeq 0, \quad (43)$$

$$\alpha_0(\alpha_0 E_1 - E_0)\left[-jY_0\left(\cot\frac{2\pi b_1}{\lambda_g} + \cot\frac{2\pi b'_1}{\lambda_g}\right) + 2Y_i\right]$$

$$+ 2E_1(Y'_f + Y'_s) - 2E_0 Y_t \simeq 0. \quad (44)$$

Solving the above coupled equations, we can determine the resonant frequency and the slot field strength ratio E_1/E_0 for the π-mode. In Fig. 7, the calculated values of the resonant frequency of the π-mode for three π-mode Laddertron cavities are shown in conjunction with the calculated values of the higher modes and with the experimental oscillation frequencies. It is clear that the calculated frequencies of the π mode agree quite well with experimental values. Also, in Fig. 8, calculated values of E_1/E_0 for the same three cavities are shown.

From (43) and (44), we can obtain an equivalent circuit representation for the π mode as shown in Fig. 9. In this figure, $Y_0^{(e)}$, $Y_1^{(e)}$, and $Y_t^{(e)}$ are, respectively, the electronic self-admittance of the longer and the shorter slot, and the electronic transfer admittance between the two slots, which are defined in Section IV.

Now, let us consider the higher modes with $n' \geq 1$. As mentioned above, we can obtain the resonant frequency and E_1/E_0 for the higher modes by solving (34) and (35). As we have interest in the higher modes only for the frequency separation with the π mode, we cut the detailed analysis on the higher modes in this paper and mention only some calculated results of their frequencies in Fig. 7. In this figure, it is interesting that the resonant frequencies of the higher modes are little affected by the cavity thickness b'_1, except when this becomes very small and the cavity wall gets very near the slotted conducting plates. This fact is easily explained by noting that the

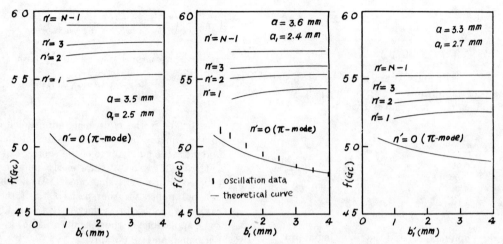

Fig. 7—The resonant frequencies of the π-mode and the higher modes of three π-mode Laddertron cavities. $h = 5.2$ mm, $b_1 = 1.5$ mm, $N = 10$.

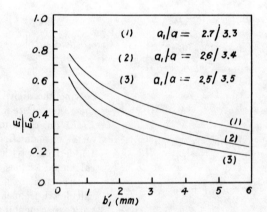

Fig. 8—Calculated values of the slot field strength ratio E_1/E_o of the three π-mode Laddertron cavities mentioned in Fig. 7.

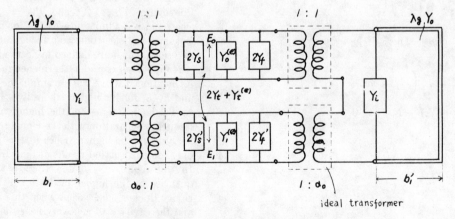

Fig. 9—The equivalent circuit of the π-mode Laddertron.

fields of the higher modes decrease exponentially away from the slots by the real propagation constants γ_{0n} and γ_{mn}, and the stored energy concentrates very near to the slots. By properly choosing the cavity dimensions such as a, h, b_1, b_1', and a_1/a, these higher modes are adequately separated from the π mode, as is seen from the examples shown in Fig. 7.

IV. THE INTERACTION OF THE SLOT FIELDS WITH THE ELECTRON BEAM

Formerly, Hechtel[2] treated the interaction of the multigaps with the electron beam for the case of a single-cavity multigap klystron under the assumption of small signals. His theory is applicable to the O-mode Laddertron, but to apply it to the π-mode Laddertron, some changes are necessary. One of the revised points is to define the electronic admittance for the slot voltage rather than for the total cavity voltage, as Hechtel did.

To develop a small signal theory of the interaction of the slots with the beam which is applicable both for the O-mode and the π-mode Laddertrons, let us introduce the following symbols:

$\bar{V}_b$ = dc beam voltage
$\bar{I}_b$ = dc beam current
$\bar{G}_b$ = $\bar{I}_b/\bar{V}_b$ = dc beam conductance
$\bar{\theta}$ = dc electronic transit angle between two adjacent slots
I_c = induced cavity current at the slot
β_0, β_1 = respective beam coupling coefficients of the longer and the shorter slots
V_0, V_1 = respective slot voltages of the longer and the shorter slots.

Under the assumption of small signals, the modulating effects of each slot and the resulting bunching action is separately calculated and linearly superposed.

Let us first consider the case of the O-mode Laddertron. The slot voltages are equal in magnitude and phase and are designated by V_0. The induced current at the second slot is caused by the velocity modulation of the beam at the first slot and is obtained by the usual bunching theory as follows:

$$I_{c2} = \tfrac{1}{2}\beta_0^2 \bar{G}_b V_0 j\bar{\theta}e^{-i\bar{\theta}}. \qquad (45)$$

The induced current at the third slot is caused by the velocity modulation of the beam at the first and the second slots, and can be obtained by summing the respective induced currents calculated separately from the velocity modulations at the first and the second slots. The results are:

$$I_{c3} = \tfrac{1}{2}\beta_0^2 \bar{G}_b V_0 (j2\bar{\theta}e^{-i2\bar{\theta}} + j\bar{\theta}e^{-i\bar{\theta}}). \qquad (46)$$

Similarly, the induced currents at all the other slots are calculated and summed up, thus giving the total induced cavity current. Dividing the total induced current by N,

[2] R. Hechtel, "Das Vielschlitzklystron, ein Generator für kurze electromagnetische Wellen," *Telefunken-Röhre* (Germany), vol. 35, pp. 5–30; September, 1958.

we obtain the mean induced current per slot, from which we have the electronic admittance as follows:

$$Y^{(e)} = \frac{1}{V_0}\frac{1}{N}\sum_{p=1}^{N} I_{cp} = \tfrac{1}{2}\beta_0^2 \bar{G}_b f(\bar{\theta}, N), \qquad (47)$$

where

$$f(\bar{\theta}, N) = \frac{N-1}{N} j\bar{\theta}e^{-i\bar{\theta}} + \frac{N-2}{N} j2\bar{\theta}e^{-i2\bar{\theta}} + \cdots$$

$$+ \frac{1}{N} j(N-1)\bar{\theta}e^{-i(N-1)\bar{\theta}} \equiv f_r + jf_i. \qquad (48)$$

The electronic admittance $Y^{(e)}$ has a large negative real part for the ranges of $2\pi(1 - 1/N) < \bar{\theta} < 2\pi$, and it has the negative maximum conductance at $\bar{\theta} = \bar{\theta}_{0p}$ as shown in Table I. Clearly, the maximum negative electronic conductance increases with the increase of the slot number N approximately proportional to $(N - 1)^2$. So, by increasing the slot number we can easily decrease the starting current for oscillation.

TABLE I

THE VALUES OF THE NORMALIZED ELECTRONIC ADMITTANCE f AND THE ELECTRON TRANSIT ANGLE $\bar{\theta}_{op}$, WHEN THE REAL PART OF f TAKES THE NEGATIVE MAXIMUM VALUES.

N	$\bar{\theta}_{op}$	$f_r(\bar{\theta}_{op}, N)$	$f_i(\bar{\theta}_{op}, N)$
2	281.5°	−2.43	0.49
4	323.1	−11.90	3.48
6	335.5	−28.02	8.33
8	341.6	−51.22	15.09
10	345.3	−81.09	23.78
12	347.7	−117.71	34.06

Now, let us consider the case of the π-mode.[3] In the following, let us put the upper suffix (1) to the quantities concerning the longer slot, and the upper suffix (2) to those concerning the shorter slot. Then, the induced currents at each slot become as follows:

$$I_{c1}^{(2)} = \tfrac{1}{2}\beta_0\beta_1 \bar{G}_b V_0 j\bar{\theta}e^{-i\bar{\theta}}$$

at the slot of $q = 1$,

$$I_{c1}^{(1)} = \tfrac{1}{2}\beta_0^2 \bar{G}_b V_0 j2\bar{\theta}e^{-i2\bar{\theta}} - \tfrac{1}{2}\beta_0\beta_1 \bar{G}_b V_1 j\bar{\theta}e^{-i\bar{\theta}}$$

at the slot of $p = 1$,

$$I_{c2}^{(2)} = \tfrac{1}{2}\beta_0\beta_1 \bar{G}_b V_0 (j\bar{\theta}e^{-i\bar{\theta}} + j3\bar{\theta}e^{-i3\bar{\theta}}) - \tfrac{1}{2}\beta_1^2 \bar{G}_b V_1 j2\bar{\theta}e^{-i2\bar{\theta}}$$

at the slot of $q = 2$,

$$I_{c2}^{(1)} = \tfrac{1}{2}\beta_0^2 \bar{G}_b V_0 (j2\bar{\theta}e^{-i2\bar{\theta}} + j4\bar{\theta}e^{-i4\bar{\theta}})$$
$$- \tfrac{1}{2}\beta_0\beta_1 \bar{G}_b V_1 (j\bar{\theta}e^{-i\bar{\theta}} + j3\bar{\theta}e^{-i3\bar{\theta}})$$

at the slot of $p = 2$.

[3] In the actual π-mode Laddertron cavity, the first and the last slots corresponding to $p = 0$ and $p = N$ have slot widths one half those of the other, as is seen in Fig. 5. So, their slot voltages are equal to $\tfrac{1}{2}V_0$. But here we consider the ideal case of same slot widths and same slot voltages on all the longer slots including the first and the last ones. It is considered that this idealization brings no essential changes to the values of the electronic admittance when the slot number is large.

Then the mean induced currents per one longer and per one shorter slot, respectively, are given by

$$\frac{1}{N} \sum_{p=0}^{N} I_{cp}^{(1)} = \tfrac{1}{2}\beta_0^2 \bar{G}_b V_0 \frac{N+1}{N} f(2\bar{\theta}, N+1)$$
$$- \tfrac{1}{2}\beta_0 \beta_1 \bar{G}_b V_1 g(\bar{\theta}, N), \quad (49)$$

$$\frac{1}{N} \sum_{q=1}^{N} I_{cq}^{(2)} = -\tfrac{1}{2}\beta_1^2 \bar{G}_b V_1 f(2\bar{\theta}, N) + \tfrac{1}{2}\beta_0 \beta_1 \bar{G}_b V_0 g(\bar{\theta}, N), \quad (50)$$

where

$$g(\bar{\theta}, N) = j\bar{\theta} e^{-j\bar{\theta}} + \frac{N-1}{N} j3\bar{\theta} e^{-j3\bar{\theta}} + \cdots$$
$$+ \frac{1}{N} j(2N-1)\bar{\theta} e^{-j(2N-1)\bar{\theta}} \equiv g_r + j g_i. \quad (51)$$

From (49) and (50), we can define the electronic self-admittances $Y_0^{(e)}$ and $Y_1^{(e)}$ of the longer and the shorter slots and the electronic transfer admittance $Y_t^{(e)}$ between the longer and the shorter slots as follows:

$$\left. \begin{array}{l} Y_0^{(e)} = \tfrac{1}{2}\beta_0^2 \bar{G}_b \dfrac{N+1}{N} f(2\bar{\theta}, N+1), \\ Y_1^{(e)} = \tfrac{1}{2}\beta_1^2 \bar{G}_b f(2\bar{\theta}, N), \\ Y_t^{(e)} = \tfrac{1}{2}\beta_0 \beta_1 \bar{G}_b g(\bar{\theta}, N). \end{array} \right\} \quad (52)$$

The function $g(\bar{\theta}, N)$ gives the normalized electronic transfer admittance, and it takes the positive maximum real part at $\bar{\theta} = \bar{\theta}'_{0p}$ as given in Table II. Using the above equations, (49) and (50) are rewritten as follows:

$$\left. \begin{array}{l} \dfrac{1}{N} \sum_{p=0}^{N} I_{cp}^{(1)} = Y_0^{(e)} V_0 - Y_t^{(e)} V_1, \\ -\dfrac{1}{N} \sum_{q=1}^{N} I_{cq}^{(2)} = -Y_t^{(e)} V_0 + Y_1^{(e)} V_1. \end{array} \right\} \quad (53)$$

Clearly these electronic admittances are inserted into the equivalent circuit, as shown in Fig. 9, by which the operation of the π-mode Laddertron can be fully explained.

TABLE II
The Values of the Normalized Electronic Transfer Admittance g and the Electron Transit Angle $\bar{\theta}_{op}$, When the Real Part of g Takes the Positive Maximum Values.

N	$\bar{\theta}_{op}$	$g_r(\bar{\theta}_{op}, N)$	$g_i(\bar{\theta}_{op}, N)$
2	149.6°	5.24	−2.17
4	163.6	17.28	−5.84
6	168.5	36.08	−11.52
8	171.3	61.70	−18.55
10	173.0	94.09	−28.67

From Tables I and II, it is seen that

$$\bar{\theta}_{0p} \simeq 2\bar{\theta}'_{0p}. \quad (54)$$

So, the electronic admittances $Y_0^{(e)}$, $Y_1^{(e)}$ and $Y_t^{(e)}$ given by (52) take their respective, approximately maximum negative and positive conductance parts, at $\bar{\theta} \simeq \bar{\theta}'_{0p}$. Accordingly, the maximum conversion power is obtained near this point, as is clear from (53).

V. Oscillation Characteristics

We have already known from Figs. 4 and 7 that the O-mode Laddertron cavity shows a much steeper resonant frequency change than the π-mode Laddertron cavity for the change of the cavity thickness b'_1. While this brings a desirable characteristic of the O-mode Laddertron in its wide mechanical tunability, it increases the dependence of the oscillation on the contact loss of the short plunger with the cavity wall, and so it increases the necessity of a precise machinery of the moving plunger. Generally speaking, for the O-mode Laddertron to exhibit good oscillation characteristics, larger beam current is needed than for the π-mode Laddertron.

In case of the π-mode Laddertron, the energy stored in the cavity is smaller than in the case of the O-mode Laddertron by the fact that the fields of the longer and the shorter slots partially cancel each other out. Accordingly, the cavity loss including the contact loss of the plunger is relatively small, and the π-mode Laddertron oscillates more easily than the O-mode Laddertron. Hence, the π-mode Laddertron can operate with a larger output coupling and lower loaded cavity. This results in the higher efficiency and the relatively wide electronic tuning range of the π-mode Laddertron.

Now let us scan the experimentally obtained oscillation characteristics of the O-mode Laddertron. In our laboratory, a 50-Gc band demountable test tube with a fixed cavity operated with a beam voltage of 2.0–2.2 kv and a beam current of 120 ma, giving the maximum output power of 7 watts and an electronic tuning range of 40 Mc. The beam transmission efficiency is about 90 per cent, and the maximum beam current density amounts to 16 a/cm². Commercially available Oki's products in the 34-Gc band operate with a beam voltage of 1.7–2.0 kv and a beam current of 100–130 ma, and have an output power of 10–15 watts, the mechanical tuning range of 2 Gc and an electronic tuning range of 50–70 Mc. The beam transmission efficiency is over 80 per cent, and the maximum beam current density amounts to 8 a/cm².

As for the oscillation characteristics of the π-mode Laddertron, the slot length ratio a_1/a of the longer and the shorter slots plays an important role. In Fig. 7, the resonant frequency distributions of the three π-mode Laddertron cavities with $a_1/a = 2.5/3.5, 2.6/3.4, 2.7/3.3$ are shown, from which it is clear that as a_1/a approaches 1, the mechanical tuning range and the separation of the π-mode with the higher modes decrease. While in Fig. 8, it is seen that as a_1/a approaches 1, the slot field strength ratio E_1/E_0 increases, and so the oscillation power also increases. Hence, to have both wide mechanical tuning

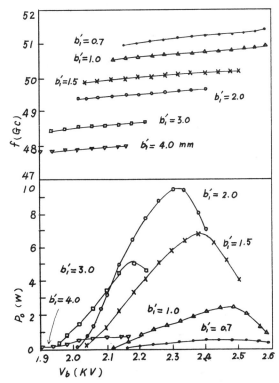

Fig. 10—The oscillation characteristics of a π-mode Laddertron in the 50-Gc band. $a = 3.4$ mm, $a_1 = 2.6$ mm, $h = 5.2$ mm, $b_1 = 1.5$ mm, $N = 10$.

range and high output power, there is an optimum value of a_1/a. In the examples shown in Fig. 7, the case of $a_1/a = 2.6/3.4$ seems best. In Fig. 10, the experimentally obtained oscillation characteristics of this case are shown. This demountable tube operated with a beam voltage of 1.9–2.6 kv and a maximum beam current of 160 ma, giving a maximum output power of 10 watts in the 50-Gc band. The mechanical tuning range exceeds 2.5 Gc and the electronic tuning range amounts to 300 Mc. The beam transmission efficiency is about 75 per cent and the maximum beam current density amounts to 17 a/cm². It is notable that the π-mode Laddertron has a much wider electronic tuning range than the O-mode Laddertron. Compared with conventional reflex klystrons in the same 50-Gc band which operate with almost the same beam voltage, this π-mode Laddertron experimental tube gives about 100 times as much power and twice as wide an electronic tuning range. By improving the design of the electron gun, the characteristics of the O-mode and the π-mode Laddertrons will be still more improved. The π-mode Laddertron is expected to operate at least up to the 150-Gc band with an output power of over 1 watt.

Acknowledgment

The author is grateful to T. Kamimura and M. Shimada for their assistance in numerical calculations and experiments. He also wishes to thank T. Kaneko and T. Nonaka for their valuable assistance in the early days of this work.

Millimeter Wave Tubes with Emphasis on High Power Gyrotrons
R.H. Chilton

(U) System designers working in the millimeter wave region of the spectrum have been severely limited in their designs by the lack of millimeter wave power. Fortunately, this critical roadblock is beginning to be removed. Simple and reliable solid state devices are now generating a few watts. In the higher power region, thermionic devices have made significant progress and there is now the feasibility for two (2) orders of magnitude increase in the power out from amplifiers.

(U) This paper will discuss high power amplifier developments at 94 GHz of the past few years, and those currently underway. Emphasis is placed on thermionic devices with the degree of specification required by complex military systems. Here the main requirement is for amplifiers, i.e., devices where the phase, amplitude, bandwidth, etc. can be controlled. Thus, old standby devices like magnetrons will not be covered other than to say that while they have high peak power with very low average power, they are difficult to pulse, and leave much to be desired in terms of reliability.

(U) A review of the state-of-the-art shows that at 94 GHz the Hughes coupled cavity TWT is probably the best tube currently available at the power and frequency. Figure 1.

(U) Before discussing the technology required to achieve 100 KW peak and high average power at 94 GHz it is meaningful to examine the coupled cavity TWT Hughes built at 94 GHz. This tube represents the best in conventional microwave design.

(U) All of the tubes that will be discussed have certain gross similarities. The electrons originate at the gun, are confined by a magnetic field as they propagate through an interaction region, and are collected. Both traveling wave tubes and klystrons operate with a slow wave structure.

(U) For Klystrons and traveling wave tubes, the diameter of the drift region is a fraction of a wavelength.

(U) The parameters of the Hughes coupled cavity TWT, the HAW-7, are summarized below:

Frequency	94 GHz
Bandwidth	4%
Power output	1 KW pk
Duty	50%
Beam voltage	40 KV
Beam current	.35 A
Cathode area compression	560
Drift tube diameter	0.5 mm
Drift tube length	90 mm

Thus, the 14 kilowatt beam must pass through the drift tube with the absolute minimum of interception on the circuit. Any interception, along with the loss due to rf skin resistance must be dissipated through the cavities which are only 0.3 mm thick.

(U) The degree of difficulty in building this tube is indicated by the fact that only one tube has been built. The next tubes will cost on the order of $60,000.

(U) Although one can scale the current HAW-7 technology to higher power, at the same frequency it is evident that physical limitations will allow modest improvements, and even then questions of tube life would remain.

(U) To go from the 1 KW level to the 100 KW level is not a matter of simple scaling. Assuming the same efficiency at 100 KW as at 1 KW would mean that a 1.4 megawatt beam would have to propagate through the slow wave structure and its drift region, which is less than a millimeter in diameter. Even if this were possible, problems would occur with voltage and rf breakdown. The cathode would be another roadblock. The cathode in the HAW-7 has a convergence ratio of over 500 which is already higher than desired for good beam formation.

(U) With this as a background, DARPA and RADC set out to explore new means of obtaining amplifiers in the 50 to 100 KW peak power, high average power range at 94 GHz. It was desired to develop an amplifier that could be operated with conventional magnets and at voltages that would be feasible in DOD systems application.

(U) The Gyrotron and the Ubitron were identified as the two most feasible approaches. Both had been demonstrated, the Ubitron as an amplifier at 54 GHz and the Gyrotron as a oscillator at several frequencies. Both are so-called fast-wave devices. They use a periodic electron beam to interact with a rf fast wave in a low-loss waveguide mode. Thus the drift tube diameter can be larger than a wavelength, thus circumventing the power handling limitations of slow wave circuits. Neither device was well understood. In early 1976 the contracts were let, one to Hughes Research Laboratories to assess the potential of the Ubitron for high power amplification at 94 GHz and the other to Varian Associates for a similar assessment of the Gyrotron. The plan was to evaluate the results of the two (2) programs and to select one device for further development.

(U) The Ubitron is a device based on the interaction of an EM wave with an undulating electron beam, hence its name. A hollow electron beam passes through the drift tube and alternating radial magnetic fields. The drift tube is, in fact, a circular waveguide propagating the TE_{01} mode. The beam is rotating as it travels down the drift tube, and its sense of rotation is periodically changed by the magnetic field. Although the beam cannot attain the same velocity as the EM wave, it is synchronized so that electron bunching and energy extraction occurs.

Phillips built a Ubitron in the early 60s and obtained 150 KW peak power at 54 GHz with a 6 percent efficiency. Figure 2 is a schematic of the Ubitron. Hughes Research Labs, under the RADC/DARPA effort, expanded the theory of Ubitrons and extended the theory for relativistically correct dc and small signal theories. They then investigated the potentials of a Ubitron for operation at 94 GHz with 50 KW of peak power at output. As a result, Hughes predicted efficiency of 1-2 percent for a wide bandwidth (50 percent) tube and 4 to 6 percent efficiency for a narrow bandwidth. The primary cause of the low efficiency predictions is the inability to obtain a high degree of undulation in the beam due to saturation effects in the pole pieces. Although these interaction efficiencies are very low, it may be possible to recover a large portion of the beam energy with a depressed collector. The Gyrotron, which is also referred to as a Cyclotron resonance device, in its simplest embodiment, is a device based on interaction of the Cyclotron resonance of an electron beam with an electromagnetic wave. Like the Ubitron and TWT it has a linear format. The Gyrotron requires a strong magnetic field which is related to the operating frequency by the Cyclotron resonance conditions.

$$\omega = n\omega_c$$

$$\omega_c = \frac{eB}{\gamma m_0}$$

ω = operating frequency
ω_c = Cyclotron frequency
n = integer (harmonic)
B = magnetic field
$\frac{e}{m_0}$ = electron charge/mass
γ = relativistic mass factor

Bunching of the electron beam is a result of the relativistic effect. A change in the kinetic energy γ causes a change in the angular velocity of the electron. In the simplest Gyrotron, the single cavity oscillator, microwave fields apply an angular velocity modulation to the electrons. As the electrons drift through the cavity, angular benching takes place. Approaching the end of the cavity, the phase between the electron benches and the microwave electric fields is adjusted so that electrons give up its kinetic energy. It is a single cavity device, one that the Russians have obtained 12 KW at 110 GHz.

(U) Several points need to be emphasized: (1) the frequency of operation is determined by the magnetic field. At 28 GHz this is 11 Kilogauss; at 94 GHz the magnetic field is 36 kilogauss. Thus, at 94 GHz a super conducting magnet is required. Conventional electromagnets are limited to a range below approximately 12 kilogauss. The alternative to superconducting magnet is to operate at a high harmonic of the Cyclotron frequency. For example, third harmonic operation would require a 11 Kilogauss magnetic field, (2) the single cavity Gyrotron is an oscillator, (3) the frequency, and especially the bandwidth, is dependent on the rf structure. For cavity Gyrotrons, the instantaneous bandwidths of 1 percent are practical. Wider bandwidths are predicted for a traveling wave Gyrotron as will be shown later.

(U) The Gyrotron program that was conducted in parallel with the Ubitron effort analyzed and experimentally demonstrated amplification from a Gyrotron operating at 10 GHz. A cavity amplifier was designed and built to operate at 10 GHz on the second harmonic of the Cyclotron frequency and is shown in Figure 2. The X-band frequency was selected because of ease and economy of experiment. The device has a magnetron injection gun, the three cavities and the output window-collector. The output/collector is a departure from conventional microwave tubes - as it was in the Ubitron. Note that the beam is allowed to expand and is collected on the side walls while the rf comes out of the on-axis window. Table I summarizes the design values and test results value of the tube.

Parameter	Design Value	Test Value
Frequency	10.35 GHz	10.35 GHz
Beam voltage	60 KV	40 KV
Gun anode voltage	31 KV	24 KV
Beam current	5A	4.5A
Main magnet field	1960 g	450 g ± 5%
Gun magnet	485 g	450 g ± 10%
Microwave gain	23-26 dB	9-10 dB
Peak body current	0	80 ma
Peak gun anode current	0	10 ma
Oscillations	None	19.5 KW at 8.33 GHz

(U) Amplification was obtained at the design value of magnetic field, however, the beam voltage had to be kept to 40 KV rather than the 60 KV design value because of oscillations which occurred at the higher voltage.

(U) There were several oscillations that could be controlled by a combination of beam power and magnetic field. At 8.33 GHz an oscillation which correlated to the TE_{211} resonance could be optimized to yield 19.5 KW peak for an efficiency of 9.8%.

(U) Although the oscillations were a detriment to amplifier performance, they were encouraging because they could be explained based on an analysis of the various modes that could exist under the test conditions.

(U) A comparison was made between the Ubitron analysis and the

Gyrotron to select the optimum approach to achieve a full spec 94 GHz tube as shown below.

Comparison of 94 GHz Devices

	Gyrotron(cavity)	Ubitron
Voltage	60-80 KV	200-300 KV
Magnetic field	36 Kg	7 Kg with field shape
Efficiency	30%	5%
Bandwidth	1%	Broad
Circuit Structure for 50-100 KW	Yes	Yes
Potential to operate at higher f	Yes	No
Additional Problems	Modes	Modes

(U) The low predicted efficiency of the Ubitron coupled with the requirement for 200-300 KV voltages were its major drawbacks. The magnetic field that can be provided by a conventional electromagnet was attractive as was the potential for very wide bandwidths.

(U) With the Gyrotron the very high CW power (12 KW) by Zaytsev of the USSR and the demonstrated amplification at Varian under the first phase of the DARPA program, coupled with the high efficiency and modest voltages, made the Gyrotron the choice for full scale development. On the negative side are the very high magnetic fields required, and the narrow bandwidths of the cavity Gyrotron.

(U) Preliminary analysis conducted by NRL and USSR literature indicated that a traveling wave interaction rather than the cavity approach could yield wider bandwidths. The analysis, when extended to harmonics of the Cyclotron resonance frequency showed that, although at some sacrifice in both bandwidth and efficiency.

(U) Because some of the potential applications involve locations that render cyrogenic magnetics impractical, and because several percent bandwidth are required for such systems as space object identification where a fm waveform is employed. The traveling wave Gyrotron operating on a high harmonic was selected for full development. The goals are listed below:

Center frequency	94 GHz
Power output	100 KW
Duty	10%
Beam voltage	80 KV
Magnetic field	Conventional magnets
Bandwidth	4%

(U) To achieve these parameters requires that the device operate on the 3rd or 4th harmonic in the traveling wave mode. Both of these areas are breaking new ground. So, to support Varian in the design of the high harmonic traveling wave Gyrotron amplifier, the Electron Beam Applications Branch of the Plasma Physics Division of NRL was funded to extend the theory and analyze various harmonic-traveling wave cases.

(U) Preliminary results of the NRL analysis is summarized below:

Harmonic	Bandwidth	Efficiency
1	13%	48 - 51%
3	5%	11 - 14%
4	4%	5.5 - 8%

(U) It therefore appears that the tube can be optimumly designed for the third harmonic. The magnetic field in this case will be approximately 12 Kg...possible to achieve with conventional electro-magnets.

(U) A two year program has just started with Varian Associates and is divided into 3 phases: Phase I - testing the current X-band Gyrotron amplifier under saturated drive power and determine performance over extended ranges of voltage and current. Phase II - Design, build and test a traveling wave Gyrotron at X-band to operate on a high harmonic. After the successful demonstration of the traveling wave approach, Phase III will develop a traveling wave Gyrotron amplifier to the above specs at 94 GHz. Once the 94 GHz tube is built, it will be tested by using the current state-of-the-art tube, the HAW-7 TWT as a driver.

(U) It should be noted that the above program is funded by DARPA and the Army's Ballistic Missile Defense Advance Technology Center, is managed by RADC and involves NRL.

(U) In other developments in fast wave interactions, Varian (working for ERDA) has recently achieved 40 dB gain in a Z cavity Gyrotron operating in the fundamental mode at 28 GHz. This tube produced 75 KW pk power at 9% efficiency. This effort is to deliver a 200 KW CW amplifier in the Spring of 1978. After that a 200 KW CW 120 GHz device will be developed.

(U) Hughes Research Lab and Northrup are currently analyzing 5-10 KW, 90 GHz devices for AFAL. Hughes is studying the Ubitron and Gyrotron with emphasis on the Traveling Wave Gyrotron. Northrup is concentrating on the cavity Gyrotron.

(U) NRL is building a 270 KW pk power traveling wave Gyrotron to operate at 35 GHz with approximately 50% efficiency and 3% bandwidth. They also are developing a Single Cavity Oscillator at 33 GHz with 200 KW pk power output for ERDA. Harry Diamond Labs, in cooperation with NRL are developing a single cavity Gyrotron oscillator at 250 GHz with 1 KW CW output. This tube will operate on the second harmonic.

(U) Now, for the first time there is the potential for giving the system engineer one of the critical missing components - a high power millimeter wave amplifier. The approaches of the Ubitron and Gyrotron overcome the fractional wavelength dependence of conventional slow wave structures and open new horizons for millimeter wave systems.

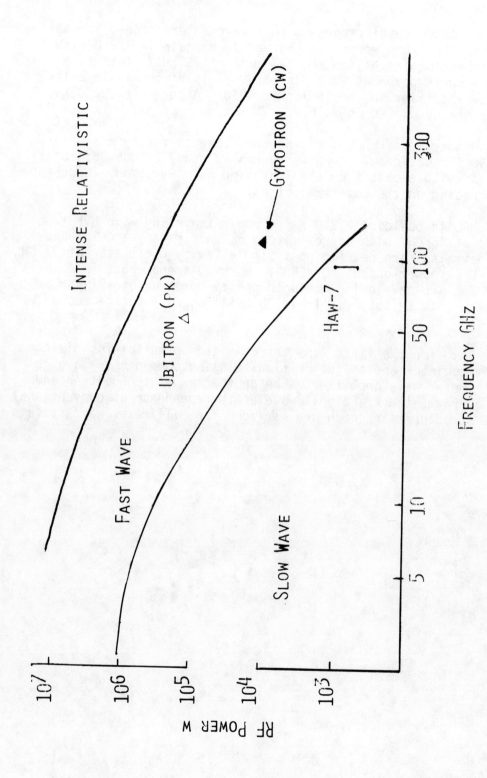

STATE OF THE ART OF MILLIMETER TUBES

FIGURE 1

Millimeter Wave Radar RF Sources and Components

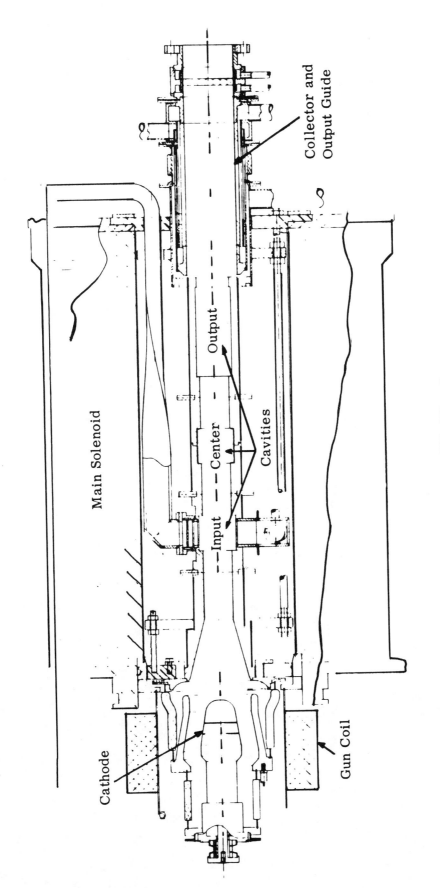

Figure 2. Layout of 3 Cavity 10.35 GHz Gyrotron

Prospects for High Power Millimeter Radar Sources and Components

T.F. Godlove, V.L. Granatstein, and J. Silverstein

Abstract

The electron cyclotron maser, or gyrotron, is under development for a variety of high-power, millimeter-wave applications, including radar. Renewed interest in this device has been spurred by experiments at NRL yielding peak power up to the gigawatt level and by the development at Gorki in the Soviet Union of efficient mm-wave sources with quasi-cw power in the 1-60 kw range. Work is in progress in the U.S. on 200 kw tubes operating in cw, long pulse, or repetetively pulsed modes. Development status is reviewed, including some technical issues relating to auxilliary components and system implications.

INTRODUCTION

The development of millimeter-wavelength sources at power levels comparable to that obtained at lower frequencies from conventional klystrons and traveling-wave tubes would be of great value to a number of important applications. For example, two significant applications are the development of clutter-free, high-resolution radar and the heating of plasma in magnetic confinement machines using the electron resonance method. Coherent cyclotron maser emission is rapidly becoming accepted as the prime candidate for device development to meet the need for such sources. The attractiveness of the electron cyclotron maser, or gyrotron, has become apparent from the results of a major long term development program at Gorki in the Soviet Union, from a research program at NRL starting in 1971, and from recent results of an ERDA sponsored program at Varian Associates. In this paper the status of device development will be reviewed along with some discussion of the outstanding technical issues and component and system implications.

DESCRIPTION OF THE PROCESS AND EARLY DEVELOPMENTS

The electron cyclotron maser employs a cloud of monoenergetic electrons in a fast wave structure such as a metallic tube or waveguide, with electron velocity transverse to an applied axial magnetic field. Such an electron ensemble can react unstably with a fast microwave signal propagating through the waveguide. Initially, the phases of the electrons in their cyclotron orbits are random, but phase bunching can occur because of the relativistic mass change of the electrons.[1,2,3] Those electrons that lose energy to the wave become lighter and accumulate phase lead while those electrons that gain energy from the wave become heavier and accumulate phase lag. This can result in a phase bunching such that the electrons radiate coherently and amplify the electromagnetic wave.

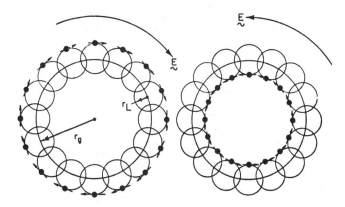

Fig. 1 - Idealized electron bunches, viewed transverse to the axis, in phase synchronism with the azimuthal electric field of the TE_{01} mode of a cylindrical waveguide. The Larmor radius is r_L and the average radius of the annular electron beam is r_g. A typical value for the waveguide radius (not shown) is $\approx 2 r_g$.

The bunched electrons can be modeled as being distributed around the large beam radius, r_g, while gyrating in circular orbits of radius $r_L < r_g$ as is depicted in Fig. 1. Such an

---

* Portions of the work described in this paper are supported by the Naval Electronic Systems Command, Task XF54-587, by the Ballistic Missile Defense Advanced Technology Center, MIPR W31RPD-73-Z787, and by the Naval Surface Weapons Center (Dahlgren), Task SF32-302-41B.

** Permanent address: Harry Diamond Laboratories, Adelphi, MD.

electron distribution can participate in an unstable interaction with waves in the drift tube, leading to exponential wave growth. In particular interaction is favored with the azimuthal electric field of the TE_{on} modes of the circular drift tube. Figure 1 shows the phase synchronism between the electron orbits and the TE_{01} field on alternate half cycles, assuming that the wave frequency is close to the cyclotron frequency. Actually, energy transfer from the electrons to the wave is optimized when the wave frequency is slightly higher than the electron cyclotron frequency (or its harmonics).

Because of the cyclotron orbit synchronism, the device emits radiation at a wavelength determined by the strength of an applied magnetic field, and not by the dimensions of some resonant structure. Thus, unlike other microwave generators, the internal dimensions of the device may be large compared to the wavelength, and high power handling capability (up to megawatts) becomes compatible with operation at millimeter wavelengths. Indeed, the highest recorded millimeter-wave power, both peak and average, have been achieved with cyclotron masers.

The first unequivocal experimental demonstration of the cyclotron maser mechanism was made in 1964 by Hirshfield and Wachtel.[4] Their 5 kV, 200 µA electron beam was passed through a combination of a "corkscrew" static magnetic field and a magnetic hill; the electrons were then injected into a high-Q cylindrical cavity resonant at 5.8 GHz with most of their energy transverse to the axial magnetic field. Later a two cavity "gyro-klystron" experiment was reported[5] which showed that an amplifier configuration was possible based on the same transverse bunching mechanism, only then the bunching was allowed to continue ballistically between input and output cavities. Other early work extended frequency into the millimeter wave and even the submillimeter wave regime; an extensive description of this early work is contained in the review by Hirshfield and Granatstein.[6] However, in all of the early experiments, the promise of exceptionally large millimeter wave power was not realized, the best result being that of Bott who generated several watts of power at $2 mm < \lambda < 4 mm$.[7]

INTENSE RELATIVISTIC ELECTRON BEAM STUDIES

New impetus to the study of the cyclotron maser mechanism itself came from research into microwave emission from intense relativistic electron beams, with beam power in the range $10^9 - 10^{11}$ W. The electron beams are generated from cold, field-emission cathodes typically as a single pulse with duration ~ 50 nsec.

Table 1 displays the maximum attained peak power levels produced with intense relativistic electron beams through the cyclotron maser process. It is especially noteworthy that these record peak powers were produced at millimeter wavelengths as well as in the more usual microwave bands.

TABLE 1 - Peak Power Levels from Cyclotron Masers Driven by Intense Relativistic Electron Beams[6]

Wavelength (cm)	Peak Power (Mw)	Diode Voltage (MV)	Diode Current (kA)
4	900	3.3	80
2	350	2.6	40
0.8	8	0.6	15
0.4	2	0.6	15

In addition to the high power levels in these intense beam experiments, it was also demonstrated[8] that the emission possessed a high degree of temporal and spatial coherence. Furthermore, the cyclotron maser was operated as a distributed-interaction amplifier[9] (8 GHz, 17 db gain, 5% bandwidth) which could be tuned magnetically. It should be noted that a distributed-interaction device has the advantage of tunability over a wide frequency range and, in addition, allows dissipation of far greater power as compared with a short resonator. Thus, its realization has considerable practical importance.

An important consequence of the intense beam experiments was the stimulation they provided for theoretical studies.[10-12] We note especially the nonlinear analysis of the saturation of the cyclotron maser instability by phase trapping,[11] and the subsequent self-consistent analysis[12] which generalized the first result to include saturation by energy depletion as well as by phase trapping. The latter work is useful not only in interpreting intense relativistic electron beam experiments but also in developing practical cyclotron maser tubes driven by electron beams with more conventional parameters.

USSR STUDIES USING MAGNETRON INJECTION GUNS

The lead in development of practical cyclotron masers using electron beams with conventional voltage and current values has been taken by a group working at the Gorki State University (USSR), where the device was given the name "gyrotron." In contrast to the cyclotron maser work in the USA after 1970, which was very much in the nature of a basic laboratory study using the relativistic beams outlined above, the Soviet work comprised an intense development effort leading to practical power tubes at millimeter and submillimeter wavelengths.[13,14,15] The key element in achieving practical devices characterized by high efficiency was in careful design of the electron gun. In the Gorki studies a crossed field, or so-called magnetron injection gun, was used to launch an annular beam with a large fraction of energy transverse to the axis and with minimum energy spread. These guns employed thermionic cathodes for cw and long-pulse operation. Experimental work on nonuniform cross section open resonators to optimize beam coupling for high efficiency has also taken place. All together, these developments have led to the announcement of the operation of two classes of devices, those operating in high (superconducting) magnetic fields, and those operating in lower conventional fields. (See Table 2 for further details.) An excellent review has been given by the Gorki group.[16]

CURRENT U.S. PROGRAM IN GYROTRON DEVELOPMENT

The success achieved in the USSR in realizing high efficiency gyrotrons has now stimulated parallel work in the USA. Under ERDA sponsorship, Varian Associates are currently developing a tube at 28 GHz with a cw power level of 200 kw for use in microwave-generated plasma studies at the Oak Ridge National Laboratory. This device is of the gyroklystron type[5] employing resonant cavities separated by drift spaces. This is in contrast to the gyromonotron oscillator, which is the type incorporated in the Soviet devices described above, involving a single oscillating cavity. Work on still a third type of device, called a gyro-traveling-wave amplifier (Gyro-TWA), is being sponsored at NRL by several DoD agencies. In this device, both an electron beam and a traveling electromagnetic wave traverse a waveguide as they interact. These three types of devices are shown schematically in Fig. 2.

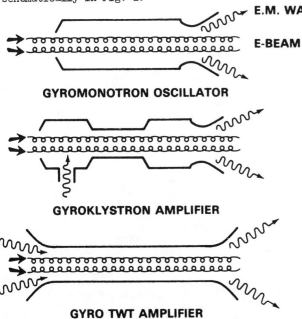

Fig. 2 - Three versions of the gyrotron.

The work at NRL is concentrating on millimeter wavelengths and on addressing scientific and technical issue at the limits of the technology. Currently studies are underway for the Navy with the aim of demonstrating efficient generation of at least 200 kw peak power at $\lambda = 8$ mm. The device configuration which is the major emphasis of this effort is the distributed, gyro-TWA because of its advantages in increased bandwidth and in handling high power. Plans are also being made at NRL to construct a gyromonotron oscillator under ERDA sponsorship; this device will also operate at $\lambda = 8$ mm and 200 kw, but will generate a long, 10 msec pulse for plasma heating. A sketch of the gyro-TWA device is shown in Fig. 3. Among the scientific issues to be addressed are the effect of self-fields of the electron beam, suppression of spurious mode generation in an overmoded waveguide, and operation at harmonics.

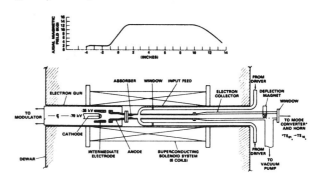

Fig. 3 - Sketch of the NRL 35 GHz gyro-traveling-wave-amplifier. A superconducting magnet is employed for future experiments at higher frequency. The field shape is shown at the top, to the same axial scale.

TABULATION OF CRM DEVICES

Table 2 provides a list of most of the cyclotron resonance masers operated to date or presently being designed that use thermionic cathode guns as the source of the electron beam. The table is meant to be representative, but not exhaustive. Further details on these devices may be found in the references cited in the column headed "Year (Ref)." Fig. 4 shows a graph of power vs. frequency for some of the gyrotron results, compared with conventional power sources.

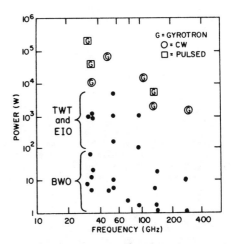

Fig. 4 - Best reported gyrotron results compared with reported results for traveling wave tubes, extended interaction oscillators, and backward wave oscillators, in the millimeter region.

TABLE 2 - CRM Devices Operated or Being Designed

λ (mm)	P(kw)[a] gain	n[b]	Tube[c] Type	Eff[d] (%)	Gun[e] Type	B (kG)	Magnet[f]	V/I (kV/A)	Mode (TE)	Year (Ref)
20	4	1	GM	50	MIG	5.4	Std	20/0.3	011	'66(13)
12	4.5	2	GM	16	MIG	4.5	Std	20/1.5	021	'66(13)
8.9	9	2	GM	40	MIG	6.0	Std	19/1	021	'73(14)
8.9	(40)	2	GM	43	MIG	6.0	Std	25/3	021	'73(14)
2.8	12	1	GM	31(36)	MIG	41	S.C.	27/1.4	021	'74(15)
1.9	2w	2	GM	10(15)	MIG	29	S.C.	18/1.4	031	'74(15)
2.0	7	2	GM	15(20)	MIG	29	S.C.	26/1.8	231	'74(15)
0.9	1.5	2	GM	6(5)	MIG	61	S.C.	27/0.9	231	'74(15)
1-2	(1µw)	1	GM		Triode	100	Std	10/0.05	over	'64(17)
2-4	(1w)	1,2	GM	2	Pierce	50	S.C.	20/	over	'65(7)
46	30µw	1	GK			2.3	Std	1-5/	TM_{010}	'66(5)
30	10dB	2	GK		MIG	2.0	Std	40/4.5	011	'77(18)
3.2	(10-100)	4	GT		MIG	9	Std			(19)
8	(300) 20dB	1	GT	(53)	MIG	13	S.C.	70/8.3	01	(20)
8	200	1	GM		MIG	13	S.C.	70/8.3	011 111	(21)
1.3	1-10	2-4	GM		MIG	<50	S.C.			(22)
11	(25) 20dB	1	GK	(44)	MIG	11	Std	80/8	01/02	(23)
11	(250)	1	GM	34(37)	MIG	11	Std	80/9	021	(23)

FOOTNOTES - a: pulsed power in parentheses; otherwise CW; b: harmonic number;
c: GM=gyromonotron, GK=gyroklystron, GT=gyro-traveling-wave amplifier;
d: experimental (theoretical); e: MIG=magnetron injection gun;
f: Std=normal electromagnet, S.C.=superconducting magnet.

TECHNICAL ISSUES

Stability to spurious oscillation is a major issue. The gain of the process is relatively high (e.g. in the current NRL design 2dB/cm) and in many cases the beam can interact with several modes of the structure. Clearly, strict attention will have to be given to all possible modes. Within limits the designer can adjust the relative transverse energy, the location of the annular beam relative to the wall, and of course the structure itself, including judicious use of resistive loading and mode suppression techniques. These techniques are common to other types of tubes, but are especially important in the gyrotron because of the high inherent gain.

Bandwidth is an important consideration for many applications. In the gyromonotron or gyroklystron versions, the cavity Q is a dominant factor; a typical bandwidth would be 0.1%. On the other hand the gyro-TWA, as noted before, is not limited in this way. From the available theory we estimate that an inherent instantaneous bandwidth of 3% should be easily achieved; perhaps with some form of contouring a bandwidth up to 10% or more may be possible. In the intense beam experiments Granatstein et al[9] measured a bandwidth of 5% for a gyro-TWA, albeit at low efficiency. Otherwise no experimental confirmation is available. At millimeter wavelengths, of course, even a 3% bandwidth amounts to several gigahertz.

Tunability is another desirable characteristic about which little is known. Assuming that stable gyro-TWA configurations can be built, we estimate that a tuning range of 5-10% should be achieved with relative ease, and that 20-30% may be possible. It is assumed that the magnetic field is tracked with the input frequency, and that minor adjustments of the gun voltages may be necessary. The speed with which the magnetic field can be tracked depends of course on the type of magnet and magnet supply.

Phase stability is a matter of concern in applications where extreme coherence is required. Preliminary estimates at NRL based on the total effective phase length of the device give approximately $4°$ phase change for 0.1% voltage variation.

SIZE AND WEIGHT

For the larger systems with power above, e.g., 5-10 kw average, the size and weight of the power supply and modulator will dominate the total. In such cases system estimates will be similar to those made for lower frequency tube systems of the same power and tube efficiency. On the other hand, for airborne or space applications the gyrotron magnet may be a dominant factor. In that event, operation of the gyrotron at a harmonic provides a clear advantage of a reduction of the required magnetic field by a factor n, the harmonic number. Interestingly enough, the highest reported efficiency at millimeter wavelengths, 43%, was obtained using the second harmonic.[14] However, gyrotron operation beyond the second harmonic has not been explored. It is clear that at least the 3rd and 4th harmonic should be also examined carefully. Operation at the 4th harmonic, for example, would produce frequencies up to 100 GHz with a magnetic field of 10 kG.

In addition to the question of harmonic operation to ease the magnet problem, the design of the magnet itself is important. The main issue is conventional electromagnet versus superconducting magnet. A conventional magnet and power supply for 10 kG is sufficiently large and heavy that serious consideration must be given to superconducting magnets and compact refrigerators even at that field level. A cursory examination of available refrigerators indicates that, while models are available for large scale cooling with MTBF measured in thousands of hours, few lightweight, compact units are available in the cooling range suitable for small superconducting solenoids. Thus some development work may be required to meet this need. With magnetic fields in the 50-100 kG range, compact gyrotron systems to 300 GHz should then be practical.

COMPONENTS

Few components are readily available to handle the 100-200 kw or more gyrotron power presently under development at millimeter wavelengths. Of primary concern are vacuum windows, mode converters, duplexers, various ferrite devices, and perhaps rotary joints. With regard to rotary joints, the use of over-moded waveguide and quasi-optical methods may be quite practical and indeed desirable. The same statement may also apply to long feed lines.

Pressurization and/or insulating gas will frequently be necessary to avoid breakdown. For cw or long pulses, the normalized power handling capacity of straight, smooth waveguides in air at atmospheric pressure, in Mw/cm^2, varies from about $0.5\ (\lambda/\lambda g)$ for several common TE modes to $0.9\ (\lambda/\lambda g)^{-1}$ for the TM_{01}^o mode, where λ_g is the guide wavelength. However, the introduction of corners and bends can easily double the electric field. A conservative point of view would dictate a reduction in the power handling capacity by perhaps a factor of four for a typical family of components. Use of insulating gas such as freon or SF_6 raises the capacity by a factor of at least 5 to 10 but introduces added complexity.

Detailed evaluations of development risk for specific components do not appear to be available. In all likelihood, components will only be developed for specific system needs. In this regard, the most immediate need is for mode converters and windows, since such components are vital to almost any application and indeed for testing of the gyrotrons themselves. For example, the TE_{on} modes, natural to many gyrotron designs, require conversion to the TE_{11} circular or the TE_{10} rectangular mode.

In summary, a high degree of confidence exists that the required components can be developed, but in general moderate development effort will be necessary for most items. In a few cases, e.g. advanced ferrite devices or very high power duplexers, the development risk may be high. In such cases, system designers may be forced to consider alternative methods using less risky components.

REFERENCES

1. R.Q. Twiss, "Radiation Transfer and the Possibility of Negative Absorption in Radio Astronomy," Aust. J. Phys. 11, 564 (1968).

2. J. Schneider, "Stimulated Emission of Radiation by Relativistic Electrons in a Magnetic Field," Phys. Rev. Lett. 2, 504 (1959).

3. A.V. Gaponov, "Interaction Between Electron Fluxes and Electromagnetic Waves in Waveguides," Radiophys. & Quantum Electronics 2, 450 & 837 (1959).

4. J.L. Hirshfield and J.M. Wachtel, "Electron Cyclotron Maser," Phys. Rev. Lett. 12, 533 (1964).

5. J.M. Wachtel and J.L. Hirshfield, "Interference Beats in Pulse-Stimulated Cyclotron Radiation," Physical Rev. Lett. 17, 348 (1966).

6. J.L. Hirshfield and V.L. Granatstein, "The Electron Cyclotron Maser - An Historical Survey," IEEE MTT 25, 522 (1977).

7. I.B. Bott, "A Powerful Source of Millimeter Wavelength Electromagnetic Radiation," Phys. Lett. 14, 292 (1965).

8. V.L. Granatstein, P. Sprangle, and M. Herndon, "An Electron Synchrotron Maser Based on an Intense Relativistic Electron Beam," J. Appl. Phys. 46, 2021 (1975).

9. V.L. Granatstein, P. Sprangle, M. Herndon, R.K. Parker, and S.P. Schlesinger, "Microwave Amplification with an Intense Relativistic Beam," J. Appl. Phys. 46, 3800 (1975).

10. E. Ott and W.M. Manheimer, "Theory of Microwave Emission by Velocity-Space Instabilities of an Intense Relativistic Electron Beam," IEEE Trans. Plasma Sci. PS-3, 1 (1975).

11. P. Sprangle and W.M. Manheimer, "Coherent Nonlinear Theory of a Cyclotron Instability," Phys. Fluids 18, 224 (1975).

12. P. Sprangle, and A. Drobot, "The Linear and Self-Consistent Nonlinear Theory of the Electron Cyclotron Maser Instability," Trans. IEEE MTT 25, 528 (1977).

13. A.V. Gaponov, A.L. Gol'denberg, D.P. Grigor'ev, T.B. Pankratova, M.I. Petelin, and V.A. Flyagin, "Experimental Investigation of Centimeter-Band Gyrotrons," Radiophys. and Quantum Electronics 18, No. 2, 280 (1975).

14. D.V. Kisel', G.S. Korablev, V.G. Navel'yev, M.I. Petelin and Sh.Ye. Tsimring, "An Experimental Study of a Gyrotron, Operating at the Second Harmonic of the Cyclotron Frequency, with Optimized Distribution of the High Frequency Field," Radio Engineering & Electronic Phys. 19, 95 (1974).

15. N.I. Zaytsev, T.B. Pankratova, M.I. Petelin and V.A. Flyagin, "Millimeter and Submillimeter-Wave Gyrotrons," Radio Engineering and Electronic Phys. 19, 103 (1974).

16. V.A. Flyagin, A.V. Gaponov, M.I. Petelin, and V.K. Yulpatov, "The Gyrotron," Trans. IEEE MTT 25, 514 (1977).

17. I.B. Bott, "Tunable Source of Millimeter and Submillimeter Electromagnetic Radiation," Proc. IEEE Corr, 330 (1964).

18. H. Jory, "Millimeter-Wave Gyrotron Development, Phase I," Final Tech. Rept., Contract F30602-76-C-0237, May 1977. Three-cavity gyroklystron built as an X-Band test model by Varian Associates under contract to the AF Rome Air Development Center and the Defense Advanced Research Projects Agency. Testing of this unit has not been completed; figures shown are preliminary. The device has also been operated as an oscillator in the third cavity. Using the TE_{211} mode, 20 kw (10% eff.) was obtained at 8.3 GHz using the 2nd harmonic.

19. "Proposal to Develop a 100 kw, Pulsed, 94 GHz Amplifier Tube - Phase II," VATP-11112, Varian Associates, Feb. 1977. Sponsored by the Rome Air Development Center, Defense Advanced Research Projects Agency, and the Ballistic Missile Defense Advanced Technology Center.

20. NRL 35 GHz test model (see text and Fig. 3 for additional description). Basic design by NRL; portions of the magnetron injection gun design and construction of the gun by Varian Associates, Palo Alto, California.

21. NRL 35 GHz gyromonotron oscillator for use in plasma heating experiments. The magnetron injection gun is identical to that described in Ref. 20. See also W.M. Manheimer and V.L. Granatstein, "Development of High-Power, Millimeter-Wave Cyclotron Masers at NRL and its Relevance to CTR," NRL Memo Report 3493, April 1977.

22. Preliminary design stage. Analysis and design shared by NRL and Harry Diamond Laboratories.

23. Varian Associates, Palo Alto, CA, "Development Program for a 200 kw, cw, 28 GHz Gyroklystron," Quarterly Rept. No. 5, July 1977, Contract No. 53X01617C. Data kindly supplied by H. Jory. Work supported by the U.S. Energy Research and Development Admin. through Oak Ridge National Laboratory. The power output, 25 kw, for the gyroklystron amplifier, is preliminary and does not represent saturation power.

Gyrotrons for High Power Millimeter Wave Generation

H.R. Jory, F. Friedlander, S.J. Hegji, J.F. Shivley, and R.S. Symons

ABSTRACT

The gyrotron is a new type of microwave tube capable of producing high power output at millimeter wavelengths. Oscillator results have been described in recent Soviet publications. This paper describes work in progress at Varian Associates, Inc. to develop an amplifier of the gyroklystron type to deliver 200 kW cw at 28 GHz. Considerable progress has been made with amplifier stability to the point that amplifier gains of up to 40 dB have been measured in a pulsed experimental amplifier. Current effort is concerned with improving efficiency. A pulsed oscillator is also described which produced 248 kW peak power at 28 GHz with 34% efficiency. A cw oscillator is under construction. Areas for future R and D are discussed. These include gyro-TWT amplifiers with increased instantaneous bandwidth (5 to 10%) and operation at harmonics of the cyclotron frequency to reduce the magnetic field requirements.

INTRODUCTION

The gyrotron is a microwave vacuum tube based on the interaction between an electron beam and microwave fields where coupling is achieved by the cyclotron resonance condition. This type of coupling allows the beam and microwave circuit dimensions to be large compared to a wavelength. Thus the power density problems encountered in conventional traveling wave tubes and klystrons at millimeter wavelengths are avoided in the gyrotron.

Rather complete histories and descriptions of the gyrotron have been published recently in the literature(1, 2). Note that the gyrotron and the cyclotron resonance maser are based on the same interaction. The most impressive results in terms of high average power at millimeter wavelengths with good device efficiency were obtained in the Soviet Union. These include 12 kW cw at 2.78 mm wavelength with 31% efficiency and 1.5 kW cw at 0.92 mm wavelength with 6.2% efficiency. The Soviet results have been limited so far to oscillator devices. However, they have discussed possible amplifier configurations.

One motivation for the Soviet work has clearly been to develop microwave power sources for heating plasmas for controlled fusion. However, once developed, these devices can clearly have other applications such as high resolution millimeter wave radar or high directivity communications.

Some of the recent work in the United States on gyrotron type devices has used short-pulse, highly relativistic beams. This work has produced very high peak power at millimeter wavelengths but beam efficiencies have been very low. An amplifier built with this type of beam demonstrated 17 dB gain (2).

The purpose of this paper is to outline the basic characteristics of the gyrotron devices and to describe work going on at Varian to develop amplifiers and oscillators using this interaction. The current Varian activity is directed toward the production of 200 kW cw at 28 GHz for heating plasmas.

BASIC CHARACTERISTICS

An important characteristic of the gyrotron is that it requires the application of a dc magnetic field which is specifically related to the operating frequency by the cyclotron resonance condition. This relationship is given by the equation

$$\omega = n\omega_c \qquad (1)$$

where ω is the operating frequency, n is an integer and ω_c is the cyclotron frequency or angular velocity of the electron given by

$$\omega_c = \frac{eB}{\gamma m_o} \qquad (2)$$

B is the dc magnetic field, e is the electron charge, m_o is the rest mass, and γ is the relativistic mass factor. Effective interaction occurs only for magnetic fields where n is near integer values. For most microwave field shapes, such as encountered in conventional waveguides and resonators, the fundamental resonance condition with n = 1 has the strongest interaction. With certain special microwave field shapes useful interaction can take place with larger integer values of n. These harmonic interactions have the advantage that the magnitude of the dc magnetic field for a given frequency can be reduced by $1/n$.

For the fundamental resonance condition, a frequency of 28 GHz requires a 10 kG magnetic field. For higher frequencies, proportionally higher fields are needed. This had led to the use of superconducting magnets in many of the Soviet experiments. Operation at the second harmonic of the cyclotron resonance has been reported (1), which allows a reduction in magnetic field by a factor of two.

In the gyrotron, bunching of the electron beam occurs as a result of a relativistic effect. This can be seen from Equation 2 where a change in electron kinetic energy, γ, results in a change in angular velocity of the electron. In the gyrotron single cavity oscillator, microwave electric fields in the early part of the cavity apply an angular velocity modulation to the electrons. As the electrons drift further through the cavity, angular bunching takes place as a result of the angular velocity modulation. Toward the end of the cavity the phase between the electron bunches and the microwave electric fields is adjusted so that the electrons give up kinetic energy. When the energy given up by the electrons exceeds the cavity losses, an oscillation results and output power is available.

Although a relativistic effect is involved in the interaction, efficient gyrotrons have been built using beam voltages as low as 18 kV. An optimum voltage range is probably 50 to 100 kV.

The frequency of the single cavity gyrotron oscillator is influenced by both the cavity resonance and the value of the dc magnetic field. In general, the output frequency is approximately linearly related to the dc magnetic field over the half power bandwidth of the cavity. Practical cavity Qs are in the range of 500 to 5000. Higher frequency stability requires tighter control on magnetic field. Oscillators with mechanically tuned cavities are quite feasible, but have not been demonstrated. Tunability of 10 to 20% should be possible with combined mechanical and magnetic tuning.

In the gyroklystron, an input cavity is used to modulate the beam and subsequent cavities are used for further amplification or energy removal. An experimental device of this type will be discussed in a later section. In the gyroklystron, instantaneous bandwidths of 1% are practical.

Another variation which should have larger bandwidth is the gyro-TWT. In this case a propagating waveguide is used for continuous interaction with the beam. Instantaneous bandwidth of 5 to 10% should be achievable and magnetic tuning should double the available bandwidth. Neither high gain nor high efficiency have yet been achieved, but they are predicted by the gyro-TWT theory.

All of the gyrotron devices require an unusual type of electron beam where most of the electron energy is transverse to the axis of the tube. This has required the development of new special electron gun configurations.

GYROKLYSTRON AMPLIFIER DEVELOPMENT PROGRAM

The goal of this program is a power source which will produce 200 kW cw at 28 GHz for heating plasma in experiments leading toward controlled fusion. An amplifier is desired in order to keep frequency and power output as constant as possible under conditions where the microwave load impedance may vary. It is also a requirement to have a design which can be scaled at a later time up to 200 kW cw at 120 GHz.

To meet this need, a gyroklystron amplifier was chosen as the preferred approach. Analytic models were developed to predict cavity coupling factors and small signal gain. A particle tracing computer code was used to predict output power and efficiency. A TE_{011} cavity of length 1.5 λ was chosen for the input cavity, and a TE_{021} cavity 2 λ long was picked for the output. The TE_{021} cavity allows reduced power density for later scaling to 120 GHz. The design values for the two-cavity amplifier are shown in Table 1.

Table 1
28 GHz GYROKLYSTRON DESIGN VALUES

Power Output	200	kW cw
Frequency	28	GHz
Cyclotron Harmonic	Fundamental	
Magnetic Field	11	kG
Beam Voltage	80	kV
Beam Current	8	A
Power Gain	30	dB
Efficiency	31	%
Bandwidth	0.2	%

The required dc magnetic field was obtained using a room temperature magnet wound with hollow core, water cooled copper conductors.

For the electron beam, a ratio of transverse to axial velocity of two to one was chosen. A digital computer code was used to determine appropriate electrode shapes in the gun to generate the beam. The gun simulation predicted a spread in transverse velocity of 5% and a corresponding axial velocity spread of 11%. The velocity spread must be minimized to obtain high efficiency.

The calculated power output for the two-cavity amplifier as a function of output cavity field strength is shown in Figure 1. The calculation assumed an ideal beam with no velocity spread. The calculated output of 280 kW and gain of 31 dB were above the design values. It was expected that velocity spread would reduce the gain and efficiency somewhat, and that space-charge effects would increase the gain and decrease efficiency.

Figure 2 shows a cross section of the first experimental amplifier. This was built to operate only pulsed in order to prove out the amplifier design at minimum cost with the expectation that modifications might be necessary. The output power for the tube is brought out through the beam collector region in 2.5 inch diameter circular guide. This vastly oversized guide is satisfactory for the plasma heating application and minimizes losses and cooling problems in transporting the output power to the plasma. The input guide is conventional WR-28 guide.

The amplifier was operated with the cathode temperature limited to control beam current. The gun anode voltage controls the ratio of transverse to axial velocity in the beam.

The gain of the first tube was found to be limited by spurious oscillations which appeared as the electron transverse energy was increased toward the design value. The reduced transverse velocity limited the gain to 13 dB. With this value of gain, saturation of the amplifier was not possible because of limited drive power. A large number of spurious oscillation frequencies were detected depending on tube parameter settings. The most troublesome ones were at 27.4 GHz and 26.9 GHz. These appeared to be associated with TM_{11} and TE_{21} type modes in the beam tunnels and in the input cavity.

The tube was modified by applying resistive loading to lower the impedances of the undesired modes. This included kanthal loading of the beam tunnels and the insertion of a lossy dielectric ring at the corner of the input cavity. The modified tube achieved stable gains of 20 dB before the onset of spurious oscillation again at 26.9 GHz. Saturation measurements were still not possible.

An additional modification was made to couple lossy dielectrics to the beam tunnel to further load the TE_{21} modes. This tube is presently in test. Initial results indicate stable gains of up to 40 dB and saturated power output of 50 kW. The reasons for the low efficiency are being explored. Possible causes are velocity spread or space-charge effects.

GYROTRON OSCILLATOR DEVELOPMENT

As a backup to the amplifier development, a single-cavity oscillator design was initiated. Power output calculations were performed for the oscillator using the same beam parameters as the amplifier. A TE_{021} cavity of length 5λ was chosen. Figure 3 shows calculated output power as a function of cavity field strength. The calculation predicted a maximum output of 260 kW. In order to make the calculation self consistent, the power given up by the beam must equal the power used by the load plus the power lost in the cavity walls. This condition can be satisfied by properly adjusting the total loaded Q of the cavity. The required Q for this condition is indicated in the figure by the dotted curve.

The horizontal line in the figure indicates the minimum Q that can be achieved by opening one end of a constant diameter cavity. Lower Q would require multiple output coupling or significant modification of the cavity geometry. Based on these considerations, a cavity loaded Q of about 400 was selected as a design value. The cross section of the pulsed oscillator was similar to that of the amplifier except for the absence of the input guide and cavity. The oscillator was designed to use the same gun, magnet, and output guide system as the amplifier.

Measured performance of the oscillator was very close to the predicted values. The maximum measured power output was 248 kW with the beam current at 9 A, slightly above the design value. This corresponded to an efficiency of 34%. The oscillator was operated up to a 5% duty factor with output of 200 kW peak and 10 kW average with a pulse length of 500 μsec. At lower duty factor the pulse length was increased to 1 msec, which was the limit of the test modulator.

Construction of a cw oscillator is now in progress. The design is similar except that the beam collector is much larger and the output waveguide is brought out to the side instead of through the collector.

The 28 GHz gyrotron amplifier and oscillator work is being conducted under subcontract to Oak Ridge National Laboratory, operated by Union Carbide Corporation for ERDA.

CONCLUSIONS

The gyrotron interaction makes it possible to achieve orders of magnitude higher power levels at millimeter wavelengths than was possible with conventional klystrons and TWTs. Figure 4 shows cw power versus frequency for the gyrotron compared to other devices. The first line on the left defines the limit of conventional production tube capability. The second line indicates what has been achieved in a few cases by pushing the linear beam devices to the limits of their capability. The next curve indicates the upper range of reported gyrotron results, and the dashed line indicates probable gyrotron results in the next few years.

Fixed frequency or narrow-band gyrotron oscillators up to the power levels indicated by the third curve can be considered available for system use in the immediate future. At the higher frequencies where super conducting magnets are currently used, the applications are probably limited to ground based or shipboard systems. Currently proven technology would limit room temperature magnet gyrotrons to 60 GHz or lower frequency using the second harmonic of the cyclotron frequency.

High gain gyroklystron amplifiers with good efficiency and bandwidth of about 1% should be available within about one year. Gyro-TWTs with instantaneous bandwidth could be available in a one to two year period. Practical amplifiers operating on high cyclotron harmonics are likely two to three years in the future.

The results reported in this paper confirm the Soviet oscillator results and design principles. The oscillator output of 248 kW peak, 10 kW average at 28 GHz with 34% efficiency is competitive with the Soviet results and represents record power output at that frequency.

Considerable work is still needed on the gyroklystron and gyro-TWT amplifiers. The feasibility of high gain in the gyroklystron has been demonstrated by the results reported here. Good progress has been made on amplifier stability, but additional work is needed on efficiency. This may require some basic investigations into beam quality, velocity spread and space-charge effects. Other development goals for the near future are gyro-TWTs with increased bandwidth, and gyro devices operating at high harmonics to minimize required dc magnetic field. With proper development the gyrotron interaction can supply a family of devices for use in high power millimeter wave systems.

REFERENCES

1. V.A. Flyagin et al, "The Gyrotron," IEEE Trans. MTT-25, No. 6, pp 514-521, June 1977

2. J.L. Hirshfield and V.L. Granatstein, "The Electron Cyclotron Maser an Historical Survey," IEEE Trans MTT-25, No. 6, pp 522-527, June 1977.

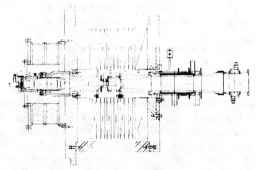

Figure 2. Layout drawing of a pulsed gyroklystron

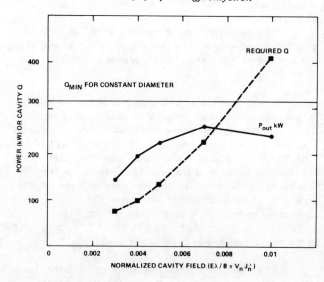

Figure 3. Calculated oscillator power output and loaded Q

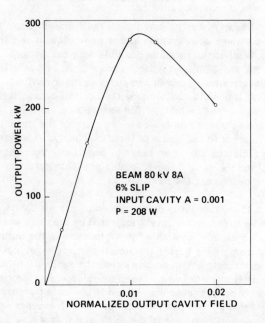

Figure 1. Two-cavity amplifier output power variation with output cavity field

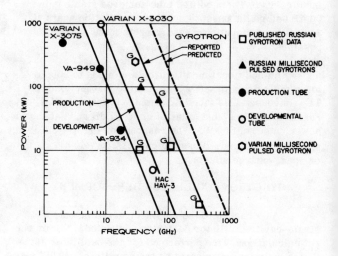

Figure 4. Microwave tube cw power state-of-the-art summary

Review of Mixers for Millimeter Wavelengths

M. McColl and D.T. Hodges

ABSTRACT

A survey of heterodyne mixers is presented for frequencies of 30 GHz to 300 GHz. The conventional Schottky diode, Josephson junction, superconductor-semiconductor (super-Schottky) diode, varactor down converter and InSb hot electron bolometric mixers are discussed in terms of their performance, physical operating principles and limitations.

The principle concerns in the choice of a heterodyne mixer are its sensitivity, bandwidth, and operating temperature. The Schottky-barrier diode mixer, operated predominately at room temperature, is a wide bandwidth device. These features plus its high sensitivity have made it the most widely used mixer element at microwave and millimeter wavelengths. The Josephson junction mixer, a cryogenic device, has yielded outstanding sensitivities well into the submillimeter. The super-Schottky diode, also a cryogenic device and the most sensitive detector of microwave radiation yet developed, may be a strong contender at millimeter wavelengths in the near future. This paper reviews the operating principles and limitations of these devices and presents an update on their sensitivities as reported in the literature to date. Also reviewed are Phillips and Jefferts' results with an InSb hot carrier bolometric mixer and Weinreb's recently reported outstanding sensitivities with a varactor-maser receiver at 115 GHz.

The sensitivity of a mixer is expressible in terms of either its minimum detectable power MDP_M or its mixer noise temperature T_M. These quantities are related by the expression

$$MDP_M = k\, T_M B_{IF} \tag{1}$$

where k is Boltzman's constant and B_{IF} is the IF bandwidth. T_M is given by

$$T_M = L_c T_d \tag{2}$$

where L_c is the conversion loss of the mixer, defined as the ratio of available power from the RF source to the power absorbed in the IF load. T_d is

---

*Supported by The Aerospace Corporation

the noise temperature of the mixer diode and is a parameter controlled by the temperature of operation, the doping of the semiconductor, and the configuration of the imbedding microwave circuit.

Conversion loss L_c of a Schottky barrier type of mixer in a tunable mount is conveniently expressed as the product of three terms[1,2,3]

$$L_c = L_0 L_1 L_2. \tag{3}$$

The intrinsic conversion loss L_0 is the loss arising from the conversion process within the nonlinear resistance of the diode and includes the impedance mismatch losses at the RF and IF ports. The L_0 of a Schottky-barrier mixer has been recently analyzed as a function of diode diameter, impedance, temperature and the Richardson constant of the semiconductor and found to increase as any one of these quantities is reduced.[4] The RF and IF parasitic losses, L_1 and L_2, respectively, are the losses associated with the parasitic elements of the diode. These losses are given by

$$L_1 = 1 + \frac{R_s}{R} + \omega^2 C^2 R R_s \tag{4}$$

$$L_2 = 1 + \frac{R_s}{R_2} \tag{5}$$

where R_s and C are the parasitic spreading resistance and junction capacitance, respectively, and ω is the signal angular frequency. R_s is the resistance which results from the crowding of the current in the semiconductor near the metal contact. R is the signal input impedance of the local oscillator pumped nonlinear resistance, and R_2 is the IF load impedance. The ω^2 dependence of the third term in Eq. (4) is responsible for the degradation in the performance of Schottky-barrier mixers at high frequencies. However, both L_1 and L_2 are geometry and material dependent, and it is by the manipulation of these dependencies in the design of the diode that one reduces L_c at high frequencies.

An epitaxial diode structure, such as shown in Fig. 1, is the standard design for high frequency Schottky-barrier mixers.[5-10] The epitaxial layer is grown very thin (as small as 1000Å) and the substrate is very heavily doped in order to reduce the diode spreading resistance. Moderate doping is chosen for the epitaxial layer so as to minimize the diode noise temperature T_d[11]. For usage at millimeter wavelengths, diode diameters with micron dimensions are used to reduce the effects of C.

The contact array diode has been proposed as a favorable alternative to the epitaxial diode as a sensitive high frequency mixer.[3] This diode achieves a low spreading resistance, and hence low L_1, by utilizing a structure consisting of a large number of small Schottky-barrier diodes connected in parallel. The number chosen is sufficiently large to maintain a low value of L_0. Arrays with more than 100 diodes, each 0.35 µm in diameter, have been fabricated but not yet tested. Figure 2a is a sketch of the cross section of the diode, and Fig. 2b shows a scanning electron microscope (SEM) photograph of several linear contact-array diodes recently fabricated.[3]

Table I summarizes heterodyne results with Schottky-barrier diodes operating at room and cryogenic temperatures. The very recent room temperature result of T_M = 390 K at 98 GHz by McMaster, et al.[14] was obtained using a subharmonically pumped two-diode hybrid integrated downconverter stripline circuit. Their result is close to those reported by Weinreb and Kerr,[13] Kerr,[2] and Schneider, et al.[8] at similar frequencies using cryogenically cooled Schottky diodes and, as such, represents a very substantial advancement in the technology of low noise receivers.

The Josephson junction, proposed in a theoretical article in 1962, was originally envisioned as consisting of two superconductors separated by a thin insulating barrier.[21] The insulator thickness was to be properly chosen to provide weak coupling for the superconducting electron pairs via tunneling. This type of coupling has since been demonstrated with a variety of barriers and structures, but the predominate configuration for high frequency mixing has consisted of a Nb point contact to a flat Nb surface.

Table II summarizes the results with Josephson junction mixers at millimeter wavelengths. The results of Taur, et al,[22] at 36 GHz are particularly outstanding because of the presence of conversion gain. Gains as large as 5 dB have been observed.[24]

The super-Schottky diode is the most sensitive detector of microwave radiation yet developed.[25-27] As a mixer at 10 GHz, it has delivered an MDP_M of 8×10^{-23} WHz^{-1} (T_M = 6K) at a liquid helium bath temperature of 1 K. The LO requirements are only 2×10^{-8} watts. A Schottky-barrier type of device, it consists of a superconducting contact to a heavily doped semiconductor. Experiments reported to date have used Pb contacts on p-type GaAs. The high solubility of GaAs for acceptor impurities fulfills the heavy doping requirement of the device, but the mobility of p-GaAs is relatively low. This low mobility results in a loss in sensitivity at high frequencies due to the effects of the parasitic loss L_1 via the third term in Eq. (4). Development efforts to extend this device to millimeter and submillimeter wavelengths are centering on techniques to lower the spreading resistance of the device. The contact array method, discussed above, and the use of Schottky-barrier contacts to semiconductors with very high mobilities, e.g., n-InSb and n-InGaSb, are being actively pursued.[3,28]

Weinreb[29] recently obtained a 160 K single-sideband receiver noise temperature at 115 GHz using a reversed bias Schottky barrier diode functioning as a varactor down converter followed by a reflected-wave type maser IF amplifier. This is the lowest noise temperature reported for an overall system at this frequency. The mixer required 3 mW of LO power and exhibited a conversion loss of 11.2 dB. The 21 GHz maser amplifier had a noise temperature of 4 K and a gain of 30 dB. The entire receiver was operated at 4.2K.

Phillips and Jefferts[30,31] working with an InSb hot electron bolometer functioning as a photoconductive mixer at 4K have achieved a double-sideband receiver noise temperature of 250 K (100 K of which was amplifier noise) at 120 GHz and 230 GHz. The LO drive required as little as 4×10^{-7} watts. However, this device has an IF bandwidth of only 2 MHz, a severe limitation for most applications.

The above discussion covers the major efforts reported on mixers operating in the 30 GHz to 300 GHz range. A summary of reported sensitivities is presented in Fig. 3 along with the quantum limit for these types of mixers $kT_M = hf$.[32] This limit is applicable only for photon energies hf much greater than kT, and hence, only for mixers cooled to cryogenic temperatures for the wavelengths under discussion.

References

[1] G. C. Messenger and C. T. McCoy, "Theory and Operation of Crystal Diodes as Mixers," Proc. IRE, vol. 45, pp. 1269-1283. 1957.

[2] A. R. Kerr, "Low-Noise Temperature and Cryogenic Mixers for 80-120 GHz," IEEE Trans. Microwave Theory Tech., vol. MTT-23, pp. 781-787. 1975.

[3] M. McColl, D. T. Hodges and W. A. Garber, "Submillimeter-wave Detection with Submicron-size Schottky-barrier Diodes," IEEE Trans. Microwave Theory Tech., vol. MTT-25, pp. 463-467, June, 1977.

[4] M. McColl, "Conversion Loss Limitations on Schottky-barrier Mixers," IEEE Trans. Microwave Theory Tech., vol. MTT-25, pp. 54-59. 1977.

[5] H. M. Leedy, H. L. Stover, H. G. Morehead, R. P. Bryan and H. L. Garvin, "Advanced Millimeter-wave Mixer Diodes, GaAs and Silicon, and a Broadband Low-noise Mixer," in Proc. 3rd Bienniel Cornell Engineering Conf. (Ithaca, N. Y.), pp. 451-462, 1971.

[6] B. J. Clifton, W. T. Lindley, R. W. Chick and R. A. Cohen, "Materials and Processing Techniques for the Fabrication of High Quality Millimeter Wave Diodes," in Proc. 3rd Bienniel Cornell Engineering Conf. (Ithaca, N. Y.), pp. 463-475. 1971.

[7] G. T. Wrixon, "Low-noise Diodes and Mixers for the 1-2 mm Wavelength Region," IEEE Trans. Microwave Theory Tech., vol. MTT-22, pp. 1159-1165. 1974.

[8] M. V. Schneider, R. A. Linke and A. Y. Cho, "Low-noise Millimeter-wave Mixer Diodes Prepared by Molecular Beam Epitaxy (MBE)" Appl. Phys. Lett. vol. 31, pp. 219-221, 1 Aug. 1977.

[9] F. Bernues, H. J. Kuno and P. A. Crandell, "GaAs or Si: What Makes a Better Mixer Diode?", Microwaves, pp. 46-55, March, 1976.

[10] J. A. Calviello and J. L. Wallace, "Performance and Reliability of an Improved High-Temperature GaAs Schottky Junction and Native-Oxide Passivation", IEEE Trans. on Electron Devices, vol. ED-24, pp. 698-704, June, 1977.

[11] T. J. Viola, Jr. and R. J. Mattauch, "Unified Theory of High-frequency Noise in Schottky Barriers", J. Appl. Phys., vol. 44, pp. 2805-2808, June 1973.

[12] L. T. Yuan, "Low Noise Octave Bandwidth Waveguide Mixer", Conf. Digest, 1977 MTT-S International Microwave Symposium, San Diego, Calif. pp. 480-482, (IEEE Cat. No. 77CH1219-5 MTT).

[13] S. Weinreb and A. R. Kerr, "Cryogenic Cooling of Mixers for Millimeter and Centimeter Wavelengths", IEEE J. Solid-State Circuits (Special Issue on Microwave Integrated Circuits), vol. SC-8, pp. 58-62, Feb. 1973.

[14] T. F. McMaster, E. R. Carlson, and M. V. Schneider, "Subharmonically Pumped Millimeter-Wave Mixers Built with Notch Front and Beam Lead Diodes", Conf. Digest, 1977 MTT-S International Microwave Symposium, San Diego, Calif. pp. 389-391 (IEEE Cat. No. 77 CH 1219-5 MTT).

[15] Hughes Aircraft, *Microwaves*, p. 63, July, 1977.

[16] A. R. Kerr, R. J. Mattauch and J. A. Grange, "A New Mixer Design for 140-220 GHz", *IEEE Trans. Microwave Theory Tech.*, vol. MTT-25, pp. 399-401, May, 1977.

[17] J. J. Gustincic, Th. de Graauw, D. T. Hodges and N. C. Luhmann, Jr., "Extension of Schottky Diode Receivers into the Submillimeter Region", 1977 MTT-S International Microwave Symposium, San Diego, Calif., (late paper).

[18] M. V. Schneider and G. T. Wrixon, "Development and Testing of a Receiver at 230 GHz," *Tech. Digest, IEEE S-MTT International Microwave Symp.*, Atlanta, Ga. (1974), pp. 120-122.

[19] P. F. Goldsmith and R. L. Plambeck, "A 230 GHz Radiometer System Employing a Second Harmonic Mixer", *IEEE Trans. Microwave Theory Tech.*, Vol. MTT-24, pp. 859-861, 1976.

[20] N. R. Erickson, private communication, 1977.

[21] B. D. Josephson, "Possible New Effects in Superconductive Tunneling", *Phys. Lett.*, vol. 1, pp. 251-253, 1962.

[22] Y. Taur, J. H. Classen, and P. L. Richards, "Josephson Junctions as Heterodyne Detectors," *IEEE Trans. Microwave Theory Tech.*, vol. MTT-22, pp. 1005-1009, 1974.

[23] J. Edrich, D. B. Sullivan, and D. G. McDonald, "Results, Potentials and Limitations of Josephson-Mixer Receivers at Millimeter and Long Submillimeter Wavelengths," *IEEE Trans. Microwave Theory Tech.*, MTT-25, pp. 476-479, 1977.

[24] Y. Taur, J. H. Classen, and P. L. Richards, "Conversion Gain and Noise in a Josephson Mixer," *Revue de Physique Applique*, vol. 9 pp. 263-268, 1974.

[25] M. McColl, R. J. Pedersen, M. F. Bottjer, M. F. Millea, A. H. Silver, and F. L. Vernon, Jr., "The super-Schottky Diode Microwave Mixer", *Appl. Phys. Lett.*, vol. 28, pp. 159-162, 1976.

[26] M. McColl, M. Millea, A. H. Silver, M. F. Bottjer, R. J. Pedersen, and F. L. Vernon, "The super-Schottky Microwave Mixer", *IEEE Trans. Magn.* Vol. MAG-13, pp. 221-227, 1977.

[27] F. L. Vernon, Jr., M. F. Millea, M. F. Bottjer, A. H. Silver, R. J. Pedersen and M. McColl, "The super-Schottky Diode", *IEEE Trans. Microwave Theory Tech.* Vol. MTT-25, pp. 286-294, 1977.

[28] M. McColl and M. F. Millea, "Schottky Barriers on InSb", *J. Electronic Materials*, vol. 5, pp. 191-207, 1976.

[29] S. Weinreb, "Millimeter Wave Varactor Down Converters", presented at *Diode Mixers at Millimeter Wavelengths Workshop* sponsored by Max Planck Institut fur Radioastronomie, Bonn, West Germany, 26-28 April 1977.

[30] T. G. Phillips and K. B. Jefferts, "A Low Temperature Bolometer Heterodyne Receiver for Millimeter Wave Astronomy", *Rev. Sci. Instrum.*, vol. 44, pp. 1009-1014, Aug. 1973.

[31] T. G. Phillips and K. B. Jefferts, "Millimeter-Wave Receivers and Their Applications in Radio Astronomy", *IEEE Trans. Microwave Theory Tech.*, vol. MTT-22, pp. 1290-1292, Dec. 1974.

[32] B. M. Oliver, "Thermal and Quantum Noise", *Proc. IEEE*, vol. 53, pp. 436-454, 1965.

Table I. Single-sideband Mixer Performance.
Using Schottky Barrier Diodes

FREQUENCY (GHz)	MDP M (WHz^{-1})	TM (K)	Lc (dB)	OPER. TEMP. (K)	REFERENCES
31	1.5×10^{-20}	1090	4.9	290	YUAN [12]
33	2.8×10^{-21}	200	5.8	15	WEINREB AND KERR [13]
35	5.8×10^{-21}	420	3.3	290	CALVIELLO AND WALLACE [10]
50	1.7×10^{-20}*	1250*		290	HUGHES AIRCRAFT [15]
75	2.4×10^{-20}*	1730*		290	
85	3.9×10^{-21}	280	7.2	18	WEINREB AND KERR [13]
85	1.1×10^{-20}	800	6.8	298	
98	5.4×10^{-21}	390	7.4	290	McMASTER, et al. [14]
102	4.35×10^{-21}	315		15	SCHNEIDER, LINKE AND CHO [8]
110	3.2×10^{-20}*	2330*		290	HUGHES AIRCRAFT [15]
115	4.1×10^{-21}	300	5.5	77	KERR [2]
115	6.9×10^{-21}	500	5.8	298	
140	4.9×10^{-20}*	3530*		290	HUGHES AIRCRAFT [15]
140	1.1×10^{-20}	800		290	WRIXON [7]
170	1.5×10^{-20}	1100	6.2	290	KERR, MATTAUCH AND GRANGE [16]
175	4.6×10^{-20}	3340	11.3	290	WRIXON [7]
186	5.5×10^{-20}	4000	10.4	290	GUSTINCIC, et al. [17]
230	1.7×10^{-19}*	12,000*		290	SCHNEIDER AND WRIXON [18]
230	1.7×10^{-19}*	12,000*		290	GOLDSMITH AND PLAMBECK [19]
244	5.0×10^{-20}	3600	10.2	290	GUSTINCIC, et al. [17]
302	6.3×10^{-20}	4600	10.9	290	
325	9.1×10^{-20}*	6600*		290	ERICKSON [20]

*Includes noise from an IF amplifier

Note: All reported double-sideband results have been increased by a factor of 2 for this table.

Table II. Single-sideband Results with Josephson Junction Mixers at 4.2 K.

FREQUENCY (GHz)	MDP$_M$ (WHz^{-1})	T$_M$ (K)	CONVERSION LOSS [gain] (dB)	REFERENCES
36	7.5×10^{-22}	54	[1]	TAUR, CLAASSEN, AND RICHARDS [22]
316	3×10^{-21}	223	9.5	EDRICH, SULIVAN, AND McDONALD [23]

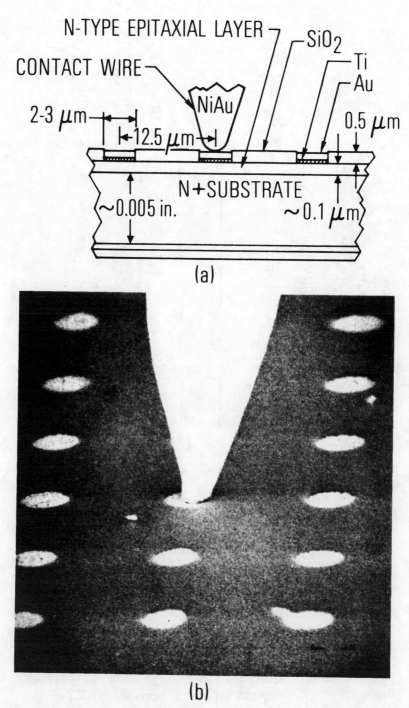

Fig. 1. Schottky-barrier diodes for millimeter-wave mixers rely on a hyper-thin epitaxially grown layer (a). Pressure holds a tapered whisker in contact with a diode in this array (b) after Bernues, et al.[9]

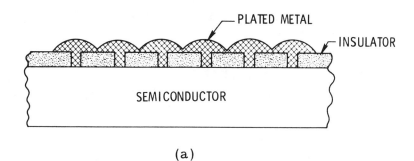

(a)

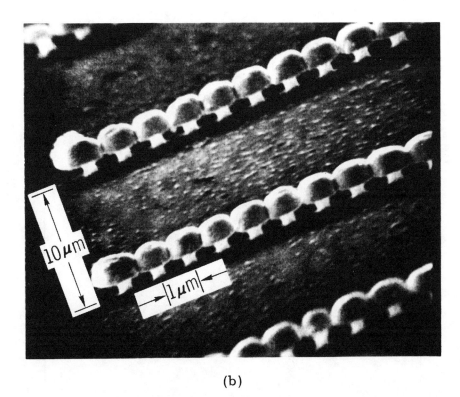

(b)

Fig. 2(a). The structure of a contact array diode.

2(b). An SEM photograph of preliminary linear contact-array diodes. The insulating layer normally present has been stripped away to permit viewing of the entire structure. To reduce skin effect losses, ohmic contacts are made on the top surface encompassing the linear array on both sides.[3]

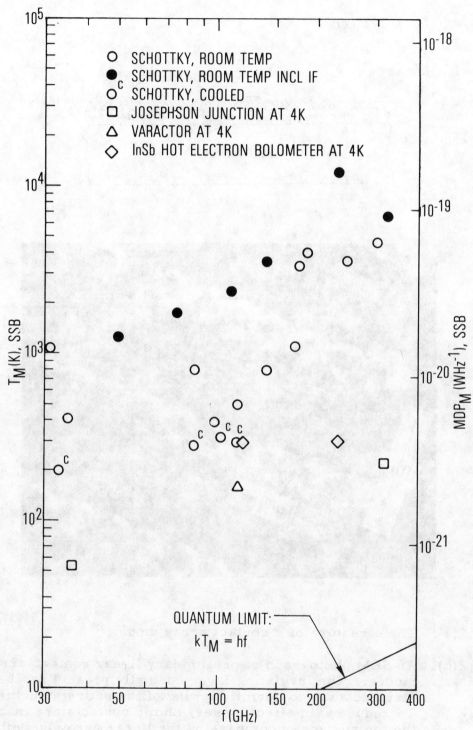

Fig. 3. Single-sideband (SSB) mixer noise temperature (T_M) and mixer minimum detectable power (MDP_M) as a function of frequency. The filled-in data points include the noise contribution of the IF amplifier.

Millimeter Wave Antennas
A.F. Kay

Abstract—Millimeter wave antennas are primarily parabolic reflectors, Cassegrain systems, and lenses. Comparative merits and the few design features peculiar to millimeter waves are discussed. Some of these antennas have larger gain and much larger gain per unit cost than any other microwave antennas. A few experimental arrays and ferrite scanners have been built. Brief mention is made of absorbers and radomes at millimeter waves.

General Characteristics

ANTENNAS for the millimeter bands continue to display the microwave trend, increasing as we move up the spectrum, in that virtually all practical antennas are based on geometrical optics designs, commonly parabolic reflectors, Cassegrain systems, and lenses. At millimeter frequencies, however, unlike lower frequencies, this gives rise to the physically small high-gain antenna. A three-inch diameter paraboloid at 100 GHz, for example, having 35 dB gain, is about the same size as other microwave components of a typical system operating at this frequency. Figure 1 shows measured patterns of such an antenna. More generally, Fig. 2 shows the gain and beamwidth vs. reflector diameter of typical quasi-optical antennas: parabolic reflectors, Cassegrains, and lenses. Nominal aperture and beam efficiencies,

$$\eta_1 = \frac{G}{\left(\frac{\pi D}{\lambda}\right)^2} \qquad \eta_2 = (3 \text{ dB full beamwidth})(D/\lambda) \quad (1)$$

of 0.55 and 75° are assumed, respectively.

Another trend has been that for the highest gain antennas which are practical, the largest antenna gain per unit cost can be obtained at the highest developed microwave frequency. Evidence accumulating now shows that this trend is continuing well into the millimeter range. This is illustrated with some well-known installations in Table I. Thus we have two interesting extremes of the millimeter antennas—the small high-gain antenna which does not appreciably increase the size of the system in which it operates and, on the other hand, physically large antennas with gain greater than hitherto seemed feasible at lower microwave frequencies. The latter are used in radio astronomy and look attractive for ground based deep space tracking and communications.

Millimeter antennas, in general, are coherent, and require close tolerance. By the Rayleigh criterion, parabolic reflector surface errors of less than $\lambda/32$ are desired and generally achieved [6]. Thickness tolerances for lenses are a factor of two more liberal. For Cassegrain systems [7], [8], the $\lambda/32$ tolerance must be divided between the two reflectors and it is generally best to do so approximately proportional to the diameters. The reflector tolerances can be achieved by the metal spinner's art up to diameters of three or four feet and frequencies up to 80 to 100 Gc/s for low cost, low volume requirements [9]. Care must be taken to use a precision mandrel, sufficiently heavy stock, and to avoid distortion in mounting to a reinforcing backup structure. Die casting or stamping is suitable for the smaller sizes for high volume production when this is required. Alternatively machined aluminum castings or weldments are attractive for somewhat better tolerance controls and the best of the present state-of-the-art [5]. A few millimeter reflectors [10] have been cast from epoxy resins flowing in a female mold rotated at a very precise angular velocity about a vertical axis. Under the influence of gravity and centrifugal force, the surface assumes a parabolic shape with focal length

$$f = \frac{38.4}{\text{rpm}} \text{ feet.} \quad (2)$$

The resin cures while at true shape and is then reflectorized by a metallic deposition. Life under temperature and weather cycling have limited the applications of these "spun-cast" reflectors. Millimeter frequencies appear to be characterized as the highest frequencies at which these standard machining and molding processes can still be used for the fabrication of reflector tolerances to the Rayleigh criterion.

Another characteristic of millimeter antennas, as opposed to microwave antennas, arises from the fact that with a few exceptions at the lower millimeter frequencies, low noise preamplifiers are not available so that limiting the antenna excess noise temperature is not required. Moreover it is often not necessary because of high atmospheric noise. Nevertheless with the usual stringent requirement for gain, energy cannot be unduly wasted in spatially extensive, far-out sidelobes or feed spillover. In radiometers this effect reduces the sensitivity and produces a requirement for noise performance little different from the lower frequencies.

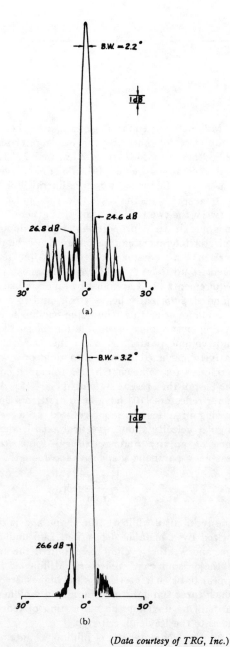

Fig. 1. (a) E-plane radiation pattern of 3-inch diameter reflector at 100 GHz. (b) H-plane radiation pattern of 3-inch diameter reflector at 100 GHz.

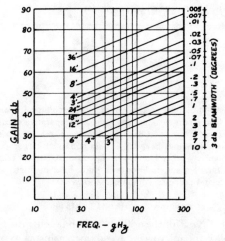

Fig. 2. Nominal gain and beamwidth of millimeter antennas.

TABLE I [1]–[5]

Facility	f, GHz	D, ft	Gain, dB	Cost $/Rel. Gain
Arecibo Radio Obs. Puerto Rico	0.43	1000	60	6.0
NRAO, Greenbank W. Va.	1	300	56.6	1.8
Haystack Hill Lincoln Lab., Mass.	18	120	73.8	0.13
Prospect Hill AFCRL, Mass.	35	29	66.5	0.15
Aerospace, Calif.	250	15	78	0.0055
NRAO, Ariz.	180*	36	83.5	0.0035

* This system was not complete at time of writing. The figure is an estimate.

Antenna Types

Figure 3 shows examples of quasi-optical millimeter antennas. The paraboloid and the Cassegrain are commonly encountered and are similar in design to the lower frequency versions [7], [8]. A horn lens is an alternative to the paraboloid which operates as follows: The waveguide throat junction excites a TE_{11} conical waveguide mode. The exterior lens surface is planar and may be grooved for better match over a narrow frequency band. The interior surface is hyperboloidal, focussing the conical guide mode energy at infinity. Figure 4 shows typical patterns of a horn lens of about the same gain as the parabolic reflector of Fig. 1.

While the dual of this antenna, the cornucopia or horn reflector, is common as a high performance point-to-point antenna at microwave frequencies, the decreased size makes the cost and weight of the horn lens more attractive at millimeter frequencies for diameters up to one or two feet. The lens may be zoned as a Fresnel lens [11] to further reduce the weight and simplify the construction. Polypenco, rexolite, and quartz are common low-loss millimeter lens materials. In these sizes, the effectiveness of the horn for improved efficiency and bandwidth is not outweighed by the size and cost of the horn. A pyramidal rather than conical horn with a properly designed rectangular aperture gives more equal E- and H-plane patterns but because of lack of symmetry can only do so with a single linear polarization, and the patterns and bandwidth are not as satisfactory as the conical horn lens. A diagonal horn [12] has been used successfully with the 15-foot Aerospace antenna.

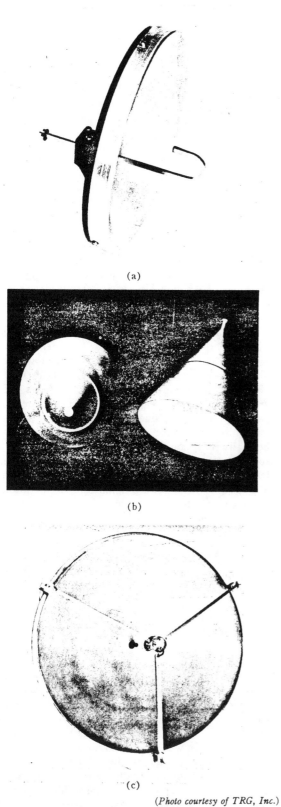

(Photo courtesy of TRG, Inc.)

Fig. 3. (a) Focal point parabolic reflector. (b) Horn lens. (c) Cassegrain fed parabolic reflector.

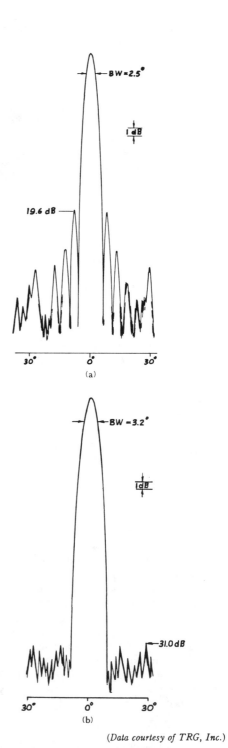

(Data courtesy of TRG, Inc.)

Fig. 4. (a) E-plane radiation pattern of 4-inch diameter horn lens at 68.5 GHz. (b) H-plane radiation pattern of 4-inch diameter horn lense at 68.5 GHz.

(Photo courtesy of TRG, Inc.)
Fig. 5. Ferrite scanning array at 50 GHz.

Not all millimeter antennas are quasi-optical. Waveguide slot arrays are feasible and can be attractive in the lower millimeter frequencies, especially for the 35 Gc/s window. Two experimental arrays of both the Purcell and leaky wave type have been built and perform well at lengths of 5 feet and frequencies up to 50 GHz [13]–[15]. Arrays of these types were especially chosen and designed to minimize the obvious tolerance and fabrication problems of long arrays at millimeter wavelengths. For example, at X band for a three-foot long resonant array of resonant slots, a tolerance of 0.001 inch must be held on slot length and cumulative slot spacing to prevent beam breakup. Such an array would be almost impossible to scale very far up into millimeter frequencies. The leaky waveguide has one wall replaced by closely spaced transverse wires whose separation is varied to force the coupling to conform to a Chebyshev distribution. Its sidelobe level at 35 GHz was -28 dB, and elsewhere in the 26 to 50 GHz band varied from -24 to -29 dB. In that band, its gain was $1\frac{1}{2}$ to 3 dB below theoretical, which was entirely attributable to the finite wall conductivity.

In a modified version which incorporated an extra section of parallel plate guide in the output, the Purcell array also worked well from 28 to 35 GHz. However, improperly taking into account the effect of waveguide losses as well as tolerance problems in generating design data produced an undercoupled array with low gain.

Figure 5 shows a ferrite scanning array having 16 circularly polarized radiating elements. This antenna was airborne for communications tests and is believed to be the highest RF electronic scanning array antenna yet developed. Specifications include:

Bandwidth:	50–52 GHz
Gain:	17 dB
Beamwidth:	3.5°×70°
Scanning Range:	±30° in plane of the array
Switching Time:	3 μs
Power Handling Capacity:	10 watts average.

Visible in the photograph are the 16 ferrite phase shifters, the corporate waveguide structure, power dividers, and hybrid with sum and tracking difference mode outputs. The size and cost performance of this system is inferior to a quasi-optical system by any standard excluding switching speed, and only a need for electronic switching justifies an antenna whose section area far exceeds its aperture area [16].

Another approach to ferrite scanning in a millimeter antenna was reported in [17], [18]. Here a ferrite slab acted like a dielectric prism when magnetized by a transversely variable magnetic field. When illuminated by a horn, a 10°-wide beam was formed. Temporal variation of the magnetic field caused the beam to shift in one plane up to ±15° with less than 13 dB sidelobe levels and 1 dB loss.

Choice of Focal Point or Cassegrain Feeds or Lenses

The trade-off considerations in the choice between focal point paraboloids and Cassegrain reflectors are similar to those found at any microwave frequency [7], [8] but at millimeter waves lead to characteristic conclusions. Waveguide losses in standard commercial practice are found to be about twice the theoretical pure copper straight waveguide loss of the same overall length as the actual feed run. This becomes very significant in large antennas and/or at higher frequencies ($\gtrsim 70$ Gc/s). This puts the focal point feed at a certain disadvantage. However tolerance, fabrication, and alignment requirements in Cassegrain feeds are also greater, thereby increasing the cost and/or causing poorer performance due to subdish diffraction, alignment, and tolerance losses. The net result is that normally focal point feeds are used first of all on reflectors less than about 60λ in diameter where Cassegrain feeds block excessively, secondly on larger sizes up to 300 or 400λ where cost is a significant consideration, and finally where the RF equipment can be placed at the focus conveniently and without excessive blocking.

The restriction on lenses is of a different character. A lens more than about 60λ in diameter cannot compete with a reflector. This is primarily because of dielectric losses but also especially at the lower millimeter frequencies because of weight and fabrication cost. Below 60λ and especially below 10λ, the lens may be preferred because of the elimination of feed blocking and the possibility of complete feed shielding.

Feed Design

In the quasi-optical antennas, only the feed design presents true microwave features. Flared rectangular or conical guide is commonly used. Often the required flare is so small as to be hardly noticeable. For focal ratios of about 0.35, none at all is required in the E plane. Since the wavelength is physically small, tapered matching sections several wavelengths long may be easily used and are physically short. The wall thickness of standard rectangular or circular guide, however, becomes an appreciable fraction of a wavelength above about 26.5 GHz, and the resulting "flange effect" tends to affect the beam in the E plane. Feed design data available at lower frequencies therefore requires modification. For example, Fig. 6 shows the 10 dB beam-

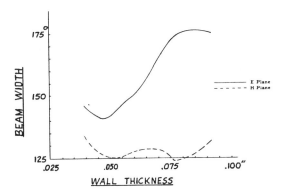

(Data courtesy of TRG, Inc.)

Fig. 6. Beamwidth vs. wall thickness of circular waveguide feed at 68.5 GHz and I.D. of 0.1015 inch. (——— E plane). (- - - - - H plane).

width of a circular waveguide feed at 68.5 GHz with ID of 0.1015 inch vs. the wall thickness. For high-gain feeds suitable for Cassegrain antennas, the wall thickness no longer has an effect. A long conical flare is often used. Figure 7(a) shows measured data on a series of such feeds having aperture diameters D/λ and lengths such that a phase error of about $\lambda/8$ existed in the aperture. A value of $\lambda/8$ is a reasonable compromise between excessive lengths and minimum phase error. Each feed consisted of a section of standard rectangular guide followed by a transition to TE_{11} mode circular guide followed by the conical flare. Five different feeds were measured at a number of frequencies, and the data lay very close to the smooth curves of Fig. 7(a). Figure 7(b) shows the continuation of these curves for smaller diameters with thin wall at the aperture and phase error less than $\lambda/8$. The data agrees fairly well with Silver [19].

Conical scanning may be obtained by all techniques used at lower frequencies, but rotating a slightly offset subdish of a Cassegrain [see Fig. 3(c)] has the advantage of not requiring a rotary joint or flexible section. Spinning the feed illuminating a lens minimizes the blocking problem which may be severe for small apertures. Tracking monopulse feeds using hybrids or magic tees have been developed up to about 96 GHz.

Near Field Antennas

Antennas which are designed to operate in the near field are rare at lower frequencies but relatively common at millimeter frequencies. The value of beam mode guide [20]-[23] rests on the fact that waveguide losses are high but diffraction losses can be made low if each antenna in the line is placed in the near field of the one preceding and refocusses to the next the energy focussed upon it (see Fig. 8). These antennas may be either lenses or elliptical reflectors.

Ability to focus to a spot which is a commonplace at optical frequencies also finds practical use in plasma diagnostic systems where millimeter waves have the further advantage of being above the opaque plasma

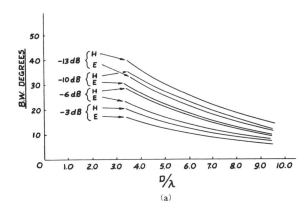

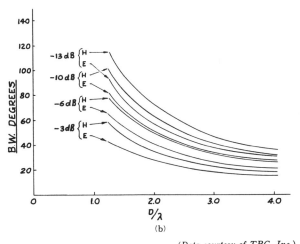

(Data courtesy of TRG, Inc.)

Fig. 7. (a) Beamwidth vs. aperture diameter—high-gain conical feed horns. (b) Beamwidth vs. aperture diameter—low-gain thin wall, conical feed horns.

resonant frequencies for electron densities of up to 10^{13}–10^{15}/cc. The plasma may be spatially probed by the spot whose 10 dB width may be quite small and is given by

$$d = \frac{2R\lambda}{D} \quad (\text{if } \lambda \leq d \leq D) \qquad (3)$$

where D/λ is the antenna diameter in wavelength and R is the range to the spot.

For this purpose, focussing horn lenses may be used. These are similar to the horn lens described under "Antenna Types" but with the change of the outer lens surface from planar to hyperbolic. An elliptical reflector is functionally equivalent and more practical for the larger sizes above 1 or 2 foot diameter. The spot may also be adjusted in range by positioning the feed. The two focal lengths are related during this adjustment by

$$f_2 = f_{20} + \frac{(f_{10} - f_1)\left(\frac{f_{20}}{f_{10}}\right)^2}{\left(1 + \frac{f_{20}}{f_{10}}\right)\left(\frac{f_1}{f_{10}} - 1\right) + 1} \qquad (4)$$

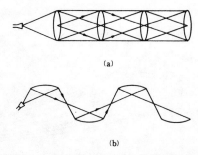

Fig. 8. (a) Beam mode waveguide. [Periodically, lenses reconstitute plane phase front, produce quasi-plane wave (TEM_{00} mode). Optically, rays would diverge and converge as shown by arrows.]
(b) Version of beam mode guide utilizing ellipsoidal reflectors.

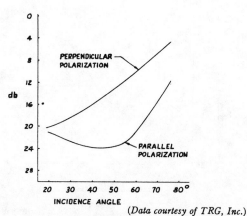

(Data courtesy of TRG, Inc.)

Fig. 10. Absorption of 96 GHz of typical microwave flat, flexible absorber.

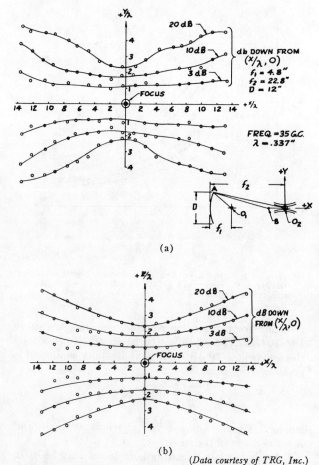

(Data courtesy of TRG, Inc.)

Fig. 9. (a) Focussing action of elliptical reflector and measured E-plane patterns of a particular case. (b) Measured H-plane patterns.

where f_{10}, f_{20} are the design focal lengths ($f_{20} > f_{10}$) and f_1, f_2 are the refocussed paraxial values.

The limits of adjustability and depth of focus may be determined approximately as follows: In Fig. 9(a), the point B defines the limit of the first Fresnel zone of the focal spot on the axis. Accordingly

$$\overline{AO_2} - \overline{AB} - \overline{BO_2} = -\lambda/2. \quad (5)$$

This expression approximately defines the 3-dB spot depth. It also approximately defines the point at which the spot is $\sqrt{2}$ times as wide as in the focal plane. Figure 9 shows measured patterns in the neighborhood of the focus illustrating this depth of focus in a particular case.

Absorbers

Commercially available physically or electrically graded absorbers, although designed for lower frequencies, appear to behave well at millimeter waves although data for the most part is sparse. Figure 10 shows data taken at 96 GHz representative of commercially available flexible planar absorber, designed for X band where its performance is only a few decibels better. Similar data could not be obtained with the common types of dentated absorber since the specularly reflected energy is so low as to be not readily detectable (more than 36 dB below that of the reference metal reflector out to incidence angles of 75 degrees). Again at 1–10 GHz such absorbers typically have −40 dB reflection.

Radomes

Very few radomes have been used at millimeter frequencies. None of the large ground antennas used in radio astronomy or space studies utilize radomes, although some have nonelectrical "astrodomes" for protection when not in use.

Airborne millimeter radomes remain to be developed especially at the higher frequencies above 70 GHz. Conventional half-wave walls and sandwiches would generally be structurally inadequate. Higher order radomes having several half wavelengths additional thickness, have poorer performance, especially transmission loss and bandwidths. Structural reinforcement of a thin wall radome using the metal space frame concept [24] appears attractive. Unlike the lower frequency counterparts which are electrically thin, the windows of such a structure would be n half-wavelengths thick with typically $n=1$ up to 70 GHz and $n=2$ up to 140 GHz.

References

[1] J. W. Findlay, "Radio telescopes," *IEEE Trans. on Antennas and Propagation*, vol. AP-12, pp. 853–864, December 1964.

[2] H. G. Weiss, "The haystack microwave research facility," *IEEE Spectrum*, pp. 50–69, February 1954.

[3] C. W. Tolbert, A. W. Straiton, and L. C. Krause, "A 16-foot diameter millimeter wavelength antenna system, its characteristics and its applications," *IEEE Trans. on Antennas and Propagation*, vol. AP-13, pp. 225–229, March 1965.

[4] E. E. Altshuler, "Earth-to-space communications at millimeter wavelengths," L. G. Hanscom Field, Bedford, Mass., AFCRL-65-566, AFCRL, August 1965.

[5] H. E. King, E. Jacobs and J. M. Stacey, "A 2.8-arc min beamwidth antenna: lunar eclipse observations at 3.2 mm," *IEEE Trans. on Antennas and Propagation*, vol. AP-14, pp. 82–90, January 1966.

[6] J. Ruze, "Antenna tolerance theory—a review," this issue, page 633.

[7] P. W. Hannon, "Microwave antennas derived from the Cassegrain telescope," *IRE Trans. on Antennas and Propagation*, vol. AP-9, pp. 140–153, March 1961.

[8] M. Viggh, "Cassegrain antennas," *Elteknik (Sweden)*, pp. 83–87, 1962.

[9] E. M. T. Jones, R. A. Folsom, Jr., and A. S. Dunbar, "Millimeter wavelength antenna studies," Supplement to Annual Rept., SRI Contract DA36-039-SC-5513, July 1952.

[10] J. W. Dawson, "28-foot liquid-spun radio reflector for millimeter wavelengths," *Proc. IRE*, vol. 50, p. 1541, June 1962.

[11] J. M. Cotton, F. Sobel, M. Cohn, and J. C. Wiltse, "Millimeter wave research," ECI, Timonium, Md., Final Rept., AF30(602)-2457, AD29654.

[12] A. W. Love, "The diagonal horn antenna," *Microwave J.*, vol. V, no. 3, pp. 117–122, March 1962.

[13] E. M. T. Jones, R. C. Honey, and R. A. Folsom, "Millimeter wavelength antenna studies," SRI Final Rept., Contract DA36-039-SC-5503, AD24043, July 1953.

[14] R. C. Honey, L. A. Robinson, and J. K. Shimizu, "Antenna design parameters," SRI Contract DA36-039-SC-73106, January 1960.

[15] R. C. Honey, "Line sources and linear arrays for millimeter wavelengths," *Proc. Symp. on Millimeter Waves*, Brooklyn Polytechnic Inst., Brooklyn, N. Y., April 1957, pp. 563–577.

[16] (a) V. Meixner, "Feasibility study and investigation of EHF communication techniques," Republic Aviation Co., 6th Quarterly (Final) Rept., AF33(657)-8855, January 1964.
(b) J. Visscher, "An EHF system for two-way aerospace communication techniques," 9th Nat'l Communication Symp., Utica, N. Y., 1963.

[17] R. E. Johnson, T. R. Schiller, and P. F. Weiss, "Ferrimagnetic beam steering at millimeter wavelengths," *1965 Internat'l Symp. IEEE Group on Antennas and Propagation*. Washington, D. C., pp. 71–77.

[18] T. R. Schiller and W. S. Heath, "Electronically steerable array," Sylvania Repts. Contract AF30(602)-3041.

[19] S. Silver, *Microwave Antenna Theory and Design*, vol. 12, Rad. Lab. Ser. New York: McGraw-Hill, 1949, p. 340.

[20] N. I. Heenan, "The beam waveguide," *IEEE Spectrum*, vol. 1, pp. 84–86, October 1964.

[21] A. F. Kay, "Near-field gain of aperture antennas," *IRE Trans. on Antennas and Propagation*, pp. 586–593, November 1960.

[22] R. G. Fellers, "Millimeter wave transmission by non-waveguide means," *Microwave J.*, vol. 5, pp. 80–87, May 1962.

[23] J. R. Christian and G. Goubeau, "Experimental studies on a beam waveguide," *IEEE Trans. on Antennas and Propagation*, vol. AP-9, pp. 256–263, May 1961.

[24] A. F. Kay, "Electrical design of metal space frame radomes," *IEEE Trans. on Antennas and Propagation*, vol. AP-13, pp. 188–202, March 1965.

[25] J. E. Degenford, "Study and investigation of millimeter and submillimeter wave receiver techniques," University of Illinois, Urbana, Contract AF30(602)-2456, 1963.

[26] M. A. Kott, "A variable beamwidth millimeter wave antenna," *IEEE Trans. on Antennas and Propagation*, vol. AP-12, pp. 662–667, November 1964.

[27] T. Sueta, "A study of antenna for millimeter wave grating spectrometer," *J. Inst. Electr. Commun. Engrs.* (Japan), vol. 42, pp. 677–683, July 1959.

[28] R. C. Honey, E. D. Sharp, C. M. Ablow, L. A. Robinson, J. L. Brenner, M. G. Andreasen, and J. Aasted, "Antenna design parameters," SRI Final Rept. DA36-039-SC-73106, May 1961.

Performance of Reflector Antennas with Absorber-Lined Tunnels

R.B. Dybdal and H.E. King

Antenna sidelobe reduction techniques find significant application in present and future system designs. Interference, whether intentional, unintentional, or radar clutter, is directly related to the antenna sidelobe level. This paper describes the sidelobe reduction achieved by adding an absorber-lined tunnel to a reflector antenna.

The initial investigation of reflector antennas with absorber-lined tunnels was conducted about 20 years ago.[1] This effort considers a configuration using a feed with a significant amount of aperture amplitude taper (≈ 27 dB) and examines several tunnel geometries.

A 6" diameter reflector with a 92 GHz diagonal horn was measured over an 80 dB dynamic range. Potential interactions between the tunnel and aperture fields were explored with 6" and 12" diameter tunnels. The tunnels were lined with 1/4" thick absorber. This antenna did not have interactions which modify main beam or peak gain characteristics. The earlier measurements[1] noted gain losses with increasing tunnel length, attributed to smaller feed amplitude taper. This antenna has a measured gain of 39.4 dB independent of tunnel length.

The tunnel lengths were chosen by a shadow boundary defined from the tunnel rim to the opposite reflector edge. Typical patterns without a tunnel are given in Fig. 1 for reference purposes. Patterns for a 12" diameter tunnel 22" long (shadow boundary 20° from the reflector axis) are shown in Fig. 2. Beyond the shadow boundary the reference pattern in Fig. 1 is significantly reduced. Similar pattern characteristics were achieved with a 6" diameter tunnel. Other tunnel lengths provided similar reduction beyond the shadow boundary.

The effectiveness of the tunnel is enhanced by terminating it with a rounded edge as Fig. 3 illustrates. The radius of the rounded edge is approximately 10λ (1-1/4") which is sufficiently large to result in significant creeping wave attenuation of the diffracted energy.

1 P. Blacksmith, W. G. Mavroides, P. J. Donovan, R. H. Beyer, "A Method of Reducing Far Out Side Lobes," AFCRC TR 57-113; July 1957, DDC No. AD 133707.

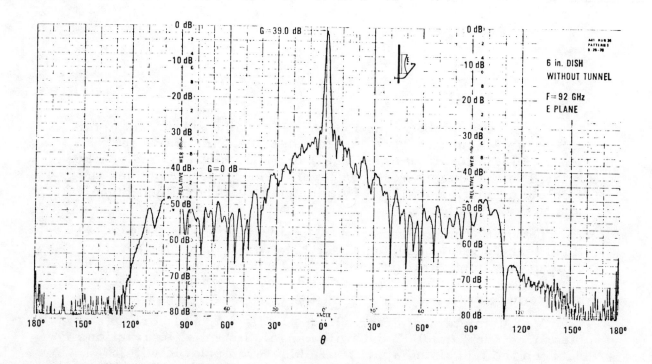
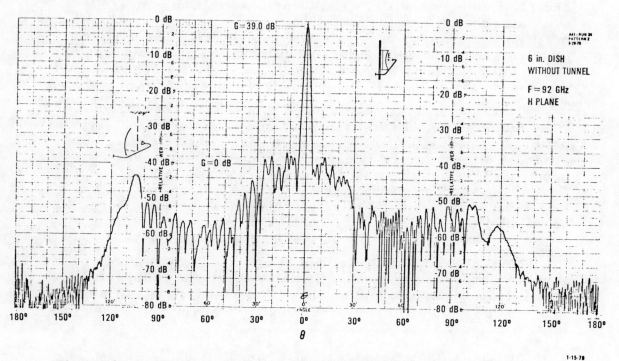

Figure 1 Pattern of 6 in. Dish Without Tunnel

Millimeter Wave Radar RF Sources and Components

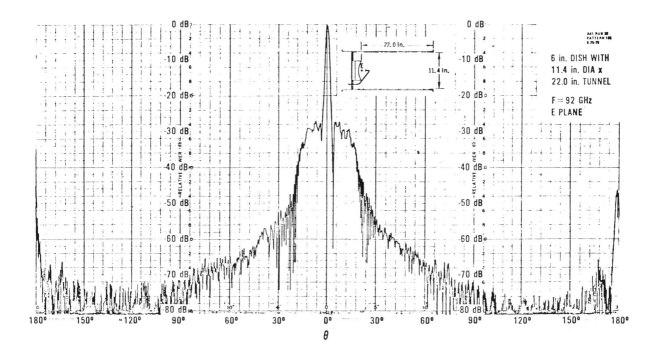

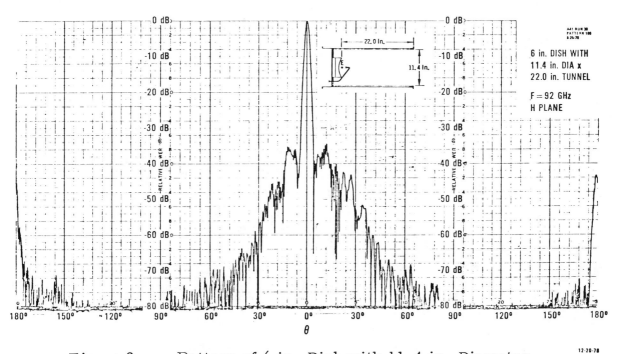

Figure 2 Pattern of 6 in. Dish with 11.4 in. Diameter x 22.0 in. Tunnel for 20° Shadow Boundary

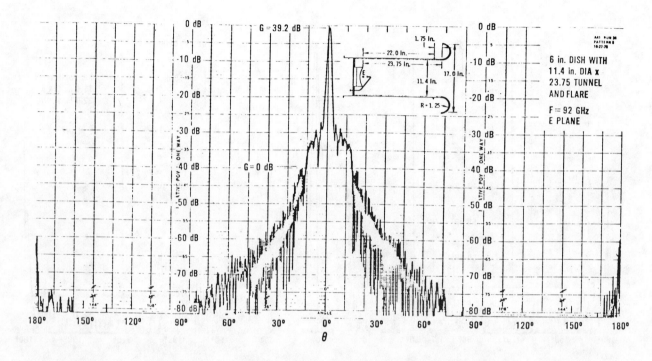
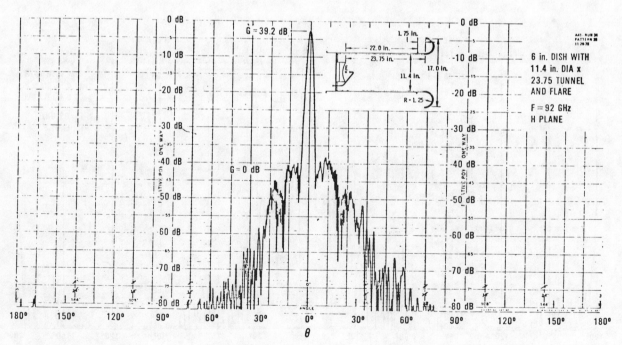

Figure 3 6 in. Dish with 11.4 in. Diameter x 23.75 in. Tunnel with Flare Section

A Millimeter-Wave Scanning Antenna For Radar Application

O.B. Kesler, W.F. Montgomery, and C.C. Liu

Proceedings of the 1979 Antenna Applications Symposium, University of Illinois/Rome Air Development Center, Sept. 1979 (Session II, first paper) DDC AD A077167

Reprinted by permission.

Introduction

During certain tactical operations, such as tracking or acquisition of a target, a high gain antenna is required to rapidly scan through a volume in space. The polarization twist reflector antenna with the hardware realizations, using photo-lithographic processes, described in this paper presents an attractive solution at millimeter-wave frequencies. The rapid scanning feature of the polarization twist antenna is achieved by scanning only a light weight flat plate rather than an entire antenna/mount assembly having high inertia. By virtue of its reflection properties the flat reflection plate need only be scanned by half the angle that the antenna pattern is scanned in comparison to a 1:1 scan to beam skew ratio of other antennas. This allows a simple spherical bearing gimbal mechanism for a 45° conical sector scan. The antenna feed remains stationary, thereby eliminating the need for a rotary joint, which at millimeter-wave frequencies is unreliable, difficult to build, and very expensive. The antennas presented in this paper are truly low cost, efficient millimeter-wave and Ka-band antennas for radar application.

Discussion of Prior Art

Prior high gain millimeter-wave radar antennas are generally of the optical generic category, such as prime focus reflectors, Cassegain reflectors, and horns. These optical type antennas, as well as planar arrays are mechanically scanned by means of a gimbal system, and as a consequence require two expensive and often band-limiting rotary joints. When monopulse direction finding is used, each of the two rotary joints have three matched channels with high tolerances, particulary at 94 GHz frequency. The beam scanning rate of these mechanically scanned antennas are limited by the inertia of the antenna, gimbal, and rotary joint mechanisms.

In the millimeter frequency band the art of electronic scanned antennas is meager. Switched beam-scanned antennas using lens or reflectors can scan only small regions due to coma aberrations. Phased arrays above 90 GHz are virtually nonexistent, although there is no scientific reason why they are not possible. However, cost and power loss are certain to restrict phased arrays for years to come.

The millimeter-wave radar antenna described in this paper yields high performance with simplicity and low cost. In particular, the described millimeter-wave radar antenna combines all of the following attributes in a single practical millimeter antenna:

1. High gain
2. Low dissipative losses
3. Four-lobe monopulse
4. Low sidelobe difference patterns
5. Wide angle scanning
6. No rotary joints
7. Spherical bearing gimbal with low inertia for acceleration
8. Innovative photolithographic fabrication techniques
    - Capacitive and inductive tuned twist reflector for polarization twist
9. Low cross-polarization
10. Wide bandwidth

Description of Antenna

Figure 1 shows a sketch of a polarization twist antenna. The polarization of the feed horn in the sketch is vertical. Rays emanating from the feed horn are reflected by the parabolic surface since the dielectric material has vertical metallic lines on the inside surface. The rays are then reflected by a flat plate which also accomplishes a 90° polarization rotation. The rotation is accomplished with 45° lines on the surface of the flat plate and 180° phase difference between the reflected electric field components parallel and perpendicular to the lines. The ray is reflected at an angle 2θ to the horizontal, as shown in Figure 2, for a plate angle of θ with the vertical. The rays pass out through the parabolic surface since the E-field is now perpendicular to the metallic lines.

The polarization twist geometry has been explored as a candidate for a scanning antenna at lower frequencies by Martin and Schwartzman [1]. Houseman [2] describes an experimental 94 GHz model. Waineo and Konienczny [3] have demonstrated a 94 GHz flight test model with scanning capability. The antenna described here was designed to simulate a missile flyable antenna, however, precise adjustment mechanisms were installed so that design data could be gathered.

Hardware Realization

A six inch millimeter polarization twist antenna constructed by Texas Instruments is shown in Figure 2. Working models were built at 35 and 94 GHz. A 45.0° conical sector of scan is possible with these antennas. Since the device pictured in Figure 2 was intended as an engineering model, many of the mechanisms were included for convenience and experimentation and could be eliminated in later models. Figure 3 shows these mechanisms. The middle micrometer moves the horn in and out relative to the parabolic surface. A screw adjustment moves the flat reflecting surface in and out relative to the horn. The micrometers were installed on the scan mechanism to accurately determine scan angle. A millimeter mixer is shown attached to the back of the antenna.

Gimbal System

Since the polarization twist reflector must only rotate by 22.5° to give 45° of beam scan, it is possible to use the spherical bearing gimbal system shown in Figure 4. This use of the spherical bearing would not be possible on conventional antennas that must be rotated 45°. Blockage is reduced by the use of the spherical bearing since the distance between the transreflector and the pivot point is minimized. The precision spherial bearing is off the shelf and inexpensive. The push rod assemblies are also low cost since there are no critical tolerances involved in their construction. In Figure 3 the micrometers that set the scan for experimental purposes would be replaced by servomechanisms in an operating system.

Horn Feed

Preliminary tests were performed using a single horn feed on both the 94 GHz and 35 GHz models. The horns were designed to give a 10 dB edge taper, excluding space loss, for lower side lobes. The 94 GHz horn was formed by milling out the E-plane walls of the waveguide at a 10° flare angle after the H-plane walls had been removed. Brass was then used to replace the H-plane walls. The 94 GHz feed horn and flat polarization twisting plate are shown in Figure 5.

An E-plane monopulse feed was constructed for the 94 GHz model. A commercially available hybrid tee was used for the comparator circuitry. The two feed apertures were brought sufficiently close together by milling off the outside of the common walls of the two feeding waveguides. The total wall thickness between the feed apertures was .020 inches prior to flaring. In order to obtain the desired compromise between sum and difference channel illumination the apertures were flared by sharpening their common wall to a knife edge and milling out the inside of the other two walls at a 10° angle. The phase was adjusted in both channels, using phase trimmers, until the maximum null depth was obtained. A phase adjustment was required because of the extremely small mechanical tolerances necessary to maintain phase and amplitude matched channels at 94 GHz. It was not, however, necessary to resort to machining the feed from a single block. Figure 6 shows the sum and difference patterns of the E-plane feed.

Polarization Twist Reflector

The 94 GHz polarization twist reflector in Figure 5 is designed such that the sum of the inductive reflection coefficient and the capacitive reflection coefficient is zero for broadside incidence at the center frequency. Changing the line width and period of the grid shown in Figure 7 changes the capacitance and inductance associated with them. Line widths and spacings were determined to satisfy the reflection coefficient requirement for a standard thickness laminate, that is close to one quarter wavelength. The use of a standard laminate thickness and an etching process make this design procedure highly desirable from the standpoints of low cost and reproducibility. Figure 8 shows the variation in line width versus period satisfying the reflection coefficient condition. The solid and dashed curves are respectively for a 20-mil thick stock piece of copper clad RT/Duroid* and a $\lambda/4$ (21.16-mil) substrate. Widths of 4.03 and 2.45 mils are respectively seen for the two cases with period A = 8 mils. The performance of the twist reflector depends upon the difference in phases of the inductive, ϕ_I, and capacitive, ϕ_C, reflection coefficients. A measure of the performance given by the ratio, in dB, of the untwisted to twisted components is $20 \log_{10} \cot|((\phi_I - \phi_C)/2)|$. The relative phase as a function of frequency at normal incidence is approximately the same for the two cases and is given in Figure 9. The figure shows that the relative phase error, deviation from 180 degrees, is less than 20 degrees over the 87 to 101 GHz band. Phase variations with scan angle in the planes parallel and perpendicular to the lines are given in Figure 10. The phase error is less than 9 degrees over twisting plate angle of 22.5 degrees or beam angle of 45 degrees.

Similar parametric design curves for a 35 GHz twist reflector are shown in Figure 11. The solid and dashed curves are respectively for 0.062 inch Duroid and $\lambda/4$ laminate. In this case a significantly wider period, A = 0.120 inches, was selected for the 35 GHz twist reflector shown in Figure 12. Figure 13 shows the relative phases of the inductive and capacitive reflection coefficients as a function of frequency. The phase variation is significantly different from the 94 GHz designs with low inductances. Here the inductance is considerably larger causing the different frequency behavior. The phase error is less than 30 degrees over the 31 to 40 GHz band for the $\lambda/4$ case. In the 0.062 inch case the band extends from 31 to 50 GHz. This shows that it can be advantageous to choose a higher inductance design and in doing so to decrease the tolerance and increase the bandwidth.

Transreflector

The parabolic dish shown in Figure 2 is a half-wavelength fiber glass structure. Epoxy resin and E-glass cloth were used for the fiber glass. The dish was formed by laying up pre-impregnated sheets on a male mold and curing in an autoclave. The dish was 29-mils thick for the 94 GHz model. A 2½-mil fiber glass laminate with copper on one side was etched with 3-mil lines and 8-mil spacings, gored, and carefully laid up on the inside of the dish. On a low cost production antenna the copper could be deposited directly on the inside of the parabolic surface and the lines could then be etched. The part of the

---

*Regulation trade mark Rogers Corporation

fiber glass structure that comes straight back toward the back of the antenna adds necessary phase shift to the rays passing through it when the beam is scanned.

As can be seen in Figure 1, the antenna aperture size is constant, independent of scan angle. Changes in beamwidth, gain, and beam steering with scan angle can be due in part to transreflector transmission effects. As a first order approximation of these effects a ray tracing computer program was written for the twist reflector geometry. Figure 14 shows the results of this program for beamwidth and gain at various scan angles. Figure 15 shows the beam steering and sidelobe results. The beamwidth increased by .02° and the gain decreased by 0.15 dB at 45° scan angle. The sidelobes increased by 1 dB and the beam steering is a maximum of 0.06° at 45° scan angle. From these results it can be seen that the effects of the transreflector should be minimal if fabrication errors do not play a significant role.

Test Results

Figure 6 demonstrates the high quality beam scanning capability of the 94 SGHz antenna out to 45 degrees. The 1.4 degree beamwidth broadens only slightly and the gain degrades less than 1 dB at 45 degrees. The measured relative gain versus scan angle is shown in Figure 17. Peak sidelobes are between -22.5 and -25 dB in the scan range. The antenna pattern integrity was maintained over a wide frequency bandwidth. The exact bandwidth determination was limited by measurement capability. Figure 18 shows the 35 GHz antenna being scanned in the H-plane. The 3.5° beamwidth varied only slightly with 45° of scan. Peak sidelobes are between -19 and -23 dB for this non-optimized engineering model. Expanded scale pattern in Figure 19 shows good pattern symmetry. The antenna has a high degree of polarization purity at 0 degree scan because the metallic lines in the parabolic surface act as a polarization filter. The cross-polarization is down more than 32 to 37 dB in the main beam. As scan is increased the polarization purity gradually degrades due to the changing orientation of the lines with scan. Figures 20 and 21 show antenna patterns at 39.89 and 30.21 GHz which demonstrate that the bandwidth is greater than 27 percent.

Figure 22 shows the 94 GHz monopulse sum and difference, H-plane patterns. The null in the difference pattern is 47.5 dB below the sum peak. The 1.8° sum beamwidth is broader than the single horn antenna beamwidth because of the non-optimum monopulse feed. The improved quality of this 94 GHz pattern when compared to the previous results reflects the aquisition of an improved measurement system.

Conclusions

The polarization twist antenna with hardware realizations described here using photolithographic processes, spherical bearing gimbal, and standard laminates is indeed an attractive low cost, rapid scanning, monopulse, millimeter wave radar antenna. It has been shown that the use of a standard copper clad laminate with etched lines is a low cost, producible design technique for the twist reflector. Various choices of line widths and periods may be selected. The specific selection of parameters may be made trading off, bandwidth, tolerance, and performance with scan angle. The polarization twist reflectors fabricated demonstrate the essentially constant gain and beamwidth with scan angle predicted by geometrical optics and an analysis of transreflector effects. Deviations from constant gain may also be explained by degraded twist reflector performance at large scan angles. The fabrication of a high quality monopulse system is possible at 94 GHz using off the shelf components.

The basic millimeter-wave antenna described here can be used in many other applications where a scanning beam is required. For example, the paraboloidal transreflector may be synthesized or spoiled to give a cosecant type fan-beam pattern. Such a pattern could have a single plane monopulse and scanning capability.

References:

1   R. W. Martin and L. Schwartzman, "A Rapid Wide Angle Scanning Antenna With Minimum Beam Distortion," Proc. of East Coast A.N.E. Conference, pp. 47-51, 1958.

2   E.O. Houseman, "A Millimeter-Wave Polarization Twist Antenna," Proc. of International A.P. Conference, pp. 51-54, 1978.

3   D.K. Waineo and J. F. Konieczny, "A Millimeter Wave Monopulse Antenna With Rapid Scan Capability," Proc. of International A.P. Conference, pp. 477-480, 1979.

Millimeter Wave Radar RF Sources and Components

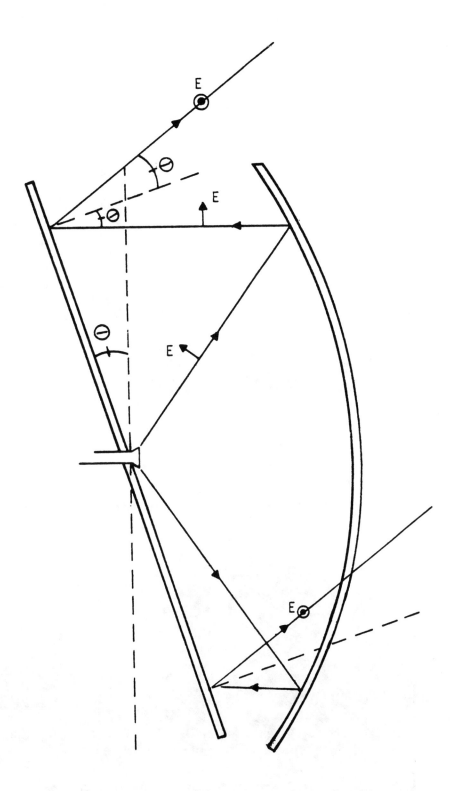

Figure 1. Polarization Twist Operation

FIGURE 2. POLARIZATION TWIST REFLECTOR

FIGURE 3. ADJUSTING MECHANISM

Millimeter Wave Radar RF Sources and Components

FIGURE 4. GIMBAL SYSTEM

FIGURE 5. 94 GHz TWIST REFLECTOR AND HORN FEED

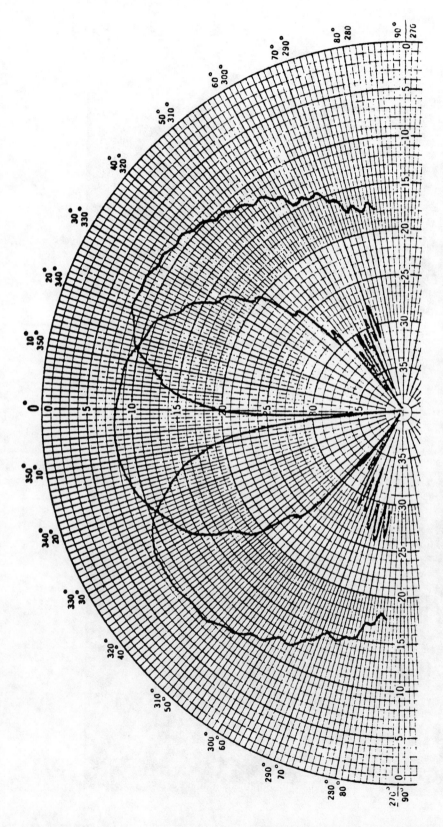

Figure 6. Monopulse Feed Pattern Sum and Difference E-Plane

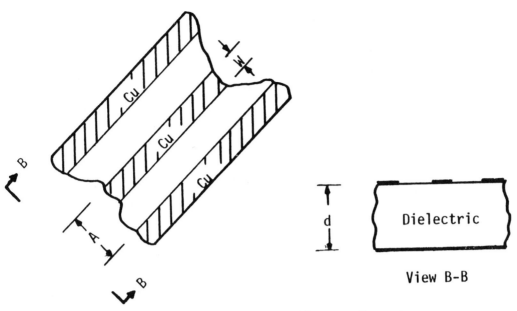

Figure 7. Twist Reflector Geometry.

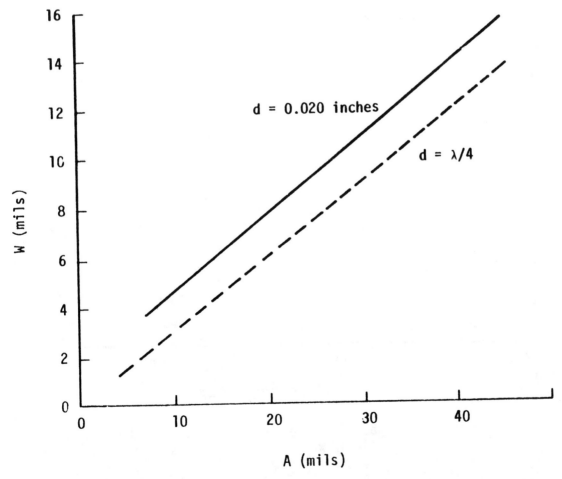

Figure 8. Line Width Versus Period for a 94 GHz Twist Reflector.

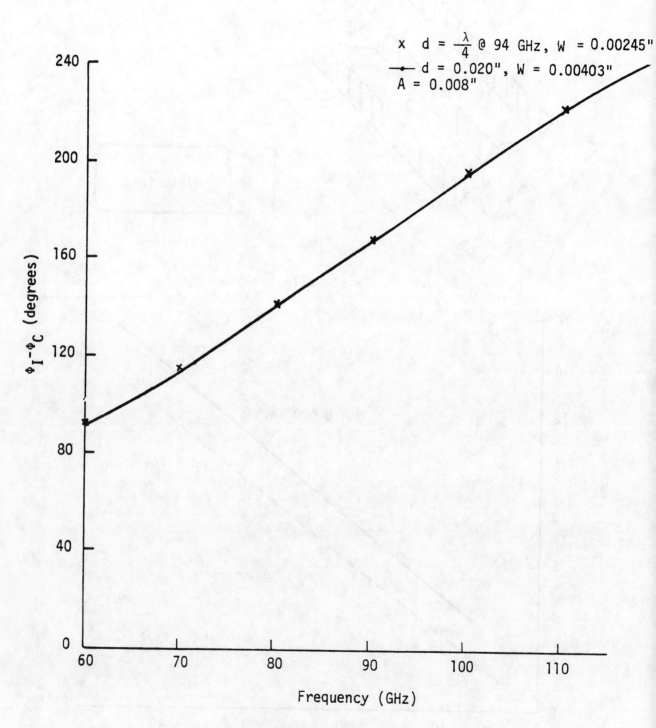

Figure 9. Phase Difference of Inductive and Capacitive Reflection Coefficients Versus Frequency.

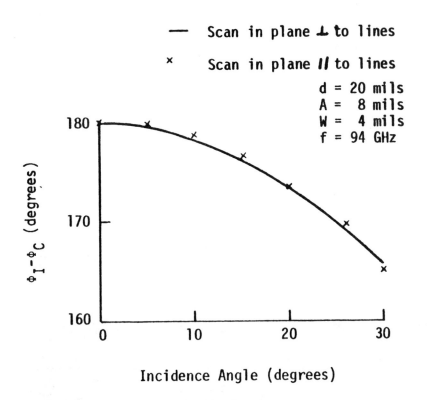

Figure 10. Phase Difference of Inductive and Capacitive Reflection Coefficients Versus Incidence Angle.

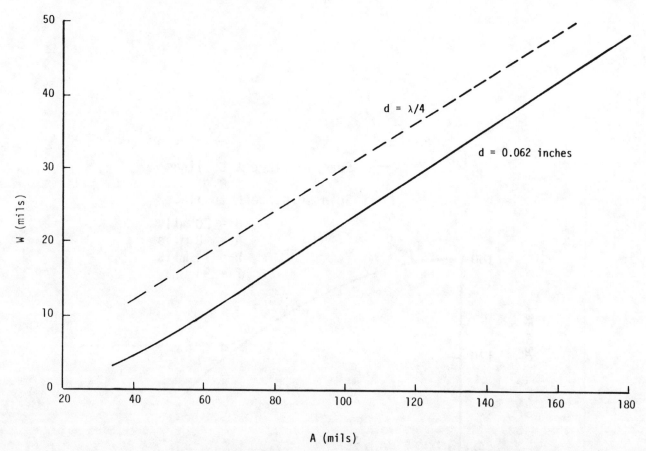

Figure 11. Line Width Versus Period for a 35 GHz Twist Reflector.

FIGURE 12. 35 GHz TWIST REFLECTOR AND HORN FEED

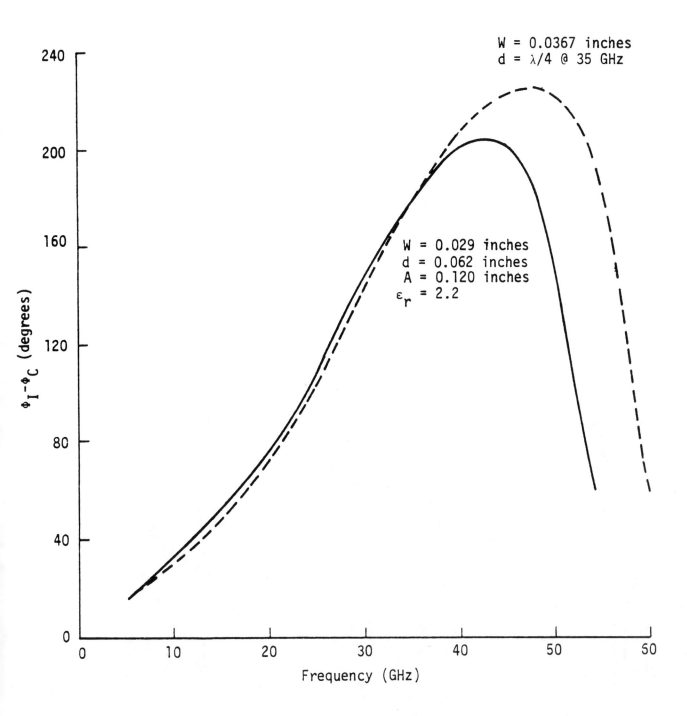

Figure 13. Phase Difference of Inductive and Capacitive Reflection Coefficients Versus Frequency.

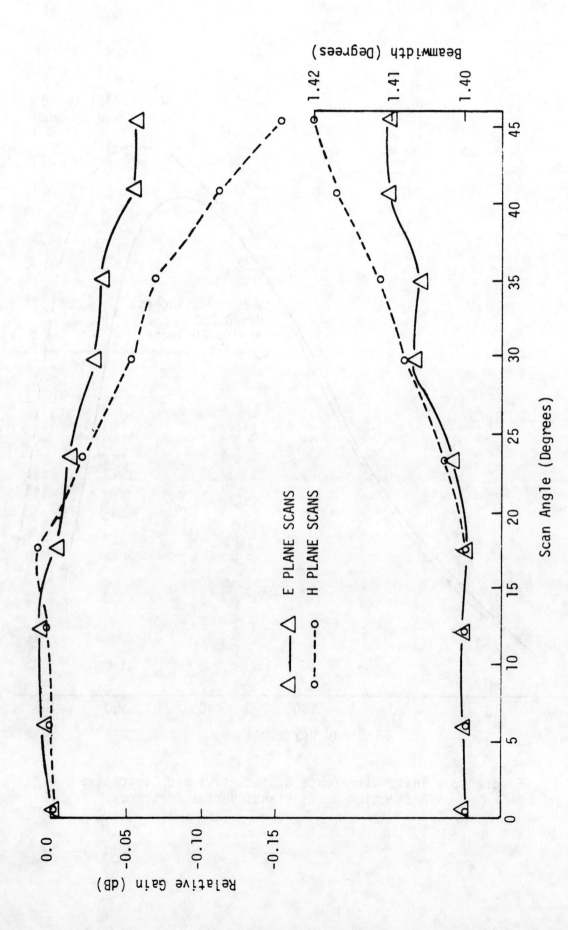

Figure 14. Calculated Transreflector Effects Gain and Beamwidth

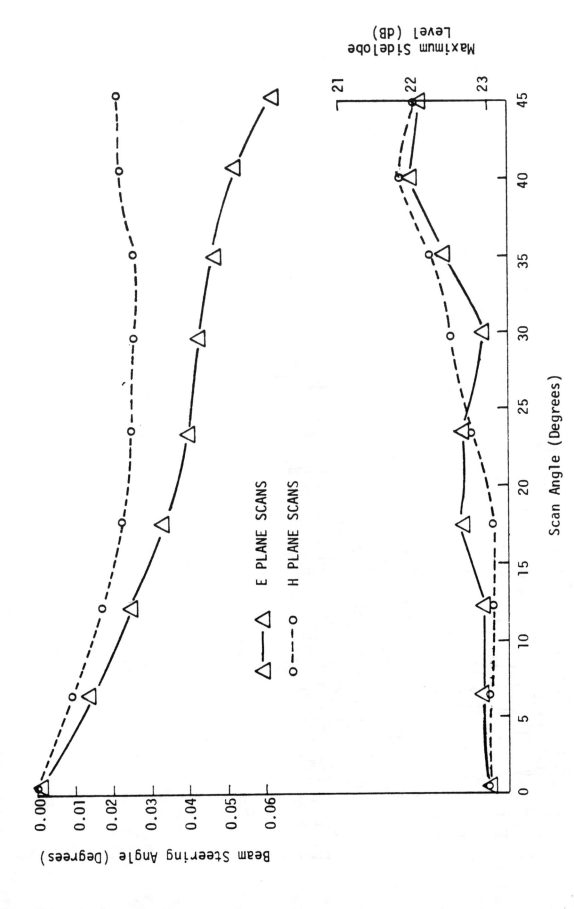

Figure 15. Calculated Transreflector Effects Beam Steering and Sidelobe Levels.

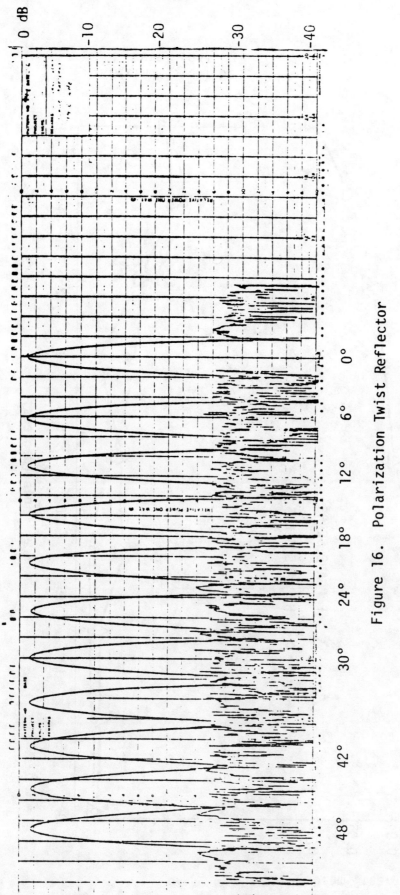

Figure 16. Polarization Twist Reflector

94.0 GHz
E-Plane Cuts
E-Plane Scans

Millimeter Wave Radar RF Sources and Components

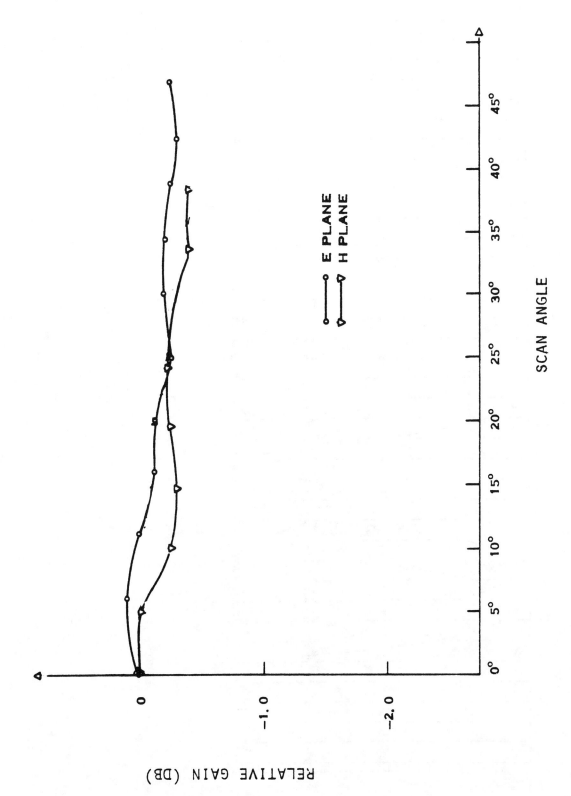

Figure 17. Measured Relative Gain (94 GHz)

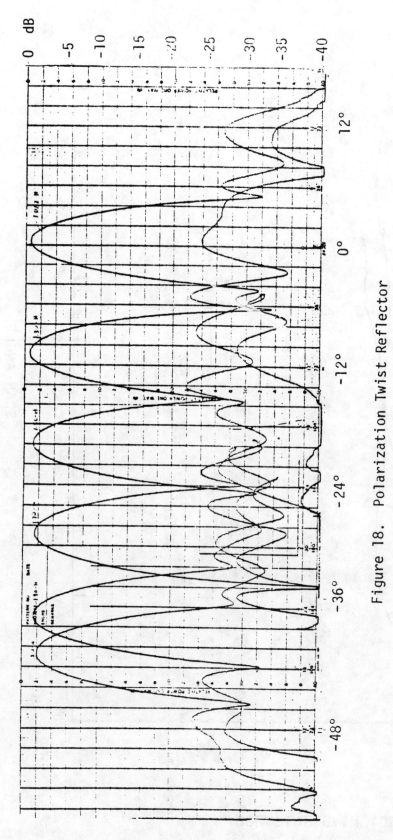

Figure 18. Polarization Twist Reflector

35.0 GHz
H-Plane Cuts
H-Plane Scans

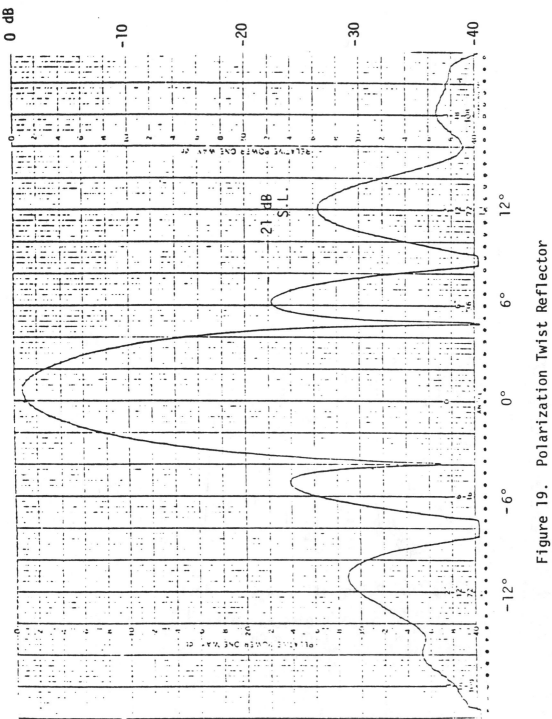

Figure 19. Polarization Twist Reflector
0° - Scan
E-Plane Cut
35.0 GHz

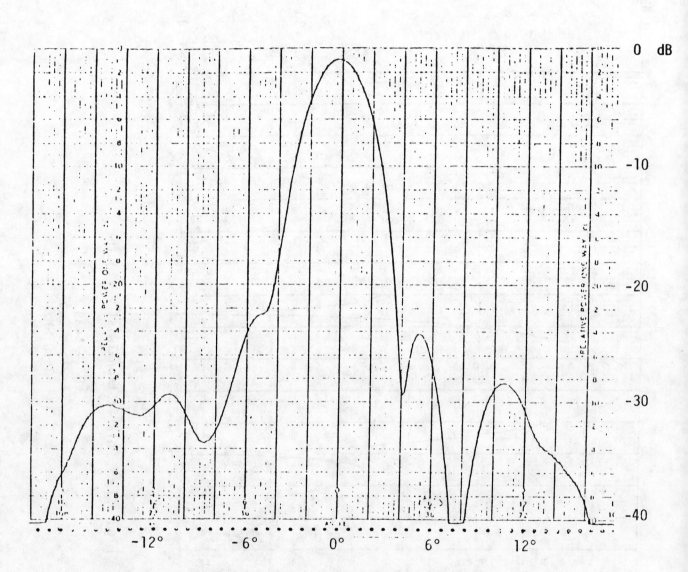

Figure 20. Polarization Twist Reflector

0° Scan
E-Plane Cut
39.89 GHz

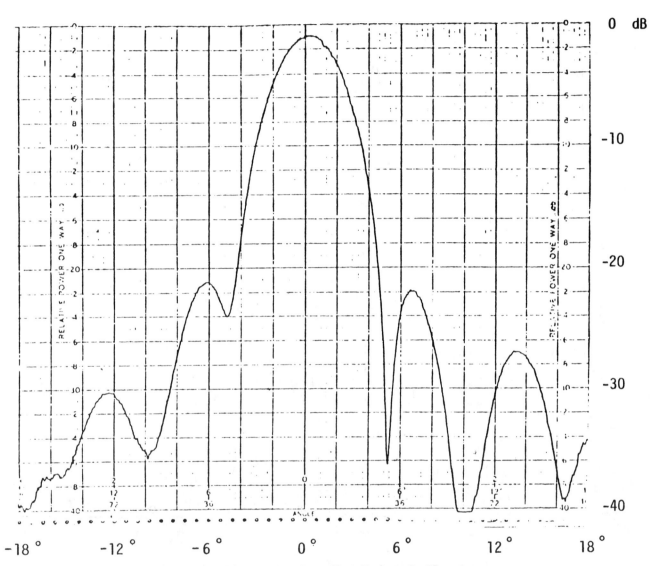

Figure 21. Polarization Twist Reflector

0° Scan
E-Plane Cut
30.21 GHz

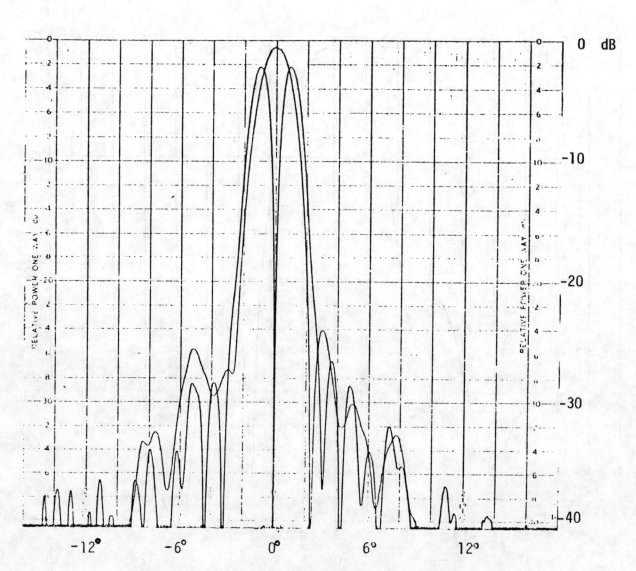

Figure 22. Polarization Twist Reflector

Sum and Difference
H-Plane Cuts
94 GHz

CHAPTER 4
K_a-BAND RADARS

Papers in this chapter will describe representative K_a-band radars. Although Tables 3, 4a, 5 and 6 in Paper 1.1 in Chapter 1 indicate that K_a-band radars have been built in production quantities for over three decades, most of the earlier K_a-band radars have been discarded. Generally, the literature on those radars is of historical interest only. Currently there are very few modern technology K_a-band radars in production, although a number have been developed recently. Accordingly, this chapter will describe several representative current developmental K_a-band radars. These will be presented in chronological order.

SYNOPSIS OF REPRINT PAPERS IN CHAPTER 4

4.1 "A Millimeter Wave Pseudorandum Coded Meteorological Radar" — Reid

K_a-band radar has been used for meteorological studies in the U.S.A. and the USSR for over three decades, as listed previously in Tables 4a and 4b of Paper 1.1 in Chapter 1. Generally these were pulsed radars. The radar described in this paper is a novel meteorological radar in that it provides three measurements: received signal amplitude, range, and Doppler frequency. Measurement of both range and Doppler frequency requires special consideration of transmitter waveform. Application of this radar is in precipitation drop size spectrum analysis and cloud studies.

After review of various considerations, a psuedo-random coded CW waveform radar is indicated as being the most suitable waveform for this purpose. Extensive description is given on methods for generation of pseudorandom noise (PN) codes. Codes, code generators, receivers and the full radar block diagram are included.

In design of the code for this radar, it is shown that carrier frequency is involved. Three wavelengths are considered: 30 mm (X-band), 8.6 mm (K_a-band) and 3.2 mm. The meteorological radar equation is used for both rainfall analysis and for cloud studies. Extensive curves are presented for expected system performance at the three wavelengths. The middle frequency is shown to be the most suitable for this application when components are considered.

This radar can also be used in a "passive" mode as a radiometer for radio astronomy, atmospheric attenuation and noise studies.

The extensive bibliography includes references both to "radar" papers, *e.g.* waveforms, and to radar meteorology.

4.2 "TRAKX: A Dual-Frequency Radar" — Cross, Howard, Lipka, Mays and Ornstein

This second K_a-band radar is a dual-frequency military range instrumentation radar which can track simultaneously at both X-band and K_a-band. The three-channel monopulse system can independently track in range, azimuth angle and elevation angle under operator control. A common Cassegrain dual-band antenna is employed. Magnetrons are used in both transmitters. A common modulator is used. A minicomputer provides real-time computation. The integral TV boresight system enables the operator to visually observe the target being tracked.

K_a-band is incorporated into this radar to permit improved tracking at low elevation angles, *i.e.* better multipath performance. Rainfall (4 mm/hr) will degrade K_a-band tracking range by a factor of 2.5 from a "normal" day.

4.3 "An FM/CW Radar With High Resolution in Range and Doppler; Application for Anti-Collision Radar For Vehicles" — Neininger

(discussed with Paper 4.4 to follow)

4.4 "MM Radar for Highway Collision Avoidance" — Wu and Tresselt

These two papers are discussed together, since they represent a European and a U.S. approach to a common radar application. The similarities and differences are very noteworthy. The first paper, by a West German author, was presented at Radar-77 in London in October 1977. The second paper, although published in November 1977, was presented at IEEE EASCON-77 in September 1977. A slightly different version of that paper is listed in the bibliography for this chapter.

Automobile collisions due to driver faults are major causes of fatalities, as shown by studies of automobile collisions in both West Germany and the U.S.A. This common problem led to different fundamental philosophies of the method of radar employment between these two papers.

The first paper uses driver visual and acoustic warning, while the second paper uses the radar for automatically applying the brakes and overriding the throttle of the automobile. The U.S. implementation also employs a sensor on the steering wheel to reduce false alarms on curves, as well as two front wheel sensors.

Both radars operate at K_a-band using Gunn oscillators, and both measure range, range rate, and Doppler sign (approach/recede). Each has a microprocessor. The German radar employs FM/CW, while the U.S. radar is a diplexed CW system with two Doppler amplifier/limiter chains in a homodyne receiver.

4.5 "Solid State mm-Wave Pulse-Compression Radar Sensor" — Winderman and Hulderman

Pulse compression techniques permit an effective increase of radar range resolution in peak power limited radars. The radar described in this paper weighs only 1.4 kg (including a 20 cm diameter Cassegrain antenna). Pulse compression ratio of the K_a-band radar is 16 dB (a factor of 40:1). A conical scan frequency of 200 Hz is used. Frequency modulation of 150 MHz of the transmitted 400 ns pulse provides an equivalent narrow pulse width of 10 ns. A surface acoustic wave (SAW) device [1] is used in the receiver. IF center frequency is 300 MHz. An unambiguous range of about 7.6 km results from the 20 KHz PRF. This PRF and the 400 ns compressed pulse width yield a duty cycle of 0.8% and an average power output of 8 mW from the IMPATT diode transmitter.

4.6 "SEATRACKS — A Millimeter Wave Radar Fire Control System" — Layman

"SEATRACKS" is an acronym standing for Small Elevation Angle TRACK and Surveillance System, an experimental K_a-band low-elevation surveillance and monopulse tracking radar. Intended shipboard applications include a fire control system for short range air defense systems, and as a navigation and surface search radar for detection of submarine periscopes.

The radar described in this paper is somewhat similar to the K_a-band portion of Paper 4.2 in this chapter. In the surveillance mode, the radar will use a digital constant false alarm rate (CFAR) receiver and an automatic target detector. A two bar raster scan pattern is used in surveillance. Method of transfer from surveillance to tracking is not described in the paper. This paper was presented at EASCON-78.

4.7 "A Millimeter Wave Radar for U.S. Army Helicopters in the 80's" — Holmes and Flick

This NAECON-80 paper offers a contrast to some other papers in this volume, and to reference 6 for Chapter 1. Studies summarized in this paper claim that K_a-band is the optimum frequency for this radar. Reference 6, Chapter 1, indicates that K_u-band is more cost-effective than K-band (and hence, by assumption, also K_a-band and 95 GHz) for U.S. Airport Surface Detection Equipment (ASDE). Paper 5.13 (Chapter 5) implies that 95 GHz is superior for a mini RPV- ground surveillance radar. Paper 5.15 (Chapter 5) indicates that K_u-band is superior to K_a-band and to 95 GHz for an *experimental* airborne radar.

Other contrasts are in the matters of target detection, clutter statistics, and differentiation between trees and targets. Consider Papers 2.A.4, 2.B.2 and 2.C.2 of Chapter 2, and Papers 5.13 and 5.16 of Chapter 5.

A more extensive radar detection computation procedure is contained in Reference 2 for this chapter.

REFERENCES FOR CHAPTER 4

[1] P.H. Carr, "The Application of SAW Technology to Radar ECCM," *Proc. DDRE Radar ECCM Symposium*, 1976. Reprinted in S.L. Johnston, *Radar Electronic Counter-Countermeasures*. Dedham, Mass.: Artech House, 1979.

[2] D.K. Barton, "Recommended Detection Computation Procedures," in D.K. Barton *Radars; Vol. 2: The Radar Equation*, Dedham, Mass.: Artech House, 1974.

BIBLIOGRAPHY TOPICS FOR CHAPTER 4

K_a-band Radars

SUPPLEMENTAL BIBLIOGRAPHY FOR CHAPTER 4

K_a-band Radars

Journal Articles

Tresselt, C.P. and Wu, Y.K., "Highway Collision Avoidance — A Potential Large-Scale Application of mm Radar," *IEEE EASCON-77 Conf. Rec.* Paper 16-7, pp. 16-7A thru G.

A Millimeter Wave Pseudorandum Coded Meteorological Radar

M.S. Reid

Abstract—This paper proposes and describes a meteorological radar for precipitation drop size spectrum analysis and cloud studies. The type of radar proposed is a millimeter wave pseudorandom coded CW radar.

This radar is analyzed on both precipitation and cloud physics studies. Its expected performance is discussed and pulse compression by pseudorandom phase coding is compared with and shown, for the given constraints, to be superior to the conventional pulsed meteorological radar. The paper describes codes for CW transmitters and pseudorandom codes in particular. It describes how the codes are generated, their structure, and their usefulness. It also describes code design and how codes are used to modulate a transmitter. The overall system design is discussed including carrier frequency, Doppler frequency spread, frequency stability, and data acquisition, recording, and reduction. Two possible locked modes of system operation are described as well as experimental work suitable for such a radar.

I. Introduction

THE AUTHOR'S project is the design, construction, and operation of a meteorological radar for precipitation drop size spectrum analysis. The radar should also be capable of contributing to a separate cloud physics study. With these given constraints this paper discusses the design of a radar system with particular reference to the transmitter signal characteristic, and proposes a millimeter (mm) wave pseudorandom coded continuous wave (CW) radar. This radar, which is presently under construction, is analyzed both on precipitation and cloud physics studies. Its expected performance is discussed, and pulse compression by pseudorandom phase coding is compared with and shown, for the given constraints, to be superior to conventional pulsed meteorological radars. The paper discusses codes for CW transmitters and pseudorandom codes in particular. It describes how the codes are generated, their structure and their usefulness. It also describes code design and how codes are used to modulate a transmitter.

II. Pulse Compression Techniques

In recent years much effort has been devoted to improving the resolving power of radars. It is now generally recognized that the matched-filter technique [1] yields the form of linear receiver characteristic that maximizes the output signal-to-noise ratio.

Manuscript received March 21, 1969.
The author is with the National Institute for Telecommunications Research, South African Council for Scientific and Industrial Research, Johannesburg, South Africa.

Pulse compression techniques [2], [3] are employed to obtain the resolution and accuracy of a short pulse while still retaining the detection capability of a long pulse. This is achieved by a special modulation within the radar pulse. This special modulation usually takes the form of a linear frequency modulation (FM), although a zigzag FM signal has also been studied [4]. The phase characteristic of the linear FM signal is quadratic. Another form of modulation has recently been investigated and this type restricts the CW carrier phase to certain discrete values as opposed to the quadratic continuum of the FM signal. This phase modulation (PM) is sometimes called "digital pulse compression." Although it is possible to phase modulate digitally by any amount between 0 and 2π, it is only steps of π that have practical application. Steps of phase modulation may then be controlled by a binary code. This type of radar is called a coded radar. Pulse compression by the use of sophisticated signals is limited by practical considerations to the above pulsed FM and CW PM types.

In general it may be said that the design of modern radar involves not only the conventional radar system design, but also the design of a suitable transmitter signal characteristic for the particular radar function [5].

III. Pulse and Coded CW Radars

In a conventional pulse radar both the pulse width and the pulse repetition frequency (PRF) are constrained [6], [7]. The pulse width must be shortened to increase range resolution, but the reciprocal of the PRF must be at least equal to the signal travel time from the maximum range otherwise range ambiguities arise. A serious limitation for certain radar situations is that short transmitter pulses require that the target Doppler frequency be determined by pulse-to-pulse phase-shift measurements instead of through an intrapulse frequency measurement. The unambiguous measurement of Doppler frequencies higher than half the PRF is then limited by the sampling theorem [8]. A further serious limitation for conventional radars is that the average transmitted power can be increased only by increasing peak power.

Thus a conventional pulse radar with short transmitter pulses and a low PRF is a good ranging radar but its average power, and therefore its detection capability,

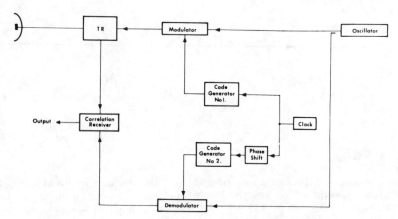

Fig. 1. Simplified block diagram of a coded radar.

is low, and it is poor with respect to unambiguous velocity measurement. Conventional CW radar, however, has satisfactory average power characteristics and satisfactory velocity measurement capability, but has poor range measurement and resolution. A coded CW radar can overcome these difficulties, as will be shown below, at the expense of search or "look" time.

Assume that in a conventional pulse radar the transmitted waveform can exist in either of two phase states, and that each pulse holds the same phase state over its width. Successive pulses are transmitted in either of the two phase states. The modulation states of the pulses could be controlled by a binary repetitive code or could be changed randomly. The latter case is complicated and has interest only for military radars. If the binary code is of period N (contains N digits) the radar's unambiguous range has been extended by a factor N, but the integration time must also be increased by the same factor.

Assume now that the binary code modulates the transmitter waveform *within* the pulse. In this case the radar's range resolution has been increased by a factor N, and the bandwidth has been increased by the same factor. In order to increase average transmitted power assume that these pulses are run together to form a CW waveform. This CW waveform will then contain contiguous pulses where the binary code determines the modulation state N times in each pulse.

CW radars using pseudorandom binary codes (see Section IV) as modulating functions thus have large maximum unambiguous ranges while still retaining fine range resolution capability. They are able to measure velocity unambiguously to high velocities with fine Doppler frequency resolution, and they have large average powers compared with the low average-to-peak power ratios of pulse radars. They have other characteristics, such as resistance to jamming, which are not of interest to the meteorologist and are ignored in this paper.

The disadvantages of CW coded radars are that their integration times are necessarily long and they are not suited to search-type or quick scanning applications. For the meteorologist interested in rainfall quantity and rainfall rate measurements over large areas the conventional scanning radar is probably the best choice, but in applications where slow scanning is advisable, such as a drop size spectrometer experiment or a cloud and atmosphere study, the coded radar is optimum.

IV. CODES, CODE GENERATORS, AND RECEIVERS

Fig. 1 shows a simplified block diagram of a coded radar. The clock set at the required code frequency drives the code generators, one of which, the reference code generator no. 2, has a variable phase shift, as shown. The output from code generator no. 1 drives the phase modulator which superimposes the code on the oscillator output. The coded CW is then transmitted through the radar TR system in the normal way. The reference code generator output, shifted in phase, drives the receiver demodulator which is fed with a small portion of the transmitter output. Each rotation of the continuously variable phase shifter delays the code output from the reference code generator by one code digit with respect to the output from code generator no. 1.

The receiver computes the correlation function between its two inputs. When the reference code has the correct delay corresponding to the range of a single target, the output from the correlation receiver is the autocorrelation function of the code. Each code digit of delay between the code generators corresponds to a range gate. If there is no target in the range gate corresponding to the setting of the phase shifter, the receiver's output is that part of the complete code autocorrelation function that lies between the peaks and that corresponds to the reference code delay.

It may be seen from the above that the most desirable autocorrelation functions are those that have high, narrow peaks and are close to zero elsewhere. Codes should be chosen accordingly. Fine range resolution can only be achieved with narrow peaks, and the separation between peaks should correspond to a target range greater than the maximum range of interest.

K_a-band Radars

The autocorrelation function $\psi(\tau)$ of a periodic time function $f(t)$ over a time period T is defined [9], [10], as

$$\psi(\tau) = \lim_{T \to \infty} \frac{1}{T} \int_0^T f(t) f(t + \tau) dt. \qquad (1)$$

It can be shown that the transformation of $f(t)$ into $\psi(\tau)$ is equivalent to taking the Fourier transform of the time function to yield a frequency function, which is squared to obtain a power function and then taking the Fourier transform of this power spectrum. $\psi(\tau)$ at $\tau = 0$ is therefore a measure of the power in the signal. A large variety of waveforms have been studied and their autocorrelation functions calculated [3], [11], [12]. It has been found that pulsed sine waves and linear FM systems ("chirp" signals) do not have optimum autocorrelation functions [3].

For a single target and Gaussian noise it has been shown that no receiver can extract more information from any received signal than the matched-filter correlation receiver [13]. Even though the optimum receiver extracts all possible information from the received signal, some transmitted waveforms have more suitable characteristics than others. One aspect of the suitability of a waveform, autocorrelation, has been discussed above. The general measure of the suitability of a waveform is Woodward's ambiguity function [13], [14]. The autocorrelation function is the zero-velocity cross section of the ambiguity function, and is, therefore, directly related to range accuracy and resolution. Studies of the radar ambiguity function have shown that a certain class of binary codes forms optimum modulating functions [15], [16]. These codes are called pseudorandom or pseudonoise (PN) codes [17], [18]. They are formed by a sequence of ones and zeros or ones and minus ones. They are periodic but have certain randomness characteristics. Fig. 2 shows a number of time waveforms with their autocorrelation functions. It will be seen that the PN code in (f) has the best autocorrelation function in figure because it has the highest, narrowest peaks and is close to zero elsewhere.

A sequence is called random if its method of generation has no characteristics or initial conditions that cause departures from true randomness. A pseudorandom sequence is one which satisfies a certain set of criteria derived from the expectation values of a random sequence for all possible methods of generating the sequence. Periodic PN sequences thus have properties close to those that true random sequences have only in a long time average sense. A true random binary sequence has approximately the same number of ones as zeros (or minus ones), whereas in a PN sequence (linear maximal sequence discussed below) the number of ones per period is always one more than the number of zeros. About one half the runs of consecutive states of the same character in the random sequence have length one, one fourth have length two, one eighth have length three, etc. Half the runs in a PN sequence have length

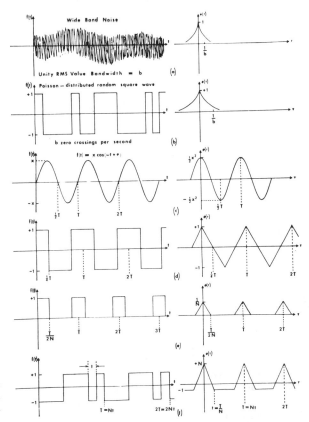

Fig. 2. Some time functions and their autocorrelation functions.

one, one fourth have length two, etc. It may be seen from the foregoing that PN sequences have a certain structure as well as certain randomness characteristics. The fact that these sequences have an underlying structure enables one to predict the properties of the resulting coded waveforms formed out of these sequences as modulating functions. The algebra of the structural properties of binary sequences has been extensively studied and is now well known [12], [19], [20]. It is possible to predict full characteristics of waveforms encoded by binary sequences and so design a suitable transmitter signal for each radar application [21], [22].

It is possible to generate a PN sequence in a relatively simple way by using shift registers. These are available today in integrated circuit form which adds to the usefulness of these sequences in coded waveforms. A shift register is basically a number of binary storage elements or a series of simple bistable circuits. The shift register has the capability of moving the digit (one of the two states of each element) stored in one element to the adjacent element by means of a shift pulse or clock pulse which is applied to each elemental circuit. A shift register generator (SRG) consists of a basic shift register to which modulo-2 adders have been added [19]. The outputs from the SRG elements form the inputs to the modulo-2 adders and the adder outputs are fed back to

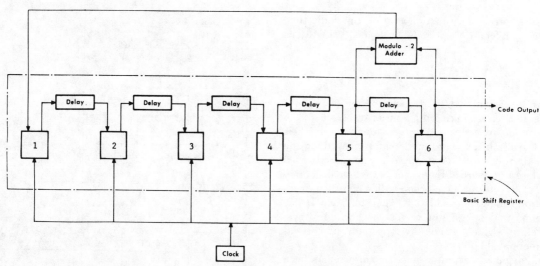

Fig. 3. A simple six-stage SRG.

some other stage of the SRG so as to form a single or multiple closed loop. In this way an SRG can produce an output much longer than the number of elements in the SRG. Fig. 3 shows a simple six-stage SRG. With each clock pulse each bistable element (1–6) is returned to zero and then given the state of the previous adjacent element. One output digit is taken from the last stage with each clock pulse and this forms the binary sequence. The feedback loop should be designed to yield the longest period output sequence (maximal length sequence) for any given number of SRG stages. This is because maximal length sequences have the optimum autocorrelation functions [23]. It may be shown that if an SRG has n stages the maximal length sequence has a period of $N = 2^n - 1$ digits [3], [7], [13], [14].

V. Code Design for Meteorological Radar

Carrier Frequency

The carrier wavelength for a precipitation drop size spectrometer should not be too short otherwise the attenuation between the antenna and the range gate becomes inconveniently large, particularly for the longer ranges. On the other hand, a short wavelength is required if the radar is to contribute significantly to cloud and atmosphere studies [24], [25], [26]. Balancing these considerations with possible antenna sizes, system complexity, and difficulty with and cost of components reduces the range of feasible carrier wavelengths to about 30–3 mm. Three wavelengths are considered in the design: 30, 8.6, and 3.2 (mm).

Code Design

If the maximum velocity to be detected is v m/s then the maximum Doppler frequency to be recorded is

$$f_d = \frac{2v}{\lambda} \qquad (2)$$

where f_d is in Hz and λ in m. If v is 20 m/s, then f_d is 1.33, 4.65, 12.5 (kHz) for wavelengths of 30, 8.6, and 3.2 (mm), respectively.

As the radar of Fig. 1 uses part of the transmitter output as local oscillator, the system mixes down to dc and thus loses sign resolution on the detected Doppler frequencies. As Doppler sign resolution should be retained a different local oscillator system should be used (see Section VI) and the range of Doppler frequencies to be recorded is twice those mentioned above, say, 3, 10, and 25 (kHz), respectively.

Assume that the SRG has n stages, a range resolution of r_r meters and a maximum unambiguous range of r_m meters. Then

$$\text{Code rate} = 2 f_d \text{ (Hz)} \qquad (3)$$

$$\text{Clock rate} = f_c = 2 f_d (2^n - 1) \text{ (Hz)} \qquad (4)$$

$$r_r = \tfrac{1}{2} c / f_c \text{ (m)} \qquad (5)$$

$$r_m = \tfrac{1}{2} c / 2 f_d \text{ (m)} \qquad (6)$$

where $\tfrac{1}{2} c$ (m/s) is half the velocity of light (radar range).

Tables I–III were drawn up from the above equations to show the range of code possibilities. In each table the maximum velocity to be detected is assumed to be 20 m/s. The maximum unambiguous range is independent of the number of stages in the SRG because r_m is a function only of the code rate and therefore depends only on the carrier wavelength and the maximum spread of Doppler frequencies to be recorded. In a practical radar the maximum unambiguous range will be somewhat greater than that shown in the tables, particularly for the longer wavelengths. This is because it is easy to distinguish between ranges of, say, 55 and 5 km in the $\lambda = 30$ mm case, for example. The tables show clearly the range resolutions that can be achieved for various maximal length SRG's.

TABLE I
CLOCK FREQUENCIES AND RANGE RESOLUTIONS FOR VARIOUS SRG FOR A CARRIER WAVELENGTH OF 30 mm AND MAX. UNAMBIGUOUS RANGE = 50 km

n	N	f_c (MHz)	r_r (m)
4	15	0.045	3333
6	63	0.19	794
8	255	0.77	196
10	1023	3.07	49
15	2767	98.30	1.5

TABLE II
CLOCK FREQUENCIES AND RANGE RESOLUTIONS FOR VARIOUS SRG FOR A CARRIER WAVELENGTH OF 8.6 mm AND MAX. UNAMBIGUOUS RANGE = 15 km

n	N	f_c (MHz)	r_r (m)
4	15	0.15	1000
6	63	0.63	238
8	255	2.55	59
10	1023	10.23	14.7
15	32767	327.7	0.46

TABLE III
CLOCK FREQUENCIES AND RANGE RESOLUTIONS FOR VARIOUS SRG FOR A CARRIER WAVELENGTH OF 3.2 mm AND Max. UNAMBIGUOUS RANGE = 6 km

n	N	f_c (MHz)	r_r (m)
4	15	0.38	400
6	63	1.58	95
8	255	6.38	23.5
10	1023	25.58	5.9
15	32767	819.18	0.2

VI. THE METEOROLOGICAL RADAR EQUATION

The meteorological radar equation is now examined for each of the three wavelengths under consideration. This will yield overall system design possibilities and give an indication of what type of cloud and atmosphere experiment will be possible with each wavelength.

The meteorological radar equation [27], [28], [29] is

$$\frac{\overline{P}_r}{P_t} = \frac{\pi^5}{48}\left(\frac{\theta\phi \cdot h \cdot A_p^2}{\lambda^6}\right) F/k/^2 \cdot \frac{Z}{r_m^2} \quad (7)$$

where

$\overline{P}_r$ = ensemble averaged received power
P_t = transmitted power
θ, ϕ = conventional horizontal and vertical beamwidth (radians)
h = depth of volume illuminated by one code word (m)
A_p = antenna aperture area (m²)
λ = carrier wavelength (m)
F = a factor to take account of the nonuniform illumination of the scatterers in the beam as well as aperture efficiency
$k = (m^2-1)/(m^2+2)$, where m = complex index of refraction of scatterers
Z = reflectivity factor, where $Z = 200 R^{1.6}$ (mm⁶/m³), where R = rainfall rate (mm/h)
r_m = maximum range (m).

Rainfall Analysis

1) $\lambda = 30$ mm.

A consideration of Table I shows that a SRG of 10 stages is adequate. A 10-stage generator is practical to build and operate. If $n=10$, then $N=1023$, the range resolution r_r is 49 m, $f_d = 5$ kHz, and $fc = 3$ MHz.

As the greatest range of interest for precipitation analysis (antenna at zenith) is probably about 20 km, r_m is set at this figure.

$$r_m = 2 \times 10^7 \text{ mm.}$$

Equation (7) then reduces to

$$\frac{\overline{P}_r}{P_t} = 3.26 \times 10^{-22}(d^2R^{1.6}) \quad (8)$$

where d = antenna aperture diameter.

With the data recording and data reduction system described below, the system bandwidth (BW) can easily be reduced to 50 kHz. With this system the BW is in effect limited only by loss in detection capability as the computer data reduction technique can theoretically simulate any BW. 50 kHz is therefore a conservative figure. With a BW of 50 kHz and a system noise figure of 8 dB, the minimum detectable signal is 1.25×10^{-15} W. Then (8) may be written as

$$P_t = \frac{0.345 \times 10^8}{d^2R^{1.6}} \text{ (W)} \quad (9)$$

for a signal-to-noise ratio of 10 dB. Another way of writing (8) is

$$S/N = 2.9 \times 10^{-7}(d^2R^{1.6}P_t) \quad (10)$$

where

d = antenna aperture diameter (mm)
R = rainfall rate (mm/h)
P_t = transmitter power (W)
S/N = signal-to-noise ratio (ratio).

2) $\lambda = 8.6$ mm.

From Table II it is seen that an SRG with $n = 10$ is also suitable for this carrier wavelength. Hence

$r_r = 15$ m
$f_c = 10$ MHz
$h = 3 \times 10^4$ mm,

and with the other parameters the same as before the radar equation reduces to

$$\frac{\overline{P}_r}{P_t} = 1.45 \times 10^{-20}(d^2R^{1.6}). \quad (11)$$

For a system BW of 50 kHz and a system noise figure of 14 dB, the minimum detectable signal is 5×10^{-15} W. Equation (11) then reduces to

$$P_t = \frac{3.4 \times 10^6}{d^2 R^{1.6}} \text{ (W)} \quad (12)$$

for a signal-to-noise ratio of 10 dB, or

$$S/N = 0.3 \times 10^{-5}(d^2 R^{1.6} P_t) \quad \text{(ratio)}. \quad (13)$$

3) $\lambda = 3.2$ mm.

The parameters chosen for this case are $n = 10$, 20 km maximum range of interest, $f_c = 25.6$ MHz, system noise figure = 17 dB. With the other parameters the same as before the radar equation reduces to

$$P_t = \frac{3 \times 10^5}{d^2 R^{1.6}} \text{ (W)} \quad (14)$$

for a signal-to-noise ratio of 10 dB, or

$$S/N = 0.33 \times 10^{-4}(d^2 R^{1.6} P_t) \quad \text{(ratio)}. \quad (15)$$

Equations (10), (13), and (15) are plotted in Figs. 4, 5, and 6, respectively. The parameters are antenna diameter and transmitter power, and all curves are plotted for a range of 20 km. Before these curves can be used to estimate system performance, attenuation by rainfall between antenna and the range gate should be taken into account.

Using a drop-size distribution due to Burrows and Attwood [30], the absorption cross sections of rain drops have been calculated [31] and used to compute attenuation for various rates of precipitation. These attenuations have been computed for a temperature of 18°C and are therefore probably pessimistic for the higher altitudes. Figs. 7–9 show the degradation in dB of the signal-to-noise ratio for various rainfall rates for the three frequencies under consideration. Table IV, from Burrows and Attwood [30] and Saxton [31], shows the values of attenuations used to compute these figures.

Cloud Studies

The radar equation (7) is used again to estimate system performance on cloud studies. In this case Z is given by

$$Z = N'D^6 \quad (16)$$

where N' is the number of particles per cubic meter. Assume a liquid water content in the cloud of M g/m³, and for the purposes of calculation, assume a uniform particle size in the range gate of D mm diameter. Then

$$N' = \frac{6 \times 10^3}{\pi}\left(\frac{M}{D^3}\right) \quad (17)$$

and

$$Z = \frac{6 \times 10^{-6}}{\pi}(MD^3) \quad (18)$$

where the reflectivity Z is in units of mm⁶/mm³.

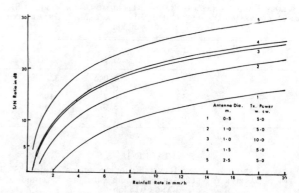

Fig. 4. Graphs of system signal-to-noise ratio versus rainfall rate for a carrier wavelength of 30 mm.

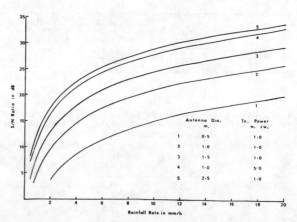

Fig. 5. Graphs of system signal-to-noise ratio versus rainfall rate for a carrier wavelength of 8.6 mm.

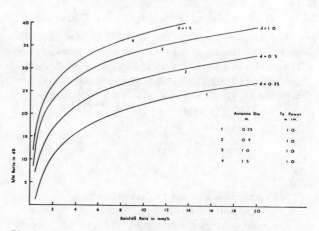

Fig. 6. Graphs of system signal-to-noise ratio versus rainfall rate for a carrier wavelength of 3.2 mm.

If the range gate is set at 5 km and using (7) and (18) and the same minimum detectable signals as before the radar equation reduces to

$$S/N = 4.4 \times 10^{-8}(d^2 M D^3 P_t) \quad \text{for } \lambda = 30 \text{ mm} \quad (19)$$

$$S/N = 4.9 \times 10^{-7}(d^2 M D^3 P_t) \quad \text{for } \lambda = 8.6 \text{ mm} \quad (20)$$

$$S/N = 5 \times 10^{-6}(d^2 M D^3 P_t) \quad \text{for } \lambda = 3.2 \text{ mm} \quad (21)$$

K_a-band Radars

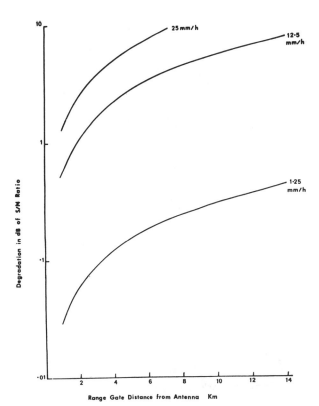

Fig. 7. Degradation in signal-to-noise ratio versus range gate distance from antenna for a carrier wavelength of 30 mm.

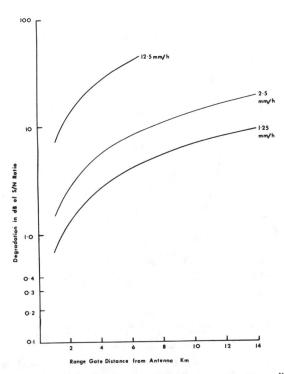

Fig. 8. Degradation in signal-to-noise ratio versus range gate distance from antenna for a carrier wavelength of 8.6 mm.

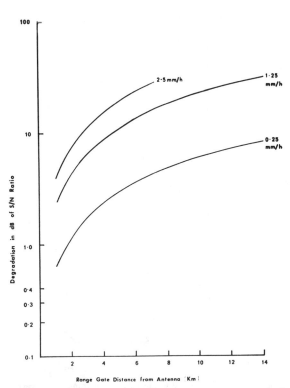

Fig. 9. Degradation in signal-to-noise ratio versus range gate distance from antenna for a carrier wavelength of 3.2 mm.

TABLE IV
ATMOSPHERIC ATTENUATIONS FOR VARIOUS RAINFALL RATES

Rainfall Rate mm/hr	Attenuation in dB/km		
	$\lambda=30$ mm	$\lambda=8.6$ mm	$\lambda=3.2$ mm
0.25			0.305
1.25	0.016	0.340	1.150
2.5		0.698	1.980
12.5	0.285	3.420	
25	0.656		

where

S/N = signal-to-noise ratio (ratio)
d = antenna aperture diameter (m)
M = liquid water content (g/m³)
D = particle diameter (microns)
P_t = transmitter power (watts).

Equations (19)–(21) are drawn in Fig. 10, which shows the ratio $S/N/P_t$ (in units of W^{-1}) plotted against MD^3 for a 1-meter diameter antenna. Two other scales show S/N ratio in dB for a 1- and a 10-watt transmitter.

VII. SYSTEM DESIGN

The carrier frequency was chosen by a study of the figures from the point of view of optimum data generation. This has to be balanced with other factors such as cost, the difficulty of working with higher frequency components and their availability. Other considerations

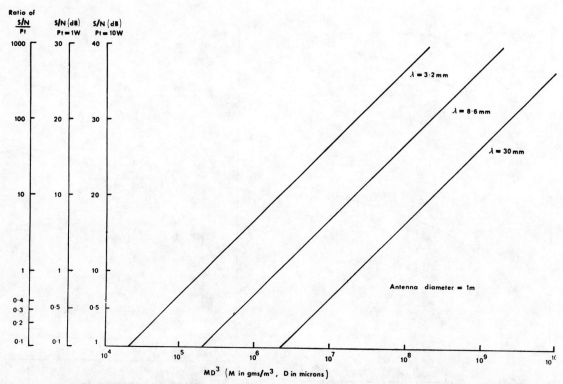

Fig. 10. Signal-to-noise ratio and transmitter power versus MD^3 for carrier wavelengths of 30, 8.6, and 3.2 mm.

are the frequency stability required for recording the wide range of Doppler frequencies proposed in this paper, and the availability of high power CW oscillators or amplifiers at the higher frequencies.

Having considered these matters a carrier wavelength of 8.6 mm was chosen. CW traveling wave tube amplifiers are commercially available today which have a gain of 60 dB and an output of 1 kW CW at 8.6 mm [32]. Figs. 4–10 show the exciting possibilities for interesting experiments on cloud base and top measurements, cloud physics including the possibility of cloud turbulence measurement [33], [34], as well as many other meteorological experiments suitable for a coded radar. It must be remembered that the design figures are conservative and that in practice the system noise figure could be a little better than the assumed value of 14 dB. With the proposed data acquisition and reduction system a bandwidth of 10 kHz should be practical. This means that the Figs. are probably pessimistic by about 6 dB.

Fig. 11 is a graph, from (20), of liquid water content M, in g/m^3, against minimum detectable droplet diameter in microns, for a carrier wavelength of 8.6 mm, an antenna diameter of 1.5 m, and a signal-to-noise ratio of 6 dB. The parameter is transmitter power, and three curves have been drawn for 10, 100, and 1000 watts. It is clear that for high transmitter powers, the minimum detectable droplet diameter is probably limited by the physics of the cloud rather than the sensitivity of the radar.

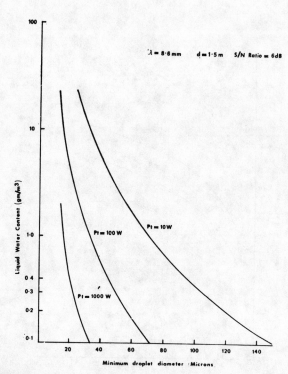

Fig. 11. Graphs of liquid water content versus minimum cloud droplet diameter for various transmitter power levels.

K_a-band Radars

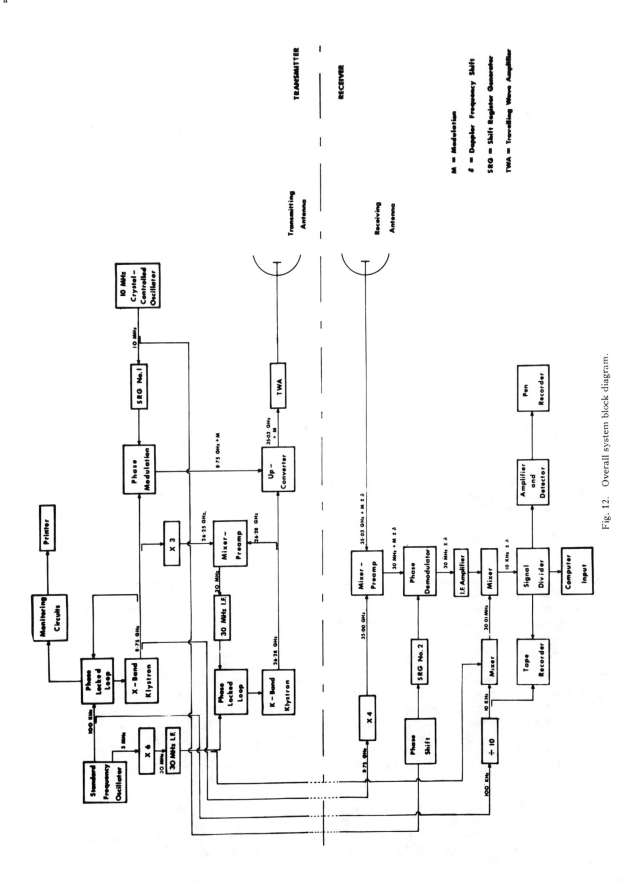

Fig. 12. Overall system block diagram.

Having chosen a carrier frequency and code parameters the detailed system design remains. Fig. 12 shows an overall block diagram of a possible system. In order to avoid the complication of a TR system at 35 GHz a bistatic radar design was chosen. This means that the receiver can be used as a radiometer for atmospheric attenuation and noise studies [35], [36] as well as radio astronomy when the equipment is not required for the meteorological program.

Equipment for calibrating the radar and for operation purely in the radiometer mode is not shown in the block diagram. This equipment consists of waveguide switches, thermal calibration loads, etc. [37].

The antennas are identical, approximately 1.5 m in diameter and mounted on modified searchlight yokes and pedestals. They are operated in the cassegrain mode. The physical separation between the antennas must be carefully considered. The further apart the antennas are the easier the shielding is between them from the point of view of transmitter leakage and sidelobes. Furthermore they cannot be too close otherwise their beams will intersect at a close vertical distance above the antennas and this may give a false echo recording if the range gate is near the maximum range and strong reflectors are present close in, in both beams. On the other hand the further apart the antennas are the greater is the problem of bringing reference signals from one to the other. This problem is particularly acute if a reference signal at the carrier frequency has to be taken across. In the system shown in Fig. 12 the highest reference signal is at X-band. The separation between antennas was chosen as 25 to 30 feet (8.5 m). The loss in this length of waveguide at X-band is negligible.

SRG's in integrated circuit form are cheap, compact and require simple, low voltage dc power supplies. As a result it is quite feasible to add a further two SRG's to both the transmitter and receiver. The radar could be switched, either manually or automatically, between these three SRG's. The number of stages could be increased in one SRG to optimize it for very short ranges (increased range resolution), and reduced in the other for maximum ranges. Without adding much complexity to the system one or more clock oscillators could be added either to the one-SRG system or to the multiple-SRG system in order to optimize the code design for the particular spread of Doppler frequencies (and velocity resolutions) required in any given experiment.

Another important consideration of the system design is overall frequency stability. This is related to the required Doppler frequency resolution. This problem is eased by the fact that the required frequency stability need be held only for a few code words. A commercial standard frequency oscillator with a short term stability of one part in 10^{11} was considered suitable as the prime system reference. This device has two outputs, 5 MHz and 100 kHz, as shown in Fig. 12.

The 100-kHz reference signal is used to phase lock the X-band klystron through a suitable locking loop with monitoring circuits. Part of this klystron's output is taken across to the receiver as a reference signal. Another portion of its output is tripled in frequency and then forms the locking signal for the K-band klystron. The major portion of its output is biphase modulated by the code generator. This X-band coded signal is then converted to the transmitter output frequency by a parametric up-converter. The up-converter is pumped by the K-band klystron which is phase-locked to the X-band reference signal, as shown.

This transmitter design avoids the problem of biphase modulating the carrier frequency at 35 GHz directly. Fast and efficient solid-stage switches are readily and cheaply available at X-band. On the other hand a suitable up-converter is neither readily nor cheaply available. This device is presently under development. This is not a major development program if one uses modern high cutoff frequency varactors. Furthermore, the efficiency of the up-converter is not critical because 600 mW are available from both the X-band and the K-band klystrons and the required up-converter output is only about 1 mW for maximum output from the traveling wave amplifier.

The X-band reference signal which is taken across to the receiver is quadrupled and then used directly as the local oscillator. The received echo signal is mixed down to 30 MHz where the demodulation is effected. After amplification the signal is mixed down to 10 kHz. 10 kHz is chosen as the recording frequency because the required Doppler spread is both plus and minus 5 kHz and in this way the sign resolution is retained.

The output signal is fed into a buffer computer where it is converted to digital form and blocked onto magnetic tape. These data tapes are analysed off-line in a conventional computer for power and frequency spectrums, etc. The 10-kHz output carrier is also amplified, detected and displayed on a pen recorder for on-line monitoring. The output signal is also recorded on a high quality tape-recorder together with a 10-kHz reference signal, to form a back-up to the computer data acquisition system.

This receiver design has the advantage that the code demodulation is carried out at the IF frequency. As the data acquisition computer has sixteen input channels, receiver and transmitter can both be monitored as well as antenna positions, ambient temperature, time, etc.

Another advantage of this design is that the conventional bank of Doppler filters is no longer necessary. The direct recording of the output signal in digital form means that filters of any suitable bandwidth can be simulated in the off-line data reduction computer. As the Doppler spectrum is determined by a mathematical transformation of the data, the analysis can be repeated on the same data for various filter resolutions.

Several variations on the block diagram are possible; for example, the quadrupler in the receiver could be replaced with a tripler and an up-converter identical to that in the transmitter. It might be possible to derive the 30-MHz reference signals in the transmitter and the receiver from local, separate, crystal controlled oscillators. The frequency stability required in the final receiver down-conversion to 10 kHz is not so stringent as it is at the higher frequencies. The design also allows a considerable freedom to adjust frequencies around those given in Fig. 12.

VIII. Conclusions

It has been shown that for those physical situations where a coded CW radar is suitable, the advantages to be gained by using such a radar in place of the conventional pulse radar are greater than simply those of cost, system stability, low power levels, ease of operation, etc. The improvements in data acquisition, sensitivity and resolution as shown in the figures are probably of prime importance. For some experimental situations it has been shown that the limiting factors are meteorological rather than instrumental.

The proposed design is capable of operation in two modes. Consider a precipitation analysis experiment. In one mode the system is locked to the particular range-gate which corresponds to the phase lag of SRG no. 2 with respect to SRG no. 1. The radar will continue to "look" at this particular range until the relative difference between SRG's 1 and 2 is changed. On the other hand the system can be operated in a loop phase-locked to the received echo. In this mode the radar will lock onto an ensemble of raindrops at a preset range and the phase difference between the SRG's will "track" this ensemble to yield a continuous history.

A wide range of experimental work is open to a meteorological radar of this type. Quite apart from rainfall analysis there is a variety of atmospheric studies which could be undertaken including, with the higher power outputs, the possibility of detecting clear air turbulence and large atmospheric aerosols. Cloud studies are important and Figs. 10 and 11 show that virtually all clouds of interest should be detectable. It would be of great interest to examine a cloud before and while precipitation starts. With this type of work in mind it must be noted that the system is such that the antennas are not restricted to a fixed zenith angle of 0°

Acknowledgment

The author wishes to thank M. C. Hodson for many hours of useful discussion, and P. J. Cabion, without whose help the project would never have been begun.

References

[1] *IRE Trans. Information Theory* (Special Issue on Matched Filters), vol. IT-6, June 1960.
[2] C. E. Cook, "Pulse compression," *Proc. IRE*, vol. 48, pp. 310–316, March 1960.
[3] J. R. Klauder, A. C. Price, S. Darlington, and W. J. Albersheim, "The theory and design of chirp radars," *Bell Sys. Tech. J.*, vol. 39, no. 4, pp. 745–808, July 1960.
[4] A. W. Rihaczek and R. L. Mitchell, "Design of zigzag FM signals," *IEEE Trans. Aerospace and Electronic Systems*, vol. AES-4, pp. 680–692, September 1968.
[5] D. E. Vakman, *Sophisticated Signals and the Uncertainty Principle in Radar*, E. Jacobs, Ed., K. N. Trirogoff, Translator. New York: Springer, 1968.
[6] M. I. Skolnik, *Introduction to Radar Systems*. New York: McGraw-Hill, 1962.
[7] *Modern Radar*, R. S. Berkowitz, Ed. New York: Wiley, 1965.
[8] M. Schwartz, *Information Transmission, Modulation, and Noise*. New York: McGraw-Hill, 1959, ch. 4.
[9] *Ibid*, ch. 7.
[10] Y. W. Lee, *Statistical Theory of Communication*. New York: Wiley, 1967, ch. 2.
[11] R. M. Lerner, "Signals with uniform ambiguity functions," *IRE Conv. Rec.*, pt. 4, p. 27, 1958.
[12] W. W. Peterson, *Error Correcting Codes*. Cambridge, Mass.: M.I.T. Press, 1961.
[13] M. H. Carpentier, *Radar, New Concepts*. New York: Gordon and Breach, 1968.
[14] P. M. Woodward, *Probability and Information Theory with Applications to Radar*. London: Pergamon, 1953.
[15] W. M. Siebert, "A radar detection philosophy," *IRE Trans. Information Theory*, vol. IT-2, pp. 204–221, September 1956.
[16] J. R. Klauder, "The design of radar signals having both high range resolution and high velocity resolution," *Bell Syst. Tech. J.*, vol. 39, p. 809, July 1960.
[17] P. L. Lindley, "The PN technique of ranging as applied in the ranging subsystem mark 1," Jet Propulsion Laboratory, Pasadena, Calif., Tech. Rept. 32-811, November 1965.
[18] L. Baumert, *et al.*, "Coding theory and its applications to communications systems," Jet Propulsion Laboratory, Pasadena, California, Tech. Rept. 32-67, March 1961.
[19] T. G. Birdsall and M. P. Ristenbatt, "Introduction to linear shift-register generated sequences," Cooley Electronics Laboratory, University of Michigan, Ann Arbor, Tech. Rept. 90, October 1958.
[20] A. A. Albert, *Fundamental Concepts of Higher Algebra*. Chicago, Ill.: University of Chicago Press, 1956.
[21] B. Elspas, "The theory of autonomous linear sequential networks," *IRE Trans. Circuit Theory*, vol. CT-6, pp. 45–60, March 1959.
[22] ——, "A note on p-nary adjacent-error-correcting codes," *IRE Trans. Information Theory*, vol. IT-6, pp. 13–15, March 1960.
[23] G. C. Bagley, "Radar pulse-compression by random phase-coding," *Radio and Elect. Engineer*, p. 5, July 1968.
[24] J. M. Colton, "Study of adaptive antenna techniques for millimeter wave applications," Advanced Technology Corp., Timonium, Md., AF 19(628)-5099, AFCRL 66–724, August 1966.
[25] V. E. Derr, "Propagation of millimeter and submillimeter waves," Martin Marietta Corp., Orlando, Fla., Rept. OR 8549, NASA CR-863, August 1967.
[26] R. L. Mitchell, "Radar meteorology at millimeter wavelengths," Aerospace Corp., El Segundo, Calif. Tech. Rept. TR-669(6230-46)-9, Air Force Rept SSD-TR-66-117, June 1966.
[27] *Ibid.*, p. 19.
[28] D. Atlas, "Advances in radar meteorology," *Advances in Geophysics*, vol. 10. New York: Academic Press, 1964, p. 317.
[29] L. J. Battan, *Radar Meteorology*. Chicago, Ill.: University of Chicago Press, 1959, ch. 4.
[30] C. R. Burrows and S. S. Attwood, *Radio Wave Propagation*. New York: Academic Press, 1949.
[31] B. R. Bean, "Attenuation of radio waves in the troposphere," *Advances in Radio Research*, J. A. Saxton, Ed., vol. 1. New York: Academic Press, 1964.
[32] B. G. James and L. T. Zitelli, "Kilowatt CW klystron amplifiers at K_a and K_a bands," *Microwave J.*, vol. 11, p. 53, November 1968.
[33] R. J. Donaldson, "A preliminary report on Doppler radar observation of turbulence in a thunderstorm," Air Force Cambridge Research Laboratories, Meterology Laboratory, Environmental Research Paper 255, January 1967.
[34] R. Wexler, "Application of Doppler radar to storm dynamics," Geophysics Div., Allied Research Associates Inc., Project 8620, January 1967.
[35] C. T. Stelzried and W. V. T. Rusch, "Improved determination of atmospheric opacity from radio astronomy measurements," *J. Geophys. Res.*, vol. 72, no. 9, p. 2445, May 1967.
[36] D. C. Hogg, "Millimeter-wave communication through the atmosphere," *Science*, vol. 159, no. 3810, p. 39, January 1968.
[37] C. T. Stelzried, "A liquid-helium-cooled coaxial termination," *Proc. IRE* (Correspondence), vol. 49, p. 1224, July 1961.

TRAKX: A Dual-Frequency Tracking Radar

D. Cross, D. Howard, M. Lipka, A. Mays, and E. Ornstein

One of the long-standing problems in the use of precision radar tracking systems has been the occurrence of multipath and clutter effects when following targets at low elevation angles. Preliminary investigation[1] into the multipath phenomenon has shown that accurate tracking can be maintained at lower elevation angles by the use of millimeter-wave trackers: targets at 100-foot altitude were successfully tracked over water with less than 0.4 mils multipath error to ranges in excess of 30,000 yards. A practical system using millimeter waves, however, must also take into account the weather and propagation limitations[2] inherent in millimeter-wave applications. The solution reached was a dual-frequency system — K_a- and X-bands — with independent tracking on any of the three radar coordinates at either of the two frequencies used.

This system, the TRAKX (Tracking Radar at K- and X-band), is designed to permit combined use of K_a-band and X-band for tracking purposes. The system allows simultaneous use of both transmitters and receivers with a single dual-feed antenna. For the required control of data flow, as well as the improved system operation available, the system is under full real-time computer control; mode control, tracking loops, data and display processing, error correction and most other system functions are done as software. Since the system is intended for precision tracking, one of the high-quality pedestals used in the Nike Hercules system was adapted for TRAKX use.

This radar is presently under construction at the Naval Research Laboratory's Chesapeake Bay Division. It is expected to be operating in the late Fall of 1976. Field testing will begin in 1977.

SYSTEM CHARACTERISTICS

The TRAKX radar transmits at 9 and 35 GHz using an 8-foot dish with combined Cassegrain and focal point feeds. Pulse lengths at both frequencies are 0.25 microsecond, with prf's from 160 to 1600. Tracking is 3-channel monopulse for both frequencies, and both echo tracking and beacon tracking are available. Ranging is done using a phase-locked frequency synthesizer, with range data provided by a time interval counter.[3]

The console contains an A-scope, displays for TV boresight optics and an alphanumeric CRT controlled by the computer. Controls are mainly push-button, with operating mode, frequency-coordinate selection and other essential functions available at the console.

The real-time computation that is the core of the system is done in a Harris 6024/4 minicomputer. The radar operating program is written in FORTRAN, as far as possible, and held strictly to modular form.

Table 1 summarizes overall TRAKX system characteristics. A simplified block diagram of the system is shown in **Figure 1**.

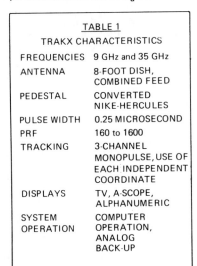

TABLE 1
TRAKX CHARACTERISTICS

FREQUENCIES	9 GHz and 35 GHz
ANTENNA	8-FOOT DISH, COMBINED FEED
PEDESTAL	CONVERTED NIKE-HERCULES
PULSE WIDTH	0.25 MICROSECOND
PRF	160 to 1600
TRACKING	3-CHANNEL MONOPULSE, USE OF EACH INDEPENDENT COORDINATE
DISPLAYS	TV, A-SCOPE, ALPHANUMERIC
SYSTEM OPERATION	COMPUTER OPERATION, ANALOG BACK-UP

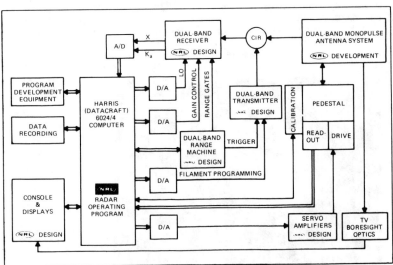

Fig. 1 TRAKX block diagram.

ANTENNA

The antenna is an 8-foot paraboloid with combined X- and K_a-feeds, shown in **Figure 2**. The K_a-band uses a 5-horn feed in the Cassegrain configuration with a dichroic sub-reflector transparent to X-band and reflective at K_a-band. Mounted behind the reflector is the X-band focal point feed, adapted from a feed used on the AN/TPQ-27. Beamwidth of the X-band radiating combination is 1°, with first sidelobes 20 dB down. The K_a-band combination offers a 0.25° beamwidth with −20.5 dB sidelobes. The gain is 45 dB at X-band, 56 dB at K_a-band.[4]

The present TRAKX system is operating at linear polarization, because of the limitations of the dichroic sub-reflector. When dichroic sub-reflectors with appropriate characteristics become available, the system can be modified to operate in the circularly-polarized mode.

PEDESTAL

The antenna has been adapted for mounting on a pedestal taken from the Nike Hercules system. The Nike-Hercules pedestal was selected because of its exceptionally high quality azimuth bearings — an important factor in radar instrumentation systems. Procured as surplus, the price was an additional attractive feature. The use of this pedestal limited dish size to eight feet, and larger antennas would require substantial mechanical modifications.

The Nike-Hercules Target Tracking Radar pedestal was stripped of its original transmitter, receiver and power supplies. A new dual-frequency transmitter and receiver were installed, with their associated power supplies. The original position package was replaced with 17-bit optical encoders, and the slip ring package was extended to accommodate the additional dual-frequency data. The drive motors were retained, but modifications to the drive motor system to improve reliability were installed.

TRANSMITTER

The transmitter uses two magnetrons, an SFD-382 tunable from 8.5 to 9.6 GHz and Litton 4064 operating at 34.85 GHz. These tubes are operated from a common line-type modulator, shown in simplified form in **Figure 3**. Both tubes transmit simultaneously, and an "X-band silence" control is provided for preventing 9 GHz radiation when appropriate. The system may be operated at prf's from 160 to 1600 in multiples of 160. Repetition rates are software controlled, and it is possible to set up prf's other than the main sequence in increments of five Hz or less if such prf's are required.

Pulse widths are 0.25 microsecond for both frequencies. The peak power available at X-band is 300 kW; 130 kW peak is available at K_a-band.

RECEIVERS

Commercial receiver, mixer and IF packages were installed in the TRAKX system. The X-band mixer operates with a 6.25 dB noise figure. The noise figure for the full rf package at K_a-band, including circulator and TR, is 13 dB.

The receiver packages furnish data for each frequency in video form to the A/D converters. Output for each channel is two range signals (early and late gates) and azimuth and elevation error signals. The digitized data is fed to the computer and handled by software for any subsequent operations.

Gain control and local oscillator tuning are integral to the receiver packages, but are under software control. Range gates are furnished by the range machine, which can be considered as a computer peripheral and is also under software control.

CONSOLE

The console is sketched in **Figure 4**. It contains positioning controls, mode controls, A-scope, TV display, and approximate displays of antenna position and AGC levels. The heart of the console, from the operator's point of view, is the TV monitor showing the picture seen by the boresight optics (a 5½" to 52" Zoomar). This occupies the center of the console and is easily seen with the operator's head in normal-reading position. A push-button matrix to the right of this display allows the operator to choose the frequency at which tracking will be done in each coordinate (range, azimuth or elevation). The operator may also choose beacon operation for any coordinate at either frequency or may choose to coast in any coordinate.

A-scopes are located above the TV display, and a slanted horizontal panel containing manual controls and controls for acquisition is located below the TV display and frequency-coordinate matrix.

A keyboard terminal to communicate with the computer is located adjacent to the main control panel. The CRT for computer alphanumeric readout is located above the keyboard, and this CRT also gives a detailed display of radar parameters during system operation.

SOFTWARE

The real-time operating software for the system is resident in a Harris 6024/4 minicomputer. The program was constructed with two primary objectives: first, the operator should not be required to deal with software during an operation; secondly, any changes that are required in the program are to be facilitated by fully documented carefully-controlled modular software and the use of FORTRAN rather than assembly language.

The radar operates in six modes: On, Standby, Calibrate, Designate, Manual, Autoacquisition and Autotrack. All modes are pushbutton selected. The pushbuttons communicate

Fig. 2 TRAKX antenna and feeds, in position on pedestal.

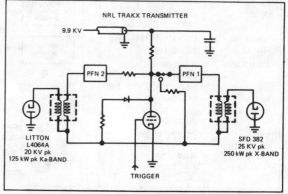
Fig. 3 Schematic of TRAKX transmitter and modulator.

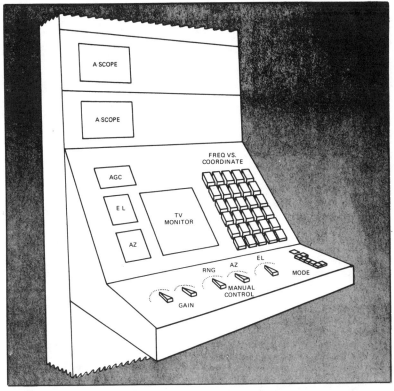

Fig. 4 Sketch of operator's console.

with the computer, and actual mode selection and set-up are done by the software at the pushbutton command. Certain items, inevitably, require operator interaction with the computer. The radar - for instance - comes up normally with a prf of 640. Changes to other prf's must be done interactively at the computer keyboard. Some calibration and error correction require keyboard use. Otherwise, the operation of the radar should not require any awareness of the computer on the part of the operator. The alphanumeric CRT will serve the operator mainly as a source of detailed data on the radar.

When changes are required or desired in the software, only module-level changes are permitted. A block of software designated as a module is usually specified by its FORTRAN coding, its input, its output and its timing effects on the system. For software changes, a full module must be replaced by another full module, and the new module must meet the timing and input/output specifications of the original.

SYSTEM PERFORMANCE

On a 1 m^2 target, the X-band portion of the system is expected to have a maximum range of 95 nautical miles for 0 dB signal-to-noise ratio. Useful tracking data usually requires a S/N of at least 13 dB, and this should be available out to 45 nautical miles.

The K_a-band ranges are more complicated. For free space, the 0 dB S/N maximum range is calculated at 93 nautical miles, the 13 dB S/N maximum range at 44 miles. If we assume a "typical" day, with an atmospheric attenuation of .28 dB/n. mi. (two-way), we can expect a 0 dB S/N maximum range of 46 n. mi. and a 13 dB S/N maximum range of 28 n. mi. For a 4 mm/hr rain, the figures drop to 14 and 11 n. mi., respectively.

In summary, the TRAKX radar system is expected to furnish good track data on a 1 m^2 target at K_a-band out to 28 n. mi. on a "normal" day, with acquisition ranges out to approximately 75 n. mi. at X-band; 4 mm/hr. rain will degrade the maximum range at which good K_a-track data can be taken to 11 n. mi.

REFERENCES

1. Kittredge, Ornstein and Licitra: Millimeter Radar for Low Angle Tracking. *EASCON Rec.*, 1974, pp 72-75.
2. Thompson and Kittredge: A Study of the Feasibility of Using 35 GHz and/or 94 GHz as a Means of Improving Low Angle Tracking Capability. *NRL Memorandum Report 2249*, May 1971, p. 8.
3. Cross: Low Jitter High Performance Electronic Range Tracker. *IEEE International Radar Conference*, April 1975, pp 408-411.
4. Final Report on Contract N00014-75-C-0500: Dual Band Monopulse Antenna Tracking System. RCA Missile and Surface Radar Division, Moorestown, NJ, March 1976.

An FM/CW Radar with High Resolution in Range and Doppler; Application for Anti-Collision Radar for Vehicles

G. Neininger

INTRODUCTION

A large portion of the traffic accidents on motorways and other high speed highways are rear-end collisions caused by improper estimation of the safety interval which is necessary between two vehicles. To determine the proper safe distance to a preceding vehicle, the measurement of distance and relative speed is necessary, in addition to other parameters such as deceleration, road condition and driver reaction time. Within the scope of the BMFT[1]-supported program, "Electronic Aids for Vehicular Traffic"[2], SEL, in cooperation with Daimler-Benz, has conceived and constructed an anti-collision system which consists essentially of a radar sensor using the FM/CW principle, and a warning computer (microprocessor). Contrary to other measurement procedures, such as infrared or ultrasonic measurement, the radar warning system is fully operational even under adverse weather conditions (rain, fog, snowfall). The parameters necessary to calculate a safe vehicular interval can be seen in the following (greatly simplified) formula:

$$a_s = \frac{v_2^2}{2b_2} - \frac{v_1^2}{2b_1} + v_2 \cdot \tau + s \quad (1)$$

a_s = safety interval
v_1 = speed $\}$ lead vehicle
b_1 = deceleration
v_2 = speed $\}$ following vehicle (radar-equipped)
b_2 = deceleration
τ = driver reaction time
s = remaining interval

The vehicles's own speed, v_2, can be derived from the vehicle's normal speedometer. The values for b_1 and b_2 vary slightly from vehicle to vehicle, vary considerably more, however, under varying road conditions (dry, wet, icy).

The speed v_1 is calculated from the speed difference v_{21} measured by the radar sensor and the vehicle's own speed, v_2. The driver reaction time amounts to approx. 0.8 to 1 sec.

SYSTEM CONFIGURATION

A block diagram of the anti-collision warning system can be seen in Fig. 1. The radar data for distance and relative speed, together with the input data for the vehicle's velocity and the road condition, are fed into the warning computer, which then calculates the necessary safety interval and compares it with the current (measured) distance. For distances which amount to less than the calculated safety interval, a visual and an acoustic warning are given to the driver.

[1] Bundesministerium für Forschung und Technologie = Federal Ministry for Research and Technology

[2] Elektronische Hilfen für den Strassenverkehr

SYSTEM REQUIREMENTS

Due to the high traffic density on motorways and highways, simple, single-target radar units cannot be employed; rather, multiple-target - distance resolving - equipment must be used.

Study of the required characteristics produced the following basic specifications:

range:	100 m, min.(approx. 360')
distance resolution:	10 m (approx. 36')
distance measuring accuracy:	± 2.5 m (approx. 9')
antenna half beam width:	2,5° (azimuth)
relative speed measurement range:	-30 to +160 km/h (approx. -18 to +100 MPH)
accuracy:	± 2.5 km/h (approx. 1.6 MPH)
sign determination:	+/-
system reaction time:	≤ 0.1 sec.

Only autonomous radar processes were considered for this system; the use of transponders could not be accepted because:

- all vehicles would have to be equipped (impractical)
- various obstacles (e.g., pedestrians, large objects on the roadway, etc.) could not be detected
- in case of an accident, which would often include destruction of the rear transponder equipment, the warning system would be inoperative.

THE FM / CW SYSTEM

Due to the fact that no available radar equipment met the specifications listed above, a radar sensor with suitable characteristics had to be developed and constructed. After thorough consideration and research, a decision in favor of an FM/CW system was made. This, however, was only possible after the theoretical basis for the distance resolution of the FM/CW method had been considerably improved.
(Raudonat and Sautter (1))

Principle

With linear frequency modulation, the following relationship is valid between the frequency deviation, ΔF, and the distance, R, in the FM/CW method:

$$f_r = \frac{4 \cdot \Delta F \cdot f_{mod} \cdot R \cdot K}{C} \quad (2)$$

ΔF = frequency deviation
f_{mod} = modulation frequency
C = speed of light
$K = \frac{T_1}{T_2}$ (for non-symmetrical linear modulation)

T_1 = modulation period
T_2 = saw-tooth signal rise time

The chronological sequence of the difference signal derived from combining the transmitted and echo signals is shown in Fig. 2. In order to enable a narrow-band evaluation of the applicable video spectrum and thereby a good distance resolution, the following measures were taken, in addition to a suitable linearization of the characteristic transmission curve:

- the interference signal f_r^* which occurs during the return of the modulation signal, is blocked.

  The target echo spectral portions lying at frequencies which are higher by the factor K are thereby suppressed, which simultaneously enables determination of the sign of the relative speed; the frequency shift of the individual spectral lines with any given relative motion (+/-) only occurs in that direction. The frequency shift of the fixed target spectral lines - which lie at whole-number multiples of the modulation frequency - is directly proportional to the relative speed, and is processed.

- Through amplitude weighting of the signal, f_r, (e.g., Hamming weighting) a strong reduction of the spectral side lobes can be achieved, so that the distance resolution capability is comparable to that of a pulse radar. Fig. 2 shows the computed spectrum of two target objects lying 10 m (36') from one another.

Block Diagram of the Present System (Figure 3)

The radar equipment consists of the following assemblies: antenna with RF section, signal processing, and the display unit. The transmission (Gunn oscillator) is frequency modulated (saw-tooth) by the modulator. A highly constant modulation frequency is important for the relative speed evaluation, and is therefore derived by division from a crystal-stabilized master oscillator, located in the frequency synthesizer.

The incoming signal is directly converted into the video frequency range and is fed through a low-noise video amplifier with subsequent amplitude weighting, to the main amplifier. There, by means of a high-pass filter, the receiving field strength for close targets is equalized. To avoid over-driving by strong echo signals, the amplification is automatically controlled. The processing of distance information is performed serially; that enables the use of a steep-flanked filter for all distance elements. The synthesizer delivers the local oscillator frequencies allocated to the individual distance intervals. After completing determination of the target distance and relative speed, the raw radar data are fed through a buffer to the microprocessor, which calculates the safety interval.

During this experimental phase, the raw radar data and the current divergence from the safety interval, Δa, are shown in the display unit; in production equipment, only Δa and /or the optical / acoustic warning are planned.

OPERATIONAL CHARACTERISTICS

The operational characteristics of the actual anti-collision warning system are summarized below:

range:	100 m (approx. 466')
distance resolution:	5...10 m (approx. 36')
distance measurement accuracy:	±2.5 m (approx. 9')
relative speed measurement range:	-50 to +160 km/h (approx. -31 to 100 MPH)
measurement accuracy:	2.5km/h (approx. 1.6 MPH)
sign:	+ / -
system reaction time:	<0.1 sec.
horizontal beam width:	≈ 2.5°
vertical beam width:	≈ 3.5°

DRIVING SIMULATION

The chance of success for an anti-collision warning system increases to the same extent that the system is successful in reducing the number of false alarms caused by irrelevant targets (e.g., guard rails, signs, etc.).

The investigations of the radar data which are necessary are aided considerably by the microprocessor. Optimizing the appropriate computer program is only possible however, if, in addition to the driving situation to be studied, the data sequence (in the form of the electrical data flow of the radar sensor) which belongs to it, is available. Therefore, in parallel with the actual radar sensor development, a magnetic tape recording procedure was developed, which meets these requirements.

The data delivered by the radar sensor for distance and relative speed are available at the signal processing unit output. The distance data measured are stored as DC voltage values using an FM recording process, after the data have undergone digital/analogue conversion. To record the relative speed amount and sign (+/-), an analogue signal (Doppler) is taken from the signal processing unit and, together with an analogue voltage representing the vehicle's own speed, is recorded on a second track.
To provide a synchronizing reference signal, the recording of an equipment-internal 15-Hz signal is adequate. During play-back, the speed signals are separated by a dividing network; the distance information runs through an A/D converter and is then fed together with the speed signal into the signal processing unit.

FALSE ALARM PROBLEM

Fig. 4 provides a partial view of the duties and data flow of the computer during real-time operation, during test drives and also when operating during tape recorder playback. The data input and processing cycle is repeated at approx. 60-msec. intervals. Bridging of individual data gaps is necessary only to a very small extent, due to the good dynamic characteristic of the radar sensor.
With tolerance testing, it is possible to eliminate individual, incorrect data or data pairs (relative speed and distance) with the aid of the proceding "history"; for example, changes in the distance or relative speed within one data measurement cycle cannot exceed certain calculable amounts due to purely physical limitations. If such specious data sequences occur, they can only be faulty measurements or a newly acquired target. In such cases, an intermediate storage of the radar data is made in parallel to the normal program. The evaluation of further data then enables a decision; either the data is eliminated (with the preceding data bridging the gap), or the da data is accepted as representing a new target object.

The suppression of false alarms is effected through a series of programmed logic and chronology comparisons. these false alarms fall primarily into the following three groups:

- momentary acquisition of objects (e.g., posts, signs, etc.) which do not represent a danger situation because they do not lie within the lane in which the vehicle is being driven

- interference target acquisition in curves

- warning indication caused by extended target objects, e.g., anti-dazzling roadway separators and guard rails.

The elimination of momentary targets can be achieved through time discrimination; a necessary minimum duration of target detection is defined through knowledge of the area of radar sensor acquisition and with consideration of the vehicle's own speed, together with the measured distance: only after this duration is exceeded, may the computer receive the data.

The measures which may be taken to suppress false alarms in curves consist principally of

- a limitation of the maximum range of the radar sensor

and/or

- slewing the antenna.

For the elimination of false alarms caused by extended target objects, the FM/CW principle offers favourable conditions because of the system's advantage of independent speed determination.

For a situation where the vehicle's own speed and the relative speed are equal (within a certain tolerance range), a special program section considers whether, after a prescribed amount of time the distance of the acquired target has changed corresponding to a value computed by integration of the relative speed ; only if this condition is fulfilled, is a warning given.

The calculation of the safety interval is performed only after checking the individual data or a data sequency has been completed. Trial driving with the anti-collision warning system has shown that it is better to indicate the difference between the safety interval and the actual measured distance (deviation from the safety interval) instead of only indicating the computed safety interval, because this allows a better and faster estimation of the danger situation by the driver.

RESULTS

The driving trials up to now (they began in the late summer of 1975), have confirmed the functional capabilities and effectiveness of the FM/CW principle. Numerous trial drives under unfavourable weather conditions (rain, fog, snowfall) have proven the all-weather capability of the system; the only limitation which should be mentioned is, that a thick, salt-saturated sludge coating on the antenna radome leads to a certain reduction of the acquisition range. This will be prevented either by a simple mechanical cleaning and/or use of a more suitable radome coating.

The preliminary specification concerning resolution capability, accuracy and reaction time are met and in some points - e.g., speed measurement - exceeded. To prevent interference - and with it, fluctuations of the echo amplitudes - caused by multi-path propagation, the antenna should be located on the vehicle as close to the road surface as possible; due to the increase in dirt on the radome face, a height of approx. 50 cm (20") has been chosen as a compromise.

The effectiveness of the microprocessor in conjunction with suitable software, has been proven.

OUTLOOK

After the German postal ministry released a frequency band at 35 GHz for interval warning system operation, the radar sensor, previously operated in the KU band (16.5 GHz), was equipped with a new antenna and RF section, which provided the considerable advantage of using a smaller antenna. Fig. 5 shows the new test vehicle.

REFERENCES

Raudonat, U. and Sautter, E. "Mehrzielfähiges FM/CW Radar zur eindeutigen Messung von Entfernung und Geschwindigkeit". Nachrichtentechnische Zeitschrift, 1977, 3, 255 - 260.

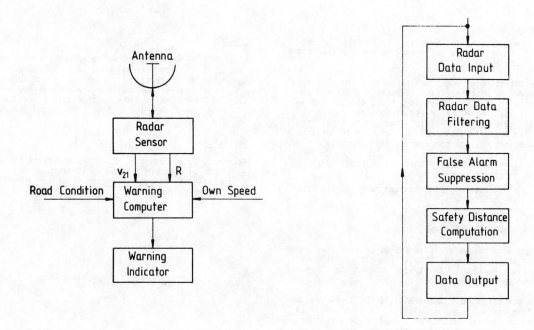

Figure 1 System block diagram

Figure 4 Microprocessor flow chart

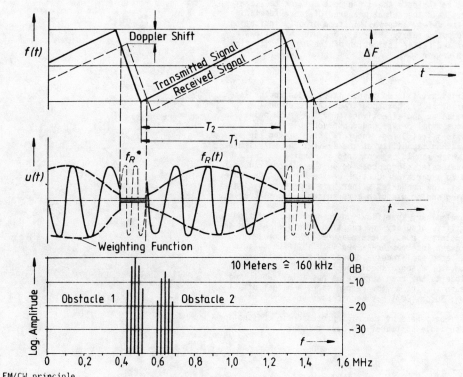

Figure 2 FM/CW principle

K_a-band Radars

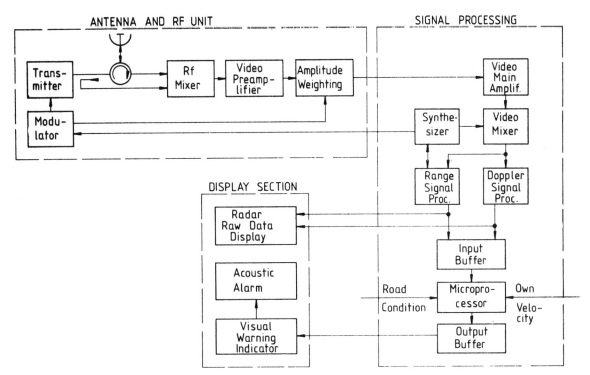

Figure 3 Overall block diagram

Figure 5 35 GHz test vehicle

mm Radar for Highway Collision Avoidance

Y.K. Wu and C.P. Tresselt

Recent innovations in microwave sources and Large Scale Integration (LSI) circuits, have made the use of radar on motor vehicles more cost acceptable. Recognizing the potential benefits of automatic braking, the National Highway Traffic Safety Administration (NHTSA) initiated research on the utilization of radar in 1974. The areas of system requirements, operational modes, and technological constraints were closely examined. Based on those results, NHTSA awarded a contract to Bendix Research Laboratories for the integration of the radar with two different braking systems, namely those on an intermediate Plymouth and a compact Ford. These are not intended to be production type systems, but rather unique test vehicles with which NHTSA will conduct a comprehensive test program. The design of the radar braking systems conform to the recommendations developed in the first two phases of the program. The radar sensors proper have been produced under subcontract by Bendix Communications Division.

PHASE I AND II STUDIES

The three-month Phase I study was initiated by NHTSA in June, 1974. Six experimental system concepts were studied: One each from the Federal Republic of Germany and the United Kingdom, and four from the United States. These system concepts were in various stages of development and represented both cooperative and non-cooperative as well as semi and fully automatic systems. The cost-benefit of a particular radar braking system was determined by analyzing traffic accident data, using a computer program designed to search accident data in several counties throughout the United States. The number of preventable accidents was determined by assuming that all automobiles were equipped with the particular radar braking system considered.

Results of the Phase I Study[1] indicated that an automatic/noncooperative radar brake system may provide a significant benefit in preventing accidents that may otherwise be caused by inattentive or tardy driver response. It further indicated that the radar braking system is technically feasible for installation in passenger automobiles subject to conditions and constraints set forth in the study guidelines. One of the major technological problems identified with the implementation of the automatic/noncooperative system was the achievement of sufficient target discrimination to allow rejection of non-hazardous objects and to maintain sufficiently low false alarm rates while retaining recognition capability on all potential hazards. A possible solution to this problem lies in reaching a viable compromise between false alarms and missed targets through performance trade-offs in the radar sensor design and sensitivity analysis of the more detailed accident data.

In May 1975 a nine-month Phase II Study was initiated by NHTSA with the following objectives:

- Conduct experimental investigation and analysis to resolve the target discrimination problem.
- Perform sensitivity analysis by means of computer simulations to determine performance trade-offs and related cost-benefits.
- Generate a preliminary performance specification for the recommended system(s).

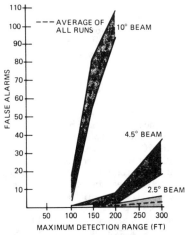

Fig. 1 False alarm rate on urban expressway.

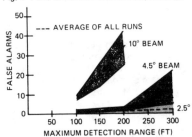

Fig. 2 False alarm rate on divided artery.

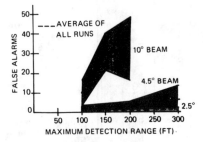

Fig. 3 False alarm rate on two-way artery.

Of particular interest were data obtained during the experimental target discrimination study. Two radar parameters, radar detection cutoff ranges and antenna beamwidths were varied to study the sensitivity of false alarms to restricted geometric coverage of the radar illumination, since many false alarms arise from objects peripheral to, but off the roadway. The test vehicle was equipped with an automatic radar braking system and four-wheel anti-lock brakes which interfaced with the radar command and control functions. Operating at a frequency of 22.125 GHz, the radar provides range and range rate data to the processing electronics which also accepts speed and steering angle information from vehicle sensors.

A summary of the test results is shown in **Figures 1, 2,** and **3** where the number of composite false alarms (Warn and Brake) are plotted for various beamwidths against maximum detection range. **Figure 1** represents urban expressway data where overpasses accounted for approximately 80% of the measured false alarms. Shaded areas for each beamwidth are bounded by maximum and minimum values recorded during 57 passes over this 20-mile course. A similar distribution for the main arteries (divided and two-way combined) is shown in **Figures 2** and **3** where the majority of false alarms (77%) were caused by roadside signs. Responses to sweep-through targets encountered during turns and from guardrails during ingress/egress maneuvers were suppressed by a steering angle inhibit function designed into the radar system.

The test results clearly indicated that false alarms could be effectively suppressed by limiting beamwidth and detection range. This was not a surprising outcome but it was significant in terms of quantifying these parameters. With a 2.5-degree beamwidth, the zero false alarm threshold detection range is between 100-150 feet.

These constraints obviously compromise system effectiveness in preventing accidents; but to reach some quantitative assessment, it was necessary to consider operational performance within a statistically accurate representation of the accident environment. Therefore, the computer simulation approach was adopted, and an accident model was generated for this purpose. Details of this study can be found in References 2 and 3. A total of 36 separate radar brake systems were evaluated for performance in a variety of accident situations.

Figure 4 summarizes the estimated savings in lives per year for both a 100 and a 200-foot range cutoff system. Clearly the longer range radar is superior for avoiding head-on crashes. However, the reported experimental test results indicated that a radar maximum range exceeding 150 feet is plagued by a false alarm problem. Since false alarms cannot be tolerated, particularly for automatic braking, the 150-foot maximum range system was selected as the prime candidate.

Other recommendations from Phase II include antenna beamwidths no greater than 2.5 to 4 degrees. Operating frequency should be raised above the 22.125 GHz of the test car in order to achieve such beamwidths with a 6 to 8 inch aperture, which can be installed on cars of the compact class. The 35 to 36 GHz region was recommended, because it provides the desired beamwidth from the aperture size stated, and microwave components at this frequency range are available from several vendors. The millimeter wave bands, defined as 40 GHz and above, can provide the beamwidth desired at an antenna size compatible with even smaller vehicles. The attenuation of desired targets in normal rainfall and fog is relatively minor at the ranges in question up through at least 60 GHz. It must be realized however, that backscatter from heavy rain increases significantly with increasing operating frequency, and presents a potential source of false alarms, which remains to be resolved. Possible solutions include the use of circular polarization, short pulse modulation, signal signature processing, deliberate attenuation of lower frequency (close-in) range returns in FM/CW radar systems, or some combination of these techniques. An alternative is to inhibit the system during heavy rain (e.g., a windshield wiper interlock).

PHASE III RADAR BRAKE SYSTEMS

The two radar systems constructed during the third phase are essentially identical to one another, and use the diplexed CW format. Phase II suggested the use of such a modulation format, partially on the basis that solid-state, 2.5W peak power sources required for a narrow pulse mm wave radar system represent a technological risk at present. The FM/CW format, however, contains no comparable technical risk and may be an acceptable alternate. The diplexed CW radar inherently forms a very accurate doppler reconstruction and hence tends to provide more accurate, noise-free velocity data than conventional FM-CW systems, where successive measurements are used to calculate velocity[4]. The system law according to which brake commands are generated uses both range and range rate data[3] with accurate velocity

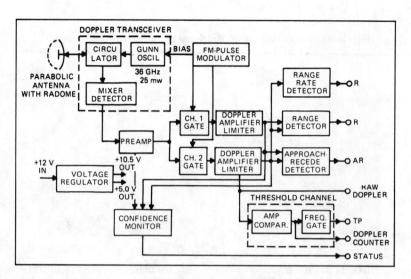

Fig. 5 Block diagram of radar sensor.

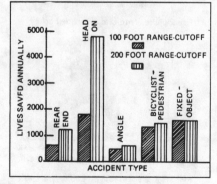

Fig. 4 Estimated annual savings in lives versus accident type.

K_a-band Radars

Fig. 6 Radar sensor electronics and antenna/radome.

Fig. 7 Installation of antenna/radome on a compact Ford.

information being at least as important as accurate range data.

The diplexed-CW system was originally devised by Boyer[5] and range cutoff features which make this system suitable for use in highway radar brake applications were added by Tresselt[6-8]. A block diagram of the radar sensor is shown in **Figure 5**. The sensor consists of two parts, an 8-inch diameter antenna/radome/transceiver assembly mounted in an opening cut in the center of the car grill, and an electronics package of roughly 3 x 6 x 10 inches mounted within a few feet of the antenna. **Figure 6** shows the sensor hardware. The electronics package is entirely solid-state and utilizes its own internal power supply. The radar transmits peak pulses of 25 milliwatts in the vicinity of 36 GHz. The normal transmission sequence is:

- RF ON: 730 ns,
  (36 GHz + 410 kHz)
- RF OFF: 1420 ns.
- RF ON: 730 ns,
  (36 GHz – 410 kHz)
- RF OFF: 1420 ns.

This pattern is repeated continuously. The homodyne video output from the mixer/detector is pulsed, corresponding to the ON intervals indicated above. Only the last 215 ns of these video pulses are gated alternately into the two doppler amplifier/limiter chains. This sampled data is filtered and amplified, forming a reconstruction of continuous doppler target information in each of the two channels. Phase comparison between the two channels provides range information, while relative velocity is determined by processing doppler frequency in one channel. The sense of velocity, approach or recede, is detected by noting the lead or lag in phase between channels. A threshold circuit rejects targets which are either too weak, or moving at too low a relative velocity.

With the 820 kHz difference in transmitter frequencies, the range comparison becomes ambiguous for targets past 300 feet. However, restriction of the duration of transmission to 730 ns eliminates responses from targets at ranges past 250 feet, as determined by measurement in an actual system against large targets. The use of a 250-foot range cutoff, instead of the 150-foot range suggested by the false alarm study, permits the addition of a warning function to the system. The various outputs from the radar are inputted to a microprocessor, using A-D conversion where necessary. The microprocessor is programmed to generate braking commands only for ranges lower than 150 or 100 feet (selectable).

Those targets which possess high enough relative velocities to pose a threat, but are located between the computed range cutoff and the absolute 250-foot range cutoff of the radar sensor, will activate an acoustical alert. Hopefully, the driver will react to this information and if a valid target is present, will initiate braking or an evasive steering maneuver. If not, and the target still fulfills the criteria for danger when range gets down to the computed cutoff, an automatic signal will be sent to a vacuum controlled, hydraulically operated actuator which operates the car's power-assisted brake pedal arm through a special linkage.

As **Figures 1** through **3** indicate, the information contained in the radar output from targets at 250 feet will definitely be contaminated by false alarms, given normal target statistics for American roads. Actual braking events, however, will only be permitted for shorter range, near-error free events. It might be noted that this form of radar possesses no inherent range resolution of separate targets. Because the doppler amplifier limits, the radar output is controlled by the target which generates the strongest return signal. Under certain infrequent multiple target geometry conditions, a desired close-in target such as a motorcycle can be made invisible by a larger target (semi-trailer) down the road. Pulsed radar with a range gate, or possibly a wide dynamic range FM/CW radar with multiple bin range-frequency-filtering, might prove capable of properly handling multiple target information, but only at the penalty of significantly greater hardware cost. If the warning function proves less than useful during road testing, the ON duration of the transmitter can by a single pot adjustment, be reduced to where range cutoff is at the 150 or 100 foot braking threshold, significantly reducing the opportunity for suppression of close-in targets noted above.

RADAR SENSOR EQUIPMENT

The antenna used in the radar sensor, which was supplied by the TRG Division of Alpha Industries, is an 8-inch diameter parabolic reflector illuminated by an open-ended waveguide feed. The reflecting surface is machined into a billet of aluminum which provides support for the bent waveguide feed located at the focus of the reflector. The integrated radome is a composite structure consisting principally of a solid formed block of styrofoam which butts against the parabolic surface. The hemispherical outer surface is covered by a single fiberglass cloth layer impregnated with

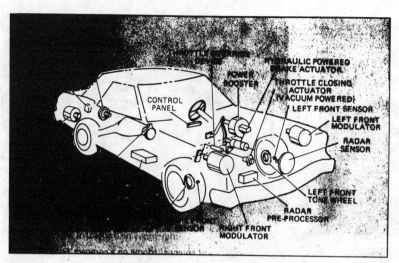

Fig. 8 Radar braking system component locations.

a mildly resilient epoxy for waterproofing.

The microwave transceiver, which is visible in **Figure 6**, is a compact waveguide block assembly made by Microwave Associates, and consists of three components: a tee circulator, Gunn oscillator, and a diode detector mount supplied with a TNC video output connection. The antenna is mounted on the car with the feed at a 45-degree angle with respect to horizontal direction. Thus two such antenna systems facing each other head-on will be cross-polarized to one another, thereby minimizing the possible effects of the transmitter of one radar "blinding" the receiver of the other, and vice-versa.

Figure 7 shows the antenna/radome mounted on the front of a compact Ford. The protrusion of the radome from the grill of this installation as well as that of an intermediate Plymouth is similar, and could be reduced in both cases by about one inch if the hood latch mechanisms were extensively modified. The bumper guards shown provide moderate protection for the radome against collisions. In the compact Ford, the electronics box is located near the passenger's side headlights, between the grill and part of the air conditioning radiator. In the intermediate Plymouth the box is in a similar area on the driver's side, but is fully out of the way of airflow to the radiator.

Two shielded leads of approximately 25 ohms connected in parallel are used to carry the pulsed Gunn diode drive to the transceiver package, while a 95 ohm shielded coax brings the detected video back to the electronics box, as shown in **Figure 6**. A multiple conductor cable carries data from this box back to the computer for warning/braking commands which is located on the floor hump between the passenger and driver. A control/status display panel is located on the passenger's side of the dashboard. Both cars have been modified to include four-wheel anti-lock brakes, to assure directional stability under heavy braking conditions. Some of the braking and anti-lock equipment for this system is located in the trunk. **Figure 8** shows the general location in an auto of the equipment associated with the radar braking system.

The 820 kHz frequency difference between pulses in the present system is provided by slightly varying the ON pulse voltage level applied to the Gunn source. An alternate approach to bias pushing for producing the desired frequency modulation of the source is to include a varactor diode in the oscillator cavity. This more than doubles the cost of sources in this frequency band at present. The Gunn bias pushing approach shows statistical variation from Gunn diode to diode which manufacturers fear might lead to a yield problem. The best and ultimately lowest cost approach can only be determined when truly large quantity production is examined in detail. An FM/CW system requires many MHz of deviation to get adequate range accuracy thus automatically necessitating a more expensive varactor tuned source.

An unusual effect was noted in the performance of the Gunn diode sources operated in the pulsed mode. In the radar, the voltage pulse levels applied during the "ON" time are about the same as the normal dc voltage suggested by the manufacturer for CW operation. In evaluation of two of the sources, FM of several megahertz within each pulse was measured, which is unacceptable for system operation. Cooperative investigation with Microwave Associates determined that diodes with a negative pushing figure exhibited the undesired FM, but that many diodes with positive pushing would exhibit excellent frequency stability during the pulse. The final radar system uses sources containing these latter diodes, which also exhibit excellent pushing factor stability with changes in ambient temperature. In common with earlier radar designs, the "ON" voltage applied to the source is increased with decreasing temperature, and heat is applied to the source case by a simple electric heater, to compensate for the tendency of the Gunn not to turn on well at reduced temperature. Operation to $-40°F$ appears feasible with careful design.

CONCLUSION

The processing which supplements the radar can function in a variety of possible operating modes. It is intended that extensive road testing of the radar brake vehicles supplied in this phase will be performed under a variety of conditions, thus increasing the knowledge base in the field of automatic braking. From the earlier testing it was learned that false alarms do not necessarily preclude mm radar from use in automated braking systems if the operating range of the radar is restricted. Ideally a radar brake system would prevent all frontal collisions. Although this system won't completely satisfy this goal, it will be an effective adjunct to safety by mitigating impact velocity in many situations, thereby reducing injury severity and saving lives. In lower speed encounters, collisions will still be entirely avoided by the system as presently envisioned. Should future work prove the desirability of radar brakes, a very large commercial market for mm radar may result.

ACKNOWLEDGMENTS

The authors wish to thank William C. Troll of Bendix Research Laboratories for supplying data on the initial two study phases of this program. This paper presents results of a study performed by The Bendix Corporation under contract DOT-HS-4-00913 sponsored by The U.S. Department of Transmportation NHTSA. The construction of two fully instrumented vehicles for additional system evaluation is being performed under contract DOT-HS-6-01450, sponsored by the same agency.

biased to operate at a higher or lower negative voltage for a lower efficiency. This technique provides an attenuation of about 1 dB per volt of fixed bias voltage. The relatively high paramp bias requires a pump power of about 35 mW. The doubler efficiency was observed to be at about 35 percent.

Figure 6 presents measured gain bandwidth data. A double tuned Tchebycheff response of a 1-dB bandwidth greater than 100 MHz at each channel was realized. The degree of double tuning was controlled independently at each channel by changing the coupling iris diameter and returning the cavity to the desired center frequency. A gain of 18 dB with a noise figure of 3.6 dB was measured for each channel.

CONCLUSION

The design criteria and performance of a switchable dual-channel parametric amplifier have been described. Two broadbanded responses of a single paramp stage separated by more than 1 GHz have been demonstrated to have an independent double tuning control. This amplifier can readily be utilized to upgrade existing communication systems to provide broadbanded dual-channel operation using a single amplifier stage.

ACKNOWLEDGMENTS

This work was supported by the Air Force Avionics Laboratory at Wright-Patterson Air Force Base, under the project direction of Mr. Roger Swanson. The basic amplifier development was done at AIL under the direction of Mr. James J. Whelehan. The advice and direction of E. Kraemer and J.J. Taub are gratefully acknowledged. Technical assistance was provided by L. Sadinsky.

REFERENCES

1. H.C. Okean, et al, "Electronically Tunable Low Noise Ka-Band Paramp-Downconverter Satellite Communication Receiver", *IEEE S-MTT,* 1975.
2. H.C. Okean and A.J. Kelly, "Low Noise Receiver Design Trends Using State of the Art Building Blocks", *IEEE Trans. MTT,* April 1977, p. 254-267.
3. M.A. Balfour, et al, "Miniaturized Non-Degenerate Ka-Band Parametric Amplifier for Earth-to-Satellite Communication Systems," *IEEE S-MTT,* 1974.
4. J.J. Whelehan, "Low-Noise Millimeter-Wave Receivers," *IEEE Trans. MTT,* April 1977, p. 268-280.
5. N. Marcuvitz, "Waveguide Handbook", Vol. 10, Radiation Laboratory Series, McGraw-Hill Book Co., New York, 1951, Sec. 6.1 and 6.5.
6. L.D. Cohen, et al, "Varactor Frequency Doublers and Triplers for the 200 to 300 GHz Range," *IEEE S-MTT,* 1975.

REFERENCES

1. Wong, R.E., D.V. Payne, W.O. Grierson, W.C. Troll, "Collision Avoidance Radar Braking Systems — Investigation, Phase I Study," *Final Report DOT-HS-801253,* October 1974.
2. Wong, R.E., W.R. Faris, W.O. Grierson, Y.M. Powell, W.C. Troll, D.V. Payne, "Collision Avoidance Radar Braking Systems — Investigation, Phase II Study," *Volume II, Technical Report, DOT-HS-802020,* September 1976.
3. Troll, W.C., R.E. Wong, Y.K. Wu, "Results from a Collision Avoidance Radar Braking System Investigation," Paper 770265, Presented at The Society of Automotive Engineers International Automotive Engineering Congress and Exposition, Detroit, February 28-March 4, 1977.
4. Belohoubek, E., J. Cusack, J. Risko, J. Rosen, "Microcomputer Controlled Radar Display for Cars," Paper 770267, Presented at The Society of Automotive Engineers International Automotive Engineering Congress and Exposition, Detroit, February 28-March 4, 1977.
5. Boyer, W.D., "Continuous Wave Radar," U.S. Patent 3,155,972, November 3, 1964.
6. Tresselt, C.P., "Multifrequency CW Radar With Range Cutoff," U.S. Patent 3,750,172, July 31, 1973.
7. Tresselt, C.P., "Range Cutoff System for Dual Frequency CW Radar," U.S. Patent 3,766,554, October 16, 1973.
8. Tresselt, C.P., "Variable Range Cutoff System for Dual Frequency CW Radar," U.S. Patent 3,898,655, August 5, 1975.

Solid State mm-Wave Pulse-Compression Radar Sensor

J.B. Winderman and G.N. Hulderman

Pulse-compression techniques provide increased range and range resolution in radars that are peak power limited. These techniques are especially applicable to solid state millimeter wave radar sensors where peak powers rarely exceed a few watts.

One such sensor, developed at Ka-band, has a 16 dB compression ratio, a 6.5 dB double sideband mixer noise figure, and greater than 100 dB dynamic range. It employs lightweight modular fabrication techniques, with all components clustered behind a 20-cm diameter Cassegrain tracking antenna which serves as an inner gimbal. The weight of the sensor, including antenna, is only 1.4 kilograms and the clustering of components results in low gimbal inertia while eliminating the need for long waveguide sections and rotating joints.

FUNCTIONAL DESCRIPTION

Figure 1 is a side view and **Figure 2** is a block diagram of the millimeter wave pulse-compression sensor. This unit contains a modulator/transmitter, receiver, antenna, duplexer, reference source, and synchronizer. The signal flow starts in the synchronizer which generates all timing functions. Specifically, the synchronizer generates the modulator trigger and controls the transmit/receive switch. It also is capable of controlling automatic gain control (AGC) blanking and providing range gate synchronization.

The synchronizer periodically triggers the pulse modulator, causing transmitter energy to be generated. This energy passes through latching switch S_1 and circulator C_1 into the antenna, where it is radiated. Simultaneously, switch S_2 routes linearly-frequency-modulated injection-locking energy from the reference oscillator to the transmitter through circulator C_2.

At all other times, switch S_1 is "off" to prevent reference oscillator energy from entering the mixer during reception of the target signal. At the same time, switch S_2 routes reference oscillator energy at the transmitter frequency to the balanced modulator where it is mixed with 225 MHz energy from the offset generator. The output of the balanced modulator passes through a filter which suppresses the reference frequency (carrier) and lower sideband. The upper sideband serves as the local oscillator.

The combination of C_2, S_2, and the reference oscillator constitutes a time-sharing reference source, for which a patent application has been made. The advantage of this type of power sharing over simple division is that full power is available for injection locking the transmitter or for local oscillator generation. A second advantage is that the receiver is non-functioning during transmission and, therefore, it is less susceptible to saturation by extraneous transmitter power at its input.

Millimeter wave energy is converted in the balanced mixer to a linearly-frequency-modulated intermediate frequency (IF). The pulses are 400 nanoseconds wide and the IF varies during each pulse from 225 through 375 MHz.

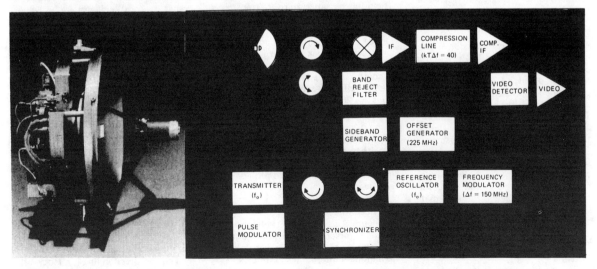

Fig. 1 Pulse-Compression Radar Sensor.

Fig. 2 Sensor Block Diagram.

TABLE 1
KEY SENSOR PARAMETERS

Frequency	Ka-band
Peak Power (W)	1
Pulse Width, Uncompressed (ns)	400
Pulse Width, Compressed (ns)	10
PRF (kHz)	20
Frequency Deviation (MHz)	150
IF Center Frequency (MHz)	300
Receiver Dynamic Range (dB)	> 100

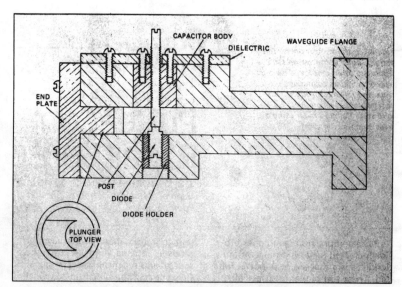

Fig. 4 Reference Oscillator, Sectional Sketch.

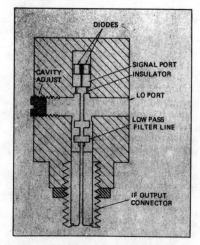

Fig. 3 Balanced Mixer, Sectional Sketch.

The compression line is a surface acoustic wave device which has a linear frequency-versus-time characteristic that is the conjugate of the frequency modulation function. Its output is a train of pulses that are taller and narrower than the input pulses, by a factor equal approximately to the time-bandwidth product (allowing for degradation caused by time sidelobe weighting). Another IF amplifier follows the compression line in order to compensate for the large surface wave transducer coupling losses.

SENSOR PARAMETERS

The chief pulse-compression sensor parameters are listed in **Table 1**. A train of one-watt, 400 nanosecond pulses is transmitted and, upon reception, the pulses are compressed in a dispersive delay line to widths of 10 nanoseconds. The ratio of the widths of the uncompressed and compressed pulses is called the compression ratio of the system.

The 20 kHz repetition frequency assures about 7.6 kilometers unambiguous range. The large receiver dynamic range reduces the possibility of saturation from echoes between the 76 meter blind range and the maximum unambiguous range.

DESCRIPTION OF COMPONENTS
Antenna
The tracking antenna is a 20 cm diameter Cassegrain design. Its subreflector is canted and spun at a 200 Hz rate by a small dc motor, thereby producing a conical angular motion of the beam.

The rear of the subreflector is divided diametrically into black (absorptive) and white (reflective) regions. Suspended above these regions and displaced 90 degrees apart are electro-optical emitter/sensor units which serve as angle reference data generators. Their output voltages are trains of orthogonal square waves with frequencies equal to the scan rate.

Duplexer
Following the antenna is a duplexer which consists of a special Microwave Associates package containing a ferrite circulator and a ferrite latching switch. The duplexer serves to route transmitter energy to the antenna and received signal energy to the mixer, and also to reduce to a negligible level any reference oscillator leakage energy to the mixer. The insertion losses through the circulator and the latching switch are each 0.4 dB, and their individual isolations are 20 dB.

A second circulator/switch package is the principal element of the time sharing reference source. It is the same type that is used as the duplexer, except that the termination of the latching switch was removed to provide two-way switching.

Balanced Mixer
Figure 3 is a sectional sketch of the balanced mixer. The 16-square-centimeter unit utilizes block construction to eliminate internal discontinuities and assure proper alignment of all parts. Two low-noise Schottky-barrier gallium arsenide diodes are stacked in series, parallel with the E-vector of the signal waveguide and separated by a thin conducting septum. The septum is an extension of a pickup probe and low pass filter which serve as the signal/local oscillator coupling post and the IF output line. Frequency adjustment of the mixer is accomplished by means of tuning screws at the closed ends of the waveguides.

The double sideband noise figure of the mixer was measured as 6.5 dB, including the 2.5 dB noise figure of the following IF amplifier. The conversion loss is 6 dB.

The sideband generator is similar in construction to the balanced mixer. It generates a carrier frequency and two sidebands, and the upper sideband becomes the local oscillator.

IF and Compensation Amplifiers
Following the mixer, six wideband gain stages provide about 75 dB of IF gain. The first stage limits the noise figure to 2.5 dB, and an interstage bandpass filter defines a passband from 225 MHz through 375 MHz. Three PIN diode attenuators are distributed among the gain stages for receiver blanking and AGC attenuation to 60 dB.

Four additional wideband gain stages following the compression line provide about 48 dB of gain, equal to the insertion loss of the compression line. A video detector and wideband video amplifier permit faithful reproduction of the compressed pulses.

Compression Line
The compression line is a Rockwell dispersive surface acoustic wave (SAW)

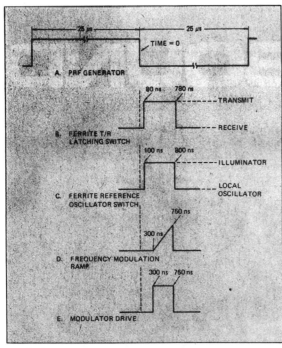

Fig. 5 Measured synchronizer timing waveforms.

TABLE 2
DISPERSIVE DELAY LINE CHARACTERISTICS

Type	Saw, Taylor Weighted
Substrate	Quartz
Case Dimensions (mm)	35.6 x 17.8 x 7.6
Center Frequency (MHz)	300
Bandwidth (MHz)	150
Input Pulse Width (ns)	400
Compression Ratio (dB)	16
Time Sidelobe Level (dB)	−25

Fig. 6 Uncompressed and Compressed Pulses.

delay device which consists of a polished quartz substrate with a pair of identical interdigital transducers displaced at opposite ends. Its characteristics are listed in **Table 2**.

The properties of the SAW delay line that make it attractive as a pulse compression filter are:

- Propagating Velocity. The propagation velocities of surface waves are five orders of magnitude lower than the velocity of electromagnetic waves in free space. Therefore, long uncompressed pulses can be processed in a device of small physical dimensions.
- Reproducibility. The method of fabrication permits reproducing identical filters from a single photographic mask.
- Frequency-Independent Delay. The acoustic delay along the substrate is independent of frequency. Hence, specially weighted interdigital transducers with frequency selective electrode pairs can be deposited to produce the desired compression effects.

An additional desirable property that applies to a delay line which employs a quartz substrate is zero temperature coefficient of delay. A change in ambient temperature causes changes in the substrate length and propagation velocity, both of which in combination produce a constant propagation delay over a wide range of temperatures.

Reference Oscillator

The reference oscillator, illustrated in **Figure 4**, employs a gallium arsenide Gunn diode to generate 200 milliwatts of CW power. This oscillator, like the balanced mixer, utilizes block construction which provides inherent heat sinking.

The diode holder is adjustable. Its position determines the depth of penetration of the source diode into the waveguide. A post extending through the cylindrical capacitor body and making contact with the top of the diode completes the dc bias path. The capacitor body and the oscillator case comprise an RF bypass capacitor.

The waveguide end plate contains a half-round tuning plunger. The plunger, together with the waveguide walls, serves as a high Q cavity at the desired operating frequency. This frequency can be adjusted by inserting thin shims between plunger end wall and the oscillator case.

The transmitter is similar in construction to the reference oscillator, except that a thin disc is employed at the diode end of the post as an impedance matching device. A silicon IMPATT diode generates 400 nanosecond pulses at 1 watt peak power.

Synchronizer

The synchronizer and the frequency modulator are housed in a common aluminum enclosure. The heart of the synchronizer is the PRF generator which consists of an astable multivibrator which produces a square wave at the pulse repetition frequency of 20 kHz.

The output of the PRF generator triggers one-shot multivibrators which produce pulses of 450 and 700 nanosecond duration. These pulses are employed to: control the latching switch drivers, control the pulse modulator through a driver stage, trigger the frequency modulator ramp generator, and synchronize a succeeding range tracker.

Figure 5 contains a set of synchronization timing diagrams illustrating the measured phases among the various waveforms. The negative-going slopes of the PRF generator waveform initiate all timing functions. Waveforms B and C should coincide in time. However, latching switch fabrication tolerances result in a 20 nanosecond difference in response time. The modulator drive pulse is 50 nanoseconds longer than the actual 400 nanosecond transmitter pulse because of the 50 nanosecond delay in transmitter turn-on.

Offset Frequency Generator

The offset frequency generator consists of a transistor oscillator stage loosely coupled to a driver and a power amplifier. The oscillator generates a 225 MHz sinusoidal waveform at 15 milliwatts power. The driver and power amplifier develop 140 milliwatts, which is of sufficient level to mix with the reference energy to produce the required local oscillator power.

Band Reject Filter

The band reject filter consists of three circular cavity resonators displaced along the centerline of a length of rectangular waveguide and coupled to the waveguide through 0.01 inch thick circular irises. The cavities are separated from one another by three-quarters of a wavelength, and they are mechanically tuned to one-half wavelength at the stop frequency.

SENSOR EVALUATION

The coherent pulse-compression radar sensor incorporates two sets of video detectors and amplifiers, one following the compensation amplifier, to facilitate performance evaluation. The signal-to-noise ratios and pulse widths were measured simultaneously at the outputs of both amplifiers while the sensor antenna was directed toward a physically small target, and the data were employed subsequently to compute an improvement factor. Use of a small target assures that the compressed pulse will not be stretched.

The improvement factor is a measure of the increases in range and range resolution that are achievable relative to those obtained using a short pulse radar transmitting the same peak power. It is expressed mathematically as:

$$I = 10 \log\left(\frac{(S/N)_c}{(S/N)_u}\right)\left(\frac{B_c}{B_u}\right) \text{ dB}$$

where S/N is the signal-to-noise ratio, B is the equivalent noise bandwidth of the system, and subscripts c and u denote "compressed" and "uncompressed".

Figure 6 is an oscilloscope display showing the uncompressed and compressed pulses riding on coincident baselines. The abscissa and ordinate calibrations are 200 nanoseconds/cm and 0.5 volts/cm, respectively.

The sensor improvement factor was calculated as 15.3 dB, which compares favorably with the design value of 16 dB. Thus, a 16 dB (factor of 40) increase in range resolution is achievable relative to that of a 1-watt, 400-nanosecond pulse sensor. Also, a 4 dB (factor of 2.5) increase in detection range is possible relative to that of a 1 watt, 10 nanosecond pulse sensor.

However, these improvements are not without cost. The minimum blind range of the pulse-compression sensor is limited to at least 61 meters by the receiver blanking circuits, and the hardware is appreciably more complex than that which comprises a short pulse sensor.

CONCLUSIONS

The performance of the millimeter wave, coherent pulse-compression radar sensor is close to what was predicted. The 16 dB compression ratio was selected to secure adequate improvements in range and range resolution while at the same time produce compressed pulses that are sufficiently wide to be processed with currently available hardware.

The widths of the uncompressed pulses are limited by the capabilities of the available transmitter diodes. Development of diodes with greater pulse width capability will permit employing higher compression ratios.

The pulse-compression scheme presented here is in no way limited to Ka-band sensors. It can be applied to radar sensors functioning at shorter wavelengths, also with good results.

BIBLIOGRAPHY

1. Cook, C. F., "Pulse Compression — Key to More Efficient Radar Transmission," *Proceedings IRE*, Vol. 48, No. 3, March 1960, pp 310-316.
2. Klauder, J. R. *et. al.*, "The Theory and Design of Chirp Radars," *BSTJ*, Vol. 39, No. 4, July 1960, pp 745-808.
3. Ohman, G. P., "Getting High Range Resolution With Pulse Compression Radar," *Electronics*, Vol. 33, No. 31, pp 53-57.

SEATRACKS — A Millimeter Wave Radar Fire Control System

G.E. Layman

Abstract

The Naval Surface Weapons Center, White Oak, MD, is investigating the potential of millimeter wave radars for shipboard fire control radar system applications. The inherent advantages of small size, narrow beamwidth, precision tracking, good performance in clutter and multipath environments, and excellent jamming immunity is being evaluated in theoretical and application studies and in the development of the SEATRACKS (Small Elevation Angle TRACK and Surveillance) feasibility radar system.

Introduction

The introduction of the latest generation of Soviet Anti-Ship Missiles, capable of a high speed, low altitude run in, has made rapid radar detection followed by continuous and accurate tracking of these targets a formidable problem. Traditional radars are confronted with excessive competing clutter, weather limitations, tracking errors and fading effects due to multipath returns. An improved performance in both surveillance and tracking are the functions the SEATRACKS (Small Elevation Angle TRACK and Surveillance) System program is evaluating. The SEATRACKS System concept combines a low-elevation surveillance system with a 35 GHz monopulse tracking radar in a single system.

Of the many possible uses of millimeter wave radars for shipboard application two particularly attractive applications have been identified. The first, alluded to above is a fire control system with low elevation surveillance feature for short range air defense systems. A second promising utilization of this frequency band is a Navigation and Surface Search Radar which includes a capability of detecting submarine periscopes. Many features of radars operating near 35 GHz are attractive, however, some undesirable characteristics are also present restricting their widespread use to short ranges when used in the lower atmosphere.

Advantages and Limitations of Millimeter Wave Radars

The most prominent characteristics of millimeter wave radars pertinent to shipboard applications will be discussed.

ANTENNA GAIN - Gain dependency on wavelength (inverse relationship) provides a relatively high gain factor at millimeter wave frequencies with physically small antennas. This feature is illustrated in Figure 1 for a typical monopulse antenna.

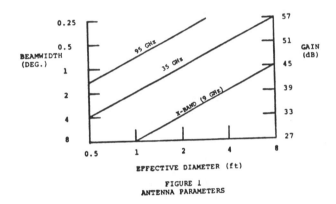

FIGURE 1
ANTENNA PARAMETERS

BEAMWIDTH - Small beamwidths are achievable with much smaller antenna apertures at millimeter waves. Figure 2 dramatically illustrates the relationship between wavelength and the antenna size for equivalent beamwidths.

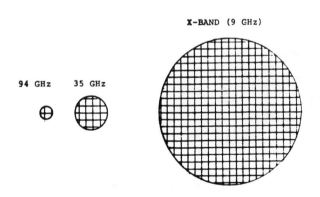

FIGURE 2
RELATIVE ANTENNA APERTURE SIZE

MULTIPATH PERFORMANCE - The specular component of the multipath reflections, being dependent upon surface roughness, are lower as frequency increases, the relative roughness being wavelength dependent. This reduction in specular reflection reduces the effects of fading and tracking errors due to multipath.

CLUTTER PERFORMANCE - Good clutter performance is obtainable at millimeter wave frequencies by having a small beamwidth (achievable with small aperture size). This allows tracking at very low angles with little clutter return by keeping most or all of the main beam out of the clutter, and by having small clutter patches in the near side lobes.

LOW PROBABILITY OF INTERCEPT - Atmospheric attenuation of the transmitted energy, although limiting the usable range, does produce the feature of severely restricting the range of detection by unfriendly sensing systems (an additional 43 dB attenuation over the free space case at 200 n. mi. in clear weather).

IMMUNITY TO ECM - A properly designed radar operating at 35 GHz will be both difficult to detect and literally impossible to jam, including main beam jamming, at stand off ranges. Even a fixed frequency, narrow spectrum radar nominally very susceptable to spot jamming is difficult. Broadening the spectrum of a 35 GHz system by any of a number of techniques produces a radar that is immune to stand off jamming. The atmospheric attenuation combined with the ability to generate only modest power levels for broadband jammers produces this effect. This feature is illustrated in Figure 3 comparing a 35 GHz radar to a similar design at X-band. The relative jamming ranges shown are those which would produce equal jamming powers into the radar receivers. Beamwidths, antenna gains, and receivers sensitivities are held equal. Six dB less transmitted power in the 35 GHz band is assumed.

VOLUME AND WEIGHT - Small volume and weights are achievable (as exemplified by the developmental model: 300 lbs. on pedestal) not only due to the small antenna size but to the scaled down version of all RF components.

FREQUENCY AGILITY - Clutter decorrelation, reduced multipath effects and enhanced detection through changing target return statistics is achieved by use of frequency agility. The agile bandwidth required is independent of transmission frequency but determined by effective pulsewidth and the number of independent returns desired. Therefore at higher frequencies the percent of bandwidth is smaller making amplitude and phase matching of multichannel systems less difficult.

ATMOSPHERIC PROPAGATION EFFECTS - Absorption of the radiated energy of millimeter wave frequencies in the lower atmosphere generally limits the usefulness to short range. The level of absorption, for clear air at sea level, is provided in the familiar curve[1] shown in Figure 4.

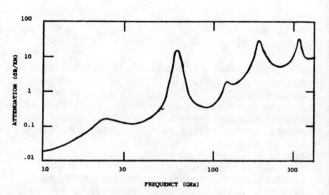

FIGURE 4
ATMOSPHERIC ATTENUATION AT SEA LEVEL

Rain absorption, and to some degree backscatter, is generally the critical range limiting factor for radar applications in the millimeter wave frequencies. The absorption[2] and backscatter[3] values are given versus rainfall rate in Figure 5.

Rain backscatter can generally be reduced to a non-limiting factor in a properly designed radar by selecting beamwidths and effective pulsewidth to produce a small clutter volume. Rain backscatter suppression on the order of 8-13 dB should also be achievable through the use of circular polarization if necessary.

The effects of fog and snow are so slight they are unnoticeable with the possibility of extremely heavy snowfall.

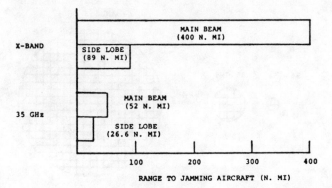

FIGURE 3
RELATIVE SUSCEPTIBILITY TO JAMMING

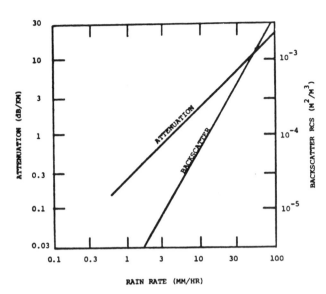

FIGURE 5
RAIN ATTENUATION AND BACKSCATTER AT 35 GHz

MTI AND PULSE DOPPLER - The short wavelength at millimeter wave frequencies introduces low velocity blind speeds in both MTI and pulse doppler processing that, in general, render these processing methods unusable.

Shipboard Application Requirements

A high performance Low Elevation Surveillance and Fire Control Radar is achievable in a 35 GHz system for short range (to the horizon) detection and tracking of air targets. Requirements to interface with specific weapons ultimately are defined by the weapon, the threat, and the tactics utilized but some guidelines may be given based upon general applications and projected radar capabilities.

General Requirements for a Low-Elevation Surveillance/Fire Control Radar System

Target Characteristics

Radar Cross Section	$1m^2$
Velocity	M 2.5
Low Altitude	30 meters
Maneuverability	5-10g turns
	5g pop up and dive in terminal phase

Radar Performance

Detection Range	20-30 KM (clear weather)
	10-20 KM (4mm rain)
Min. Range	0.3 KM
Reaction Time	6-10 seconds to firm track and handoff
Sea Conditions	up to sea state 5
Coverage	0°-3° elevation (surveillance)
	0°-90° elevation (track)
	360° azimuth
Update Rate (surveillance)	2 seconds
Automatic Detection	CFAR, ATD, automatic handoff
Tracking Accuracy	< 1 mrad, 6 meter range
Radar Height	15-20 meters
On Pedestal Weight	225 lbs. (including transmitter & receiver)
Radome	5 ft. (max.) diameter sphere
Pedestal Performance	> 30°/sec slew
	> 30°/sec^2 acceleration

The second application identified for shipboard usage of millimeter waves is a Navigation and Surface Search Radar with submarine periscope detection capability. This system should be capable of performing all the functions required of existing systems plus be capable of detecting and tracking (track-while-scan) a periscope, with one square meter radar cross section, at 16 KM in in clear air and sea state 3 and 8 KM in 4 mm of rain and sea state 5.

The SEATRACKS System

A basic block diagram of the SEATRACKS System is shown in Figure 6 and photographs of the system (less the mobile trailer) in Figures 7 and 8. A general description of the individual blocks follows.

MONOPULSE ANTENNA AND COMPARATOR

The antenna is a two-foot diameter cassegrain reflector with a four-horn monopulse feed. Polarization is linear vertical with capability built in to change to linear horizontal or circular. Peak gain of the sum channel is 43 dBi with a one degree 3 dB beamwidth. Side lobes are -18 dB nominal. Phase shifters are incorporated into the comparator and tuned to equalize the electrical length of all three channels (required for future accommodation of frequency agility). The insertion loss of any one channel is 0.75 dB maximum. Power handling capability of the antenna/comparator is 100 KW peak, 60 watts average.

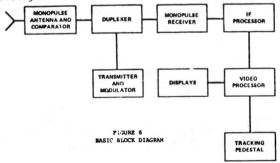

FIGURE 6
BASIC BLOCK DIAGRAM

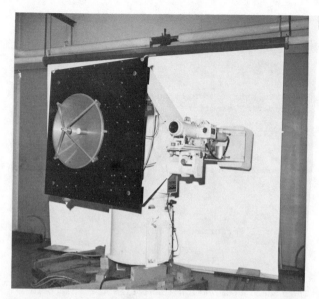

FIGURE 7
PEDESTAL MOUNTED RADAR

FIGURE 8
DISPLAYS AND CONTROL PANELS

TRANSMITTER AND MODULATOR

The heart of the transmitter system is 100 kw peak Varian SFD 319 coaxial magnetron. The tube operates at a maximum duty cycle of 0.00055 to provide 55 watts average power. The lightweight solid state modulator provides selectable PRF's of 5.5, 1.1, and 0.55 KHz with pulse widths of 0.1, 0.5, and 1.0 microseconds respectively. A power reduction network allows reduction of power by 10, 20, 30, or 40 dB. A coupler at the output provides power to the AFC network. The modulator control panel is mounted remotely. The system is pressurized with Sulfurhexafluoride (SF6).

DUPLEXER

The duplexer, developed for this system, consists of three latching ferrite switches per channel and one differential phase shift circulator per channel. Although, from a circuit protection standpoint, the differential phase shift circulators are unnecessary in the difference channels, they provide better channel to channel dynamic phase tracking. Voltage sensing interlock devices in the latching switches prevents the modulator from being triggered in the event of latching circuit failures. Total receiver isolation during the transmit pulse is 80 dB.

THREE-CHANNEL MONOPULSE RECEIVER

Each channel of the monopulse receiver consists of a balanced mixer/preamp with a 5.1 dB noise figure and 6 dB conversion loss. The 1 dB compression point is +3 dB and the dynamic range is approximately 65 dB with a total RF to IF gain of 32 dB. The local oscillator is a Gunn Oscillator with 50 mw of output power. Phase trimming is done at the L. O. arms of the difference channels to adjust the relative phase of the IF difference signals with respect to the IF sum signal. Pin-diode attenuators provide gain trim in all three channels.

The AFC circuit consists of a gated frequency discriminator followed by an integrator and amplifier. The AFC voltage is directly applied to the Varactor tuning diode inside the L. O. source. The AFC circuit acts independently of the IF strip. As a result, the IF frequency of the AFC mixer can be chosen to be different from 60 MHz to allow for a combined worst-case drift of the magnetron and Gunn L. O. higher than 60 MHz. The AFC error signal is maintained at all times by means of a sample-and-hold circuit even in the absence of target return.

IF PROCESSOR

A Varian MIF 8394 Monopulse IF subsystem as modified by Varian and NSWC, accepts the three outputs of the monopulse receiver (Σ, Δ_E, and Δ_{Az}) and generates an error signal for both Azimuth and Elevation and also puts out a linearly amplified sum channel video. Simplicity and reliability are built into the IF system by using insensitive passive components to convert the amplitude/phase information, form the three input channels, into phase only information.

VIDEO PROCESSOR

The video processor consists of circuitry that performs the following functions: acquisition, range tracking, angle processing, synchronizer, AGC processing, display preprocessing, servo loops and display symbol generation.

Each of these subsystems incorporates recent technological advances in integrated circuit and solid state devices technology to attain high performance levels while simultaneously reducing the system size and complexity, thus retaining high reliability and high serviceability.

K_a-band Radars

A digital Constant False Alarm Rate Receiver and an Automatic Target Detector has been developed at the Naval Surface Weapons Center, for automated operation in the Surveillance Mode, but has not yet been integrated into the system. The latter subsystem is built around a high speed, 16 bit microprocessor.

TRACKING PEDESTAL

The entire radar, which weighs about 300 lbs., is mounted on a Scientific-Atlanta 3100 series tracking pedestal. This unit is capable of slew rates of up to $30°/sec$ and slew accelerations of $30°/sec^2$. It can travel ± 370 in azimuth and $-5°$ to $+185°$ in elevation. The TV camera is mounted on the pedestal. The entire radar and pedestal is mounted on a Scientific Atlanta trailer designed to eliminate mechanical resonances for this system.

DISPLAYS AND CONTROLS

A dual beam "A" Scope (simultaneous full scale and expanded section of range) is provided for manual target tracking acquisition. A "B" Scope (azimuth-range) is included for radar display in the Low Elevation Surveillance Mode. The target is also displayed on a closed circuit TV monitor. Both wide angle and Telephoto lenses are remotely selectable on the pedestal mounted camera providing $10°$ and $1°$ fields-of-view respectively. Composite video superimposing a radar track error symbol onto the TV image provides a means of recording transient fluctuations in radar aimpoint without the integrating effects of the pedestal servo response.

Controls include all functions required for surveillance (performed with a two bar raster scan pattern), acquisition, and tracking.

A summary of the system features is given in the following chart.

System Description Summary

Frequency	34.67 GHz
Power	100 KW peak (controlable in 10 dB steps to 10 watts)
PW	0.1, 0.5, 1.0 µsec selectable
PRF	500 - 5500 PPS
Beamwidth	$1°$
Polarizations	Vertical, horizontal, circular
Gain Control	IAGC, Range Gated AGC, Manual
Frequency Control	Transmitted pulse sampled AFC
RF Processing	3 channel Σ, Δ_{az}, Δ_{el}
IF Processing	Monopulse IF converts amplitude to phase
Transmitter Source	100 KW Coaxial Magnetron
Modulator	Solid State
Radar Weight	300 lbs. (exclusive of pedestal weight)
Circuit Protection	Latching Ferrite Switches
Displays	A-Scope, B-Scope
Video Processing	Precision Range Tracker, synchronizer, Az and El error channels, interlock protection circuits, control circuits
Detection	Digital Constant False Alarm Rate Receiver, Microprocessor Automatic Target Detector (planned)
Pedestal	$30°/sec$ scan rate, $30°/sec$ acceleration
Auxiliary Features	TV System, digitized position and range readout, two bar raster scan, Joy Stick position control

Performance and Test Program

The feasibility model SEATRACKS is estimated to operate with the detection range and basic radar tracking accuracy (good S/N, exclusive of pedestal errors, resolver errors, etc.) shown in Figures 9 and 10. The test program is designed to verify these estimations.

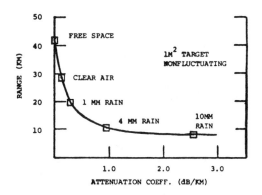

FIGURE 9
PREDICTED DETECTION RANGE

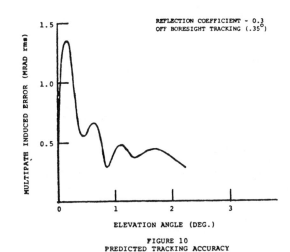

FIGURE 10
PREDICTED TRACKING ACCURACY

One of the primary intents of the test program is to obtain accurate data on the ability of a 35 GHz radar to accurately track a low elevation target over water with small tracking errors. The radar will be located at a height of 40 feet above water surface with a stationary target located at variable heights at locations of 3-4 thousand feet. The geometry will be varied to accurately controlled dimensions to simulate a low flying missile passing through multipath enhancement and cancellation regions.

A less rigid but possibly more conclusive set of tests will be performed with the 35 GHz SEATRACKS jointly sharing a pedestal with an X-Band radar having an equivalent beamwidth. Direct comparison of accuracies will be observed in tracking low flying aircraft with each system.

Conclusions

State-of-the-art development of component technology in the lower millimeter wave frequencies have made the feasibility of a radar in this band quite obtainable. Although widespread use of millimeter wave radars for shipboard applications is not anticipated, they do have unique characteristics for several uses. For application requiring accuracy, covertness, and/or small size (with short ranges acceptable) the millimeter wave frequencies are a strong candidate.

Acknowledgements

Key NSWC personnel participating on this program include L. Black, N. DeMinco, D. Kirkpatrick, J. McCorkle and C. Payne. Specific developed components, subsystems, or consulting contributions were made by the following companies:

TRG	Monopulse Antenna
Microwave Associates	Latching Ferrite Switches
Hughes Aircraft	RF Receiver, Low Power Transmitter
Varian	Coaxial Magnetron, Monopulse IF
Scientific Atlanta	Trailer
Technology Service Corp.	Technical Consulting, Analysis
Ritter Corp.	Modulator

References

1. Rosenblum, E. S., Microwave Journal, March 1961, pp. 91-96
2. Kosowsky, L. H. et al, "A 70-GHz Monopulse Track Radar", Proceedings of the SOUTHEASTCON, 4 April 1977.
3. Richard, V. W. and Kranamerer, J. E., NELC Millimeter Waves Techniques Conference, 1974.

A Millimeter Wave Radar for the U.S. Army Helicopters in the 80's

T.R. Holmes and E.A. Flick

ABSTRACT

U.S. Army tactics in repelling a hypothetical ground attack by Soviet and Warsaw Pact forces rely heavily upon highly mobile, fast striking attack helicopters. Current weapons systems use highly accurate infrared, laser and optically aimed and guided weapons to detect, track and destroy hostile threats. Radar will greatly enhance the capability of helicopters in periods of inclement weather and reduced visibility due to battlefield smoke and haze. At such time optical, infrared and laser systems suffer significantly reduced capability.

This problem has been studied in detail. Atmospheric, weather and ground clutter models have been analyzed and simulated. Various tradeoff studies have been performed to obtain a set of parameters for use in the synthesis of a system achievable within the state-of-the-art. Trade studies have included operating frequency band, transmit signal format, baseband signal processing techniques, antenna design characteristics, packaging configuration and display requirements.

1. THE THREAT

Current scenarios for a hypothetical conflict between Soviet and Warsaw Pact forces and the U.S. and its NATO allies envision massive assault by armor and infantry supported by Anti aircraft weapons and fixed and rotary wing aircraft. Tactics envisioned will incorporate the Barbarossa type assault used by the German army of World War II. This type of attack features multiple echelons with one echelon spearheading the attack while the second leapfrogs ahead to exploit the penetration by the initial assault. A third echelon provides support and replacement forces.

U.S. Army tactics in repelling a hypothetical ground attack by Soviet and Warsaw Pact forces rely heavily on highly mobile, fast striking attack helicopters. These helicopters, operating "Nap of the Earth" in a target rich environment, use highly accurate infrared laser and optically aimed and guided weapons to detect, track and destroy hostile threats, blunting and turning back the assault. It has been recognized that the addition of radar will greatly enhance the capability of helicopters in periods of inclement weather and reduced visibility due to battlefield smoke and haze. At such times optical infrared and laser systems suffer significantly reduced capability.

The WARSAW PACT forces are equipped with an array of weapons systems. Soviet weapons likely to be primary targets for U.S. Army helicopters include T-62 and T-72 tanks as well as BMP armored personnel carriers. The primary threats to U.S. helicopters include:

1. ZSU 23-4, 23 mm quad barreled anti-aircraft system.
2. S-60, 57 mm, dual barreled anti-aircraft system.
3. SA-6 surface to air missile.
4. Mi-24 Hind helicopter
5. MiG 21 Fishbed aircraft.
6. MiG 23 Flogger aircraft.

The ZSU 23 with its associated Gundish radar is capable of detecting and tracking U.S. attack helicopters at long ranges and delivering devastating firepower. The Mi 24 Hind helicopter is heavily armed and fast moving. The Hind is a formidable threat. Typical range-altitude profiles for ground based anti-aircraft threats are shown in Figure 1.

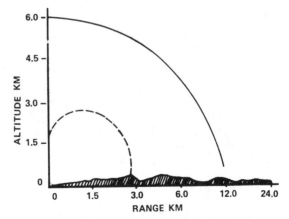

Fig. 1. Representative Threat Profiles

*This work has been supported in part by U.S. Army ARRADCOM, Mr. John Spangler, Project Manager.

The need for cover and concealment provided by nap of earth tactics as well as capability to detect threats at greatest possible range is self evident.

2. ENVIRONMENT AND ENVIRONMENTAL MODELS

The hypothesized Soviet assault is expected to occur during the October-November time frame when weather is expected to be poor. A summary of statistical weather data for a portion of Eastern Europe is presented in Table 1. It is readily apparent that some form of visibility limiting weather (fog or rain) is expected with high frequency during this time frame.

Table 1. Adverse Weather Summary

VISIBILITY LESS THAN 3 KM 43% OF SEP MORNINGS

AVERAGE FOR DURATION 5 HRS. IN FALL, 6 HRS IN WINTER

CEILING LESS THAN 1000 FT. 28% OF SEP DAYS, 34% OCT DAYS

VISIBILITY LESS THAN 3 KM 38% OF ALL DEC HRS.

MEASURABLE RAIN 132 DAYS OF YR. (AVG DURATION 8.5 HRS.)

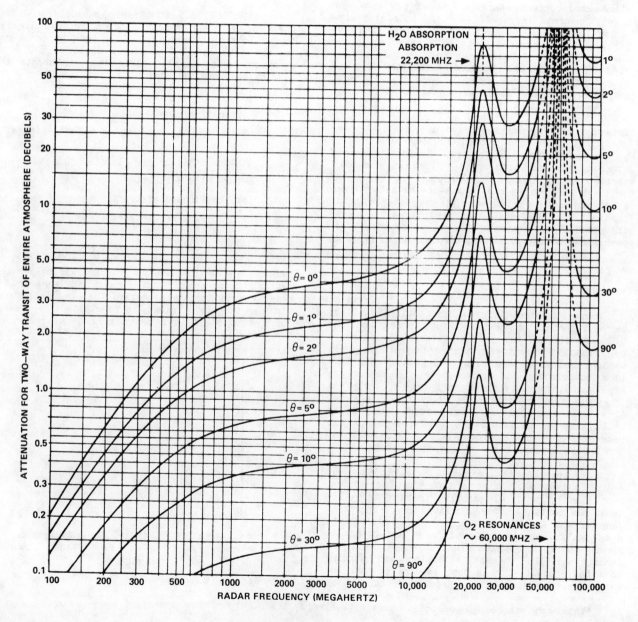

Fig. 2. Atmospheric Attenuation Curves

Additionally Soviet T-62 and T-72 tanks (and other vehicles) are routinely outfitted with smoke generating equipment. It is standard Soviet doctrine to lay down smoke screens during military operations. A full scale conflict is likely to result in various additional smokes (such as burning oil), aerosols and debris.

In this type environment optical and IR systems suffer decreased range and accuracy. Figure 2 shows measurements made of atmospheric attenuation as a function of frequency. (Reference 1.) From the curves one sees three frequency bands of interest for a possible radar system. These bands are:

1. Around 15 GHZ (K_u band) and lower.

2. Around 35 GHZ (K_a band).

3. Around 96 GHZ (W-band).

A major part of the environment includes the ground clutter background. Extensive measurements have been made for various types of ground clutter. (Reference 2,3.) A model for prediction of crossection as a function of grazing angle and frequency (or wavelength) has been formulated. The basic form of the model is presented in Table 2. The units of crossection are square meters per square meter of surface illuminated. The constants C_1, C_2, C_3 are empirically determined.

Table 2. Ground Clutter Crossection Model

$$\overset{\bullet}{\sigma} = -C_1 + C_2 (\psi/\psi_o) + C_3 \log(f/f_o)$$

DONE BY DYER & HAYES OF GIT
FOR FRANKFORD ARSENAL

Models for backscatter crossection, absorption and scattering losses due to raindrops have been developed using measured data. (Reference 4.) These models are summarized in Table 3.

In addition the clear air absorption shown in Figure 2 may be modeled by using the results of Van Vleck. (Reference 5.)

3. SYSTEM REQUIREMENTS AND ANALYSIS

An analysis of the previously discussed tactics, threats and operational environment lead to the conclusion that a potential radar system must have the following characteristics:

1. Detect and track moving and fixed ground targets in heavy tree clutter.

2. Detect and track rotary wing (moving and hovering) and fixed wing aircraft.

3. Detect and track at standoff ranges.

4. Detect and track targets in moderate rain (nominally 4 mm/hr.) and dense fog (100 meter visibility).

Table 3. Rain and Fog Models

VOLUMETRIC BACKSCATTER DUE TO RAINFALL IN dB METERS^2/METERS^3

$$N(r) = A(f) r^{B(f)}$$

WHERE:

$$A(f) = 1.782 \times 10^{-15} f^{5.74}$$

$$B(f) = 2.108 - 0.135f + 0.00688 f^2 - 0.000118 f^3 + 0.620 \times 10^{-6} f^4$$

ONE-WAY EXCESS ATTENUATION IN db/KILOMETER RESULT FROM RAINFALL AND FOG

$$\alpha(r) = 6.28 \times 10^{-6} \left(\frac{f}{1+\frac{f}{17}}\right)^{5.37} (r)^{m(f)}$$

WHERE $m(f) = 1.182 - 4.622 \times 10^{-3} f^{1.018}$

$$\alpha(f) = f^{1.95} \times 1666 V^{-1.43} e^{-6.866} (1 + 0.0045T)$$

WHERE r IS RAINFALL RATE IN MILLIMETERS/HOUR AND
V IS ONE-WAY VISIBILITY IN FEET.
f IS FREQUENCY IN MHZ.

In order to assess the ability of a candidate system to meet these requirements, the models previously cited may be used. The radar range equation may be used to predict target, clutter, weather and noise power. The expressions become:

Clutter power:

$$P_c = \frac{P_T G^2 \lambda^2 c\tau\theta_B}{(4\pi)^3 R^3 \cos\psi L_T} \frac{\sigma_o}{\epsilon} \qquad (1)$$

Weather Power:

$$P_w = \frac{P_T G^2 \lambda^2 (1-\cos\theta_B) E}{(4\pi)^3 L_T R^4} \frac{\eta}{\epsilon} \qquad (2)$$

Target Power:

$$P_S = \frac{P_T G^2 \lambda^2}{(4\pi)^3 L_T R^4} \frac{\sigma_t}{\epsilon} \quad (3)$$

Noise Power:

$$P_N = KT \left(\frac{1.2}{\tau}\right) F \quad (4)$$

Where:

P_T	=	Transmitter power
G	=	Antenna gain
c	=	Speed of light = $f\lambda$
θ_B	=	Antenna beamwidth
R	=	Range
KT	=	Noise spectral density
σ_t	=	Target radar crossection
F	=	Receiver noise figure
L_T	=	Equipment loss
ψ	=	Grazing angle
σ_o	=	Predicted clutter crossection
η	=	Predicted weather backscatter crossection
ϵ	=	Predicted composite atmospheric absorption
$\frac{1.2}{\tau}$	=	IF bandwidth
τ	=	Radar pulsewidth
E	=	$(R + \frac{c\tau}{2})^3 - (R - \frac{c\tau}{2})^3$

Analysis of candidate modulation techniques revealed that for the power available from state of the art coherent millimeter wave sources, the cost and complexity of a fully coherent, pulse compression system was prohibitive for this application. A more cost effective technique is presented by use of a frequency agile system with noncoherent detection. The composite signal to interference ratio obtained for such a noncoherent, frequency agile system is

$$S/I = \frac{N^\rho (SNO)}{1 + \left(\frac{N}{B\tau}\right)^\rho (CNO) + \left(\frac{N}{B\tau}\right)^\rho (WNO)} \quad (5)$$

Where

SNO	=	Signal (target)-to-noise ratio out of a noncoherent envelope detector
CNO	=	Clutter to noise ratio
WNO	=	Weather to noise ratio
N	=	Number of pulses processed
B	=	Frequency agile bandwidth
ρ	=	Processor efficiency relative to a perfect, coherent integrator

Given a set of constraints in the form of required detection range, probability of detection, false alarm rate, size and weight, tracking accuracy, etc., one is now in a position to optimize specific system parameters.

Table 4 shows the result of a trade-off analysis of operating frequency band subject to a constraint on the allowable size of antenna which can be supported. It can be seen that although K_u band appears attractive from the standpoint of weather performance, system resolution is poor. While 96 GHZ provides good system resolution, weather and atmospheric absorption performance is poor. The analysis favors the 35 GHZ band for implementation.

Table 4. Frequency Tradeoff

PARAMETER	16 GHZ	35 GHZ	96 GHZ
ATTENUATION[1], dB/km	0.62	1.6	5.1
BACKSCATTER CROSSECTION[2] dBm^2/m^3	-60	-49	-38
WAVELENGTH, mm	17.7	8.6	3
BEAMWIDTH[3], DEG	2.7	1.3	0.5
RESOLUTION CELL WIDTH[4], m	188	90	35

1 ONE-WAY 4mm/HR, RAINFALL PLUS NORMAL ATMOSPHERIC O_2, H_2O. 400 FT VISIBILITY FOG

2 ONE-WAY 4mm/HR RAINFALL RATE

3 18" ANTENNA DIAMETER

4 4000M RANGE

Antenna design for the helicopter borne radar presents a challenge. The desire for air and ground search as well as track capability dictate a dual beam shape. In addition, since ground clutter and weather clutter are somewhat polarization sensitive, the antenna should be capable of radiating selectable polarization, horizontal, vertical or circular. The antenna will have an 18 inch aperture and be capable of switching between a spoiled beam and pencil beam in microseconds. A full aperture 1/4 wave plate polarizer will be used to obtain selectable polarization. The spoiled beam will support full monopulse tracking.

Moving target detection is achieved using an all digital MTI. The spectrum of land clutter has been studied (Reference 3) and can be modeled by a function of the form

$$c(f) = \frac{1}{1 + \left(\frac{f}{f_c}\right)^x} \qquad (6)$$

Where f_c = 17 Hz AT 35 GHZ
x = 2.5 AT 35 GHZ

After extensive modeling and simulation, it was found that a 5-pole elliptic recursive digital filter implementation can achieve 26 dB MTI improvement for moving targets in this type clutter background. The filter passband is adaptively controlled to compensate for clutter spread while scanning off the ground track.

Stationary targets are detected by processing the frequency agile returns using conventional techniques. The all digital signal processor employs a digital integrator, adaptive threshold (CFAR) circuit and a double threshold (M of N) detector.

4. SUMMARY

After 4 years of study and research a radar system for use by U.S. Army helicopters has been designed. The radar will incorporate the following features:

1. Operate in the 35 GHZ band.

2. Provide search/track capability in pencil or spoiled beam using full monopulse tracking.

3. Provide fixed or frequency agile operation for detection of moving or stationary targets.

4. Provide moving target detection with all digital, adaptive MTI.

5. The radar system will also provide capability not presently in existance for navigation, obstacle avoidance and ground mapping. These capabilities will provide significant enhancement of the ability of helicopters to fly to and from the battlefield in this adverse visibility environment.

The system described is currently under development and will be fielded in the early to mid 80's time frame.

REFERENCES

(1) Blake, L.V., "A Guide to Basic Pulse Radar Maximum Range Calculation", AD 701 321, Dec. 1969.

(2) Hayes, R.D. and F.B. Dyer, "Land Clutter Characteristics for Computer Modeling of Fire Control Radar Systems", AD 912 490 May 1973.

(3) Hayes, R.D. and F.B. Dyer, "Radar Land Clutter Measurements at Frequencies of 9.5, 16, 35 and 95 GHZ", AD A012 709 March 1975.

(4) Hayes, R.D., F.B. Dyer, and N.C. Currie, "Radar Backscatter from Rain at 9.375, 35, 70 and 95 GHZ", AD A007254, Feb. 1975.

(5) Van Vleck, J.H., "The Absorption of Microwaves by Oxygen" and "The Absorption of Microwaves by Uncondensed Water Vapor", Physical Reviews, 71 (No.7): 413-424, 425-433, Apr. 1947.

CHAPTER 5

MILLIMETER WAVE RADARS

It will be recalled from Chapter 1 that the current IEEE Radar Standard for letter-bands defines millimeter waves for radar as the region of 40 - 300 GHz. Accordingly, representative radars operating above K_a-band will be described in this chapter. Inspection of Tables 5, 6, and 7 of Paper 1.1 in Chapter 1 indicates that most of the reported millimeter radars are for military or military-related activities. Accordingly, only one civilian application of millimeter wave radar (Paper 5.5) will be described in this chapter.

Comparison of papers in this chapter with those of the preceding chapter indicates that most of the radars in Chapter 5 are intended to operate at very short ranges, as compared to, for example, Papers 4.2 and 4.6. There is a general similarity between Papers 4.3, 4.4 and 4.5 in Chapter 4 and the many solid-state radars in this chapter.

Radar applications of papers in this chapter include ground-based land surveillance, airborne land surveillance, missile seekers, instrumentation radars, ballistic missile defense radars, and space object identification. Generally, 70 GHz radar work has been discontinued; hence, most of the papers in this chapter operate at 95 GHz. Papers on a few 70 GHz radars are included to illustrate certain applications and/or radar techniques.

Papers will be presented in chronological order to give an appreciation for the approximate technological maturity represented therein.

SYNOPSIS OF REPRINT PAPERS IN CHAPTER 5

5.1 "Combat Surveillance Radar AN/MPS-29 (XE-1)" — Long, Rivers and Butterworth

Radar Set AN/MPS(XE-1), designed by the Georgia Institute of Technology Engineering Experiment Station in 1957-9 for the U.S. Army Signal Research and Development Laboratory, was the first working millimeter radar to receive AN nomenclature. Its intended application was combat surveillance of vehicles, weapons, projectiles and personnel. Extremely high resolution is achieved by use of a narrow (.2 deg azimuth) beam and narrow pulse duration (50 ns). Scan rate of 20 per second over a 30 degree sector is achieved by use of a geodesic Luneburg lens antenna. Displays include A and B scopes and aural for non-coherent Doppler. Operational capability is both scanning and searchlight modes. Special means are provided for transfer from scanning to searchlight mode within 2 seconds of command.

An extensive description of the antenna, ring switch, mode transfer mechanism, and limited test results are contained in the paper. It is to be noted that although better mixer diodes (*cf.* Paper 3.D.1, Chapter 3) and solid state local oscillators (*cf.* papers in Section 3A, Chapter 3) are available now, performance of this radar is comparable to that of current radars, although it was built almost 25 years ago.

This paper was presented at the Sixth Tri-Service Radar Symposium, 1960.

5.2 "A 94-GHz Radar for Space Object Identification" — Hoffman, Hurlbut, Kind, Wintroub

This paper, presented at the 1969 IEEE International Microwave Symposium, is noteworthy in several ways: 1) it is one of the earliest 95 GHz radars reported in the literature; 2) the intended target range (about 1000 nmi, maximum) is much greater than the "typical" mm wave radar in this chapter; and 3) the techniques employed represent a very difficult task.

The radar is a linear FM pulse compression system with a 1 ms pulse width and 1 GHz frequency modulation for a time-bandwidth product of 10^6 and a pulse compression ratio of 10^6 (60 dB) (*cf.* Papers 3.A.5 and 4.5), resulting in a range resolution of six inches.

Utilization of this very long pulse width requires extreme system linearity. A phase-locked loop is used to provide frequency sweep linearity (*cf.* papers in Chapter 3).

Although not mentioned in this paper, antenna sidelobes could present a problem. Consider the radar cross section patterns for various satellite materials of Paper 2.B.1 in Chapter 2. If the antenna main lobe were to be pointed at a "sidelobe" of a radar cross section pattern, an antenna sidelobe could conceivably be pointed at an RCS pattern "main lobe" at the same time. (Of course, this situation could be reversed.) Possibly the two signals might be at different range gate positions; if in the same range gate, the signals would add vectorially. A false return would then result. If the two signals were in different range gates, moving the gate selector could cause display of the erroneous antenna sidelobe received signal.

This paper is extended and updated by Paper 5.11 later in this chapter.

5.3 "Vehicle-Mounted Millimeter Radar" – Dyer and Goodman

Experience gained in the design and testing of Radar Set AN/MPS-29 (XE-1), described in Paper 5.1 of this chapter, led to design of the Vehicle-Mounted Millimeter Radar (informally called VEH).

VEH was designed for the U.S. Army Harry Diamond Laboratories by the Georgia Institute of Technology Engineering Experiment Station in the mid-1960s. Portions were designed by HDL personnel; the indicator was built by Beta Instrument Corp., with participation by both HDL and Georgia Tech in its design.

Application of the VEH is that of target identification and ranging with capability of tracking outbound shells in order to provide firing commands and thereby improve first-round kill probability. Subsequent testing revealed a capability for blind, *i.e.*, non-visual vehicle navigation (*e.g.*, obstacle avoidance) by this radar.

Installation in a U.S. Army M 113 Personnel Carrier dictated many changes from that of the AN/MPS-29, although both utilize a rapid sector scanning geodesic Luneburg lens antenna. The VEH uses a folded geodesic Luneburg lens antenna and has a higher sector scan rate (up to 70 per sec). A periscope provides an optical view of the radar illuminated area. Freedom of antenna sector from vehicle orientation is provided by rotating the radar console and antenna about a common axis such that the radar operator rotates with the antenna. As in the AN/MPS-29, displays include A and B scopes and aural non-coherent Doppler.

Extensive discussion of the helmet type folded geodesic Luneburg lens antenna scanner, radar electronics, and test results are contained in the paper. The table of measured radar cross sections of various targets, although not given as RCS patterns, is noteworthy. The wide range of values shown, although not stated, is due primarily to target RCS pattern lobes and indicates the impropriety of using a single value of RCS for radar design (*cf.* Paper 2.B.1, Chapter 2).

This paper was presented at the 18th Tri-Service Radar Symposium in 1972.

5.4 "Development and Test of a 95 GHz Terrain Imaging Radar" – Wilcox

Narrow beamwidth and short pulse width achievable at millimeter wavelengths make this frequency region attractive for terrain imaging radar. The radar described in this paper operates at 95 GHz and has a ten foot range resolution resulting from a twenty ns second pulse duration. Azimuth resolution of 40 ft. per nautical mile (6 dB points) results from the 2 ft. diameter Cassegrain fed parabolic antenna. Antenna polarization can be either circular or linear. Radar display is a PPI. Maximum display range is 8 nmi. Transmitter employs a magnetron with a klystron local oscillator in the receiver. IF amplifier center frequency is 335 MHz.

Photographs of scope display of this radar are very impressive. Individual utility poles can be detected at a range of 6 nmi. Although individual automobiles can be detected at considerable range, it is not possible to identify them; hence, alternate means of target identification would be necessary.

This paper was presented at the Naval Electronic Laboratory Center 1974 Millimeter Wave Techniques Conference.

5.5 "A Solid-State 94 GHz Doppler Radar" – Bernues, Kuno and McIntosh

Solid-state millimeter-wave technology as described by several papers in Chapter 3 has made possible the use of homodyne systems for Doppler radars. A homodyne receiver is a superheterodyne where the local oscillator is at the same frequency as the transmitter frequency, making the IF at zero frequency, *i.e.* video frequency. In most homodyne radars, the local oscillator signal is obtained by leaking a part of the transmitter signal to the mixer. Such a radar has several advantages, *e.g.* elimination of local oscillator frequency drift, the system is inherently fully coherent and a lighter system can result from elimination of a separate local oscillator.

The very compact homodyne radar described in this paper has been used as a harbor ship speedometer and as a hand-held police radar speedometer.

Photograph of the latter is shown in Paper 1.1 (Chapter 1), Figure 13. Detection range is approximately 1 - 3 km, depending on the target type and aspect angle. The IMPATT diode transmitter has a CW power output of about 100 mW.

5.6 "Millimeter Wave Monopulse Track Radar" — Kosowsky, Koester and Graziano

The radar described in this paper was probably the first millimeter wave monopulse tracking radar developed. It was actually developed several years earlier, this being a later and more complete paper describing the radar.

Topics included in this paper are propagation factors (clear air attenuation, rain, fog and clouds, snow) frequency selection, monopulse concepts, performance predictions, system description, and preliminary experimental results.

Target acquisition is accomplished visually using a helmet-mounted optical sight worn by the radar operator. For actual system employment, it would be necessary to use other target acquisition techniques such as that of Paper 3.E.3 in Chapter 3. Other papers on this radar are listed in the bibliography for this chapter.

5.7 "Combined Electro-Optical/Millimeter Wave Radar Sensor System" — Holm, Foster, Loefer

It will be recalled from the comparison of millimeter waves and optical systems in Table I, Paper 1.5, Chapter 1, that the two systems are complementary in capabilities.

Papers 5.1 and 5.3 in this chapter describe millimeter wave combat surveillance radars. These radars have limited capability for target identification, however.

This paper describes the marriage of the radar of Paper 5.3 with an electro-optical sensor. Target search/acquisition is provided by the radar. Target identification is then accomplished by the boresighted EO system.

In the phase I system, a TV system is used as the EO system; however, this will be replaced by an active laser radar later in the program.

This paper describes the millimeter wave radar, its modified geodesic lens scanning antenna, integration of the TV system, limited test results and plans for the laser radar.

Although not specifically identified in the bibliography, the radar is that of Paper 5.3 in this chapter. The radar is novel in its ability to scan both vertically and horizontally, as well as in its non-coherent Doppler aural display. It will be noted that the radar uses a klystron local oscillator and does not use Schottky barrier diode mixers. These could improve radar performance; however, present radar performance is adequate to demonstrate the concept of a combined system.

This paper was presented at the 6th DARPA-Tri-Service Millimeter Wave Conference in 1977.

5.8 "Radar Tracking of an M-48 Tank at 94 and 140 GHz" — McGee and Loomis

A more descriptive title for this DARPA-Tri-Services Millimeter Wave Conference paper could be "A Dual Band Millimeter Wave Automatic Tracking Radar". The radar described in this paper consists of two interchangeable RF assemblies with a common receiver and automatic angle and range tracking system. Currently the radar can operate at 94 or at 140 GHz. Capability to operate at 217 GHz was initiated.

RF source at both frequencies is a pulsed IMPATT diode with inherent frequency chirp during the pulse. Local oscillator is a Gunn diode for 95 GHz and a klystron at 140 GHz. PIN diode attenuators are used to reduce transmitter spike leakage through the circulator. Conical scan with a 60 Hz scan frequency is used on the Cassegrain antenna.

A photograph of this radar appears in Reference 10 of Paper 2.A.1, Chapter 1.

5.9 "94 GHz Active Missile Seeker Captive Flight Test Results" — Yoshitani and Beebe

This interesting DARPA-Tri-Service Millimeter Wave Conference paper describes results obtained from the captive flight tests of a 94 GHz tracking radar used as an active missile seeker. The radar is similar to that of Paper 5.5 of Chapter 5 of this volume, except that this automatic range and angle tracking radar uses a conically scanned Cassegrain antenna and the transmitter is chirped.

The term "aimpoint wander" used in this paper and by some millimeter wave radar practitioners is actually angle noise, in accordance with the generally used IEEE Standard Radar Definitions [1, 2]. Angle noise is due to both glint and scintillation error. An English language translation of reference [2] of this paper will be found in reference [3] in the bibliography for this chapter. An extensive bibliography of angle noise is in reference [4]. Papers treating effects of frequency agility are cited in reference [5].

As stated by the authors, results obtained from these tests are at odds with experience of others at X-band. Accordingly, the results reported here require further study and analysis. This is the first open publication of results of tracking tests of mm wave radars (*cf.* the preceding paper).

Extension of these tracking test results to the design of a missile system requires caution. It is not proper to assume a direct correspondence of these results to missile miss distance. Characteristics of the missile autopilot hardware must be included. This requires use of a six degree of freedom missile simulation.

5.10 "Potential Applications of Millimeter Wave Radar to Ballistic Missile Defense" — Jones

Millimeter-waves offer a potential improvement over microwaves for ballistic missile defense radars. This *IEEE EASCON-77* paper surveys potential BMD applications of millimeter waves. Applications cited are both ground based and missile borne. Operating ranges vary from 200 m (fuzing) to 2500 km (exoatmospheric range tracking). Accompanying antenna diameters are 5 cm to 10 m. Average powers vary between less than 1 W to 100 kW. Truly these applications cover a wide spectrum.

The problem of obtaining required performance from ground based mm radars during inclement weather is not addressed in this paper. A BMD radar cannot choose its operating weather.

5.11 "Millimeter Radar Application to SOI" — Dybdal

This *IEEE EASCON-77* paper is an extension and update of Paper 5.2 in this chapter, with emphasis on system considerations, *e.g.* resolution, radar requirements, millimeter component technology and target characteristics. Characteristics of two rather ambitious systems are discussed. Presumably such a system might not have to operate in inclement weather, as a BMD system of the preceding paper might have to.

A useful bibliography on millimeter wave components and various radar techniques is included.

5.12 "mm-Wave Instrumentation Radar Systems" — Currie, Scheer and Holm

Instrumentation radars operating at 35, 70 and 95 GHz developed by the Georgia Institute of Technology Engineering Experiment Station are described in this paper. This is an extension of an *IEEE EASCON-77* paper (listed in the bibliography for this chapter).

These radars are intended as research tools for applications such as measurements of terrain backscatter, atmospheric attenuation, etc. As described, the radars do not have capability of target tracking. However, these radars are very versatile in frequency agility, polarization agility, dual polarization and coherent transmitter-receivers. Five video receiving channels are provided: H and V amplitude and phase as well as pseudo-coherent phase video.

5.13 "A Millimeter Wave Radar for the Mini-RPV" — Kosowsky, Koester, Gelernter and Johnson

The mini-RPV (remotely piloted vehicle, *e.g.* a drone aircraft) offers an attractive potential for ground surveillance through use of a millimeter wave radar. A radar for this application must be extremely light, and operate in severe ground clutter and in adverse weather. A 95 GHz brassboard radar described in this *AIAA/DARPA Conference on Smart Sensors* paper has been developed for this purpose. This polarization agile radar has several unusual features for a mm-wave system, *e.g.* an antenna having $\cos^2 \cos^{1/2}$ elevation pattern, a magnetron transmitter, and a digital signal processor. The fixed target enhancement (FTE) using polarization diversity discrimination is described. Operation in HRGM (high resolution ground mapping) and moving target indicator (MTI) modes are also possible with this radar. Further information on this radar is contained in other papers listed in the bibliography for this chapter.

Data contained in this paper regarding smoke propagation attenuation supplements that of Paper 2.A.7 in Chapter 2.

5.14 "Model Simulation of Target Characteristics and Employing Millimetre Wave Radar Systems" — Gabsdil and Jacobi

Fuzing radars for military weapon systems operate in a very unique target signature environment, *i.e.*, at very short distances. Usual target far field radar cross section patterns are not usable for such short range operation. This *AGARD* paper describes use of millimeter-wave radars for scaled frequency simulation of fuze operation. Frequency scaled radar cross section measurement techniques using microwaves are well known, *e.g.* [6]. Millimeter waves have also been used for scale model RCS measurements as described in reference [41] of Paper 1.6, Chapter 1. That reference describes a unique method for the *simultaneous* measurement of target radar cross section *and* glint (*cf.* Paper 5.9).

Proper investigation of fuze operation is very difficult. This paper primarily treats the application rather than the apparatus. Characteristics of the apparatus would be of interest, however. Laws for scaling of parameters in Table 1 should have wide use for various scaling activities. Measurement results with spectra at selected intervals demonstrate data which can be obtained.

References cited above should be useful to readers of this paper in view of the absence of references by the authors.

5.15 "HOWLS Radar Development" – Lynn

This paper from an AGARD Avionics/Guidance and Control Symposium is somewhat complementary to Paper 5.13. The specific problem here is the location and classification of hostile artillery by use of mini-RPV radar. Representative radars at four frequencies: 10, 16, 35 and 95 GHz (X-, K_u-, K_a-, and mm, respectively) are considered. The antenna has a csc^2 pattern (*cf.* Paper 5.13).

An extensive discussion is included on signal processing techniques: fixed target detection and target classification. An experimental K_u-band system was built as a result of its greater range than K_a- or mm. Although possibly only coincidental, it is interesting to note that K_u-band was selected over K- and K_a-band for the ASDE application discussed in the introduction to Chapter 1. This choice provides an interesting contrast to Paper 5.13 in this regard.

5.16 "Millimeter Airborne Radar Target Detection and Selection Techniques" – Novak and Vote

Typical military ground targets will exist in a different environment from aircraft targets. An aircraft may fly at low altitudes against a clutter background. Fortunately, aircraft motion permits use of moving target indicator (MTI) techniques to detect the aircraft. Ground targets present a very difficult detection problem. In addition to severe ground clutter, the ground target may be motionless, *e.g.* a weapon or parked vehicle. Another paper in this chapter (5.13) proposes polarization properties as a means of target-clutter differentiation. This paper addresses the evaluation of several digital computer target detection algorithms for this purpose. Radar data used in this work were obtained from the K_u-band HOWLS radar described in the preceding paper.

The principal detection algorithm used involves the lead/lag CFAR windows technique. Clutter distribution models considered include Rayleigh, Ricean, Weibull, and log-normal. It is found that log-normal is the most appropriate for characterizing amplitude variations within a given cell. This is reflected in the clutter papers in Chapter 2 of this volume.

Target selection algorithms are also investigated. The algorithm found to be the most promising involves examination of the spread-to-mean ratio and target-to-clutter ratio of each target cell.

This paper is informative as to the problems of automatic detection and selection of military ground targets by airborne mm wave search radars.

5.17 "USAF Millimeter-Wave Seeker Development" – Disalvio

Millimeter wave radar offers a potential for guidance of air-to-surface missiles. This *IEEE NAECON-79* paper is a review of the U.S. Air Force Armament Laboratory millimeter wave "contrast" guidance demonstration program.

Although referred to in the paper as "all weather", effects of rain are not considered here.

Work by a number of contractors on this program are described. Interestingly, scale model RCS measurements of a battle tank are being made using the facility cited in the previous summary of Paper 5.14 in this chapter. Glint measurements would be very useful in a better understanding of the experimental tracking test results of Paper 5.9 in this chapter. Another version of this paper is listed in the bibliography for this chapter.

5.18 "A Terminal Guidance Simulator for Evaluation of Millimeter Wave Seekers" – Wismer and Witsmeer

In the discussion of Paper 5.9 previously in this Chapter, it is indicated that a full missile simulation is necessary in order to obtain target miss distance; tracking accuracy is not a direct indicator of miss distance. This paper describes a unique active and passive terminal guidance seeker RF chamber simulator. A minicomputer is used to control the illuminator array. Primary emphasis is given to the use of fluorescent lights as a novel source of radiation from a simulated target for passive seeker evaluation. The authors properly indicate that the role of a simulator is restricted to tracking; the vital function of target acquisition (*cf.* Paper 1.1, Chapter 1) must be investigated by other means.

This paper was presented at MM-80 (Military Microwave-80) London, October 1980.

REFERENCES FOR CHAPTER 5

[1] "IEEE Standard Radar Definitions," *IEEE Standard 686-1977*, November 1977.

[2] "Standard Dictionary of Electrical and Electronic Terms," *IEEE Std. 100-1977*, IEEE, N.Y.

[3] R.V. Ostrovityanov, "On the Problems of Angular Noise," *Radio Engrg. and Electron. Phys.* Vol. 11 No. 4, April 1966, pp. 507-515.

[4] D.K. Barton, "International Cumulative Index on Radar Systems," 1978, *IEEE Publication Nr JH 4675.5.* See "Glint," pp. 29-30, especially J.H. Dunn and D.D. Howard, "Target Noise," Chapter 28 in *Radar Handbook*.

[5] Papers by Lind and by Sims and Graf in Barton, *loc cit.*

[6] C.G. Bachman, H.E. King, and R.C. Hansen, "Techniques for Measurement of Reduced Radar Cross Sections," *Microwave Jnl.* Vol. 6 No. 2, Feb. 1963, pp. 61-7; No. 3, Mar. 1963, pp. 95-101; No. 4, April 1963, pp. 80-86.

BIBLIOGRAPHY TOPICS FOR CHAPTER 5

mm-Wave Radars

SUPPLEMENTAL BIBLIOGRAPHY FOR CHAPTER 5

mm-Wave Radars

Journal Articles

Kosowsky, L.H., *et al.*, "A Millimeter Wave Surveillance Radar for RPV's," *Proc. IEEE SOUTHEASCON-77*, April 1977, pp. 238-43.

Kosowsky, L.H., Koester, R.L. and Carulli, J., "A 70 GHz Monopulse Track Radar," *Proc. IEEE SOUTHEASCON-77*, April 1977, pp. 331-8.

Scheer, J.A., Eaves, J.L. and Currie, N.C. "Modern Millimeter Wave Instrumentation Radar Development and Research Methodology," *IEEE EASCON-77 Rec.*, pp. 16-6A through H.

Disalvio, A.N., "USAF Millimetre-Wave Seeker Developments," *International Def. Ref.* Vol. 13 No. 1, 1980, pp. 42-6.

Pratt, D.H., Kosowsky, L.H., Koester, K. and Johnson, W., "Mini-RPV Radar Test Program," *Proc. Symp. of the Assn. for Unmanned Vehicle Systems* (AUVS-80), June 1980, Dayton, OH.

Barrett, C.R., Jr., and Ryan, D.A., "Surveillance and Target Acquisition Radar for Tank Location and Engagement: STARTLE," *Proc. Military Microwaves Conf.*, 1980 (MM-80).

Woolcock, S.C., and Plaster, M.W., "Backscatter Measurement Radar Systems Operating at 140 GHz and 280 GHz," *Proc. Military Microwaves Conf.*, 1980 (MM-80).

Combat Surveillance Radar AN/MPS-29 (XE-1)

M.W. Long, W.K. Rivers and J.C. Butterworth

6th Tri-Service Radar Symposium Record, 1960

At the request of the Signal Corps, development of an experimental 4.3 millimeter ground surveillance radar was undertaken in July 1957, and the radar was delivered to Fort Huachuca for field observations in August 1959. The use of short wavelengths provides the potentiality of developing very narrow beam antennas of reasonable size. Difficulties associated with these wavelengths include high atmospheric attenuation, increased waveguide losses, and increased phase errors because of dimensional errors. This paper describes a narrow-beam, rapid-scanning radar and results of field operation.

Radar System

The system was designed for the high resolution display of ground targets at short ranges; use is made of non-coherent doppler to assist in identifying moving targets. A photograph of the experimental system prepared for operation is shown in Figure 1. The system is operable in a scanning mode and is capable of being switched on command to a searchlight mode for observing targets at any desired azimuth and range. Figure 2 is a photograph of the display console. Visible are the B-scope, A-scope, and speaker for non-coherent doppler return. A wobble-stick for selecting range and azimuth coordinates in the searchlight mode is located between the A- and B-scopes.

System parameters are given in Table 1.

-----

*This work was sponsored by the United States Army Signal Research and Development Laboratory under Contract No. DA 36-039 SC-74870.

Figure 1. System Prepared for Operation

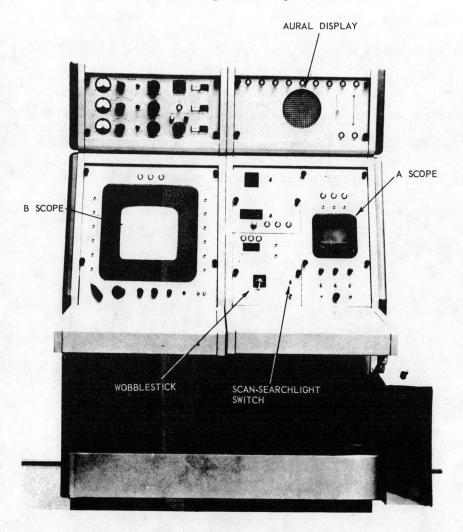

Figure 2. Display Console

Table 1. System Parameters

Resolution Characteristics

Azimuth beamwidth	0.2 degree (3.5 mils)
Radar pulse width	0.05 microsecond (7.5 meters)
Elevation beamwidth	0.3 degree (shaped to -4 degrees of elevation)
Polarization	vertical
Scan rate	20 scans per second
Scan sector	30 degrees (150 beamwidths)
P.R.F.	10,000 pulses per second

Sensitivity Characteristics

Antenna gain (including losses)	54.7 db
Antenna losses	2.8 db
Waveguide losses	0.4 db
TR-ATR losses (BL-24)	4.6 db max (2-way)
Transmitter (BL-221)	15 kw
Receiver noise figure (Philco L-5506 crystals and 4 db IF N.F., QK-369 klystron)	18 db

Scanner

One of the novel features of the radar is the scanning antenna. The antenna consists of a geodesic Luneberg lens for azimuthal collimation and a modified parabolic cylinder for vertical collimation and beam-shaping. The geodesic lens is fed by a rotary feed mechanism which utilizes a ring switch of new design.

Figure 3 is an exploded drawing of the scanner. The geodesic lens consists of a pair of closely nested conducting surfaces which are shaped to transform a point source of TEM mode radiation at its feed arc into a collimated line source at the output of the lens diametrically opposite the point source. This type of lens is well known and is described in the literature[1,2]

The mechanism employed for feeding the lens is of the ring switch type. It is based on a broadband, low-loss switch junction that was developed for this purpose. The ring switch, which acts as a commutator, consists of a stationary input section and a rotary output section which has eleven sectoral horns for feeding the geodesic lens. Energy is transferred by the input junction to a split rectangular waveguide ring formed by the stator and rotor of the ring switch. Energy leaves the ring through the output junction and the feed horn which are passing across the 30-degree feed arc. This energy spreads throughout the lens and then passes through the parallel

- - - - -

1. F. G. R. Warren, "The Tin Hat, A Modified Luneberg Lens," *Proceedings of the Fourth Symposium on Scanning Antennas*, Naval Research Laboratory, Washington, D. C., April 21-22, 1952, p. 104. CONFIDENTIAL.

2. J. S. Hollis and M. W. Long, "A Luneberg Lens Scanning System," *IRE Transactions on Antennas and Propagation*, Vol. AP-5, No. 1, January 1957.

Millimeter Wave Radars

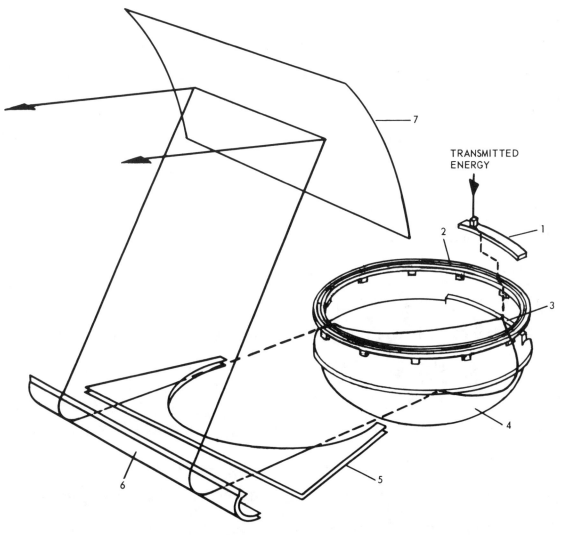

Figure 3. Exploded View of Scanner

plate extensions to the focal line of the modified parabolic cylinder reflector. In principle, the shape of the lens is such that there is no degradation of the beam as it scans over the 30-degree azimuth sector.

The diameter of the lens is 60 inches, or 350λ, and the spacing between the conducting surfaces is 1-3/8λ. The measured sidelobe level of the lens-reflector combination does not exceed 22 db for any of the various feed horns and positions of scan. The lens was spun of mild steel, machined, copper plated, and protected with a thin coating of Irilac, a low loss plastic surface coating. Templates for the conducting surfaces were made on a digitally controlled profiler, and the surfaces were machined on a tracer controlled boring mill.

A picture of the complete antenna assembly is shown in Figure 4. The antenna is mounted on a pedestal and trailer of one of the early AN/MPQ-4 radars.

The specification for the vertical beam pattern was that the system give a nominally constant return from targets of equal size on a plane 35 meters below the radar lying between elevation angles of minus 4 degrees and minus 0.2 degree. The corresponding range interval on this plane is 500 meters to 10,000 meters. The reflector is a modified parabolic cylinder; it has a projected height of 4 feet and a length of 7 feet.

Figure 5 is a vertical beam pattern of the reflector. The dashed line is the desired contour based on a 0.6 db per kilometer one-way atmospheric attenuation. The beam shaping calculations were made by the Fourier inte-

Figure 4. Scanner Set Up for Operation

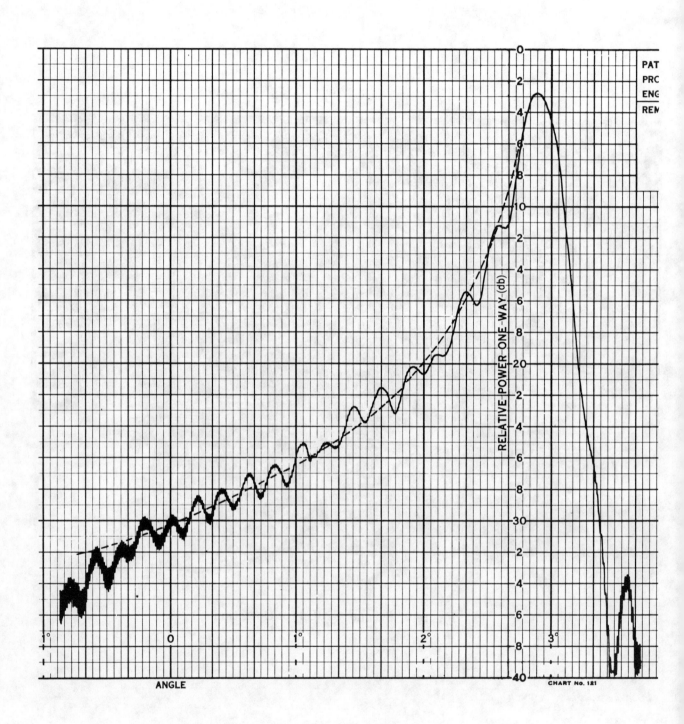

Figure 5. Elevation Pattern

gral method. An analog computer was employed to determine the far-zone beam pattern which would be generated by assumed aperture phase and amplitude distributions. A digital computer was programmed to determine the coordinates of the reflector which would give the desired aperture phase distribution.

The path taken by the microwave energy in going through the ring switch was shown in Figure 3. Figure 6 is a drawing of two junctions which were developed for feeding energy into and out of the switch. The basic junction geometry consists of a cavity formed by two septa in the E-plane, which is normal to the plane of the figure. The cavity couples two rectangular waveguides, one of which is split longitudinally in the E-plane. Figure 7 is a photograph of a ring switch rotor junction. The junction has a simple geometrical configuration and requires no additional matching devices. The scanner VSWR does not exceed 1.13 over the 3 percent design bandwidth of this system. The useful bandwidth of the type of junction used exceeds 25 percent.

The E-dimension of the ring switch guide and junction is $1\text{-}3/8\lambda$, corresponding to the conducting surface spacing of the lens. The other dimension corresponds to that of RG-98/U which is $7/8\lambda$. This height was chosen as a compromise between loss due to dissipation and that due to mode conversion caused by bending in the E-plane. The advantage gained by using this height over a spacing of $\lambda/2$ is shown in Table 2. The estimated data were obtained by comparing measured losses of similarly formed straight waveguide sections having heights of $\lambda/2$ and $1\text{-}3/8\lambda$.

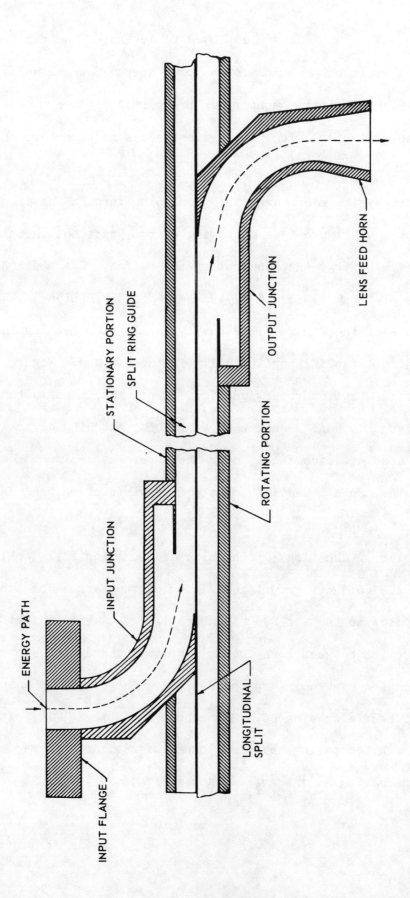

Figure 6. Cross Section Through Ring Switch

Millimeter Wave Radars

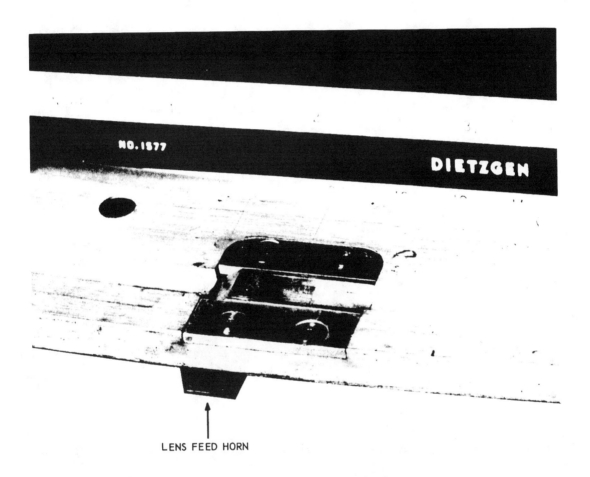

Figure 7. Ring Switch Rotor Junction

Table 2. Scanner Losses

	$\lambda/2$ Height	$1\text{-}3/8\lambda$ Height
Ring switch - start scan	0.8 db*	0.5 db
center	1.3 db*	0.8 db
finish scan	1.7 db*	1.0 db
Lens	3.5 db*	1.3 db
Reflector feed assembly	0.5 db	0.5 db
Maximum scanner loss (one way)	5.7 db	2.8 db

*Estimated value

The radar has the capability of going from a scan rate of 20 scans per second to a searchlight mode in less than 2 seconds after command. The beam searchlights targets along an azimuth selected by the operator prior to command and can be moved at will to follow a moving target. In order to stop the scan and position the beam to the accuracy required and in the time required, a unique mechanism was devised.

In the scanning mode the ring switch shown in Figure 8 is driven in a clockwise direction by a hydraulic motor. After an operator positions a B-scope cursor (generated by a magnetic pickup on the cursor arm) on a target and gives the command to searchlight, rotation of the ring gear and switch is stopped by the hydraulic brakes on the brake arm. Part of the energy of

Millimeter Wave Radars

Figure 8. Searchlight Mechanism

rotation is stored in the liquid spring, which is attached between the brake arm and the frame of the scanner. The spring moves the ring backwards until one of eleven cam followers is caught by an extended hook on the cursor arm. The excess energy of the spring is dissipated by the hydraulic damper also attached to the brake arm. Changes from the scan to the searchlight mode are made with pointing errors of less than 1/2 mil. A hydraulic positioner attached to the cursor arm can move the ring, and hence the beam, to any position within the 30-degree scan sector. This positioning is controlled by the operator by means of the wobble-stick at the display console.

Field Observations

This experimental radar was operated at Fort Huachuca, Arizona, for 6 months in 1959 and 1960. During that time, limited data were obtained on maximum ranges for detecting men and vehicles. Some of these results are shown in Table 3.

Table 3. Maximum Range for Detection of Targets

B-Display	
Walking man	5 km
Light vehicles	10 - 15 km
2-1/2 ton truck	18 km
Helicopter (H-19)	15 km
Aural Display	
Walking man	8 km
8 walking men	10 km

Maximum range measurements were also made for low and high angle ground bursts and for low angle air bursts. Bursts from an 8-inch howitzer and a 105-mm howitzer were reliably detected at 5500 meters and 2200 meters, respectively. Only about 50 percent of the rounds for an 81-mm mortar were detected, even for a range as short as 3000 meters.

Photographs of the B-display are shown in Figures 9 and 10. The first is a display of the town of Sierra Vista, Arizona, as seen from the test area of Fort Huachuca. The lowest range marker visible is 4 kilometers, and the range markers are 1 kilometer apart. Some of the principal targets are identified. The second display is of a section of improved dirt road. The peak of the beam is depressed to an elevation angle of minus 25 mils, and it intersects the target area at about 1000 meters range. Visible is return from 10 to 20-inch high sage grass and shadowing due to 10-foot mesquite trees. The banded shape of the illuminated area is due to the contour of the terrain, which rises above the antenna beam toward the upper left of the display. The fullness of the B-display with target echoes when the radar observed areas containing only sage grass and mesquite trees was striking. Based on impressions of target density as viewed optically, only a few targets would be expected on the display. This may be caused by limited color contrast between vegetation and ground in the Fort Huachuca area.

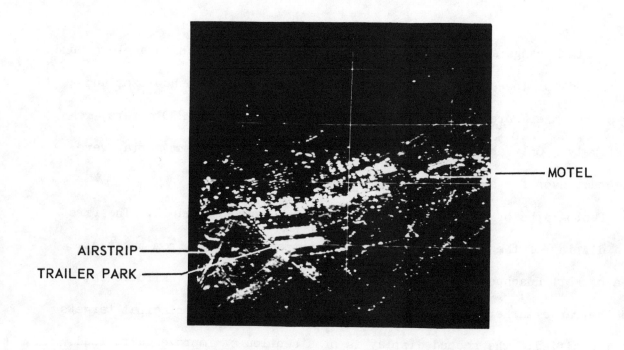

Figure 9. B-Display of Sierra Vista, Arizona

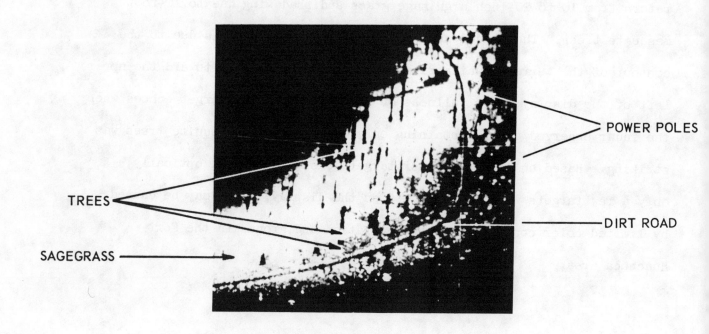

Figure 10. B-Display of Post Road, Ft. Huachuca, Arizona

It is interesting that unidentified moving targets were observed on the B-display at ranges of 500 meters and less. By sending an observer into the area, it was concluded that individual 2- to 3-inch flying grasshoppers were being resolved. Grasshoppers were frequently seen on the display during the field observations.

Discussion

The major problems which had to be overcome in developing the 4.3 millimeter system were associated with ohmic losses, phase errors caused by mechanical errors, and rapid positioning the 0.2 degree azimuthal beam.

The ohmic losses were minimized by using $1-3/8\lambda$ spacing for the conducting surfaces of the geodesic lens, by using $1-3/8\lambda$ E-dimension for the TE_{01} mode ring switch, and by holding transmission path lengths to a minimum.

The lens, which provides the narrow azimuthal beamwidth, has a loss of only 1.3 db. At the center of scan the ring switch loss is 0.8 db, and the loss does not exceed 1.0 db at any scan position. The entire antenna loss at the center of scan is 2.6 db; the loss varies by about $\pm 1/4$ db with scan because of the varying path length of the ring switch.

The severity of mechanical tolerances associated with such a short wavelength was reduced by using only components having simple geometrical shapes. A by-product of the simplicity existing in the ring switch junction which was developed to relax difficulties of machining is a 25 percent bandwidth for a 1.2 VSWR. The use of electroforming enhanced the accuracy with which the numerous feed horns and switch junctions were reproduced.

By carefully machining the lens spinnings and the reflector surface, phase errors were minimized to the extent that the azimuthal sidelobe level does not extend 22 db over the entire 30-degree scan sector.

The requirement to rapidly change the 0.2 degree beam from the 20 scans per second mode to the searchlight mode within 2 seconds represented a challenge. Development of a unique mechanism permitted this to be accomplished with a pointing error of less than 1/2 mil.

Limited measurements were made with the system under relatively dry weather conditions, for which the one-way atmospheric attenuation is expected to be 0.6 db per kilometer. The results indicate that vehicle echoes not masked by ground clutter can be reliably detected on the B-display for ranges out to 10,000 meters. With the aural display, a small group of walking men can also be detected at 10,000 meters.

Acknowledgements

Development of the radar was made possible by the effort and cooperation of a large number of people in the Radar Division, U. S. Army Signal Research and Development Laboratory, and at the Engineering Experiment Station, Georgia Institute of Technology.

A 94-GHz Radar for Space Object Identification

L.A. Hoffman, K.H. Hurlburt, D.E. Kind and H.J. Wintroub

Abstract—A feasibility demonstration radar system that operates at 94 GHz is outlined. A major goal of the program is the obtaining of radar echoes from orbiting objects for space object identification purposes. The radar is a linear FM pulse compression system with an eventual pulse time (1 ms)–bandwidth (1 GHz) product of 10^6. Experimental results of short range tests are discussed as well as details of the phase-locked loop linearizing system employed.

Introduction

THEORETICAL studies [1]–[12] over the past several years of the general radar resolution problem have been the stimulus for an experimental program [13] aimed at demonstrating the practicability of utilizing the high inherent resolution potential of millimeter-wavelength radar. In brief, the large bandwidth available at 94 GHz (several thousand MHz) and the very short wavelength should enable a radar to obtain a "range profile" of a satellite that shows more physical or structural details than range profiles at lower frequencies. The high Doppler sensitivity at 94 GHz should permit a precise measurement of the spin rate for spinning or tumbling satellites; the high carrier frequency offers the possibility of using synthetic aperture processing to "compress" the antenna beam along the track, so that two-dimensional resolution in range and along the track is obtained for better determination of satellite properties.

Significant progress has been achieved on this feasibility demonstration program that has as its major goal the obtaining of radar echoes from orbiting objects that can be suitably processed for satellite identification purposes.

System Considerations

Because the RF power available at millimeter-wave frequencies is limited, a long pulse is necessary to obtain sufficient energy "on target" to realize a usable signal-to-noise ratio. Therefore, pulse compression must be used to obtain the desired range resolution. The system configuration under development for obtaining radar echoes from space objects is shown in simplified block diagram form in Fig. 1. The transmitted signal is a frequency-modulated 1-ms pulse. The modulation produces a linear frequency sweep of 1 GHz, thus the time–bandwidth product is 10^6. The return radar echo is compared with a reference pulse that is identical to the transmitted pulse, for the case of zero Doppler shift, but offset in time. The difference frequency output derived from the mixer is therefore a constant frequency during the correlation period for a single, fixed target. An actual target will, however, return a multiplicity of echoes as well as Doppler data, and the target return signals will be analyzed by a bank of contiguous matched filters having a time–bandwidth product of 1, or a bandwidth of 1 kHz. This results in a pulse compression ratio of 10^6 and a theoretical range resolution of 6 inches.

Two of the major components in the experimental radar transmitter are the millimeter-wave tubes operating at 94 GHz. The driver is a 3-watt backward-wave oscillator (BWO), and the power amplifier is a 1-kW peak traveling-wave-tube amplifier (TWTA), the Hughes Research Laboratories LOW-1 and HAW-3, respectively. The BWO is operated in the CW mode and has a control voltage-frequency output scale factor of ~ 2 MHz/V. The frequency sweep range is 3 GHz within a power output level change of 6 dB. The maximum allowable duty cycle ratio of the TWTA is 10 percent; thus the pulse duration establishes the maximum PRF at 100 pps with the tube energized during the transmit period only.

For satellite echo tests, the Aerospace Corporation's precision millimeter-wave 15-foot dish, which has a gain of 70 dB at 94 GHz, will be programmed to follow a chosen space object by means of stored ephemeris data fed into the digital computer employed for the antenna pointing and control requirements. The practicability of tracking space objects in this manner has been tested satisfactorily on such satellites as Echo and Pegasus, wherein pointing confirmation was obtained by use of optical monitoring techniques.

The system parameters outlined were dictated by the constraints of the available components and the objectives of the radar. Fig. 2 is a graph of the expected received signal strength and S/N for targets of various sizes. It is expected that typical satellite targets will have scatterers that will appear larger than 1 m². Preliminary evaluation of the system reported herein has been done by the use of an overland 10-nmi test range using two 2-foot dish antennas in bistatic configuration. A variety of different sized corner reflectors were used for targets, with one mounted on a track and moveable over a range of 25 feet. Fig. 3 shows the 15-foot precision millimeter-wave dish, the pair of 2-foot dishes used in the 10-nmi test range, and the components of the 866-ns propagation delay line path.

The very difficult problem of processing extremely large time–bandwidth products has led to the choice of the linear ramp FM pulse as a means of making this processing practicable. Repeatability of the sweep is mandatory since there are no means by which the transmitted signal can be delayed so that the delayed version can be used in a correlator.

The linearity requirements are difficult to meet. Only if

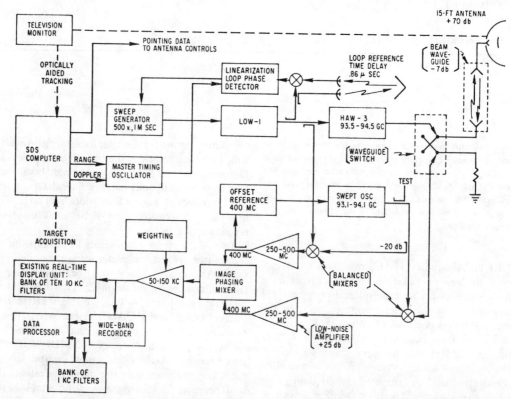

Fig. 1. Millimeter-wave radar configuration.

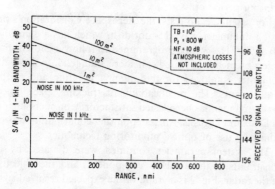

Fig. 2. Performance characteristics.

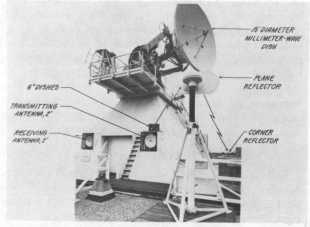

Fig. 3. Antenna tower facility.

the sweep is extremely linear will target echoes from different ranges (or individual scatterers) be properly correlated with the repeated reference pulse so that the range can be determined by examining the range filters; nonlinearities will tend to smear the signal output over the filter bank. However, since it is necessary to investigate only a small range interval about a satellite position (its equivalent range extent), the system is feasible.

Linearization of the frequency-sweep signal is accomplished utilizing a phase-locked loop. As shown in Fig. 4, part of the BWO RF output is tapped off and launched over a short air path via the 6-inch dishes and corner reflector to produce a nondispersive delay of 866 ns. The delayed sweep signal is mixed with the undelayed signal in a homodyne system that would produce a constant-frequency difference signal if the RF sweep were linear. This difference signal is compared in a phase detector with a stable reference oscillator output, and the error is applied as a correction to the BWO through the modulator, thus completing the loop. Linearization to within a 6-deg rms phase error has been achieved, and it is expected that this can be improved. An analysis concerning the required degree of linearization is presented in Appendix I. In Appendix II, optimization of the short-range delay is considered.

Millimeter Wave Radars

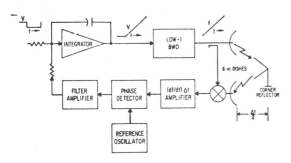

Fig. 4. Block diagram of linearization loop.

Major critical components required for the millimeter-wave radar, in addition to the tubes and linearizer, include low-loss transmission lines, mixers, and a duplexer. These components have been part of the laboratory investigations. Of these items, the low-loss transmission line is probably the most difficult implementation. Placing the radar on the antenna structure is undesirable because of weight restrictions (other equipment is placed on the antenna, which is in constant use for other research activities). A low-loss system of oversize waveguides and quasi-optical waveguide transmission lines (beam waveguide) is being designed so that both the radar receiver and transmitter can be placed at the bottom of the antenna tower, some 30 feet below the feed horn.

Experimental Results

Much of the preliminary evaluation of system performance has been completed over the 10-nmi test range. Due to the test-range round trip propagation time of approximately 124 μs, the pulse length employed for evaluation-experimentation has been limited to 100 μs. The transmitted signal consists of a 100-MHz frequency sweep in this pulse period. This pulse length in turn establishes requirements on the matched filters in the range discriminator filter bank of 10-kHz each bandwidth. Consequently, the pulse compression ratio is 10^4 and the range resolution becomes ~ 5 feet. The tests thus far on the 10-nmi range have resulted in radar returns in which the adjacent channel responses are down 9.5 dB from the desired central responses (Fig. 5). These adjacent channel responses result from sweep nonlinearities and the $\sin x/x$ spectral distribution of received pulse signals. Tests with simulated pulse returns from an idealized test oscillator indicate that most of the adjacent channel response may be attributed to the $\sin x/x$ spectral distribution. Considerable improvement in the reduction of adjacent channel responses can be obtained with appropriate pulse shaping and weighting, but generally at the expense of resolution.

The TWT amplifier output employed was ~ 800 watts. This power level, in conjunction with 2-foot corner reflectors and the short 10-nmi range, results in a very high signal-to-noise power ratio and thus provides a good means of studying sweep linearity and range sidelobe effects and other possible system anomalies.

A focused-wave quasi-optical transmission line (beam waveguide) is being developed to transmit the millimeter-wavelength power from the transmitter/receiver duplexer

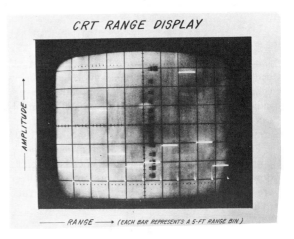

Fig. 5. CRT range display (10-nmi range).

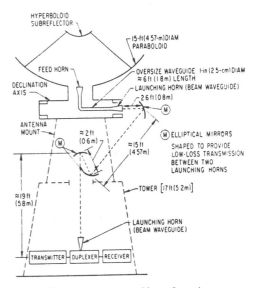

Fig. 6. Beam-waveguide configuration.

in the antenna tower base to the feed horn of the 15-foot-diam antenna (Fig. 6). The beam waveguide portion of the complete RF transmission line consists of three concave reflectors and two beam launching horns. (A second version under development will employ a special angle feed and only two reflectors.) The beam waveguide transmission path, ~ 20 feet long, performs the function of two rotary joints in addition to providing a minimum of insertion loss. Spherical mirrors have been designed by computer, best-fitting the contours to the theoretically optimum ellipsoidal surfaces. Swept-frequency tests of the beam waveguide in the laboratory have indicated a total insertion loss (including the beam launching efficiency) of 1.8 dB from 93.5 to 94.5 GHz. The total two-way loss for the complete RF transmission system will be ~ 7 dB.

Conclusions

The expected detection performance for the space radar on a single-pulse basis is adequate despite atmospheric attenuation. The attenuation, due primarily to water vapor density, is dependent on antenna look angle and weather conditions. Past experience in propagation research [14] indicates that considerable investigation of potential targets can be conducted under conditions wherein the two-way atmospheric attenuation will range from 1 to <6 dB.

Present interference levels in adjacent range bins are high (although expectedly so), the first ones being only 9.5 dB down from the desired central response level. Techniques are being investigated that can be employed to reduce these to lower values for a feasibility demonstration.

Appendix I

A linearity requirement estimate for repeatable FM sweeps of a large time–bandwidth radar correlator is made as follows. Let the reference pulse be defined as

$$e_r(t) = p(t) \cos\left\{2\pi\left[f_0 t + \frac{K}{2}t^2 + \int_0^t F(t')dt'\right] + \theta\right\} \quad (1)$$

where

$p(t)$ = envelope or pulsing function (ideal)
f_0 = unmodulated carrier frequency, Hz
K = frequency slope of the FM waveform, Hz s
$F(t')$ = deviation from linear frequency sweep, Hz
θ = phase constant, rad.

The mixing product for this reference pulse and a normalized target return pulse displaced by τ seconds is (eliminating the sum frequencies)

$$e_m(t) = \frac{p(t)p(t-\tau)}{2}\cos\left[2\pi\left(f_0\tau + K\tau t - \frac{K\tau^2}{2}\right) + 2\pi\int_0^t F(t')dt' - 2\pi\int_0^{t-\tau} F(t')dt'\right]. \quad (2)$$

The expression $2\pi f_0\tau - 2\pi K\tau^2/2$ represents a phase constant that will be denoted as θ_0. The expression $2\pi K\tau t$ is representative of the desired constant beat frequency $K\tau$ and will be expressed as ωt, where ω is $2\pi K\tau$. The integral expressions, representing the undesired phase deviation (noise spectrum), may be combined as

$$\phi(t) = 2\pi \int_{t-\tau}^t F(t')dt'. \quad (3)$$

Thus we have

$$e_m(t) = g(t,\tau)\cos\left[\omega t + \theta_0 + \phi(t)\right] \quad (4)$$

where

$$g(t,\tau) = [p(t)p(t-\tau)/2].$$

Expanding (4), we obtain

$$e_m(t) = g(t,\tau)[\cos\phi(t)\cos(\omega t + \theta_0) - \sin\phi(t)\sin(\omega t + \theta_0)] \quad (5)$$

and from (4) it is seen that $\phi(t)$ must be small to maintain the main response to noise sideband power ratio at a reasonable value. Then, (5) may be written

$$e_m(t) = g(t,\tau)\{[1 - (\phi^2(t)/2)]\cos(\omega t + \theta_0) - \phi(t)\sin(\omega t + \theta_0)\}. \quad (6)$$

The matched filter response is, to the first approximation, proportional to $1-[\phi^2(t)/2]$ since the sine term results in a broad-band low-level spectrum compared to the narrow-band lobe response. Thus, the peak of the filter response is down to $1-[\phi^2(t)/2]$ from unity.

The power in the main lobe response relative to unity is then given approximately by

$$(1 - [\overline{\phi^2}/2])^2 \simeq 1 - \overline{\phi^2} \quad (7)$$

where $\overline{\phi^2}$ is the rms value of $\phi^2(t)$. The power in the sidelobes due to sweep nonlinearities is approximately $\overline{\phi^2}$ on the same relative scale. The exact distribution of sidelobe energy depends on $\phi(t)$. The noise of the nonlinearities of the sweep is

$$\phi(t,\tau) = 2\pi \int_{t-\tau}^t F(t')dt' \approx 2\pi\tau F(t). \quad (8)$$

If $F(t)$ may be represented by its rms value,

$$\bar{\phi} \approx 2\pi\tau F_{\text{rms}}. \quad (9)$$

Using the frequency deviation of 18.4-kHz rms (Appendix II) and a τ of 10^{-7} second that corresponds to a range window centered at 50 feet, the rms for $\bar{\phi}$ is 0.01156 rad. Thus, the total noise power $\overline{\phi^2}$ is 38.7 dB below the desired signal.

It should be noted that this analysis has considered only the repeatable nonlinearities in the sweep.

Appendix II

It is necessary that the bandwidth of the phase-locked loop be sufficiently broad to permit tracking of the sweep nonlinearities without losing lock. It is also desirable to make the frequency error of the linearized sweep as small as possible. It is useful to determine how these parameters vary as a function of the selected propagation delay and to determine the optimum delay.

The bandwidth of the phase-locked loop must be maintained at a value less than the frequency at which phase shift around the loop is 90 deg. Otherwise, the loop will oscillate at the frequency at which the phase shift becomes 90 deg. The total phase shift is composed of amplifier and filter phase shifts and phase components due to propagation delay τ_p and phase detector delay τ_ϕ. It is assumed that amplifier and filter phase shifts are small compared to the phase shifts due to time delay. Then it becomes desirable to optimize the system for minimum frequency error while tracking non linearities. Let the loop time delay be defined as

$$\tau = \tau_p + \tau_\phi \quad (10)$$

where

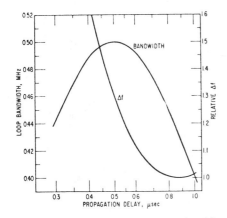

Fig. 7. Loop bandwidth versus propagation delay.

τ = total time delay around the loop
τ_p = range propagation delay
τ_ϕ = phase detector storage time.

For the proposed radar, an FM sweep rate of 1000 MHz/ms (i.e., 10^{12} Hz/s) has been selected. Thus, the beat frequency f_b from the linearizing homodyne receiver can be expressed as

$$f_b = 10^{12} \tau_p. \qquad (11)$$

A sample-and-hold phase detector has been developed for use in the loop because of its inherent broad bandwidth characteristics. The phase detector storage time is expressed as

$$\tau_\phi = 1/nf_b \qquad (12)$$

where n is the number of samples per cycle of the reference frequency. Expressed in terms of propagation delay

$$\tau_\phi = 10^{-12}/n\tau_p \qquad (13)$$

$$\tau = \tau_p + 10^{-12}/n\tau_p. \qquad (14)$$

If we assume that all amplifier and filter phase shifts are small, the maximum attainable loop bandwidth B is the frequency at which the phase shift contribution due to delay is 90 deg:

$$B = (1/4)[1/(\tau/2)] \qquad (15)$$

$$B = 1/2(\tau_p + 10^{-12}/n\tau_p). \qquad (16)$$

The loop bandwidth for the case where n is 4 has been plotted as a function of propagation delay in Fig. 7. The loop bandwidth is maximum when the propagation delay and the phase detector storage time are both 0.5 μs.

The broad-band phase-locked loop functions as a modulation following loop as it tracks the nonlinearities in the sweep. It is desirable to optimize the loop parameters for minimum frequency error Δf. The frequency error may be expressed as

$$\Delta f = \phi/2\pi\tau_p. \qquad (17)$$

Also, the phase error ϕ is inversely proportional to B^2, i.e.,

$$\phi = \kappa/B^2. \qquad (18)$$

Substitution gives

$$\Delta f = \kappa/2\pi\tau_p B^2 \qquad (19)$$

$$\Delta f = (\kappa/2\pi\tau_p)[2(\tau_p + 10^{-12}/4\tau_p)]^2. \qquad (20)$$

The optimum τ_p for minimum Δf, can be determined by setting $d(\Delta f)/d\tau_p = 0$ and solving for τ_p:

$$\tau_p = (\sqrt{3}/2)10^{-6} = 866 \text{ ns} \qquad (21)$$

$$\Delta f = \phi/2\pi\tau_p = 184 \text{ kHz/rad}. \qquad (22)$$

The relative Δf and loop bandwidth have been plotted as a function of propagation delay in Fig. 7. In the linearization loop, phase deviations on the order of 0.1 rad rms have been observed. This corresponds to a frequency deviation of 18.4 kHz rms.

References

[1] A. W. Rihaczek and R. L. Mitchell, "A high resolution approach to satellite identification," *Proc. 2nd Meeting on Ground Identification of Satellites (GISAT)* (Mitre, Bedford, Mass., October 2–4, 1967).
[2] A. W. Rihaczek, "Delay-Doppler ambiguity function for wideband signals," *IEEE Trans. Aerospace and Electronic Systems*, vol. AES-3, pp. 705–711, July 1967.
[3] R. L. Mitchell and A. W. Rihaczek, "Linear FM radar with clutter suppression capabilities," patent disclosure, April 21, 1966.
[4] ——, "Matched-filter responses of the linear FM waveform," *IEEE Trans. Aerospace and Electronic Systems*, vol. AES-4, pp. 417–432, May 1968.
[5] A. W. Rihaczek, "Measurement properties of the chirp signal," Aerospace Corp., El Segundo, Calif., Tech. Rept. TDR-469(5230-43)-2, August 1965.
[6] ——, "Method of generated pulse compression signal of extremely large time-bandwidth product," patent disclosure, June 8, 1965.
[7] ——, "Millimeter wave radar for BMD," Report of the IDA RESD Radar Summer Study 1965, Institute for Defense Analyses, Study S-213, December 1965 (S).
[8] D. D. King and A. W. Rihaczek. "Radar performance prediction from test range measurements," *IEEE Trans. Aerospace and Electronic Systems*, vol. AES-2, pp. 462–463 July 1966.
[9] A. W. Rihaczek, "Radar resolution of moving targets," *IEEE Trans. Information Theory*, vol. IT-13, pp. 51–56, January 1967.
[10] ——, "Radar signal design for target resolution," *Proc. IEEE*, vol. 53, pp. 116–128, February 1965.
[11] ——, "Range accuracy of chirp signals," *Proc. IEEE*, vol. 53, pp. 412–413, April 1965.
[12] ——, "Target resolution: capabilities of modern radar and fundamental limits," in *Radar Techniques for Detection, Tracking and Navigation*, W. T. Blackband, Ed. New York: Gordon and Breach, 1966, AGARDograph 100.
[13] L. A. Hoffman, H. J. Wintroub, and D. E. Kind, "Large time-bandwidth product modulation experiments at 94 GHz," *Proc. 2nd Meeting on Ground Identification of Satellites (GISAT)* (Mitre, Bedford, Mass., October 2–4, 1967).
[14] L. A. Hoffman, W. A. Garber, and H. J. Wintroub, "Propagation observations at 3.2 millimeter," *Proc. IEEE*, vol. 54, pp. 449–454, April 1966.

Vehicle Mounted Millimeter Radar
F.B. Dyer and R.M. Goodman, Jr.

18th Tri-Service Radar Symposium Record, 1972

(U) An experimental, vehicle-mounted, 4.3-mm radar has been constructed to investigate the potential of rapid-scan millimeter radar in a battlefield environment. The radar system has capabilities for ranging and for target acquisition and identification, and has shown good reliability with state-of-the-art millimeter components. The system has been operated both stationary and on a tracked vehicle running at top speed. Targets investigated have included tanks and other large vehicles, personnel in clutter, helicopters, wires, and other manmade and natural objects.

(U) Many of the radar parameters may be varied--for example, pulse lengths of 15, 30, and 45 nanoseconds are available, and the 0.5° beam may be scanned over a 45° sector at any rate between 1 scan/minute and 70 scans/second. A quick-change mechanism allows the radar to be searchlighted along the boresight of an optical system and aural Doppler to then be used for identification of any selected target. The increase in the silhouette of an M113 armored personnel carrier was minimized through use of a unique folded geodesic Luneberg lens antenna which exhibited first-sidelobe levels lower than 28 dB at all scan angles. Either an offset PPI or a micro B-scope may be used for the rapid-scan display. Key features of construction of the modular, solid-state radar are described.

(U) Besides the system flexibility and key engineering aspects, also discussed is the ease with which inexperienced personnel were able to usefully operate the system after only a minimum orientation period. It is considered significant that the operators quickly obtained sufficient confidence to actually navigate the moving vehicle by radar while accomplishing their primary duties. Qualitative discussions of such factors as propagation and target characteristics at 4.3 mm, including semi-quantitative data from measurements made during the evaluation, are also presented. Other potential applications of the techniques developed under this program are discussed.

1. INTRODUCTION

(U) Effective utilization in a practical radar system of the higher millimeter frequencies (i.e., above 35 GHz) is a goal that has been pursued by many organizations for nearly twenty years. While a certain degree of technical success has been achieved, none of the efforts has resulted in a "production" system. The purpose of this paper is to discuss some of the factors involved in the development of a useful millimeter radar in the context of a specific vehicle-mounted, 4.3-mm radar. The intent is to bring into clearer focus the need to match the system to the mission, and to point out the impact of certain limitations and advantages of millimeter radar on its use.

(C) The investigations* reported here result from several years of effort at both Harry Diamond Laboratories (HDL) and Georgia Tech in the development of millimeter wave radars for ranging, command fuzing, and fire-control use in battlefield situations, especially for flat-trajectory ordnance such as that used by tanks. Several radars were designed and constructed using a ruggedized 70-GHz (4.3-mm) magnetron (the Bomac BL 234) developed for Harry Diamond Laboratories for use in a high-shock environment such as exists in an armored vehicle. The most successful antenna system was based on an adaptation of the geodesic lens scanner which had been developed at Georgia Tech [1,2,3]. Although the primary goal of this program was to assess the utility of the 70-GHz radar as a ranging and fuzing system for armored vehicles, the investigation included a variety of tasks spanning many of the requirements for target

---
*Under Contract DA-49-186-AMC-275(A).

acquisition, identification, and navigation, as well as a limited, but necessary, target cross-section measurement program. The description here is focused on the vehicle-mounted version of the millimeter radar, not only because of its success in the original context, but because of its pertinence to current vehicle radar requirements.

(C) Very early in the program it was determined that the various concepts of the ranging and fuzing system could best be evaluated if installed on an armored vehicle. The M113 Armored Personnel Carrier was chosen for its availability and relative convenience as a test bed for such an evaluation program. The system concept included a high degree of modularity of subsystems and a sophisticated antenna system with a relatively wide-angle, rapid-scan capability in a compact, rugged configuration suitable for the space limitations of the vehicle.

(U) Of the various antenna approaches considered, the geodesic lens with a ring-switch scanner most nearly met the stringent microwave and physical requirements which were considered necessary for a practical radar. While a conventional geodesic Luneberg lens normally has a height-to-diameter ratio of approximately 1/3, this was considered to present an objectionable increase in vehicle silhouette. This problem was minimized by the successful design and fabrication of a new type of folded geodesic lens which has a height-to-diameter ratio of 1/6, and, as an unanticipated bonus, remarkably low sidelobes. The complete antenna, including lens and two reflecting surface, has sidelobes below 28 dB in the azimuth plane (the elevation beam is deliberately broadened for ground mapping) [4,5].

(U) An initial version of the radar was fabricated in 1966; however, the vehicle-mounted system was not completed until mid-1968. This system has been used in a series of measurements at various test sites since that time with a remarkable reliability record. Although most of the data reported herein were obtained either at Aberdeen Proving Grounds, Maryland, or at the HDL Gaithersburg, Maryland, Field Site, similar results were observed at other sites. The radar is currently in use at White Sands, New Mexico.

2. SYSTEM CONSIDERATIONS

(C) At the outset, the program goal was to develop a radar system which combined target identification and ranging with the capability of tracking outbound shells in order to provide firing commands and thereby improve first-round kill probability. In order to investigate the feasibility of this concept and goal, two areas were of paramount importance in terms of new investigations: (1) the determination of target discrimination methods for vehicle targets not moving but with engines idling, and (2) determination of the practical problems of tracking an outbound shell with a radar and providing command fuzing information to the shell from the radar. Both of these areas were investigated concurrently at Georgia Tech and at Harry Diamond Laboratories in the initial phase of the program.

(C) The system design conceived to meet the program goals envisioned using the advantages of millimeter wavelengths to provide reasonable system performance in a compact, lightweight equipment. A short wavelength was considered mandatory for development of a flush-mounted antenna of reasonable size and silhouette height without compromising the high resolution required for accurate bearing information. The choice of a short wavelength also facilitates generation of short pulse lengths which are desirable for meeting precision ranging and target acquisition and identification requirements. The flat trajectory fire and relatively short range requirements of the tank mission are also compatible with use of millimeter wavelengths, in that atmospheric attenuation and other weather losses do not severely degrade performance over ranges of a few kilometers. The choice of a short wavelength also is potentially advantageous, in that the high carrier frequency results in a relatively large frequency shift versus velocity, thus allowing for simple Doppler processing and easy recognition by the operator of relatively slowly moving targets.

(C) The rational behind this radar system was based on the desire to address several key problems through use of the radar by relatively untrained battlefield personnel. Key aspects considered are simple operation, high confidence by the operator in the quality of data as well as close coupling of operator and system, and data-rate/range performance specifications commensurate with the intended goal of tracking shells and other relatively small, rapidly moving targets.

(C) Results of the initial investigations at Georgia Tech indicated that it was entirely feasible to identify light civilian vehicles, jeeps, and trucks by their idling engine characteristics using a 70-GHz radar. The surface motion of an armored vehicle was found to be too small when stationary with only an idling engine for reliable identification at 70 GHz. However, a characteristic impulse was observed in the Doppler signal due to the initial crank-up of the main engine of a tank, thus potentially providing a way to discriminate between an "inert" target and a "live" target under certain conditions. Doppler signatures and amplitude distributions from various classes of vehicles in motion appear to be sufficiently distinctive to allow effective discrimination between vehicle classes.

Millimeter Wave Radars

(C) A number of measurements were made of the effective radar cross-section of various caliber ordnance ranging from 7.62-mm small-arms projectiles up to a 105-mm shell (see Section 4). Investigations were made of methods for implementing the fuzing command system. The conclusions resulting from these investigations were that: (1) it is feasible to track the larger shells to ranges appropriate for many battlefield situations, but not at the longest ranges at which armored vehicles might be engaged; and (2) the mechanics of providing for fuzing commands and fabrication of the fuzes were within the state-of-the-art. Unfortunately, the command-fuzing work was not pursued beyond preliminary investigations due to funding constraints in the latter stages of the program.

(U) Investigations were made early in the program of target cross-sections and ranging accuracy which might be anticipated under a variety of field conditions. The results of the cross-section measurements are discussed in Section 4; in general, results were encouraging for the use of a millimeter ranging system. Advantages of the high frequency were demonstrated by the reduced effect of multipath signals on ranging accuracy. Unanticipated benefits of the combined use of 70 GHz and a rapid scan rate were a remarkable ability for detection of small-cross-section obstacles (such as wires) and a significant potential for blind navigation by use of only radar information, even at the top speed of the vehicle. Ranging accuracy of better than 1 meter was consistently achieved in this system.

(U) Figure 1 shows the geodesic Luneberg-type folded-lens antenna mounted on the M113 armored personnel carrier. Figure 2 shows the interior of the M113 in which the radar console plotting table and driver's position are visible. The radar console and antenna are linked together and revolve about a common axis. The operator faces the direction illuminated by the radar and a periscope provides him with an optical view of the illuminated area. The operator may position the 45° scan sector in any azimuth direction, thus covering the full 360°. Controls are available for changing antenna scan rate, azimuth position, and elevation, and for switching between scan and searchlight modes. The radar is powered by 24 volts DC, and a switch is provided to select either the vehicle power system or a 400 ampere-hour battery mounted under the personnel benches. Principal system parameters are given in Table I.

(U) In order to meet the constraints imposed by the requirements of Military Potential Tests, the radar was designed to mate with a specific vehicle (the M113) rather than being developed as a general add-on device. Thus, basic radar parameters and configurations were tailored for the M113 only. While the M113 is relatively spacious, the requirement for access to certain components and the desire to retain capability for as much of the vehicle's primary mission as possible constrained the location and size of various units. It was decided that the best location for the radar was within the commander's cupola, with the antenna replacing the turret top and the console replacing the normal commander's position in the M113. Some of the accessory equipment was located in the side compartments of the vehicle or under the personnel benches, as shown in Figure 2.

(U) It was felt to be very important to minimize the training time required of potential radar operators, particularly to reduce the time required for them to acquire confidence in the utility of the radar. This goal, along with the more commonly considered human factors in the design of man-operated equipment, resulted in the concept of the rotatable operator's position coupled with an optical view. In training situations and in circumstances of reasonable optical visibility, the optical view (either telescope-aided or unaided) can be used to improve the coupling of the operator to the radar display. Availability of an optical view coordinated with the radar display was found to be valuable to many of the personnel who have used this radar. Operators have ranged from relatively experienced engineers to military personnel who had never viewed a radar scope. In every case, only a few minutes were required to initiate these operators in use of this radar in all of its various modes. The television-like presentation which results from scan rates of 30 to 60 per second was found to be of great utility in the recognition of targets and in the general orientation of all operators; the steady display was of particular benefit to relatively untrained operators.

(U) Because the M113 vehicle was not intended as a firing platform, actual use of the radar for weapon directing and ranging was not evaluated. Careful ranging tests resulted in a probable ranging error of less than 1 meter for targets of 15-dB signal-to-noise or greater, and probable errors of 3 meters or less for any target detectable in noise. The stability of the ranging circuit was good to approximately 0.2 meter; the bulk of the error was due to uncertainty in positioning the range cursor on the target.

3. MAJOR SUBSYSTEMS

3.1 (U) ANTENNA AND SCANNER

(U) The antenna and scanner assembly is composed of a helmet-type folded geodesic lens, a ring switch mechanism, and a reflector system. A cross-section view of these elements (shaded portion of pictorial) is shown in Figure 3 along with other components located in the turret area. Each of these elements is discussed below.

Figure 1. Folded geodesic lens antenna installation. (U)
(Unclassified)

Millimeter Wave Radars

Figure 2. Radar installation. (U)
(Unclassified)

TABLE I. SYSTEM PARAMETERS (U)
(Unclassified)

1. Antenna: Folded, Geodesic Luneberg Lens feeding a parabolic cylinder reflector.

   Dimensions: 24-inch diameter by 3.5-inch height.
   0.55° azimuth BW; 3.5° vertical BW, positionable in elevation from -10° to 20°
   Gain: 43.2 dB (corrected for loss).
   Polarization: Vertical

2. Scanning by a seven-section ring switch which allows a scan rate from approximately 1 scan/min to 70 scans/sec with a scan angle of 45° ($\pm$ 22 1/2° about the boresight). Boresight may be varied over 360° of azimuth.

3. Transmitter: Magnetron: Bomac BL 234C
   Peak Power: 500 watts (nominal)
   Frequency: 69 to 71 GHz (fixed)
   RF Pulse Width: 15, 30, 45 nanoseconds (selectable)
   Pulse Repetition Frequency: 5 kHz to 25 kHz (selectable)
   Modulator: Triggered Blocking Oscillator
   Duplexer: Bomac BL-P-017D

4. RF Component Loss: 6 dB (total excluding loss in antenna)

5. Local Oscillator: Varian VA 250 Klystron

6. Mixer: Philco IN2792 mixer crystal with a conversion loss of 15 dB. Equivalent noise figure of 18 dB minimum. Equivalent noise figure including IF amp of 25 dB maximum.

7. IF: Gain: 70 dB
   Center Frequency: 400 MHz
   Bandwidth: 60 MHz
   Noise Figure: 5 dB

8. Video Amp: Bandwidth: 60 MHz
   Gain: 150 (voltage)

9. Displays:

   A. Main Indicator Unit:
   (1) Sector-Scan Mode - Azimuth Angle: 45°
   Maximum Range Displayed: 1000, 2000, 3000, or 4000 meters

   (2) B-Scan Mode - Azimuth Angle: 45°
   Maximum Range Displayed: 75, 150, 225, or 300 meters; positionable from 0 to 4000 meters

   B. Slave Unit: Hewlett-Packard 180A Oscilloscope: A-Scope, B-Scan, or Sector-Scan Modes.

   C. Aural Display: Noncoherent Doppler from any resolution cell - 360° azimuth, range from 0 meters to 4000 meters.

   D. Aural Display: Sampling gate width of 40 ns. Readout by headset, speaker, or intercom system.

10. System Power Requirements: Basic Radar: 24 VDC at 8 amp
    Display: 24V at 15 amp
    Scanner: 24V at 5 amp nominal

11. Power Source: Main Battery: 24V, 400 amp-hour
    Vehicle Battery: 24V, 100 amp-hour. Run radar with alternator only.
    Battery Charger: 220 VAC, 20 amp, 60 Hz
    Radar can operate for 12 hours without charge of main battery.

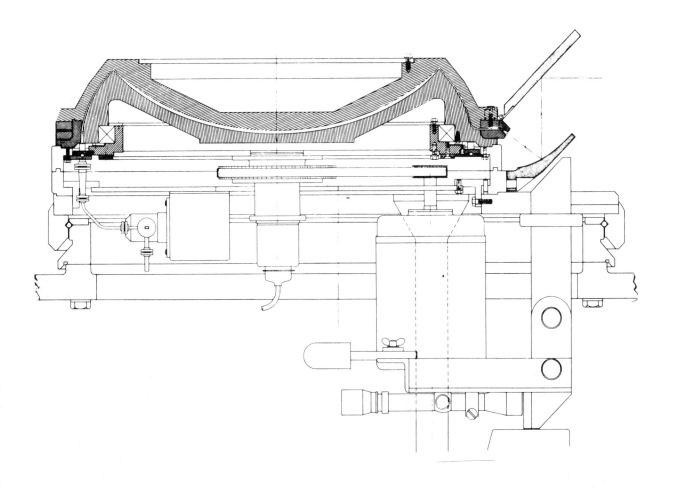

Figure 3. Pictorial cross-section of folded geodesic lens antenna. (U)
(Unclassified)

3.1.1 (U) THE HELMET-TYPE FOLDED GEODESIC LENS

(U) The geodesic Luneberg lens is a microwave focusing device consisting of a pair of closely nested conducting surfaces of revolution which are parallel in the sense that their normal separation is everywhere constant [6,7]. Assume the electromagnetic energy to be in the TEM mode and to follow this mean surface between the conducting plates. The velocity of this mode is independent of the plate spacing and is equal to the propagation velocity in an unbounded medium. The physical lengths of rays in a geodesic lens are equivalent to the optical lengths of rays in a dielectric lens; therefore, collimation of the rays is accomplished by curving the mean surface in a manner which yields equal path lengths from a point on the periphery of the lip to a line diametrically opposite and tangent to the periphery. Figures 4 and 5 show typical ray paths through an unfolded geodesic lens.

(U) Geodesic lenses in the past have had a typical height-to-diameter ratio of approximately 1/3. In order to reduce this ratio, the lens for this system is folded to obtain a height-to-diameter ratio of approximately 1/6. It has long been recognized that, in principle, the mean surface of a normal geodesic lens can be folded without changing the equal-path-length characteristics of the geodesic ray paths. However, folding the surface must not cause a sharp discontinuity at the fold lines which would prevent operation of a microwave device. There is, furthermore, a limitation on the minimum radius of curvature which can be used in a curved parallel-plate device of normal spacing without excessive mode conversion [1]. A major achievement in this program was the successful development of a fold design to solve these problems [1].

(U) Ohmic losses in waveguide can be reduced by increasing the transverse dimensions. Because the path lengths in geodesic lenses are relatively long, it is advantageous at millimeter wavelengths to increase the spacing of the parallel plates as much as possible. As the spacing of curved plates is widened, two effects become important. (1) The TEM propagation constant is increased slightly and energy is converted from the TEM mode to higher order modes. (2) The propagation constant perturbation and mode coupling are increased as the spacing is increased and as the radius of curvature is decreased. Equations and curves describing these effects are included in Reference 1.

(U) The folded geodesic lens used here has a nominal diameter of 24 inches. It is characterized as a helmet-type lens by the shape of the input and output lip configuration, having a vertical input lip to permit feeding with a waveguide ring switch and a horizontal output lip, and thus resembling a soldier's helmet. Separation between the conducting surfaces is 0.185 inch at the input and tapers to a spacing of 0.070 inch between input lip and fold. The narrow (0.070-inch) spacing continues through the fold. The spacing is increased through a taper to 0.186 inch at the central portion of the lens. A typical azimuth antenna pattern for midscan with 0° beam elevation is illustrated in Figure 6. Loss measurements indicated one-way attenuation of approximately 1.4 dB in the geodesic lens and its flat-plate extensions. The geodesic mean surface was calculated by the conical approximation method [1].

3.1.2 (U) RING-SWITCH MECHANISM

(U) A waveguide ring switch is a device for sequentially switching from a stationary waveguide into one or more circularly moving waveguides. The basic structure consists of a ring waveguide split longitudinally to enable one portion to rotate with respect to the other. On transmission, energy is coupled into the ring guide and constrained to propagate in one direction by an input junction attached to the stationary part of the switch. The energy flows down the waveguide until it reaches an output junction attached to the rotating part; here it is coupled out of the ring into a waveguide. The ring switch for this system has a mean diameter of 24 inches; it has one input junction and seven equally spaced output junctions. Each of the output junctions is terminated with a feedhorn designed to feed the geodesic lens. The junctions are of an offset ring-switch type and were originally designed at Georgia Tech for the AN/MPS-29 radar developed for the Army Signal Corps [3]. The various components of this ring switch make extensive use of electroforming techniques because of the dimensional accuracies required. A typical offset ring-switch junction and lens feedhorn is shown in Figure 7 which shows dimensions as well as the basic configuration.

(U) The ring switch is driven by a 24-volt DC series traction motor through a timing-belt pulley system having a 3.75 to 1 reduction ratio to achieve the high scan rate required. Speed control of the drive is obtained by varying the duty factor of the 24-volt DC power source with a switching transistor regulator using multiple feedback loops to compensate for drive-level variation, temperature effects, and load changes. Final scan speed is referenced to a tachometer directly connected to the ring switch. This arrangement allows precise control of scan speed over a very wide range and also provides the various timing signals used to synchronize the display sweeps.

(U) A portion of the ring-switch drive system is coordinated with a searchlighting mechanism capable of rapidly positioning, upon command, one of the seven feedhorns of the ring-switch rotor at the center of the lens feed sector. This allows the antenna beam to be positioned along the antenna boresight coincident with the vertical cross hair of the telescopic

Millimeter Wave Radars

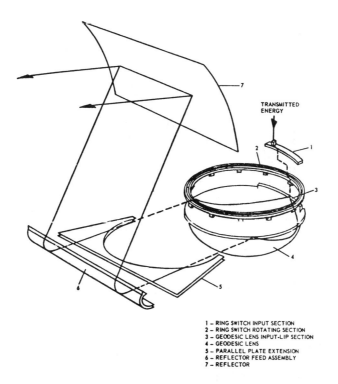

1 – RING SWITCH INPUT SECTION
2 – RING SWITCH ROTATING SECTION
3 – GEODESIC LENS INPUT-LIP SECTION
4 – GEODESIC LENS
5 – PARALLEL PLATE EXTENSION
6 – REFLECTOR FEED ASSEMBLY
7 – REFLECTOR

Figure 4. Antenna assembly exploded view showing two energy ray paths. (U)
(Unclassified)

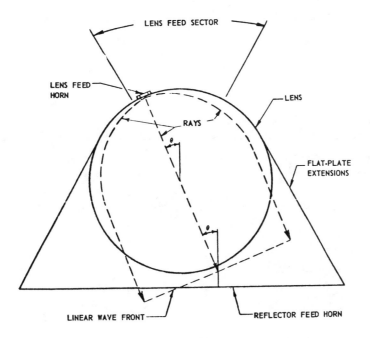

Figure 5. Ray paths through a lens when the lens feed horn is an angular
distance from the center of the lens feed sector. (U)
(Unclassified)

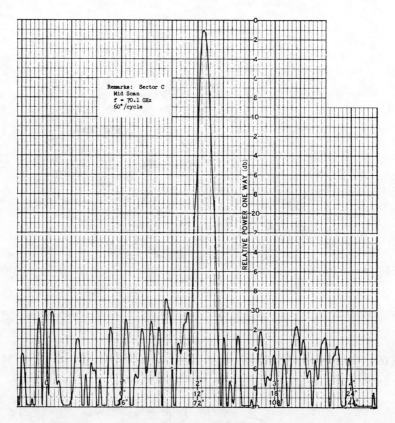

Figure 6. Typical antenna radiation pattern. (U)
(Unclassified)

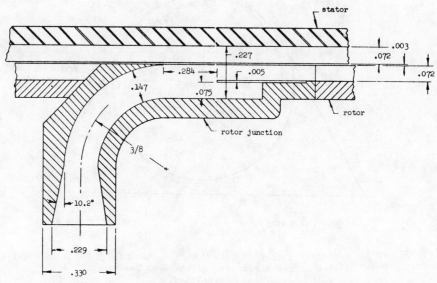

Figure 7. Rotor junction and lens feedhorn. (U)
(Unclassified)

sight. Searchlighting is achieved in a fraction of a second and allows the radar to be used in the noncoherent Doppler mode for investigating the motion signature of a target within the selected cell. The system can easily be returned to the scan mode simply by pressing the scan button.

3.1.3 (U) REFLECTOR SYSTEM

(U) A low-silhouette installation, adjustment of elevation angle of the radar beam, and minimum blind area around the vehicle were the criteria considered in design of the reflector system. The low-silhouette requirement dictated an antenna with only azimuth positioning capability. It was concluded that the best system for providing vertical collimation and beam-tilt adjustment in elevation consisted of a fixed parabolic cylinder illuminated by the line-source lens output horn. The cylinder reflector is positioned to direct the collimated 1/2° x 3-1/2° beam vertically and it is then reflected forward by the tiltable plane reflector (see Figure 4). The tilt of this secondary reflector, which can be adjusted by the operator, positions the beam in elevation; a 45° tilt of this reflector produces a 0° elevation angle of the beam.

(U) The parabolic-cylinder reflector is of conventional design and was machined to its final contour with tolerances better than 0.005 inch. The flat-plane reflector was made from a piece of 1/2-inch aluminum plate, selected and hand-finished for flatness. The final plate was flat to within a 0.005-inch total indicator reading; across most of the surface the variation was less than 0.003 inch.

3.2 (U) ELECTRONICS

(U) The philosophy followed in design and construction of the electronic units for this system was based on the desire to achieve an experimental system to fulfill the performance objectives with minimum maintenance problems and with minimum cost and complexity. The objective was to provide a realistic test of the millimeter radar concept with a minimum of the development problems which may result from state-of-the-art design. It was not considered practical to design and construct the electronic units to strictly military standards or to specify military standard on all components. Figures 8 and 9 illustrate the basic assembly and construction techniques used. Figure 10 shows a simplified block diagram of the radar which illustrates the relative simplicity of this system.

(U) The principal components of transmitter/receiver are shown in Figure 8. The basic design of the transmitter and receiver used well-proven concepts and is unusual only in the special nature of certain components. For example, the modulator is a parallel-triggered blocking oscillator that provides the 3,500-volt pulse needed by the Bomac BL-234C magnetron. The pulse voltage is adjustable in amplitude to allow matching of the drive to individual magnetrons. The transmitted pulse width is somewhat a function of the particular magnetron used; however, nominal pulse widths of 15, 30, and 45 nanoseconds are operator-selectable by changing the RC timing network in the blocking oscillator modulator. The prf is manually adjustable from 10 to 25 kHz; however, it is normally coupled to the pulse width in order to provide constant duty factor.

(U) The prf output of the magnetron is coupled to the ring switch through a Bomac BL-P-017D duplexer and normally a 1N2792 detector is provided for monitoring power and pulse shape. The returning rf is directed by the duplexer through a TRG precision attenuator and through a directional coupler in which the local-oscillator signal is injected. A single 1N2792 mixer was chosen because the 400-MHz IF center frequency reduces the effect of local-oscillator noise.

(U) The local oscillator is based on a Varian VA-250 klystron with a highly stabilized power supply and temperature compensation. The rest of the transmitter/receiver assembly is essentially conventional microwave design and makes use of good mechanical engineering practices to meet the stringent vibration and shock requirements.

(U) The main indicator was built by Beta Instrument Corporation under separate contract from HDL, with Georgia Tech participating in its design. The enclosure for the indicator was fabricated as an integral but removable part of the console in order to reduce interface problems. The design of the indicator is conventional (see Figure 11) except in two areas. (1) Moderately high resolution was required because of the short pulse length, and (2) the requirement for continuously variable antenna scan rate put severe limitations on the sweep and timing circuits. The indicator successfully met the design goals in both areas.

(U) The indicator has display scales of 4000, 3000, 2000, or 1000 meters in the PPI mode, and 300, 225, 150, or 75 meters in the micro-B mode. An azimuth angle of 45° was displayed in all cases. The sweep-resolving circuits used a combination of diode function generators and operational amplifiers to produce the resolved sweep waveforms. This approach was chosen at the time because of its reduced risk and lower cost; current integrated-circuit technology would lend itself readily to fully digital sweep generation.

Figure 8. Transmitter and receiver components. (U)
(Unclassified)

Millimeter Wave Radars

Figure 9. Radar console construction. (U)
(Unclassified)

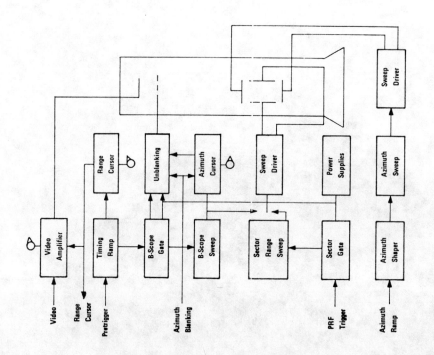

Figure 11. Indicator block diagram. (U) (Unclassified)

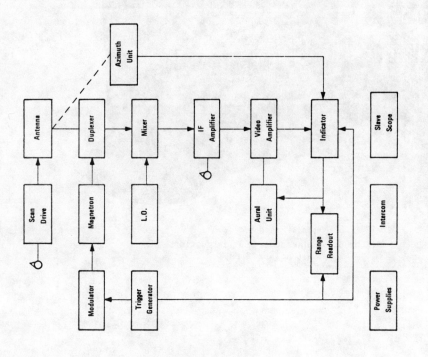

Figure 10. Block diagram of radar system (U) (Unclassified)

(U) The principal timing relations of the synchronization circuitry are shown in Figure 13. The indicator range sweep triggers and the various prf-related timing pulses are generated from an oscillator which is manually variable but stable. The stable delays between the various timing pulses were chosen in order to achieve an accurate ranging capability in the system. The ranging unit used in this system was developed at HDL and is based on a vernier time measuring technique similar to that described by Baron [8].

(U) The azimuth synchronizing waveforms shown in figure 12 are generated in the azimuth unit which is operated by impulses from a multi-track shaft encoder. Seven blanking vanes are spaced about the periphery of the lower driven disk of the scanner and divide the 360° into seven equal segments. These vanes are sensed optically to provide start-stop triggers for the transmitter. Also coupled to this disk is the incremental shaft encoder which produces 3600 pulses per revolution; these pulses are summed to form a staircase waveform with 0.1° equivalent angular resolution steps as illustrated in Figure 13. The staircase generator is synchronized with the blanking pulses to form a sawtooth waveform with 450 pulses per staircase, with a period determined by the rotational rate of the scanner; the resulting waveform is direct coupled to the azimuth sweep circuits in the indicator. The staircase waveform which is produced is independent of the rotational rate of the scanner in both amplitude and linearity.

(U) The principal components of the aural unit are shown in the simplified diagram in Figure 13. The pulse-to-pulse amplitude modulation which results from the relative motion of radar scatterers within a resolution cell is boxcar detected and processed in a manner referred to as incoherent Doppler detection. Because of the relatively high carrier frequency, the Doppler shifts from most moving targets of interest fall in the audible region, typically 100 Hz to 10,000 Hz. For example, a vehicle moving radially at 30 miles per hour produces a Doppler shift of about 7 kHz.

(U) The boxcar output is filtered to provide emphasis to the lower frequencies and to produce the 20-dB-per-decade rolloff above 3 kHz which is necessary to reduce the objectionable note from the switching regulators of the power supply. A special dynamic-range-compression audio amplifier (adjustable) is used to set the signal voltage level; the audio output is distributed through a power amplifier to speakers and to the various intercom stations in the vehicle. The combination of preemphasis and audio compression serves to reduce the dynamic range of the signal to a level acceptable to the ear for the variety of listening conditions which exist in the noisy vehicle and to enhance the detectability of the signal from slowly moving targets.

4. TEST RESULTS

(U) Tests and experimental investigations with the radar can be divided into two categories: (1) familiarization of potential operators with the various operational features of the radar and qualitative assessment of military potential, and (2) semiquantitative measurements of target cross sections, both in the clear and in simulated detection situations. Some of the more pertinent of these observations are reported in this section.

(U) Qaulitative assessments were made of the detectability of various targets during the course of the qualitative measurements, particularly at Aberdeen, and, in addition, in a few instances the targets were moved about through foliage and other obscuring clutter in order to assess their detectability. One of the most dramatic effects noted in these observations was that the very narrow antenna beam (0.5°) resulted in radar detections of targets obscured behind tree lines and easy visual observation. For example, a series of observations were made of a helicopter (HU-1B) moving behind a line of trees approximately 1000 meters from the radar which was located on the east side of the Munson Test Area at Aberdeen. The radar operator easily followed the helicopter as it moved behind the line of trees and was able to quickly place the Doppler tracking gate on him each time he rose above the trees. Although the helicopter was essentially hidden from view optically most of the time, it was never lost to the radar operator.

(U) The ease with which operators were able to learn to use the radar is attributed, in large measure, to the mapping capability of the combination of rapid scan and high resolution. Also foliage returns were observed to fluctuate in a characteristic fashion such that operators were able to quickly locate point targets in extended foliage clutter. The photographs shown in Figure 14 and 15 illustrate the mapping capability of the radar but do not adequately demonstrate all of the information actually available to the radar operator. Figure 14 also illustrates the wire detection capability of the radar. Although the particular utility line shown in Figure 14 is at a range of less than 100 meters, similar wires were detected at ranges beyond 400 meters.

(U) A significant aspect of detection with a high-resolution radar which is sometimes overlooked is that the concepts of target-to-clutter ratio and $\sigma°$ of clutter do not apply

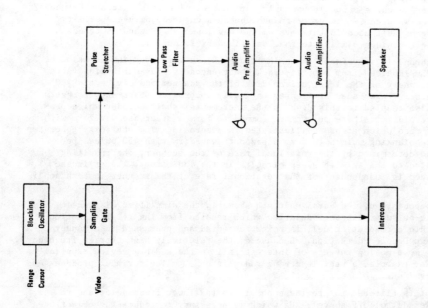

Figure 13. Aural unit block diagram. (U)
(Unclassified)

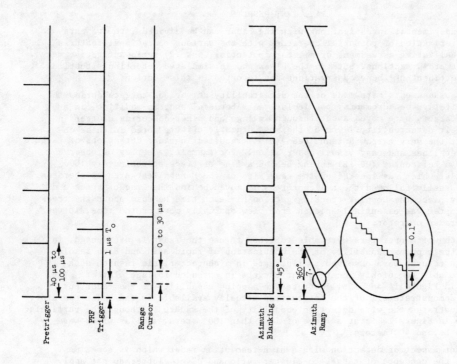

Figure 12. Range and azimuth synchronizing waveforms. (U)
(Unclassified)

Millimeter Wave Radars

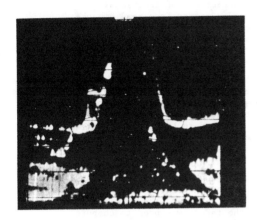

14a. Normal illumination of intersection. (U)

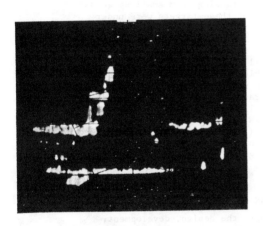

14b. Vertical beam elevated to show height effects. (Note return from utility wire on near side of intersection) (U)

Figure 14. Photographs of the B-display. (U)
(Unclassified)

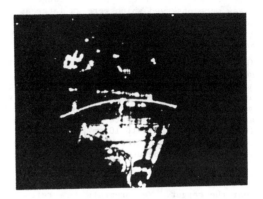

15a. Looking south across the Munson Test Area. (U)

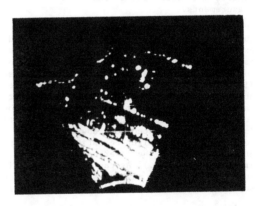

15b. Looking east across the Munson Test Area. (U)

Figure 15. Portions of the Munson Test Area. (U)
(Unclassified)

without some qualifications. That is, the detection problem becomes either: (1) detection and classification of a target in a background of discrete clutter returns, or (2) detection of a target with marginal signal-to-noise ratio. With the above limitations in mind, observations of area-extensive ground clutter, such as grass and other low foliage coverage, gave values of $\sigma°$ which typically ranged from 15 to 20 dB. Cross-sections of individual trees were observed which were as large as 10 to 50 square meters. Returns from road embankments, rocks, and other ground returns exhibited diffuse scattering and thus returns at all incidence angles, but still gave effective cross-section values consistent with the $\sigma°$ measurements reported in Reference 9.

(U) Only a few observations were made while appreciable precipitation was occurring. One series of measurements were made in rain which was categorized by observers as "moderate" although no quantitative measurements were made of the actual rainfall. The radar suffered an effective loss of range performance equivalent to an increase in noise figure of approximately 15 dB when observing targets at 1000 to 1500 meters. Observations reported by Lin and Ishimaru [10] indicate that attenuation at 4.3 mm might be expected to be in excess of 10 dB in moderate (4 to 10 mm/hr) rainfall. These are experimental values and are higher than generally predicted analytically. The return from the rain was observed to be noise-like but range dependent unlike normal receiver noise.

(U) Since convenient signal-generator standards do not exist at 4.3 mm, all of the cross-section measurements were made relative to calibrated standard corner reflectors located near the target. Measurements were referenced to a calibrated rf attenuator; the dB difference between the return from the standard corner reflector and the target was used to compute effective cross-section in the conventional manner. Careful measurements of returns from accurate spheres and corner reflectors were made to determine the repeatability of the measurements and absolute accuracy. Day-to-day variations in calibration were less than 3 dB (total). Careful standarization measurements were made at the beginning of each measurement sequence in order to ensure that the receiver noise figure had not changed; when poor noise figures were observed, alignment or component replacement was undertaken prior to making measurements.

(U) Measurements were made on a variety of potential targets ranging from small objects such as projectiles through large trucks and buildings. Some of these targets were under experimental control, others were made as targets of opportunity appeared such as in the case of the observations of helicopters, trains, and aircraft. A brief summary of the measurements which were made in the initial tests of the vehicle radar is given below in Table II. Other similar measurements are contained in References 11, 12, and 13. The principal observations are summarized in graphical form in Figure 16 which shows tangential range versus target cross-section at 4.3 mm. The two sloping lines on the graph represent estimates of reasonable extremes in performance for practical millimeter radars. The radar described in this paper operated just above the lower line. The upper line illustrates the gains in performance which might be anticipated by use of an 8-to-10 kW magnetron such as the Bomac BL 246 and use of the lower-noise mixers currently available (noise figures of 12 dB or better are being reported.) As can be seen from Figure 16, detection ranges on personnel of as much as 1500 meters might reasonably be expected. Larger targets, such as tanks, might reasonably be expected to be detectable at ranges of 8 to 10 km.

5. SOME IMPLICATIONS

(C) The intent of this paper is to illustrate an approach to the design, development, and installation of a millimeter radar on a military vehicle. The requirements of the vehicle and of using relatively untrained military personnel as operators were strong factors in the basic design and in details of the fabrication. System constraints imposed by requirement for a rugged, low-silhouette antenna dictated in large measure the system performance which might be obtained, as did the availability of components suitable for use in the high-shock environment. Component availability lead to the choice of 4.3 mm rather than 3 mm which might have been a slightly better choice from system considerations. The ultimate goal of the program of providing the Army with a millimeter radar for ranging and command fuzing of tank ordnance (with the vehicle radar described here as an intermediate step) was not reached because of other program considerations. No technical problems were encountered which would have prevented completion of the program as originally conceived.

(U) One of the largest bonuses to come out of this program was the remarkable facility with which untrained operators were able to use the combined optical and radar data and the improved coupling of information to the operator by the use of a high-resolution, rapid-scan radar. The proven ability to navigate at moderately high speed, and to see small, low-height obstacles with this system shows the considerable potential that millimeter radar has for the navigation and obstacle-avoidance roles required in high-performance vehicles currently

TABLE II. SUMMARY OF TARGET CROSS-SECTIONS (U)
(Unclassified)

Category	Observation Ranges (meters)	Cross-Section (square meters)
I. Projectiles (tail aspect)		
A. 30 mm	20	.02 to 0.1
B. 90 mm HEAT	20	.003 to 0.15
C. 90 mm	calculated RCS	0.1 to 10
D. 105 mm	calculated RCS	0.5 to 18
E. 152 mm	calculated RCS	1 to 86
II. Personnel		
A. Man, unobscured	100 (path over grass)	0.1 to 1.2
B. Man, obscured	100	0.1 to 0.4
C. Man, obscured	200	0.08 to 0.1
III. Vehicles		
A. Light (jeep, automobile)	500	10 to 100
B. Trucks	500	10 to 1000
C. Helicopter (HU-1B)	1000	100 to 500
D. Tank (M48)	500	5 to 300
1. ±45° about front	500	5 to 50
2. ±45° about rear	500	10 to 300
E. Tank (M48)	100 (path over grass)	5 to 100
F. Aircraft (C-124)	2000	100
G. Train	3000	1000

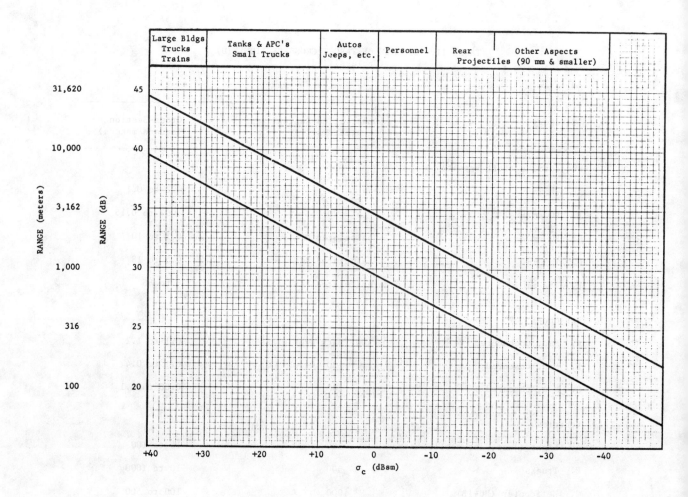

Figure 16. Tangential signal level range versus radar cross-section. (U)
(Unclassified)

being considered such as the surface effect vehicle (SEV). Other investigators have also recognized this need [14].

(U)   Millimeter-wave radar is still potentially attractive as an alternative to optical devices for ranging and fire-control applications. An example of one such concept which was considered briefly on the program reported here was a compact ranging device (estimated to weigh 25 lbs., complete) which would have a rapidly scanned 8° azimuth sector and to be displayed to match the optical view of a telescope. Such a compact package could be man-portable or mounted on a variety of vehicles. Figure 17 shows a sketch of such a system. Versions of this system, not including the rapid-scan feature, have been fabricated at Harry Diamond Laboratories and have demonstrated utility for target ranging.

(U) A number of areas connected with millimeter radar need further investigation; it is recommended that this be recognized when consideration is given to the application of millimeter radar. Particular attention should be given to evidence which indicates that backscatter and attenuation due to rain are appreciably greater than has been theoretically predicted. Additional controlled experiments are needed in the area of millimeter propagation. Additional measurements are needed of effective target cross-sections, particularly in realistic clutter environments. Detailed measurements are needed of clutter cross-section and temporal behavior in order to develop a basis for the application of sophisticated signal-processing techniques. Additional effort is needed in the development of components, particularly duplexers and mixers and high-power transmitting sources, in order to increase the options available to the systems designer and in order that systems with high reliability capabilities can be fabricated for use in military environments.

6. ACKNOWLEDGEMENTS

(U) The work reported here has involved the efforts of many people at Georgia Tech and Harry Diamond Laboratories, and we gratefully acknowledge their aid. Particularly important contributions were made by W. S. Foster, R. C. Johnson, and W. K. Rivers at Georgia Tech. We also thank the many participants at Harry Diamond Laboratories for their interest and support. Special note is made of the roles of C. D. Hardin, W. J. Moore, F. Weiss, C. E. H. Edward, E. Naess, and J. Salerno in the millimeter program.

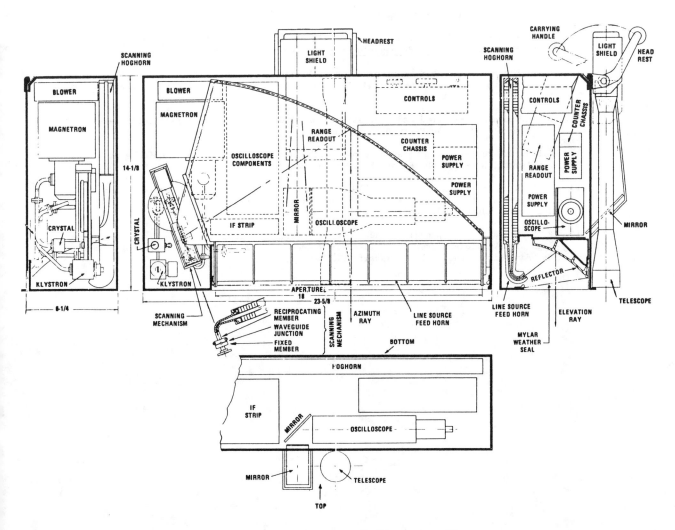

Figure 17. Proposed radar range finding system. (U)
(Unclassified)

7. REFERENCES
(Unclassified)

1. R. M. Goodman and F. B. Dyer, "Millimeter Wave Radar (U)," Final Report on Contract DA-49-186 AMC-275(A), Georgia Institute of Technology, 29 February 1968, Confidential.

2. M. W. Long, W. K. Rivers, and J. C. Butterworth, "Combat Surveillance Radar AN/MPS-29 (XE-1) (U)," Sixth Annual Radar Symposium Record, University of Michigan, June 1960, Confidential Paper, Secret Report, AD 318 072.

3. M. W. Long, G. E. Allen, Jr., et al., "Combat Surveillance Radar," Final Report on Contract DA-36-029 SC-74870, Georgia Institute of Technology, 30 June 1960.

4. R. M. Goodman, Jr., R. C. Johnson, et al., "A Folded Geodesic Luneberg Lens Antenna," Seventeenth Annual Symposium on USAF Antenna Research and Development, University of Illinois, November 1967, Unclassified.

5. R. C. Johnson and R. M. Goodman, Jr. "Geodesic Lenses for Radar Antennas," EASCON '68 Record, IEEE Trans. Aerospace and Electronic Systems, 64-68 (September 1968).

6. J. S. Hollis and M. W. Long, "Luneberg Lens Scanning System," IRE Trans. Antennas and Propagation AP-5, 21-25 (January 1957).

7. R. C. Johnson, "The Geodesic Luneberg Lens," Microwave Journal 5, 76-85 (August 1962).

8. R. G. Baron, "The Vernier Time-Measuring Technique," Proc. IRE 45, 25-30 (1957).

9. H. E. King and C. J. Zamites, Jr., "Terrain Backscatter Measurements at 40 to 90 GHz," IEEE Trans. Antennas and Propagation AP-18, 780-784 (November 1970).

10. J. C. Lin and A. Ishimaru, "Propagation of Millimeter Waves in Rain," Scientific Report No. 4 on Contract F19628-69-C-0123 (AFCRL-71-0310, T.R. 144), University of Washington, May 1971, Unclassified.

11. C.E.H. Edward, "Expected Radar Cross-Section of Ground Targets at Millimeter Wavelengths," HDL Report R-620-65-4, Harry Diamond Laboratories, 19 July 1965.

12. C.E.H. Edward, et al., "Notes on Cross-Section Measurements of Small Objects at 4 mm Wavelength, Note 1," HDL Report M-410-68-3, Harry Diamond Laboratories, May 1968.

13. J. E. Kammerer and K. A. Richer, "70 Gc Radar Characteristic Measurements for American and Russian Vehicles (U)," BRL MR 1683, Ballistic Research Laboratories, Aberdeen Proving Ground, June 1965, Confidential, AD 369 082.

14. P. Hoekstra and D. Spanogle, "Radar Cross-Section Measurements of Snow and Ice for Design of SEV Pilotage System," Cold Regions Research and Engineering Laboratory, Hanover, New Hampshire, June 1971, Unclassified.

Development and Test of a 95 GHz Terrain Imaging Radar

INTRODUCTION

This paper describes Goodyear Aerospace Corporation's millimeter-wave radar system and preliminary test results. The transmitted wavelength of the radar is approximately 3.15 millimeters, which corresponds to a carrier frequency of 95 GHz. The radar system consists of a pencil beam antenna, narrow pulse transmitter, crystal mixer superheterodyne receiver, and a plan-position indicator (PPI) display.

The millimeter-wave radar offers significant advantages over longer wavelength systems because relatively high resolution can be obtained with physically small and comparatively low-cost equipment. High resolution at X-band with present-day radar technology is obtained by using physically large antennas, pulse compression, and/or synthetic aperture techniques which lead to large, heavy, complex systems that are expensive to manufacture. At short ranges of the order of five miles, a real aperture millimeter-wave radar transmitting a narrow pulse provides a high-resolution, real-time display. Unlike an optical sensor, a high-resolution airborne, millimeter-wave radar theoretically offers an excellent MTI capability with the ability to operate at night and will contine to perform with reduced range in inclement weather.

The carrier frequency for a millimeter-wave radar should be chosen after considering component availability, atmospheric attenuation, and rainfall backscatter effects. Foul weather radar performance is directly related to atmospheric attenuation and rain backscatter. Backscatter or rain clutter increases the minimum noise level, and attenuation reduces the signal level returned from a target. Thus, the overall signal-to-noise ratio is reduced and results in reduced detection capability. After all these facts were considered, 95 GHz was selected because

1. Atmospheric attenuation becomes more severe with increasing frequency, but windows, or low attenuation regions, occur in the frequency spectrum. One window is centered at about 95 GHz.

2. Hardware at 95 GHz is becoming readily available.

3. Although <u>attenuation</u> caused by rain is more severe at 95 GHz than at 70 GHz, this effect is compensated for because the <u>backscatter</u> from the rain is lower at 95 GHz than at 70 GHz. The lower backscatter at the higher frequency can be attributed to the following:

    a. The volume of rain from which the signals can reflect is proportional to λ^2 for a fixed-aperture-size pencil beam antenna, and λ for an equivalent elevation shaped beam antenna

    b. The index of refraction of water varies directly with λ. Thus, there is less mismatch at the raindrop surface at 95 GHz than at 70 GHz and therefore less scatter.

The analysis used to predict the foul weather performance of millimeter-wave radar has been summarized in a Goodyear Aerospace paper entitled "Millimeter-Wave Weather Performance Projections," presented at the 1974 Millimeter-Wave Conference.

SYSTEM DESCRIPTION

The millimeter-wave radar system can be broken down into three basic blocks: antenna, transmitter-receiver, and display. Figure 1 is a simplified block diagram of the millimeter radar system. The antenna is hard mounted to the transmitter-receiver. Scanning is accomplished by rotating the antenna, transmitter-receiver package. Figures 2 and 3 are photographs of the antenna, transmitter-receiver package showing how the radar is mounted on the pedestal and the microwave hardware assembly. Figure 4 is a photograph of the transmitter tube and the high-voltage pulse modulator.

System parameters are summarized as follows:

Operational frequency	95 GHz
Display type	PPI
Range resolution	10 ft (τ = 20 ns)
Azimuth resolution	40 ft/NMI (-6 dB)
Range (maximum displayed)	8 NMI
Azimuth scan angle	360 deg
Primary power requirement	115 VAC, 3 phase, 400 Hz, 700 watts total.

The antenna is a two-foot parabolic reflector with a Cassegrain feed. A switchable polarizer on the antenna feed provides the capability of transmitting either left-hand circular, right-hand circular, or linear polarization. Both azimuth and elevation antenna position signals are supplied to the display. The gain of the antenna is 52 dB, and the 3-dB beamwidth is 0.38 degree.

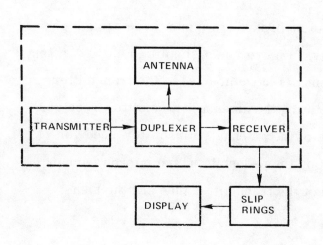

Figure 1 - Millimeter-Wave Radar Simplified System Diagram

Figure 2 - Antenna and Transmitter-Receiver Mounted on Pedestal

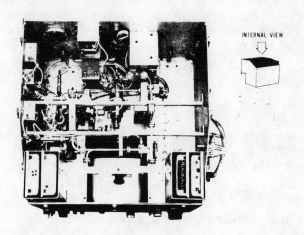

Figure 3 - Transmitter-Receiver Internal View

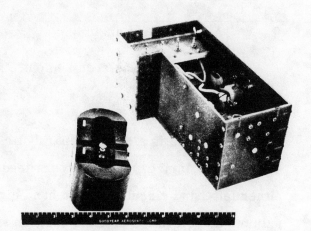

Figure 4 - Magnetron and Modulator

The transmitter includes the high-power RF tube, high-power waveguide components, modulator, and the necessary timing and protective circuitry. Low-voltage power supplies are common to the transmitter-receiver. An Amperex DX423 magnetron, capable of generating 8 kW of RF power, is used as the high-power tube. The modulator circuit is a thyratron line type with a Darlington pulse forming network (PFN) driving the magnetron. The high voltage required to charge the PFN is developed by a solid-state magnetic modulator rather than a high-voltage DC supply. This magnetic modulator reduces the requirements of the DC power supply from approximately 2 kV down to about 250 volts. The RF pulse width is 20 nanoseconds.

The receiver contains an RF local oscillator, mixer, IF amplifiers, and detector. A single-ended crystal mixer is used as the receiver front end. The local oscillator is a low-power klystron.

The preamplifier-IF amplifier section has a center frequency of 335 MHz and a passband of 80 MHz with an overall gain of greater than 100 dB. The IF amplifiers have built-in gain controls. The gain can be electronically controlled and provides a 60-dB range. After amplification at IF, the received signal is detected, amplified, and fed to the display unit.

The display unit is a PPI utilizing a five-inch cathode-ray tube (CRT). The image on the display is built up by scanning the antenna in both elevation and azimuth. A camera is used to photograph the display as the antenna scans, thus producing a photograph of the radar image.

Three display ranges are provided: 1.5, 5, and 8 miles. Range marks can be superimposed on the display in 0.1-mile steps throughout the entire range. Also, a single range mark to a 0.1-mile accuracy may be positioned on the display at the desired range.

TEST RESULTS

All tests were performed at Goodyear Aerospace, Arizona Division. During the test phase of the program, the radar was placed atop the radar tower on Building 13 (see Figure 5). Note that radar transmission from the roof of the radar tower on Building 13 is blocked by Buildings 6 and 26. The palm trees, signs, water tower, and telephone poles surrounding the test site all cast long radar shadows, as is apparent on the imagery.

The radar platform is only 54 feet above the ground, presenting a problem in that the incidence angles are very shallow. Figure 6 is photographs taken from the roof of the radar tower. These photographs have been superimposed on a plan view of the plant area. Note the radar shadows cast by the various buildings surrounding the radar tower. Figure 6 accents the shadow problem and shows some of the areas with very small radar cross-section that are clearly visible in the radar imagery. From an elevation of 54 feet, the incidence angle or grazing angle of the radar beam is approximately 6 deg at 500 feet range and decreases to 0.5 deg at 1 mile.

Figure 7 is an aerial view, or plan position photograph, and a sample of radar imagery made from the roof. The aerial view has the same geometrical terrain relationships as the radar imagery, while Figure 6 illustrates the limited field of view of an optical sensor located at the radar site.

Figures 8 through 11 are examples of the radar imagery made with the millimeter-wave radar. Imagery of the same area made at different times was

Millimeter Wave Radars

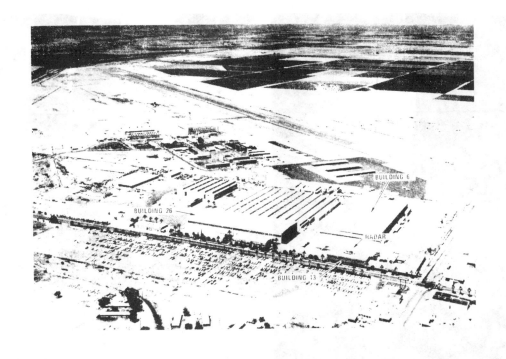

Figure 5 - Aerial View of Goodyear Aerospace, Arizona Division

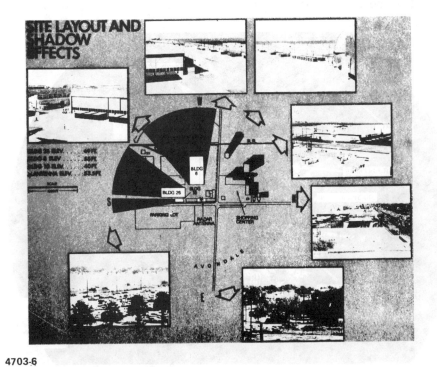

Figure 6 - Site Layout and Shadow Effects

Figure 7 - Radar Imagery versus Aerial Photography

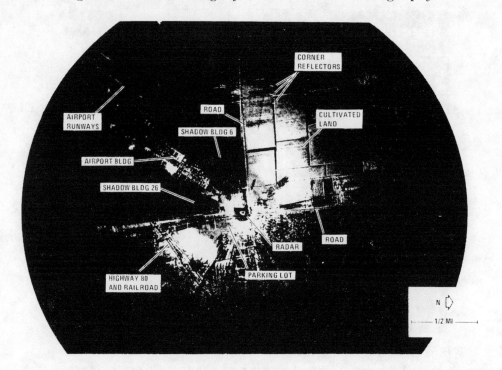

Figure 8 - Goodyear Aerospace Plant and Vicinity 20 November 1973 (PM)

Millimeter Wave Radars

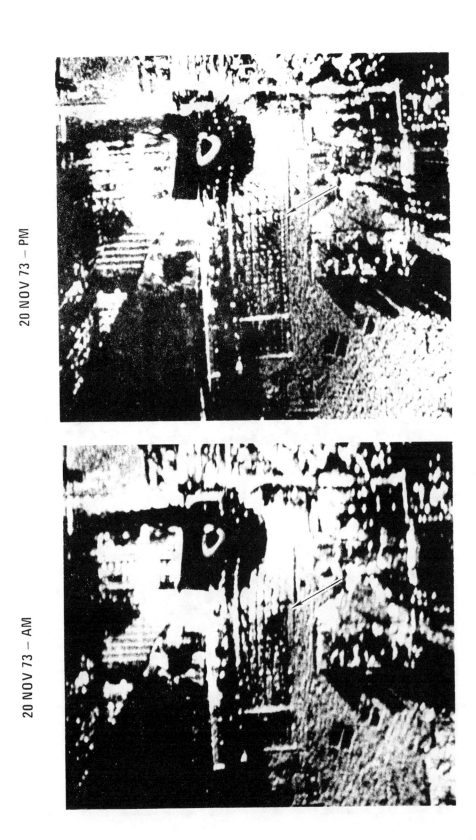

Figure 9 - Change Detection - Parking Lot

Figure 10 - Change Detection - Cultivated Areas

Millimeter Wave Radars

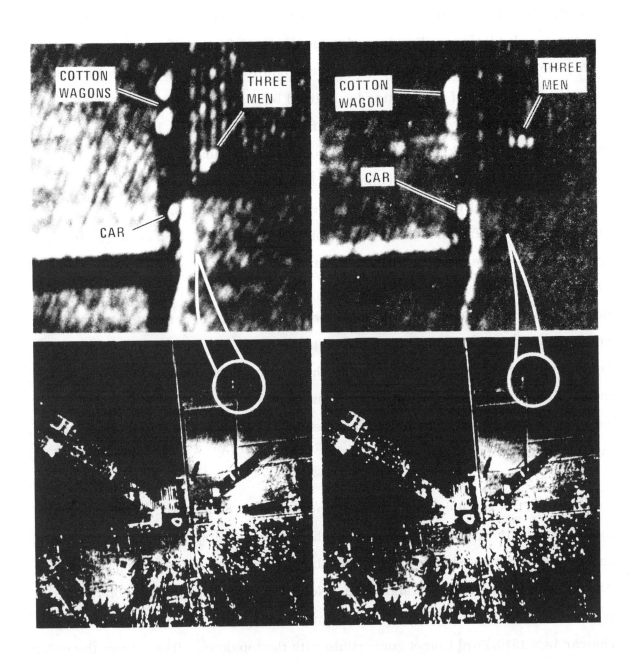

Figure 11 - Change Detection - Men

compared to that of Figure 8. Some very interesting change detection can be done with three radar images. For example, close scrutinization of the Goodyear Aerospace parking lot reveals that there are significant differences. In Figure 8 the eastern boundary of the parking lot is lined by 55-gallon drums, but in another sample of imagery, it was noted that the drums were missing. Figure 9 is an enlargement of the parking lot areas of the images and clearly reveals the difference. Inspection of the farmland in the northwest corner of Figure 8 and two other samples of imagery also discloses detail differences. Figure 10 is an enlargement of the farmland area in the northwest corner imaged at three different times. In the earliest image on the left-hand side, it can be seen that a farmer is just starting to disk under the cotton vines in preparation for plowing, but imagery made approximately four hours later (center) shows that about three-fourths of the cotton plants in one field have been disked under. The right-hand side, made nine days later, shows that the field in the northwest corner has been plowed, leveled, and the dirt has been mounded in rows in preparation for raising hay. Note that the rows of cotton were running north and south and the rows in the newly planted hay field are running east and west.

Figure 11 is two pieces of imagery made about an hour apart. The enlarged sections of the imagery reveal images of three men standing in the field. In one image, the men were standing in a triangular pattern, and in the other, they were standing side-by-side in a straight line. Note that in one image, two cotton wagons are parked in the field, and in the other image, only one cotton wagon is present. The car is a 1970 Ford Comet convertible with the top down. Range from the radar to the men was approximately 3100 feet.

Imagery demonstrating the long range capability of the radar has been made. With the display range set at eight nautical miles, point targets are visible out to the edge. Power poles are visible out to a range of six miles. The near-range detail in the imagery made on the eight-mile scale is comparable to that of Figure 8.

CONCLUSIONS

Millimeter-wave radar can satisfy the requirement for a relatively all-weather, lightweight, compact, inexpensive sensor for various operational applications. Some hardware obstacles do exist, e.g., high-gain, shaped vertical beam antennas; inexpensive, long-life RF power sources, and less expensive RF mixers are not yet available. Increased interest in millimeter waves for solving tactical problems will stimulate component development. Reduced complexity is the key to producing an inexpensive system. Millimeter waves provide a tool for obtaining high performance with simplified equipment.

It has been demonstrated that a real aperture, short pulse millimeter-wave radar has an excellent terrain and cultural target imaging capability even at very low depression angles. Although no direct quantification of the radar parameters has been made, the imagery results demonstrate that an obvious adequacy of the parameter combination exists. In particular, cultural targets such as building dihedral and trihedral corners, that normally cause receiver saturation problems at X-band, are effectively suppressed in the millimeter-wave radar imagery. Because of the very limited first Fresnel zone dimensions at millimeter wavelengths, multipath interference also appears to be virtually nonexistent at 95 GHz.

Assuming that the altitude of the radar was 1000 feet rather than 54 feet, the present system would map out to 3 nautical miles. If the altitude were increased to 2000 feet, the range would be extended to 3.5 nautical miles. Using techniques described in "Millimeter-Wave Weather Performance Projections," the range as a function of rainfall rate has been projected. Figure 12 is a curve showing the relationship of the radar range to rainfall rate. It is interesting to note that greater than one-mile range is predicted for rainfall rates lower than 15 millimeters per hour.

A major impediment to operating in a forward area appears to be the lack of confidence which both the pilots and engineers have exhibited in the performance of a self-contained airborne system for instrument landing. Contributing to this reluctance is the inability of a pilot to detect potential hazards such as aircraft, vehicles, personnel, debris, or other obstructions on a runway during a low-visibility landing. The test results made with a millimeter-wave radar operating on a stationary platform supply the necessary detail, but experiments must be performed to determine if the necessary detail can be obtained from a moving platform.

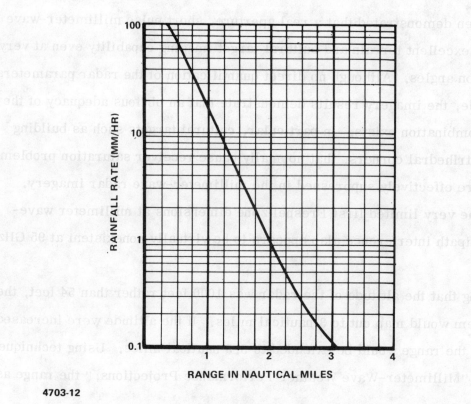

Figure 12 - Radar Range versus Rainfall Rate (1000-Ft Altitude)

A Solid State 94 GHz Doppler Radar

F.B. Bernues, H.J. Kuno, and J. McIntosh
IEEE 1975 MTTS International Symposium Digest
Pages 258-260

© 1975 by the IEEE. Reprinted by permission.

Abstract

The design, construction and performance of a small portable all solid-state Millimeter-Wave Doppler Radar operating at 94 GHz are described. Design trade-offs are discussed. Laboratory and field performance data are also presented.

Introduction

In recent years there have appeared a variety of solid-state Doppler sensors operating at frequencies up to V-band. Recent advances in millimeter-wave solid state sources and components have made it possible to benefit from the advantages inherent in higher frequency operation, such as:

1. Higher gain and smaller beamwidth for the same antenna size. The higher gain (increasing by a factor proportional to λ^2) results in longer range, whereas the smaller beamwidth results in improved spatial resolution, i.e. ability to discriminate between adjacent targets, as well as improved clutter rejection capability.

2. Higher Doppler frequency for the same target speeds. This in turn translates into higher sensitivity (due to the reduction of 1/f noise) and higher frequency resolution.

This paper describes the design principles and the performance of a Solid-State Doppler Radar operating in the low atmospheric attenuation window centered around 94 GHz.

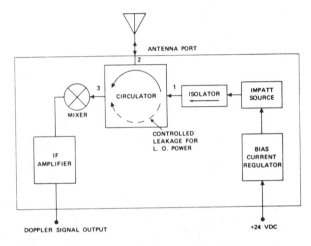

Figure 1 Block diagram of solid-state 94 GHz Doppler Radar.

Design Considerations

The radar is of a homodyne type, consisting of an RF source, a circulator, a single-ended mixer, an IF preamplifier and a parabolic antenna (see the block diagram and photograph in Figures 1 and 2 respectively). The RF source is an IMPATT oscillator (with an integral isolator) capable of delivering over 100 mW CW output power. The mixer uses a silicon Schottky barrier diode with an extremely small junction (3 μm diameter). The circulator has the dual purpose of separating the transmit and receive signals and leaking a small fraction of the signal power which serves as local oscillator (L.O.) power for the mixer.

The design goal is to achieve maximum sensitivity. The theoretical limit of tangential sensitivity is given by

$$S = -140 + 10 \text{Lg}_{10} B + NF$$

where
  S = tangential sensitivity in dBm
  B = IF preamplifier bandwidth in Hz
  NF = Noise figure of mixer-preamp in dB

This includes thermal and 1/f noise[1] but not noise from the transmitter. It is assumed that the preamplifier has a low frequency cutoff of at least 2 KHz, corresponding to a minimum detectable target speed of about 10 mph.

Noise from the transmitter can not be neglected, however, because the isolation of the circulator is carefully controlled so that a portion of the trans-

Figure 2 Photograph of Doppler Radar.

mitter power is used as L.O. power for the mixer. FM noise is of no importance for target distances below 2 miles. AM noise could limit the sensitivity of the radar if a large amount of power is leaked. To determine the optimum leakage, it is necessary to know the AM noise of a typical IMPATT diode at this frequency. Measurements of this noise have shown[2] that a typical value is 120 dB below carrier per 100 Hz bandwidth, for frequencies at least 1 KHz off the carrier frequency. Figure 3 shows the ratio $P_{received}/P_{L.O. noise}$ (in dB) versus L.O. power leaked to the mixer (in dBm), assuming $P_{transmitted}$ = 100 mW and a preamplifier bandwidth of 20 KHz. $P_{received}$ is the power arriving to the mixer from a target with $\sigma = 1\ m^2$, and $P_{l.o.\ noise}$ is the noise arriving at the mixer from the IMPATT source.

To minimize noise from the transmitter, leakage should be as small as possible, but as the available L.O. power is reduced, the conversion loss of the mixer deteriorates. Figure 4 shows a typical plot of conversion loss versus available L.O. power. By using forward-bias, the conversion loss can be kept reasonably low when the available L.O. power is reduced to low value (1 mW or less). This in turn reduces L.O. noise contribution to the receiver.

The maximum range of the radar depends mainly on target cross-section (σ) and transmitted power. Figure 5 shows a plot obtained from the classical radar equation, including 4 dB roundtrip atmospheric attenuation[3]. Measurements[4] of σ at X-band for cars give values of about 10 m^2. Measurements at W-band are not common; as an example,[5] a square aluminum plate of 0.66" side has a value of $\sigma = 0.1\ m^2$. Consequently, we can predict a worst case of $\sigma = 1\ m^2$ for targets such as cars, boats, etc.: typical values will be between 10 m^2 and 1000 m^2, depending on angle, etc.

With a transmitter power of +20 dBm and an antenna gain of 52 dB (24" parabola), the maximum range will vary between 1000 and 3000 meters, depending on the target.

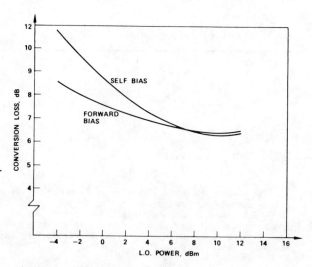

Figure 4 Shows a typical plot of conversion loss vs. available L.O. power.

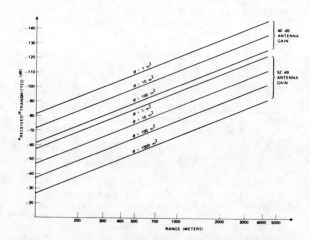

Figure 5 A plot obtained from the classical radar equation.

Construction and Performance

The unit was tested in the laboratory by using a small rotating wheel as a target. The speed of the wheel could be varied so that the Doppler frequencies fall within the intended range.

It was found that the maximum sensitivity could be obtained by carefully adjusting the isolation of the circulator (between 15 and 20 dB), the amount of forward bias and, to some extent, the input impedance of the preamplifier.

Figure 6 shows the measured output voltage of the preamplifier (in volts, peak to peak), versus the power returned to the mixer by the target. It can be seen that the tangential sensitivity is about -75 dBm.

Figure 7 shows an oscilloscope display of the Doppler signal output, with the rotating wheel as a target, when the ratio $P_{received}/P_{transmitted}$ is -70 dBm.

In field tests, targets such as cars at a distance of about 1.5 miles gave Doppler signals of about 350 mV peak; at a distance of 0.5 miles, the voltage output varies between 1 to 2 volts, and the radar could clearly

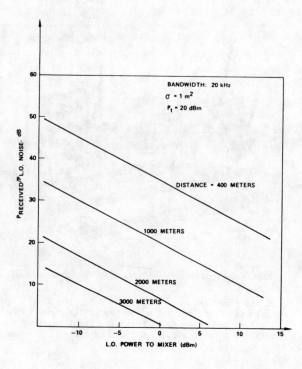

Figure 3 Shows the ratio $P_{received}/P_{L.O.\ noise}$ vs. L.O. power leaked to the mixer.

discriminate between two cars in adjacent traffic lanes.

Figure 8 shows the spectrum of the Doppler return from a boat. The Doppler signal was recorded on an ordinary cassette tape recorder and then played back into an audio spectrum analyzer. The radar was "viewing" the stern of the boat as the range was increasing from 1/4 mile to 3/4 mile. Further field tests will be conducted to gather cross-section and sea-clutter data under several meteorological conditions, as well as to determine maximum usable range.

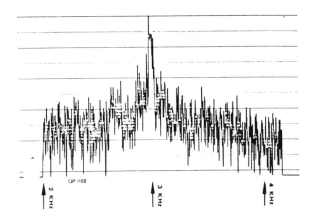

Figure 8 Spectrum of Doppler return from small boat.

Conclusions

A small solid-state 94 GHz CW Doppler Radar has been constructed. The unit demonstrates the advantages of millimeter-wave operation to reduce clutter (specially sea-clutter) and to achieve high spatial and speed resolution. Sensitivity and range can be improved by using appropriate signal processing circuitry.

References

1. *Microwave Semiconductor Devices and Their Circuit Applications*, H.A. Watson, Ed., McGraw Hill Book Co., p 371.

2. H.J. Kuno, K.P. Weller, D.L. English, "Tunable Millimeter-Wave Packaged Impatt Diode Oscillators", 1974 MTT Symposium Digest, p. 320.

3. D.M. Grimes and T.O. Jones, "Automative Radar: A Brief Review," Proceedings IEEE, vol. 62, No. 6, pp. 804-822, June 1974.

4. T.O. Jones, D.M. Grimes and R.A. Dork, "A Critical Review of Radar as a Predictive Crash Sensor," SAE Paper 720424, May 1972.

5. R.B. Dybdal and C.O. Yowell, "Millimeter Wavelength Radar Cross-Section Measurements," G-AP International Symposium, Boulder, CO, pp. 49-52, August 1973.

Acknowledgements

This work was supported in part by the U.S. Coast Guard Research and Development Center, Groton, Connecticut, under Contract No. DOT-CG-81-74-1082. The opinions or assertions contained herein are the private one of the authors and are not to be construed as official or reflecting the views of the Commandant or the Coast Guard at large.

The authors wish to thank N. B. Kramer and Y. W. Chang for their valuable suggestions and help, and L. D. Thomas for his technical assistance in the construction of this radar.

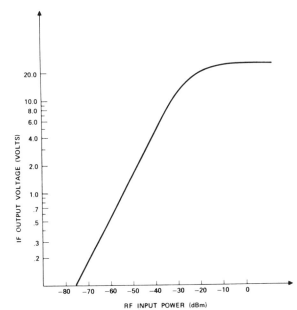

Figure 6 Measured output voltage of Doppler signal volts, peak-to-peak) as a function of power returned from target to antenna port (in dBm).

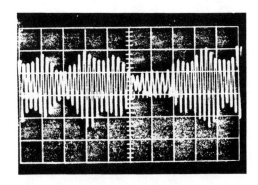

Figure 7 Doppler return from rotating wheel. $P_{received}/P_{transmitted}$ is -70 dBm. Horizontal scale: 1msec/div. Vertical scale: .5 Volts/division.

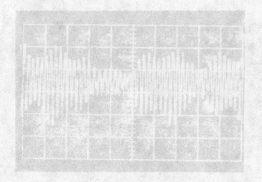

Millimeter Wave Monopulse Track Radar

L.H. Kosowsky, K.L. Koester, and R.S. Graziano

SUMMARY

A high accuracy millimeter wave monopulse radar has been developed for use in high resolution tracking applications. The system utilizes a real aperture millimeter wave antenna transmitting a medium power noncoherent narrow pulse.

An analysis of attenuation and backscatter in adverse weather conditions indicates that against a 10 square meter target, a range of 10 km may be obtained in clear air and 4.8 km in 4 mm/hr or rain. A description of the system configuration is given together with currently available experimental data. Accuracies of better than 1.0 mr have been achieved.

1. INTRODUCTION

The utilization of millimeter wavelengths for high resolution, high accuracy applications is becoming increasingly significant as components and technology develop rapidly. For a given size aperture, there are advantages to the use of millimeter wave radiation namely, high resolution and ECM immunity. As a consequence, a millimeter monopulse radar is an attractive sensor for short range, adverse weather tracking applications such as terminal guidance and fire control.

Because adverse weather plays a significant role in determining the performance limits of any millimeter radar, a careful examination and measurement of the absorption and backscatter from fog, rain, and snow has been made. It has been shown, both theoretically and experimentally that fog and snow produce little effect on system performance, whereas rain contributes significant absorption and backscatter and must be carefully considered in system design. The effects of weather and clutter on system performance have been examined in detail. Curves depicting detection range are presented for a variety of weather conditions.

The tracking of complex targets requires knowledge of the glint characteristics of the target as well as the clutter in which it is immersed. Because of the complexity in analytically deriving the glint statistics of targets and clutter, an experimental program was undertaken to develop a four lobe amplitude monopulse track radar which could be utilized to obtain experimental data and to provide a "brassboard" configuration for use in demonstration tracking systems.

The millimeter track radar which has been configured, operates at 70 GHz, utilizes an 18 inch antenna, and has a static tracking accuracy of less than one milliradian. The system transmits a 100 nanosecond, 10 kW pulse at a PRF of 4 KHz, utilizing a pulsed magnetron source. The receiver achieves a 10 dB noise figure with Schottky Barrier mixer crystals. Experimental tracking data are presented for isolated targets and targets in clutter.

2. THE TRACK PROBLEM

Many tracking problems require accurate tracking of a relatively small target in close proximity to a large competitive clutter area. This is particularly true in the case of tracking low flying aircraft and missiles from a ground location. As shown in Figure 1, we shall consider cases in which the aircraft is located 30 meters above a clutter area at ranges out to 10 kilometers. The solution to this difficult tracking problem has proceeded classically along a number of different paths, namely:

Clutter suppression by statistical data processing, e.g. MTI

Coherent techniques e.g. pulse compression, synthetic aperture, doppler discrimination

Spatial discrimination e.g. narrow beamwidths

Of the techniques enumerated, the direct generation of narrow beamwidths by multiwavelength apertures provides the simplest most direct method of resolving small targets in the presence of clutter. In order to utilize reasonable size antennas, the designer is forced to high frequencies of operation where propagation becomes a significant parameter in limiting system performance.

3. PROPAGATION FACTORS IN THE MILLIMETER REGION (60-100 GHz)

The design of millimeter wave systems is extremely sensitive to propagation losses in the atmosphere. Although a detailed analysis of the factors involved in producing energy loss is beyond the scope of the present paper, several of the most significant propagation characteristics are summarized below.

3.1 Clear Air Attenuation

The attenuation of millimeter wave energy in clear air has been the subject of intensive investigation by a number of researchers (VanVleck, J.H., 1964, Rosenblum, E.S., 1961, Straiton, A.W. and C.W. Tolbert, 1960, and Koester, K.L., Sept 1971) and has been confirmed by careful measurement (Tolbert, C.W. and A.W. Straiton, 1957, and Hogg, D.C., 1968). A summary of available data is given by Rosenblum, and is shown in Figure 2. Examination of the M-band region, 60-100 GHz, shows that the clear air attenuation varies from a peak of 15 dB/km at the oxygen line at 60 GHz, to a minimum of 0.39 dB/km at the "window" at 94 GHz. It should be noted that the propagation loss is strongly dependent on the amount of oxygen in the atmosphere and thus varies considerably as a function of altitude and path length through the atmosphere.

3.2 Rain

In addition to the energy loss resulting from atmospheric gases, the performance of a radar operating in rain is degraded by the absorption and scattering of the energy by the rain drops. The propagation of millimeter waves in rain has been investigated in depth (Koester, K.L., Sept 1972, and Wilcox, F.P. and R.S. Graziano, 1974). The attenuation as a function of rain rate is shown in Figure 3 for 70 and 94 GHz. It can be seen that for a moderate rainfall rate of 4 millimeters per hour, the attenuation is 2.3 dB/km at 70 GHz and 2.8 dB/km at 94 GHz (Wilcox, F.P. and R.S. Graziano, 1974).

Accompanying the absorption loss in rain is the microwave energy reflected by the rain drops which contributes additional noise to the system. The backscatter coefficients for rain have been measured (Richard, V.W. and J.E. Kammerer, 1974), and are shown in Figure 4. The use of circularly polarized antennas results in a 15 dB reduction of rain backscatter at 70 GHz and 94 GHz as shown by Richard.

3.3 Fog and Clouds

Of great significance to the performance of a track radar is its ability to follow aircraft and missiles through cloud cover and fog. The attenuation coefficient for millimeter wave propagation of fog has been investigated in detail (Koester, K.L., March 1971, Koester, K.L. and L.H. Kosowsky, Nov 1970, and Koester, K.L. and L.H. Kosowsky, Sept 1970) and has been shown to be linearity dependent on the liquid water content of the fog. In addition, there is a significant dependence of attenuation on the temperature of the fog. The attenuation is shown in Figure 5 as a function of liquid water content. The attenuation coefficient for a liquid water content of 0.1 g/m^3 is 0.35 dB/km at 70 GHz and 0.47 dB/km at 94 GHz at 0°C.

The backscatter coefficient of fog in the millimeter band is more than two orders of magnitude smaller than that of rain and has a negligible effect on radar system performance (Koester, K.L., March 1971).

3.4 Snow

In cold climates, the presence of snow severely influences optical detection of targets. The attenuation in snow as a function of snowfall rate (Skolnik, M.I., 1962) is shown in Figure 6 for frequencies of 70 and 94 GHz. In a moderate snow (accumulation rate of 30 millimeters per hour) the attenuation is 0.62 dB/km at 70 GHz and 1.95 dB/km at 94 GHz. Experiments conducted by Norden at 70 GHz indicate that the backscatter from very heavy snow is negligible.

4. FREQUENCY SELECTION

The theoretical analysis and experimental work performed by Norden has demonstrated that millimeter radar systems will provide adequate adverse weather performance.

During the course of Norden's internally funded programs, a careful comparison was made of cost and availability of components at both 70 and 94 GHz. Greater availability and lower cost of components led to the choice of 70 GHz as a representative frequency for experimental work in the millimeter band. The work at 70 GHz is compatible with 94 GHz requirements permitting design anywhere within the band.

5. MONOPULSE CONCEPTS

Monopulse, as defined by Rhodes, "....is a concept of precision direction finding of a pulsed source of radiation. The direction of the pulsed source is determined by comparing the signals received on two or more antenna patterns simultaenously." (Rhodes, D.R., 1959). The two basic types of systems which are commonly used, phase or amplitude, are shown in Figure 7. The type of system derives from the antenna configuration since phase information received in a phase monopulse can be readily converted to amplitude information and vice versa.

In either case, by forming the sum and difference, the information on target location is contained in the relative amplitudes of the two signals.

In the system discussed in this paper, an amplitude monopulse system is used as the antenna sensor in which four squinted beams are produced from a single aperture to yield an azimuth difference pattern, an elevation difference pattern, and a common sum pattern.

Figure 8 shows the feed/comparator for a single coordinate amplitude monopulse antenna. In this configuration, the sum pattern is used for transmission, while on reception, both the sum and difference patterns are obtained. The sum signal provides target intensity information as well as range to the target. The difference signal is processed together with the sum signal to yield the error signal.

Of significance to the designer is the sensitivity of the error signals and the discrimination of scatterers in the vicinity of the target. Both of these conditions are optimized with narrow highly directional antenna beams (Skolnik, M.I., 1962). Thus, the smaller the beamwidth of the antenna, the larger the error slope and the better the tracking accuracy.

Since, for a given aperture, the beamwidth decreases directly with the wavelength, the sensitivity of a monopulse antenna operating at 70 GHz can be expected to be seven times greater than a comparable aperture operating at 10 GHz.

6. PERFORMANCE PREDICTIONS

A block diagram of the monopulse system currently being evaluated at Norden is shown in Figure 9. As indicated, the system consists of the following major subsystems: transmitter/modulator, microwave package, antenna, receiver, data processing unit, gimbal and servo electronics unit. (A description of each subsystem is contained in the following sections.) Parameters of the radar are summarized in Table 1.

Table 1. Millimeter Monopulse Track Radar Parameters

TRANSMITTER		
Frequency	70 GHz	
Peak Power	10 kW	
Pulse Width	100 ns	
Pulse Repetition Rate	4 kHz	
ANTENNA		
Type	Parabola	Cassegrain
Sum Beamwidth (Pencil)	0.66°	0.67°
Gain	47.2 dB	45.5 dB
Sidelobes	-18 dB	-25 dB (Lower)
Null Depth	32 dB	35 dB
Polarization	Linear	Circular
RECEIVER		
Type	Linear, Lin/Log	
IF Frequency	60 MHz	
Bandwidth	20 MHz	
Noise Figure	10 dB	

The range performance of the tracking radar can be determined using the radar range equation. The single pulse signal to noise ratio is given by

$$(S/N)_{SP} = \frac{P_T G^2 \lambda^2 \sigma_T}{(4\pi)^3 \, KTB \, \overline{NF} \, R^4 \, L_S L_A} \quad (1)$$

where

P_T is the peak transmit power

G is the peak antenna gain

λ is the transmit wavelength

σ_T is the target cross section

K is Boltzmann's constant

T is the absolute temperature of the receiver

B is the receiver bandwidth

$\overline{NF}$ is the system noise figure

R is the range to the target

L_S is the system loss

L_A is the atmospheric loss

Using the radar parameters in Table 1, system performance against a 10 square meter target was calculated using Equation 1. The single pulse signal to noise ratio is shown in Figure 10 as a function of range for clear air and 100 meter visibility fog.

For the noise only case, the single pulse signal to noise required to obtain a 90% probability of detection can be determined using the curves presented by Rubin and DiFranco (Rubin, W.L. and J.V. DiFranco, 1964). Assuming a 10^{-6} probability of false alarm, a Swerling 1 target, and the integration of 100 pulses, the $(S/N)_{SP}$ required for a 90% Pd is approximately 6 dB. Thus for the clear air and fog cases radar range in excess of 10 km is achieved.

During rainfall, rain backscatter must be considered to determine the system performance. The rain backscatter introduces a clutter return which is proportional to the rain clutter cross section given by

$$\sigma_\omega = V \sigma_r' \quad (2)$$

where V is the volume of the radar pulse packet and σ_r is the normalized rain backscatter cross section.

The volume of the pulse packet for a pencil beam antenna pattern is approximately

$$V = \left(\frac{\pi}{4}\right) (R\theta)^2 \, \Delta R \quad (3)$$

where

R is the radar range

θ is the antenna half power beamwidth

ΔR is the length of the pulse packet

Using the normalized rain backscatter cross section measured by Richard, the rain clutter cross section was determined as a function of range using Equations (2) and (3), (Richard, V.W. and J.E. Kammerer, 1974). The results are summarized in Table 2. A 15 dB reduction in rain backscatter through the use of circular polarization has been assumed.

Table 2. Rain Clutter Cross Section

Range (KM)	Rainfall Rate (mm/hr)	Rain Cross Section (σ_ω)* (m^2)
1	1	.005
	4	.01
5	1	.125
	4	.27
10	1	.54
	4	1.08

*A 15 dB reduction in σ_ω due to the use of circular polarization has been assumed.

To determine system performance during rainfall, the integration gain must be evaluated by first examining the decorrelation time of rain. The standard deviation of the rain clutter frequency spectrum is due primarily to wind turbulence. If we assume a fluctuating wind speed (Ws) of 1 m/sec, then the standard deviation of the rain clutter spectrum is given by

$$\sigma_c = \frac{2\ Ws}{\lambda} = 465\ Hz$$

The number of independent clutter samples is approximately

$$N_I = \frac{2.5\ \sigma_c\ N}{f_r}$$

where

f_r is the radar PRF

N is the number of radar pulses

For the 100 pulse case,

$$N_I = 29\ \text{pulses}$$

Thus the effective number of pulses integrated for signal to rain clutter improvement is 29.

As the target observation time increases to

$$\sigma_c > .4\ f_r$$

then

$$N = N_I$$

may be employed. At the radar range of interest, the signal to clutter plus noise ratio is limited by thermal noise. Thus, the signal to clutter plus noise required for a 90% Probability of detection remains approximately 6 dB. At shorter ranges, where the rain backscatter limits system performance, the effective number of radar pulses must be determined. As shown in Figure 10, the radar range for a 90% Probability of detection for 1 mm/hr and 4 mm/hr of rain is 9.3 km and 4.8 km respectively. A summary of the system range performance predictions is contained in Table 3.

Table 3. System Range Performance Predictions

Environment	Attenuation (dB/km)	Backscatter (m^2/m^3)	Target Detection Range (km)
Clear Air	.4	-	>10 km
Fog (100m Visibility)	.7	-	10.0
Light Rain (1mm/hr)	.8	1×10^{-4}	9.3
Moderate Rain (4mm/hr)	2.7	2×10^{-4}	4.8

Target Cross Section = 10 square meters

Target Model = Swerling Case 1

Probability of detection = 90%

Probability of false alarm = 10^{-6}

7. SYSTEM DESCRIPTION

The Norden experimental millimeter track radar is shown in Figure 11. A description of each of the major subsystems follows.

7.1 Transmitter/Modulator

The transmitter/modulator employs a conventional line-type thyratron modulator with a solid state driver to produce the grid trigger and a pulse-to-pulse regulator to reduce output power amplitude variations due to fluctuating input voltage. The magnetron, which operates at 70 GHz, provides 10 kW peak power at an 0.0004 duty ratio. The nominal transmitted pulse width is 100 nanoseconds. A ferrite isolator prevents magnetron pulling due to load VSWR. A pressurization unit provides 31 psia of dry air pressure to prevent arcing in the waveguide and pulse transformer housing.

7.2 Antenna

Two monopulse antennas are available for use with the system. The antenna used in the experiments described in this paper is an 18-inch parabola with an f/D ratio of 0.67. The microwave comparator and four-horn feed are constructed as an integral unit to minimize precomparator phase shift. A manual polarization selection capability allows the choice of either vertical or horizontal polarization depending on antenna orientation.

Sum channel gain at the comparator is 47.2 dB. The radiation pattern is a pencil beam with a 3 dB beamwidth of 0.66 degree and first sidelobe levels of -18 dB with respect to the sum peak. The null depth of the difference pattern is -32 dB. The four lobe antenna was designed for use between 69 and 71 GHz and meets its performance specifications over this region.

A Cassegrain monopulse antenna was also developed for use with the system. The 18-inch plastic reflector provides a sum channel gain of 45.5 dB and a 3-dB beamwidth of 0.67 degrees. The lower sidelobe level is -25 dB with respect to the sum peak. The null depth of the difference pattern exceeds 35 dB. The isolation between sum and difference ports exceeds 30 dB.

The microwave comparator and four-horn feed were constructed as an integral unit to minimize pre-comparator phase shift. The antenna provides circularly polarized energy (axial ratio < 1.2 dB) to minimize rain clutter signals.

7.3 Microwave

The microwave package incorporates specially designed waveguide assemblies to minimize insertion loss. The sum channel loss is 2.8 dB while losses for the difference channels are less than 1.0 dB each. The duplexing function is accomplished in a ferrite circulator which provides more than 22 dB isolation.

Translation to IF is accomplished by mixer/preamplifier assemblies using Schottky Barrier diodes in a balanced mixer configuration. The units provide 25 dB minimum gain (RF to IF) with a noise figure of 10 dB and a 20 MHz bandwidth centered at 60 MHz, resulting in a receiver sensitivity of -87 dBm. The sensitive Schottky crystals are protected by a switchable ferrite attenuator inserted before the mixer in each channel.

An IMPATT diode oscillator is used as the system local oscillator. A voltage-controlled current source drives the IMPATT which, used in conjunction with an AFC module, maintains the IF within 1 MHz of the center frequency.

7.4 Receiver

Because of the attenuation in the atmosphere, significant variations in signal return can be expected as a function of target range. Thus for amplitude monopulse systems, in which AGC is an important design consideration, it becomes advantageous to convert to angle processing circuitry to minimize the effects of the large dynamic range.

Amplitude monopulse information contained in the IF signals is converted to phase information by the Monopulse IF (MIF) Processor and phase detected to determine the angle-off-boresight in each difference channel. The sum channel signal is split with one output acting as a reference for the phase detector and the other output processed through a log amplifier for information and display purposes. The overall receiver dynamic range is 68 dB.

The angle-error output voltages of the MIF Processor are a function of the phase difference between the sum and difference channel signals and range from zero volts for the on-boresight condition to a maximum value of about 45 mV. The polarity or sense of the error signal depends on the location of the target with respect to the antenna boresight axis. For targets above or to the left of the axis, the sense is positive. For targets below or to the right of the axis, the sense is negative.

Of particular interest is the slope of the angle-error curve, especially in the linear region around boresight, since it is this signal that will ultimately provide the control signals for the servo system. For the millimeter track radar, the slope (or scale factor) at the output of the MIF Processor is 10 mV/mr in the linear region. Scaling amplifiers with variable gains are provided for interface with the processing circuitry.

7.5 Data Processing Unit (DPU)

The data processing unit provides the essential circuitry needed to process the received signals. In addition, the system timing trigger of 4 KHz and all synchronization signals are developed in the unit. The MIF and AFC modules are also contained in the DPU.

The pulsed sum video, azimuth and elevation error signals are received from the MIF and processed.

The sum video generates a position-correcting servo loop for range tracking a single range-gated target. A range swath of 4 nautical miles is searched in a period of 1 second.

Clutter rejection logic is also incorporated in the DPU. The logic, which uses pulsewidth discrimination, allows lock to be broken if the range gate locks onto broad-based clutter. As soon as lock is broken the range gate continues searching for a hard target at a closer range.

Range and range rate readouts are also provided in the system.

7.6 Pedestal

The Antenna/Receiver is mounted on a pedestal which allows tracking the azimuth and elevation. The transmitter/modulator and data processing unit are mounted in the base of the pedestal. Rotary joints are utilized to send energy to the antenna/receiver.

The antenna can track a target over a ± 45 azimuth sector and -5° to +30° in elevation.

8. PRELIMINARY EXPERIMENTAL RESULTS

8.1 Target Acquisition

To demonstrate the acquisition and track capability of the millimeter track radar, the system was integrated with a helmet-mounted optical sight. The sight, provided by General Electric, Burlington, Vermont, is used for rapid acquisition of a potential target and hand-off to the track radar. The helmet sight, operator and track radar are shown in Figure 12.

The helmet sight is comprised of a reticle sight mounted on a standard flight helmet and an electromechanical linkage from the helmet to the common mount of the radar antenna. The reticle sight has a birefringent optical element that creates a pattern of concentric circles. Once the operator aligns the sight to the target, the radar antenna will follow his head motion.

As the operator moves the helmet sight to maintain the target in the sight, the slaved beam is maintained on the target. The automatic range acquisition is engaged and the radar locks onto the target.

When this condition is met the radar system takes over the angle track function and drives the antenna in azimuth and elevation. If the radar should lose range lock, the antenna will automatically track the helmet providing a back-up safety feature.

8.2 Tracking Tests

All tests of the monopulse radar have thus far been conducted as the Norden facility in Norwalk, Connecticut, USA. The radar was placed in a tower facility which is 18.3 meters above ground level. A view from the laboratory is shown in Figure 13.

A corner reflector, shown in Figure 14, is mounted on a pole 17.7 meters in height, at a range of 487.7 meters from the tower. Clutter, at the base of the support pole, is 36 milliradians below the reflector.

Figure 15 shows an A-scope presentation of radar returns as a function of range. Examination of the figure indicates that when the radar is tracking the reflector, the clutter returns in the vicinity of the pole are well below the minimum detectable signal. Since the pole subtends about 36 milliradians at this range, it can be seen that targets can readily be tracked to within 2° of the clutter. The ground clutter shown in the figure is 365.8 meters behind the corner reflector.

Initial observations indicate that no measurable wander occurs, even during windy conditions. This phenomena will be investigated in detail in comprehensive glint studies to be conducted during Calendar 1976.

A series of dynamic tracking experiments were conducted using a Bell 47G helicopter. Figure 13 shows the helicopter flying in the vicinity of the Norden test facility.

The target aircraft was acquired using the helmet sight and tracked. Results of the tests were being evaluated at the time this paper was prepared.

9. CONCLUSIONS

A high accuracy millimeter wave monopulse radar has been developed for use in low angle tracking applications. Predictions of system performance indicate that adverse weather performance permits system operation in dense fog and rainfall rates of 4 mm/hr. Tracking tests are in progress using a corner reflector and helicopter.

Application of the millimeter track radar to aircraft and missiles for weapon delivery and terminal guidance is being investigated.

10. ACKNOWLEDGEMENT

The authors wish to acknowledge the assistance of John Carulli who was responsible for the assembly and test of the brassboard radar and to John Fitzgerald, Tom Ikeda and Albert Sokolowski for their assistance in the development of the radar system.

11. REFERENCES

Hogg, D.C., January 1968, "Millimeter Wave Communication through the Atmosphere," Science, pp. 39-46

Koester, K.L. and L.H. Kosowksy, November 1970, "Attenuation of Millimeter Waves in Fog," presented at the Fourteenth Radar Meteorology Conference

Koester, K.L., March 1971, "Propagation of 70 GHz Energy through Rain and Fog," Norden Technical Report 4302 R 0001, Rev. A

Koester, K.L. and L.H. Kosowsky, September 1971, "Millimeter Wave Propagation in Fog," presented at the IEEE International Antennas and Propagation Symposium

Koester, K.L., September 1971, "Clear Air Propagation in the 4 mm Band," Norden Technical Report 4337 R 0009

Koester, K.L., September 1972, "Millimeter Wave Propagation," Norden Technical Report 4392 R 0005

Richard, V.W. and J.E. Kammerer, 1974, "Millimeter Wave Rain Backscatter Measurements," presented at NELC Millimeter Waves Techniques Conference

Rhodes, Donald R., 1959, Introduction to Monopulse, McGraw-Hill, New York

Rosenblum, E.S., March 1961, "Atmospheric Absorption of 10 to 400 KMCPS Radiation," Microwave Journal, pp. 91-96

Rubin, W.L. and J.V. DiFranco, April 1964, "Radar Detection," Electro Technology, pp. 61-90

Skolnik, Merrill I., 1962, Introduction to Radar Systems, McGraw-Hill, New York pp. 164-197

Straiton, A.W. and C.W. Tolbert, May 1960, "Anomalies in the Absorption of Radio Waves by Atmospheric Gases," Proceedings of the IRE, Vol. 48, pp 898-903

Tolbert, C.W. and A.W. Straiton, April 1957, "Experimental Measurement of the Absorption of Millimeter Radio Waves over Extended Ranges," IRE Transactions on Antennas and Propagation, pp 239-241

Van Vleck, J.H., 1964, "Theory of Absorption by Uncondensed Gases" in D.E. Kerr (ed.), Propagation of Short Radio Waves, pp. 646-664, Boston Technical Publishers, Massachusetts

Wilcox, F.P. and R.S. Graziano, March 1974 "Millimeter Wave Weather Performance Projections," presented at the NELC Millimeter Waves Techniques Conferences

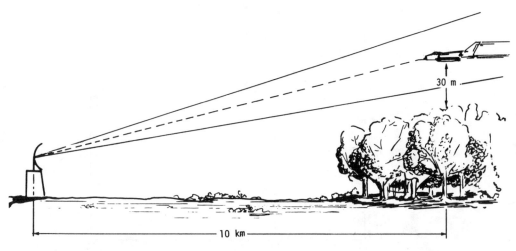

Figure 1. Low Angle Track Scenario

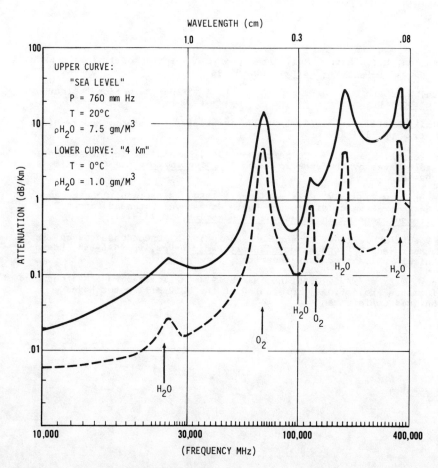

Figure 2. One-Way Attenuation for Horizontal Propagation in Clear Air (E.S. Rosenblum 1961)

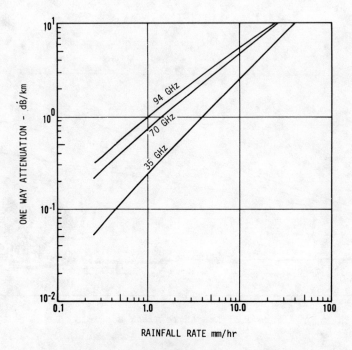

Figure 3. One-Way Attenuation in Rain

Millimeter Wave Radars

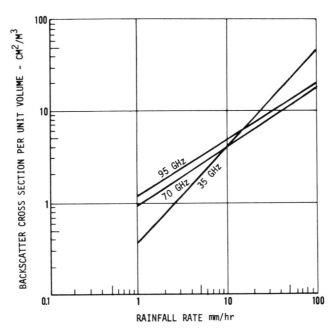

Figure 4. Measured Backscatter Cross Section of Rain (V.W. Richard and J.E. Kammerer - March 1974)

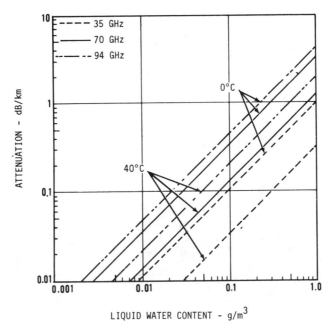

Figure 5. One-Way Attenuation in Fog

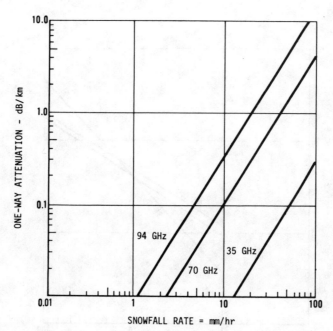

Figure 6. One-Way Attenuation in Snow

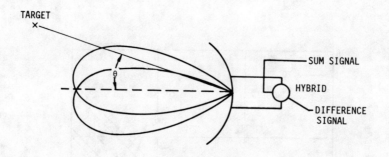

A) AMPLITUDE MONOPULSE

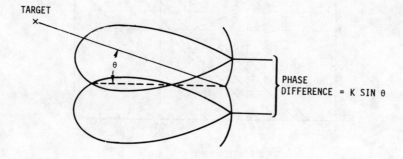

B) PHASE MONOPULSE

Figure 7. Monopulse Antenna Configuration

Millimeter Wave Radars

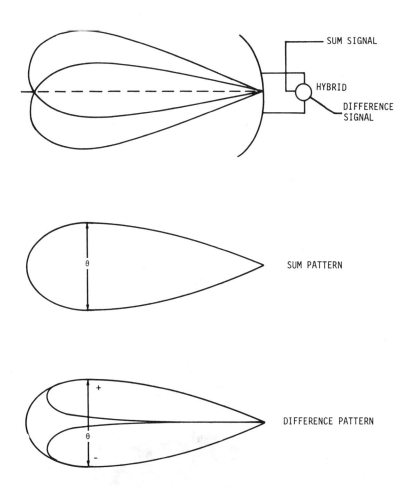

Figure 8. Single Coordinate Amplitude Monopulse Antenna Configuration

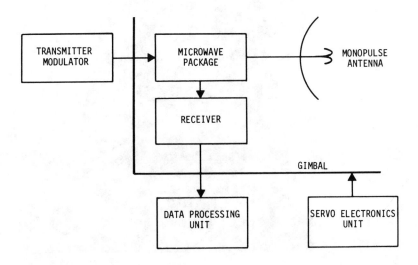

Figure 9. Block Diagram of Millimeter Monopulse Track Radar

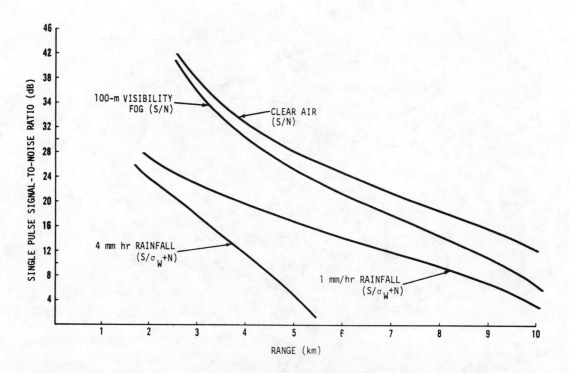

Figure 10. Predicted Performance of Millimeter Track Radar

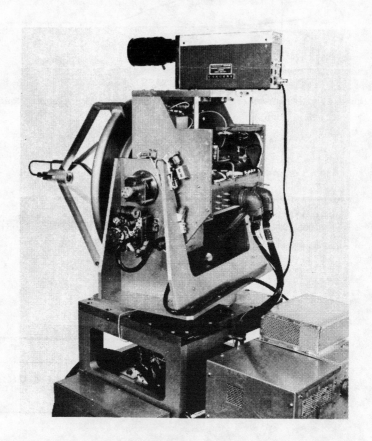

Figure 11. Norden Experimental Millimeter Track Radar

Millimeter Wave Radars

Figure 12. Norden Millimeter Track Radar with Operator and Helmet Sight

Figure 13. View from Tower Facility

Figure 14. Corner Reflector Target

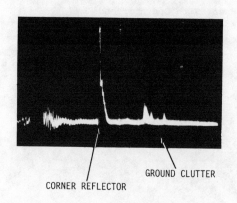

VERTICAL SCALE 0.5 V/cm
HORIZONTAL SCALE 1.0 μs/cm

CORNER REFLECTOR   GROUND CLUTTER

Figure 15. A-Scope Display of Sum Channel Video

DISCUSSION

H SITTROP: Have you made any very low-angle tracker experiments over water, in particular to investigate the "Beckmann Spizzichino" specular reflection coefficient over water? If not, are you planning to do so?

L H KOSOWSKY: No tracking experiments have been made over water and none is presently planned unless sponsorship is obtained. The monopulse radar would be capable of making such measurements.

J SNIEDER: Has attention been given to the attenuation due to rain where the drop size distribution has been taken into account? This distribution gives a better relationship to attenuation than does total rainfall.

L H KOSOWSKY: We have not taken into account the drop size distributions of raindrops in computing the attenuation coefficients, and we have not considered multiple scattering in the rain volume. Various drop size distributions in different parts of the world would also affect the calculations. The published curves appear to be valid in the US.

J FREEDMAN: 1) Does the gain of 15 dB in backscatter attenuation, using circular polarisation on rain, depend upon theoretical or experimental work?

2) How dependent is the result on the sphericity of the raindrops?

L H KOSOWSKY: 1) The 15 dB reduction in rain backscatter was measured at 70 GHz using a circularly polarised 18 inch antenna.

2) The dependence of polarisation sensitivity on raindrop deformation has been studied by Oguchi in Japan. Our results indicate that 15 dB improvement can be achieved at 4 mm/hr of rain, but I do not know of data at higher rain fall rates.

Combined Electro-Optical/Millimeter Wave Radar Sensor System

W.A. Holm, W.S. Foster, and G.R. Loefer

A combined electro-optical/millimeter wave radar sensor system is described and the results of preliminary field tests with this system are presented. The electro-optical sensor is a conventional television camera. Future modifications of this system, which includes installing a pulsed laser radar as the electro-optical sensor and future plans for this program are discussed.

I. INTRODUCTION

Tactical surveillance and weapon guidance systems depend critically on the Army's ability to not only acquire, but also to identify and to precisely locate targets of military significance. Conventional microwave radars have traditionally fulfilled the target acquisition role for the Army. However, due to resolution limitations inherent in the microwave frequency domain, these radars have met with limited success in target identification and precise target location required for accurate weapons delivery. To overcome these inherent inadequacies in radar systems, there has recently been a rapid emergence of electro-optical (EO) sensors, such as low-light-level television, forward looking and other infrared sensors and laser radar. Since these devices operate at, or are sensitive to, frequencies in the infrared to visible region of the electromagnetic spectrum, they have a much greater resolution than that of microwave radars and thus are able to more adequately fulfill the target identification and precise target location roles.

Two basic limitations of EO sensors prevent these sensors from completely replacing the radar sensor in tactical operations. First of all, EO sensors are inadequate in a search and target acquisition role due to the length of time needed for these extremely high resolution sensors to scan a given spatial volume. Secondly, EO sensors suffer much higher atmospheric attenuation losses than microwave sensors, especially in degraded weather conditions or in a smoke, fog or aerosol environment. Under these atmospheric conditions and without a microwave radar aboard, an Army vehicle, e.g., tank, would literally be "blind" to its surroundings.

To overcome the individual inadequacies of the two sensor systems and at the same time take advantage of their respective capabilities and strengths, both sensors can be utilized together in an augmenting fashion to form a combined electro-optical/radar sensor system. With this in mind, the Engineering Experiment Station (EES) at the Georgia Institute of Technology recently began a multi-phase program under contract with the U. S. Army ERADCOM to demonstrate the feasibility of such a combined EO/millimeter wave radar sensor system. In this dual sensor system, the millimeter wave radar system performs its conventional role of searching large spatial volumes in order to acquire and crudely locate a target. Once the target is acquired, the higher resolution EO sensor is directed toward the target for identification, accurate and precise target location and weapons delivery.

The ultimate goal of the program is to have for the EO sensor a pulsed, heterodyned CO_2 laser radar suitable for weapon guidance. In Phase I of the program, which is nearing completion, basic interfacing problems between the two sensors are being investigated. A conventional television camera (vidicon) and monitor are being used to simulate the laser radar. In this paper, the preliminary results of this initial phase of the program are discussed and the plans for future phases are reviewed. In Section II, the combined EO/millimeter wave sensor system itself is discussed. In Section III, Phase I of the program is discussed, the results of the preliminary field tests with the combined sensor system are presented and some of the EO/Radar interfacing problems discussed. Finally, in Section IV, future plans for the program, including those involving the laser radar sensor, are reviewed.

II. EO/MILLIMETER WAVE RADAR SENSOR SYSTEM

The radar system used in the combined dual sensor system is Georgia Tech's 70 Ghz (4.3 mm), high-resolution, rapid scan radar originally built by EES for Harry Diamond Laboratories. This radar was made mobile by installing it in a M-109 shop van. The radar console and antenna are linked together and revolve about a common axis. The operator faces the direction illuminated by the radar and a periscope, whose optical axis is aligned with the antenna electromagnetic axis, provides him with an optical view of the illuminated area. The basic system parameters of this radar are given in Table I.

The radar antenna assembly was recently modified by the addition of two microwave reflectors (see Figure 1). When in use, these reflectors essentially rotate the microwave beam through 90° upon leaving the antenna, thus enabling the radar to operate in a vertical scan mode. To activate the vertical scan mode, the tiltable plane reflector-mirror assembly (see Figure 2) must be folded back toward the antenna so that the microwaves can reach the vertical scan reflectors. While in the vertical scan mode, the EO sensor is inoperative.

The EO sensor is a conventional television (TV) camera (vidicon) which is mounted in the van and aligned with the periscope so that the EO optical axis and radar microwave axis are mechanically co-boresighted in azimuth (see Figures 2 and 3). Attached to the camera is a lens with variable magnification from x15 to x60 corresponding to a field-of-view from approximately 14.2 mrad to 3.5 mrad. Another TV camera is focused on the radar display and the outputs from the two cameras are fed into a special effects generator which in turn is connected to a video tape recorder and monitor (see Figure 4). Thus, the radar display and optical imaging can be displayed together in a "split-screen" effect.

III. PHASE I OF THE PROGRAM AND RESULTS OF THE PRELIMINARY FIELD TESTS

A. Phase I of the Program

In Phase I of the program, basic interfacing problems between the two sensors are being investigated. These problems include:

1. Determination of the minimum angular uncertainty in target location achievable with the radar sensor alone, and

2. Determination of the best method of coupling the two sensors together.

The basic target locating accuracy of the radar in a clutter and multipath environment is of critical importance in determining the field-of-view (FOV) to be scanned by the laser radar. Because of the large amount of time required by the laser to scan large spatial volumes, the FOV must be kept to a minimum. Therefore, a determination must be made of the minimum angular uncertainty in target location achievable with the radar.

There are several ways to "handoff" from the radar to the EO sensor, i.e., several methods of coupling the two sensors together. These methods vary from a system where the laser sensor is on a set of gimbals and is electronically coupled to the radar, to a system where the coupling is done both electronically and mechanically, to a totally mechanically coupled system. As was mentioned in Section II, for feasibility demonstration purposes the EO sensor is currently mechanically coupled to the radar in azimuth with no coupling in elevation.

B. Results of the Preliminary Field Test

Preliminary calibration/shake-down field tests of the combined EO/Radar sensor were conducted at Ft. Gillem, Georgia on 20-30 Sept., 1977. The tests were conducted in a field with relatively flat terrain bounded with trees on either side and with a range of over 1200 meters.

The beamsplitting experiments consisted of simply having the radar operator center the azimuth or elevation cursor on the target return as displayed on a B-scope. The angular placement of the cursor was then compared to the true angular position of the target as measured with a theodolite located in the antenna assembly. No attempt was made on these initial series of field tests to eliminate the human factor and its affect on the test results. Targets included both corner reflectors and a small pick-up truck. Results of these tests are summarized in Table II. Within one standard deviation, a target could be located in both azimuth and elevation to an accuracy of approximately 1 mrad. Given the radar's 9.6 mrad beamwidth, this represents roughly a 9:1 beamsplit.

Finally, to demonstrate the overall feasibility of combining an EO sensor with a radar sensor, several "handoff" experiments were performed. The EO and radar boresights were aligned and then several targets were acquired by the radar. Once a target had been acquired and aligned on the radar's boresight, the EO sensor was activated to determine whether or not the target was in the EO sensor's FOV, and if so, to identify the target. During these tests, the EO sensor was kept at its maximum FOV. The combined sensor system performed very well during these tests with the target being well centered on the monitor. Several video tape recordings were taken from which two still frames are shown in Figures 5 and 6.

IV. FUTURE PLANS FOR THE PROGRAM

In the remainder of Phase I of the program, which will terminate on 31 Jan. 1978, extensive field tests will be conducted in which the minimum angular uncertainty in target location achievable with the radar will be precisely determined. This data will be operationally checked by actually conducting various "handoff" experiments in which the FOV of the TV camera will be varied in order to simulate different possible FOV's being scanned by the laser sensor. In this way, the actual FOV to be scanned by the laser sensor can be determined. Based on preliminary results, a 3 mrad FOV is currently being planned. In addition, a determination will be made in this phase of the program as to the best method of coupling the two sensors together.

In Phase II of the program, which is already underway, a CW CO_2 laser sensor system will be constructed and integrated with the radar. Within the next three months this integration will be completed, with a He-Ne laser simulating the CO_2 laser. In four to six months, a CW direct detection system will be operative with:

1. a 5 watt CW CO_2 laser,
2. a resolution cell of 0.1 mrad, total FOV 1 mrad
3. a frame rate of 20 hz

In 10 to 18 months, a CW heterodyned receiver will be operative with a second laser as L.O.

In Phase III of the program, a pulsed, heterodyned CO_2 laser radar will be developed as the EO sensor. This sensor will have a 5 KW or greater peak power, 18 Khz PRF and a CW laser as L.O.

ACKNOWLEDGMENTS

The authors wish to acknowledge the following people for their contributions to this program: N. T. Alexander, J. E. Davidson, J. A. McKenzie and S. Y. Willis.

TABLE I. BASIC RADAR SYSTEM PARAMETERS

1. Antenna: Folded, geodesic lens feeding a parabolic cylinder reflector. The antenna is 25 inch diameter by 3.5 inch height. Gain: 43.2 dB (corrected for loss)

2. Scanning Modes: Scanning by a seven ring switch which allows a scan rate of approximately 1 RPM to 50 RPS

   <u>Azimuth Scan Mode:</u>

   Scan Angle: 45° (±22.5° about boresight)
   Azimuth beamwidth: 0.55° (positionable through 360°)
   Vertical Beamwidth: 3.5° (positionable from -10° to 20°)

   <u>Vertical Scan Mode</u>

   Scan Angle: -5° to 10° (approximately)
   Azimuth beamwidth: 3.5° (positionable through 360°)
   Vertical Beamwidth: 0.55° (positionable through ±5°)

3. Transmitter:

   Magnetron: Bomac BL 234C
   Peak Power: 500 watts
   Frequency: 69 to 71 GHz
   RF Pulse Width: 20 and 45 nanoseconds (adjustable)
   Pulse Repetition Frequency: 5 KHz to 25 KHz (variable)
   Modulator: Triggered blocking oscillator
   Duplexer: Bomac BL-P-017D

4. RF Component Loss: 6 dB (Total)

5. Local Oscillator: Varian VA 250 Klystron

6. Mixer: Philco IN2792 Mixer crystal; conversion loss 15 dB
   Equivalent noise figure 18 dB minimum
   Equivalent noise figure with IF 25 dB maximum

7. IF: Gain: 70 dB
   Center Frequency: 400 MHz
   Bandwidth: 60 MHz
   Noise Figure: 5 dB

8. Video Amp:

   Bandwidth: 50 MHz
   Gain: 150 (voltage)

9. Displays-indicator units:

   Sector scan
   B-scan mode
   Non-coherent Doppler aural display

Millimeter Wave Radars

Figure 1. 70-Ghz Antenna Assembly With Vertical Scan Reflectors Mounted On Top Of Geodesic Lens Antenna

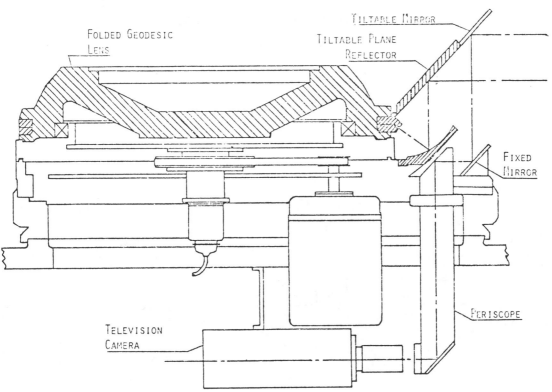

Figure 2. 70-GHz Folded Geodesic Lens Antenna With TV Camera Mounted For Viewing Through Periscope

Figure 3. TV Camera Aligned With Periscope And Radar Display And Controls (2nd TV Camera Not Shown)

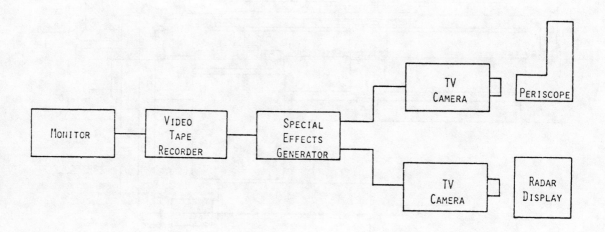

Figure 4. Schematic Drawing of EO/Radar Monitoring and Recording System

TABLE II. BEAMSPLITTING RESULTS FROM PRELIMINARY FIELD TESTS

Elevation Beamsplitting

Target	No. of Trials	Range (m)	Resolution (One Standard Deviation)
Corner Reflector	30	168	1.047 mrad
Corner Reflector	30	168	1.066 mrad
Truck	6	112-408	0.544 mrad
Truck	8	455-1026	1.009 mrad

Azimuth Beamsplitting

Target	No. of Trials	Range (m)	Resolution (One Standard Deviation)
Corner Reflector	30	356	1.117 mrad

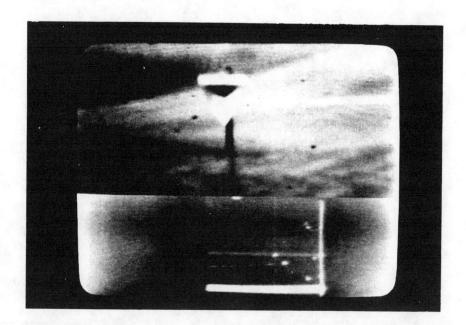

Figure 5. Still Frame Photo Of Video Tape Recording Showing Split Screen Of Corner Reflector As Simultaneously Imaged By EO Sensor (Top) And Displayed By Radar On B-Scope

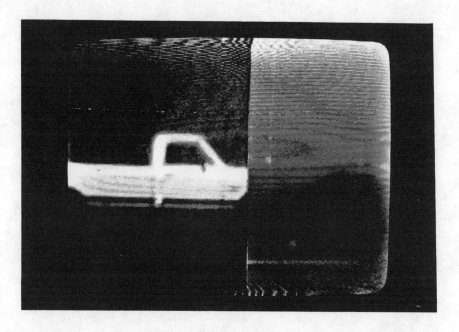

Figure 6. Still Frame Photo Of Video Tape Recording Showing Split Screen Of Truck As Simultaneously Imaged By EO Sensor (Left) And Displayed By Radar On B-Scope

Radar Tracking of an M-48 Tank at 94 and 140 GHz

R.A. McGee and J.M. Loomis

ABSTRACT

This report provides a description of the 94-GHz and 140-GHz radar systems used in the ground-based tracking of M-48 tanks.

1. INTRODUCTION

For years, the Ballistic Research Laboratory (BRL) of the US Army Armament Research and Development Command, has been actively investigating the effects of various obscurants and near-earth propagation phenomena on millimeter-wave target detection and tracking. Atmospheric obscurants, target signatures, multipath, and clutter have been examined at frequencies above X-band. In order to better understand the difficulties in millimeter low angle tracking, in particular the detection and tracking of ground targets, a mobile radar system was constructed. Utilizing components developed previously for the BRL and various other on-hand items, a tracking system was built to operate at two frequencies - 94 and 140 GHz. Currently, work is continuing to extend this to 217 GHz.

The tracking radar system was constructed on an 8x12 ft low-bed trailer for mobility. It consists of a precision tracking mount, an instrumentation compartment containing signal processing, radar control, and recording apparatus, and two RF assemblies which can be interchanged. Since both radar assemblies operate with the same power and control circuits, they can be interchanged in approximately 10 min.

2. RADAR ASSEMBLY

Figure 1 is a diagram of the radar without signal processing. Since the 94- and 140-GHz radars have the same basic component configurations, only the 94-GHz radar will be described and the differences between the two will follow as shown in Table I.

The tracking antenna is a 0.915-m Cassegrain consisting of a parabolic main reflector and a conically scanning hyperbolic subreflector mounted on a DC motor. The conical scan frequency may be adjusted remotely from 0-85 Hz by varying the DC drive to the motor. For tracking, the conical scan frequency was set at 60 Hz.

The receiver/transmitter is an all solid-state assembly using pulsed transmitter waveform and a single conversion superheterodyne receiver. A passive three port circulator provides isolation between the pulsed IMPATT transmitter and the mixer, and it couples the radar to the antenna through a circular feed horn.

The outstanding characteristic of the tracking radar is the pulsed IMPATT source. This source was developed for the BRL by the Electron Dynamics Division of the Hughes Aircraft Company under Contract DAAD05-74-C-0748. It is a single drift device mounted in a reduced height waveguide cavity and provides, at 94 GHz, a peak power of 1.68 W into the circulator. Because of the rise in junction temperature of the device during the 50-ns transmitted pulse, the frequency decreases (chirps) nonlinearly. This provides a transmitted frequency spread of approximately 0.5% of the center frequency.

The mixer is double-balanced and uses self-biased Schottky barrier diodes. With an integral IF preamplifier having a gain of 17 db and the mixer having a conversion loss of 5 db, the RF to IF gain is 12 db and the single sideband (SSB) noise figure is 10 db. Local oscillator power is provided by a Varian GaAs Gunn diode oscillator, which is capable of delivering 17 mW at the 94-GHz center frequency. The LO drive power to the mixer is dropped to 5 mW through a variable attenuator. Since no T/R switch is provided in the radar, the mixer diodes are saturated by the transmitter pulse leaking back through the circulator. Although this leakage power, approximately 13 mW peak, is not sufficient to damage the mixer diodes, it will overload the IF amplifiers and produce a "ringing" in the A-scan. A PIN diode SPDT IF switch (the second input is terminated with a 50-ohm load) is positioned after the IF preamplifier to provide 60 db of isolation during the transmit pulse. This switch can also be used to gate out the returns from large targets outside the range of interest. This prevents "ringing" the IF but, because of the characteristics of the PIN diode, switching spikes may be introduced into the IF. These are removed by shaping the switch drive current and placing a high pass filter (100 MHz) after the switch. Using this switching mechanism, a minimum range of 150 m is realized.

After the switch, a PIN diode voltage controlled attenuator is placed for gain control. Up to 60 db of attenuation can be provided, either manually or automatically, to maintain the target return signal within the tracking loop range. Automatic gain control is provided by an external circuit which monitors the return pulse video output. An additional IF amplifier is used to bring the return signal up to the sensitivity of the video detector-amplifier (-40 dbm). The final IF bandwidth is 500 MHz centered at 350 MHz.

The video detector has a 3-db video bandwidth of 16 MHz but, because of the slew rate of the video amplifier, this is reduced to 12 MHz for large returns. Total power required by the radar is 15 W.

There are three major differences between the component parts of the 94- and 140-GHz radars. First, the 140-GHz circulator is a passive four port with the extra port terminated with a waveguide load. Second, local oscillator power is provided by a Varian 140-GHz Klystron which requires a cooling fan. Third, the mixer is a single diode mounted in a Sharpless package, and requires a DC bias. This single-ended mixer requires the local oscillator power be injected through a 10-db directional coupler in line with the signal input. An extra amplifier is needed as an IF pre-amplifier; the SSB NF, including preamplifier, is 18 db. The overall IF bandwidth is 700 MHz with a center frequency of 1.35 GHz.

3. SIGNAL PROCESSING

The raw video signal from the radar is first gated synchronously with the IF switch, and then passed to three boxcar integrators. One boxcar integrator pulse intergrates the signal from the target of interest, and then low pass filters the signal down to 200 Hz. The other two boxcars integrators are used in the automatic range tracker. These integrators provide error signals to the range tracker by differencing the returns from "early" and "late" range gates. The automatic range tracker then provides a trigger to the first integrator to keep it on the target. This integrated video is fed to the AGC circuit and to the hard limiter. The AGC has a time constant of 30 ms and is capable of removing most signal fluctuations. Because of the large specular returns from the sides of most vehicles, the hard limiter was added to prevent the tracking loop dynamic range from being exceeded on returns that the AGC could not smooth.

The resultant video signal (30 Hz to 200 Hz) is fed to a Quadrant Gated Detector which was developed by the BRL* for tracking radar systems. Th quadrant detector, using the synchronous signal from the antenna conical scan motor, derives the error signals to drive the tracking mount servo-amplifiers. The servo-amplifiers and tracking mount are component parts of two different systems which were modified for this application. The overall precision of the auto-tracking system is shown in Table II.

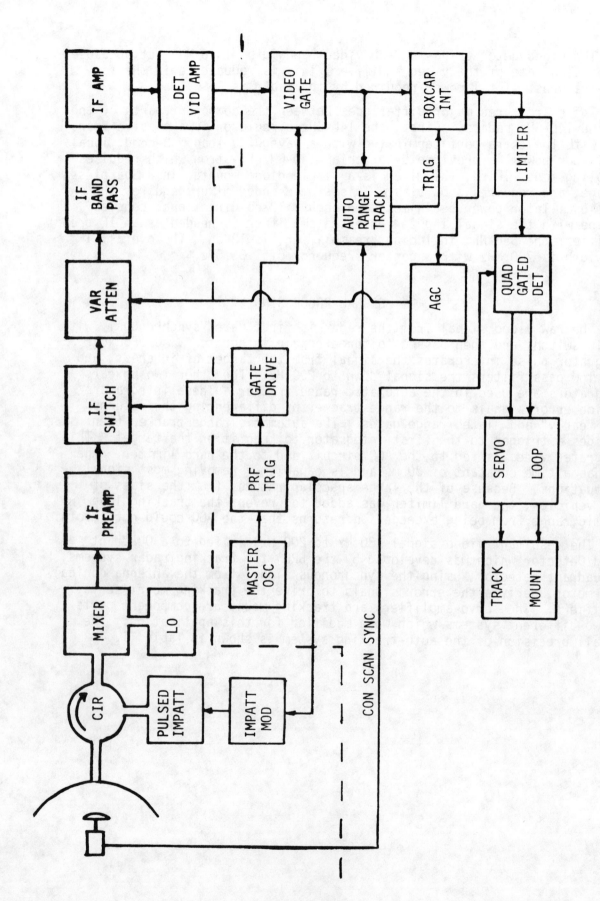

FIGURE 1. 94- AND 140-GHz TRACKING RADAR

TABLE I. RADAR CHARACTERISTICS

Component	94 GHz	140 GHz
Antenna		
Cross Over	3 db	3 db
$\theta_{3\,db}$	4.7 mrad	3.1 mrad
Gain	56 db	59 db
Loss	$\cong$3 db	$\cong$3.5 db
Circulator		
Type	Passive 3 port	Passive 4 port
Isolation	21 db	20 db
Loss	0.5 db	3 db
Mixer	Double-balanced	Single-ended with 10-db coupler for LO injection
NF (SSB) (includes preamp)	10 db	18 db
Pulsed IMPATT		
Pulse Width	50 ns	70 ns
Peak Power	1.68 W	0.77 W
Chirp	300 MHz	1 GHz
Center Frequency	94.35 GHz	138.5 GHz
Local Oscillator		
Type	Gunn	Klystron
Power to Mixer	5 mW (into Hybrid)	17 mW (into 10-db coupler)
Frequency	94 GHz	140 GHz
IF Amplifiers		
Frequency	100-600 MHz (100-600 MHz bandpass)	0.8-2.5 GHz (1-1.7 GHz bandpass)
Gain	70 db	70 db
Variable Attenuator	Pin diode voltage Controlled 0-60 db	Pin diode voltage Controlled 0-60 db
Video Detector Amplifier		
TSS	-40 dbm	-40 dbm
3 db Bandwidth	16 MHz-1 kHz	16 MHz-1 kHz
Gain	30 db	30 db
Power Out of Antenna		
Peak	0.75 W	0.17 W
Average	0.675 mW	0.214 mW
SSB NF + Losses	13.5 db	24.5 db

TABLE II. PROCESSING AND TRACKING CHARACTERISTICS

SNIR (through pulse integration and bandwidth reduction)	36.75 db at 94 GHz 36.0 db at 140 GHz
Conical Scan Frequency	60 Hz
AGC Time Constant	30 ms
Bandpass into Quadrant Gated Detector	30-200 Hz
Bandwidth into Tracking Loop	10 Hz Az + EL
Maximum Tracking Slew Rate	20°/s
Tracking Accuracy @ S/N 13 db	$\cong$0.31 mrad
Mount Position Accuracy	15 bit or 0.19 mrad
Optical Boresight Resolution	$\cong$±0.1 mrad (calm day)

94 GHz Active Missile Seeker Captive Flight Test Results
R. Yoshitani and M.E. Beebe
6th DARPA-Tri Service Millimeter Waves Conference, 1977

ABSTRACT

A captive flight program was conducted to test the tracking behavior of a 94 GHz active radar seeker against armored vehicles. The test program is described and its significant results discussed. Excellent tracking results were obtained and they are explained in terms of two kinds of averaging, one because of wideband frequency chirp and the other because of phase front averaging by the antenna aperture.

1. DESCRIPTION OF TEST PROGRAM

A series of captive flight tests of a 94-GHz missile seeker test bed was conducted during Winter 1977 at the Naval Weapons Center, China Lake, California. Captive flight tests were supported by a joint USN-USAF contract. The seeker was carried in a helicopter and tracking runs were made against tanks and trucks, in benign as well as severe clutter background and in snow. The purpose of the experiment was to test whether active radar tracking at 94 GHz can provide sufficiently small aimpoint wander for missile homing. Although both target and clutter signature measurements were made, only the tracking run results will be discussed.

The parameters of the active radar are shown in Table I. It is an incoherent pulsed radar with conical scan tracking. It transmits a wideband (1 GHz) chirp to smooth the clutter and target return. The incoming RF signal was homodyned to 500 MHz IF and then incoherently detected. The video was processed through five range gates, with the outside four gates forming an adaptive threshold for the center gate. The video in the center gate was used for both ranging and angle estimation.

The seeker assembly consisting of a AGM-65 TV seeker and the 94 GHz seeker was carried on a UH-1N helicopter as shown in Figure 1. Either seeker could be operated in a Master-Slave mode. The TV seeker was used as an acquisition aid and a tracking reference for the MMW seeker. After target acquisition with the Master TV seeker, the operator manually searched in range for the MMW signal return, commanding automatic track after acquisi-

TABLE I. HUGHES 94-GHz ACTIVE SEEKER PARAMETERS

ANTENNA	
Type	Cassegrain - 10-inch diameter conical scan
Scan Frequency	120 Hz
Gain	45 dB (excluding feed and duplexer losses)
Beamwidth	0.9 degree
Sidelobes	-16 dB
Polarization	Vertical - LHCP - RHCP
Squint Angle	0.35 degree
TRANSMITTER	
Type	Duplexed chirped pulsed IMPATT
Power	Nominal 1 watt
Pulsewidth	0.1 μs
PRF	35 KHz
Chirp Bandwidth	Nominal 1 GHz
RECEIVER	
Type	Homodyne, Balanced Mixer
IF	50 to 500 MHz, linear with delayed AGC
NF (DSB)	8.5 dB
SYSTEM	
NF (SSB)	12.5 dB
Integration	80 ms
Sensitivity	-107 dBm
Dynamic Range	84 dB

FIGURE 1. 94 GHz SEEKER HELICOPTER INSTALLATION

tion was achieved. Lockon generally occurred at ranges between 1200M and 2000M. After lockon, the helicopter flew a constant aspect and grazing angle (10 degrees or 20 degrees) approach to the target.

2. TRACKING ERROR MEASUREMENT PROCEDURE

Tracking errors were measured using the TV tracker aimpoint as reference. Before takeoff, the TV aimpoint and the radar aimpoint were adjusted to coincide by tracking a 20M^2 corner reflector. The tracking error characteristic did not change when the TV reference was removed and the error was measured solely from the radar tracking loop.

To determine the measurement accuracy of the tracking system, runs were made against a 200M^2 corner reflector. Because of its small physical size and large cross section, the 200M^2 corner reflector well approximates a point target. Its tracking error, particularly at close range where the effect of

TABLE II. TRACKING DATA ON 200M^2 CORNER REFLECTOR

Run No.	Channel	RMS Error (mr)
Mugu 006-1	Az	0.234
Mugu 006-1	El	0.378
Mugu 005-1-1	Az	0.212
Mugu 005-1-2	Az	0.143
Mugu 005-1-1	El	0.450
Mugu 005-1-2	El	0.277

the clutter is negligible, is therefore the system error. Table II shows the rms tracking error for the 200M^2 corner reflector. Note that the upperbound is about 0.45 mr, which is about three percent of the beamwidth. Note also that the elevation channel error is always larger than that of the azimuth channel. The reason is probably the reflector image on the ground which would appear to the seeker as a second reflector.

3. TRACKING ERROR MEASUREMENT RESULTS

Targets on which tracking data were taken included an M41 tank, an M53 self-propelled gun and a 2-1/2 ton truck. On viewing the video tapes made during tracking runs of the TV seeker output, some qualitative statements can be made regarding the 94 GHz tracking. There appeared to be three distinct regions of tracking behavior. At very long ranges, usually just after lockon, the aimpoint wandered over the target and its surrounding area. Once the lockon was secure, however, neither the type of clutter (sand, vegetation, snow) or its roughness seemed to make much difference. With further range closure as the target angular extent became comparable to the beamwidth (∼800M) aimpoint usually wandered slowly over the target but seldom veered off the target. The aimpoint motion became smaller and slower with range closure, particularly when the range was within 300M.

The rms tracking errors for an M41 tank are shown in Table III. They are typically of the order of 1 milliradian, which is surprisingly small. There appears to be no consistent difference between the elevation and azimuth channel nor between the tank and truck.

Also, the magnitude of the angular error remained nearly constant with range closure. This means that the aimpoint excursion on the target became smaller as the range closed. This behavior appeared to contradict the theory of angle scintillation that shows the rms error increasing with decreasing range. As will be seen, the constant rms angle error can be explained in terms of aperture averaging.

TABLE III. RMS TRACKING ERROR FOR TANK

Run No.	Target	Channel	RMS Error (mr)
Mugu 006-2	Tank (in lagoon)	Az	0.78
Mugu 006-2	Tank (in lagoon)	El	1.00
Mugu 006-4A	Tank Front	Az	0.60
Mugu 006-4A	Tank Front	El	0.55
Mugu 006-4B	Tank Front	Az	0.88
Mugu 006-4B	Tank Front	El	0.83
Mugu 006-4C	Tank Front (MM Master)	Az	0.61
Mugu 006-4C	Tank Front (MM Master)	El	1.30
Mugu 006-3	Tank Side (MM Master)	Az	0.59
Mugu 006-3	Tank Side (MM Master)	El	0.49

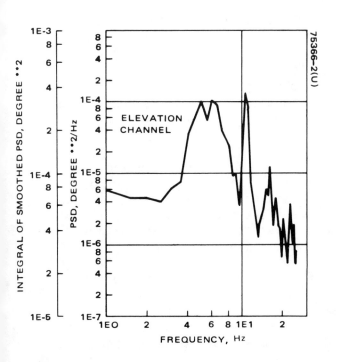

FIGURE 2. POWER SPECTRUM OF TRACKING ERROR FROM 200 m² CORNER REFLECTOR

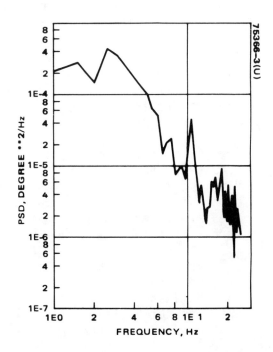

FIGURE 3. POWER SPECTRUM OF TRACKING ERROR FROM TANK

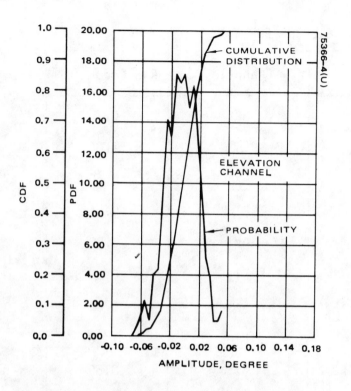

FIGURE 4. PROBABILITY DENSITY FUNCTION OF TRACKING ERROR FROM 200 m² CORNER REFLECTOR

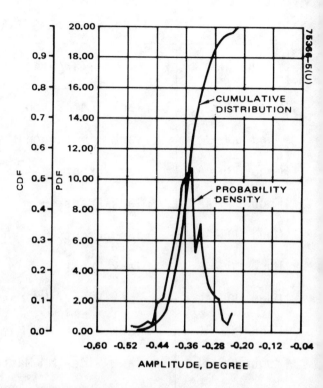

FIGURE 5. PROBABILITY DENSITY FUNCTION OF TRACKING ERROR FROM TANK

Typical power spectra of the tracking error are shown in Figures 2 and 3. Figure 2 is the spectrum from the 200M² corner reflector and Figure 3 is from a tank. Note the distinct difference in shape, the spectrum from the 200M² reflector is a bandpass shape, whereas that from the tank has more of a lowpass shape, indicating that for the corner reflector the aimpoint motion had a distinct frequency whereas for the tank the aimpoint apparently executed a slow wander. It is interesting to note, however, that the probability density function of the tracking error has similar shapes, both nearly Gaussian, as shown in Figure 4 and 5.

Another view of the tracking error can be obtained from analyzing the target amplitude return. Tracking error and amplitude are generally negatively correlated, i.e., when the amplitude is large, the associated tracking error is small and vice versa. Figure 6 shows the typical power spectrum of the target amplitude fluctuation. Note that the spectrum is lowpass with the corner frequency less than 1 Hz.

The corresponding probability density function of the target cross section is given in Figure 7. The measured density function has a strong main peak due to a dominant cross section. For this reason the density chosen to fit the data was not Rayleigh but Ricean, which assumes a strong deterministic component plus a small random component. The Ricean density in terms of power is

$$f_\sigma(x) = \frac{1+p^2}{\bar{\sigma}} \exp\left\{-p^2 + (1+p^2)\frac{x}{\bar{\sigma}}\right\} I_0 \left[2p \sqrt{(1+p^2)\left(\frac{x}{\bar{\sigma}}\right)}\right], \quad x \geq 0$$

Millimeter Wave Radars

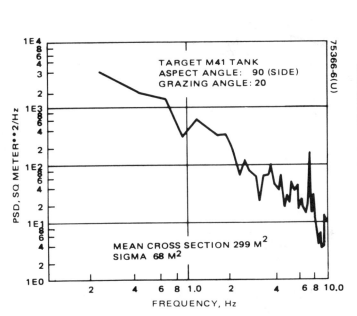

FIGURE 6. POWER SPECTRUM OF TARGET AMPLITUDE FLUCTUATION

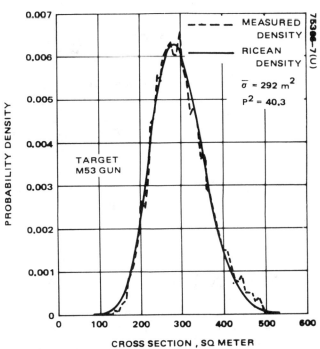

FIGURE 7. PROBABILITY DENSITY FUNCTION OF TARGET CROSS SECTION

where $\bar{\sigma}$ is the mean target cross section, p^2 is the ratio of the power in the deterministic to the random component, and $I_0(.)$ is the modified Bessel function of the first kind, zeroth order. Note the excellent fit of the Ricean density to the measured density. The mean cross section $\bar{\sigma}$ varied from 70M^2 to 1000M^2 and the power ratio p^2 was in the 25 to 50 range.

Much of the above data seem to be at odds with our experience at X-band and lower frequency radars. Two parameters of the 94 GHz seeker are distinctly different from those of the low frequency radars: one is of course the wavelength and the other is the unusually wide (1 GHz) frequency excursion of the transmitter. To quantify the effects of these two parameters on tracking, the simplest model of a target, one consisting of two point reflectors was analyzed.

4. TWO-REFLECTOR TARGET TRACKING MODEL

The two-reflector target is a useful model in understanding the tracking error phenomena because it is simple enough to allow some direct analysis and yet complex enough to retain the basis electromagnetic interference effect. For an active radar, the aimpoint displacement from the center of reflectors (see Figure 8) is given by [1] - [3],

$$e = \frac{d}{2} \frac{(1-a^2)\cos\theta}{1+a^2 + 2a\cos\left(\frac{4\pi d}{\lambda}\sin\theta + \phi\right)} \qquad (1)$$

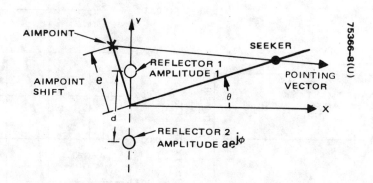

FIGURE 8. TWO-REFLECTOR TARGET AND ITS AIMPOINT DISPLACEMENT

where λ is the wavelength, θ is the target angle, d is the separation between the two reflectors, a is the ratio of the lower reflector amplitude to that of the upper (a < 1) and ϕ is the phase of the lower reflector relative to that of the upper. Figure 9 shows a plot of e vs θ for a 94 GHz radar without the frequency chirp. Note that because $d/\lambda_0 >> 1$ at millimeter wavelengths, a small change in θ can cause a large oscillation in e, particularly when a is large. On the basis of such an analysis it was thought that glint at 94 GHz would be too large for terminal homing.

Such an analysis, however, neglects the smoothing effect of the transmitter frequency chirp and of aperture averaging. Note that e depends on wavelength λ, hence, on frequency. If the transmitter frequency is linearly swept over a bandwidth Δf, the aimpoint shift e must be averaged over Δf. Moreover, e as given by (1) is the instantaneous slope of the phase front at a particular θ. For an antenna subtending a finite angle α, the phase front must be averaged over the antenna aperture. Thus, for a frequency chirp of Δf during a pulse duration T and phase front averaging over antenna angular extent α, the averaged aimpoint shift is [4],

$$\bar{e}(\theta) = \frac{1}{\alpha} \int_{\theta - \frac{\alpha}{2}}^{\theta + \frac{\alpha}{2}} d\theta' \frac{1}{T} \int_0^T e(\theta', \lambda(t)) \, dt$$

$$= \frac{d}{2} \cos \theta \left\{ \frac{\sin \frac{\alpha}{2}}{\frac{\alpha}{2}} + 2 \sum_{k=1}^{\infty} (-a)^k A_k(\theta) \cos \left[k \left(\frac{4\pi d}{\lambda_0} \sin \theta + \phi \right) \right] \right\}$$

(2)

where

$$A_k(\theta) = \frac{\sin \left[\frac{2\pi d}{\Delta \lambda} k \sin \theta \right]}{\frac{2\pi d}{\Delta \lambda} k \sin \theta} \frac{\sin \left[\frac{2\pi d \alpha}{\lambda_0} k \cos \theta \right]}{\frac{2\pi d \alpha}{\lambda_0} k \cos \theta}$$

(3)

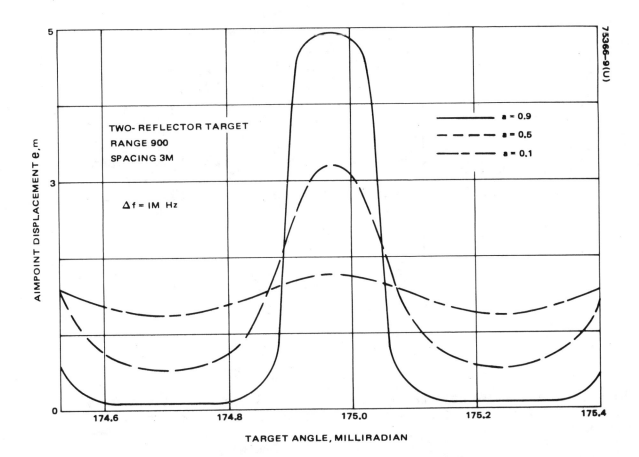

FIGURE 9. PLOT OF e versus θ

and where λ_0 is the center wavelength, $\Delta\lambda = c/\Delta f$, and c is the speed of light. In deriving (2) and (3), it was assumed that $\Delta f/f_0 \ll 1$ where $f_0 = c/\lambda_0$ and cos $\alpha \simeq 1$ and sin $\alpha \simeq \alpha$. Figure 10 shows a plot of $\bar{e}$ vs θ corresponding to the angle range of Figure 9. Note the reduction in amplitude of $\bar{e}$ due to aperture averaging at 300M. Frequency averaging produces a similar effect.

It is easy to see from (2) that the aimpoint motion is given by the sum of sinusoids whose amplitudes are reduced by $A_k(\theta)$. The term $A_k(\theta)$ consists of two sin x/x factors, the first due to frequency averaging and the second due to aperture averaging. Note that sin x/x becomes negligibly small when $x \geq \pi$. For frequency averaging this criterion leads to

$$d \sin \theta \geq \frac{c}{2\Delta f} \qquad (4)$$

Since d sin θ is the range differential between the reflectors, (4) is the familiar range decorrelation criterion.

For aperture averaging, noting that $\alpha = D/R$ where D is the antenna dimension and R the range, the sin x/x criterion leads to

$$\frac{d \cos \theta}{R} \geq \frac{\lambda_0}{2D} \qquad (5)$$

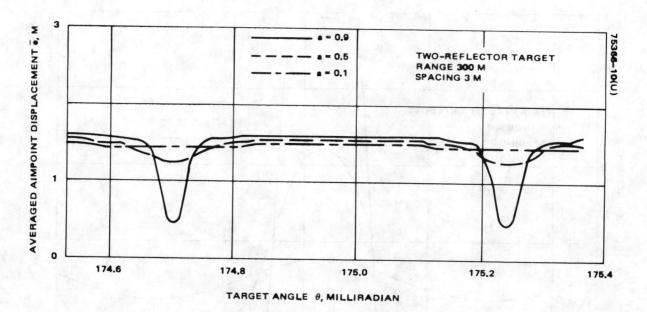

FIGURE 10. PLOT OF $\bar{e}$ versus θ

which gives the angle decorrelation criterion. Since the antenna beamwidth is given by $1.2 \lambda_0/D$, it follows from (5) that the aimpoint motion is significantly reduced when

$$\frac{\text{target subtense}}{\text{beamwidth}} \geq 0.417$$

In this form the aperture averaging has been experimentally observed earlier on aircrafts [3].

In terms of range, the aperture averaging would significantly limit the aimpoint wander when

$$R \leq \frac{2DS}{\lambda} \triangleq R_g \qquad (6)$$

where S is the target extent $d \cos \theta$. The right hand side of (6) may be called the glint reduction range. For the Captive Flight Test with a 25 cm antenna, assuming a 2M mean target spacing, R_g = 312M. This appears to be consistent with the observed aimpoint motions.

When either (5) or (6) hold, the sum of (2) collapses to one term, and the angular error is

$$\eta = \frac{\bar{e}}{R} \simeq \frac{d \cos \theta}{R} \left\{ 1 - 2aA_1(\theta) \cos \left(\frac{4\pi d}{\lambda_0} \sin \theta + \phi \right) \right\} \qquad (7)$$

Thus, even though the static part of the error must increase with range closure, the rapidly varying part, hence, the rms error, essentially remains constant because $A_1(\theta)/R$ is nearly constant because of the fact that $d/\lambda_0 \gg 1$.

The foregoing analyses of the 94-GHz seeker captive flight tracking test indicate some tradeoffs among seeker parameters. In general, the frequency chirp bandwidth should be as wide as possible for reduced terminal tracking error but wide bandwidth also penalizes acquisition range by introducing more noise. For accurate missile homing, a long glint reduction range is desirable because it would give a missile more time to damp out the error incurred prior to entering this region. From (6) it is clear that missile homing can be improved by increasing D/λ, and since the antenna diameter D is usually fixed by external constraints, the only way to increase D/λ is by going to a higher frequency. For example, for a target spacing of 2M, the glint reduction range for a 12.7 cm (5 inch) diameter antenna is 60M for a 35-GHz seeker and 160M for a 94-GHz seeker. The glint reduction range for 35 GHz is probably too short to be useful for missile homing but that for 94 GHz may be long enough to reduce the homing miss. Even if the seeker terminal mode is to be passive, a long glint reduction range would seem to be advantageous because it would reduce the transient effect of the active-to-passive handover.

5. CONCLUSION

A captive flight program was conducted to measure the tracking error of a 94 GHz active radar seeker against armored vehicles. The measured rms error was found to be surprisingly small and nearly constant with range closure. It was suggested from an analysis of a two-reflector target that these phenomena were due to two kinds of averaging, one caused by wideband frequency chirp and the other by phase front averaging by the antenna aperture. Quantitative criteria were obtained that show when each averaging became effective in reducing the rms tracking error, and implications of these criteria for seeker design were discussed.

REFERENCES

1. Meade, J.E., "Target Considerations," Chapter in Guidance, A.A. Locke, ed., New York: Van Nostrand, 1955, p. 440-441.

2. Ostrovityanov, "Angular Noise," Radiotekhnika i Elektronica, Vol. 4, 1966, p. 507-515.

3. Dunn, J.H. and Howard, D.D., "Radar Target Amplitude, Angle and Doppler Scintillation from Analysis of the Echo Signal Propagating in Space," IEEE Trans on MTT, Sept. 1968, p. 715-728.

4. R. Yoshitani, Hughes Internal Memo 5743/328, November 1977.

Potential Applications of Millimeter Wave Radar to Ballistic Missile Defense
G.B. Jones
IEEE EASCON-77 Conference Record, 1977
Pages 16-3A — C.
© 1977 by the IEEE. Reprinted by permission.

ABSTRACT

Sensors operating at millimeter wavelengths offer a potential payoff in many areas of BMD. Because of their excellent resolution and measurement accuracy, they are particularly suitable for target and interceptor tracking, homing, and fuzing, in both natural and hostile environments. Short wavelength scattering is sensitive to small details in the target configuration providing enhanced discrimination possibilities. The narrow beamwidths and wide operating bandwidth possible at millimeter wavelengths give these sensors significant ECM resistance relative to microwave radars.

In this paper, we survey potential BMD applications of millimeter waves emphasizing those applications which make use of their advantages and are relatively insensitive to their disadvantages.

GENERAL

The field of microwave radar has traditionally provided the sensors for past and current terminal defense and midcourse ballistic missile defense (BMD) concepts. Recent developments in millimeter wave components and some ideas that have come from recent BMD concept studies suggest that there might be a substantial payoff in BMD if certain functions were performed with millimeter wave sensors. The advantages and disadvantages of millimeter wave sensors when compared to microwave sensors that perform the same or similar functions are shown in Table 1.

Several of the major performance parameters that were considered in arriving at Table 1 were based on the curves shown in Figure 1. Many of the arguments for use of millimeter wave sensors in general missile systems apply equally in BMD (Reference 1), although the restrictions imposed by weather and propagation effects can be minimized in BMD systems that operate in the exosphere. The frequencies of current BMD interest are those which fall between 30 and 100 GHz. The transmission windows in the atmosphere at 35 and 94 GHz are attractive for ground-based systems and the atmospheric opacity at 60 GHz can act as a shield from ground-based RFI and ECM in spaceborne sensors.

For most of the millimeter wave applications to be discussed in this paper, the millimeter wave sensor must operate in conjunction with another sensor, such as microwave radar or passive optics, to provide an acquisition or handover volume which the millimeter wave sensor is capable of searching in a reasonable time. Two significant areas are the system performance payoff and technology advances required for each millimeter wave application.

To provide a simple framework for evaluating various BMD applications of millimeter wave sensors, Table 2 describes the application considered (detection, track, etc.) in terms of the operating regime (boost phase, midcourse, etc.).

	MILLIMETER WAVE	MICROWAVE
ADVANTAGES	• Good Angular, Radar, Doppler Accuracy • Good ECM Resistance • Small Size and Weight	• Good Search Performance • Good Beam Agility • All-Weather Capability
DISADVANTAGES	• Poor Search Performance • Limited Traffic Capacity	• Large Size and Weight • Poor ECM Resistance

TABLE 1

COMPARISON OF MILLIMETER WAVE AND MICROWAVE SENSORS

It is not reasonable at this time to consider applications using boost phase intercept or autonomous millimeter wave search in midcourse; thus, these blocks in the Table were omitted. Concepts using passive millimeter wave and sub-millimeter radiation have been proposed for detection in re-entry and boost phase, respectively. These will be mentioned in this section, but the emphasis of the paper will be on use of active sensors. One possible way to implement the concept of combining a millimeter wave sensor with passive optics is shown in Figure 2.

APPLICATIONS

Based on the considerations above, several interesting areas of application of millimeter wave sensors are early re-entry track and discrimination, exo-atmospheric range tracking, and re-entry homing. The sensor requirements for those applications present some interesting challenges when compared with current and projected millimeter wave sensor capabilities. (References 2 and 3) Important areas for further hardware development include low-cost antenna elements for larger arrays; small, light-weight power sources for missile-based applications; antennas and radomes configured for missileborne sensors. There are a number of uncertainties as well. For all the applications considered, target signature data is needed for sensor sizing as well as evaluating discrimination performance. There is considerable uncertainty in most target Radar Cross Section (RCS) levels as well as in the detailed signatures needed for imaging and other discrimination techniques.

One of the major questions in utilizing millimeter wave sensors is what configuration the sensor antenna will take. The two obvious ones are the parabola and the phased array. To be compatible with other system requirements, some sort of electronic beam steering will probably be required, so a single horn-fed parabola is not considered here. What may work for applications where a limited scan is required is an electronically steerable feed. Some possible configurations are shown in Figure 3. Which one would be best for a given system would be dependent on frequency, physical space, and power level. This variation in power level is principally dependent on application and physical constraints. Figure 4 shows how the power of a millimeter wave system might vary as the physical space available to accommodate a sensor aperture is varied. The figure presents the information for the four most likely applications of millimeter wave sensors to BMD systems. It can be seen that a fair amount of latitude is available.

The cyclotron maser, or gyrotron, may very well be able to generate the levels of power required for millimeter wave ground-based and possibly large space-based sensors. Power of 100 kilowatts average and up to multimegawatt peak levels is possible using these devices. (Reference 4) For distributed array configurations such as missileborne homing and fuzing, peak powers in the 1 to 10 watt range can do the job.

A number of other applications exist whose feasibility depends on the resolution of signature uncertainties or advances in hardware capability. Should these feasibility issues be resolved favorably, the spectrum of BMD applications could be broadened considerably. However, even with the current state of millimeter wave technology, there are a number of areas where such sensors could have significant payoff to BMD.

	BOOST	MIDCOURSE	RE-ENTRY
DETECTION	Passive (Sub-millimeter)	Active	Passive Active
TRACKING	Active	Active	Active
DESIGNATION, DISCRIMINATION	Active	Active	Passive Active
GUIDANCE, HOMING, FUZING		Active	Active

TABLE 2

MILLIMETER APPLICATIONS SUMMARY

Millimeter Wave Radars

REFERENCES

1. "Applications of Millimeter Wave Technology to Army Missile Systems," by A. H. Green, EASCON 77, September 1977.

2. "Some Comments on the Present Status of Millimeter Waves," by Dr. L. R. Whicker, Microwave Journal, November 1974, page 41.

3. "Evolution of Millimeter Waves," by Dr. Benjamin Lax, Microwave Journal, May 1976, pages 26-30.

4. "Prospects for High Power Millimeter Radar Sources and Devices," by T. R. Godlove, et.al., EASCON 77, September 1977.

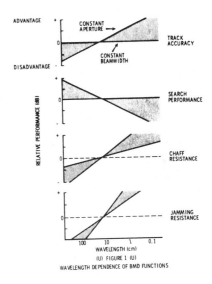

(U) FIGURE 1 (U)
WAVELENGTH DEPENDENCE OF BMD FUNCTIONS

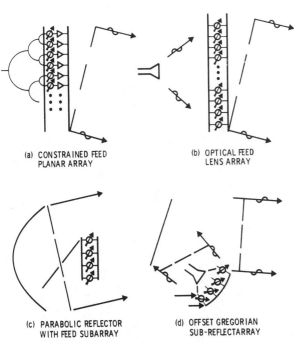

(U) FIGURE 3 (U)
mm-WAVE SENSOR CONFIGURATIONS (U)

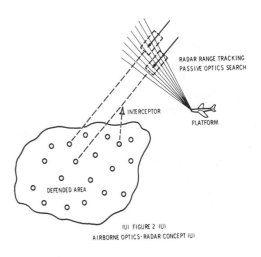

(U) FIGURE 2 (U)
AIRBORNE OPTICS-RADAR CONCEPT (U)

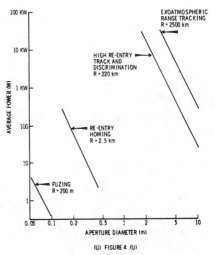

(U) FIGURE 4 (U)
NOMINAL BMD MILLIMETER WAVE RADAR REQUIREMENTS (U)

Millimeter Radar Application to SOI

R.B. Dybdal

Abstract

The millimeter wavelength regime offers significant potential advantages for SOI (Space Object Identification) applications. These advantages result from the ability to achieve good angular resolution from apertures of modest physical size, the capability to derive accurate orbital parameters from high doppler sensitivity, and the imaging capability commensurate with spacecraft dimensions which result from wide bandwidth waveforms inherently achievable at these frequencies. The present technology capability allows serious consideration of developing millimeter sensors for SOI applications.

This paper describes the system requirements for SOI radar systems utilizing some recently measured RCS (radar cross section) characteristics of spacecraft elements. These system requirements are compared with available and projected near-term component technology. Exemplary system designs are decribed.

*This work was supported by the U.S. Air Force under Contract F04701-76-C-0077.

I. Introduction

Applications for millimeter wave systems are being more seriously considered as the technology in this wavelength regime matures. The progress in this technology development has been followed in recent symposiums (Refs. 1, 2). For SOI (Space Object Identification) applications, millimeter wave systems offer potential improvements in resolution capability stemming from both wideband radar waveforms for range resolution and high doppler sensitivity required for cross range imaging. The achievable resolution for millimeter wave systems is commensurate with the dimensions of typical satellite component parts, and for this reason, considerable interest exists for the development of such systems. The purpose of this paper is to identify potential operating characteristics and to summarize the radar requirements for SOI applications.

While millimeter wavelength systems offer potential benefits in target resolution, limitations exist in their operation. Propagation in adverse weather conditions may result in outages in operation; however, such outages may be tolerable in SOI applications since target information is derived from several satellite passes. As the operating frequency increases, radar search capability suffers and becomes more expensive. As a consequence, the millimeter radar systems described here are directed to particular satellite positions by pointing information derived from other search systems.

While a significant amount of component research and development is being performed in both the millimeter and submillimeter wavelength regimes, component technology is presently available in the two window frequencies, 35 and 94 GHz, and technology for the next window frequency, 140 GHz, soon reaching feasibility. The development and demonstration of suitable high resolution radar electronics has already taken place at 94 GHz (Refs. 3, 4, and 5). The available technology performance is sufficient to contemplate systems with the capability of imaging satellites at altitudes as high as 10,000 mi.

II. System Considerations

The development of a millimeter imaging radar for SOI applications involves the exploration of many different tradeoffs. Component technology for such systems has been developed largely on an individual basis; i.e., large antennas have been developed as have low noise receivers, signal processing equipment, etc. The integration of this equipment into an operational radar has to be completed. At the same time, other studies have been performed to investigate target RCS characteristics, coherent integration time limitations, etc. The development of an operational facility therefore requires consideration of both component technology as well as the derivation of appropriate system design parameters.

A. Resolution Performance

The desired resolution performance of an imaging radar is an important parameter. High resolution radar waveforms and their characteristics are described in great detail in Ref. 6. While true short pulse waveforms can be generated, for the SOI application and its associated average power requirements for detection purposes, long pulse waveforms with pulse compression and integration are appropriate and provide synthetic short pulse responses. Resolution in the range direction δ_R is given by

$$\delta_R = \frac{Kc}{2B} \qquad (1)$$

where c is the velocity of light and B is the bandwidth of the radar waveform. The constant K is ideally unity; however, in order to maintain low range sidelobe performance, a value of 1.3 is more typical. The range resolution as a function of radar waveform bandwidth is given in Fig. 1. To achieve range resolution commensurate with the detail of typical satellite dimensions, a waveform bandwidth on the order of 1 GHz is required. It is desirable to achieve this waveform bandwidth as a small percentage bandwidth of the radar operating frequency. If the waveform bandwidth represents a significant percentage bandwidth, dispersion effects degrade the image performance. The millimeter regime allows one to achieve a large waveform bandwidth which is a small percentage bandwidth. For example, a 1 GHz bandwidth radar waveform represents only a 1% bandwidth at 100 GHz.

Resolution may also be achieved in the cross range direction and depends on target rotation with respect to the radar. The target rotation can either result from satellite rotation within the orbit or from rotation in aspect angle caused by translation of the satellite along its orbit. Cross range resolution is achieved only in the plane of rotation and is derived in an analogous fashion to synthetic aperture processing (Ref. 7). The cross range resolution δ_{CR} is given by

$$\delta_{CR} = \frac{K\lambda}{2\Delta\theta} \quad (2)$$

where λ is the operating wavelength, $\Delta\theta$ is the differential angular change in target aspect angle, and again K depends on the weighting applied to the samples with a typical value being 1.3. The resolution achieved in the cross range dimension is again favored by increasing the frequency to reduce the operating wavelength. Stated in another manner, the high doppler sensitivity at millimeter wavelengths enhances cross range resolution.

The resolution limits described thus far should receive additional comment. These limits are defined on the basis of separating two equal amplitude returns with different range or doppler values. Returns having different level returns will, of course, have resolution distances degraded from the ideal values. With the processing required to achieve the synthetic responses, the range or doppler sidelobe levels become important; the sidelobe level is the residual signal surrounding the central return. The sidelobe contributions from large returns can mask the returns from smaller target detail. The relative number of significant scattering components at a given aspect angle affects the overall image quality. As will be shown later, operation at higher frequencies results in narrower specular returns; such behavior allows the components of the target to be separated spatially and viewed on an individual basis by the imaging radar.

B. Radar Requirements

Analysis of radar requirements are derived in many varying forms for different applications. These analysis are based on the radar range equations. This paper uses the following form

$$\frac{P_{AVE} G_r G_t}{k T_s} = \frac{(4\pi)^3 R^4 (S/N)}{\lambda^2 \sigma L T_c} \quad (3)$$

where
P_{AVE} = average transmitter power
$G_{r,t}$ = receive and transmit antenna gain
k = Boltzmann's constant
T_s = total system noise temperature
R = radar to satellite range
S/N = signal-to-noise ratio
λ = operating wavelength
σ = RCS (radar cross section)
L = total system loss
T_c = coherent integration time

This form of the range equation is valuable for several reasons. The left side of the equation is associated with the radar parameters, average transmit power, antenna gain, and system noise performance. The right side of the equation is associated with the link and target parameters. This separation is desirable since the required radar parameters may be easily determined for a particular link and target configuration.

This equation could easily be rewritten in terms of the usual tracking radar parameter, the power aperture gain product. However, system noise temperature varies greatly with technology, and antenna gain, rather than effective aperture, seems to be a more appropriate parameter in this wavelength regime. For millimeter systems, physically small antennas lead to large gain values, and antenna gain is ultimately limited by atmospheric phase decorrelation as well as achievable mechanical precision. In addition, narrower antenna beamwidths place more stringent requirements on the antenna mount and pointing systems; these requirements affect system costs and tractibility.

C. Millimeter Component Technology

This section will discuss millimeter component performance as related to the SOI application. The basic quantities are: achievable antenna gain, average transmitter power, and system noise temperature. In addition, the system must have sufficient bandwidth capability to achieve the desired range resolution requirements. These quantities correspond to the left side of Eq. 3.

The development of large, high gain antennas capable of millimeter operation has been actively pursued with much of the impetus coming from the radio astronomy community. The limitations on gain performance arise in two areas, mechanical precision and atmospheric phase decorrelation. At present, the gain performance of millimeter antennas is bounded by achievable mechanical precision. A tutorial review on the effects of mechanical precision is given in Ref. 8. A common rating for precision antennas is the precision of manufacture D/ϵ where D is the aperture diameter and ϵ is the rms surface error. For a 1 dB gain loss, the corresponding rms error is roughly $\lambda/26$ where λ is the operating wavelength based on the approximate Ruze exponential form for the tolerance loss. At the present time, antennas with D/ϵ values of 40,000 to 120,000 have been constructed; proposed systems are contemplating values up to 400,000.

One of the earliest (1963) precision millimeter antennas is the 15 ft diameter parabola in operation at The Aerospace Corporation (Ref. 9). This antenna has been operated in a radiometric mode as high as 220 GHz (Ref. 10). At 93 GHz its measured gain is 70.5 dB which corresponds to an overall 55% aperture efficiency. Its precision of manufacture is 100,000 at zenith and degrades with gravitation deformation at 60° zenith to a value of 66,700. A survey of millimeter antennas may be found in Ref. 11. A 45 ft diameter antenna has been recently constructed for the University of Massachusetts (Ref. 12). This antenna has a precision of manufacture of 118,200, which illustrates present state-of-the-art. Figure 2 illustrates typical antenna diameters required to achieve the necessary gain for the SOI application.

Suitable duplexers for millimeter wavelength operation should receive more development effort. The requirements for the duplexer are high power capability, sufficient isolation, and low loss. The loss in the duplexer is particularly important for systems using cooled receivers which achieve very low noise temperature performance. An alternative to the duplexer is to use separate antennas for receiving and transmitting and operate the radar in a quasi-monostatic mode. This approach avoids duplexer losses, simplifies the feed configuration, and allows a minimal system temperature with a cooled receiver, at the expense of a second antenna. This type of operation is also used at lower frequencies in the JPL Goldstone Radar (Ref. 13).

The development of high power transmitters has already been surveyed in this session (Ref. 14). The transmitter requirements for an imaging radar include high average power levels, bandwidth commensurate with the desired resolution performance, and a minimal amount of dispersion. The dispersion in the transmitter degrades the high resolution waveform; a feedback technique to maintain the waveform performance is described in Ref. 3. Tutorial articles on millimeter transmitters may be found in Refs. 15 and 16. The recent development of the gyrotron has particular significance for this application. The gyrotron is capable of kilowatt average power levels and is capable of 3 to 4% bandwidth, which is on the order of 1 GHz even at Ka band frequencies.

The total system noise temperature is the other technology related factor. The total system temperature, as used in Eq. 3, is made up of several components. The antenna temperature at millimeter wavelengths results from atmospheric attenuation caused by water vapor and oxygen absorption. The antenna temperature depends on frequency and antenna elevation angle. Typical values at the earth's surface are given in Ref. 17. These values may be decreased by locating the radar at a high altitude site, preferably a site which is relatively dry to minimize rainfall outages.

The performance of low noise millimeter wave receivers has been reported in several recent articles (Refs. 18, 19, 20, and 21). Recent progress in this area has led to the development of extremely low noise receivers, particularly for cryogenically cooled devices. With cooling, noise temperature values as low as 100°K can be achieved.

Millimeter receiver technology includes room temperature and cooled mixers, cooled paramps and super-Schottky diodes. For mixer performance, Kerr (Ref. 18) reports 500°K room temperature and 300°K cooled performance with gallium arsenide material at 115 GHz. Edrich (Ref. 20) reports the performance of a cooled paramp at 47 GHz and projects the performance of this technology at 100 GHz as 250°K for a system cooled to 20°K and having a 1.5 GHz bandwidth. The super-Schottky diode (Ref. 21) utilizes superconductivity, and the projected performance at 100 GHz for an all superconducting receiver is 100°K (Ref. 22). With such noise levels, particular attention must be paid to all components of the total system temperature. The bandwidth requirements for the receiver also merits some comment. If stretch processing is used with a chirp waveform (Ref. 3) the LO is swept and the receiver is required to tune over the bandwidth needed for the desired resolution performance with an instantaneous requirement dictated by sweep rate, pulse duration and doppler margin. This factor allows some additional latitude in optimizing noise performance.

D. Link and Target Properties

The right hand side of Eq. 3 is dependent on the link geometry and target properties. Two of the quantities on this side of the equation, the target RCS and coherent integration time require particular discussion. It should also be pointed out that the system loss term, as it is used here, includes atmospheric attenuation which is frequency dependent. Typical values for this loss component may be found in Ref. 23.

The RCS characteristics of a satellite for an imaging radar application merit some comment. For conventional radar systems whose resolution performance is much larger than the physical extent of the target, the appropriate RCS value for system sizing represents the total contribution from all components of the target. For an imaging radar, the system sizing depends on the RCS level of the individual components of the target within the achievable resolution cell.

RCS measurements on typical satellite hardware and components have been performed at 93 GHz (Ref. 24) and represent returns which may be located in distinct resolution cells. These measurements were performed in an anechoic chamber (Ref. 25) and the maximum target dimensions as limited by the far field condition was 6 in. Range resolution corresponding to 6 in. requires approximately 1.3 GHz radar waveform bandwidth which may be achieved at 93 GHz.

Solar array cells and thermal blanket material are found in abundance on typical satellites. The measured results shown in Figs. 3 and 4 indicate the specular nature of millimeter scattering. The return from the thermal blanket material having a somewhat wrinkled surface indicates the reduction in peak RCS return by comparison of the thermal blanket side and the smooth aluminum backing side. The return from structural components such as angle stock segments and from sensors such as antennas is given in Figs. 5 and 6 illustrating satellite structure that, typically, is visible to an interrogating radar.

These measurements illustrate the RCS at millimeter wavelengths are characterized by narrow specular returns, high dynamic range, and surface roughness dependence. The specular nature of the scattering is an advantage for an imaging radar. The narrow lobe widths result in spatial separation of target detail; i.e., as the target aspect angle changes, the specular levels show up only for a limited angular region. The target details are therefore seen by the radar on an "individual" basis. Such spatial separation enhances image definition.

In terms of system sizing, a minimum RCS level is required to determine radar performance requirements. This choice is somewhat subjective and obviously depends on the particular target but, based on these measurements, a value of -20dBsm will probably encompass most returns from specular regions. The lower level scattering will probably be buried in the sidelobes of the processed image.

The limitations on coherent integration time depend on the target motion. Two components of motion should be considered. The first component is due to the orbital characteristics and the second component is due to motion of the satellite within its orbit, e.g., rotation of a spin stabilized satellite. For the first component, the limiting factor in coherent processing is the ability to compensate doppler variation in orbit. Recent investigations by Lincoln Laboratory (Ref. 26) have successfully demonstrated coherent integration improvements by correcting for the acceleration terms for synchronous altitude targets. The application of such corrections to other orbits should be investigated in future efforts. Target motion within orbit provides a more restrictive limitation for an imaging radar. The limitation in this case is the amount of time a scattering center return is visible to the radar. In this case the scattering center may move through resolution cells and the processing becomes more complex (Ref. 27).

III. Example System Characteristics

With the above background, example systems and their requirements can be examined. The requirements as given by the right hand side of Eq. (3) can be plotted as a function of frequency with satellite-to-target range as a variable. This is done in Fig. 7 with the following assumptions. A 10 dB signal-to-noise ratio and a minimum RCS level of -20 dBsm are assumed. The system loss is equal to 5 dB plus the two way propagation loss at a 30° elevation angle given in Ref. 23. These losses include such factors as deviations from a matched filter response; other losses are reflected by using actual antenna gain values etc. For reference purposes, a one second coherent integration time is assumed; other integration times may be easily adjusted by scale changes.

The requirements, as shown in Fig. 7 indicate the effects of the water vapor line at 22 GHz and the oxygen absorption region at 60 GHz. Two frequencies in the millimeter region, the two lower window frequencies 35 and 94 GHz, will be considered for example systems. The parameters chosen to characterize these systems will be based on reported performance; the values selected do not represent systems under actual development.

For a Ka-band (35 GHz) system, consider an antenna such as the 120 ft. diameter Haystack antenna (Ref. 11). This antenna has a measured gain value of 73 dB corresponding to a 13% aperture efficiency. This performance is due in part to a 2.8 dB radome loss, and with a more suitable radome design, a gain value of 75 dB can be utilized. Assume a transmitter with an average power of 10 KW is used; this level is achievable with TWT technology. For noise performance the zenith antenna temperature is 17° and a cooled paramp based on Ref. 20 should have a 200°K performance; a system with an overall temperature of 300°K should be achievable. Using these component values for the radar requirement parameter, $\frac{P_{AVE} G_r G_t}{k T_s}$, a performance level of 394.8 dB-sec is obtained. This level of performance allows a satellite beyond 10,000 mi. to be imaged on the basis of the values given in Fig. 7.

For a 94 GHz system, consider a 45 ft. diameter antenna as constructed for the University of Massachusetts (Ref. 12). With its 0.11 mm rms tolerance, the achievable gain should be approximately 80 dB. For a transmitter, a gyrotron with a 20 KW level will be assumed; this power level is consistent with values reported in Ref. 16. The zenith antenna temperature at 94 GHz is 53°K, and a cooled paramp with a 250°K noise level has been projected (Ref. 20); allowing some margin an overall 600°K system temperature can be projected. Using these component values for the radar requirement parameter, $\frac{P_{AVE} G_r G_t}{k T_s}$, a performance level of 403.8 dB-sec is achieved. Referring to Fig. 7, such a system should be capable of imaging satellites out to 5 to 10 thousand miles.

The performance levels described are rather ambitious, yet they illustrate the potential for millimeter SOI radars. Obviously, many tradeoffs exist in establishing system parameters, economics not being the least of these considerations. The system requirements can be relaxed somewhat if images of lower altitude satellites are the system objectives. The presentation of the system requirements, as given in Fig. 7, allows an easy evaluation in the tradeoff in postulated system parameters.

IV. Summary and Conclusion

This paper has surveyed millimeter wave SOI radar systems. Such systems are capable of achieving resolution commensurate with typical satellite structural detail. The technology in this wavelength regime has undergone significant development in recent years, and component technology applicable to SOI systems is briefly described. The requirements for an imaging radar have been addressed and measured RCS data of component parts has been utilized in establishing these requirement. Finally, it was shown that current technology could be used to configure systems capable of imaging satellites in 5,000 to 10,000 mi. ranges, encompassing a large portion of the satellite population.

References

1. ..., Proceedings of the "1974 Millimeter Waves Techniques Conference," Vols. 1, 2, and 3, Naval Electronics Laboratory Center, March 26-28, 1974.

2. ..., "Report of the ARPA/Tri-Service Millimeter Wave Workshop," Applied Physics Laboratory/Johns Hopkins University Report APL/JHU QM-75-009, January 1975.

3. L. A. Hoffman, K. H. Hurlbut, D. E. Kind, and H. J. Wintroub, "A 94 GHz Radar for Space Object Identification," IEEE Trans. Microwave Theory and Techniques, Vol. MTT-17, pp. 1145-1149, December 1969.

4. H. R. Anderson and J. F. Heney, "Millimeter-Wave Imaging Radar for Space-Object Identification," 23rd Annual Tri-Service Radar Symposium, U.S. Military Academy, West Point, NY, July 12-14, 1977.

5. H. E. King, R. B. Dybdal, E. E. Epstein, J. D. Michaelson, J. W. Montgomery, and F. I. Shimabukuro, "Millimeter Wave Technology Program," Aerospace Corporation Tech. Rept. TR-0075(5650)-1, April 1975.

6. A. W. Rihaczek, Principles of High-Resolution Radar, McGraw-Hill (1969).

7. W. M. Brown and L. J. Porcello, "An Introduction to Synthetic-Aperture Radar," IEEE Spectrum, pp. 52-62, September 1969.

8. J. Ruze, "Antenna Tolerance Theory- A Review," Proc. IEEE, Vol. 54, pp. 633-640, April 1966.

9. H. E. King, E. Jacobs, and J. M. Stacey, "A 2.8 Arc-Minute Beamwidth Antenna: Lunar Eclipse Observations at 3.2 mm," IEEE Trans. Antennas and Propagation, AP-14, pp. 82-91, January 1966.

10. W. A. Johnson, T. T. Mori, and F. I. Shimabukuro, "Design, Development and Initial Measurements of a 1.4-mm Radiometric System," IEEE Trans Antennas and Propagation, AP-18, pp. 485-490, July 1970.

11. J. R. Cogdell, J. J. C. McCue, P. D. Kalachev, A. E. Salomonovich, J. G. Moisew, J. M. Stacey, E. E. Epstein, E. E. Altshuler, G. Feix, J. W. B. Day, H. Hvatum, W. J. Welch, and F. T. Barath, "High Resolution Millimeter Reflector Antennas," IEEE Trans Antennas and Propagation, AP-18, pp. 515-529, July 1970.

12. S. L. Hensel and L. E. Rhoades, "The Development and Manufacture of ESSCO Precision mm Wave Radio Telescopes," Electronic Space Systems Corp. Tech. Rept. MTP-77-0008, May 1977.

13. M. I. Skolink, Radar Handbook, McGraw-Hill (1970) p. 33-16.

14. T. F. Godlove, "Prospects for High Power Millimeter Radar Sources and Devices," EASCON 77, see paper this session.

15. D. C. Forster, "High Power Sources at Millimeter Wavelengths" Proc IEEE 54, pp. 532-539, April 1966.

16. V. A. Flyagin, A. V. Gaponov, M. I. Petelin, and V. K. Yulpatov, "The Gyrotron," IEEE Trans Microwave Theory and Techniques, MTT-25, pp. 514-521, June 1977.

17. M. I. Skolnik, Radar Handbook, McGraw-Hill (1970) p. 2-32.

18. A. R. Kerr, "Low-Noise Room Temperature and Cryogenic Mixers for 80-120 GHz," IEEE Trans Microwave Theory and Techniques, MTT-23, pp. 781-787, October 1975.

19. J. J. Whelehan, "Low-Noise Millimeter-Wave Receivers," IEEE Trans Microwave Theory and Techniques, MTT-25, pp. 268-280, April 1977.

20. J. Edrich, "A Cryogenically Cooled Two-Channel Paramp Radiometer for 47 GHz," ibid., pp. 280-285.

21. F. L. Vernon, Jr., M. F. Millea, M. F. Bottjer, A. H. Silver, R. J. Pedersen, and M. McColl, "The Super-Schottky Diode," ibid., pp. 286-294.

22. A. H. Silver, "Superconducting Low Noise Receivers," IEEE Trans Magnetics, MAG-11, pp. 794-797, March 1975.

23. M. I. Skolnik, Radar Handbook, McGraw-Hill (1970), p. 2-51.

24. R. B. Dybdal and H. E. King, "93-GHz Radar Cross Section Measurements of Satellite Elemental Scatterers," IEEE Trans Antennas and Propagation, AP-25, pp. 396-402, May 1977.

25. R. B. Dybdal and C. O. Yowell, "VHF to EHF Performance of a 90-ft. Quasi-Tapered Anechoic Chamber," IEEE Trans Antennas and Propagation, AP-21, pp. 579-581, July 1973.

26. L. B. Spence and G. P. Banner, "Observations of Synchronous Satellite ATS-3 with Three Coherent Radars," Lincoln Laboratory Tech Note 1975-36, June 1975.

27. W. M. Brown and D. J. Fredricks, "Range-Doppler Imaging with Motion Through Resolution Cells," IEEE Trans Aerospace and Electronic Systems AES-5, pp. 98-102, January 1969.

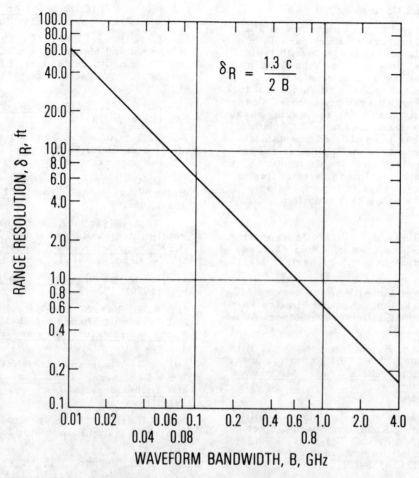

Fig. 1 Bandwidth Requirements for Range Resolution

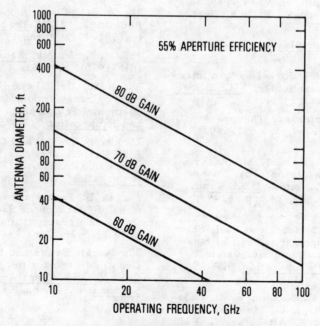

Fig. 2 Millimeter Antenna Characteristics

Millimeter Wave Radars

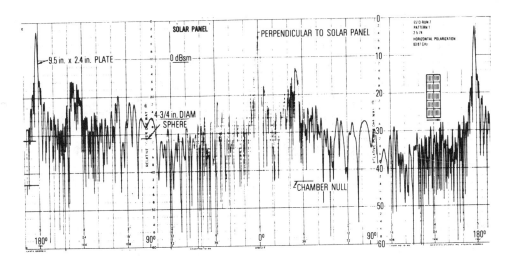

Fig. 3(a)   93 GHz Response of Solar Array Sample

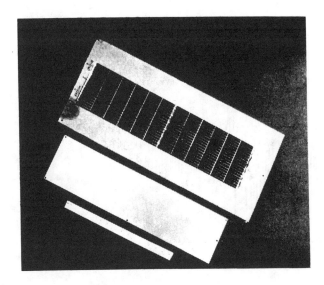

Fig. 3(b)   Solar Array Sample

Fig. 4(a)   Thermal Blanket Sample

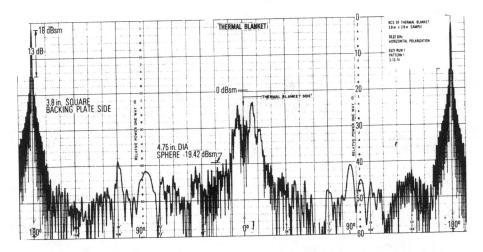

Fig. 4(b)   93 GHz Response of Thermal Blanket Sample

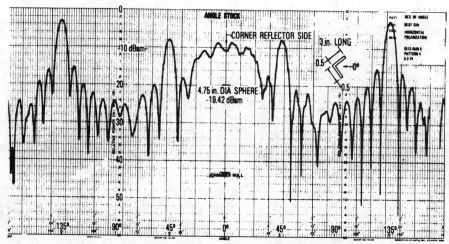

Fig. 5   93 GHz Response of Angle Stock

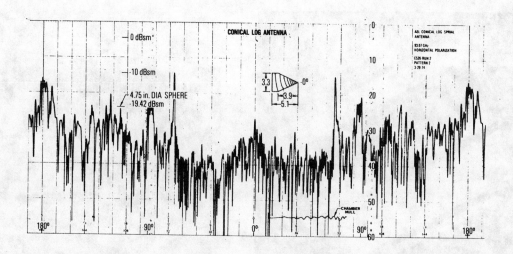

Fig. 6(a)   93 GHz Response of Conical Log Spiral

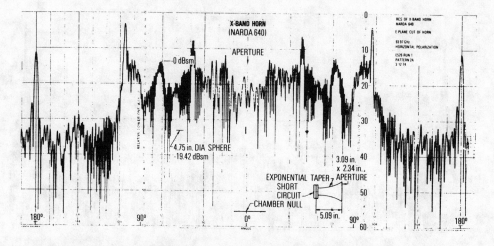

Fig. 6(b)   93 GHz Response of X-Band Standard Gain Horn

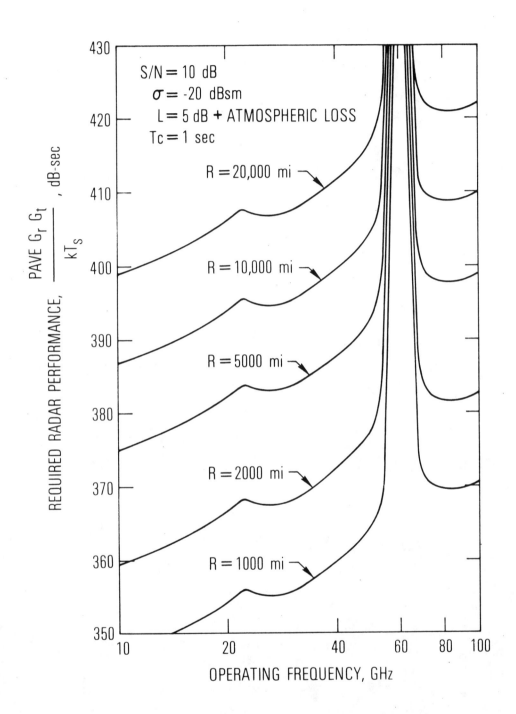

Fig. 7 Radar Requirements for Satellite Detection

mm-Wave Instrumentation Radar Systems

N.C. Currie, J.A. Scheer, and W.A. Holm

INTRODUCTION

The Georgia Institute of Technology Engineering Experiment Station has conducted research in millimeter wavelength radar techniques for more than 25 years. This research is continuing to characterize environmental effects, clutter, and tactical targets; to develop and evaluate detection, discrimination, and classification techniques; and to assess the potential and feasibility of millimeter waves for general and specific applications. Numerous system design philosophies and concepts have evolved from the applied research, but few systems have been built because of the lack of sufficient design data. Although the system designs and concepts may evolve from systems developed for operation at microwave frequencies, the propagation and reflectivity characteristics for millimeter waves are considerably different.

Fortunately, state-of-the-art advances in components and solid state technology for millimeter waves have evolved in parallel with the applied research. This made it possible to continuously improve instrumentation systems to further enhance the quality of research. Recent technology advances now make it practical to consider building operational radar systems, but further design data is needed to assess system feasibility for particular applications.

The lack of appropriate data at millimeter wave frequencies for use in developing system concepts[1] indicates a need for extensive applied research to characterize the reflectivity of targets and background at millimeter wave frequencies. High-quality instrumentation-grade millimeter-wave radar sensors are required for reflectivity measurements. These sensors must have the integrity similar to that associated with test equipment for measuring the required target and clutter characteristics for the system designer's use.

A comprehensive measurements program must be complemented by appropriate data analyses. The measurement program procedures must evolve from the data requirements and the data collection program must be tailored according to the plan to guide its activities. The subsequent analysis of the data must be consistent with the system designer's needs.

ELEMENTS OF AN INSTRUMENTATION SYSTEM

Design Philosophy

Typically, it is desirable that the measurement accuracy of test equipment be ten times as great as the allowable error in the measurement being made. For the same reasons, it is desirable that the integrity of a measurement radar system used to collect system design data be considerably higher than typical operational systems. The integrity of instrumentation systems is measured in terms of various parameters; e.g., amplitude stability for non-coherent systems, frequency and phase stability for coherent systems, and polarization isolation for dual polarized systems. A ten-to-one improvement in the integrity of an instrumentation radar over an operational radar is not always easy to realize, particularly when the operational system is using state-of-the-art technology. The instrumentation system, however, must be at least as good as, if not better than, the radar system for which data are being collected and analyzed. This requirement sets a lower bound on the measurement system performance.

Calibration

In addition to the integrity of an instrumentation system, one element which sets it apart from an operational system is the calibration capability built into the radar system. Typically, in a measurements program, the instrumentation system is operated for a variety of tests at various times of the day and night. The data are stored on either analog instrumentation recorders or digital recorders. Due to the long term drift characteristics of the instrumentation recording system, periodic calibration is required. Calibration signals must be recorded just prior to or directly following the actual measurement. Non-coherent systems may be calibrated by either of two methods. One method consists of injecting the output of an RF signal generator signal into either the antenna connection or a directional coupler located in the receiver waveguide path and incrementally varying the signal generator output level over the dynamic range of the receiver. This procedure provides a record of the receiver output voltage as a function of input power level. The other method is to boresight the radar antenna on a standard calibrated radar corner reflector at a known distance and incrementally insert attenuation in the receiver section until the reflector return is attenuated down to the receiver noise level. This procedure provides a record of the receiver output voltage as a function of radar cross section into the receiver at a given range. It is desirable that both methods be combined whenever possible to permit correlation, by means of the radar equation, of the signal generator calibration with the corner reflector calibration. Agreement between these two independent procedures indicates a high integrity and confidence factor in the operation of the instrumentation system. A dual polarized receiver in the radar requires that the calibration procedure be performed for each polarization.

If the radar receiver has a phase channel, a coherent receive capability, or a pseudo-coherent detection capability, the phase must be calibrated as well as the amplitude. The phase is calibrated by incrementally inserting different phase shifters in the receiver channel and recording the output of the phase detector. This provides a record of the phase detector output voltage as a function of input signal phase. As with the amplitude calibration, the phase must be calibrated for each channel of a multichannel receiver.

Adaptability

The instrumentation radar system must be adaptable to permit its use in diverse measurements programs since the particular data needed for a specific application determine the required operational configuration of the instrumentation sensor. Such adaptability requires that the instrumentation sensor have a variety of operation modes. Variable operation mode parameters typically incorporated in millimeter wave instrumentation radars include frequency agility, polarization, pulse length and pulse repetition frequency, and signal processing techniques. Pulse lengths typically of interest in developing modern operational systems range from 10 nanoseconds to 1 microsecond. Pulse repetition frequencies of interest typically range from 1 kHz to 1 MHz. Frequency agility modes of interest are intrapulse frequency agility or chirp mode, pulse-to-pulse frequency agility and FMCW. Polarization modes of interest are linear, including horizontal and/or vertical on transmit and dual polarized horizontal and vertical on receive, and circular polarization on transmit and receive. The use of polarization agility is sometimes of interest in developing modern system

tor control of these parameters. The state-of-the-art in the millimeter wave component technology from 35 GHz up through at least 95 GHz is such that most of these operating parameters can be implemented in instrumentation radar systems. Proper design of the timing and control circuitry allows for remote control of these parameters.

Components — State-of-the-Art

Instrumentation grade radar systems are generally not developed around state-of-the-art componentry since calibration parameter flexibility and accuracy are more important than maximum detection range, minimum sensitivity, or physical size. In the millimeter region, however, component availability has necessitated that state-of-the-art components be used in many instrumentation radars to obtain acceptable measurement performance.

Available magnetrons can provide transmitter peak powers from 50 kW (70 GHz, 95 GHz) to 150 kW (35 GHz); they may also be frequency agile with tuning bandwidths of 500 MHz (35 GHz, 70 GHz). Most magnetrons enable operation with fixed pulse widths ranging from 20 to 500 nanoseconds, but some are capable of short pulse, 7 nanosecond (95 GHz), operation.

Solid state transmitters using pulsed IMPATT or Gunn devices can provide

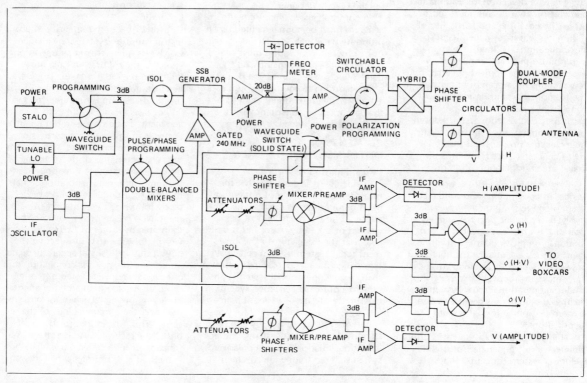

Fig. 1 Block Diagram of Georgia Tech 35 GHz Instrumentation Radar.

Millimeter Wave Radars

output peak power levels of 10 watts (35 GHz), 1 watt (70 GHz) or 5 watts (95 GHz) with a chirp bandwidth of approximately 500 MHz. Pulse widths are typically 10 to 100 nanoseconds and may be variable over a limited range. Gunn diode amplifiers can also deliver up to 3 watts peak power at 35 GHz for pulse widths ranging from 10 to 250 nanoseconds and bandwidths of 1 GHz.

Millimeter receiver components are now being developed with specifications similar to their lower frequency counterparts. Integrated mixer/preamplifier noise figures (dsb) of 4.0 dB (35 GHz), 6.0 dB (70 GHz) and 8.0 dB (95 GHz) are available, along with low noise Gunn diode local oscillators. Ferrite circulators are available with isolations as high as 40 dB (35 GHz) and 20 dB (70 GHz and 95 GHz), and insertion losses of 0.1 dB (35 GHz) and 0.7 dB (70 GHz and 95 GHz). Duplexers and conventional receiver protectors are not readily available, although TR limiters with typical insertion losses of 2 dB at 35 GHz are available for 100 kW power. Ferrite switches are commonly used in the receiver line to provide 35 dB isolation during the transmission period, but have high insertion losses of 1-to-2 dB. Where high speed recovery times are required (short range operation), pin diode switches are available which provide isolations of 30 dB (35 GHz) and 15 dB (95 GHz), and insertion losses of 1 dB (35 GHz) and 2-to-3 dB (95 GHz).

Taking into account the devices currently available and the resultant system losses, practical instrumentation radar sensitivities for 20 MHz IF bandwidths are −90 dBm (35 GHz), −84 dBm (70 GHz), and −82 dBm (95 GHz). The sensitivities are worse than at lower frequencies, but the increase in antenna gain for a given aperture size results in approximately the same range performance as at lower frequencies for a given transmitted power and target cross-section.

EXAMPLES OF MILLIMETER INSTRUMENTATION RADAR SYSTEMS

Georgia Tech recently developed measurement radar systems operating at 35, 70, and 95 GHz for comprehensive millimeter wave reflectivity measurements, thus enhancing their millimeter research capabilities. Brief descriptions of these systems follow.

35 GHz Instrumentation System

The salient features of the Georgia Tech K_a-band solid state measurement radar are:

- Coherent
- Portable-Self Contained
- Dual Polarized
- Polarization Agile
- Frequency Agile — Pulse-to-Pulse and Intrapulse (Chirp)
- Remote Operation and Recording Capability.

RF section

Figure 1 shows the block diagram of the RF and IF portion of the radar, and **Table 1** gives the operating parameters. In the coherent mode, the local oscillator signal is generated in a highly-stable 35-GHz phase-locked Gunn-diode source. The 35-GHz signal is applied to the single sideband (SSB) generator where it is offset by the gated 240 MHz signal to produce the pulsed RF at 35.24 GHz. The limited power output of the SSB generator requires that two solid state amplifiers be used to increase the signal to a useable transmit level. The first amplifier is a two-stage Gunn-diode amplifier having about 30 dB gain to provide 200 mW, and the second is a three-stage Gunn-diode amplifier increasing that level to about 3 watts for pulse widths up to 500 nsec. The power is limited to 200 mW for longer pulse widths. When the phase-locked Gunn source is replaced with the varactor-tuned source, pulse-to-pulse and intrapulse frequency agility is achieved by tuning the oscillator in a programmed mode. The amount of agility as well as the number of pulses between frequency jumps can be selected by the operator at the radar control panel.

TABLE 1
OPERATING PARAMETERS OF GEORGIA TECH 35, 70, AND 95 GHz INSTRUMENTATION RADARS

	35 GHz		70 GHz	95 GHz	
Transmitter					
RF Frequency:	35.24 ± 0.40 GHz (fixed or agile)		69-71 GHz	95 GHz (nominal)	
Peak Power	200 mW (10 nsec to CW) 3 watts (10-500 nsec)		500 watts	1 kW	
Pulse Width:	10 nsec to CW		20, 45 nsec	20 nsec	
PRF:	variable to 40 kHz		5-25 kHz	0-4000 Hz	
Antennas					
Type:	18" Dish	5" horn/lens	24" diameter folded geodesic lens/reflected	12" Cassegrain	3" horn/lens
Beamwidth:	1.25	4.5°	.55° x 3.5° (Az x El)	0.7°	3°
Gain:	43 dB	31 dB	43.2 dB	47 dB	35 dB
Sidelobes:	−20 dB	−20 dB	.28 dB	−20 dB	−20 dB
Polarization:	Dual Polarized		Vertical	Dual Polarized	
Receiver:	H,V			H,V	
Transmit:	H,V,RC,LC (agile or fixed)			H,V,RC,LC (agile or fixed)	
Receiver					
Type:	Coherent, Integrated Mixer/Preamps Linear	Logarithmic	Linear	Logarithmic	
Bandwidth:	500 MHz	160 MHz	60 MHz	100 MHz	
Dynamic Range:	30 dB	60 dB	30 dB	70 dB	
Sensitivity:	−80 dBm	−85 dBm	−78 dBm	−82 dBm	
Detection:	Square Law Amplitude Coherent (phase) Pseudo-Coherent (phase)		Amplitude	Amplitude and Pseudo-Coherent (phase)	
IF:	240 MHz		400 MHz	160 MHz	

The microwave assembly is configured to produce either linear, circular, or elliptical polarization at the antenna. In the linear polarization mode, either horizontal (H), or vertical (V), polarization is generated, depending on the state of the Faraday rotational switch. The Faraday switch can be switched in an inter-pulse period, providing pulse-to-pulse H and V polarization diversity. When the quadrature hybrid is inserted between the Faraday switch and the circulators, right and left elliptical polarization is generated, depending on the state of the Faraday switch. Circular polarization is generated by appropriate settings of the phase shifters and can likewise be switched between right and left circular polarization on a pulse-to-pulse basis.

Receiver Section

In all transmitting modes, the H and V channels are received simultaneously. In the coherent mode, both amplitude and absolute phase are measured. In the noncoherent mode, the relative phase between H and V is measured, which is called pseudo-coherent detection. This detection scheme has been shown to provide a powerful target/clutter discriminant technique at the microwave frequencies.[2]

The dual receiver section consists of two solid-state wide-band (10-500 MHz) mixer/preamplifiers and high-output (500 MHz) linear IF amplifiers, one channel for H and the other for V. Two IF amplifiers are used in each channel, one to drive the phase detectors and one to drive the amplitude detectors. A phase detector is placed between the H and V channels to measure relative phase for the pseudo-coherent detection.

The local oscillator signal for the mixers is derived from the stable (or tuned) 35-GHz source ahead of the single sideband generator. This provides the CW signal, 240 MHz below the RF signal, needed to produce the 240 MHz IF out of the mixer.

The IF reference signal is generated in a highly stable crystal oscillator. This signal is controlled to provide the pulsed and phase-coded IF signals and applied to two double-balanced mixers. The pulsed IF signal is used in the single-sideband generator to provide the pulsed 35.24 GHz RF signal applied to the solid-state amplifiers. A CW sample of the oscillator signal serves as the IF reference for the receiver phase detectors, making the radar coherent on receive.

Recording and Display

Five video channels enable recording of the amplitude and phase for both the H and V channels as well as the pseudo-coherent phase video. All five wideband video signals are simultaneously box-carred (sampled and held) at the pulse repetition rate at some slant range determined by the placement of the range gate. The signal of one of the amplitude channels, selected by the operator, is converted from analog to digital. The signal in the selected range bin is averaged over a selectable number of pulses, and the average signal level is displayed by light emitting diodes (LEDs) as well as on a microammeter mounted on the control box to facilitate range gate placement.

Control

With the exception of the control of waveguide phase shifters and attenuators, all radar control is accomplished remotely from the radar chassis itself via a multiconductor cable. With the exception of one DC control voltage, all the control commands are digital

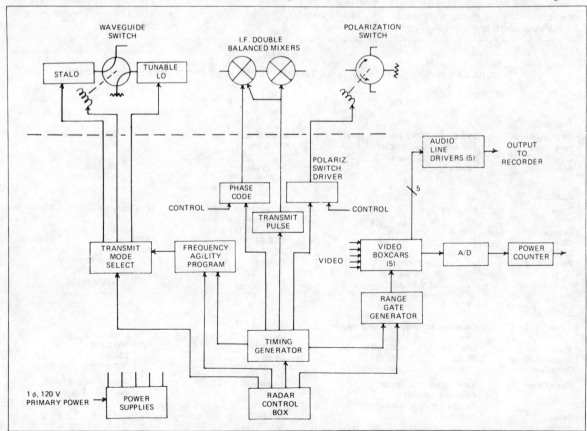

Fig. 2 Diagram of Control Circuits for Georgia Tech 35 GHz System.

Millimeter Wave Radars

logic levels to avoid the effect of noise on the control box cable. Operating controls include those listed below:

- PRF
- Pulsewidth
- Pulse-to-Pulse Frequency Agility
- Intra-Pulse Frequency Agility
- Polarization Mode
- Coherent/Agile
- Range Gate Control
- Power Averager Control
- Integration Time Control

These controls provide logic signals to the various control circuits associated with radar operation. **Figure 2** shows these control and regulator circuits which include:

- Radar Timing
- Frequency Agility Programming
- Polarization Switch Driver
- Phase Code and Transmit Waveform Generation
- Range Gate Generation
- Boxcars
- Linedrivers
- Power Counter

Power Supplies

All radar dc power is supplied by the separate power supply assembly. The various supply voltages are all derived from a single phase, 47 Hz to 420 Hz, 120 volt source.

70 GHz Mobile Radar System

Georgia Tech's 70 GHz, high resolution, rapid scan radar system was originally built by Georgia Tech for Harry Diamond Laboratories.[3] This radar was installed in a M-109 shop van in order to create a mobile facility. The radar console and antenna are linked together and revolve about a common axis. The operator faces the direction illuminated by the radar; a periscope, whose optical axis is aligned with the antenna electromagnetic axis, enables the operator to view the illuminated area.

While not originally an instrumentation radar, this radar system has been modified for use in several research programs conducted at Georgia Tech.[4] Currently, it is an integral part of a research program being conducted for the U.S. Army Electronic Research and Development Command (ERADCOM) in which the feasibility of a combined laser/millimeter wave radar sensor system is being investigated.[5]

Many of the radar parameters are variable, e.g., the RF pulse width is selectable from 20 to 45 nanoseconds, the PRF is variable from 5 to 25 kHz, and the antenna scan rate is variable from approximately one scan/minute to 50 scans/second. Scanning is accomplished by a seven-section ring switch which revolves around the antenna. The antenna is a 24-inch diameter helmet-type folded geodesic lens. This lens produces a fan-shaped beam with a 0.55° azimuth beamwidth and a 3.5° elevation beamwidth. The radar antenna was recently modified to give the radar even more versatility. Two microwave reflectors were added to rotate the beam through 90°, thus enabling the radar to operate in a vertical scan mode as well as horizontal. The basic system parameters of this radar are given in **Table 1**, and a block diagram is shown in **Figure 3**.

95 GHz Instrumentation Radar System

The Georgia Tech GT-M radar is an instrumented, calibrated, short-pulse, measurement radar operating at approximately 95 GHz. Major parameters of the radar are summarized in **Table 1**. The radar is housed in a small protective container which consists of two separate compartments; one containing the Extended Interaction Oscillator and modulator, and the other containing the receiver. The packaging approach combines isolation to minimize interaction and interference problems with good portability and accessability. Sufficient space is provided in the package to allow the radar to be used in a number of different experiments.

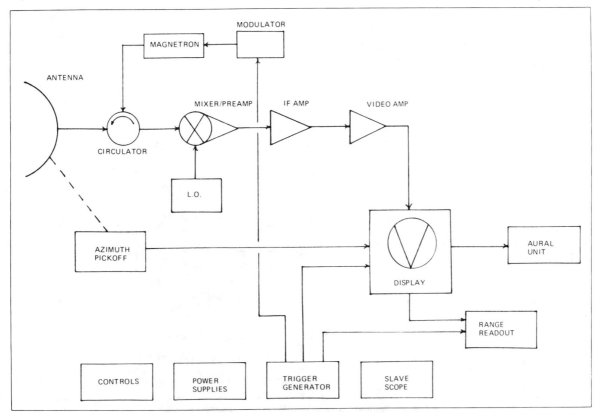

Fig. 3 Block Diagram of Georgia Tech 70 GHz System.

The system block diagram is shown in **Figure 4**. The radar is configured to allow vertical, horizontal, left circular (LC), right circular (RC), or ±45° linear polarized transmissions and simultaneous receptions of vertically and horizontally polarized components of the reflected signal. Pulse-to-pulse polarization agility for LC/RC or ±45° polarizations can be transmitted. The receiver detects the amplitude and relative phase of the H and V components to provide pseudo-coherent detection.

The system is configured to permit accurate acquisition of carefully calibrated pulse-to-pulse radar reflectivity data. Logarithmic receivers of large dynamic range (typically 70 dB) are normally used, although linear receivers are also available.

Amplitude calibration of the system is accomplished either by injecting a known signal into the receiver front end through a coupler or by using precision calibrated attenuators with a standard target. In the latter method, the precision variable attenuators are used to establish known power levels at the receiver input. Either approach yields calibrations essentially independent of the transfer function of the data acquisition and recording equipment. Phase calibration is achieved by boresighting the antenna on a reference target with a known phase relationship between the reflected components (for example, a diplane) and varying the mechanical phase shifter between the transmitted components from 0° to 180°.

The 95 GHz radar system has been used as a radar cross-section measurement tool on several measurement programs in the recent past for the Army and Air Force[6-8] and is currently being used on a joint program for ECOM/MIRADCOM to study millimeter target/clutter discrimination techniques.

MILLIMETER MEASUREMENTS METHODOLOGY

The results and recommendations of research and development programs are generally viewed as accurate and complete when used to form the basis for system design. Since such systems, with the large quantities normally involved in production, involve a large investment, it is important that the research data represent a comprehensive and realistic operational environment on which to have systems design and development.

The procedures for millimeter measurements are similar to those at lower frequencies; however, there are several problem areas encountered for millimeter measurements which must be given special attention. These problem areas include absolute calibration, atmospheric propagation effects, and environmental characterization.

Calibration

Accurate calibration of radar measurements is always a difficult task, but is even more of a problem in the millimeter wave region than at lower frequencies. Millimeter system components are not as stable as lower frequency components so that changes in transmitted power or receiver drift are more likely. The only solution to this type of problem is to calibrate often; of course, this slows down the measurement process. Another problem encountered is that absolute-calibration reference targets tend to have measured radar cross-sections significantly

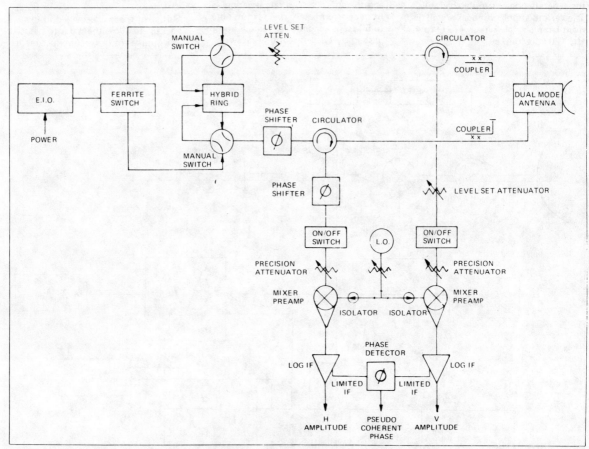

Fig. 4 Block Diagram of Georgia Tech 95 GHz Instrumentation System.

lower than their theoretical cross-sections. The solution to this problem is to make calibration targets as accurate as possible, and then to calibrate them independently before beginning field tests. This, then, requires that large expenditures be made on calibration targets and their mounts since the pointing accuracy of calibration targets such as diplane and trihedral reflectors become much more important in the millimeter region. One other minor nuisance is the difficulty in pointing the radar antennas for calibration; since millimeter calibration targets tend to be smaller than at microwave frequencies, they are harder to see.

One helpful characteristic of millimeter wave measurements is that multipath tends to be less of a problem since antenna beamwidths tend to be smaller and the terrain along the propagation path appears rougher so that the effects of multipath lobing on calibration are reduced.

Propagation Effects

At millimeter frequencies, radiation propagating through the atmosphere can be strongly affected by meteorological conditions. For clear air, appreciable attenuation occurs for electromagnetic energy traveling through the medium even in the so-called transmission windows near 35 GHz, 95 GHz, 140 GHz, and 300 GHz. For the ranges which are presently being considered for millimeter radar applications, however, clear air attenuation can be neglected. But the effects of the precipital atmosphere cannot be neglected. Heavy fog, rain, snow, and hail can have significant effects on propagating millimeter wave energy through both attenuation and scattering mechanisms. Anomalous propagation is present at millimeter frequencies as well as at lower frequencies. When measurements are made under these conditions, these atmospheric effects must be accounted for.

The best method for accounting for atmospheric effects on a radar signal is to periodically inject a known power level into the receiver while continuously monitoring a calibration target. The power from the radar calibration target can be related to the injected power by using the radar equation and known system parameters. Changes in power from the calibration reflector can then be readily detected and accounted for. In addition, measurement of meteorological parameters such as rain rate or drop-size distribution can confirm the meteorological effects detected by the above method since previous experiments have related such parameters to attenuation and backscatter.[9]

Environmental Characterization

Because the environment has a greater effect on millimeter wave propagation and reflectivity than at lower frequencies, it is more important to adequately characterize the environment. It has been determined that reflectivity from rain depends more strongly on drop-size distribution than rain rate.[9] Snow backscatter is very strongly affected by the surface roughness and the free water content,[8] backscatter from foliage depends on the surface moisture present,[10] and sea return is strongly affected by wind generated droplets and small facets on the ocean surface.[11]

Thus, these types of parameters must be accurately quantified if the interaction mechanisms are to be understood. Unfortunately, the instrumentation and techniques for such measurements are in most cases not yet fully developed. A number of rain drop-size spectrometers, including both optical and hydrophonic types, are now available, but they are expensive and measure the drop-size for a very small area. Snow characterization is still a process of collecting snow samples, freezing and weighing them for determination of the free water content, and using pictures to estimate surface roughness. However, Jones[12] recently developed a recursion formula which allows the free water content of a snowpack to be determined from snow temperature and density. Sea characterization is still confined to wave gauges or to stereo-photography. However, neither technique accurately characterizes airborne droplets which may have strong effects on the backscatter properties. Finally, no practical technique has been developed for sensing surface moisture on foliage on the ground. Infrared sensors can give relative ground moisture readings, but they are hard to calibrate since the readings depend on soil type as well as moisture content. Thus, environmental characterization is as difficult as it is important.

SUMMARY

The unique problems associated with building millimeter wave systems and performing millimeter wave measurement programs include the availability of millimeter wave components, the influence of environmental effects, and the need to fully characterize the environment to understand the physical mechanisms involved. These problems have necessitated the use of state-of-the-art components, both for the sensor and environmental characterization hardware and techniques. As a result a number of problems remain to be solved, as demonstrated in this article. With the present rapid increase in the state-of-the-art in both millimeter wave technology and environmental characterization technology, these problems should soon be solved, and millimeter wave measurements will become as routine as microwave measurements are presently.

REFERENCES

1. Dyer, F. B., Ewell, G. W., and Hayes, R. D., "Data Requirements for the Millimeter Seeker Problem," *Fourth DoD Millimeter Workshop*, November 1976.
2. Gelernter, B., Johnson, W., Reedy, E. K., Eaves, J. L., Piper, S. O., Parks, W. K., and Brookshire, S. P., "Stationary Target Discrimination and Classification Studies," *23rd Annual Tri-Service Radar Symposium Record*, West Point, N.Y., July, 1977.
3. Goodman, R. M., Jr., and Dyer, F. B., "Millimeter Wave Radar," Final Report on Contract DA-49-186-AMC-275(A), Engineering Experiment Station, Georgia Institute of Technology, February, 1968.
4. Alexander, N. T., Craven, T. S., and Foster, W. S., "Target Acquisition Experiments Employing a Millimeter Wave Rapid Scan Radar," Final Report on Contract N00014-76-C-0860, Mod. P00001, Engineering Experiment Station, Georgia Institute of Technology, October, 1976.
5. Holm, W. A., Foster, W. S., Loefer, G. R., "Combined Electro-Optical/Millimeter Wave Radar Sensor System," *Proc of the Sixth DARPA/Tri-Service Millimeter Wave Conference*, Adelphi, MD, November, 1977.
6. Currie, N. C., Dyer, F. B., and Hayes, R. D., "Radar Land Clutter Measurements at Frequencies of 9.5, 16, 35, and 95 GHz," Technical Report No. 3 on Contract DAAA 25-73-C-0256, Engineering Experiment Station, Georgia Institute of Technology, April, 1975, ADA01279.
7. Currie, N. C., Martin, E. E., and Dyer, F. B., "Radar Foliage Penetration Measurements at Millimeter Wavelengths," Technical Report No. 4 on Contract DAAA 25-76-C-0256, Engineering Experiment Station, Georgia Institute of Technology, December, 1975, ADA023838.
8. Currie, N. C., Dyer, F. B., and Ewell, G. W., "Characteristics of Snow at Millimeter Wavelengths," *Digest of the 1976 IEEE AP-S Int'l Symp.*, Amherst, MA, October, 1976.
9. Currie, N. C., Dyer, F. B., and Hayes, R. D., "Some Properties of Radar Returns from Rain at 9.375, 35, 70, and 95 GHz," *Proc of the IEEE 1975 Int'l Radar Conference*, Arlington, VA, April, 1975.
10. Currie, N. C., et al., "Millimeter Radar Backscatter Measurements, Part I. Snow and Vegetation," Final Report on Contract F08635-76-C-0221, AFATL-TR-77-92, July, 1977.
11. Trebits, R. N., et al., "Millimeter Radar Sea Return Study," Technical Report No. 1 on Contract N60921-77-C-A168, Engineering Experiment Station, Georgia Institute of Technology, to be published.
12. Jones, E. B., Private Communications with N. C. Currie, December, 1977.

A Millimeter Wave Radar for the Mini-RPV

L. Kosowsky, K. Koester, B. Gelernter, and W. Johnson
AIAA/DARPA Conference on Smart Sensors, Nov. 1978
Reprinted by permission.

(U) Abstract -- A brassboard millimeter wave radar operating at 95 GHz has been developed to evaluate state-of-the-art components and techniques for use in the mini-RPV. The radar is designed for tactical surveillance in adverse weather. Modes of operation include high resolution ground map, fixed target enhancement and moving target indication.

(U) During the course of the radar development, two technical areas have been shown to be critical to the development of an operational system, namely, high power sources and polarization signatures. This paper provides a status report on the brassboard radar and the results of measurements on targets and clutter using the fixed target enhancement mode.

1. Introduction (U)

(U) The mini-RPV offers a unique potential as a surveillance and intelligence gathering tool. Because they are unmanned and relatively expendable, mini-RPVs provide ideal platforms for missions beyond the forward edge of the battle area (FEBA) which makes it possible to observe the disposition of enemy forces with a high resolution sensor.

(U) Norden Systems, Inc. a subsidiary of United Technologies Corporation has successfully completed two portions of a development program to demonstrate the ability of a 95 GHz radar sensor to perform real time battlefield surveillance and targeting. The development program was apportioned into three parts: (a) design and fabrication, (b) ground testing, and (c) flight testing. Part (a) has been completed, part (b) is a continuing effort, and part (c) is being planned.

(U) In a tactical situation, the millimeter radar is flown at low altitudes beyond the FEBA into a highly vulnerable combat zone by an unmanned, ground controlled mini-RPV, as shown in Figure 1. The prime responsibility of the radar is to recognize hard fixed and moving targets such as tanks, truck convoys, gun emplacements, or boats. The collected radar data is processed in-flight and relayed back to the control point for real time tactical evaluation.

---

*(U) Work sponsored by PM for RPV, AVRADCOM via ERADCOM under contract No. DAAB07-76-C-0843.

†(U) Dr. L. Kosowsky and K. Koester are with Norden Systems, Inc. Dr. B. Gelernter and W. Johnson are with the U.S. Army Electronics Research and Development Command.

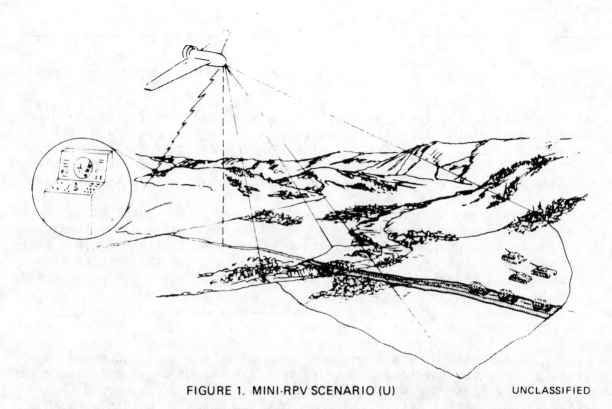

FIGURE 1. MINI-RPV SCENARIO (U)   UNCLASSIFIED

(U) The radar modes of operation include fixed target enhancement (FTE) for the detection of fixed targets in clutter, moving target indication (MTI) for the detection of hard moving targets surrounded by ground clutter, and high resolution ground mapping (HRGM) for target orientation and damage assessment.

(U) Since the excellent angular resolution obtainable with millimeter wavelength radars of moderate dimensions is not fully adequate for target recognition, there is a need for techniques to distinguish the radar returns emanating from threat targets from those generated by other scatterers. The army has recognized this problem and has expended work, in this and other programs, to meet this need. The problem of discriminating between targets and clutter is being addressed in this program with the FTE mode. Other techniques for the discrimination and automatic classification of targets from their radar returns are being investigated under ERADCOM sponsorship.

2. Battlefield Environment (U)

(U) There are various technical problems associated with the development of an airborne platform including launch, control and retrieval. As these challenges are successfully met, the problem of providing a

suitable sensor assumes increased importance. While falling short of
providing the resolution and imaging capability available in E/O devices, a radar sensor operating at millimeter wavelengths has the great
advantage over E/O devices of being operable in the presence of severe
weather conditions including fog, haze, relatively dry snow and moderate amounts of rain (Ref. 1).

(U) The study of the transmission properties of millimeter, IR and optical wavelengths in the presence of various man-made obscurants such as
smoke, battlefield debris and dust, is a subject of increasing current
interest. Figure 2, extracted from a recent Ballistic Research Laboratories report (Ref. 2), shows results obtained in smoke measurements
conducted by BRL at Ft. Hood, Texas. According to the BRL report "four
salvos of 2.75 inch WP-wick rockets were fired in groups of 14 rockets
per salvo at times zero, 5 minutes, 14 minutes and 18 minutes". The
figure shows the measured transmittance as a function of time for the
indicated millimeter, IR and optical wavelengths. Although the smoke
concentration and path length were not measured and quantified during
this test, there can be little argument with the author's evaluation
that the "... WP - smoke reduced the infrared and visual transmission
to near zero for extended periods of time while producing no measurable
attenuation at millimeter wave frequencies". Figure 3, extracted from
the same report, shows the transmittance in the presence of an unquantified amount of dust as measured at Aberdeen Proving Grounds. The report reads in part, "the dust was created by a jeep running parallel to
the line of sight with the wind carrying the dust across the measurement
path. Again there was no measured attenuation at millimeter wave frequencies...". Further and more controlled tests held at Dugway Proving
Ground during smoke week (Ref. 2, 3), again indicated that transmission
at millimeter wavelengths was unaffected by dust and the man-made obscurants used in these tests, while transmission in the IR and optical bands
was significantly affected for varying durations.

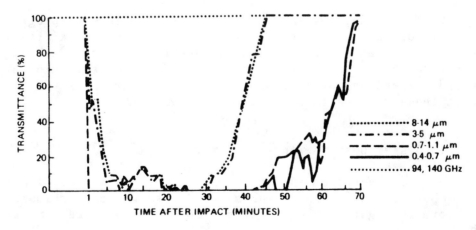

FIGURE 2. TRANSMISSION THROUGH WP SMOKE (U) (Ref. 2)   UNCLASSIFIED

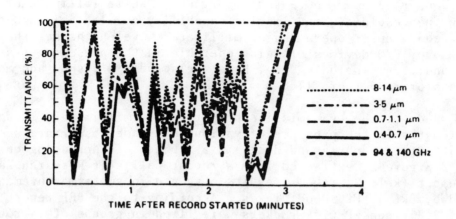

FIGURE 3. TRANSMISSION THROUGH DUST (U) (Ref. 2)   UNCLASSIFIED

3. **RPV Brassboard Radar** (U)

   a. **Background** (U)

   (U) The RPV program was conceived as a means for assessing the utility of millimeter wave radar to fill the need for an all-weather, battlefield surveillance system. The radar is configured to serve as a test-bed for a variety of waveforms and data processing techniques and to evaluate the capability of the radar to locate and identify potential targets. Norden integrated newly designed and modified proven equipment with comprehensive instrumentation into a flyable brassboard radar. The brassboard was designed for flight testing in a CH-47 helicopter test bed. Digital video recording and playback instrumentation was included in the millimeter radar design to permit post operation data evaluation.

   (U) The ground testing portion of the development program consisted of laboratory and range testing at Norden and functional testing at the Fort Huachuca test site in Arizona. Additional ground testing will continue with the quantitative measurement of the polarization signatures of hard targets and assessment of the ability of the 95 GHz system to detect these targets in a battlefield environment.

   b. **System Description**

   (U) The RPV system is configured as a forward-looking surveillance radar operating at 95 GHz. The system is shown in functional block diagram form in Figure 4. Critical system parameters are summarized in Table 1.

(U) Table 1. RPV Radar Parameters (U)

FREQUENCY	95 GHz (NOM)
PEAK GAIN	36 dB
AZIMUTH BEAMWIDTH	0.33° (WITH 63 CM REFLECTING SURFACE)
	0.44° (WITH 42 CM REFLECTING SURFACE)
ELEVATION PATTERN	$csc^2 \, cos^{1/2}$
POLARIZATION	HORIZONTAL } (PULSE TO PULSE SWITCHABLE)
	VERTICAL
PEAK POWER	2 kW
PULSEWIDTH	20, 50 NS
PRF	4 kHz
IF FREQUENCY	160 MHz
BANDWIDTH	24, 60 MHz
SCAN WIDTH	40, 30, 20 DEG
SCAN SPEED	40, 30, 20 DEG/SEC
	MANUAL POSITIONING
PITCH LIMITS	+20, -30 DEG
ROLL LIMITS	±45 DEG

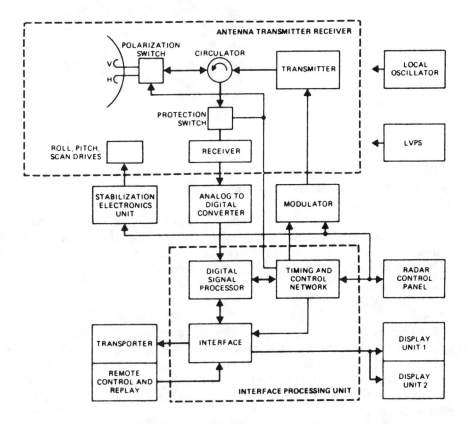

FIGURE 4. RPV SURVEILLANCE RADAR BLOCK DIAGRAM (U)   UNCLASSIFIED

(U) The Antenna Transmitter Receiver (ATR) brassboard is shown in Figure 5. The major portions of the ATR are contained on the scanning, stabilized portion of the antenna structure. The reflector and feedhorn are supported at the front of the balanced scanning structure, the receiver assemblies at the top, and the transmitter, magnetron, and pulse transformer at the mid and lower sections, respectively. Waveguide sections functionally join the RF sections.

FIGURE 5. ANTENNA TRANSMITTER RECEIVER (U)   UNCLASSIFIED

(U) A detailed description of the various units comprising the system can be found in references 4-8. The following sections contain a description of the polarization diverse csc^2 antenna developed by Norden as well as the 94 GHz magnetron developed under Norden sponsorship.

c. <u>Antenna</u> (U)

(U) <u>Reflector</u> -- The reflector is a modified AN/APQ-148 sandwich structure (aluminum honeycomb sandwiched between fiberglass skins) 63 cm wide and 43 cm high with a focal length of 36 cm. The reflecting surface is sprayed with high conducting silver paint to minimize ohmic losses. A surface tolerance of ±0.015 cm was maintained to achieve good sidelobe levels.

(U) The elevation contour is shaped to provide a $\csc^2\theta \cos^{1/2}\theta$ elevation pattern extending from 12° to 45° below the horizon. The peak of the elevation beam, with a gain of 36 dB, is depressed 12° from the horizon. The azimuth contour is parabolic to yield a pencil beamwidth of 0.33° or 0.44° with microwave absorber applied to the reflector to reduce the reflecting surface width.

(U) Feedhorn -- The feedhorn, which provides vertical and horizontal polarization is offset from the reflector. Horizontal polarization is used in the high resolution ground map and moving target indication modes of operation. Polarization diversity, in which the polarization is changed from vertical to horizontal on a pulse-to-pulse basis, is used in the fixed target enhancement mode.

(U) The offset eliminates secondary rays from passing through the feedhorn or support structure, improving sidelobe levels. In addition, greater structural rigidity is provided.

(U) Antenna Tests -- Antenna tests were conducted to determine the actual elevation pattern, azimuth beamwidth, sidelobes, and gain of the brassboard antenna. The tests were conducted at 95.5 GHz using both vertical and horizontal polarization. Complete patterns were taken for the two reflector configurations. The antenna parameters are summarized in Table 2.

(U) Table 2. Antenna Parameters (U)

PARAMETER	DESIGN GOAL	MEASURED	
ELEVATION PATTERN	$\csc^2\theta \cos^{1/2}\theta$ -12° TO -45°	$\csc^2\theta \cos^{1/2}\theta$ -12° TO -52°	
ELEVATION SIDE LOBES	-20 dB	-25 dB	
AZIMUTH BEAMWIDTH	0.43°	0.33°	0.44°*
AZIMUTH SIDE LOBES	-18 dB	-26 dB	-18 dB*
GAIN	37 dB	36 dB	35 dB*

*WITH MICROWAVE ABSORBER SECTIONS

(U) The millimeter brassboard radar is designed to operate at an altitude of 610 m. At that altitude, elevation coverage from -11.7° to -37.6° is required to illuminate the 1 to 3 km ground patch. The elevation patterns for the two azimuth beamwidths are shown in Figures 6 and 7. The peak of the beam is located 12° below the horizon for both configurations.

(U) The $\csc^2\theta \cos^{1/2}\theta$ design curve is shown in Figures 6 and 7 for comparison. The patterns follow the $\csc^2\theta\cos^{1/2}\theta$ curve from -12° to -42°. For both azimuth beamwidths, the difference in gain for vertical and horizontal polarization is less than 0.36 dB for the -12 to -38° portion of the pattern.

(U) The elevation sidelobes, which are not shown on the patterns are more than 25 dB below the peak of the beam. The design goal was -20 dB.

(U) The azimuth patterns for the two antenna configurations are shown in Figures 8 and 9. Half power beamwidths of 0.33° and 0.44° were achieved. The design goal was 0.43°. Test cuts were made at -12°, -24°, -36°, and -48° in elevation.

(U) Azimuth sidelobes of -26 dB and -18 dB were achieved for the 0.33° and 0.44° configurations, respectively. The sidelobe specification was -18 dB.

(U) The measured peak gain of the antenna is 36 dB for the 0.33° beamwidth configuration and 35 dB for the 0.44° beamwidth configuration.

d. Magnetron (U)

(U) Introduction -- During the development and test of the 95 GHz surveillance radar for the mini-RPV, several difficulties with the Amperex magnetron were identified. The tube exhibited a wide frequency spectrum and a significant negative frequency drift with aging. In addition, the tube had a relatively short operational life. In order to evaluate alternative high power sources, Norden Systems sponsored the development of a 95 GHz magnetron by the English Electric Valve Company, Ltd.

Millimeter Wave Radars

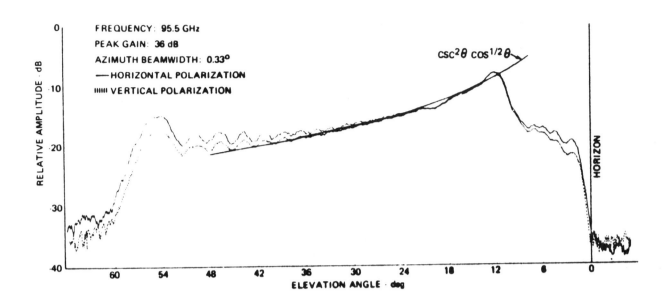

FIGURE 6. ELEVATION PATTERNS FOR 0.33° BEAMWIDTH CONFIGURATION (U)

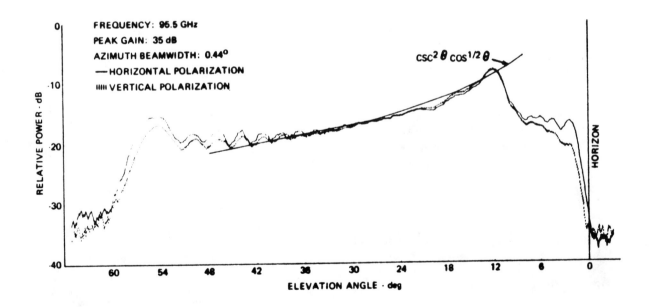

FIGURE 7. ELEVATION PATTERNS FOR 0.44° BEAMWIDTH CONFIGURATION (U)

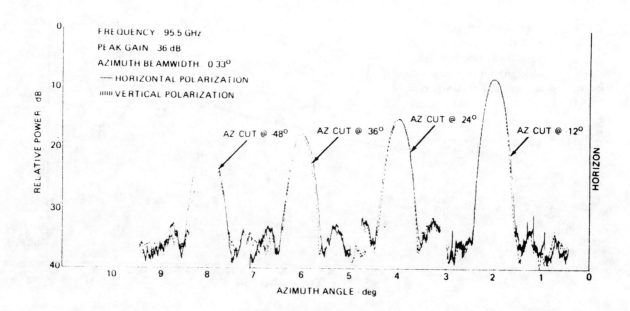

FIGURE 8. AZIMUTH PATTERNS FOR 0.33° BEAMWIDTH CONFIGURATION (U)

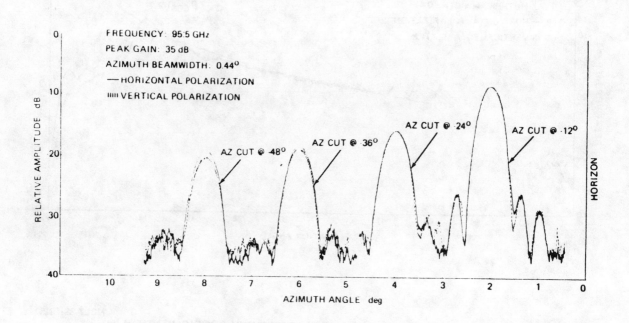

FIGURE 9. AZIMUTH PATTERNS FOR 0.44° BEAMWIDTH CONFIGURATION (U)

(U) **EEV Magnetron** -- The 95 GHz tube design was based on an 80 GHz magnetron which is being produced in quantity by EEV. That tube has an expected life of 750 hours. Life tests have demonstrated less than a 10% degradation in power after 500 hours of operation. The frequency change with use is 1 MHz/hr max. and 0.5 MHz/hr typical. This frequency shift is usually negligible after a hundred hours of operation.

(U) The 95 GHz tube shown in Figure 10, has an expected life of 750 hours. Since only a few of the tubes have been built, there is not enough data to establish operational characteristics. They are, however, expected to be comparable to the 80 GHz tube.

(U) **System Modifications** -- In order to increase the performance and reliability of the mini-RPV Surveillance Radar, the newly developed magnetron has been incorporated into the brassboard radar system. Modifications to the modulator have been completed and system testing with the magnetron is underway. Preliminary magnetron performance characteristics are described in the following sections.

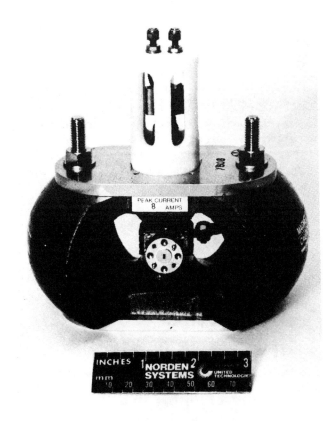

FIGURE 10. EEV 95 GHz MAGNETRON (U)

(U) Magnetron Tests -- The RF detected pulse from the EEV magnetron is shown in Figure 11 for the 20 and 50 nanosecond pulsewidth modes. The pulse rise time is approximately 10 nanoseconds and has less than 2 nanoseconds of pulse jitter.

(U) The spectra for each pulsewidth are shown in Figure 12. The spectra have a mainlobe width of approximately $2/\tau$ with a sidelobe level of 8 dB. The actual mainlobe widths are 100 and 60 MHz for the 20 and 50 ns pulses respectively.

4. Fixed Target Enhancement (FTE) Techniques (U)

   a. Polarization Diversity (U)

(U) Polarization diversity may be used to enhance the detectability of targets in ground clutter since hard targets are less sensitive to transmitted polarization states than natural or terrain scatters (9). Target returns tend to exhibit approximately the same amplitude when illuminated alternately by horizontally and vertically polarized pulses. Ground clutter, however, tends to exhibit an amplitude variation when illuminated with alternate orthogonal polarizations. By transmitting and receiving alternating orthogonal linear polarizations, and sensing the amplitude modulation on the radar return, targets may be discriminated from ground clutter.

(U) This technique has been successfully tested at lower than millimeter radar frequencies where excellent clutter rejection and interpulse subclutter visibility were achieved.

(U) Consider two consecutive radar returns, E^V and E^H. The amplitude variation, ΔE, is given by:

$$\Delta E = E^V - E^H$$

and the average amplitude, E_{ave}, of these two returns is

$$E_{ave} = 1/2 \, (E^V + E^H)$$

Therefore, the amount of the amplitude modulation is given by:

$$P = \frac{|\Delta E|}{E_{ave}} = \frac{|E^V - E^H|}{1/2(E^V + E^H)} \tag{1}$$

Millimeter Wave Radars

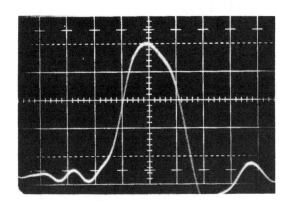 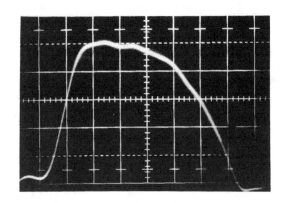

20 ns Pulsewidth
Vertical Scale: 50 mV/div
Horizontal Scale: 10 nS/div

50 ns Pulsewidth
Vertical Scale: 50 mV/div
Horizontal Scale: 10 nS/div

FIGURE 11. RF DETECTED PULSE (U)   UNCLASSIFIED

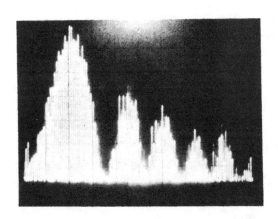 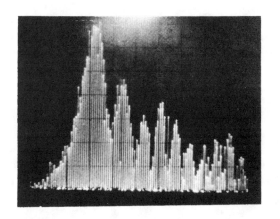

20 ns Pulsewidth
Vertical Scale: 2 mV/div
Horizontal Scale: 30 MHz/div

50 ns Pulsewidth
Vertical Scale: 2 mV/div
Horizontal Scale: 30 MHz/div

FIGURE 12. RF SPECTRA (U)   UNCLASSIFIED

(U) In the FTE processor, equation (1) is evaluated for two consecutive pulses and compared with a preselected threshold level K2. If P > K2, a clutter target is declared, otherwise a hard target is declared. The discrimination logic is shown in block diagram form in Figure 13. The threshold value is selected according to the nature of the clutter, the allowed probability of error, etc.

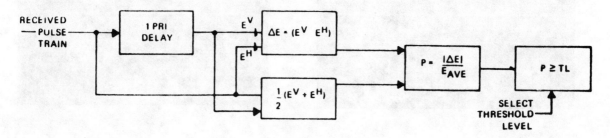

UNCLASSIFIED

FIGURE 13. POLARIZATION DIVERSITY DISCRIMINATION LOGIC BLOCK DIAGRAM (U)

b. Experimental Results (U)

(U) Introduction -- In order to evaluate the FTE mode of operation, the test site shown in Figure 14 was mapped by the radar operating in the high resolution ground map mode.

(U) A typical B-scan of the area is shown in Figure 15. The display was taken with the radar scanning a 30 degree sector at a scan rate of 30 deg/sec.

(U) FTE Mode -- The clutter cancellation capability of the FTE mode was evaluated at the Norden facility test site. Two 15.2 cm reflectors and a 7.6 cm reflector were aligned with the radar and a large tree at the edge of a grassy field.

(U) The processed video return in the HRGM mode is shown in Figure 16-a. The returns from the 7.6 cm reflector and tree are approximately one-third the amplitude of the return from the 15.2 cm reflector. Processed video returns in the FTE mode are shown in Figure 16-b, 16-c, and 16-d. With the wide acceptance threshold (K2=5.3), the return from each of the targets is doubled in amplitude. This is shown in Figure 16-b. With the narrow acceptance threshold (K2=32) the returns from the three trihedrals are doubled while the tree return was rejected as shown in Figure 16-c.

Millimeter Wave Radars

FIGURE 14. VIEW FROM NEST (U)   UNCLASSIFIED

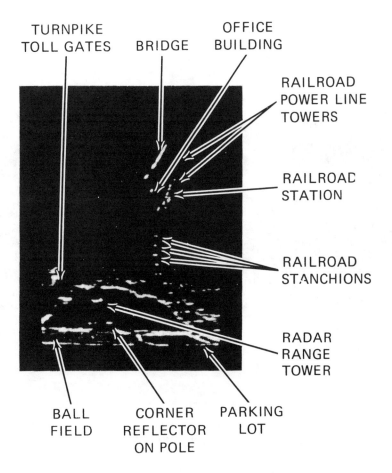

FIGURE 15. HRGM MODE DISPLAY (U)   UNCLASSIFIED

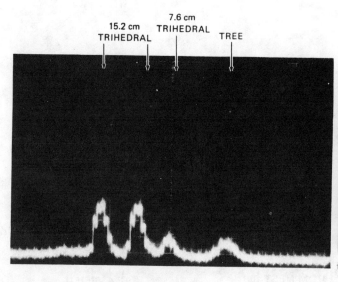
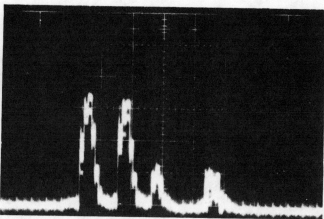

a. MODE: HRGM
GAIN: -1.5 V
WIND VELOCITY: 8 km/hr

b. MODE: FTE
K2: 5.3
WIND VELOCITY: 8 km/hr

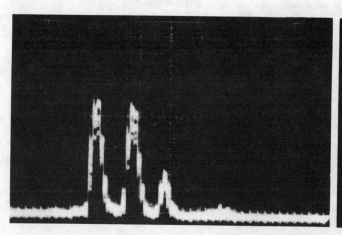
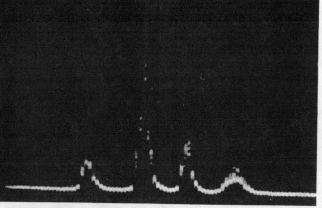

c. MODE: FTE
K2: 32
WIND VELOCITY: 8 km/hr

d. MODE: FTE
K2: 32
WIND VELOCITY: 50 km/hr

SCALE:
VERTICAL: 0.5 V/div
HORIZONTAL: 2 μs/div

FIGURE 16. FTE MODE CLUTTER CANCELLATION (U)   UNCLASSIFIED

(U) In each of the above cases, the wind velocity was less than 8 km/hr. It was found that the polarization dependence of the tree return decreases as the wind velocity increases. This effect is shown in Figure 16-d when the winds gusted as high as 50 to 70 km/hr. While the tree return is partially rejected, the cancellation is not as large as in Figure 16-c. This condition occurs because of the cross modulation between the doppler spectrum of the clutter and the induced pulse-to-pulse polarization variation of the clutter. This condition should not occur in an operational system with a high prf.

(U) An FTE display is shown in Figure 17 for a threshold level of 10.7. As can be seen, the returns from the clutter are almost completely rejected. (A HRGM display of the same terrain is shown on Figure 15).

5. Conclusions (U)

(U) The mini-RPV radar brassboard is moving into the measurements phase of development. Utilization of the radar for the measurement of polarization signatures of military vehicles will be undertaken in the spring of 1979 using the brassboard radar which has been modified with an EEV magnetron and an expanded data gathering system. Of prime importance is the demonstration of target detection and clutter reduction in a realistic battlefield environment.

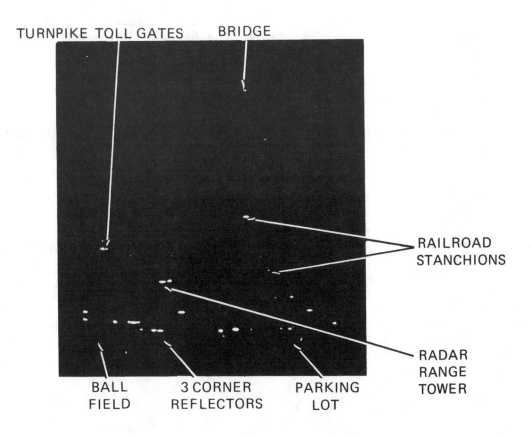

FIGURE 17. FTE MODE DISPLAY (K = 10.7) (U)   UNCLASSIFIED

6. Acknowledgement (U)

(U) The authors wish to acknowledge the support of D. Dunlap, (Office of RPV, AVRADCOM), and J. Carulli, J. Kricek, J. Gambardella, R. Ferraro of Norden Systems for their efforts in the design and test of the millimeter brassboard radar.

7. References (U)

1. Gamble W. L., and A. D. Hodgens "Propagation of Millimeter and Submillimeter Waves" U.S. Army MIRADCOM Technical Report TE-77-14, June 1977.

2. LaGrange, A., "Effect of Obscurants on Millimeter Wave Systems", presented at the Smoke Symposium, February 1978.

3. Shelton, H.R., "Recent Developments in Smoke Test Methodology", presented at the Smoke Symposium, February 1978.

4. Kosowsky, L.H., et al "A Millimeter Wave Radar for RPV's" presented at the Army Aviation Electronics Symposium, April 1976.

5. Pearce, R.H., et al, "3.2 Millimeter Surveillance Radar for the U.S. Army Mini-RPV" presented at the NATO AGARD Symposium, October, 1976.

6. Kosowsky, L.H. et al "A Millimeter Wave Surveillance Radar for RPV's", presented at the IEEE South East Conference, April 1977.

7. Wagner, R.H., et al "Millimeter Wave Radar for RPV Applications", presented at the 23rd Annual Tri-Service Radar Symposium", July 1977.

8. Zizzo, E.A., et al "Final Technical Report - Millimeter Surveillance Radar for the Mini-RPV, U.S. Army Research and Development", Technical Report ECOM-76-0843-F, June 1977.

9. Appling, B.C., and J.L. Eaves, "Design Analysis for Implementation of Polarization Agility for Target Recognition", Georgia Institute of Technology Report EES/GIT A-1723.

10. Soong, A. et al "Target Discrimination Measurements at Millimeter Wavelengths" presented at the 23rd Annual Tri-Service Radar Symposium, July 1977.

1   Introduction

In the development of radar proximity fuzes - which is done at AEG-TELEFUNKEN since about 1960 - one has on principle to deal with a radar system where the received signal is used to measure the encounter geometry between fuze and target and to determine from this the optimum detonation point.

This measurement is carried out in the extreme near-field of the target - normally aircraft - thus the distance between target and sensor is about maximum one order of magnitude greater than the target dimension while the radar operating wavelength is about a hundredth off.

Thus the normally used radar equation is no longer valid, the radar does not see a single fluctuating point target, but a number of reflection points whose amplitudes and locations are changing during the engagement, summing up or interfering to form a really complex signal.

On the other hand, these signals are needed first to optimize the system as well as to determine the effectiveness of the entire system and this has to be done for the whole set of possible encounter geometries.

On principle there are three solutions to this problem

- theoretical computation
- use of original targets in real flight tests
- use of down-scaled model measurements.

The first solution, the theoretical computation of these signals, leads to extreme difficulties although the different possible approaches - wire grid model, plate model and physical optics - are already approximations.

It is, nevertheless, worth to note that a very simple geometric optics-model leads in some special cases to satisfactory results. In this model, the complex target is represented by a few - typically 3 to 10 - point sources but the amplitudes and exact locations of these point sources have to be determined in general first by measurements.

The use of original targets in real flight tests is a possible solution. Figure 1 shows such a flight test where the radar fuze is on top of the mast and the marks on the left side are used for measuring the flight geometry.

Despite of the fact that such experiments may be dangerous, especially for small misdistances, they are of course very expensive and extremely difficult to control. Therefore they can at the most be used as limited, selected basic data but not as the required comprehensive data set for all interesting encounter geometries.

Therefore only the third solution, model measurement, remains valid and has in fact given a satisfactory solution for our problem.

2   Model Simulation

The possibility to use down-scaled model simulation is given by the linearity of Maxwell's equations. If one excludes non-linear media as ferromagnetics or ionized media with magnetic fields, the valid transformation equations are given in Figure 2. Here the index "O" stands for "original" and "m" for "model", respectively.

There are two basic scale factors

a) one of which governs the basic electromagnetic properties as well as target dimension

b) while the other one takes into account the scaling of time dependent properties.

Since the interesting targets are metallic ones and provided that the conductivity of model metallisation is high enough - the received target signal is only dependent on the ratio of target dimension to radar wavelength. Furthermore, the signal amplitude is independent of the relative engagement velocity which only determines the frequency speed of the doppler signal. The original signal can then be easily obtained by time transformation using different recording and reproducing scale factors.

3 Experimental set-up

In our model simulations at AEG-TELEFUNKEN in Ulm, we use radar systems with frequencies up to 90 GHz, usually with emphasis at 33 GHz and 86 GHz, corresponding to scaling factors of about 10 - 30. The radar sensors used are normal CW systems with the block diagram shown in Figure 3.

The transmitted signal provided by a gun oscillator at the lower and an impatt-oscillator at the higher frequency is fed via an antenna duplexer to the antenna. The received signal is mixed with a part of the transmitter signal supplied via a coupler or directly via the reverse circulator path. The resulting video signal is then amplified and recorded.

The RF output power of these radars is of about 100 mW at the lower frequencies down to 20 mW at the higher frequencies. The antennas used in the model simulation should show the same gain and beamwidth as the original ones, a requirement which may lead to problems which will be discussed later.

The next Figure 4 shows our simulation facility. The whole measuring set-up is housed in an anechoic chamber with the dimensions 15 by 4 meters. The target model, here a MIG 21 (a Fishbed), is mounted on top of a plastic rod which itself is mounted on a movable platform. This platform is driven on rails past the fixed radar at a constant speed of about 0.5 m/s.

Platform and rails are under microwave absorber plates which are tilted to avoid multipath problems. This measurement method, fixed radar and moving target, is in contrast to methods used in similar facilities in the UK and France where the radar is moved.

Therefore some comments why we proceed in this manner:

When moving our CW radars which do not have range resolution and low antenna gains the chamber walls, even when covered with best absorber material, would give high radar returns at target doppler frequencies. In our set-up only the moved platform and the target mounting can introduce error signals but this can be overcome by proper construction and the use of absorber material.

The recording set-up is shown in Figure 5. Instead of recording the measured signals with analog tape recorders we now use a 32 k, 8 bit transient recorder for intermediate storage. This has the advantage that the recording as well as the reproducing of the signal can be done at any desired speed.

For immediate inspection a display is used and last not least the signals are recorded in digital form on normal commercial tape cassettes.

Further data processing including fuze logics are up to now introduced off-range but we are just integrating a small computer in the system which, as we hope, will at least do a part of this on-line.

4 Results

Figure 6 shows - as an example - the measured signal for a sphere with a radius of 10 centimeters which is normally used for calibration purposes.

The four traces show the signal recorded with for different gain factors at the top distance marks are shown. The decrease of the doppler frequency when approaching the point of nearest distance as well as the jump in phase at this point is easily seen.

The next Figure 7 shows the signal of the real aircraft target, here a FIAT G 91, in a nearly head-on geometry. This picture shows clearly the "interference packets", typical for these complex targets as well as a general feature, that is decreasing frequency with decreasing distance, valid for the point target.

This fact is in detail shown in Figure 8 and Figure 9 where for this signal the power spectral densities are given for different time intervals. Time interval a) shows the small peak given by the relative target velocity while interval c) shows the broad low frequency spectrum when the sensor is near the part of closest approach.

5 Modelling problems

The models of aircraft, missiles and helicopters are made of wood which is then silver sprayed sufficiently to give about 5 skin depth at the corresponding model frequency. For the higher scale factors we also use commercial plastic model kits with good results. These models have the advantage of being light that means giving no mounting problems as well as become very cheap ones in contrast to the wood modules.

The choice of a scale factor as high as possible has the advantage that this leads to small dimensions of the targets as well as to small dimensions of the whole measuring set-up. Especially from a practical commercial point of view the last point is a very important one. Also the corresponding higher frequencies lead, for given relative velocities, to higher doppler frequencies. This means in general higher signal to noise ratio by avoiding frequency noise problems such as 50 Hz power lines, amplifier drift and so on as well as lower requirement on spectral purity of the RF oscillator.

Nevertheless there are some practical limits for the scale factor. First the lower sensor sensitivities at the higher frequencies at least for a given cost level. Also target dimension may reach a level where the modelling of the target details will be a problem.

Therefore we feel that scale factors in the order of 10 to 30 will be an optimum choice.

A special problem arises from the fact that at least in our case, in the original systems, the antennas are smaller than the wavelength. Therefore it is normally not possible to scale down these antennas and therefore new antenna design at the model frequency is necessary.

The next Figure 10 shows, for example, a low gain 90 GHz antenna which was developed in our central millimeter laboratory by Dr. Rembold an his staff.

6 Conclusion

Millimeter wave systems have been very successfully used for model simulation measurements to determine the near-field characteristics of complex targets.

These data are used for system optimisation, for the proof of system effectivity and are also used now for quality control in the production of the systems.

Fig.1 Fly-over trials

Length	$l_m = \dfrac{1}{q} l_o$
Frequency	$f_m = q f_o$
Conductivity	$\sigma_m = q \sigma_o$
Cross Section	$\sigma_{Rm} = \dfrac{1}{q^2} \sigma_{Ro}$
Antenna gain	$g_m = g_o$
Velocity	$v_m = \dfrac{1}{k} v_o$
Doppler frequency	$f_{Dm} = \dfrac{q}{k} f_{Do}$
Reprod. factor	$p = \dfrac{k}{q}$

Fig.2 Scaling of parameters

Millimeter Wave Radars

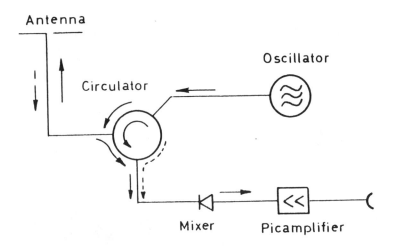

Fig.3 Blockdiagram of radar sensors

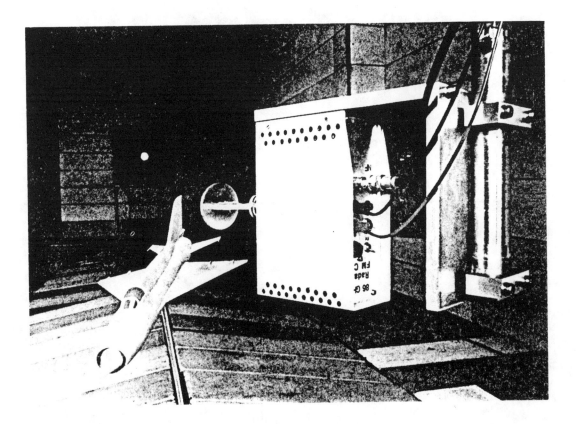

Fig.4 Simulation facility

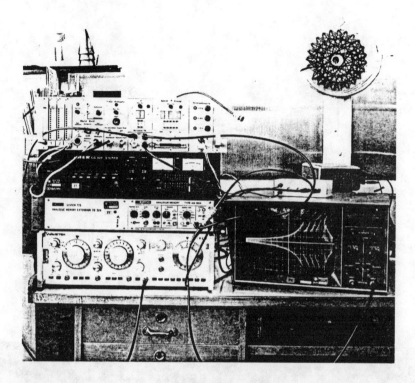

Fig.5 Recording set-up

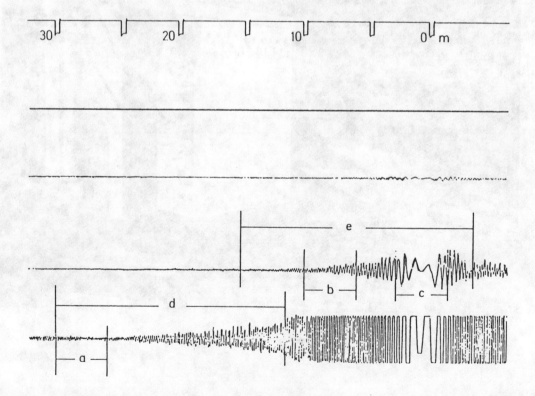

Fig.6 Signal of a sphere

Millimeter Wave Radars

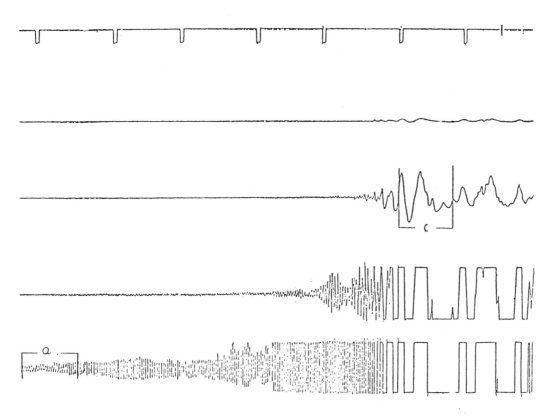

Fig.7 Signal of an aircraft

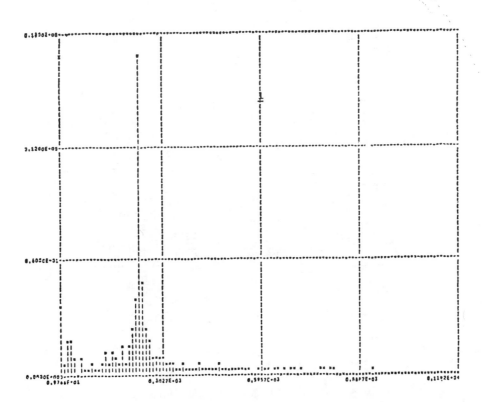

Fig.8 Spectrum of Figure 7. Time interval a

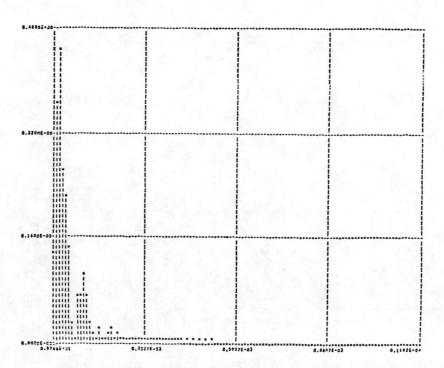

Fig.9   Spectrum of Figure 7. Time interval c

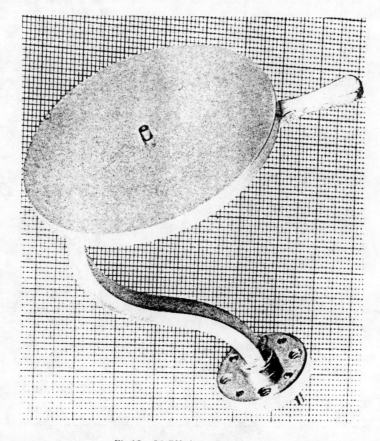

Fig.10   86 GHz low gain antenna

DISCUSSION

M.Carter, UK
    How good is the Radar absorptive material used in the anechoic chamber.

Author's Reply
    Approximately 5–10%.

HOWLS Radar Development

V.L. Lynn

SUMMARY

The use of airborne radars for locating and classifying hostile artillery is being investigated. At shorter operating ranges, approximately 2-10 km, advances in component and antenna technology have made possible consideration of radars with substantial performance and with weights and costs suitable for use in a mini-RPV. The selection of radar parameters for this application is discussed.

The principal uncertainty in such a radar is the feasibility and performance of techniques for detecting and classifying stationary targets. To pursue the development of these techniques, an experimental radar system has been constructed. The parameters are consistent with those which could be implemented in a mini-RPV although the system will be installed in a manned aircraft for data collection and experiments. The techniques being considered, the experimental system, and the development program are described along with preliminary results.

1. INTRODUCTION

The joint DARPA-Army Hostile Weapons Location System (HOWLS) Program was initiated to develop new or improved means for locating and identifying enemy indirect-fire weapons. One all-weather approach which is being investigated is the use of airborne radar at ranges of 2-50 km.[1] At least at the shorter ranges, advances in component and antenna technology have made possible the consideration of radars with substantial performance and with weights and costs suitable for use in a mini-RPV.

Early in the program, design studies were undertaken to assess the applicability of airborne radar to the location of weapons, to examine alternate implementations, and to define the developments required. General Electric and Hughes Aircraft participated in parallel efforts of four months duration. The principal conclusions which could be drawn from these studies were as follows:

(a) The utility of airborne radars for this purpose depends on the development of techniques for the detection and classification of stationary targets.

(b) Implementation of a short-range radar for this purpose in a mini-RPV appears generally to be straightforward within the state-of-the-art. The most significant area where component development is required is that of lightweight, low-cost antennas.

(c) A mini-RPV radar was designed for detection and identification of artillery targets (10 m^2) in clutter at ranges of 2-3 km. The airborne portion of this system was estimated to weigh 30 pounds and to cost approximately $20,000 (FY 74 dollars) in quantity production, exclusive of airframe, flight controls, and data links.

The dominant need, then, is the development and understanding of target detection and classification techniques. From an examination of the existing data base, it was clear that an extensive measurement program would have to be undertaken to support the development and, later, the testing of such techniques. Existing facilities and radars were investigated to satisfy these needs and several were utilized in specific measurements. However, none could fulfill the overall requirements, and the construction of an experimental system was initiated in early 1975. The principal organizations in this effort are the M.I.T. Lincoln Laboratory and the General Electric Company.

This experimental system has been constructed, integrated, and utilized in preliminary data collection, both from ground sites and in airborne operation. The data processing capabilities of the system have been used to examine the early data from this radar and also to

---

*AVP/GCP Joint Symposium - Avionics/Guidance and Control for RPV's - Florence, Italy 4-8 October 1976.

process data available from a number of other sources. Although the experimental effort is just getting under way, some preliminary results are available.

The following sections describe the program in more detail. Some possible applications and implementations are discussed to provide an understanding of the principal parametric tradeoffs which were considered. The techniques that are under investigation are outlined along with some of the preliminary results. Finally, the critical issues and consequent experimental requirements are summarized, and the system which has been constructed to fulfill these requirements is described.

2. RPV APPLICATIONS AND SYSTEM PARAMETERS

The focus of the HOWLS Program is on detection and classification of indirect-fire weapons. Although it is unlikely that a radar would be deployed to deal only with these targets, even though they represent an important problem, it is useful to examine the applications and the performance-implementation tradeoffs in this constrained area. In actual practice, the radar capabilities required to solve this problem will be sufficient as well to provide a new and powerful capability for general battlefield surveillance and target acquisition.

Airborne radars offer the potential for surveillance of large areas in a short period and for all-weather target detection and classification. Some of the airborne radar capabilities and observations which might be considered in surveillance of weapons are as follows:

Fixed target detection or target enhancement describes a basic capability for processing the complete radar returns to identify a limited number of resolution cells which contain potential targets. Several techniques have been proposed or tried, but the capability is generally embryonic and achievable performance cannot be quantified.

Target identification or classification depends on observation and understanding of the signatures of both targets of interest and others. In the case of indirect-fire weapons the signatures which appear to be most unique are those associated with the actual firing.

Context mapping: A moderate resolution ground map can provide an important augmenting capability for identification and imposes few additional requirements on the radar beyond those for fixed target detection. For example, artillery is emplaced according to fairly well-defined, operationally-driven rules, and therefore, the combination of fixed target detection and a mapping of the terrain can provide a substantial basis for deductive identification.

Moving Target Indication (MTI) is an important and well-understood technique. In general, given a radar to perform the above functions, inclusion of MTI represents only a small additional step.

One can view the use of such capabilities in two ways. The first represents a two-stage process in which all fixed, man-made targets are located, and then the tools of identification and contextual mapping are applied to classify the targets. In the second, the initial detection and identification are simultaneous, based on observation of a unique signature, and then, fixed target enhancement and MTI are required only in a limited area to provide continuous track of this detected target. This second case may place much less stringent requirements on the fixed target enhancement and simplify the implementations.

Figure 1 illustrates some of the currently available mechanisms for surveillance of hostile weapons. Where, then, can an airborne radar contribute to the solution? It can certainly provide good detection and identification of the firing weapon but is best viewed in this role as a highly mobile extension or backup of the ground-based sensors which also can do this. The capabilities for medium-range MTI and for all-weather location of fixed, non-firing weapons are important, and these are areas where improved airborne radar can contribute significantly. However, the "fire and move" targets appear to represent a unique role for airborne radars because of the ability to observe both fixed and moving targets at some distance in all weather. This condition is particularly well matched to the potential radar capabilities and is difficult or impossible to handle adequately with other sensors. It is clearly a case of increasing importance on the modern battlefield.

Millimeter Wave Radars

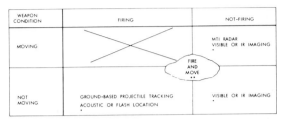

Fig. 1. Surveillance of hostile indirect-fire weapons.

A simple case of a multiple rocket launcher illustrates the point. The weapon is easily located and identified by its firing signature but may quickly leave this position before counterfire can be brought to bear. As the launcher proceeds to a rendezvous with its resupply vehicles, it can be tracked, whether moving or fixed at any given time, by the same radar which originally observed the firing. The identification and track can be maintained until the target is brought under fire.

The essential but undeveloped elements of performance in these applications are the fixed target detection and target identification capabilities. Nonetheless, recognizing these uncertainties, it is reasonable to assume some likely performance values and to take a first-order look at some of the important system tradeoffs. This paper will examine only mini-RPV implementations although the HOWLS Program is equally directed at radars for larger platforms.

The first question which generally comes to mind is the frequency of the radar. For RPV implementations in these kinds of applications, it is likely to be chosen as high as possible to minimize the antenna beamwidth and size while remaining consistent with the limits of component technology and tolerable propagation attenuation. Systems at 10, 16, 35 and 94 GHz might be considered as spanning the range of interest.

Fig. 2a. Aquila RPV with electronic scan array.

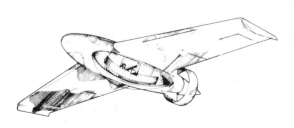

Fig. 2b. Aquila RPV with mechanical scan dish.

Particularly in the case of a mini-RPV, the antenna form is a critical variable. Figure 2 is an artist's sketch of the two principal forms of antenna as they might be implemented in the U. S. Army Aquila RPV. Factors which must be considered include the following:

-- A phased array can generally utilize a larger aperture since it does not physically move; this directly relates to increased signal-to-clutter ratio.

-- The additional gain of the larger array is offset by increased losses incurred in the corporate feed structure and the phase shifters. For example, at Ku-band, the use of a 2-meter array would introduce a net loss of sensitivity of about 4 dB over a 1-meter dish; this is based on existing componentry, and losses certainly can be reduced somewhat.

-- A dish will probably have a lower cost and can provide 360° coverage if this is needed, but may introduce problems as a result of mechanical scan motions.

-- The array provides an agile beam which may be quite important in the types of multimode system needed and also provides an inherent growth to incorporate jammer nulling if this proves useful.

-- Current technology will not support a phased array at the highest frequencies.

The choice between these forms depends on the operating frequency and the intended application, but, in the general surveillance case, the array probably has the edge.

One other important antenna form deserves mention but will not be further considered in this paper. This is the frequency-scanned array which has the potential virtues of very low cost and very light weight with losses lower than the phased array. A Ku-band, frequency-scanned array could be constructed for instantaneous signal bandwidths of less than approximately 25 MHz.[2] Whether such an antenna is practical in these applications depends on the requirements for signal bandwidth which, in turn, are driven by the needs of the signal processing techniques to be employed.

For mini-RPV radars, the transmitter power represents a real constraint on performance. On the Aquila, for example, approximately 500 watts of prime power are available, and it seems reasonable to allocate about half to the transmitter. If 20% overall transmitter efficiency can be achieved, this limits the RF output power to the order of 50 watts. The choice of device is further restricted by the requirements that it also be light, say less than 5 pounds, and air cooled. Up to 20 GHz, TWT's can be found which approximate these requirements. With some development, a TWT might be available at 35 GHz, but at 94 GHz it appears necessary to use low-duty cycle magnetrons. Based on the same restrictions, it might be reasonable to anticipate output powers of 5 w and 2 w respectively for 35 and 94 GHz.

To provide insight into the parameter tradeoffs, it is useful to define some systems which might be representative in these applications and to examine the ranges at which each might achieve various capabilities. Table 1 describes four such representative radars and conditions. These systems are dominated by current technology and fairly stringent weather. This tends to bias the performance in favor of lower frequencies, so, to test the sensitivity of these assumptions, it is reasonable to look also at transmitters which provide the same power levels for all frequencies and at clear weather propagation.

Table 1
Representative Radars and Conditions

Frequency (GHz)	10	16	35	94
Antenna	array	array	dish	dish
Azimuth dimension (m)	2	2	1	1
Transmitter average power (watts)	50	50	5	2
Receiver				
Bandwidth (MHz)	50	50	50	50
Noise Temperature (°K)	750	1200	1900	3000
System losses; two-way (dB)	18	21	11	11
Propagation loss; two-way (dB/km)				
2 mm/hr. rain (99% weather)	0.04	0.2	1.4	3.7

Elevation beam	Fixed; 20° CSC^2
Azimuth scan rate	20°/sec
Clutter	Moderate vegetation; σ_o = -16 dB

Table 2 presents some performance results for these conditions. In drawing conclusions, it is important to bear in mind the limitations of this comparison. The clutter and signature models and fixed target enhancement performance are not well quantified and may be significantly in error. The calculations are based on a constant-angle search although, in practice, either a fixed beam or a constant-area search might be more applicable in some modes; in a relative comparison, these tend to favor the lower frequencies.

Some of the more general conclusions that can be drawn from Table 2, recognizing these limitations and that many system variations are straightforward, are as follows:

(a) The maximum ranges of millimeter systems (35 and 94 GHz) will be of the order of 5-10 km. This pretty much requires that they penetrate hostile territory but is adequate to permit operation at an altitude of 3 km which will minimize data link masking problems and proximity to smaller ground-based air defenses. At these ranges, the millimeter radars:

--can probably achieve fixed target detection at or near maximum range and have no need for synthetic aperture techniques.

--could achieve useful ranges for target signatures as low as about -10 dBsm.

(b) The maximum ranges of X and Ku-band systems can be 30 km or greater and these radars can perform in either a standoff or penetration mode. In a standoff role at ranges of 20 km, these radars:

--can perform fixed target detection only if modest (approximately 20:1) doppler beam sharpening is feasible in low-cost, mini-RPV implementation.

--can probably observe target signatures as low as -10 dBsm.

However, the same radars in a penetration role:

--can probably perform fixed target detection with real aperture techniques.

--can observe target signatures as low as about -30 dBsm.

(c) There should be no difficulties in implementing MTI for the maximum range of any of these systems.

Table 2
Performance for Various Capabilities

Capability	10 GHz	16 GHz	35 GHz	95 GHz
Maximum range for S/N > 12 dB on 10 m^2 target (km) Representative system and conditions	58	31	11	5
Variations:				
Clear weather (Table 1 power)	64	42	23	12
50 watts (Table 1 propagation)	58	31	15	7
Clutter suppression to achieve S/C > 12 dB at maximum range (dB)	20	15	11	2
Fixed Target Detection maximum range (km)				
Real-aperture; S/C > 10 dB after 6 dB enhancement	4	6	6	(18)*
Doppler beam sharpening (20:1); S/C > 10 dB without enhancement	18	30	(30)*	(90)*
Signature observations; maximum range for S/N > 12 dB (km)				
Cross section of 0 dBsm	30	20	7	4
Cross section of -10 dBsm	20	12	6	3
Cross section of -30 dBsm	7	4	2	1

*S/N Limited

The overall choice of system parameters is not entirely clear because the uncertainties tend to dominate the conclusions. Nonetheless, a flexible mixed-mode system which can provide substantial capabilities in both standoff and penetrating roles has much to offer, and this also favors the lower frequencies.

The HOWLS development work is being undertaken at Ku-band on the basis of a rationale similar to that outlined above. The choice of frequency at this stage of the development, however, does not appear to be at all critical. The results of the processing technique development work should readily extrapolate to any of the bands of interest with a minimum of additional uncertainty and, to the extent practicable, effort has been undertaken to obtain a data base over the entire band.

3. SIGNAL PROCESSING TECHNIQUES

The dominating uncertainty in the use of airborne radar for these applications is the development of techniques for fixed target detection and for target classification. There is evidence to indicate that both are possible, but there is no quantitative understanding of the performance and limitations. Table 3 summarizes the techniques which are currently under consideration. As indicated earlier, MTI and Context Mapping are considered to be well understood and straightforward to implement, and these will not be discussed further.

Table 3
Signal Processing Techniques

Target Detection
    Fixed Targets
        Clutter decorrelation (frequency or polarization agility)
        Target extent estimation (range, angle)
        Reduced resolution element (doppler beam sharpening)
    Moving Targets
        MTI
Target Classification
    Contextual mapping (target grouping or flow)
    Projectile detection/tracking
    Spectral signatures
    RCS analysis

Fixed Target Detection

Fixed target detection, the separation of man-made targets from the natural background clutter, relies on the statistical scattering differences of these classes of returns. Natural clutter is generally characterized by multiple scatterers, randomly distributed and oriented within a resolution cell while man-made targets usually have regular geometric properties.

Adaptive thresholding and Constant False Alarm Rate (CFAR) techniques are basic to any automatic target detection processes. Modified versions of conventional CFAR detectors[3,4] have been implemented and tested against data from the FLAMR radar. These estimate the clutter power in each cell by examining the surrounding cells and setting the threshold based on the estimated average clutter power thus obtained.

In a clutter-limited situation, the target must be enhanced by exploiting the statistical differences between targets and clutter before this adaptive thresholding can be applied. Two general approaches can be considered which differ primarily in perspective--decorrelation of the clutter with appropriate radar signal structure and estimation of target extent.

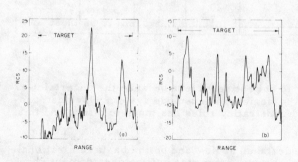

Fig. 3. High resolution data.

Consider frequency agility as one radar signal structure mechanism for clutter decorrelation. The performance depends on the relative decorrelation of the target and the clutter as the frequency is varied. Figure 3 shows high resolution data taken on two different aspects of a typical target. Figure 3a shows a case in which a single scatterer dominates the target return. In this instance, as the radar frequency is varied to decorrelate the clutter, the target would be expected to remain coherent and hence integrate, over a wide band. Figure 3b on the other hand, shows another case, that in which the target is represented by a multiplicity of more or less equal scatterers which can reinforce or cancel each other, as the relative phase varies. These target returns would be expected to decorrelate as the frequency is shifted over a band roughly commensurate with the target length. Based on the limited data examined to date, it appears that Figure 3b represents the usual case and Figure 3a the exception.

Data from narrow band, multifrequency measurements have been used to examine these frequency agile effects. Figure 4 indicates some of the results for a typical target and aspect.

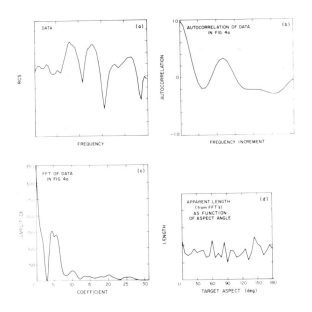

Fig. 4. Multifrequency data; typical target.

Figure 4a shows the cross section as a function of frequency. Figure 4b, the autocorrelation of these data, indicates that the target completely decorrelates in a frequency shift which is consistent with the target length. Another similar view is given in Figure 4c, the FFT of the data of Figure 4a; this plot can be scaled in apparent target range extent, and, if a threshold is established to estimate this target coherence length, the result for a range of aspects is given in Figure 4d where the values are again consistent with the actual target dimensions. It is apparent, at least for the cases examined, that bandwidths significantly greater than that appropriate to the target dimensions, will not further enhance the average signal-to-clutter ratio.

Clutter decorrelation techniques will likely rely principally on variable frequency and polarization observations. Time and aspect diversity might be considered but probably impose such stringent requirements on the system as to be impractical as the primary mechanism. They are, however, of interest as corollary techniques in, for example, frame-to-frame processing of filtered images.

Since the detection problem is dominated by the signal-to-clutter ratio, techniques in the azimuth domain are also important. The most obvious example being pursued is doppler beam sharpening which can provide azimuthal resolutions approximating the target dimensions. In a manner analagous to that in range, monopulse techniques might be considered for estimation of the target extent in angle.

Optimum fixed target detection can be achieved, then, by matching in one way or another, the radar signal in all domains (range, angle, doppler, polarization) to the characteristics of the targets and adaptively comparing each cell with its neighborhood.

Target Classification

The techniques which appear to be of most interest for identification of hostile weapons are indicated in Table 3. The most powerful classification mechanisms seem to be those associated with the firing weapon signatures.

The detection of the projectile is obvious and well understood. Artillery projectiles have rather small cross sections, and therein lies the limitation in this technique. It should be noted, however, that unlike conventional ground-based projectile tracking radars which require high signal-to-noise to establish a precise track for location, the airborne radar need only provide a crude history of the projectile sufficient to correlate it with a particular fixed ground target.

Spectral signatures may be the most useful but are certainly the least understood in any degree of detail. In the case of the firing weapon, these may include any of the blast-induced motions such as the gun recoil, acoustic shock effects or effluent discharge.

These are the more promising techniques which are being examined in detail. Others such as context mapping, cross-section analysis and associative sensors (e.g., acoustic, flash) are under investigation as well.

4. EXPERIMENTAL REQUIREMENTS

The critical issue, then, in the development of airborne radar for these applications is the development and understanding of the signal processing techniques. The primary experimental requirements are therefore dictated by the need for a high-quality data base with sufficient breadth to allow examination of all potential approaches.

Initially, the raw radar data must be recorded in a form suitable for use on a large, general-purpose computer so that extensive processing can be brought to bear without undue concerns for computational efficiencies and limitations and with the full power of a large-machine system aiding the developers. These data will be examined in detail and then utilized to determine the relative effectiveness of techniques and variations in non-real-time operation. At this stage, the airborne radar receiver will act only as a front end with essentially all the radar signal and data processing accomplished by the off-line computer.

As an understanding of the techniques develops, the algorithms which evolve off-line will be implemented in a real-time form. This is required, not only to provide the practicalities, constraints, and approximations of real-time versions, but, more importantly, to allow extension of the techniques to incorporate interactive or adaptive use of the radar.

In the later stages of development, specific algorithms or parts of algorithms will be incorporated in special-purpose hardware implementation as a part of the airborne system. Clearly, in any operational system, the maximum of processing consistent with performance must be accomplished on board the airborne platform to minimize the data link bandwidths.

This describes a philosophy of development which has proven efficient and effective in similar efforts. Its implications in the requirements on the experimental system are access to the unprocessed radar data and recording of these data in digital form. The radar itself must be designed with an extreme flexibility of control and growth. The requirements for a well-calibrated system with carefully controlled analog signal path with, for example, high spectral purity, low sidelobes, and wide dynamic range, are obvious.

Some of the more important, specific requirements are as follows:

Resolution: Sufficient range and cross-range (azimuth) resolutions are required for contextual mapping and for fixed target detection. These require resolutions on the order of 3 to 15 m in each dimension with fixed target detection probably the more demanding case.

Real-Aperture: The requisite azimuthal resolution could be achieved with real apertures at relatively short ranges or with synthetic aperture techniques at longer ranges. Since the latter requires coherent processing to achieve resolution, it tends to seriously complicate or compromise coherent processing for other purposes such as target classification or fixed target detection. For this basic reason, but also because of the other limitations imposed by synthetic aperture techniques, it is clear that the experimental system should utilize real aperture for azimuthal resolution. Short ranges impose no limitations on experimental technique development and extrapolation of capabilities and implementations to longer ranges, and more complex processing is straightforward.

Coherence: A coherent system is required for spectral signature studies and for some moving-target techniques. This capability is also required to establish the practicality of doppler beam sharpening in implementations suitable for mini-RPV use.

Sensitivity: In the experimental environment, high signal-to-noise should be available to permit well-controlled experiments and analysis. For reasonable apertures, a minimum average power of about 10 w is required even for the short ranges where most tests are planned for the experimental system.

Waveform and Repetition Rates: Range and doppler resolution and ambiguities must be sufficiently flexible to accommodate the uncertainties in target characteristics and signal processing techniques. Pulse trains with a wide variety of repetition frequencies are required with pulse coding to provide compatible resolutions and energies.

Beam Agility: This capability is required to permit simultaneous and coordinated use of contextual mapping, MTI and fixed target detection and classification techniques.

Flexible combinations of these with their varying dependence on signature durations may require a pointing agility which can only be achieved with electronic steering of the antenna beam.

<u>Control Flexibility</u>: In conjunction with the antenna beam agility requirement, all radar parameters must be computer controlled to permit rapid changes, mode interleaving, and to allow later inclusion of adaptive real-time processing.

<u>Data Manipulation</u>: One of the more important requirements of the experimental system is the provision of facilities for easy editing and manipulation of large volumes of data. The importance of developing adequate tools for analysis and technique experimentation cannot be over emphasized and may represent the difference between ultimate success and just another set of experiments.

Two other important areas are not addressed above but may, in fact, represent critical requirements. These are the provisions for polarization agility and for range resolutions smaller than 3 meters. Both are of interest in detection and classification techniques. However, each represents a significant additional implementation cost for an experimental system and each can be investigated adequately, at least initially, with other existing facilities. Therefore, it was believed more appropriate to require only that the experimental system be designed and constructed to allow easy incorporation of these capabilities at a later time.

5. EXPERIMENTAL SYSTEM

The parameters and architecture of the experimental system reflect the philosophy and requirements outlined above. It has been designed as a powerful instrument for measurement and analysis and for off-line technique development which can evolve to a real-time implementation and eventually a demonstration.

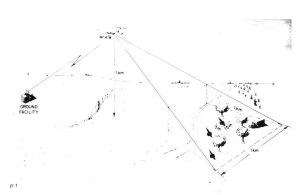

Fig. 5. Experimental geometry.

Figure 5 indicates the approximate geometry which will be used for much of the experimental work. At these shorter ranges of 2-3 km, the real-aperture beam of less than a half degree will provide as much azimuthal resolution as necessary, and the experiments are kept as simple as possible.

The overall experimental system architecture is depicted in Figure 6. Physically, this is a three-part system. The airborne radar was built by General Electric. The van-mounted Ground Support Facility was designed by Lincoln Laboratory and centers on a Modcomp IV-25 computer system. The Data Processing System was constructed by Lincoln in conjunction with the Laboratory's IBM 370/168 Computer Facility.

Initially, raw radar data, in the form of complex digital samples, are telemetered to the Ground Facility where they are recorded on high-density, computer-compatible tapes. These tapes are then processed on the Data Processing System. The parameters of the radar can be controlled by the Ground Facility Computer or, alternatively, by a local control panel in the aircraft. As algorithms are developed and tested, there are provisions in the Ground Facility for real-time implementation and test and for adaptive control of the radar by the algorithms.

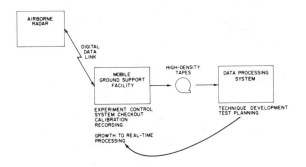

Fig. 6. Experimental radar system architecture.

The parameters of the experimental radar are summarized in Table 4. These parameters are consistent with those which could be implemented in a mini-RPV, but no effort has been made to reduce the weight and size to approach such an implementation. In the experimental system, the radar is installed on a manned, light aircraft.

Table 4
Experimental Radar Parameters

Frequency	16.0 to 16.5 GHz
Antenna gain (two way at antenna flange)	56 dB (+ 30° scan)
Transmitter	TWT 1.0 kw peak, 20 w avg.
Losses (excluding antenna)	7 dB
Noise figure	6 dB
Single-pulse S/N (1 m^2 target at 2.3 km)	+13 dB
Frequency agility band	500 MHz in 64 steps
Azimuth resolution (2.5 m antenna)	7 mrad
Range resolution modes	3, 5, 10, 75 m
Projectile detection (0.01 m^2 target)	2.5 km
Dual polarization	Not initially
Azimuth monopulse	Not initially
Overlapping azimuth beams	Up to 8
Beam agility	256 μsec repoint time
Airborne A/D converters	8 bits I and Q at 15 MHz or 6 bits I and Q at 60 MHz
Airborne processing	PRF buffering only
Maximum PRF	40 KHz
Maximum two-way sidelobes near	-40 dB
far	-50 dB

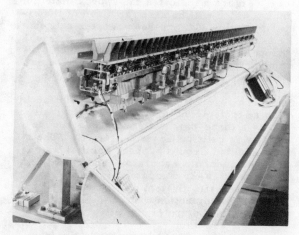

Fig. 7. Antenna in pod.

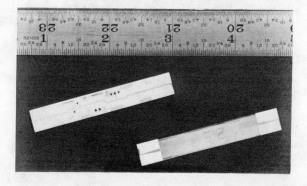

Fig. 8. Phase shifter substrate.

The antenna developed by General Electric for this radar is unique. It was designed to be amenable to very lightweight construction, although again, in the interests of economy and simplicity for a one-of-a-kind system, more conventional construction was used with no special regard for actual weight. The 2.5 meter-long, linear array is shown in Figure 7 as it is partially assembled in a hinged, cylindrical, protective pod for installation on the test aircraft. A transmissive window covered with fiberglass radome material is located in front of the horn flare but is not visible in the picture. Each of the 208 radiating elements is constructed on a microstrip substrate measuring 1 x 6 cm which includes a 3-bit diode phase shifter, the radiating element, and a probe which is inserted into the feed waveguide.[5] A close-up photograph of the substrate is shown in Figure 8. The elements are arranged such that one end projects into the horn flare radiating aperture, and the other couples into the waveguide of the corporate feed structure. The elements and phase shifters in the antenna of this experimental system are relatively lossy, having an average one-way loss of 3.2 dB, due to the microstrip construction which was chosen in a tradeoff between loss and cost. This loss is expected to be reduced to approximately 2 dB in future versions.

Figure 9 is a block diagram of this airborne radar. It is configured around a microprogrammed computer which provides both control and a capability for formatting the

Millimeter Wave Radars

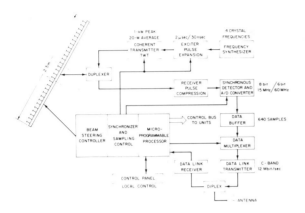

Fig. 9. Airborne radar block diagram.

radar data and auxiliary information in a general message structure so as to simplify subsequent data retrieval and processing. Radar control commands are distributed to the various subsystems over a control bus using double rank buffering in each subsystem so that transfer to a new operating state can be accomplished simultaneously throughout the radar on the occurrence of an execute command. This permits controlled transitions to be made from one operating mode to another so that mode interleaving can be employed. For example, an MTI mode for detection of artillery projectiles can be interleaved with a fixed target detection operation so that the association of the backward extrapolation of a crude track of the projectile with a fixed target can provide identification of that target as a gun.

Both pulse compression and unmodulated pulse waveforms are available with pulse lengths of 20 nsec to 1.5 μsec and with repetition rates (PRF) variable from 60-40,000 Hz. The frequency can be changed from pulse to pulse within a 500 MHz band divided into 8 MHz steps. For coherent modes, however, only four stable frequencies are provided in the initial configuration.

The data link PRF buffering circuitry is designed to operate at 60 MHz. The hardware is configured to accommodate various A/D converter rates and number of bits up to 8 bits at 60 MHz when such units become available. Present plans are to use 6-bit, 60-MHz A/D's for the highest resolution modes and 8-bit, 15-MHz A/D's for applications requiring larger dynamic range. Bandwidth-limiting filters are provided to match the system bandwidth to that of the A/D converter. The range swath which can be sampled, telemetered, and recorded for each pulse repetition interval obviously depends on many factors. However, as an example, 8-bit complex samples with 5 m resolution can be recorded for a range swath of approximately 1 km at PRF's as high as 3000 pps, providing an ambiguous range of 40 km and ambiguous velocity of 100 km/hr.

The airborne radar has been constructed so it is easily removed from the aircraft as two units--a pallet of electronic equipment and the antenna pod which is readily detached. A good bit of the early data collection will be accomplished from ground environments which can be more controlled, and this facilitates the transition between airborne and ground-based experimentation.

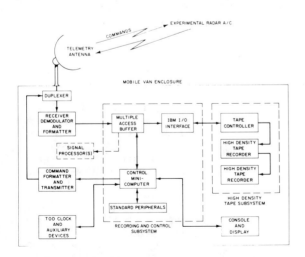

Fig. 10. Ground Support Facility block diagram.

The Ground Support Facility block diagram is shown in Figure 10 as it is currently installed in a large van. The principal functions of this van are to provide the following:

(a) Data recording.

(b) Experiment and radar control.

(c) Complete system checkout and calibration.

(d) Limited real-time processing and displays necessary to control functions.

(e) Growth to full, real-time processing and adaptive, closed-loop control of the radar.

(f) Limited off-line processing and editing of data to verify performance.

The functions of this facility center around a Modcomp IV-25 minicomputer with 4-port, multiple-access buffer memory. The memory is configured to simultaneously accept telemetered data, output data to the high-density tape recorders, and exchange data with a special-purpose, real-time processor to be added later, each at approximately 10 Mbps rates. The fourth port is used for communication with the computer system controller and displays.

Because of the processing speeds involved, it is impractical to consider doing extensive real-time processing with a small general-purpose computer; the bulk of the real-time operation in later stages of development will be accomplished by a specially configured processor under control of the minicomputer.

A recording rate on the order of 10 Mbps is adequate to meet the requirements for the various experimental situations. To obviate the need for a transcription capability and facility, 6250 bpi, computer-compatible tape recorders were selected over the alternative of continuously running, multi-channel instrumentation recorders. Data rates approaching 10 Mbps can be realized when processing large records; the maximum data rate is 9.4 Mbps for the 32 K byte records which are used.

The third element of the Experimental System is the Data Processing System. This serves as a powerful tool for analysis of data from the HOWLS Airborne Radar and from other sources and for the development and evaluation of algorithms. The intent of this off-line facility is not to implement a specific set of algorithms but rather to provide a modular system where individual processing techniques can easily be installed or modified as dictated by experimental requirements. With the exception of specific display subsystem functions, all processing is accomplished with software using the IBM 370/168 computer at Lincoln Laboratory.

The functional operations of the software system are depicted in Figure 11. These functions have been implemented as a series of utility routines under the flexible control of an interactive executive program in such a way that any function may easily be modified or replaced and new routines can be added.

Some examples can best convey the capabilities of this system as an analytic tool. Figure 12a displays in a PPI format the single-frequency-return, radar amplitudes over a 60° sector with ranges shown to about 4 km, taken with a single-pulse-per-beam position. These data were collected with the radar at the General Electric building looking out over an area where test targets have been set up in a controlled location at a range of 3.5-4.0 km. The linear amplitudes are shown in "grey scale" with

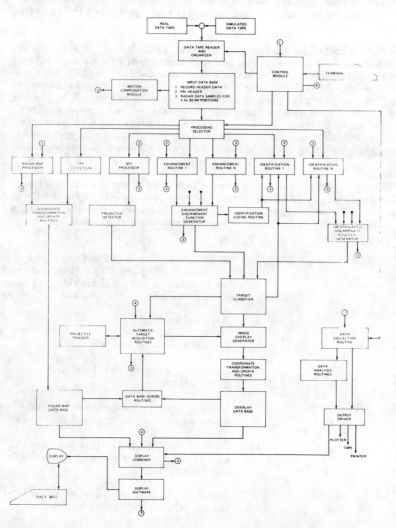

Fig. 11. HOWLS data processing software--data flow functional diagram.

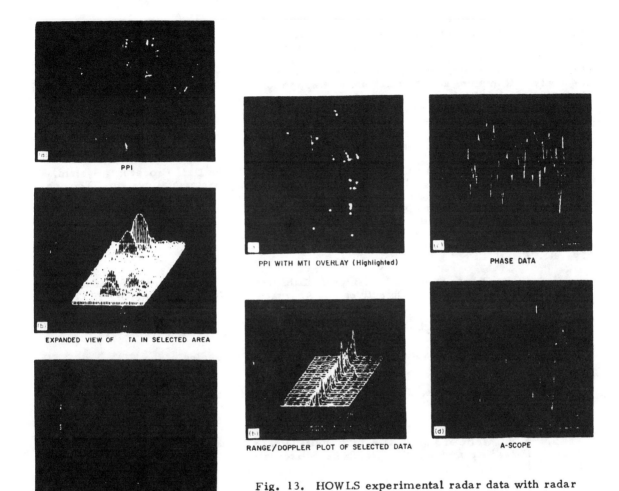

Fig. 12. HOWLS experimental radar data with radar mounted on building.

Fig. 13. HOWLS experimental radar data with radar airborne.

most of the low cross-section clutter suppressed and with all values representing cross sections greater than about 10 m² highlighted. The box was positioned and sized by the operator to encompass the test target area of interest for further analysis. Figure 12b is then an expansion of the data in that box in three-dimensional format with, in this case, the logarithm of amplitude as the vertical scale. The statistics of clutter and target returns, of particular interest in fixed target detection, are shown in Figure 12c as a histogram of the amplitudes observed within this same area. On this scale, the large targets occur in the range of amplitudes of 30-70 but are difficult to discern because of their low occurrence.

Figure 13 displays data taken with the radar airborne. Figure 13a shows the map image similar to that before but, in this case, with more of the clutter visible and with moving targets indicated by white squares. The azimuth sector is 32° and the range displayed is approximately 3 km. The MTI in this example was a simple doppler filter bank with a sliding window adaptive threshold. The operator's selection for further examination in this case is indicated by a line in range at a single azimuth position, and Figure 13b is a range/doppler plot of these data. Since the motion sensing for the aircraft has not been installed yet, there is no motion correction, and ground clutter appears slightly to the right of zero velocity. There is a DC component (zero-velocity) perceptible which results from slight imbalance in the quadrature video detectors. A target moving at about +6 m/sec. can be observed at approximately the midpoint in range. Figure 13c represents the phase history of the clutter in the selected sample and shows the phase change between pulses, for the 64 pulses integrated by the FFT, averaged for the range elements in the selected sample. Since a fixed target would be expected to be a horizontal line intersecting the vertical axis consistent with aircraft velocity, the variance of data in this plot is a measure of the limit of cancellation due to aircraft motions and the intrinsic clutter motion and is another representation of the clutter spectral width in Figure 13b. As a comparison with Figure 13b, Figure 13d shows the conventional A-scope format for the same data, where the 64 pulses have been non-coherently averaged. The range scales are the same as in Figure 13b; it is clear from a comparison that the large return at 2.75 km results from clutter which is about 30 m (three resolution elements) beyond the moving target.

In addition to illustrating the analysis aids inherent in the Data Processing System, the examples of Figures 12 and 13 convey the status of the experimental system. The construction and integration of the three main elements were completed this summer, and the photos show data as collected by the radar, telemetered to and recorded by the Ground Facility, and processed on the off-line system.

6. REFERENCES

1. D. R. Olsen, et al, M.I.T. Lincoln Laboratory, "HOWLS Airborne Radar Program: Background, Objectives and Approach", 1976, Project Report TT-10.

2. F. G. Willwerth and J. A. Weiss, M.I.T. Lincoln Laboratory, "Frequency Scan Antenna Design for RPV Radar Sensors", 1975, Project Report TT-5.

3. H. M. Finn, "Adaptive Detection in Clutter", Proceeding of the NEC, Vol. XXII, 1966.

4. H. M. Finn and R. S. Johnson, "Adaptive Detection Mode with Threshold Control as a Function of Spatially Sampled Clutter-Level Estimates", RCA Review, September 1968.

5. A. R. Wolfe and M. E. Davis, "Digital Phase Shifter Elements for a Ku-Band Phased Array Radar", 1976, MTT-S International Microwave Symposium, June 1976.

7. ACKNOWLEDGMENT

This paper describes the work of many people at the M.I.T. Lincoln Laboratory and at the General Electric Company. The significant contributors are too numerous to list. The work is sponsored primarily by the Defense Advanced Research Projects Agency and has enjoyed the guidance and support of DARPA and the Army.

Millimeter Airborne Radar Target Detection and Selection Techniques

L.M. Novak and F.W. Vote

ABSTRACT

Results of a study of candidate target detection and selection techniques for use in millimeter airborne radar systems are presented. Improved target and ground clutter models are developed and the implications of these mathematical models on target detection and selection performance is discussed.

INTRODUCTION

This paper presents the results of a study of several candidate target detection and selection techniques for use in a millimeter airborne radar system. A primary goal of these studies has been to develop algorithms which are computationally simple, provide reliable detection and acquisition performance against typical ground targets (high probability of detection, P_D, and low probability of false alarm, P_{FA}) and are compatible with the constraints of a millimeter airborne radar system. The algorithms must perform in a severe ground clutter environment, hence, must perform well in a non-homogeneous, non-stationary ground clutter background containing clutter edges, discontinuities, shadows, etc.

Figure 1 depicts one of the typical HOWLS[1] radar images used in the target detection/selection studies (note that the image display is 3-D with image intensity proportional to amplitude as shown). With reference to the Figure, several comments of importance are given. First, the scenes have been augmented with numerous corner reflectors for purposes of providing ground truth and to provide for calibration of the radar data. Thus, target and clutter cross-sections can be accurately estimated. Note also the bright forward edge between the meadow and tree areas (also, shadows exist within and immediately behind the tree regions). Some targets are positioned within the edge of the tree area and visible. This scene might be typical of a real world clutter environment which includes targets located in a non-homogeneous clutter background.

Since fixed target detection based upon target amplitude only is easily decoyed by simple corner reflectors, selection and acquisition of a target of opportunity from a set of target detections is of importance. Thus, the paper presents results of a study of one simple, easily implementable target selection algorithm.

FIXED TARGET DETECTION ISSUES

The millimeter radar target detection studies to be reported in this paper have focussed on development of fixed target detection algorithms which are computationally simple and provide reliable detection performance against typical targets located in a real ground clutter environment. The performance of numerous target detection algorithms was evaluated empirically using HOWLS radar images. It was found that detection performance was essentially dominated by several important effects including the effects of clutter discontinuities (bright clutter edges, shadows, etc.), clutter non-homogeneities, and variations of target cross-section with aspect angle as depicted in Figure 1. As a result of these initial studies, a simple baseline detection algorithm (a line CFAR algorithm) was selected for further evaluation. It was found that the effects of clutter discontinuities could be minimized by the use of lead/lag CFAR windows[2] which sense the presence of a clutter discontinuity and adapt the detection threshold accordingly to inhibit false alarms along clutter discontinuities without a significant degradation of detection probability. Typical performance results using this lead/lag CFAR approach are illustrated in Figure 2. For the HOWLS radar image shown in the figure, without the lead-lag CFAR logic implemented, detection performance was found to be P_D = 0.76 and P_{FA} = 3.4 x 10^{-3}. A number of false alarms are seen to have occurred along two typical clutter edges in the scene. With the lead-lag logic implemented it was found that P_D was essentially unchanged whereas most of the clutter discontinuity false alarms have been eliminated, resulting in the P_{FA} = 1.1 x 10^{-3}. As a result of this and similar analyses, the lead/lag CFAR window approach was found to be a simple yet viable solution to the clutter discontinuity problem and was therefore included as a part of a baseline algorithm for the remaining target detection studies. It is expected that the effects discussed will be greatly depression angle dependent. The depression angle range in the images used varied between about 20 and 50 degrees. For ground sensors (low depression angle), the discontinuity effects may be more pronounced.

The target detection studies focused next on the two other major issues, namely the effects of clutter non-homogeneities and the effects of target cross-section variation with radar viewing angle. It was found that the non-homogeneous spatial variation of ground clutter causes a reduction or loss in achievable detection performance from that predicted by the classical detection theory of Marcum[3] and Swerling[4]. It was also found that the variation in target backscatter with aspect angle significantly effects system detection performance.

The basic underlying problem is summarized in Figure 3 which shows typical data distributions obtained from several HOWLS radar images utilized in these studies. In the top figure, distributions of single frequency clutter samples and target-plus-clutter samples are shown, indicating for the given data set, standard deviations of approximately 5 dB and 6 dB for the clutter and target-plus-clutter samples, respectively. Target-to-clutter ratio was estimated to be about 6 dB. Clearly there is considerable overlap of the two distributions making it difficult to separate targets from clutter. Averaging of independent frequency samples was utilized to achieve a narrowing of the data distributions. As indicated in the bottom of Figure 3, considerable narrowing of both the clutter and target-plus-clutter distributions (i.e., variance reduction) has resulted, making it easier to separate targets from clutter. Note, however, that if an ideal variance reduction of 1/N is assumed (where N is the number of independent frequency samples averaged) then the clutter and target plus clutter distributions should have narrowed even further, as indicated in the figure. This implies the achievable detection performance will not be as good as that predicted for ideal clutter and target models. We remark that the ideal variance reduction of 1/N is based upon the assumption that the data being averaged is statistically homogeneous.

Analysis of HOWLS clutter data shows that the clutter average backscatter coefficient has an inherent spatial variation (i.e., the assumption of spatial homogeneity for clutter is not valid) and this limits the achievable reduction in clutter variance that can be achieved through frequency averaging. Narrowing of the target-plus-clutter distribution is also limited by the variation in the average target cross-section with viewing angle. In the remaining sections we shall develop improved target and clutter models based upon HOWLS data and determine the degradation in detection performance relative to that predicted by classical approaches. Also, it will be demonstrated that detection performance predictions based on these improved target and clutter models agree well with experimental results obtained using HOWLS radar images.

CLUTTER DISTRIBUTIONS AND MODELING OF GROUND CLUTTER

Since detection performance is dependent upon the clutter background in the local neighborhood of the target cell, considerable attention has been given towards development of an improved mathematical model for ground clutter. Basically we have chosen to characterize ground clutter with two probability distributions, one distribution characterizes the clutter amplitudes within a given clutter cell and the second distribution characterizes the average clutter power spatially over a given set of clutter cells.

In characterizing clutter amplitudes within a given clutter cell one has several fluctuation models to choose from. The simplest approach is to select clutter amplitudes from a Rayleigh distribution. Mathematically this corresponds to a clutter cell containing a large number of point scatterers randomly distributed within the cell.[5] A given proportion of the clutter cells could be modeled with a Ricean distribution if each cell contains a large discrete clutter scatterer plus many smaller scatterers. The Ricean distribution is a two-parameter distribution, and one must specify the ratio of DC or non fluctuating component to the AC or fluctuating component, as well as an average total clutter power for the given clutter cell. Since this is difficult to do, we have chosen the simpler Rayleigh scintillation model for characterizing ground clutter within a given cell. A comparison with experimental data obtained from HOWLS radar images is given below.

Other researchers (Booth[6] and Goldstein[7]) have considered modeling the amplitude of ground clutter within a given cell with other two-parameter distributions. In reference [6], Booth proposed a Weibull clutter model and in reference [7], Goldstein studied both Weibull and Log-Normal clutter models and their effect on target detection performance. These models, however, are more appropriate for characterizing the mean clutter backscatter from a set of spatially distributed clutter cells rather than for characterizing amplitude variations within a given cell. Researchers at Georgia Institute of Technology[8] have also suggested that the Log-Normal distribution is a good approximation to measured ground clutter. In reference [8] a Log-Normal representation for clutter was obtained from a set of clutter measurements gathered over a large set of spatially distributed clutter cells. For our analysis studies, we have chosen to represent the spatial variation of clutter with a Log-Normal distribution since this model appears most consistent with the existing literature on clutter measurements and also agrees with clutter data obtained from the HOWLS images.

One final remark about our mathematical modeling of ground clutter is appropriate. We have not yet done a complete study of the spatial correlation properties of the HOWLS clutter data. This property has received considerable attention in the literature. The concept is simply that if a clutter cell has a given average backscatter, σ_o, one would not expect the next adjacent clutter cell to have a completely different (statistically independent) average backscatter, i.e., the next cell may have a somewhat different value of σ_o but it should be "close to" the previous cell in value (statistically correlated). This implies there is a finite spatial correlation distance or length which characterizes the clutter backscatter spatial variation. In References [9], [10], [11], a simple exponential correlation function was used to constrain the variation of clutter backscatter spatially. We remark that this parameter may also be a function of the system implementation, since, for example, if the antenna beam illuminates overlapping areas of the ground or if range is over-sampled (range gates overlap spatially) there will be spatial correlation of the average clutter backscatter.

TEST OF CLUTTER MODEL USING HOWLS DATA

From the discussion of the previous paragraphs it is reasonable to characterize ground clutter as Rayleigh within a given cell with a Log-Normal spatially distributed backscatter coefficient. To test this assumption, experiments were performed using a patch of real ground clutter obtained from one of the HOWLS images. The clutter patch was a fairly uniform patch selected from the meadow region of the scene depicted in Figure 1 (the size of the area was 10 range gates by 25 azimuth positions, yielding 250 clutter samples). Thirty-two single frequency images were available for this experiment, each of the images corresponds to one of the available HOWLS frequencies. The reduction in clutter standard-deviation obtained through the averaging of independent samples was determined. Clutter patches were first constructed by linearly averaging independent single frequency images as implied in Figure 4. Clutter patch images comprised of $\{1,2,3,4,\ldots,32\}$ averaged frequencies were obtained and the standard deviation of each clutter patch was computed. Note that the standard deviation versus N is normalized to the N=1 case. Figure 4 shows the results achieved using this real set of clutter data. If clutter were spatially homogeneous, one would expect the reduction in standard deviation to follow the theoretical curve shown on Figure 4 as $1/\sqrt{N}$, where N represents the number of independent samples or pulses averaged.

From the figure it is observed that for the real clutter data the reduction in standard deviation reached a lower bound equivalent to that obtained with approximately 12-15 independent pulses. This is due to the spatial variation of clutter, i.e., one may reduce the Rayleigh scintillation within any given clutter cell by averaging many independent frequency samples within that cell, however, once this is accomplished, the spatial variation of the average clutter backscatter remains. It is this clutter variation which effects the performance of the CFAR detection algorithm.

Also shown in Figure 4 is a set of curves showing reduction in standard deviation versus N for computer generated clutter data which has been mathematically modeled as described above. For any given clutter cell we model the amplitude of the radar pulse returns as having a Rayleigh distribution. Frequency diverse pulses are assumed to provide independent or decorrelated samples from the given Rayleigh distribution. The frequency separation of each transmitted pulse used from the HOWLS data was sufficient to reduce the pulse to pulse correlation to a negligible value. Finally,

32 independent Rayleigh amplitudes were generated for each of the 250 clutter cells, and each cell was then scaled to have the corresponding mean of the real clutter data. In this way the spatial correlation of the average clutter backscatter is present in the synthetically generated clutter. From the curves of Figure 4 it can be seen that the simulated clutter exhibits variance reduction properties similar to that exhibited by the real clutter data. Although it is likely that the match can be somewhat improved by including in the model a small frequency to frequency correlation between Rayleigh samples, this was not pursued since it is felt that this would be a second order effect.

The probability distribution of the mean clutter cross-section over the 250 cells is shown in Figure 5, and indicates the distribution to be approximately Log-Normal. To test the assumption of Rayleigh amplitudes within each given clutter cell, the curve of Figure 6 is given. For the given real clutter data, each clutter cell was first normalized by its cell average thereby removing the spatial variation across the clutter patch. Then the cumulative distribution of the spatially normalized single frequency data was obtained as plotted in Figure 6. The figure implies the frequency samples may be approximated quite accurately with a Rayleigh distribution for this HOWLS scene.

TARGET DISTRIBUTIONS AND MODELING OF TARGETS

The signature of typical targets considered in these studies was characterized with an appropriately selected statistical target backscatter model for purposes of predicting average detection probability. The approach provides a statistical characterization of typical ground targets. The radar cross-section of targets is characterized as having a Log-Normal probability distribution with parameters x and ρ selected to fit measured data (x is the distribution mean and ρ is the ratio of mean to median). The statistical model assumes the aspect angle of approach to the target is "equally likely" in all possible directions. Thus, the target backscatter is modeled as a random variable which depends upon the radar aspect angle relative to the target. Detection probability is then obtained by evaluating an average P_D over all possible equally weighted aspect angles, which mathematically is equivalent to averaging over the probability distribution of target backscatter.

The cumulative probability distribution of mean target cross-section for the set of targets considered in these studies has been plotted in Figure 5. This distribution comprises the total target data set taken at 8 different radar viewing angles, and includes the variation of target cross-section with aspect angle. The distribution is seen to be approximately Log-Normal.

A study to determine the probability distribution of the single frequency samples of target returns was also performed. This study was similar to that described previously for clutter data. Specifically, each target cell was normalized by its cell mean, resulting in a normalized set of target data. The cumulative distribution of the normalized single frequency target samples was then obtained and plotted (see Figure 6). It was concluded that the target returns are, on the average, brighter than the clutter returns (recall the target-to-clutter ratio was estimated to be about 6 dB), and that the individual frequency returns for this class of targets are also, to a good approximation, Rayleigh distributed.

For purposes of evaluating detection performance, the target was modeled as a Swerling II target with N independent pulses integrated. That is, given the mean target cross-section at one specific aspect angle, the detection probability is computed using the well known results of Swerling.[4] Variation of the target cross section is then accounted for by averaging detection probability over all possible aspect angles, i.e., by averaging detection performance over the target Log-Normal spatial distribution. The approach is similar to that used in Reference [10].

To obtain the desired performance predictions one must first determine the CFAR coefficient required to provide a desired P_{FA}. This was achieved through a Monte Carlo simulation of Log-Normal clutter samples (as per the previously discussed clutter model) and computing a large sample of independent adaptive thresholds. An average false alarm probability (e.g., $P_{FA} = 10^{-3}$) was then determined using theoretical Swerling formulas. The adaptive thresholds were adjusted to give $P_{FA} = 10^{-3}$ by proper selection of CFAR scale factor. This mixed theoretical/Monte-Carlo approach permits one to easily evaluate P_{FA} accurately with a small number of Monte Carlo trials (500).

Probability of detection was then evaluated by standard Monte Carlo techniques. Target cell amplitudes (actually, target-plus-clutter amplitudes) were simulated according to the previously described Log-Normal target-plus-clutter model. These target-plus-clutter amplitudes were compared with the CFAR threshold amplitudes and the number of detections were tabulated. This approach has several advantages. First, the technique can be applied to various CFAR implementations including single parameter or two parameter algorithms, and P_{FA} may be accurately determined using relatively few Monte-Carlo trials. In the case of a two-parameter CFAR, correlation between estimates of clutter mean and clutter standard deviation is exactly accounted for in the adaptive threshold. Next, the spatial variation of the target cross-section is easily included, and measured data distributions could be utilized directly instead of the theoretical Log-Normal distribution.

Some analytical solutions to the detection of a Log-Normal target in Log-Normal clutter may be found in the literature, however, the effects of the CFAR detector are not included in these previous analyses. An analytical solution to the problem including the effects of the CFAR detector is being attempted.[12]

DETECTION PERFORMANCE PREDICTIONS IN NON-HOMOGENEOUS LOG-NORMAL CLUTTER

Figure 7 presents the predicted theoretical detection performance for a Swerling II, N=32 pulses, target in both a homogeneous clutter background (or a receiver noise background) and in a Log-Normal (non-homogeneous) clutter background. To provide a meaningful comparison of various target/clutter situations, all curves in Figure 7 correspond to a false alarm probability of $P_{FA} = 10^{-3}$. Curve No. 1 is the ideal detection performance for a Swerling II, N=32 pulses, target in receiver noise.[4] This curve contains no CFAR loss, i.e., it assumes the clutter backscatter coefficient is constant over all clutter cells, is known exactly, and that a fixed detection threshold may be set to achieve $P_{FA} = 10^{-3}$. The effect of using a finite number of clutter samples to estimate the mean of the clutter distribution results in a loss (CFAR loss) as implied by Curve No. 2 which corresponds to detection performance of the baseline algorithm which averages 8 clutter cells, four on each side of the target cell. From these curves it appears as though a detection probability of $P_D = 0.5$ can be obtained with input target-to-clutter ratio of only -2 dB. However, this performance is highly optimistic since these curves do not model the ground environment or the spatial variation of target cross-section very well.

In order to assess the effect of target cross-section variation on detection performance, Curve No. 3 has been included in Figure 7. This curve presents the expected detection probability for a Swerling II, N=32 pulses, target in a homogeneous clutter background (receiver noise). The curve includes CFAR loss. In this case, however, the target cross-section has been given a 3 dB variation with aspect angle. Interestingly, for low detection probabilities there is a slight improvement in detectability due to the fact that an occasional target glint will improve P_D at low target-to-clutter ratios. At detection probabilities above 0.3 the result of target spatial cross-section variation is a loss in detectability.

Finally, Curve No. 4 shows the predicted detection performance of a Swerling II, N=32 pulses, target in a Log-Normal non-homogeneous clutter background. This curve also includes the CFAR algorithm loss. Thus, it is seen that to achieve $P_D = 0.5$ and $P_{FA} = 10^{-3}$ in a ground clutter background, one requires about 6 dB input target-to-clutter ratio. Curve No. 4 includes a composite of CFAR loss, and losses due to clutter non-homogeneity and target cross-section spatial variation.

The previous paragraphs have described the method used to evaluate expected detection performance for Swerling type targets embedded in a spatially non-homogeneous Log-Normal clutter model. This approach has been used to predict detection performance against targets in ground clutter and validation of the method has been provided using HOWLS radar data.

Figure 8 summarizes the results obtained. As implied in the figure, a set of targets were located in a moderate clutter background (meadow-scrub clutter) and a set of 8 images were obtained at various viewing angles. Detection performance was evaluated using the baseline CFAR algorithm. The experimental average detection performance (P_D versus P_{FA}) is shown plotted in Figure 8. Analysis of the experimental data resulted in an estimated target-to-clutter ratio of approximately 6 dB.

Superimposed on the experimentally derived performance results are theoretical performance predictions for both homogeneous and non-homogeneous clutter models. As seen in the figure, the theoretical performance curves for a Swerling II, N=32 pulses, target ($\sigma_t = 3$ dB) in non-homogeneous Log-Normal clutter bound the experimental data points quite well. We remark that the experimental target-to-clutter ratio of 6 dB provides quite good agreement with the theoretical predictions. In particular, the slope of the theoretical non-homogeneous curve matches the experimental data much better than the theoretical homogeneous curve. The slope of the experimental data (a relative measurement between data points with one instrument) is expected to be experimentally more significant than the absolute value of any point. A calibration of the system to 1 dB absolute value is difficult; the good agreement between theory and experiment is considered most significant in the P_{FA} region 10^{-1} to 10^{-3} where the number of data samples is reasonable statistically. Finally, note that theoretical performance predictions based on the classical Swerling target (with $\sigma_t = 3$ dB) in homogeneous clutter (or receiver noise) are highly optimistic as indicated in the figure.

TARGET SELECTION STUDIES

This section describes the study of a target selection algorithm designed to select the "most probable targets" from a set of detections obtained by a CFAR detection algorithm. The algorithm is a two step sequential procedure and the rationale behind the algorithm is briefly described as follows.

The radar is assumed to examine an acquisition area containing one or more targets. The total acquisition area is scanned by the radar and the CFAR algorithm obtains a set of detections (including targets, corner reflectors and false alarms). For each CFAR detection, the spread-to-mean ratio of the target cell data is computed and stored in memory along with the observed target-to-clutter ratio (T/C ratio is simply the ratio of target amplitude to local clutter mean estimated by the CFAR), and also the coordinates of the observed detections. Detections are then ranked in descending order according to observed target-to-clutter ratios. A hypothetical ranking would look as follows:

Target	T/C Ratio	S/M Ratio	Remarks
1	20 dB	0.07	Corner Reflector
2	20 dB	0.07	Corner Reflector
3	10 dB	0.22	Corner Reflector
4	10 dB	0.22	Corner Reflector
5	7 dB	0.29	Corner Reflector
6	7 dB	0.52	"Most Probable Target"
7	5 dB	0.52	Target
8	5 dB	0.35	Corner Reflector

The spread-to-mean ratios shown in the above hypothetical table correspond to a point type target in a homogeneous clutter cell (modeled mathematically with a Ricean distribution). The target is mathematically modeled with a Rayleigh distribution since it is comprised of many scatterers. Note in the table that there are no detections below a target-to-clutter ratio of 5 dB since the CFAR threshold is assumed to be 5 dB (corresponding to, for example, a single-parameter CFAR with a scale factor of K = 1.75).

The target selection algorithm is implemented as described. The table is searched in descending fashion. Target No. 1 is automatically rejected since its spread-to-mean ratio is only 0.07, clearly a point target (corner reflector). Target Nos. 2, 3, 4 and 5 are also rejected on this basis. In this example the spread-to-mean ratio must be at least 0.45, since as discussed below, the expected spread-to-mean ratio for a multi-scatterer target is 0.52. Target No. 6 is selected as being the most probable target. This second threshold is selected to provide high probability of rejecting point targets and low probability of rejecting multiscatter targets.

TARGET SIGNATURES

Figure 9 presents typical curves of amplitude versus frequency for a corner reflector and several target detections obtained from a HOWLS radar image. These curves show the amplitude of each pulse versus frequency and have been normalized to the mean amplitude. Visual comparison of the curves of the target signatures with that of the corner reflector shows there is a significant difference between the corner and the multi-scatterer target signatures. This provides a means for discrimination between the corner and targets. One method for achieving this discrimination which has been studied to date is the spread-to-mean ratio test (actually the ratio of the standard deviation-to-mean). Performance results for this algorithm are presented below.

THRESHOLD SELECTION FOR SPREAD-TO-MEAN ALGORITHM

Some theoretical results have been obtained for the selection of the spread-to-mean threshold (i.e., the minimum acceptable spread-to-mean ratio). The curves shown in Figure 10 are the results of a Monte-Carlo simulation of ideal corner reflectors and a typical multi-scatterer target. These curves present histograms of the estimated spread-to-mean ratios for simulated corner reflectors and multi-scatterer targets. The mean and standard deviation estimates are based on 32 independent samples (corresponding to 32 frequency diverse pulse returns). Theoretically, excellent discrimination between corners and multi-scatterer targets can be achieved due to the large separation in the average spread-to-mean ratios. For example, as indicated in the figure, a corner reflector having 10 dB target-to-clutter ratio exhibits an average spread-to-mean ratio of 0.22 whereas the multi-scatterer target exhibits an average spread-to-mean ratio of about 0.52. Note in Figure 10 that data is shown for corner reflectors having signal-to-clutter ratios of 3 dB (worst case) and 10 dB. Setting the decision threshold at 0.45 as indicated in the figure (corresponding to the crossover point of the histograms) yields a minimum probability of error under worst case conditions.

Figure 11 shows the theoretical performance (P_D, P_{FA}) of the spread-to-mean algorithm. For the worst case corner reflector (3 dB signal-to-clutter) using a decision threshold of 0.45 we find for the spread-to-mean discriminant that

$$P_D \triangleq P(\text{target} > .45) = 0.87$$

$$P_{FA} \triangleq P(\text{corner} > .45) = 0.15$$

With a 6 dB signal-to-clutter ratio and the decision threshold of 0.45 we find for the spread-to-mean discriminant that

$$P_D \triangleq P(\text{target} > .45) = 0.87$$

$$P_{FA} \triangleq P(\text{corner} > .45) = 0.00$$

The corresponding curve for S/C = 10 dB exhibited essentially perfect discrimination and thus was not visible when plotted.

TARGET SELECTION STUDIES USING HOWLS IMAGES

Results using this target selection algorithm have been obtained using eight HOWLS radar images. These images comprise eight different views of the target set used in this study. In each image the targets are at a different aspect angle with respect to the radar. Summaries of the target detection and target selection (i.e., target acquisition) results achieved using the proposed target selection algorithm are presented next. Basically, in each scene an acquisition area of 400 m x 400 m was processed with the targets approximately centered in the surveillance area. The baseline CFAR algorithm was run over the area and the appropriate data collected. Figure 12 provides a summary of the number of detections obtained, etc., for one typical scene. A tabulation of all detections achieved along with the corresponding target-to-clutter and spread-to-mean ratios is shown. In the table, detection No. 8 is pointed out as a "most probable target". Most probable targets are defined to be those detections having the largest target-to-clutter ratios and an acceptable (greater than 0.45) spread-to-mean ratio. Other targets are also indicated (e.g., detections No. 9, 10, 11) as well as rejected corner reflectors and false alarms. The radar image shown in the accompanying figure indicates detected objects and corresponds to the data in the table. Note the detections have been labeled in the figure to correspond with the data in the table.

A summary of the results of these studies is shown in the scatter plot data of Figure 13. This figure summarizes experimental data for all detected targets and corner reflectors in the eight HOWLS images processed.

CONCLUSIONS

Target and clutter models are developed which fit experimentally obtained radar data. Target detection performance predictions based on these improved target and clutter models were shown to agree well with experimental results. It is also shown that target acquisition performance may be improved by the use of target selection techniques. For example, a simple algorithm for achieving selection of multi-scatterer targets from point targets (corner reflectors) was described. Other discriminants may be used to provide enhanced target selection performance.

REFERENCES

1. V. L. Lynn, "HOWLS Radar Development", AVP/GCP Joint Symposium - Avionics/Guidance and Control for RPV's - Florence, Italy, Paper No. 31, 4-6 October 1976.

2. V. G. Hansen, "Constant False Alarm Rate Processing in Search Radars", Proc. Int. Conf. on Radar - Present and Future, October 1973.

3. J. I. Marcum, "Studies of Target Detection by Pulsed Radar", IRE Trans., Vol. IT-6, April 1960.

4. P. Swerling, "Studies of Target Detection by Pulsed Radar", IRE Trans., Vol. IT-6, April 1960.

5. W. S. Burdic, Radar Signal Analysis, McGraw Hill

6. R. R. Booth, "The Weibull Distribution Applied to the Ground Clutter Backscatter Coefficient", U.S. Army Missile Command, Redstone Arsenal, RE-TR-69-15, June 1969.

7. G. B. Goldstein, "False Alarm Regulation in Log Normal and Weibull Clutter", IEEE Trans. AES, Jan. 1973.

8. N. C. Currie, et al., "Radar Land Clutter Measurements at Frequencies of 9.5, 16, 35 and 95-GHz", Tech. Report No. 3, Georgia Institute of Technology, 2 April 1975.

9. F. E. Nathanson, "Spatial Correlation of Land Clutter", TSC Memo No. TSC-LB-jab, July 1973.

10. L. M. Novak, "On the Detection Performance of a Cell-Averaging CFAR in Nonstationary Weibull Clutter", Proc. 1974 IEEE Symp. Info. Theory, Oct. 31, 1974.

11. L. J. Greenstein, et al., "A Comprehensive Ground Clutter Model for Airborne Radars", IIT Research Inst. Report, Chicago, Ill., Sept. 1969.

12. R. E. Stovall, MIT Lincoln Laboratory, Private Communication.

ACKNOWLEDGMENT

We wish to express our appreciation to V. L. Lynn, Lincoln Laboratory, for his enthusiastic and tireless guidance, to our colleagues working on Millimeter Terminal Homing and the HOWLS programs, to the Air Force Armament Development and Test Center for many discussions and information, and to DARPA for the use of the HOWLS facilities and data.

This work was sponsored by the Department of the Air Force, the Defense Advanced Research Projects Agency, and the U.S. Army.

"The views and conclusions contained in this document are those of the contractor and should not be interpreted as necessarily representing the official policies, either expressed or implied, of the United States Government."

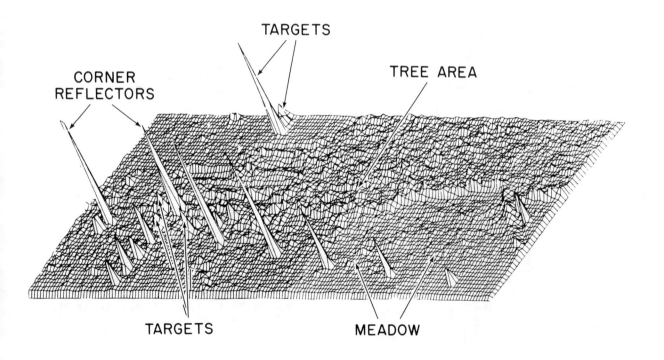

FIGURE 1. Typical Radar Image (Image Intensity Proportional to Amplitude)

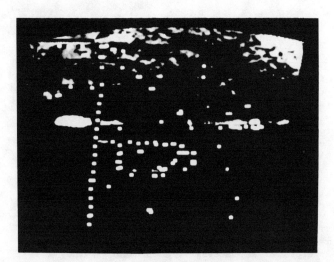
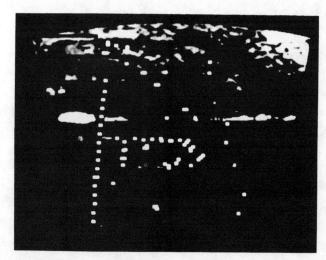

FIGURE 2. Radar Images With Detections Superimposed (White Boxes)

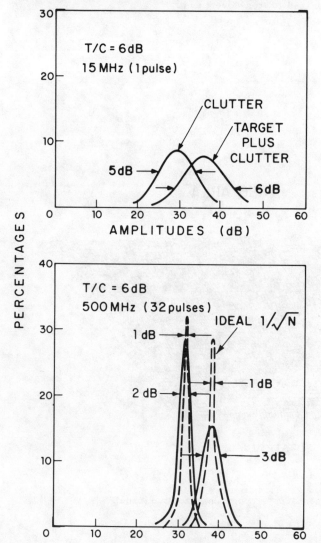

FIGURE 3. Typical Data Distributions of Clutter and Target-plus-Clutter

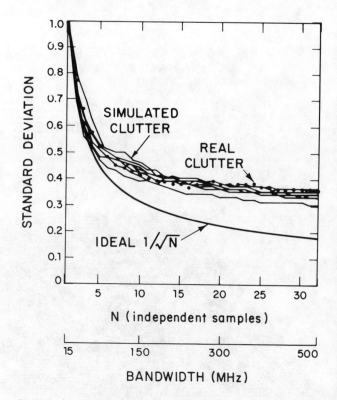

FIGURE 4. Reduction in Standard Deviation of Clutter Versus Number (N) of Independent Pulses Averaged

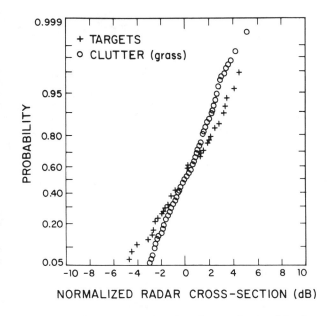

FIGURE 5. Cumulative Distributions of Normalized Target and Clutter Data

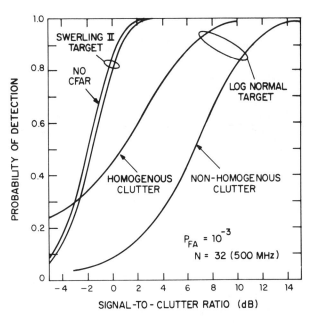

FIGURE 7. Theoretical Detection Performance Curves

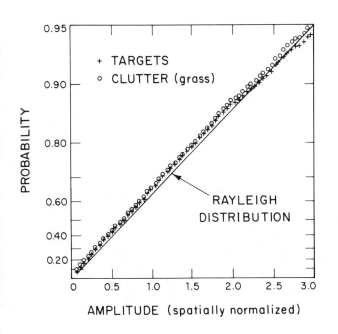

FIGURE 6. Cumulative Distributions of Spatially Normalized Target and Clutter Data

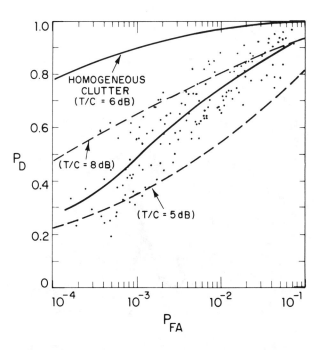

FIGURE 8. Experimental Detection Performance Curves (Theoretical Performance Curves ---)

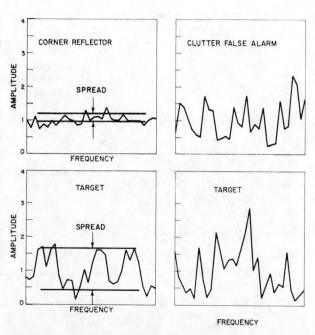

FIGURE 9. Typical Target and Clutter Signatures (Normalized)

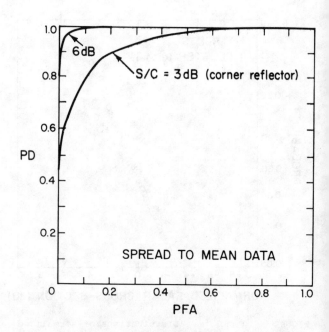

FIGURE 11. Theoretical Performance of Spread-to-Mean Algorithm

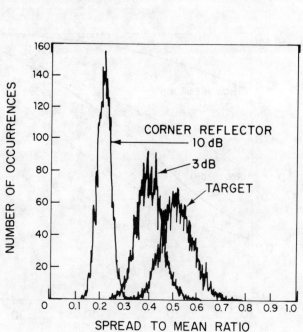

FIGURE 10. Distributions of Spread-to-Mean Ratio (Monte Carlo Simulation)

	DETECTIONS	T/C RATIO (dB)	S/M RATIO
1	CORNER REFLECTOR	24.0	0.137
⋮	⋮	⋮	⋮
7	CORNER REFLECTOR	11.8	0.273
8	TARGET	11.1	0.479*
9	TARGET	9.8	0.453
10	TARGET	7.6	0.674
11	TARGET	7.6	0.538
12	FALSE ALARM	7.2	0.574
⋮	⋮	⋮	⋮
15	TARGET	6.4	0.773
⋮	⋮	⋮	⋮
19	FALSE ALARM	5.1	0.553

MOST PROBABLE TARGET*

FIGURE 12. Target Selection Algorithm Performance Results

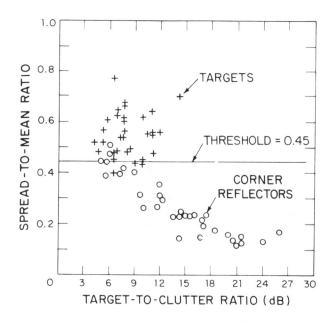

FIGURE 13. Scatter Plot of Experimental Data

USAF Millimeter-Wave Seeker Development
A.N. Disalvio

ABSTRACT

The Millimeter-Wave Contrast Guidance Demonstration (MCGD) program is evaluating millimeter-wave (30-40 and 90-100 GHz) radar for guidance of standoff, all weather, air-to-surface missiles against tank targets in the European scenario. The Air Force Armament Laboratory is pursuing four diverse seeker approaches to solve the lock-on-after-launch detection and track problem. Four seekers have been designed and developed, and are now being exhaustively tested through laboratory, tower and captive flight tests. The results will be exercised via detailed seeker, target and clutter models and missile simulations to determine worst case delivery envelopes, detection capabilities, and weapon accuracies against armored targets for various scenarios.

INTRODUCTION

Due to advances in the technology of microminiaturization coupled with state-of-the-art advances of millimeter-wave frequency (30-300 GHz) components, it has now become feasible to investigate radar for small, air-to-surface terminally guided missiles. Until now, radar has not been used for autonomously guiding small air-to-ground tactical missiles due to the need for high resolution to detect targets in background clutter.

Operating radar at the highest frequency possible permits use of wideband transmitted waveforms to obtain high resolution which results in high signal-to-noise ratios even with small targets such as armored vehicles in a natural clutter environment. Although millimeter-wave frequencies extend throughout the 30-300 GHz band, component availability for development programs are only readily attainable up through 100 GHz. Figure 1, taken from Reference 1, shows that there are two windows in the MMW band at 25 to 40 GHz and 80 to 100 GHz. Dropping below 25 GHz results in degraded resolution.

The 30 to 100 GHz region provides low weight through small component size (i.e., waveguides) compared to X-band and better adverse weather efficiency when compared to infrared and electro-optical systems. TV systems cannot operate in fog or haze and are degraded by nighttime operation, while MMW transmittance is two orders of magnitude better than IR in light fog (Fig 2). The advantage of small missile size and weight is required due to aircraft carriage limitations. To obtain small weapon diameters, the seeker antenna must be correspondingly sized. The antenna size, however, is inversly proportional to the beamwidth (Beamwidth = λ/D, where λ=wavelength and D = aperature diameter)[2]. The beamwidth, in turn, is the footprint on the ground which is proportional to the resolution--or measure of the ability of the radar to separate targets, or separate a target from competing clutter. Thus, the smaller the weapon diameter, the higher the frequency must be to maintain the correct resolution. This is why MMW is being investigated to solve the adverse weather antiarmor problem of the European scenario.

Standoff attack in adverse weather with a launch-and-leave capability requires a seeker which is; (1) able to search undulating variegated terrain, (2) detect tank targets, (3) track the tanks and (4) guide the missile to impact. The problems associated with meeting all of these objectives are difficult to define, much less determine their solutions.

(1) Search. Terrain search is initiated when the missile is dropped from the aircraft, flies to the target and the seeker acquires the ground. The ground may be dry, wet, contain heavy vegetation, have a rough contour, and/or be covered with snow. For example, the radar return from apparently smooth snow was reported from Georgia Tech[3] to have 25 dB reflectivity variations in only seven percent range changes at 30 degrees to 45 degrees depression angles. Another major problem is that tree lines form an abrupt amplitude change which have the appearance of an extended target.

(2) Detection. Separating the target from clutter requires the seeker to have a large RF bandwidth, high resolution, and signal processing or discrimination capability. A large RF bandwidth decorrelates the clutter return from different ranges by generating fluctuations of the clutter interference pattern. Range decorrelation is given by $\Delta r = \frac{c}{\Delta f}$ where Δf is the

width of the frequency chirp and c is speed of light. Resolution can be broken into range and azimuth components. Range resolution is attainable through very narrow pulsewidths, while azimuth resolution is achievable through either small beamwidths or synthetic beams via doppler techniques. Small beamwidths are obtained by making the aperture size large with respect to the wavelength. Synthetic beams through doppler beam sharpening techniques are not yet within the present state-of-the-art in small low-cost missiles. Signal processing techniques, such as spatial discrimination and constant false alarm rate (CFAR), are used to discriminate between extended clutter in range and azimuth (i.e., tree lines, long buildings, etc.), and returns obviously not from tanks such as excessive amplitude returns from manmade structures.

(3) Transition and Track. Following acquisition of a target, the seeker must transition to track. Depending on the beamwidth or ground footprint size, speed of scan, missile velocity and gimbal stopping time, it is usually necessary to stop the antenna, reverse the direction of scan, and reacquire the target. Tracking the target after missile pitchover (assuming horizontal flight) becomes less difficult with decreasing range due to the narrowing of the beamwidth (increasing resolution).

(4) Terminal Tracking. Tracking to impact becomes a problem when the seeker closes to a range where scintillation affects tracking accuracy. Scintillation is caused by different scatterers on the target becoming dominant reflectors or adding in phase to cause a large change in the apparent centroid of the target. Amplitude scintillation is the effect of amplitude of the target return changing with time. Angle scintillation or glint is the shifting of the center of reflectivity with time. There are many discussions in the open literature [4,5] which show that the instantaneous aimpoint, due to glint, can actually appear to be outside the angular extent of the target. There are two ways to reduce the effect of glint: (a) use a wideband transmitter to decorrelate the glint and smear its spectrum where $\Delta f = c/2d_r$ (Δf = change in transmitted frequency, d_r = length of target in range dimension) (Ref Fig 3), and (b) switch to passive at a range of between 200m and 300m from the target, depending on the sky overcast conditions.

The Air Force Armament Laboratory (AFATL) initiated a MMW seeker design, development, evaluation and demonstration program by awarding four parallel contracts to Hughes, Rockwell, Boeing, and Honeywell. The first three were awarded in June 1978; Honeywell's was awarded in July 1978. Boeing is teamed with Sperry Microwave. All four contractors are required to accomplish identical tasks, but have selected four diverse approaches to the MMW target detection and track problem. Two companies' seekers operate in the lower MMW region of 30-40 GHz, while the other two, Hughes and Rockwell, are in the 90-100 GHz window. They also use different transmit waveforms, antenna configurations, tracking schemes, and processing techniques.

The program is designated the Millimeter-Wave Contrast Guidance Demonstration (MCGD). It is divided into two phases. Phase I was scheduled to be four months in duration. Each contractor designed, fabricated, and instrumented a MMW seeker system. They also developed automatic data reduction and analysis routines, detailed seeker computer models and two six-degree-of-freedom (DOF) missile simulations; one simulation of probable profiles of a minimissile, the other of a GBU-5 test missile airframe. Also, during Phase I the contractors prepared detailed tower and captive flight test plans for Air Force approval.

Phase II was initiated only after all data was submitted and the contractors successfully completed the requirements of Phase I. Honeywell started Phase II on 20 November 1978 and is conducting tower tests at Minneapolis MN. Boeing and Hughes began Phase II on 26 and 29 January 1979, respectively. Boeing will do Tower Tests at Kennedy Space Flight Center, while Hughes will perform theirs at Eglin Air Force Base. Rockwell entered Phase II on 15 February 1979 and will perform their Tower Tests at Ft Hunter-Liggett CA. The Phase II tasks consist of tower and captive flight testing. The tower tests are being performed to determine detailed seeker operation, characterization, and parameter sensitivities. The tower and captive flight tests will determine seeker/target and seeker/clutter interactions at realistic depression angles and under dynamic conditions. Automatic data reduction will be used to obtain rapid printouts of glint and scintillation effects, target and seeker generated noise, aimpoint distribution of the seeker vs armored targets and point sources, blind range effects, and adverse weather effects. These results will be programmed into the missile simulations to perform predictive analyses of the MCGD seeker and missiles to successfully complete possible free-flight testing under a follow-on Phase III program.

BACKGROUND

The Millimeter-Wave Contrast Guidance Demonstration (MCGD) program was initiated to evaluate the MMW potential for guidance of air-to-surface missiles against tank targets in the European scenario. In a non-nuclear engagement the NATO countries are at a distinct disadvantage. The Warsaw Pack countries have arranged a 2.5:1 to 4.5:1 tank advantage and a 1.5:1 airplane advantage against NATO forces. [6,7] Along with this disparity is the distance of the United States to Central Europe compared to that for the Soviet Union for rapid resupply and reinforcement. In one scenario for an invasion of Europe a "...Group of Soviet Forces, Germany Front would be expected to defeat NATO forces designated for the defense of central Europe and reach English channel objectives in fourteen days or less."[8] Therefore, NATO ground and airborne forces would have to stop large numbers of conventional armored vehicles moving forward at high speed. Additionally, a capability to destroy multiple tanks in a massed armored thrust in adverse weather with one pass of an aircraft would, in itself, prove to be a deterrent to either a non-nuclear conflict or the massing of tanks should a conflict be engaged. Moreover, if an aircraft can stand off from the enemy's defenses and fire multiple missiles into an area where enemy forces are deployed, the aircraft would be less vulnerable to ground fire.

Numerous MMW technology efforts have been performed both under Government contract and independent research, which have contributed to provide a technology base sufficiently broad to support a MMW seeker development and test program. Martin Marietta, Sperry, General Dynamics, Hughes and Honeywell have all demonstrated a capability to design, fabricate and test MMW seekers. The United States Army, Navy and Air Force performed feasibility demonstrations of noise illumination MMW contrast seekers, the results of which showed potential for employing MMW in antiarmor seekers. The Air Force Armament Laboratory contracted with Honeywell, Inc to develop and demonstrate terminal accuracy of a MMW seeker in a benign environment. This was a helicopter-borne captive flight test program which was completed in February 1978. The Remotely Piloted Vehicle (RPV) Program office at Wright-Patterson Air Force Base OH has a test program to guide an RPV with a MMW seeker built by General Dynamics. The Army had contracts with Sperry and Martin Marietta, while the Navy had a comprehensive test effort with Hughes Aircraft Corporation.[9] Rockwell became a contender in the MMW seeker area through their work on a Ku-band seeker designed, fabricated and tested in 1977.[10]

In spite of all these efforts, the critical objectives of lock-on-after-launch target detection in realistic clutter had not been demonstrated. The Air Force Armament Laboratory, thus, initiated the MCGD program to solve the MMW seeker problems, but due to the high risk involved decided to investigate four different approaches. Of the six proposals submitted, four were selected.

In addition to the above programs which directly led to the MCGD effort, there are numerous other tasks which contribute to MCGD through the accumulation of evidence and data on background and targets. The following section will deal only with those associated with AFATL.

SUPPORTIVE PROGRAMS

The Armament Laboratory is pursuing several exploratory development programs to perform background measurements and develop simulations for investigating lock-on-after-launch adverse-weather terminal guidance capabilities. They are directed toward measuring target signatures, obtaining clutter distribution statistics, and determining the capabilities and limitations of MMW seekers. These results are then used to develop mathematical models of the target (three-dimensional), clutter and seeker characteristics. Their overall goal will be to provide inputs into seeker development programs such as the MCGD effort; to enable the seekers to incorporate improved target detection and discrimination techniques.

There are five measurement programs, either conducted in 1978, or are presently on-going in AFATL. (1) Two tasks by Georgia Tech are to accomplish active and passive tower-based measurements. The active-mode measurement effort uses a 35 GHz radar capable of operating in a pulsed or FM mode with polarization diversity and PRF pulsewidth flexibility. Backgrounds consist of snow-covered, frozen, and wet terrain. An M-60 tank is the primary target. The passive measurements are being performed at 35 and 94 GHz on several different target vehicles, at several depression angles, extensive target aspect angles and two polarizations.[5] (2) A University of Kansas measurement effort is being conducted to obtain backscatter measurements, ground truth data and radiometric temperature measurements of a variety of snow. This test has been performed in conjunction with the Army (MICOM). (3) A target signature measurement program at 35 GHz of various targets is being performed by the Environmental Research Institute of Michigan. The goal is to identify the physical location and characteristics of scattering phenomena. These scatterers will then be mapped into a three-dimensional target signature model. (4) An effort is being made to build a scale model of a tank, obtain active mode target signatures at scaled frequencies and compare the results with other test programs. This work is being performed by EMI Electronics in England.

There are three simulation efforts being conducted: (1) Lincoln Laboratories, Inc is developing models of seekers, targets and clutter. They are defining the detection probability as a function of the environment and target model. The data base is predicated on data gathered in the AFATL/Honeywell, Inc test program completed in February 1978 and the Army's HOWLS program. (2) A program with Systems Control, Inc to develop models of target, clutter, and seeker characteristics and determine MCGD seeker performance is on-going. Each of the four MCGD contractors' simulations and models will be investigated. (3) In 1978, Honeywell, Inc completed a study to determine electronic countermeasure (ECM) effects on the detection and tracking of the Honeywell MMW seeker.

MCGD EQUIPMENT DESCRIPTION

The MCGD system (Fig 4) is made up of a MMW seeker, an independent reference system (IRS) and a wide field-of-view (WFOV) TV. The IRS in three of the four MCGD systems are narrow field-of-view (NFOV) TVs, while the fourth is an automatic lock-on laser seeker. The IRS is used as an independent measurement system which is locked onto the geometric centroid of the target and is used to measure the difference between it and the MMW seeker's pointing directions. This difference is recorded and analyzed for tracking and line-of-sight pointing errors. The IRS is locked onto the center of the target through target enhancement techniques such as lights, cloth, and laser corner reflectors. The WFOV TV is used to monitor surrounding terrain and clutter for ground truth.

The MMW seekers (See Fig 5) were built in response to AFATL defined requirements covering antenna size, search area, missile velocity, clutter level and variability, expected target signature characteristics, probability of target detection over the search area, false detection rate, and terminal accuracy. The seeker's design consists mainly of three basic major subassemblies:

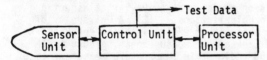

The sensor unit consists of the gimballed antenna, radome, transmitter/receiver, IF processing modules and range tracking circuits. The seeker control unit performs video processing, automatic gain control, search error signal generation and instrumentation interface. The heart of the seeker is the signal processor which performs system timing, acquisition and detection processing, and clutter capture processing.

Computer programs for rapid and accurate test data reduction and analysis were prepared for use on the Armament Development and Test Center (ADTC) computer. The objective of computer-aided data analysis is to efficiently process large volumes of test data in a timely manner and obtain an unbiased assessment of the test results. The volume and complexity of the test data in the MCGD program dictated the requirement (Ref Fig 6). The data is generated from the MMW seeker, the IRS, instrumentation, test equipment, video tapes, test log sheets, and support elements such as tracking radar sites. All this data is put onto a common merge tape for processing by the ADTC Math Lab.

Honeywell's task involves a modification of the pulsed seeker operating in the 30-40 GHz range which they built and tested for AFATL in 1975-1978. It employs a chirped pulse waveform and has a conscan parabolic antenna with a ring focus feed. The detection is performed by a matched filter. The IRS is a NFOV TV. It incorporates an optical discrimination system that places a video track box on the centroid of brightness and maintains track during motions of the test aircraft and target. The MMW seeker and IRS are mounted on the same base structure and are coaligned. The pointing direction of the MCGD seekers is indicated through seeker gimbal angle transducers and is recorded and compared with the optical tracker pointing direction data to determine the tracking performance of the MMW seeker to within one milliradian.

Boeing, in conjunction with Sperry Microwave, has built a 30-40 GHz FM/CW seeker. The Boeing/Sperry design makes excellent use of state-of-art MMW components to achieve required performance (i.e., exceptionally low-noise receiver). Boeing uses a frequency modulated/continuous-wave transmission and has a conscan lens antenna. Their detection is also performed by a matched filter. The IRS is an active 5mw CW laser tracking system used in conjunction with a small target-mounted laser reflector to provide automatic centroid tracking with an accuracy of better than 1mr. The laser IRS features fully automatic target detection, acquisition and tracking at ranges of 5 Km in clear weather and 3.5 Km in light rain, which is better than a NFOV TV. A WFOV TV (19°) enables the system operators to view the entire target area. The laser employs a low-inertia tracking system having twice the tracking rate of the MMW seeker.

Rockwell, Inc has taken an approach which is pushing the state-of-the-art in antenna, gimbal, and tracking design at MMW frequencies. The seeker's operating frequency is between 90-100 GHz and was considered a high risk due to component availability and the very tight Phase I schedule. The physical characteristics of the seeker make it directly compatible with the mini-missile size, mass and power restrictions without requiring significant new development and testing. Rockwell utilizes a modified chirp pulse with frequency agility and has a monopulse inverted cassegrain with twist reflector antenna. Unlike the previous two efforts, Rockwell's detection is accomplished through CFAR and spatial discrimination. The instrumentation allows easy flight line or in-flight calibration and quick-look data display. A 14-channel tape recorder is used to record signals from the MMW seeker, narrow and wide FOV TV system, time code generator, and intra-aircraft communications. The 3.5 degree NFOV TV reference system can either independently track an optically enhanced target or can be slaved to the MMW seeker. The latter capability was added for poor visibility conditions when the TV tracker would have a short acquisition range or have difficulty maintaining track. The IRS has a slew rate in excess of 100 degrees/second and a maximum off-boresight capability of ± 70 degrees in both axes.

Hughes Aircraft Company's seeker (Fig 5) operates in the 90-100 GHz frequency spectrum, for which Hughes claims is the highest operating frequency for which adequate adverse weather performance can be obtained, and which will provide a viable low-cost production base for missile developments in the near future.[11] Hughes uses a chirped pulse waveform and a conscan cassegrain antenna. Their detection is performed through CFAR and spatial discrimination. Hughes addresses the problem of glint, multipath and scintillation through the aperture averaging phenomena.9 Hughes feels that glint reduction occurs when the seeker closes in range such that the antenna subtends a significant angle and averages the slope of the reflected phase front. Aperture averaging theory thus predicts that glint at terminal ranges is reduced with increasing frequency and/or aperture. The IRS is a 2.5 degree FOV Maverick TV guidance unit. They also employ a 15 degree FOV fixed TV camera.

MCGD SIMULATION AND TEST

The MCGD test portions of the program include a completely automatic and autonomous search, detection, acquisition and tracking accuracy measurement program in both fair and adverse weather conditions. The output of these tests will define system performance as a function of range, velocity, and environmental conditions and will accomplish three main objectives: (1) establish that the seeker and supporting instrumentation meet or exceed the Air Force nominal requirements and qualify for continued testing, (2) demonstrate seeker performance capabilities in natural clutter and realistic flight environments, and (3) provide empirical data to support and validate the system simulations.

Although each contractor has approached the modeling effort in somewhat different ways, in general, the simulations consist of separate, highly detailed seeker, target and clutter models as well as two system simulations. The system simulations (SS) which include their own generalized seeker, target and clutter models, demonstrate autonomous automatic search, lock-on-after-launch, and terminal guidance (Fig 7). One SS is a contractor-developed replica of his own minimissile concept in sufficient complexity to allow derivation of a complete flight geometry time-history from launch to target impact. The second SS is the GBU-5 airframe, which may be used as the test vehicle in the planned follow-on free flight test program. (Fig 8 depicts the GBU-5.) These two predictive simulations are based upon measurements taken during the acceptance, tower and (eventually) captive flight tests as well as extrapolated results. The simulations are modular, with six degrees-of-freedom, which can be varied parametrically to determine the sensitivity or limitations imposed by the flight path angle, velocity, search techniques, and maneuverability. The GBU-5 simulation will specifically be used to determine the probability of a free-flight test at Eglin Air Force Base as well as predicting terminal guidance accuracies. The detailed seeker, clutter, and target models include the theoretical operation as well as the physical limitations and non-linearities that realistically describe the seeker modes of operation. The seekers are primarily modeled on a one-for-one digital emulation of the hardware design with over twenty seconds of computer time for each second of real time. The targets are represented by n-point scatters with variable values for each point providing real time glint and scintillation effects. The clutter models include discontinuities, adjustable ground cell dimensions (accounts for increasing resolution with decreasing range to target during the track segment), and weather effects. The seeker and clutter interactions are, in at least one case, performed by convolving the antenna pattern with the ground scene to generate a new pattern which provides a good representation of the antenna/radome interaction with the ground and the target. All simulations and models have been documented by the contractors with full card-deck and software comments for delivery to the Air Force.

The MCGD test program is broken up into three major tests, each of which is a milestone that the contractor must successfully pass before proceeding onto the next phase of the program. The initial test was the Acceptance Test performed at the contractor's facility. Honeywell completed theirs in November 1978, Hughes and Boeing in January 1979, and Rockwell in February 1979. The objective of the Acceptance Test is to provide assurance that the MMW seeker will perform as specified to actively search for, detect, acquire and track designated targets in benign background and in clear and adverse weather. Acceptance Testing was performed in three major categories: (1) system integration and calibration tests, (2) environmental tests, and (3) formal demonstration tests.

The second major test is the Tower Test which will take approximately six to eight weeks to complete, and requires 60 hours of test. The basic objectives are to investigate seeker effectiveness in adverse weather; evaluate seeker/software performance in automatic search, detection, acquisition and tracking; and obtain engineering data on target and clutter signatures for updating and validating the MMW seeker/target clutter model. Data is being accumulated on radar cross-section of M-60 tanks; clutter amplitude distribution, variation, internal motion, and emissivity; false target density; target detections during search; and accuracies. All tower testing will be completed by March 1979.

The Captive Flight Tests will be conducted during the Spring/Summer of 1979 for a period of four months at Eglin Air Force Base, Florida, Rome Air Development Center, New York, and Ft Drum, New York. The basic requirement is for 100 productive hours of testing. The captive flights will be flown in T-39 Sabreliners except for Boeing who will use a T-33. The purposes of these tests are to obtain engineering data on target and clutter signatures for updating and validating the MMW seeker/target/clutter models, to evaluate seeker performance with the dynamic conditions of flight and a variety of target/clutter situations, and to determine the effects of adverse weather conditions on seeker operation. Targets will include tanks, self-propelled guns and trucks. This task will be completed on 31 July 1979. Allowing for data reduction, the final report should be published in September 1979.

CONCLUSIONS

The results of the MCGD captive flight tests coupled with the respective simulations will provide the Government with the first concrete evidence that MMW radar is a viable candidate for a standoff, adverse-weather, lock-on-after-launch minimissile seeker against mobile armored targets. Additionally, the results will be examined in detail for use in engineering development programs, such as the Air Force's Wide Area Anti-armor Munitions (WAAM) program. Following the successful conclusion of Phase II on 31 July 1979, there will still be much more work to be accomplished. A first priority requirement is the determination of the susceptibility of the MCGD seekers to ECM and what can cost-effectively be done to thwart it. The capability to detect targets in or bordering on urban area, to separate actual targets from man-made clutter, and to prioritize targets must be accomplished through sophisticated signal processing techniques which will be investigated in lateral programs by AFATL. Phase III, which is not yet contracted, will include GBU-5 drop tests at Eglin Air Force Base, Florida. Phase III will provide the terminal accuracy data necessary to complete the Laboratory's role in this generation of MMW seekers.

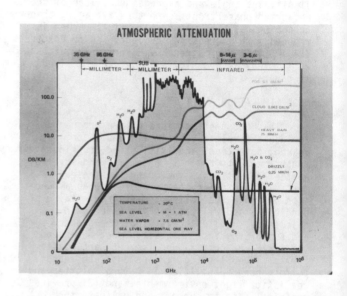

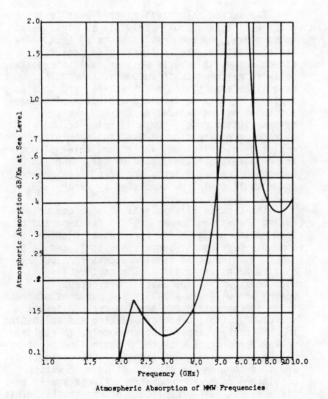

Atmospheric Absorption of MMW Frequencies

Fig 1

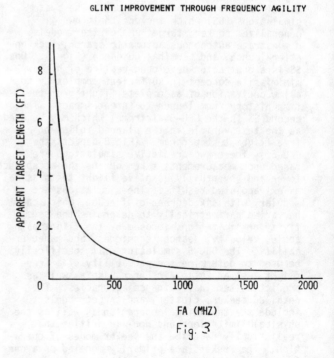

Fig. 3

Millimeter Wave Radars

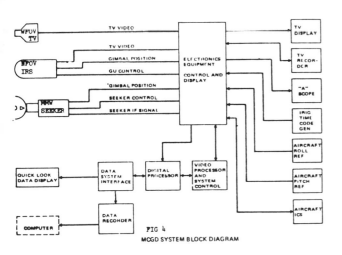

FIG 4
MCGD SYSTEM BLOCK DIAGRAM

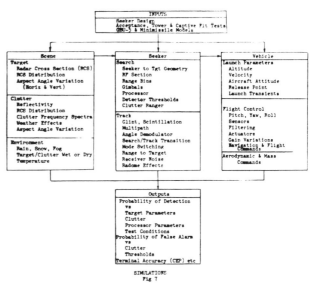

SIMULATIONS
Fig 7

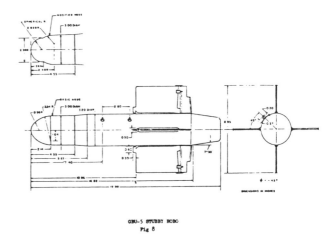

GBU-5 STUBBY HOBO
Fig 8

Fig. 5

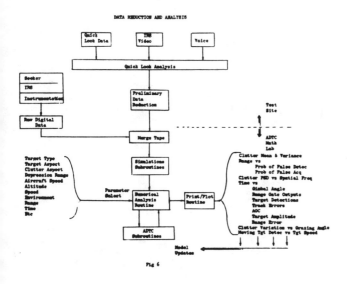

DATA REDUCTION AND ANALYSIS

Fig 6

BIBLIOGRAPHY

1. Rosenblum, E. S.; "Atmospheric Absorption of 10-400 kcps Radiation," The Microwave Journal, March 1961.

2. M. Skolnik; Radar Handbook, 1970.

3. AFATL-TR-77-92, Vol I, "Radar Millimeter Backscatter Measurements," July 1977, Georgia Institute of Technology.

4. Dunn, J. H. and Howard P. D.; "Radar Target Amplitude, Angle and Doppler Scintillation from Analysis of the Echo Signal Propagating in Space," IEEE Trans MTT, Sept 1968, Pg 715.

5. APL Report TG-1047, "Near Field Glint at X and Ka Bands," Dec 1968.

6. Air Force, Vol 60, No 12, Dec 77, "Tables of Comparative Strength." Pgs 123 & 124.

7. International Defense Review, Vol 10, No 3,

A Terminal Guidance Simulator for Evaluation of Millimeter Wave Seekers
K.L. Wismer and A.J. Witsmeer
Proc. Military Microwaves Conference, MM-80, London, Oct. 1980
Reprinted by permission.

ABSTRACT

This paper contains a theoretical and physical description of the simulator which has been built as part of the Boeing Aerospace Company's research laboratories. The simulator is presently limited to evaluation of passive terminal guidance seekers operating in the millimeter region 10-100 GHz. The paper will also cover the expansion of the present facility into the active regime. This particular chamber is the only one of its kind in the USA and probably in the world. Its unique feature is the use of aluminum foil to reflect the excess microwave energy out of the room through a skylight. This principal in effect makes the walls appear cold like the sky background so that the seeker sees only the array with its picture of the target scene.

INTRODUCTION

The past three decades have witnessed development of missiles which can sense the target and modify their path during flight to achieve intercept. These terminal guidance systems normally include: a sensor, measuring the line-of-sight angles and sometimes range to the target; a gimbal system, permitting the sensor in the case of a null seeking device to physically track the line-of-sight independent of missile motion; pickoffs, reading the gimbal position and rates; steering algorithms or guidance laws, directing the missile to change its flight path in response to sensed information; and a control mechanism, forcing the missile physically to alter its course. A slight variation occurs where the sensor is capable of tracking targets off center and may be fixed to the missile body. We often refer to this as electronic gimballing.

As a major United States missile builder and weapon system integrator, Boeing has long been interested in terminal guidance systems. Early in their development when we were building BOMARC, the first radar guided anti-aircraft missile, our engineers identified, design of the track loop, as the key to minimizing miss distance. The error sources contributing to miss distance included: angle noise, a statistical error caused by variations in the centroid of the energy returned by the target (glint) plus the non range dependent servo and receiver noise; angle bias, caused by pickoff misalignment and random distortions, and dynamic coupling between the azimuth and elevation of the seeker itself or between the seeker motion and the missile rotational dynamics. Some difficulty was experienced in measuring these errors in a dynamic environment and correctly assessing their impact on miss distance. Flight tests for BOMARC were very costly and could be conducted only with a high probability of success. The statistical nature of the errors required a very large number of test cases to predict system performance within a reasonable confidence interval. These circumstances combined to percipitate the need for a terminal guidance laboratory capable of running a large number of simulated test flights with actual hardware under repeatable controlled conditions. The first such facility was completed in 1960 and has been expanding ever since to incorporate other wavelengths. In addition to our own facility we have designed and build simulators for various agencies of our Department of Defense. Terminal Guidance Laboratories have long established their value by providing early identification of design problems in the Sparrow, Hawk and Patriot missile systems. The specific subject of this paper is the recent expansion of this laboratory capability into the millimeter wave region.

Before describing our millimeter wave terminal guidance simulator, I feel advised to review with you some general concepts. A terminal guidance laboratory is usually made up of several chambers each designed to accommodate a prescribed portion of the frequency spectrum and all connected to a central computer facility capable of supporting real-time operations with high-speed processing. The individual chambers are all similar containing the basic elements shown in Figure 1 including a target generator

Figure 1. Boeing Millimeter Wave Terminal Guidance Simulator

to simulate the target and background as they appear to the sensor and at the opposite end of the room a three-axis platform for mounting the seeker and simulating missile rotational motion. The chamber must be sized to keep the target generator outside of the sensor near field and its sides should not reflect energy back into the seeker or its sidelobes to create false targets nor should it add significantly to the background modifying the signal-to-noise ratio. The target generator can take several forms depending on the wavelength and type of sensor (i.e., imaging, correlation, or spot contrast). In the emitter array variety multiple elements of the array are used simultaneously to generate the target images using either synthesized signals or delayed transmissions from the seeker itself.

The target generator must have the ability to expand the target and background image to simulate range closure. The central computer receives the seeker output, integrates the equations of motion and commands both the three-axis table and the target generator to new positions. More detail on terminal guidance simulators can be found in the literature [1].

One characteristic which should be mentioned early in this discussion involves the focus of these facilities on the tracking and accuracy evaluation of terminally guided missiles rather than the target acquisition problem. These facilities do not in any way obviate the separate requirement for tower and captive flight testing of seekers to establish their acquisition probabilities in the field with real targets. The targets simulated in the laboratory only represent our best knowledge of the actual target structure and as such can only provide a controlled calibration of a seekers capability to acquire in a given background.

TWO UNIQUE DESIGN APPROACHES

The rapid development of millimeter wave component technology in the last few years has generated its application to terminal guidance seekers to capitalize on the adverse weather feature of radar along with the greater resolution at shorter wavelengths. This greater resolution has made practical the use of these seekers against ground targets despite their high clutter background and small size. A new twist was added when several candidate seekers were offered with a passive mode during the final critical phase of flight where the seeker ceased to radiate and measured only the radiation received from the objects within its beam. This approach significantly reduced target glint which becomes more pronounced at higher frequencies. Since system accuracy was our objective it was crucial to simulate at least the passive portion of the flight.

Several chambers have been built at longer wavelengths using anechoic material to absorb the stray radiation and reduce side lobe interference. While it has been demonstrated that this technique could be used at higher frequencies the emissivity of good absorbers is by definition close to unity and its blackbody radiation would saturate a passive receiver. The only known method for reducing the effective background temperature was to cryogenically cool the entire room -- an obviously impractical solution. A unique solution to this problem was obtained by covering the walls, ceiling and floor with highly reflective foil and cutting an opening in the ceiling to provide the cool sky background temperature. The opening was placed directly above the seeker under test which was mounted in a sloping surface as shown in Figure 1. Thus whenever the sensor looked in the chamber it would see the reflected sky background temperature and a target generator could now be added to create the desired increases above the cool background. As will be shown later this concept proved good enough to be used as an "effective" anechoid chamber for the active mode simulation.

The second major problem was how to build a multiple element array of millimeter wave energy emitters sufficient to cover the seeker field-of-view and spaced close enough to accommodate the expected seeker resolution capability. Millimeter wave components are costly and usually restricted in bandwidth. It was observed that plasma devices such as ordinary flourescent bulbs generate energy over a broad band and tests indicated that their range, linearity and response times were more than adequate for our purposes. This allowed us to economically construct a very low cost array of 1280 elements each individually driven under computer control. The fluorescent tubes were mounted in a household funnel to collimate the energy. A ball and socket arrangement using a plastic ball and aluminum holders facilitates alignment of each element individually to allow the forming of a spherical shaped array. Both the target array elements and the foil lined chamber are unique patented Boeing concepts that have made the development of such a facility possible.

A prototype passive millimeter wave simulator was constructed during 1977-78 to verify the concept. The prototype facility consisted of a small chamber (4x4x3 meters) whose surfaces were lined with aluminum foil, a 96 element array mounted on a flat surface and non-gimballed 35 GHz radiometer positioned on a two-axis mount. Modifications, improvements and component feasibility tests were conducted during the following year. The analyses and tests provided the technology necessary to develop the fully operational facility shown in Figure 2 with the 1280 element array located at

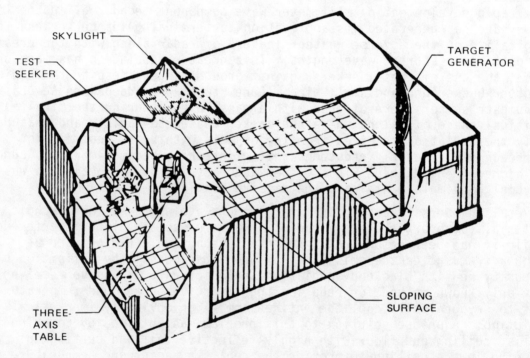

Figure 2 Millimeter Wave Guidance Simulator Elements

one end of the chamber and the sloping surface around the three-axis flight table 6.7 meters across the room. A 35 GHz radiometer was developed in parallel to evaluate the performance of the facility. A rainshield fabricated of aluminum and lexan plastic was placed over the opening permitting facility operation in all weather as experienced in actual terminal engagements.

TESTS AND TEST RESULTS

Tests were performed at the component and system level to verify design parameters and determine the overall capability of the facility. At the component level, the noise source generators (fluorescent lamps) were evaluated for their response time, linearity and stability of the output radiation as well as their power output in the millimeter region. These fluorescent lamps are commercially manufactured and their electrical and physical tolerances are not closely controlled. Consequently, it was necessary to set up a test procedure to select lamps that performed within specific tolerances. The rejection rate was nearly 30 percent still a considerable savings over factory control of specific parameters. The command current required versus the radiometric temperature at 35 GHz was measured on each lamp. Another component level investigation determined the optimum collimation angle and collimator for the 100mm lamp source given the designed chamber geometry. Collimation of the lamps required that the energy radiated had to fall on the diamond shaped surface and be totally reflected through the sky doors. Six funnel shaped collimators with solid angles varying from 40° to 90° were tested with several lamp sources. The combinations were mounted on a Leitz Head angular adjustable mount and scanned angularly across the 35 GHz radiometric sensor. Evaluation of the data indicated a 70° collimator provided the optimum forward projection of the radiated energy. Commercially available aluminum funnels provided the proper collimation angle and an opening of proper diameter to insert the lamp at the lowest cost. Cost of the array components was one of the key considerations in the facility development since there are 1280 elements on the array. System level tests were performed to evaluate the performance of the array and measure the background radiation levels at 35 and 94 GHz with two available radiometers. Horizontal scans of the 35 GHz radiometer across the array and adjacent walls in the chamber for five lamp current levels are plotted in Figure 3 with the lowest temperature (120°K) measured at the lowest current level possible without extinguishing the lamps. Varying the target size (number of array elements) from 2 by 2, to 10 by 10 and 20 by 20 lamps at maximum intensity again with horizontal scans gave the results shown in Figure 4. A 2 by 2 element

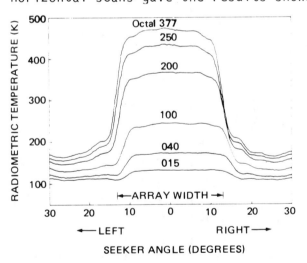

Figure 3. Varying Intensity (Current) Level on Entire Array

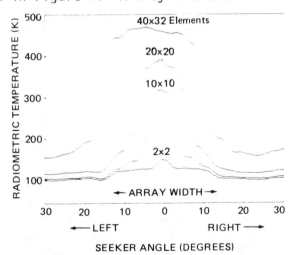

Figure 4. Target Size Variations on Array With Constant Intensity

target measured 20°K above the background. The array was again scanned while turning on different columns of elements at a reduced energy level to verify the uniformity of lamp radiation with the results shown in Figure 5. A plot of the computer command to the array elements versus radiometric temperature with the antenna beam filled is shown in Figure 6. The linear region of computer command versus radiometric temperature extends from 120°K to 440°K. Sky temperatures measured over the millimeter

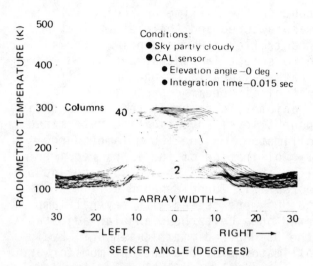
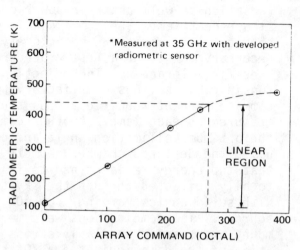

Figure 5. Varying Columns of Elements on Array With Constant Intensity

Figure 6. Radiometric Temperature Measured on Array Versus Computer Command

wave frequency region [2] are compared with temperatures measured in the test chamber at 35 and 94 GHz in Figure 7. The data at 35 GHz reflects the temperature measured with the radiometer mounted 6.7 meters away and pointed at the array. The upper reading was taken with clouds, the lower reading without. At 94 GHz, the higher reading was measured with the radiometer pointing at the center of the array while the lower reading was taken looking directly into an overcast sky.

ACTIVE SEEKER TEST CAPABILITY

As stated earlier our present operational facility is limited to the more critical final flight phase where passive operation is anticipated. We are now in the process of adding the capability to evaluate active seekers in the same chamber. A dual chamber approach (anechoid and reflective) was considered and abandoned when test data verified that the reflective chamber provides adequate attenuation of the alternate signal paths. The rejection ratio of the main reflected signal was measured at more than 25 db. Using the same chamber for both active and passive modes allows us to evaluate the hand-over transition in the laboratory considered to be critical to system performance.

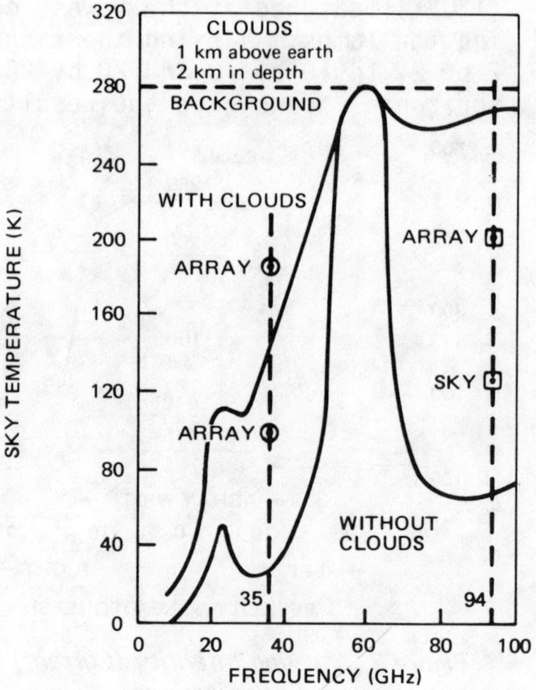

Figure 7. Radiometric Sky Temperature Versus Operating Frequency

Initially a set of three small narrow beamwidth antennas will be mounted in a triad configuration near the center of the array and interspersed between the present broadband elements. These antennas will be driven by signals taken either from the active seeker and delayed to account for the range variations or generated independently synchronous with the seeker internal timing. The amplitude and phase of each radiator will be adjusted under computer control with attenuators and phase shifters so as to provide the combined wave front necessary to accurately position the target. After completion of an initial test phase more elements will be added to the array to accommodate larger changes in target missile relative geometries and alternate frequency bands. The initial active capability will be limited to frequency-modulated continuous-wave systems within a narrow range of frequencies. Even this limited early capability will significantly improve our support of millimeter wave terminal guidance for future missile applications.

REFERENCES

1. Swetnam, G. D. and Belrose, F. M. "Radio Frequency (RF) Homing Missile Guidance and Control Simulation Techniques, Facilities and Experiences" Proc. of NATO/AGARD Conference No. 257, May, 1978.

2. Seashore, C. R., Miley, J. E. and Kearns, B. C. "MM-Wave Radar and Radiometer Sensors for Guidance Systems" Microwave Journal, p. 47, August, 1979.